Super Field Theories

NATO ASI Series

Advanced Science Institutes Series

A series presenting the results of activities sponsored by the NATO Science Committee, which aims at the dissemination of advanced scientific and technological knowledge, with a view to strengthening links between scientific communities.

The series is published by an international board of publishers in conjunction with the NATO Scientific Affairs Division

A	**Life Sciences**	Plenum Publishing Corporation
B	**Physics**	New York and London
C	**Mathematical and Physical Sciences**	D. Reidel Publishing Company Dordrecht, Boston, and Lancaster
D	**Behavioral and Social Sciences**	Martinus Nijhoff Publishers
E	**Engineering and Materials Sciences**	The Hague, Boston, Dordrecht, and Lancaster
F	**Computer and Systems Sciences**	Springer-Verlag
G	**Ecological Sciences**	Berlin, Heidelberg, New York, London,
H	**Cell Biology**	Paris, and Tokyo

Recent Volumes in this Series

Volume 153—Physics of Strong Fields
edited by Walter Greiner

Volume 154—Strongly Coupled Plasma Physics
edited by Forrest J. Rogers and Hugh E. Dewitt

Volume 155—Low-Dimensional Conductors and Superconductors
edited by D. Jérome and L. G. Caron

Volume 156—Gravitation in Astrophysics: *Cargèse 1986*
edited by B. Carter and J. B. Hartle

Volume 157—The Physics of the Two-Dimensional Electron Gas
edited by J. T. Devreese and F. M. Peters

Volume 158—Physics and Chemistry of Small Clusters
edited by P. Jena, B. K. Rao, and S. N. Khanna

Volume 159—Lattice Gauge Theory '86
edited by Helmut Satz, Isabel Harrity, and Jean Potvin

Volume 160—Super Field Theories
edited by H. C. Lee, V. Elias, G. Kunstatter, R. B. Mann, and K. S. Viswanathan

Series B: Physics

NATO Advanced Research Workshop ...

Super Field Theories

Edited by

H. C. Lee

Chalk River Nuclear Laboratories
Chalk River, Ontario, Canada

V. Elias

University of Western Ontario
London, Ontario, Canada

G. Kunstatter

University of Winnipeg
Winnipeg, Manitoba, Canada

R. B. Mann

University of Toronto
Toronto, Ontario, Canada

and

K. S. Viswanathan

Simon Fraser University
Burnaby, British Columbia, Canada

Plenum Press
New York and London
Published in cooperation with NATO Scientific Affairs Division

Proceedings of a NATO Advanced Research Workshop on
Super Field Theory,
held July 25-August 5, 1986,
in Vancouver, Canada

Library of Congress Cataloging in Publication Data

NATO Advanced Research Workshop on Super Field Theory (1986: Vancouver, B.C.)
Super field theories.

NATO ASI series. Series B, Physics; vol. 160)
"Proceedings of a NATO Advanced Research Workshop on Super Field Theory, held July 25-August 5, 1986, in Vancouver, Canada"—T.p. verso.
Includes bibliographical references and index.
1. Field theory (Physics)—Congresses. 2. Kaluza-Klein theories—Congresses. 3. Superstring theories—Congresses. I. Lee, H. C. (Hoong-Chien), 1941- . II. Title. III. Series: NATO ASI series. Series B Physics; v. 160.
QC793.3.F5N365 1986 530.1'4 87-14157
ISBN 0-306-42660-9

A Division of Plenum Publishing Corporation
233 Spring Street, New York, N.Y. 10013

Printed in the United States of America

PREFACE

The Super Field Theory Workshop, held at Simon Fraser University, Vancouver, Canada July 25 - August 5, 1986 was originally intended to be a sequel to the 1983 Chalk River Workshop on Kaluza-Klein Theories and the 1985 Workshop on Quantum Field Theories held at the University of Western Ontario. The scope of the workshop was therefore not to be very big, with a program of about 20 papers, an anticipated 30 to 45 participants, and with much time scheduled for discussion and personal contact. These goals were soon changed in the face of wide interest in the workshop, both for participation and for giving talks, so that the workshop materialized with about 90 participants and 40 talks. This volume contains the texts, some considerably expanded from the oral version, of most of the talks presented at the workshop. Not included are a few talks whose manuscripts were not made available to the editors.

In the last few years the subject of particle physics and unified field theory has developed in a way not witnessed in the last fifty years: a confluence with mathematics, especially in geometry, topology and algebra at an advanced level. This has vastly expanded the horizon of the discipline and heightened the expectation that a true understanding of the fundamental laws of physics may soon be within reach. Most aspects of this new development are covered, many in pedagogical detail, by articles in this volume. Topics discussed include light-cone physics, supergravity, anomalies, Kaluza-Klein theory, nonlinear sigma-model, strings and superstrings, Kac-Moody and Virasoro algebras, BRST algebra, superconformal algebra, supermanifolds and super Riemann surfaces.

We thank all the speakers and participants for making the workshop interesting. We are especially thankful to those lecturers who have taken the extra effort in making their contributions pedagogical. Chris Hull won (in a fashion that violated causality) the Frozen BC Salmon Award for delivering the manuscript that is the longest, most detailed and containing the most original material not published elsewhere; it was also delivered to us three months later than any other manuscript. We are glad it was worth waiting for.

The scientific organizers of the workshop were the editors of this volume. K.S. Viswanathan and R.M. Woloshyn were the local organizers. The Physics Department of Simon Fraser University graciously hosted the workshop. In particular we thank Jacque Link and her colleagues from the Conference Service at SFU for providing efficient support on site. Ed Levinson was the photographer.

The workshop was primarily supported by a NATO Advance Research Workshop award; we thank the NATO Science Committee for providing this fund and other infrastructure assistance. We acknowledge generous grants from Natural Science & Engineering Research Council of Canada, TRIUMF, Simon Fraser University, Atomic Energy of Canada Research Company, University of Winnipeg and the Canadian Institute of Particle Physics.

Finally we are very appreciative of Margaret Carey for secretarial assistance carried out, quite often beyond the call of duty, during the entire course of the workshop, and for preparing a number of manuscripts in this volume.

Chalk River
March 1, 1987

H.C. Lee
V. Elias
G. Kunstatter
R.B. Mann
K.S. Viswanathan

CONTENTS

LECTURES

SEMINARS

LIGHT-CONE PHYSICS

Lars Brink

Institute of Theoretical Physics
S-412 96 Göteborg
Sweden

1. INTRODUCTION

In classical physics differential equations are used to describe the evolution of a physical system. A well-defined problem must then include a set of initial values. In the quantum case, the corresponding information is provided when canonical commutators are specified. In a non-relativistic case these are specified on a surface of equal time, since constant time surfaces are the only ones that every particle trajectory crosses exactly once. In the relativistic case the existence of an upper bound to velocities means that the concept of simultaneity becomes ambiguous. Through each space-time point we can draw a hypercone (the light-cone); events occuring outside the light-cone cannot influence, or be influenced by an event at the tip of the cone. It is then a matter of definition how the surface of simultaneity is drawn. Any space-like surface may be used to set the initial data (or canonical commutators). Dirac[1)] observed that initial data can be given on various surfaces, leading to different forms of dynamics. The conventional formulation is to specify a surface at $x^0 = 0$, but here I will use another choice of Dirac, namely a hypersurface tangent to the light-cone, defined by x^+, where we use the notation $x^{\pm} = \frac{1}{\sqrt{2}}(x^0 \pm x^{d-1})$. Dirac called this choice the "front form". For historic reasons I will use the slightly inappropriate name "the light-cone formalism". Here x^+ is the evolution parameter, i.e. the "time". The choice is a singular case of a space-like surface. It is a limiting one for a surface to set initial data on. This will be reflected when we specify these data.

The initial value problem is a characteristic initial value problem rather than the ordinary Cauchy problem. A well-behaved case is depicted in fig. 1,

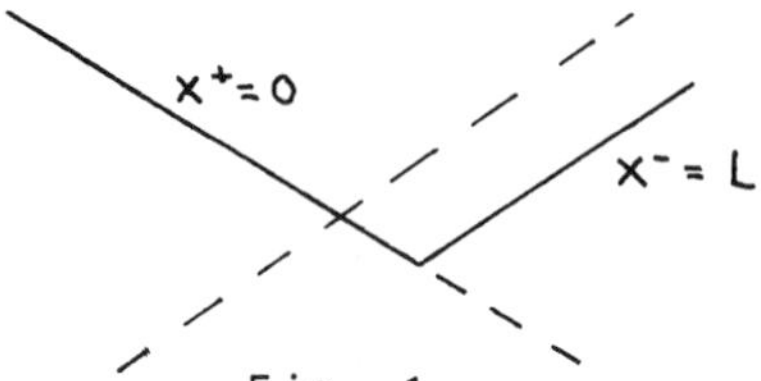

Fig. 1

where initial data are given on a pair of intersecting null-planes. This one has been thoroughly studied in general relativity [2]. However, we would like to perform the limit $L \to -\infty$. This is in fact possible for spaces with d>2 provided that the fields go to zero when $x^- \to \pm\infty$ and that no time-derivatives are present in the interaction terms[3]. We must keep this result in mind. It is quite plausible that the light-cone dynamics we will set up could miss some non-trivial solutions. Let us also note that the initial data surface contains the null-vector x^-. Therefore non-localities in x^- are expected.

Light-cone dynamics became popular in current algebra, which was seen to simplify for this case[4]. It was then often referred to as "the infinite momentum frame". However, I must stress that it does not involve a choice of Lorentz frame. It was then observed by Weinberg[5] that the old-fashioned time-ordered perturbation theory also simplifies, acquiring a sort of non-relativistic structure. The relation between covariant perturbation theory and canonical quantization on the light-cone was studied by a number of authors[6] and the equivalence was, somewhat formally, asserted [7].

Another arena for light-cone dynamics came with the advent of string theory. In order to fully fix a gauge when describing a free string, one was led into "the light-cone gauge"[8]. In fact, in the case of the superstring[9], the apparent impossibility to find a covariant supersymmetric formulation spurred us to investigate supersymmetric theories using light-cone dynamics[10]. This investigation was quite successful. It provided the first proof that N = 4 Yang-Mills theory is a finite quantum theory[11], in fact, the first non-trivial 4-dimensional such a theory. It also provided a method to construct field theories for the superstring theories[12], showing the non-existence of local counterterms[13] (other than the actions themselves) providing yet another indication that these theories are perturbatively consistent.

There is one point in our programme which is logically separate from our use of light-cone dynamics. We will only deal with the physical degrees of freedom, which means that we have to accept that the dynamical (super-)Poincaré transformations, i.e. the ones that transform the system forward in time, are non-linearly and non-locally realized on the fields. This is often an awkward feature, but there are practical advantages for supersymmetric theories (as we will see) which in some cases are bigger than the obvious disadvantages.

In these lectures I will describe two ways to arrive at light-cone field theories. For point-like particles I will use our knowledge of covariant field theory to find the light-cone versions of the interesting field theories. In the case of strings I will first describe first-quantized strings and proceed in an algebraic way to define a second-quantized theory involving interactions. This way could, of course, also be used for point-particles, but might seem too round-about with the vast knowledge of covariant field theory which has been acquired in recent years.

2. FREE FIELDS FOR POINT-PARTICLES ON THE LIGHT-CONE

Consider the action for a scalar, complex free field.

$$S = \int d^4x \, \bar{\phi} \, \Box \, \phi \quad . \tag{2.1}$$

If we introduce light-cone coordinates we write it as

$$S = \int d^4x \, \bar{\phi}(-2\partial_+\partial_- + \partial^i\partial^i)\phi \tag{2.2}$$

with $\partial_\pm = \frac{1}{\sqrt{2}} \left(\frac{\partial}{\partial x^0} \pm \frac{\partial}{\partial x^3}\right) = \frac{\partial}{\partial x^\pm}$

and $i = 1,2$. (We use space-like metric).

Let us formally partially integrate the "space"-derivative ∂_- to obtain

$$S = 2\int d^4x \, \partial_-\bar{\phi} \, \left(\partial_+ - \frac{\partial^i\partial^i}{2\partial_-}\right)\phi \quad . \tag{2.3}$$

In this expression we have allowed the use of the non-local operator ∂_-^{-1}. We will discuss it more thoroughly later, but here we point out that for trivial boundary conditions it can be defined by

$$\frac{1}{\partial_-} \phi(x^-) = \int_{-\infty}^{\infty} \varepsilon(x^- - y^-)\phi(y^-)dy^- \quad . \tag{2.4}$$

Since signals can propagate "instantaneously" along the x^--direction, the non-locality in this direction is unavoidable on physical grounds. As we will see, the non-locality will be confined to the x^--direction also in the interacting theories, which is an important advantage of light-cone dynamics.

It is straightforward to perform a canonical analysis on the action (2.3). The momentum conjugate to ϕ is

$$\Pi_\phi = 2\partial_-\bar{\phi} \tag{2.5}$$

and the hamiltonian is

$$H = \int d^3x \, \partial_-\bar{\phi} \, \frac{\partial^i\partial^i}{\partial_-} \, \phi \quad . \tag{2.6}$$

The equation of motion is

$$\partial_+\phi = \frac{\partial^i\partial^i}{2\partial_-} \phi \quad . \tag{2.7}$$

We can directly second-quantize the theory by

$$[\Pi_\phi(x), \phi(x')]_{x^+=x'^+} = -i\delta^3(x-x') \quad , \tag{2.8}$$

or using (2.5)

$$[\bar{\phi}(x),\phi(x')]_{x^+=x'^+} = \frac{-i}{2\partial_-^x}\,\delta^3(x-x') \quad , \tag{2.9}$$

where we again light-heartedly have divided by ∂_-. Note that the commutator only involves fields and no momenta. This is a difference with covariant treatments and stems from the fact that the action is linear in light-cone time. (The eq. (2.5) is a second class constraint in Dirac's terminology[14]*).

It is now straightforward to read off the canonical representation of the Poincaré generators (either by the Noether procedure or by guessing them directly)

$$P^\mu = 2\int d^3x\; \partial_-\bar{\phi}\;\partial^\mu\phi \tag{2.10}$$

$$J^{\mu\nu} = 2\int d^3x\; \partial_-\bar{\phi}\;(x^\mu\partial^\nu - x^\nu\partial^\mu)\phi \quad . \tag{2.11}$$

On mass-shell we can use (2.7) and make the substitution $\partial_+ \rightarrow \frac{\partial^i\partial^i}{2\partial_-}$. The remarkable fact is then that the algebra still closes off the mass-shell. (If we had started from the dynamics of the free particles this fact would have followed directly). We also note that now

$$P_+ = H \quad . \tag{2.12}$$

It is quite easy to see that the representation (2.10) and (2.11) with the substitution above can be generalized. We can add a helicity term to $J^{12} = J_z$ and make additions also to J^{i-} and write

$$J^{12} = 2\int d^3x\; \partial_-\bar{\phi}\;(x^1\partial^2 - x^2\partial^1 - \lambda)\phi \tag{2.13}$$

$$J^{+-} = 2\int d^3x\; [\partial_-\bar{\phi}\; x^-\partial_-\phi - x^+H] \tag{2.14}$$

$$J^{+i} = 2\int d^3x\; [\partial_-\bar{\phi}(x^+\partial^i - x^i\partial^+)\phi] \tag{2.15}$$

$$J^{-i} = 2\int d^3x\; [\partial_-\bar{\phi}\;(x^-\partial^i - \lambda\,\frac{\partial^i}{\partial_-})\phi + x^iH] \quad . \tag{2.16}$$

This is the representation for a massless spin-λ field. (This can be checked by constructing the Pauli-Lubanski operator). Since such a field only carries two degrees of freedom it can be represented by a complex field. In fact, I must stress that the representation can be used for any spin-λ field. Honestly speaking in this formalism we do not yet even see any difference between half-integer or integer spin fields!

In the case of spin-1 we could have arrived at the representation (2.13)-(2.16) by starting with the action

$$S = -\frac{1}{4}\int d^4x\; F_{\mu\nu}\,F^{\mu\nu} \quad . \tag{2.17}$$

By choosing a gauge $A^+ = 0$ (light-cone gauge), eliminating A^- via its equation of motion (which is algebraic) and linearly combining $\phi = \frac{1}{\sqrt{2}}\,(A^1 + iA^2)$ we obtain the action

---

* In principle one should be more careful when quantizing. The constraint implies that one must construct Dirac brackets and then quantize. This amounts to a factor 1/2 in (2.8), which, however, here can be scaled away.

(2.1). The representation (2.13) - (2.16) is then obtained from the original covariant generators by making compensating gauge transformations to stay in the gauge and introducing the solution of A^-.

In the interacting case for a scalar field one can add say a ϕ^3 term. The only effect will be a ϕ^3 term in H and hence in all the dynamical generators. Note that now the dynamical Poincaré transformations are non-linear. For spin-1 we could start with a non-abelian gauge theory and perform the steps sketched above[10]. An alternative way is to guess the interaction terms in the dynamical generators and check until one has a closing algebra. In fact the latter method is probably the quickest way to get the result. The remarkable thing is that one can find 3-point couplings for any spin-λ at one go[15]. By guessing an arbitrary form and grinding through the algebra one finds the following action to work

$$S = S_{kin} + \alpha \sum_{n=0}^{\lambda} (-1)^n \binom{\lambda}{n} \int d^4x \, [\partial_-^{\lambda} \bar{\phi} \, \frac{\bar{\partial}^{\lambda-n}}{\partial_-^{\lambda-n}} \phi \, \frac{\bar{\partial}^n}{\partial_-^n} \phi + c.c.]$$

$$+ O(\alpha^2) \quad , \tag{2.18}$$

with $\partial = \frac{1}{\sqrt{2}} (\partial^1 + i\partial^2)$ and $\bar{\partial} = \frac{1}{\sqrt{2}} (\partial^1 - i\partial^2)$. For odd λ the fields must transform as the adjoint representation of a Lie group and the corresponding group multiplication is understood. The dimension of the coupling constant α is $(\text{length})^{\lambda-1}$. In the case of $\lambda = 1$ the action closes after having added a 4-point coupling, while for higher λ an infinite power series in the fields is needed. In the case of λ = 2, this series could, in principle, be obtained from the Einstein-Hilbert action by choosing a light-cone gauge and eliminating all auxiliary field components. For spins greater than 2 there are no covariant actions constructed so far and it is unclear if such self-interacting theories exist.

In the case of spin-1 the action constructed can be shown to be unique. There are no possible local counterterms other than the action itself lending credence to the renormalizability of non-abelian gauge theories. In the case of gravity the situation is quite different. The dimension dependent coupling constant (α = κ Newton's constant) makes higher derivative terms possible. The most general kinetic term is[13]

$$L = \bar{\phi} \sum_{n=1}^{N} a_n \Box^n \phi \quad , \quad a_1 = 1 \tag{2.19}$$

with N arbitrary. This action represents in fact N degrees of freedom, one for each pole in the propagator $(\Sigma\, a_n\, {}^n)^{-1}$. We find that the Poincaré algebra can be realized on a field vector with one massless spin-2 field and N-1 massive spinless fields, some of which are tachyons.

Also for the 3-point coupling, the term (2.18) is not unique and the following terms are possible[16]

$$S^{(3)} = \sum_{k=0}^{p} \sum_{n=0}^{2+p} \kappa^{2p+1} (-1)^{k+n} \binom{p}{k} \binom{2+p}{n} \int d^4x$$

$$[\partial_-^2 \bar{\phi}\ \frac{\partial^{p-k}\bar{\partial}^{p+\lambda-n}}{\partial_-^{p+\lambda-k-n}}\ \phi\ \partial_-^{p-k-n}\partial^k\bar{\partial}^n\phi + c.c] \quad , \qquad (2.20)$$

where p is an arbitrary integer.

To really prove that there are infinity many new counterterms in gravity we should close the algebra to all orders. This is practically impossible and is in fact unnecessary, since we know the result to be true from covariant studies.

We have so far considered only Bose fields. Since the 3-point couplings only work for integer spin, we have another proof of the spin-statistics theorem[17].

For fermions we use anticommuting fields. If we start with the Dirac equation for a Majorana (or Weyl) field we can divide up the spinor in its two light-cone components through

$$\psi = \frac{1}{2}(\gamma_+\gamma_- + \gamma_-\gamma_+)\psi \equiv \psi_+ + \psi_- \quad . \qquad (2.21)$$

The spinor ψ_- will then satisfy an algebraic equation and can be eliminated. ψ_+ has two real components and can be written as a complex anticommuting field.

In principle one can now write any known field theory in light-cone dynamics. However, there seem to be few advantages for most theories compared to covariant methods unless the theory is supersymmetric. (For an ambitious programme for computing loop graphs in the light-cone gauge see ref 18). Hence we now turn to this case.

3. SUPERSYMMETRIC FIELD THEORIES

In light-cone dynamics in d dimensions it is only the transverse symmetry group SO(d-2) that is covariantly described. That is yet another reason why a spinor is divided up into its two "light-cone spinors". The same decomposition is then natural for the supersymmetry generator Q. In 4 dimensions we can use a complex notation and write the generators as Q_+^m and Q_-^m letting them transform as N under SU(N)[10]. The supersymmetry algebra is then

$$\{Q_+^m, \bar{Q}_{+n}\} = \sqrt{2}\ \delta_n^m\ P_- \qquad (3.1)$$

$$\{Q_-^m, \bar{Q}_{-n}\} = \sqrt{2}\ \delta_n^m\ H \qquad (3.2)$$

$$\{Q_+^m, \bar{Q}_{-n}\} = -2\ \delta_n^m (P^1 + iP^2) \quad . \qquad (3.3)$$

From this algebra we conclude that Q_+ is a kinematical (i.e a linearly realized) generator and Q_- a dynamical generator.

The light-cone approach is essentially perturbative, in the sense that we construct the theories order by order in the coupling constants. Hence it is natural to seek a representation of the super-Poincaré algebra first for the free case. If we represent the generators first as operators on x-dependent fields we let $P^\mu \to -i\frac{\partial}{\partial x^\mu}$. We see then that the Q's can be constructed if we also let the fields depend on anticommuting coordinates θ^m and $\bar{\theta}_n$. This space being bigger than necessary allows us also to define covariant derivatives d^m and $\bar{d}_n$ anticommuting with the q's. To find an irreducible representation of the supersymmetry, we first impose a "chirality" condition on the superfields

$$d^m \phi(x,\theta,\bar{\theta}) = 0 \quad . \tag{3.4}$$

An alternative representation is to just use θ and not $\bar{\theta}$. This one has some advantages, but makes complex conjugation much more intricate. When we consider 10-dimensional theories we will use this representation.

When N = 4 × integer we can further impose the condition.

$$\bar{\phi}(x,\theta,\bar{\theta}) = \frac{1}{2^{N/4}\, N!} \frac{\bar{d}^n}{\partial_-^{N/2}} \phi(x,\theta,\bar{\theta}) \quad , \tag{3.5}$$

where we have defined

$$\bar{d}^N = \varepsilon^{m_1 \ldots m_N} \bar{d}_{m_1} \ldots \bar{d}_{m_N} \tag{3.6}$$

$$d^N = \varepsilon_{m_1 \ldots m_N} d^{m_1} \ldots d^{m_N} \quad . \tag{3.7}$$

In this way the minimal supermultiplet is contained in an index-free superfield with helicities λ, $\lambda-1/2, \ldots, -\lambda$, where $\lambda = N/4$. Note that all components of the superfield are physical fields. The action for a free superfield is straightforwardly seen to be[19)]

$$S_0 = \frac{1}{4} \int d^4x d^N\theta d^N\bar{\theta}\; \bar{\phi} \frac{\Box}{\partial_-^{N/2}} \phi \quad . \tag{3.8}$$

The dimension of ϕ is $(\text{length})^{-1+N/4}$. We interpret this to mean that the superfield starts out as

$$\phi(x,\theta,\bar{\theta}) = \frac{1}{\partial_-^{\lambda}} \phi(x) + \ldots \quad . \tag{3.9}$$

A canonical procedure leads to

$$[\bar{\phi}(z),\phi(z')]_{\text{equal } x^+} = -\frac{i}{2} \frac{\bar{d}_z{}^N d_{z'}^N}{(N!)^4 \partial_{-z}^N} \delta^{3+2N}(z-z') \tag{3.10}$$

with $z = (x^\mu, \theta^m, \bar{\theta}_m)$. (The reason why covariant derivatives occur is that functional derivatives w.r.t ϕ is not straightforward. See ref 19). We can then write the super-Poincaré generators as

$$G = 2i \int d^3x d^N\theta d^N\bar{\theta}\; \partial_- \bar{\phi} \frac{1}{\partial_-^{N/2}} g\, \phi \tag{3.11}$$

where the resp. g's are

$$d^m = -\frac{\partial}{\partial\bar{\theta}_m} + \frac{i}{\sqrt{2}} \theta^m \partial_-, \quad \bar{d}_n = \frac{\partial}{\partial\theta^n} - \frac{i}{\sqrt{2}} \bar{\theta}_n \partial_- \quad ,$$

$$q_+^{\ m} = -\frac{\partial}{\partial\bar\theta_m} - \frac{i}{\sqrt2}\,\theta^m\partial_- \ , \quad \bar q_{+n} = \frac{\partial}{\partial\theta^n} + \frac{i}{\sqrt2}\,\bar\theta_n\partial_- \quad ,$$

$$q_-^{\ m} = -\frac{\bar\partial}{\partial_-}\,q_+^{\ m} \ , \quad \bar q_{-n} = -\frac{\partial}{\partial_-}\,\bar q_{+n}$$

$$p_+ = -i\,\frac{\partial^i\partial^i}{2\partial_-} \ , \quad p_- = -i\partial_- \ , \quad p^i = -i\partial^i$$

$$j^{12} = -i(x^1\partial^2 - x^2\partial^1) + \frac{1}{2}\,\theta\cdot\frac{\partial}{\partial\theta} - \frac{1}{2}\,\bar\theta\cdot\frac{\partial}{\partial\bar\theta}$$

$$j^{+i} = -i(x^+\partial^i + x^i\partial_-) \qquad (3.12)$$

$$j^{+-} = i(x^-\partial_- - x^+\,\frac{\partial^i\partial^i}{2\partial_-} + \frac{1}{2}\,\theta\cdot\frac{\partial}{\partial\theta} + \frac{1}{2}\,\bar\theta\cdot\frac{\partial}{\partial\bar\theta}\,)$$

$$j^{-i} = -i(x^-\partial^i - x^i\,\frac{\partial^i\partial^i}{2\partial_-} - \theta\cdot\frac{\partial}{\partial\theta}\,\frac{\partial^1}{\partial_-} - \bar\theta\cdot\frac{\partial}{\partial\bar\theta}\,\frac{\partial^2}{\partial_-}\,) \quad .$$

This algebra can be truncated to N's not being divisible by 4. One then must relax (3.5) and use both ϕ and $\bar\phi$ as independent fields.

To construct interacting theories we can again try to enlarge the representation (3.12) by adding 3-point couplings to the dynamical generators[19]. Here the anticommutator (3.2) is quite suitable as a starting-point. It is in fact easier to guess a 3-point coupling for Q_- than for H. Starting with a general term grinding through the super-Poincaré algebra again we can find a solution for any case with $\lambda_{max} = \frac{1}{4}$ N. The result for Q_- is

$$\bar Q_{-m} = -2\alpha \int d^3x d^N\theta d^N\bar\theta\, \frac{1}{\partial_-^{\,N/2}}\,\bar\phi \sum_{n=0}^{\lambda-1} (-1)^n \binom{\lambda-1}{n}$$

$$\bar\partial^{(\lambda-1-n)}\,\partial_-^{\,n}\,\frac{\partial}{\partial\theta^m}\,\phi\,\bar\partial^n\,\partial^{(\lambda-n)}\,\phi \qquad (3.13)$$

and $Q_-^{\ m}$ being the complex conjugate. The hamiltonian to this order is

$$H^\alpha = \alpha \int d^3x d^N\theta d^N\bar\theta\, \frac{1}{\partial_-^{\,N/2}}\,\bar\phi\,\Big[\sum_{n=0}^{\lambda-1} (-1)^n \binom{\lambda}{n}$$

$$\bar\partial^{(\lambda-1)}\,\partial_-^{\,n}\phi\,\bar\partial^n\partial_-^{\,(\lambda-n)}\phi + h.c\Big] \quad . \qquad (3.14)$$

Again we do not see any limit for N = 8 (which corresponds to N = 8 supergravity). For λ_{max} odd the fields must transform as the adjoint representation of a Lie group and group multiplication among the fields is implied.

To show that the theories exist we have to construct the generators to all orders in the coupling constant α. For N=4 this is, in fact, simple. Checking $\bar Q_{-m}$ for this case it is easy to convince oneself that because of dimensional reasons no 4-point coupling is possible. Hence the complete solution to the anticommutator (3.2) gives the full hamiltonian with a 3-point and a 4-point coupling (as we expect for a Yang-Mills theory). This theory was first constructed from the covariant theory choosing the light-cone gauge[10].

Higher N's must have an infinite series of interaction terms with rising powers of α as in the non-supersymmetric case. To construct higher order terms by just going through the algebra is practically impossible. We have also, so far, found no way of finding the complete sum by indirect methods.

For the case of N = 1,2,3 and 4 Yang-Mills theories, the complete actions have been constructed[20]. It is also straightforward to include matter in the cases, where such multiplets exist. The great advantage in this formalism is that each multiplet is described by an index-free superfield. Feynman rules can be derived and the perturbation expansion can be studied. Non-covariant gauges such as the light-cone gauge have been intensively studied in the literature and one can use these results in studying higher order corrections[18]. Our form of the light-cone gauge is particularly suited since we always use index-free fields. It is straightforward, for example, to invoke dimensional regularization in the transverse space keeping supersymmetry and hence renormalize in a supersymmetric way.

In the case of N = 4 Yang-Mills, the maximal such multiplet, one can even prove that the theory is completely finite in a specific gauge[11]. If we perform naîve power counting, we can conclude that any Feynman graph is finite, if power counting is a legitimate procedure. According to Weinberg's theorem it is, if we are allowed to perform a Wick rotation. This is not necessarily the case in the light-cone gauge. The occurence of the non-local factor p_-^{-1} in amplitudes introduces new poles, whose exact situation in the complex plane must be decided upon. If we go back to Eq. (2.4) we have defined $\frac{1}{\partial_-}\phi(x^-)$ such that

$$\partial_-\left(\frac{1}{\partial_-}\phi(x^-)\right) = \phi(x^-) \qquad . \tag{3.15}$$

This means that there is a freedom in the choice of $\frac{1}{\partial_-}\phi(x^-)$. We can add to the RHS of (2.4) any function not depending on x^-. One can regard this freedom as a remaining gauge degree of freedom. Only by specifying p_-^{-1} completely, we specify the gauge completely. The choice in (2.4), "the principle value prescription" amounts to

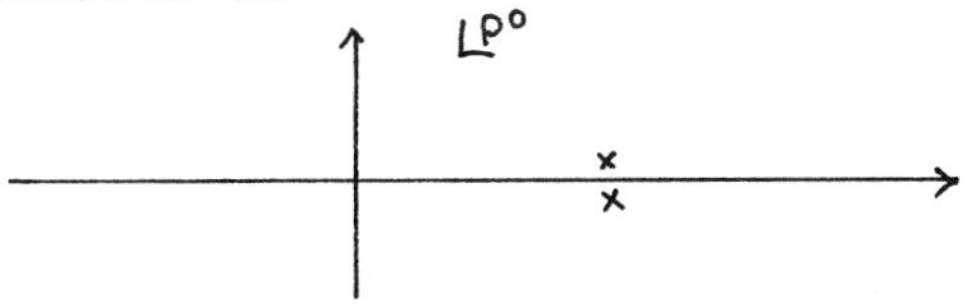

and we cannot perform the Wick rotation. However, there exist another prescription, "the Mandelstam prescription"[11], where

$$\frac{1}{p^+} \to \frac{1}{p^+ + i\varepsilon p^-} \qquad . \tag{3.16}$$

Here the pole structure is the following

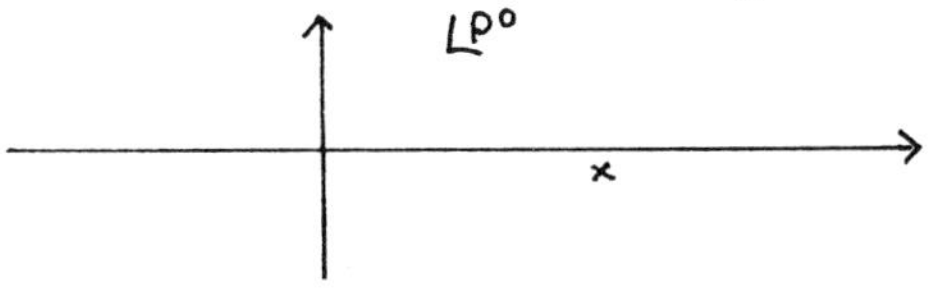

and a Wick rotation can be performed. For this choice, the whole perturbation expansion is finite term by term. This is the first non-trivial 4-dimensional field theory for which the above property has been established. It is a remarkable result, which still awaits its use in physics. It can very well be accidental. The importance of this Yang-Mills model is so far that its 10-dimensional counterpart is an integral part of superstring theory[9].

One may ask if the convergence is even better than what powercounting indicates. The answer is yes. Marcus and Sagnotti[21] checked the N = 4 theory in 6 dimensions and found the two-loop graphs to be finite although light-cone gauge power counting indicates the opposite. There are, in fact, more finite theories in four dimensions. It has been shown that the N = 2 theory has only one-loop contributions to the β-function[22]. By adding matter appropriately one can cancel this divergence for a class of models[23]. In the light-cone gauge formulation this finiteness is due to cancellations between different graphs and it takes a calculation of the relevant divergent parts to establish this fact.

For supergravity one is hampered, as said above, by the fact that no one has been able to give a closed form for the full action. One can, however, use the formalism to look for possible new counterterms, which would ruin the quantum theory. One finds that for $N > 3$ one cannot construct kinetic counterterms as in (2.19)[13]. However, one should be able to construct new interaction terms with more derivatives. From covariant considerations we know such terms to exist[24].

Also supergravity theories in higher dimensions can be discussed in the light-cone gauge[25]. Since the d = 10 theories are the massless sectors of superstring theories we defer the presentation of them to the next section. Let me just point out that for those theories possible kinetic counterterms can easily be constructed[13].

The real conclusion we can draw from the various studies of supergravity theories is that the quantum theories diverge in an uncontrollable way. We need something new to solve the quantum gravity problem. String theories are the plausible candidates to solve this problem and we now turn to them.

4. STRINGS

4.1 Bosonic Strings

So far we have considered theories based on point-particles. Let us now instead consider theories based on strings. Usually one starts by describing the free strings with a geometric covariant action. This is the natural starting-point, and the one which probably would lead to a deeper understanding of string theory. However, here we will instead stick to light-cone dynamics forgetting about geometry. I will, in fact, follow a rather extreme line; a line that will lead quickly to results, but will obscure the geometry. We start by considering a bosonic string with coordinates $x^{\mu}(\sigma,\tau)$. The world-sheet is spanned by a time-like variable τ and a space-like one σ which runs between 0 and π, and μ runs over the d-dimensional Minkowski space. Although we will start by considering a free string, we must construct a free such theory which could interact in a natural way. (In fact, one lesson we have learnt from string theory is that in a consistent theory, interactions are essentially built in if we just allow the free strings to split and join[26]).

The interacting theories that we aim at should satisfy

(i) Poincaré invariance
(ii) causality
(iii) unitarity.

In the standard treatment the Poincaré invariance is automatic; the unitarity is ensured by the gauge invariance of the theory, the reparametrization invariance, and causality is implemented by choosing the correct description of the interactions. Here we will start in another end. In order for a string to be able to interact causally, we demand that each point along the string carry the same time, i.e. the time coordinate should not depend on σ. This is certainly the first requirement to impose in order to have consistent interactions for extended objects. In a gauge invariant theory, it is enough that the points carry the same time up to a gauge transformation. We do have a choice here, however. We can either choose x^0, or one of the light-cone directions $x^{\pm}$ as the evolution parameter. Since we are going to describe a massless string, i.e. a string with no explicit mass term in the free action, it is natural to choose x^+ as the "time". Our first assumption is then*

$$x'^{+} = 0 \quad , \tag{4.1}$$

i.e.

$$x^{+}(\sigma,\tau) = x^{+} + p^{+}\tau \quad , \tag{4.2}$$

where x^+ and p^+ are integration constants. We will interpret p^+ as the total momentum in the + -direction. In order to do so we have put a dimensionful constant to one. We will soon correct this fact.

To implement the third requirement above, we assume that all the propagating degrees of freedom lie in the transverse part of $x^{\mu}(\sigma,\tau)$, i.e. in $x^{i}(\sigma,\tau)$ where $i=1,\ldots,d-2$. Since we are going to describe freely propagating strings, it is natural to choose an action

$$S = -\frac{T}{2}\int_{\tau_i}^{\tau_f} d\tau \int_0^{\pi} d\sigma \; \eta^{\alpha\beta}\partial_\alpha x^i \partial_\beta x^i \quad , \tag{4.3}$$

with α and β taking the values τ and σ and the metric $\eta^{\alpha\beta}$ is chosen space-like. T is a proportionality factor (which ensures that x^i is a length) and will turn out to be the string tension.

The possible flaw with this action is that it is not explicitly Poincaré invariant. The invariance can only be proven by an explicit construction of the Poincaré generatons. Before performing the analysis let us check the dyna-

---

*We use the standard notation that $\dot{x} \equiv \frac{\partial x}{\partial \tau}$ and $x' \equiv \frac{\partial x}{\partial \sigma}$.

mical content of the action. When varying it to obtain the equations of motion some care has to be exercised because of the finite σ-interval. Disregarding possible surface terms, one obtains the equations of motion

$$\ddot{x}^i - x''^i = 0 \quad . \qquad (4.4)$$

The canonical conjugate momentum density is

$$p^i(\sigma,\tau) = T\dot{x}^i(\sigma,\tau) \quad , \qquad (4.5)$$

and the total momentum is

$$p^i = T\int_0^\pi d\sigma\ \dot{x}^i(\sigma,\tau) \quad . \qquad (4.6)$$

By comparing with (4.2) we find that by choosing $T = \frac{1}{\pi}$, also p^+ defined by (4.2) satisfy (4.6). For the rest of the lectures we will make this choice.

We can now quantize the theory by the canonical commutators with $\hbar=1$).

$$[x^i(\sigma,\tau),\ p^j(\sigma',\tau)] = i\ \delta^{ij}\delta(\sigma-\sigma') \quad . \qquad (4.7)$$

Returning to the surface terms we find two sets of boundary conditions which make the surface terms vanish:

(i) $x^i(\sigma=0,\pi) = 0$, open strings:

$$x^i(\sigma,\tau) = x^i + p^i\tau + i\sum_{n\neq 0}\frac{1}{n}\alpha_n{}^i \cos n\sigma\ e^{in\tau} \qquad (4.8)$$

(ii) x^i periodic , closed strings:

$$x^i(\sigma,\tau) = x^i + p^i\tau + \frac{i}{2}\sum_{n\neq 0}\frac{1}{n}(\alpha_n{}^i e^{-2in(\tau-\sigma)}$$

$$+ \tilde{\alpha}_n{}^i e^{-2in(\tau+\sigma)}) \quad . \qquad (4.9)$$

The canonical commutators are satisfied if

$$[\alpha_n{}^i,\alpha_m{}^j] = n\ \delta_{n+m,0}\ \delta^{ij} \qquad (4.10)$$

$$[\tilde{\alpha}_n{}^i,\tilde{\alpha}_m{}^j] = n\ \delta_{n+m,0}\ \delta^{ij} \quad . \qquad (4.11)$$

The Hilbert space of states can be constructed by introducing a vacuum $|0\rangle$ and demanding

$$\alpha_n{}^i\ |0\rangle = 0 \qquad n>0$$

$$\tilde{\alpha}_n{}^i\ |0\rangle = 0 \qquad n>0 \quad . \qquad (4.12)$$

The space of states is then constructed by letting α^i_{-n} and $\tilde{\alpha}^i_{-n}$, $n>0$, act as creation operators building up an explicitly positive-norm Hilbert space.

It remains to check the Poincaré invariance. The generators must be constructed out of $x^i(\sigma)$, $p^i(\sigma)$, x^+, p^+ and possibly a zero mode x^-conjugate to p^+. Since there is no such linear Lorentz transformation, they must be <u>non-linear</u>.

We will first construct the representation of the generators classically and use Poisson brackets instead of the commutators (2.7). We will also use a notation with lower case letters for these generators and use capital letters eventually for second-quantized representations. In the light-cone frame the generator p_+ is the Hamiltonian. Now

$$p_+ \sim i \frac{\partial}{\partial x^+} = i \frac{\partial \tau}{\partial x^+} \frac{\partial}{\partial \tau} = \frac{1}{p_-}(i \frac{\partial}{\partial \tau}) \sim \frac{1}{p_-} H \quad , \qquad (4.13)$$

where H is the Hamiltonian connected to the Lagrangian (4.3). The Hamiltonian density is

$$\mathcal{H} = \dot{x}^i p^i - \mathcal{L} = \frac{1}{2\pi}[\pi^2 p^{i\,2} + x'^2] \quad . \qquad (4.14)$$

Hence the translation part of the algebra is

$$p_+ = \frac{1}{2\pi p_-} \int_0^\pi d\sigma [\pi^2 p^{i\,2} + x'^{i\,2}] \qquad (4.15)$$

$$p_- = p_- \qquad (4.16)$$

$$p^i = \int_0^\pi p^i(\sigma) d\sigma \quad . \qquad (4.17)$$

The Lorentz part we write as

$$j^{ij} = \int_0^\pi d\sigma (x^i p^j - x^j p^i) \qquad (4.18)$$

$$j^{+i} = \int_0^\pi d\sigma (x^+ p^i + x^i p_-) \qquad (4.19)$$

$$j^{+-} = -x^+ p_+ + x^- p_- \qquad (4.20)$$

$$j^{-i} = \int_0^\pi d\sigma [x^-(\sigma) p^i(\sigma) - x^i(\sigma) p^-(\sigma)] \quad . \qquad (4.21)$$

Here we have introduced $x^-(\sigma)$ which should be determined in terms of x^i and p^i apart from its zero mode x^-. We have introduced the generators one by one. The generator j^{ij} follows from the explicit transverse invariance in (4.3). The generators j^{+i} and j^{+-} are natural and are easily seen to be correct. Finally we determine j^{-i} first classically (to avoid ordering problems). A tedious calculation shows that the algebra is indeed satisfied if we choose

$$x'^-(\sigma) = \frac{-\pi}{p_-} p^i x'^i \quad . \qquad (4.22)$$

If we so check the quantum algebra we symmetrize the generators in order to keep the hermiticity. This affects the checking of the commutator $[j^{i-}, j^{j-}] = 0$, since j^{i-} is cubic. <u>A detailed computation shows that the commutator is satisfied only if d=26[8] (the critical dimension)</u>.

Inserting the solution (4.8) into p_+ (4.15) we obtain

$$p_+ = \frac{p^{i\,2}}{2p_-} + \frac{1}{p_-} \sum_{n=1}^{\infty} \alpha_{-n}^i \alpha_n^i \quad , \qquad (4.23)$$

i.e.

$$p^2 = -2 \sum_{n=1}^{\infty} \alpha_{-n}^i \alpha_n^i = -m^2 \quad . \tag{4.24}$$

Reintroduce the dimensionful constant as $\alpha' = \frac{1}{2\pi T}$, the Regge slope. Then the correct expression is

$$\alpha' m^2 = \sum_{n=1}^{\infty} \alpha_{-n}^i \alpha_n^i \quad . \tag{4.25}$$

The mass-squared is hence built up by an infinite set of "harmonic oscillator energies" [27]. For the quantum case each oscillator will contribute an energy due to zero-point fluctuations (which is due to the symmetrization above). The lowest mass level will then be

$$\alpha' m_0^2 = \frac{d-2}{2} \sum_{n=1}^{\infty} n \quad . \tag{4.26}$$

This is clearly a divergent sum which must be regularized. We do so by comparing the sum to the Riemann ζ-function[28]

$$\zeta(s) = \sum_{n=1}^{\infty} n^{-s} \quad . \quad \mathrm{Re}\ s > 1 \quad . \tag{4.27}$$

This is a function that can be analytically continued to $s = -1$ and

$$\zeta(-1) = -\frac{1}{12} \quad . \tag{4.28}$$

In this way the infinite series has been regularized into

$$\alpha' m_0^2 = -\frac{d-2}{24} \quad . \tag{4.29}$$

Hence this term should be included in p_+.

There is also a more standard way[26] of obtaining this result by adding counterterms to the action (4.3), which amounts to renormalizing the speed of light, (which, of course, is another parameter of the theory). It is quite attractive that it is the quantum theory that demands a finite speed of light.

The really important consequence of Eq. (4.29) is that the lowest state is a tachyon. This really means that an interacting theory based on this bosonic string will not make sense. Also since the tachyon is the scalar state with no excitations of the higher modes, I find it hard to believe that there exists a consistent truncation in which the tachyon is left out.

Let us now redo the last analysis for closed strings. Inserting the solution (4.9) we find

$$\frac{\alpha'}{2} m^2 = \sum_{n=1}^{\infty} (\alpha_{-n}^i \alpha_n^i + \tilde{\alpha}_{-n}^i \tilde{\alpha}_n^i) \equiv N + \tilde{N} \quad . \tag{4.30}$$

Also in this sector we find a tachyon. For closed strings we get a further constraint. Consider again Eq. (4.22). Integrating it between 0 and σ we find $x^-(\sigma)$. Although $x^-(\sigma)$ is depending on the x^i's we must demand that it represent a component of $x^\mu(\sigma)$ and hence must be periodic. Then

$$\int_0^{\pi} d\sigma\, x'^- = 0 = \frac{1}{p^+}\int_0^{\pi} d\sigma\, \dot{x}^i x'^i = \frac{\pi}{p^+}(N - \tilde{N}) \quad , \qquad (4.31)$$

i.e. classically $N=\tilde{N}$ and quantum mechanically we impose this condition on the physical states.

The formulation derived here is the "light-cone gauge" formulation of the Nambu-Hara-Goto string action[29]. In the usual description[8] one starts with a reparametrization and Poincaré invariant action. By specifying a gauge, one arrives at the formulation above. Such an approach which certainly is more general starts with a geometric action. This is quite different from the approach here, where we start with a non-covariant action which, however, explicitly describes free positive-norm states.

The Poincaré algebra spanned by the generators (4.15)-(4.21) with the constraint (4.22) contains all the information about the strings. This is typical for the light-cone gauge. By finding the non-linear representation we know the complete dynamics of the system, since the Hamiltonian is one of the generators. The regrettable thing is that so far we have not found a deductive way to find the generators, but have had to allow some guesswork and then check.

By describing the dynamics of free bosonic strings we have found that such a theory is much more constrained than a corresponding theory for point-particles. This is certainly a most wanted property, since one lesson we have learnt from modern gauge field theories for point particles is that seemingly whole classes of theories are theoretically consistent and only experiments can tell which theories Nature is using. In string theories we can entertain the hope that only one model is consistent and this then should be the theory of Nature!

4.2 Spinning Strings

The representation of the Poincaré algebra found in the last section was found to lead to an inconsistent theory. To obtain a consistent one, we need to change the expression for p_+ (4.15) so as to avoid tachyons. The most natural thing is to introduce a set of anticommuting harmonic oscillators such that their zero-point fluctuations compensate the ones from the commuting oscillators.

The first problem to solve is to determine in which representation of the transverse symmetry group SO(d-2), the new oscillators should be chosen. The x-coordinates belong to the vector representation. We can always try this representation also for the new set, which we shall do first, but we should keep in mind, that for certain values of d-2, there are other representations with the same dimension as the vector one.

Consider hence a set of anticommuting harmonic oscillators $d_n{}^i$ satisfying

$$\{d_m{}^i, d_n{}^j\} = \delta_{n+m,0}\delta^{ij} \quad . \qquad (4.32)$$

If the relevant mass formula for open strings is

$$\alpha' p^2 = -\left(\sum_{n=1}^{\infty} \alpha_{-n}^i \alpha_n{}^i + \sum_{n=1}^{\infty} n\, d_{-n}^i d_n{}^i\right) \quad , \qquad (4.33)$$

we can deduce p_+ from this expression. To write it in a coordinate basis we introduce two normal-mode expansions

$$\lambda^{1i} = \sum_{n=-\infty}^{\infty} d_n{}^i \, e^{-in(\tau-\sigma)} \tag{4.34}$$

$$\lambda^{2i} = \sum_{n=-\infty}^{\infty} d_n{}^i \, e^{-in(\tau+\sigma)} \quad , \tag{4.35}$$

such that

$$\{\lambda^{Ai}(\sigma,\tau),\ \lambda^{Bj}(\sigma',\tau)\} = \pi\delta^{AB}\delta^{ij}\delta(\sigma-\sigma') \quad . \tag{4.36}$$

The generator p_+ can then be written (classically) as

$$p_+ = \frac{-1}{2\pi p^+} \int_0^\pi d\sigma(\pi^2 p^{i\,2} + {x'}^{i\,2} - i\lambda^{1i}{\lambda'}^{1i} + i\lambda^{2i}{\lambda'}^{2i}) \quad . \tag{4.37}$$

Before trying to construct the remaining generators let us consider the dynamics following from (4.37). Since $p_+ = \frac{1}{p_-}$ H, the corresponding action is

$$S = -\frac{1}{2\pi}\int d\tau \int_0^\pi d\sigma[\eta^{\alpha\beta}\partial_\alpha x^i \partial_\beta x^i + i\bar{\lambda}^i\rho^\alpha\partial_\alpha\lambda^i], \tag{4.38}$$

where we combine the two λ^i's into two-dimensional 2-component spinors and use the Majorana representation for the 2×2 Dirac matrices, here called ρ^α. To get a consistent theory

the surface terms obtained upon variation of (4.38) must be zero. In the case of open strings the boundary conditions for the λ's are

$$\lambda^{1i}(0,\tau) = \lambda^{2i}(0,\tau) \tag{4.39}$$

$$\lambda^{1i}(\pi,\tau) = \begin{cases} \lambda^{2i}(\pi,\tau) & (4.40a) \\ -\lambda^{2i}(\pi,\tau) & (4.40b) \end{cases} \quad .$$

In fact we have two choices. The first choice together with equations of motion gives the solutions (4.34) and (4.35). In this sector we know that there are no tachyons.

The other choice results in expansions

$$\lambda^{1i} = \sum_r b_r{}^i e^{-ir(\tau-\sigma)} \tag{4.41}$$

$$\lambda^{2i} = \sum_r b_r{}^i e^{-ir(\tau+\sigma)} \quad , \tag{4.42}$$

where the index r takes all half-integer values. The b's satisfy the anticommutators

$$\{b_r{}^i, b_s{}^j\} = \delta_{r+s,0}\ \delta^{ij} \quad . \tag{4.43}$$

The (classical) mass-shell condition now reads

$$\alpha' m^2 = \sum_{n=1}^{\infty} \alpha_{-n}^i\alpha_n{}^i + \sum_{r=1/2}^{\infty} r b_{-r}^i b_r{}^i \quad . \tag{4.44}$$

Computing the contributions from the zero-point fluctuations we find

$$\alpha' m_0^2 = \frac{d-2}{2}\left[\sum_{n=1}^{\infty} n - \sum_{n=1}^{\infty}(n-1/2)\right]$$

$$= \frac{d-2}{2}\sum_{n=1}^{\infty} n(1-\tfrac{1}{2}+1)$$

$$\rightarrow -\frac{d-2}{16} \quad , \qquad (4.45)$$

when the sum is renormalized. Again we find tachyons! Hence this sector of the model is unphysical and basing an interacting theory upon it would lead to inconsistencies.

The states we have discovered spanned by the d- and b-oscillators together with the α's are in fact the spectrum of the Ramond[30]-Neveu-Schwarz[31] model. The states constructed out of d-oscillators all have to transform as fermions and constitute the states of the Ramond sector, while the ones constructed out of b-modes which are bosonic constitute the Neveu-Schwarz sector. Both sectors are needed in order to have a model with both fermions and bosons.

There is, in fact, a way to truncate the spectrum to avoid tachyons, which can be proven to be consistent with interactions[32]. Consider the projector

$$P = \tfrac{1}{2}(1+(-1)^{\Sigma b_{-r} b_r}) \quad . \qquad (4.46)$$

By demanding it be zero on physical states

$$P|\text{phys}\rangle = 0 \qquad (4.47)$$

we obtain a tachyon free spectrum. A consistent interaction can be set up, if also the spinors are chosen to be of Majorana-Weyl type (the critical dimension is 10). In fact, this projection is necessary for the quantum corrections to be consistent. This leads to the superstrings, which we will describe in the next section.

The Poincaré generators can be constructed and in the quantum case the algebra only works in d=10. I will not give them here, but will discuss them in the next section.

In the case of closed strings there are also two sectors depending on what boundary conditions are chosen. One sector is obtained if the λ's are periodic in σ. This leads to 2 sets of integer moded oscillators $d_n{}^i$ and $\tilde{d}_n{}^i$. The other sector is obtained by choosing the λ's antiperiodic, which leads to 2 sets of half-integer moded oscillators $b_r{}^i$ and $\tilde{b}_r{}^i$. This sector has a tachyon. With the projection above it disappears.

We have discussed the spinning string completely in the light-cone gauge. Also for this string theory, there exists a covariant reparametrization invariant action[33], which when gauge fixed leads to the representation above.

4.3 Superstrings

In the last section we added anticommuting degrees of freedom to cancel zero-point fluctuation. They transformed as the vector representation of SO(d-2). For d=3,4,6 and 10 we could also choose the lowest spinor representation, since it has the same dimension as the vector one. Since SO(d-2) is a compact group, the scalar product of two spinors is just the contracted sum as for vectors (which is the only product used in sect. 4.2). We can, then, take over all work in sect. 4.2. We only make the substitution

$$\lambda^{Ai} \to S^{Aa} \quad , \tag{4.48}$$

where A still is a 2-component spinor index and a is a d-2-component spinor index. This leads to the superstring theory, which was discussed in another formulation in the last section.

The action for the superstring theory is then

$$S = -\frac{1}{2\pi} \int d\tau \int_0^\pi d\sigma [\eta^{\alpha\beta}\partial_\alpha x^i \partial_\beta x^i + i\, \bar{S}^a \rho^\alpha \partial_\alpha S^a] \ . \tag{4.49}$$

We know that this action leads to a sector (for open strings) which starts with massless particles as the lowest lying states. In this sector the solution to the equations of motion is

$$S_a^{\ 1} = \sum_{n=-\infty}^{\infty} S_n^{\ a}\, e^{-in(\tau-\sigma)} \tag{4.50}$$

$$S_a^{\ 2} = \sum_{n=-\infty}^{\infty} S_n^{\ a}\, e^{-in(\tau+\sigma)} \tag{4.51}$$

with the anticommutation rule

$$\{S_a^{\ A}(\sigma,\tau),\ S_b^{\ B}(\sigma',\tau)\} = \pi\ \delta_{ab}\delta^{AB}\ \delta(\sigma-\sigma') \tag{4.52}$$

$$\{S_n^{\ a},\ S_m^{\ b}\} = \delta_{n+m,0}\delta^{ab} \quad . \tag{4.53}$$

The operators $S_{-n}^{\ a}$ with n positive are creation operators. They will take a bosonic state that it acts on to a fermionic one. Hence this sector will contain both bosons and fermions, in fact equally many of each kind at each mass level, building up supermultiplets at each level.

The other sector which follows by using the other set of boundary conditions corresponding to (4.39) and (4.40b), we know has tachyons. Furthermore the fermionic oscillators will be half-integer moded and there will not be an equal number of bosons and fermions at each mass level, thus ruining the possibility to have a supersymmetry. Since the first sector contains all we want, we simply decree, that we only use the boundary conditions (4.39) and (3.40a).

Similarly for closed strings we decree that we only use periodic boundary conditions. This leads to the following solutions to the equations of motion

$$S_a^{\ 1} = \sum_{n=-\infty}^{\infty} S_n^{\ 1a}\, e^{-2in(\tau-\sigma)} \tag{4.54}$$

$$S_a^{\ 2} = \sum_{n=-\infty}^{\infty} S_n^{\ 2a}\, e^{-2in(\tau+\sigma)} \quad . \tag{4.55}$$

The arduous task now is to check if there is a representation of the Poincaré algebra spanned on this string theory. In fact there is[35], and the marvellous fact is that it can also be extended to a super-Poincaré algebra! Let me so write down the representation of the super-Poincaré algebra. For notations, see Appendix. Again I stress that there is some guesswork behind the construction of it. For the case of closed strings the algebra turns out to be an N=2 super-Poincaré algebra. The most general algebra is (we here anticipate the result that the quantum theory only works for d=10)

$$p_- = p_- \tag{4.57a}$$

$$p_i = \int_0^\pi d\sigma\; p_i(\sigma,\tau) \tag{4.57b}$$

$$p_+ = \frac{1}{2\pi p_-} \int_0^\pi d\sigma\; [\pi^2 p^{i\,2} + x'^{i\,2} - i(S^1 S'^1 - S^2 S'^2)] \tag{4.57c}$$

$$q_1^{+a} = \sqrt{\frac{2p^+}{\pi}} \int_0^\pi d\sigma\; S_1^a \tag{4.58a}$$

$$q_2^{+a} = \sqrt{\frac{2p^+}{\pi}} \int_0^\pi d\sigma\; S_2^a \tag{4.58b}$$

$$q_1^{-\dot{a}} = \frac{1}{\pi\sqrt{p^+}} \int_0^\pi d\sigma (\gamma^i S_1)^{\dot{a}} (\pi p^i - x'^i) \tag{4.58c}$$

$$q_2^{-\dot{a}} = \frac{1}{\pi\sqrt{p^+}} \int_0^\pi d\sigma (\gamma^i S_2)^{\dot{a}} (\pi p^i + x'^i) \tag{4.58d}$$

$$j^{ij} = \int_0^\pi d\sigma [x^i p^j - x^j p^i + \frac{1}{4\pi}(S^1 \gamma^{ij} S^1 + S^2 \gamma^{ij} S^2)] \tag{4.59a}$$

$$j^{+i} = \int_0^\pi d\sigma (x^+ p^i - x^i p^+) \tag{4.59b}$$

$$j^{+-} = x^+ p^- - x^- p^+ \tag{4.59c}$$

$$j^{-i} = \frac{1}{2} \int_0^\pi d\sigma [\{x^-(\sigma), p^i\} - \{x^i, p^-(\sigma)\} \tag{4.59d}$$

$$- \frac{i}{4\pi\sqrt{\pi} p^+} (S^1 \gamma^{ij} S^1 (\pi p^j - x'^j) + S^2 \gamma^{ij} S^2 (\pi p^j + x'^j)) + 4 \frac{p^i}{p^+}] , \tag{4.59e}$$

where

$$x'(\sigma) = \frac{\pi}{p^+} p^i x'^i + \frac{i}{2p^+}(S^1 S^1 + S^2 S^2) \quad . \tag{4.60}$$

$p^+ = -p_-$ and we perform all calculations for $p^+ > 0$. In fact this algebra is enough to cover all known string models apart from a d=2 model[36].

(i) Type_IIb_superstrings: This is the full algebra (4.57)-(4.59) with periodic boundary conditions for the coordinates. Note that this is a chiral model, since the creation operators S_{-n}^{1a} and S_{-n}^{2a} create spinors of only one chirality.

(ii) Type IIa superstrings: The anticommuting coordinate S_2 can instead be chosen to transform as the other spinor representation, $S_2^{\dot{a}}$. Nothing is affected in the algebra since S_1 and S_2 are never contracted with each other. S_1 and S_2 cannot be put together as a 2-component spinor, but who cares? This model is not chiral since the spinor states can be combined to Majorana states.

(iii) Type I superstrings: For open strings we must use the boundary conditions corresponding to (4.8) and (4.39) and (4.40a). Then $q_1^+=q_2^+$ and $q_1^-=q_2^-$ and the supersymmetry is reduced to an N=1 one. One can also perform this truncation for closed strings.

(iv) The bosonic strings: Put $S^1=S^2=0$. No supersymmetry, of course.

(v) The heterotic superstring[37]: Consider first the bosonic closed string. The generators can be written as a sum of two pieces, one built from the right-moving part of x and p in terms of α-oscillators and one in terms of the left-moving part in terms of the $\tilde{\alpha}$-oscillators. Both parts separately satisfy the algebra, and one can in principle set one part to zero. Consider so the full algebra (4.57)-(4.59). It is straightforward to see that in the terms where S^A couple to x and p, S^1 couple to the right-moving part and S^2 to the left-moving. (They are right-moving and left-moving resp.) A consistent truncation can now be made by putting, say the left-moving parts to zero. This reduces the algebra to an N=1 supersymmetry. The heterotic superstring is now constructed by putting together one right-moving superstring constructed as above and a 26-dimensional bosonic left-moving string. The 16 internal coordinates span a hypertorus and carry either $E_8 \times E_8$ or Spin $(32)/Z_2$ as internal symmetry group.

(vi) The spinning string: By performing triality transformations back to vectors λ^i such as $S^{1a}S^{1a} \rightarrow \lambda^{1i}\lambda^{1i}$ and $S^1\gamma^{ij}S^1 \rightarrow \lambda^{1i}\lambda^{1j}$ and similarly for S^2 one can easily read off the representation for the spinning string. No supersymmetry survives, of course.

It is now straightforward to construct representations in the various cases. I refer to the literature[9] for a detailed study, and here I just remind that for closed strings the massless sector always contains a graviton field. When we construct field representations it will be an enormous simplification if we can use scalar fields. In a first-quantized formulation this amounts to be able to formulate a scalar vacuum to let the operators act on. For type IIb this is possible within the SO(8) formalism. In the other cases one has to break SO(8) into SU(4) x U(1). I will not go into details with the heterotic string here. There is a vast literature on the subject[37]. Let me just point out that it is tachyon-free although the vacuum fluctuations do not cancel. The argument that they have to cancel, which was our guiding principle, is only necessary for open strings. In the case of open strings we see from (4.30) and (4.31) in the bosonic case that

$$\frac{\alpha'}{2} m^2 = 2N \quad , \tag{4.61}$$

i.e. the mass spectrum for a closed string is twice the one from the right- (or left-) moving modes. This is generally true and it is enough that one of the two sectors (right-moving or left-moving) has vanishing vacuum fluctuations to avoid tachyons.

5. STRING FIELD THEORY

The formulation so far has only dealt with free strings. One can introduce interactions by allowing the strings to split and integrate functionally over world-surfaces with all the possible topologies[26]. This might be the most expedient way to compute multiloops. It is, however, intrinsically perturbative in nature. An alternative is to set up a full-fledged field theory[12]. It is dubious that it will simplify loop calculations, but it might offer us better chances to investigate non-perturbative aspects. I will exemplify here by setting up a field theory for the heterotic string[38]. I will be quite detailed in defining the field. the procedure will be completely algebraic as I alluded to in the introduction.

5.1 Coordinates and Symmetry

The light-cone gauge formulation of the heterotic superstring theory[37] is based on eight transverse coordinates, $X^i(\tau,\sigma)(i = 1,\dots,8)$, which satisfy the two-dimensional wave equation with solutions

$$X^i(\tau,\sigma) = X_R^i(\tau-\sigma) + X_L^i(\tau+\sigma) \quad , \tag{5.1}$$

where $X_R(\eta)$ and $X_L(\eta)$ are arbitrary functions which are periodic in η (with L and R denoting left and right moving coordinates). The conjugate momentum is similarly of the form

$$P^i(\tau,\sigma) = P_R^i(\tau-\sigma) + P_L^i(\tau+\sigma) \quad . \tag{5.2}$$

The fermionic coordinates $S^a(\tau,\sigma)$ are SO(8) spinors ($a = 1,\dots,8$) as well as being chiral (right-moving) spinors on the world sheet. This means that for a given value of a S^a has only one independent component satisfying the chiral Dirac equation

$$\left(\frac{\partial}{\partial\tau} + \frac{\partial}{\partial\sigma}\right) S^a = 0 \quad , \tag{5.3}$$

which implies

$$S^a = S^a(\tau-\sigma) \quad . \tag{5.4}$$

The internal symmetry is described by $X^I(\tau+\sigma)$ (with $I = 1,\ldots,16$) (and we need not indicate the subscript L since these coordinates are always left-moving).

As with any Majorana fermion S^a is a self-conjugate field so that it satisfies the anticommutation relations

$$\{S^a(\tau,\sigma), S^b(\tau,\sigma')\} = \pi\, \delta^{ab}\delta(\sigma-\sigma') \quad . \tag{5.5}$$

Similarly it is an easy exercise to deduce that if the left-moving and right-moving space-time coordinates are decoupled

$$[X_L^i(\tau+\sigma), X_L^j(\tau+\sigma')] = -\frac{i}{2}\,\pi\,\delta^{ij}\,\frac{1}{\partial_\sigma}\,\delta(\sigma-\sigma') \quad , \tag{5.6}$$

$$[X_R^i(\tau-\sigma), X_R^j(\tau-\sigma')] = \frac{i}{2}\,\pi\,\delta^{ij}\,\frac{1}{\partial_\sigma}\,\delta(\sigma-\sigma') \quad . \tag{5.7}$$

Notice that adding (5.6) and (5.7) gives the usual commutation relation

$$[X^i(\tau,\sigma), X^j(\tau,\sigma')] = 0 \quad . \tag{5.8}$$

In similar fashion the internal bosons satisfy

$$[X^I(\tau+\sigma), X^J(\tau+\sigma')] = -\frac{i}{2}\,\pi\,\delta^{IJ}\,\frac{1}{\partial_\sigma}\,\delta(\sigma-\sigma') \quad . \tag{5.9}$$

To prepare for the discussion of the field theory we normalize the σ parameter so that

$$-\pi|\alpha| \leq \sigma \leq \pi|\alpha| \quad , \tag{5.10}$$

where

$$\alpha = 2p^+ \quad , \tag{5.11}$$

and p^+ is defined to be positive for incoming strings and negative for outgoing strings. We have defined the integrated δ-function by

$$\frac{1}{\partial_\sigma}\,\delta(\sigma-\sigma') \equiv [\sigma-\sigma'] + \frac{1}{2} = \frac{\sigma-\sigma'}{2\pi|\alpha|} + \sum_{n\neq 0}^{\infty} \frac{1}{2\pi i n}\, e^{in(\sigma-\sigma')/|\alpha|}, \tag{5.12}$$

($[\sigma-\sigma']$ means the integer part of $\sigma-\sigma'$.)

In order to describe a string field theory we need to represent the super-Poincaré symmetry and the internal symmetry on fields which are functionals of the classical coordinates. We therefore need to define the phase space in terms of a maximal commuting set of coordinates and their conjugate momenta. Since the $X^I(\tau-\sigma)$ do not commute and the $S^a(\tau,\sigma)$ do not anticommute they cannot be used as independent coordinates in the argument of the functional string field. As far as the $S^a(\tau,\sigma)$ are concerned the resolution is to define new Grassmann coordinates, $\theta^A(\tau-\sigma)$, where

$$\theta^A(\tau-\sigma) \sim S^A(\tau-\sigma) + iS^{A+4}(\tau-\sigma) \quad , \tag{5.13}$$

for $A = 1,\ldots,4$. The other combination of the S^a's are denoted $\lambda_A(\tau-\sigma)$ where

$$\lambda_A(\tau-\sigma) \sim S^A(\tau-\sigma) - iS^{A+4}(\tau-\sigma) \quad . \tag{5.14}$$

The symbol ~ denotes an arbitrary normalization which is determined by convention so that

$$\{\lambda_A(\tau,\sigma),\Theta^B(\tau,\sigma')\} = \delta_A^B\delta(\sigma-\sigma') \tag{5.15}$$

while

$$\{\Theta^A(\tau,\sigma),\Theta^B(\tau,\sigma')\} = 0 = \{\lambda_A(\tau,\sigma),\lambda_B(\tau,\sigma')\} \quad . \tag{5.16}$$

The Θ^A's form an SU(4) spinor Grassmann coordinate while the λ_A's form an antispinor which plays the rôle of the momentum conjugate to Θ^A. The heterotic string superfield will be chosen a functional of $\Theta^A(\tau-\sigma)$ but not $\lambda_A(\tau-\sigma)$ (while for the type II theories there is a second, left-moving spinor: $\Theta^A(\tau+\sigma)$ for type IIb and $\Theta_A(\tau+\sigma)$ for type IIa). In this formulation the SO(8) symmetry of the transverse space is no longer manifest since it is represented by coordinates belonging to a SU(4) x U(1) subgroup. The U(1) charge of Θ^A is $\frac{1}{2}$ while λ_A has a U(1) charge of $-\frac{1}{2}$. The eight X^i's form an SO(8) vector which decomposes under SU(4) x U(1) as

$$X^I \rightarrow X^l,\ X^{\hat{\imath}},\ X^r \quad , \tag{5.17}$$

where $\hat{\imath} = 1,\ldots,6$ is a SO(6) ~ SU(4) vector index and r and l denote the combinations

$$X^r = (X^7 - iX^8)/\sqrt{2} \tag{5.18}$$

and

$$X^l = (X^7 + iX^8)/\sqrt{2} \quad . \tag{5.19}$$

The U(1) charges are +1 for X^l, 0 for $X^{\hat{\imath}}$ and -1 for X^r.

We shall treat the internal coordinates, $X^I(\tau+\sigma)$ in a manner analogous to the fermion coordinates by defining the linear combinations

$$Y^I(\tau+\sigma) = X^I(\tau+\sigma) + iX^{I+8}(\tau+\sigma) \tag{5.20}$$

and

$$\bar{Y}^I(\tau+\sigma) = X^I(\tau+\sigma) - iX^{I+8}(\tau+\sigma) \quad , \tag{5.21}$$

where I = 1,...,8 in these expressions. It is easy to see that $\bar{Y}$ is simply related to the momentum, Π^I, conjugate to the coordinates Y^I by

$$\Pi^I(\tau+\sigma) = \dot{\bar{Y}}^I(\tau+\sigma) \quad , \tag{5.22}$$

so that

$$[\Pi^I(\tau+\sigma),Y^J(\tau+\sigma')] = -i\delta^{IJ}\delta(\sigma-\sigma') \tag{5.23}$$

and

$$[\Pi^I(\tau+\sigma),\Pi^J(\tau+\sigma')] = 0 = [Y^I(\tau+\sigma),Y^J(\tau+\sigma')] \tag{5.24}$$

The supercharge algebra of the heterotic string is the same as for the right-moving modes of the type II theories. The sixteen components of a single ten-dimensional chiral supercharge decompose into four SU(4) spinors and anti-spinors ($\varepsilon(\alpha) = 1$ for $\alpha>0$ and -1 for $\alpha<0$)

$$q^{+A} = \frac{\varepsilon(\alpha)}{2\sqrt{\pi}} \int_{-\pi|\alpha|}^{\pi|\alpha|} d\sigma \; \Theta^A(\tau,\sigma) \quad , \tag{5.25}$$

$$q^{+}_{A} = \frac{1}{2\sqrt{\pi}} \int_{-\pi|\alpha|}^{\pi|\alpha|} d\sigma \; \lambda_A(\tau,\sigma) \quad , \tag{5.26}$$

$$q^{-A} = \frac{1}{2} \int_{-\pi|\alpha|}^{\pi|\alpha|} d\sigma \; [\sqrt{2}\rho^{\hat{i}AB}\Theta_B(\tau,\sigma)\acute{X}^{i}_{R}(\tau,\sigma) + 2\pi\varepsilon(\alpha)\lambda^A(\tau,\sigma)\; \acute{X}^{1}_{R}(\tau,\sigma)] \tag{5.27}$$

$$q^{-}_{A} = \frac{1}{2} \int_{-\pi|\alpha|}^{\pi|\alpha|} d\sigma \; [2\; \acute{X}^{r}_{R}(\tau,\sigma)\Theta_A(\tau,\sigma) + \sqrt{2}\pi\varepsilon(\alpha)\rho^{\hat{i}}_{AB}\lambda^B(\tau,\sigma)\; \acute{X}^{i}_{R}(\tau,\sigma)] \tag{5.28}$$

The matrices $\rho^{\hat{i}AB}$ and $\rho^{\hat{i}}_{AB}$ are SO(6) gamma matrices defined in Appendix. The q^- operators anticommute to give the hamiltonian

$$\{q^{-A},q^{-}_{B}\} = \delta^A_B\, h = 2\delta^A_B \int_{-\pi|\alpha|}^{\pi|\alpha|} d\sigma \; \varepsilon(\alpha)\{\pi(P^i - \tfrac{1}{\pi}\acute{X}^i)^2 + 2i\acute{\Theta}^A\lambda_A\} \tag{5.29}$$

Although the left-moving coordinates are inert under supersymmetry transformations they are related to the right-moving. coordinates by the constraint that the field must be invariant under rigid shifts in σ generated by the operator

$$k = \int_{-\pi\alpha}^{\pi\alpha} d\sigma\; \acute{X}^-(\tau,\sigma) = 4\pi \int_{-\pi\alpha}^{\pi\alpha} d\sigma\; \varepsilon(\alpha)(\acute{X}^i P^i - i\acute{\Theta}^A\lambda_A + \acute{Y}^I\Pi^I) \quad . \tag{5.30}$$

This is a remnant of the reparametrization invariance of the covariant theory. By requiring the constraint

$$k = 0 \tag{5.31}$$

on physical states we can rewrite h in a form that is more symmetric between left- and right-movers

$$h = 2\int_{-\pi\alpha}^{\pi\alpha} d\sigma\; \varepsilon(\alpha)\{(\pi P^i)^2 + \tfrac{1}{\pi}(\acute{X}^i)^2 + 2i\acute{\Theta}^A\lambda_A + 2\acute{Y}^I\Pi^I\} \quad . \tag{5.32}$$

The operators h and k should be normal ordered but we shall leave a discussion of that point until the next section. The fact that the left-moving coordinates are inert under supersymmetry is a novel feature of superstrings - in point particle theories all coordinates must transform under supersymmetry. It is clear, that the hamiltonian for variables moving in <u>one</u> direction must have this form which identifies h (τ-translations) with a σ-translation. The other generators of super-Poincaré transformations can so be written in analogous fashion. One important fact is that j^{ij} has no spin term.

Lastly, the internal symmetry generators, which can represent $E_8 \times E_8$ or Spin(32)/Z_2[37], are obtained by inserting the definition of Y^I and Π^I ((5.20) and (5.22)) into the expressions for the generators of the symmetry groups in terms of emission vertices

$$E(K) = \int_{-\pi|\alpha|}^{\pi|\alpha|} d\sigma :e^{i\bar{R}^I Y^I(\tau+\sigma)+iK^I \bar{Y}^I(\tau+\sigma)}: C(K,\bar{R}) \quad , \quad (5.33)$$

where $K^I = k^I + ik^{I+8}$ ($I = 1,\ldots,8$) and k^I is a real 16-dimensional lattice vector with the condition

$$\sum_{I=1}^{16} (k^I)^2 = 2 \qquad (2.34)$$

required for the consistency of the theory at the level of the free theory. The E's, together with

$$R^I = \int_{\pi|\alpha|}^{\pi|\alpha|} d\sigma \quad \acute{Y}^I(\tau+\sigma) \qquad (5.35)$$

and

$$\bar{R}^I = \int_{-\pi|\alpha|}^{\pi|\alpha|} d\sigma \ \acute{\bar{Y}}^I(\tau+\sigma) \quad , \qquad (5.36)$$

generate the internal symmetry group. The term $C(K,\bar{R})$ is an operator 1-cocycle chosen such that the generators generate a Lie algebra. At the level of the one-loop diagrams the restriction that the theory is unitary is known to force the lattice to be self-dual as well as satisfy (5.35) which results in one or other of the lattices Γ_{16} or $\Gamma_8 \times \Gamma_8$.

5.2 Wave Functionals and Measures

The string wave functional, Ψ, satisfies the Schrödinger equation

$$h\Psi = i\frac{\partial\Psi}{\partial x^+} \quad , \qquad (5.37)$$

where h is the operator defined in the last section. In passing to the string field theory the wave functional, Ψ, will be identified with the string field, Φ, and (5.37) will then be the free field equation of motion. The wave function can be expanded in a complete set of eigenfunctions of (5.37). Since h is a sum over hamiltonians for each mode of the coordinates X^i, θ^A and Y^I each eigenfunction is an infinite product of eigenfunctions of the individual hamiltonians for each mode with a given occupation number.

The wave functions for the X^i coordinates are simply the eigenfunctions of harmonic oscillators and are well understood. Here I will discuss the wave functions associated with the modes of Y^I and of θ^A. It will be instructive to see the close parallel between the treatment of the left-moving bosonic variables and the right-moving Grassmann variables.

5.2.1 Bosonic Left-Moving Modes

The Fourier expansions of the coordinates and momenta are given by (setting $\tau = 0$)

$$Y^I(\sigma) = \frac{\Lambda^I \sigma}{2|\alpha|} + \frac{1}{\sqrt{2\pi|\alpha|}} \sum_{n=-\infty}^{\infty} y_n^I e^{in\sigma/\alpha} \quad , \tag{5.38}$$

and

$$\Pi^I(\sigma) = -\frac{i}{\sqrt{2\pi|\alpha|}} \sum_{n=-\infty}^{\infty} \frac{\partial}{\partial y_{-n}^I} e^{in\sigma/|\alpha|} \quad . \tag{5.39}$$

Since the momentum zero modes, p^I, lie on a discrete lattice the parameters

$$\Lambda^I = p^I + ip^{I+8} \tag{5.40}$$

are complex combinations of the discrete variables.

From (5.20)-(5.22) we see that

$$\frac{\partial}{\partial y_n^I} = \frac{n}{|\alpha|} y_n^{I+} \quad . \tag{5.41}$$

We can therefore make the correspondence in terms of creation and annihilation operators

$$a_n^I = i\sqrt{\frac{n}{|\alpha|}}\, y_n^I \qquad a_n^{\bar{I}} = -\frac{i}{\sqrt{\frac{n}{|\alpha|}}} \frac{\partial}{\partial y_n^I} \tag{5.42}$$

and

$$a_{-n}^{\bar{I}} = \frac{i}{\sqrt{\frac{n}{|\alpha|}}} \frac{\partial}{\partial y_n^I} \qquad a_{-n}^I = i\sqrt{\frac{n}{|\alpha|}}\, y_{-n}^I \quad . \tag{5.43}$$

The contribution to the $(\text{mass})^2$ operator from the X^I coordinates can be written as

$$M_{int}^2 = 4 \int_{-\pi|\alpha|}^{\pi|\alpha|} d\sigma\, \acute{Y}^I \Pi^I =$$

$$= -i\, 2\sqrt{2}\pi|\alpha|\, \Lambda^I \frac{\partial}{\partial y_0^I} + \sum_{n \neq 0} 4n\, y_n^I \frac{\partial}{\partial y_n^I} \quad . \tag{5.44}$$

Normal ordering this expression gives a contribution to the total $(\text{mass})^2$ of the ground state

$$\frac{1}{4} M_{0,int}^2 = 8 \sum_{n=1}^{\infty} n = -\frac{2}{3} \quad , \tag{5.45}$$

where the last step involves subtracting an infinite constant by the usual string rule $\sum n = -1/12$[27].

The wave functions corresponding to the states created by the modes in (3.6) and (3.7) are easily written in the y_n^I representation. Thus, corresponding to a state

$$|m\rangle_{n,I} = \frac{1}{\sqrt{m!}} (a_n^{+I})^m |0\rangle \tag{5.46}$$

the wave function is

$$\psi_{n,m}(y_n^I) = \frac{1}{\sqrt{m!}} \left(i\sqrt{\frac{n}{|\alpha|}}\, y_n^I\right)^m \quad , \tag{5.47}$$

while for the state

$$|m\rangle_{-n,I} = \frac{1}{\sqrt{m!}} (a_{-n}^{+I})^m |0\rangle \tag{5.48}$$

the wave function is

$$\psi_{-n,m}(y^I_{-n}) = \sqrt{m!}\left(i\sqrt{\frac{n}{|\alpha|}}\, y^I_{-n}\right)^m / y^I_{-n} \quad , \tag{5.49}$$

when $n>0$. Here the vacuum has been assigned a wavefunction $\psi_{-n,0}(y^I_{-n}) = (y^I_{-n})^{-1}$ since this is the highest power not annihilated by any number of derivatives (creation operators). It is easily verified that the eigenfunctionals of the hamiltonian really are products of these wavefunctions, and that they give the correct vacuum energy. The zero mode part of the wave functional is formally written as

$$\Psi_{0,K}\,(\Lambda, y_0) = e^{i\,\bar{K}^I y_0{}^I}\,\delta_{\Lambda,K} \quad , \tag{5.50}$$

where $\delta_{\Lambda,K}$ is an eight-dimensional lattice site delta function. There is no principal problem with this treatment, if just the exponential is supposed to fourier transform into $\delta_{\bar{\Lambda},\bar{K}}$ which combines with $\delta_{\Lambda,K}$ to a delta function in the 16-dimensional lattice. We will return to discuss the zero-modes more thoroughly elsewhere.

Scalar products are defined by introducing conjugate wave functions,

$$\tilde{\psi}_{n,m}(y_n^I) = \sqrt{m!}\left(i\sqrt{\frac{n}{|\alpha|}}\, y_n^I\right)^{-m} / y_n^I = -\psi_{-n,m} \tag{5.51}$$

and

$$\tilde{\psi}_{-n,m}(y^I_{-n,m}) = \frac{1}{\sqrt{m!}}\left(-in\sqrt{\frac{n}{|\alpha|}}\, y^I_{-n}\right)^m = \psi_{n,m}(-y^I_{-n}) \quad , \tag{5.52}$$

so that

$$\int \frac{dy_n^I}{2i\pi}\, \psi_{n,m}(y_n^I)\tilde{\psi}_{n,k}(y_n^I) = \delta_{mk} \quad . \tag{5.53}$$

The integration contour is defined to encircle the origin. Note that the 'δ function' is defined by

$$\delta(y_n^I, y_n'^I) = \sum_{m\geq 0} \tilde{\psi}_{n,m}(y_n^I)\psi_{n,m}(y_n'^I) = \frac{1}{y_n^I - y_n'^I} \quad \text{if } n>0$$

$$= \frac{1}{y_n'^I - y_n^I} \quad \text{if } n<0 \quad . \tag{5.54}$$

This has the property that

$$\int \frac{dy_n^I}{2i\pi}\, f(y_n^I)\delta(y_n^I, y_n'^I) = f(y_n'^I) \quad . \tag{5.55}$$

In this expression the integration contour is taken so that $|y_n^I| > |y_n'^I|$ for $n>0$ and $|y_n'^I| > |y_n^I|$ for $n<0$. The part of the functional measure involving the y^I coordinates is given by

$$\int D^8Y^I(\sigma) = \sum_{\Lambda} \prod_{I=1}^{8} \int dy_0 \prod_{n\neq 0} \frac{dy_n^I}{2i\pi} \quad . \tag{5.56}$$

5.2.2 Fermionic Right-Moving Modes

The fermionic right-moving modes can be treated in a way completely analogous to the bosonic ones. We start by Fourier expanding the coordinate Θ^A and the momentum λ_A

$$\Theta^A(\sigma) = \frac{1}{\sqrt{2\pi|\alpha|}} \sum_{n=-\infty}^{\infty} \Theta_n{}^A \, e^{in\sigma/|\alpha|} \tag{5.57}$$

$$\lambda_A(\sigma) = \frac{1}{\sqrt{2\pi|\alpha|}} \sum_{n=-\infty}^{\infty} \frac{\partial}{\partial\Theta_{-n}{}^A} \, e^{in\sigma/|\alpha|} \quad . \tag{5.58}$$

Comparing with (5.13) and (5.14) we find

$$\frac{\partial}{\partial\Theta_n^A} = (\Theta_n^A)^+ \quad . \tag{5.59}$$

For fermionic modes it is advantageous to use creation and annihilation operators Q_n^A and Q_{nA} satisfying

$$Q_{-n}^{\ A} = Q_n^{+A} \quad , \qquad Q_{-nA} = Q_{nA}^+ \tag{5.60}$$

satisfying

$$\{Q_m^A, Q_{nB}\} = \alpha\, \delta_{m+n}\delta_B^A \tag{5.61}$$

$$\{Q_m^A, Q_n^B\} = \{Q_{mA}, Q_{n\beta}\} = 0 \tag{5.62}$$

We can therefore make the identification

$$Q_n^{+A} = \sqrt{\alpha}\, \Theta_{-n}^A \qquad Q_{nA} = \sqrt{\alpha}\, \frac{\partial}{\partial\Theta_n{}^A} \tag{5.63}$$

$$Q_{nA}^+ = \sqrt{\alpha}\, \frac{\partial}{\partial\Theta_n} \qquad Q_n{}^A = \sqrt{\alpha}\, \Theta_n{}^A \tag{5.64}$$

Corresponding to the state

$$|m>_n{}^A = \begin{cases} |0>_n{}^A & m = 0 \\ \frac{1}{\sqrt{\alpha}}\, Q_m^{+A}|0>_n{}^A & m = 1 \end{cases} \quad . \tag{5.65}$$

the wave functions is then

$$\chi_{n,m}(\Theta_n^A) = \begin{cases} 1 & m = 0 \\ \Theta_n{}^A & m = 1 \end{cases} \tag{5.66}$$

and similarly

$$\chi_{-n,m}(\Theta_{-n}{}^A) = \begin{cases} \Theta_{-n}{}^A & m = 0 \\ 1 & m = 1 \end{cases} \quad . \tag{5.67}$$

The scalar product is defined by introducing the conjugate wave functions

$$\tilde{\chi}_{n,m}(\Theta_n{}^A) = \chi_{-n,m}(\Theta_n{}^A) \quad . \tag{5.68}$$

Recalling that the Berezin rules for integration over Grassmann variables we see that

$$\int d\Theta_n{}^A\ \tilde{\chi}_{n,m}\ \chi_{n,\ell} = \delta_{m\ell} \quad . \tag{5.69}$$

Finally we notice that a δ-function in Grassmann variables is

$$\delta(\Theta_n^A - \Theta_n^{\prime A}) = \Theta_n{}^A - \Theta_n^{\prime A} \quad . \tag{5.70}$$

We want to stress the similarities between the bosonic and fermionic coordinates in wave functions, conjugations, integrations and δ-functions.

5.3 FREE FIELD THEORY

The string field is a scalar functional of the coordinates, i.e.

$$\Phi \equiv \Phi[X^i(\sigma), \Theta^A(\sigma), Y^I(\sigma), x^+, p^+] \equiv \Phi[\Sigma, x^+, p^+] \quad , \tag{5.71}$$

where Σ denotes the collection of coordinates, $\Sigma = (X^i(\sigma), \Theta^A(\sigma), Y^I(\sigma))$. In the expansion of the string field in a complete set of eigenfunctions of the non-zero modes there is one term for each of the states of occupation of every oscillator level. This results in an infinite series of terms with coefficients which are functions of the zero modes of the coordinates, x^i, Θ_0^A, y_0^I and Λ. These coefficient functions are the component fields. In analogy with previous light-cone theories we introduce a 2nd quantized representation by demanding commutation relations.

$$[\Phi[\Sigma_1, p_1^+, x^+], \tilde{\Phi}[\Sigma_2, p_2^+, x^+]] = \frac{1}{2p_1^+}\ \delta(p_1^+ + p_2^+)$$

$$\int d\sigma_0 \Delta^8\ [X_2^i(\sigma) - X_2^i(\sigma+\sigma_0)]$$

$$\times\ \Delta^4[\Theta_1^A(\sigma) - \Theta_2^A(\sigma+\sigma_0)]\Delta^8[Y^I(\sigma) - Y^I(\sigma+\sigma_0)]$$

$$\equiv \frac{1}{2p_1^+}\ \delta(p_1^+ + p_2^+)\Delta^{20}[\Sigma_1 - \Sigma_2] \quad , \tag{5.72}$$

where $\tilde{\Phi}$ denotes a field defined in terms of the conjugate wave functionals.

Expanding the fields in (5.72) in terms of their components reproduces the usual light-cone commutation relations for the point fields.

In the free theory the generators of the (super-Poincaré) $\times$ (internal symmetry) algebra is represented by linear transformations of the fields. If we denote the representation of a generator in terms of the fields by G where g is the same generator represented on the coordinates, then the free generators are of the form

$$G_2 = 2 \int dp^+ \int D^{20}\Sigma\ \tilde{\Phi}[\Sigma, p^+, x^+] g\ \Phi[\Sigma, -p^+, x^+] \quad , \tag{5.73}$$

where the subscript 2 indicates an expression that is quadratic in fields. It is easy to check, using (5.72) that

$$[G_2, \Phi] = g\Phi \quad . \tag{5.74}$$

For example, the free field hamiltonian, H_2, is of this form with g replaced by h. This results in the expected sum of components free-field hamiltonians. In obtaining this it is necessary to do the y_n^I integrals using the rules given earlier. The Grassmann integrals over the modes of θ^A are likewise simple to evaluate using the orthonormality of the component wave functions contained in a string field and its conjugate.

5.4 INTERACTING FIELD THEORY FOR STRINGS

In the point-particle case we saw that interactions could be introduced by finding higher-point contributions to the dynamical generators in the super-Poincaré algebra. Here we will attempt the same procedure in the string case.

To obtain the precise form of the interaction terms we will work in two steps. Firstly, we will represent the generators in a functional form. A general form for the dynamical generators will be assumed and then the closure of the algebra has to be imposed. Then to give the precise meaning of this form, we must go over to a mode basis to check that each coupling is properly defined.

We write a general 3-string contribution to a generator as

$$G_3 = \kappa \int D\Sigma_1 D\Sigma_2 dp_1{}^+ dp_2{}^+ \mu(p_1{}^+,p_2{}^+)\tilde{\Phi}[x^+,-p_1{}^+-p_2{}^+,\Sigma_1+\Sigma_2]$$

$$\times\, g(\sigma_1,\sigma_2)\{\Phi[x^+,p_1{}^+,\Sigma_1]\Phi[x^+,p_2{}^+,\Sigma_2]\} + \text{c.c.}. \qquad (5.75)$$

Here $\Sigma_1+\Sigma_2$ denotes the string configuration obtained as the union of the two curves described by Σ_1 and Σ_2. $g(\sigma_1,\sigma_2)$ is an operator that has to be determined as well as the factor $\mu(p_1{}^+,p_2{}^+)$. To make the interaction local, this operator acts at points σ_1 and σ_2 infinitesimally close to the point common to Σ_1 and Σ_2, the interaction point. When we transform to the mode basis we will find that convergence factors have to be inserted to damp singular behaviour near this point, but they are of no importance for the closure of the algebra.

When trying to close the algebra we will deal with commutators between two- and three-string operators. These calculations are tedious, and abundant use of partial functional integrations have to be used. It turns out to be enough to consider the supersymmetry algebra to find a unique answer for the hamiltonian (up to p^+- factors). Again the heterotic string is the prototype. Since $Q^-{}_2$ does not involve any y_I^I-dependence the factor g in (5.75) will not depend on y^I or $\frac{\delta}{\delta y^I}$ and the computation only takes into account the right-moving superstring part. Hence this computation can also be used for the other types of superstrings. Furthermore, we see immediately that the 3-string hamiltonian for a bosonic string has g = 1. The result after a lengthy calculation is

$$g_n = \acute{\underline{x}}^R - \sqrt{2}\, \rho^{\hat{\imath}}_{AB}\, \underline{\lambda}^A \underline{\lambda}^B \acute{\underline{x}}^{\hat{\imath}} + \frac{2}{3}\, \varepsilon_{ABCD} \underline{\lambda}^A \underline{\lambda}^B \underline{\lambda}^C \underline{\lambda}^D \acute{\underline{x}}^L \quad , \qquad (5.76)$$

where

$$\underline{x} = x_1 - x_2 \quad \text{and} \quad \underline{\lambda} = \lambda_1 - \lambda_2 \quad . \tag{5.77}$$

To complete the construction of the representation the remaining dynamical generators should be constructed. This will determine the missing factor μ in (5.75). This is an arduous task and Lorentz invariance has been checked and μ constructed only by other methods (such as checking a four-point amplitude).

A remarkable fact is that although Q^- has a 3-string term, the commutator $\{Q^-,Q^-\}$ does not produce a 4-string contribution to the hamiltonian[39]. The full hamiltonian consists of only the kinetic and the 3-string term! This is very different from point-particle gravity.

Finally let me point out again that no local counter-terms other than the action itself is possible. This is a strong indication that no divergencies occur in the higher loops.

5.5 THE VERTEX IN THE OSCILLATOR BASIS

The vertex defined so far is somewhat formal (to say the least) and to make its definition precise we will treat it in the oscillator basis, where we can check that each three-point coupling among states is well-behaved. In principle we should treat the two different kinds of bosonic coordinates separately, but it is enough to describe the internal ones, since the space-time coordinates can be decomposed into a pair of such coordinates.

5.5.1 The Internal Bosonic Coordinates

An eigenstate of definite y_n^I is given by

$$\begin{aligned} |y_n^I\rangle &= \sum_{m=0}^{\infty} \langle m|y_n^I\rangle |m\rangle \\ &= \sum_{m=0}^{\infty} \psi_{n,m}(y_n^I)\frac{1}{\sqrt{m!}}(a_n^{I+})^m|0\rangle \end{aligned} \tag{5.78}$$

so that $\langle m|y_n^I\rangle = \psi_{n,m}(y_n^I)$. Correspondingly, the bra vector $\langle y_n^I|$ is defined by

$$\langle y_n^I| = \langle 0|\sum_{m=0}^{\infty} \frac{1}{\sqrt{m!}}(a_n^I)^m \psi_{n,m}(y_n^I) \quad . \tag{5.79}$$

Performing the summations in these equations explicitly gives

$$|y_n^I\rangle = \frac{1}{y^I{}_n + iK_n^{-1}a_n^{I+}}|0\rangle \quad , \qquad \langle y_n^I| = \langle 0|e^{ia_n^I K_n y_n^I} \quad ,$$

$$|y_{-n}^I\rangle = e^{-ia_{-n}^{I+}K_n y_{-n}^I}|0\rangle \quad , \qquad \langle y_{-n}^I| = \langle 0|\frac{1}{y_{-n}^I - iK_n^{-1}a_{-n}^I} \quad , \tag{5.80}$$

where $K_n = \sqrt{\frac{n}{T\alpha T}}$. These states respect the identifications

(5.42-5.43)

$$a^I_{-n}|y^I_{-n}> = -iK_n y^I_{-n}|y^I_{-n}> \qquad (5.81)$$

etc.

The single mode states $|y^I_n>$ are multiplied together to a string state $|\Sigma>$ (here Σ denotes only the internal 8-dimensional configuration). Since there is one Fock space for each point in the lattice, the zero-mode wave functions are also included in $|\Sigma>$. We then form a three-string state

$$|V>>> = \mu_{int} \int D\Sigma_1 D\Sigma_2 |\Sigma_1>|\Sigma_2>|\widetilde{\Sigma_1+\Sigma_2}> \quad . \qquad (5.82)$$

$|\tilde{}>$ means again conjugation of the occuring eigenfunctions. The coupling of three mass eigenstates is obtained by taking the scalar product of these with $|V>>>$. As we pointed out in sect. 5.4, the functional integration $\int D\Sigma = \int D\Sigma_1 D\Sigma_2$ has to be carefully performed to pick up contributions from all poles.

This vertex contains the same information as a 3-string interaction with no prefactor (5.76). The prefactor must be treated separately afterwards.

To evaluate (5.82) we introduce infinite products over states (5.80) and identify the configuration Σ_3 with $\Sigma_1+\Sigma_2$ for all values of σ. This calculation is quite lengthy and I must refer to the literature for details[38]. The final result is

$$|V>>> = \exp\{ \frac{1}{2} \sum_{r,s=1}^{3} \sum_{n,m=1}^{\infty} a^{I+}_{rn} \sqrt{n}\, N^{rs}_{nm} \sqrt{m}\, a^{I+}_{sm}$$

$$+ \Delta^I \sum_{r=1}^{3} \sum_{n=1}^{\infty} N^r_n \sqrt{n}\, a^{I+}_{rn} + \tau_0 \left(\sum_{r=1}^{3} \frac{1}{\alpha_1} + \frac{\Lambda^{I2}_r}{2\alpha_r} \right)\}|0>|0>|0> \quad , \qquad (5.83)$$

where

$$N^{rs}_{mn} = \frac{mn\alpha}{n\alpha_r+m\alpha_s} N^r_m N^s_n \qquad (5.84)$$

$$N^r_m = \frac{1}{\alpha_r} \frac{(-1)^{m+1}}{m!} \frac{\Gamma(m(1+\frac{\alpha_{r+1}}{\alpha_r}))}{\Gamma(1-m\frac{\alpha_{r+1}}{\alpha_r})} \qquad (5.85)$$

$$\alpha_r = 2p^+_r \qquad (5.86)$$

$$\alpha = \alpha_1\alpha_2\alpha_3 \qquad (5.87)$$

$$\tau_0 = \sum_{r=1}^{3} \alpha_r \ln\alpha_r \qquad (5.88)$$

$$\Delta^I = \alpha_1\Lambda_2{}^I - \alpha_2\Lambda_1{}^I \quad . \qquad (5.89)$$

Λ^I is the zero-mode in x^I, which in a space-time case is p^I.

5.5.2 The Grassmann Coordinates

The discussion of the Grassmann integrations parallels that of the integrations over the y^A_n modes quite closely. The eigenstates of the modes Θ^A_n and Θ^A_{-n} are given by

$$|\Theta^A_n\rangle = (1 + \frac{1}{\sqrt{\alpha}}\,\Theta^A_n Q_{-nA})|0\rangle = \exp(\frac{1}{\sqrt{\alpha}}\,\Theta^A_n Q_{-nA})|0\rangle \qquad (5.90)$$

and

$$|\Theta^A_{-n}\rangle = (\Theta^A_{-n} + \frac{1}{\sqrt{\alpha}}\,Q^A_{-n})|0\rangle \quad . \qquad (5.91)$$

The conjugate eigenstates are given by

$$\langle\Theta^A_n| = \langle 0|(\Theta^A_n - \frac{1}{\sqrt{\alpha}}\,Q^A_n) \qquad (5.92)$$

and

$$\langle\Theta_{-n}^{A}| = \langle 0|\exp\,(-\frac{1}{\sqrt{\alpha}}\,\Theta_{-n}^{A}\,Q_{nA}) \quad . \qquad (5.93)$$

It is easy to check that

$$\Theta^A_n|\Theta^A_n\rangle = \Theta^A_n|\Theta^A_n\rangle \quad , \qquad (5.94)$$

where $\Theta^A_n = Q^A_{-n}/\sqrt{\alpha}$ is the operator for the mode. A similar relation holds for the modes Θ^A_{-n}.

The normalization has been chosen in (5.90)-(5.93) so that

$$\langle\Theta^A_n|\Theta'^A_n\rangle = \pm\delta(\Theta^A_n - \Theta'^A_n) \quad , \qquad (5.95)$$

with the plus-sign for positive and minus-sign for negative n (and use has been made of the relation $\delta(\eta) = \eta$ for any Grassmann variable η):

The mode expansion for the coordinate $\Theta^A(\sigma)$ resembles that of the Bose coordinate $Y^I(\sigma)$. There is also great similarity between the expressions for the eigenstates of Θ^A_n in (5.90) and 5.91) and the eigenstates of y^I_n in (5.80) and (5.82). Furthermore the rules for Berezin integration are very similar to the rules for the contour integration over the y^I_n's. Taken together these facts imply that there is a close parallel between the calculation of the mode expansion of the functional overlap that ensures the continuity of $Y(\sigma)$ at the interaction time and the one that ensures the continuity of $\Theta^A(\sigma)$. There are only slight differences in the placement of factors of α_r and in certain normalizations. Since this is the case we shall not repeat the algebra of the last subsection but merely quote the result for the factors in $|V\rangle\rangle\rangle$ which involve the fermion modes which we shall denote by $|V\rangle\rangle\rangle_Q$. This is given by

$$|V\rangle\rangle\rangle_Q = \mu_{int}\,(\det\frac{\Gamma}{2})^8 \exp\{\frac{1}{2}\sum_{r,s=1}^{3}\sum_{n,m=1}^{\infty} Q^{(r)A}_{-m}U^{rs}_{mn}Q^{(s)}_{-nA}$$

$$+ \sum_{r=1}^{3}\sum_{m=1}^{\infty} V^r_m Q^{(r)A}_{-m}\Theta_A\} \quad , \qquad (5.96)$$

where

$$U^{rs}_{mn} = \frac{m}{\alpha_r}\,\bar{N}^{rs}_{mn} \qquad (5.97)$$

$$V^r_m = -\alpha_1\alpha_2\alpha_3\,\sqrt{2}\,\frac{m}{\alpha_r}\,\bar{N}^r_m \qquad (5.98)$$

$$\Theta_A = \frac{1}{\alpha_3}\,(\Theta_{01A} - \Theta_{02A}) \quad . \qquad (5.99)$$

Finally we have to consider the prefactor in H_3, (5.75) and (5.76). It is easy to convince oneself that it corresponds to a mode-basis vertex operator

$$|H\rangle\rangle\rangle = g_h(\sigma_1,\sigma_2,\sigma_3)|V\rangle\rangle\rangle \quad . \tag{5.100}$$

Letting the operator g_h commute through the exponential in $|V\rangle\rangle\rangle$ one finds that g_h (5.76) is well-defined (and hence H_3) if (5.77) is changed into

$$\underline{x} = \lim_{\varepsilon\to 0} \sqrt{\varepsilon}\,(x_1-x_2);\; \underline{\lambda} = \lim_{\varepsilon\to 0} \sqrt{\varepsilon}\,(\lambda_1-\lambda_2) \quad , \tag{5.101}$$

where subscripts 1 and 2 indicate action at points σ_1 and σ_2 on strings 1 and 2 resp. at a distance ε from the interaction point.

6. DISCUSSION

In the preceding sections we have seen how to construct light-cone field theories for point-particles and strings. The construction is relatively straightforward but involves always a bit of guesswork. The method is quite useful, especially for strings, in order to construct a perturbation expansion. It is perhaps also the simplest way to construct new models and to check the limits of various assumptions. The real drawback is that the method is completely algebraic. It does not give us any intuition about underlying geometry. This is usually all right for theories not involving gravity. However, we seek a unified model for all interactions and superstring theory is a candidate one. Hence, we would like to have a geometric principle to build the theory from. This should really not be coordinate dependent (as our formalism is). In an ultimate theory the light-cone formalism is probably a certain gauge choice, with the advantages and disadvantages that follows from a specific choice. I do hope I have convinced the reader that there are quite a few advantages.

APPENDIX

Some Notations and Conventions

The algebra of SO(8) has three inequivalent real eight-dimensional representations, one vector and two spinors. We use 8-valued indices i,j, ... corresponding to the vector, a,b, ... corresponding to one spinor and ȧ, ḃ, ..., corresponding to the other spinor. Dirac matrices $\gamma^i{}_{a\dot{a}}$ may be regarded as Clebsch-Gordan coefficients for combining the three eights into a singlet. A second set of matrices $\tilde{\gamma}^i{}_{\dot{a}a}$ is also introduced. We choose

$$\tilde{\gamma} = \gamma^T \tag{A.1}$$

$$\{\gamma^i, \tilde{\gamma}^j\} = 2\,\delta^{ij} \quad . \tag{A.2}$$

The 16 × 16 matrices

$$\begin{vmatrix} 0 & \gamma^{i}{}_{a\dot{a}} \\ \tilde{\gamma}^{i}{}_{\dot{b}b} & 0 \end{vmatrix}$$

form a Clifford algebra. We also define

$$\gamma^{ij}_{ab} = \tfrac{1}{2}[\gamma^{i}_{a\dot{a}}\ \tilde{\gamma}^{j}_{\dot{a}b} - \gamma^{j}_{a\dot{a}}\ \tilde{\gamma}^{i}_{\dot{a}b}] \quad . \tag{A.3}$$

These matrices are seen to be antisymmetric in a and b using (A.1).

We can also define

$$\gamma^{ij}_{\dot{a}\dot{b}} = \tfrac{1}{2}[\tilde{\gamma}^{i}_{\dot{a}a}\ \gamma^{j}_{a\dot{b}} - \tilde{\gamma}^{j}_{\dot{a}a}\ \gamma^{i}_{a\dot{b}}] \quad , \tag{A.4}$$

which in a similar fashion is antisymmetric in ȧ and ḃ.

To span the whole 8 × 8 dimensional matrix spaces we also define

$$\gamma^{ijkl}_{ab} \equiv (\gamma^{[i}\ \tilde{\gamma}^{j}\ \gamma^{k}\ \tilde{\gamma}^{l]})_{ab} \tag{A.5}$$

$$\gamma^{ijkl}_{\dot{a}\dot{b}} \equiv (\tilde{\gamma}^{[i}\ \gamma^{j}\ \tilde{\gamma}^{k}\ \gamma^{l]})_{\dot{a}\dot{b}} \tag{A.6}$$

These matrices are symmetric.

The general Fierz formula is

$$M_{ab} = \tfrac{1}{8}\,\delta_{ab}\ \mathrm{tr}\ M - \tfrac{1}{16}\,\gamma^{ij}_{ab}\ \mathrm{tr}(\gamma^{ij}M)$$

$$+ \tfrac{1}{384}\,\gamma^{ijkl}_{ab}\ \mathrm{tr}(\gamma^{ijkl}M) \quad . \tag{A.7}$$

In the case of SU(4), the six-vector can be obtained as the antisymmetric tensor product of two 4's or two $\bar{4}$'s. The corresponding Clebsch-Gordan coefficients (or Dirac matrices) are denoted $\rho^{I}{}_{AB}$ and ρ^{IAB}. They are normalized as usual so that

$$\rho^{IAB}\ \rho^{J}{}_{BC} + \rho^{JAB}\ \rho^{I}{}_{BC} = 2\ \delta^{A}{}_{C}\ \delta^{IJ} \quad . \tag{A.8}$$

We also define

$$\rho^{IJ}{}_{A}{}^{B} = \tfrac{1}{2}(\rho^{I}{}_{AC}\ \rho^{JCB} - \rho^{J}{}_{AC}\ \rho^{ICB}) \quad . \tag{A.9}$$

REFERENCES

1. P.A.M. Dirac, Rev. Mod. Phys. 26 (1949) 392.
2. R.K. Sachs, J. Math. Phys. 3 (1962) 908.
   H. Müller zum Hagen and H.J. Seifert, Gen. Rel. Grav. 8 (1977).
3. R.A. Neville and F. Rohrlich, Nuovo Cim. 1A (1971) 625.
4. S. Fubini and G. Furlan, Physics 1 (1965) 229.
5. S. Weinberg, Phys. Rev. 150 (1966) 1313.
6. S.-J. Chang and S.-K. Ma, Phys. Rev. 180 (1969) 1506,
   J.B. Kogut and D.E. Soper, Phys. Rev. D1 (1970) 2901.
7. S.J. Brodsky, R. Roskies and R. Suaya, Phys. Rev. D8 (1973) 4574,
   J.H. Ten Eyck and F. Rohrlich, Phys. Rev. D9 (1974) 2237,
   J.M. Cornwall, Phys. Rev. D10 (1974) 500,
   C.B. Thorn, Phys. Rev. D19 (1979), 639, 1934.

8. P. Goddard, J. Goldstone, C. Rebbi and C.B. Thorn, Nucl, Phys. B56 (1973) 109.
9. For recent reviews, see
J.H. Schwarz, Lectures at Scottish Summer School 1985.
M.B. Green in "Workshop on Unified String Theories", eds. M.B. Green and D.J. Gross (World Scientific 1986).
L. Brink, Lectures at "Ecole d'été de physique theorique", Les Houches 1985.
10. L. Brink, O. Lindgren and B.E.W. Nilsson, Nucl. Phys. B212 (1983) 401.
11. S. Mandelstam, Nucl. Phys. B213 (1983) 149,
L. Brink, O. Lindgren and B.E.W. Nilsson, Phys. Lett. 123B (1983) 323.
12. M.B. Green and J.H. Schwarz, Nucl. Phys. B218 (1983), 43
M.B. Green, J.H. Schwarz and L. Brink, Nucl. Phys. B219 (1983) 437,
M.B. Green and J.H. Schwarz, Nucl. Phys. B243 (1984) 475
13. A.K.H. Bengtsson, L. Brink, M. Cederwall and M. Ögren, Nucl. Phys. B254 (1985) 625.
14. P.A.M. Dirac, Can. J. Math. 2 (1980) 129.
15. A.K.H. Bengtsson, I. Bengtsson and L. Brink, Nucl. Phys. B227 (1983) 31.
16. A.K.H. Bengtsson, thesis ITP-Göteborg (1984).
17. I. Bengtsson, Phys. Rev. D31 (1985) 2525.
18. H.C. Lee and M.S. Milgram, Phys. Lett. 133B (1983) 320, Nucl. Phys. 268 (1986)543.
19. A.K.H. Bengtsson, I. Bengtsson and L. Brink, Nucl. Phys. B227 (1983) 41.
20. L. Brink and A. Tollstén, Nucl. Phys. B249 (1984) 244.
21. N. Marcus and A. Sagnotti, Phys. Lett. 135B (1984) 85.
22. M.T. Grisaru and W. Siegel, Nucl. Phys. B201 (1982) 292.
23. P.S. Howe, K.S. Stelle and P.C. West, Phys. Lett. 124B (1983) 55.
24. S. Deser, J.H. Kay and K.S. Stelle, Phys. Rev. Lett. 38 (1977) 527,
S. Deser and J.H. Kay, Phys. Lett. 76B (1978) 573,
S. Ferrara and P. Van Nieuwenhuizen, Phys. Lett. 78B (1978), 573.
P.S. Howe and U. Lindström, Nucl. Phys. B181 (1981), 487,
R.E. Kallosh, Phys. Lett. 99B, 122.
25. M.B. Green and J.H. Schwarz, Phys. Lett. 122B (1983), 143.
26. S. Mandelstam, Nucl. Phys. B64, (1973) 205.
27. L. Brink and H.B. Nielsen, Phys. Lett. 45B (1973) 332.
28. F. Gliozzi, unpublished (1976).
29. Y, Nambu, Lectures at Copenhagen Symposium, unpublished (1970),
O. Hara, Progr. Theor. Phys. 46 (1971) 1549.
T. Goto, Progr. Theor. Phys. 46 (1971) 1560.
30. P.M. Ramond, Phys. Rev. D3 (1971) 2415.
31. A. Neveu and J.H. Schwarz, Nucl. Phys. B31 (1971) 86; Phys. Rev. D4 (1971) 1109.
32. F. Gliozzi, J. Scherk and D.I. Olive, Phys. Lett. 65B, (1976) 282; Nucl. Phys. B122 (1977) 253.
33. L. Brink, P. Di Vecchia and P.S. Howe, Phys. Lett. 65B (1976) 471,
S. Deser and B. Zumino, Phys. Lett. 65B (1976) 369.
34. M.B. Green and J.H. Schwarz, Phys. Lett. 109B (1982) 444.

35. M.B. Green and J.H. Schwarz, Nucl. Phys. B181 (1981) 502.
36. M. Ademollo, L. Brink, A. D'Adda, R. d'Auria, E. Napolitano, S. Sciuto, E. Del Giudice, P. Di Vecchia, S. Ferrara, F. Gliozzi, R. Musto, R. Pettorino and J.H. Schwarz, Nucl. Phys. B111 (1976) 77.
37. D.J. Gross, J.A. Harvey, E. Martinec and R. Rohm, Phys. Rev. Lett. 54 (1985) 502; Nucl. Phys. B256 (1985), 253; Nucl. Phys.
38. I follow here L. Brink, M. Cederwall and M.B. Green, Institute for Theoretical PHysics, Göteborg (1986) to be published.
39. The last reference in 12.

SUPERGRAVITY

B. de Wit

Institute for Theoretical Physics
Princetonplein 5, P.O. Box 80.006
3508 TA Utrecht, The Netherlands

1. INTRODUCTION

In these lectures we shall describe three topics in supergravity. We start by discussing supergravity in 10 space-time dimensions, whose importance stems from the fact that theories of this type arise in the zero-slope limit of superstrings. We emphasize the structure of the underlying supermultiplets, and devote particular attention to the off-shell features of Einstein-Yang-Mills supergravity. Subsequently we discuss the issue of consistent truncations of theories with extra compactified dimensions. Most of this discussion is in the context of Kaluza-Klein supergravity, and we summarize some of the results obtained in d=11 supergravity with the extra dimensions compactified to the seven-sphere. Finally we consider some new topological terms that can be present in gauge invariant actions, which have been found in the context of d=4, N=2 Einstein-Yang-Mills supergravity. These terms have a surprising mathematical structure which is related to Lie-algebra cohomology.

2. INTRODUCTION TO d=10 SUPERGRAVITY

In 10 space-time dimensions the supersymmetry generators Q_α satisfy the well-known anticommutation relation

$$\{Q_\alpha, \bar{Q}_\beta\} = -2i(\not{P})_{\alpha\beta} , \tag{2.1}$$

where Q_α transforms as a spinor and P_μ denotes the energy-momentum operators. The charges Q_α can be restricted to Majorana-Weyl spinors, satisfying

$$C^{-1}\,\bar{Q}^T = Q \;, \tag{2.2}$$

$$\Gamma^{(11)}\;Q = -\,Q \;, \tag{2.3}$$

where

$$\Gamma^{(11)} \equiv i\,\Gamma_1\,\Gamma_2\,\Gamma_3\,\ldots\,\Gamma_{10} \tag{2.4}$$

is the analogue of γ_5 in 4 dimensions, and C is the d=10 charge conjugation matrix. Because of (2.2) and (2.3) Q is restricted to 16 real components.

We are mainly interested in massless representations of (2.1), so that the energy-momentum vector P_μ is lightlike. Choosing the spatial components of P_μ in the 9-th direction yields

$$P_\mu = (o,o,\ldots,\omega,i\omega) \quad , \quad (\omega > o) \tag{2.5}$$

which is invariant under SO(8) rotations of the transverse momenta. Because $P^2 = 0$, it is easy to see that $\not{P}\Gamma_{10}$ is proportional to a projection operator,

$$(\not{P}\Gamma_{10})(\not{P}\Gamma_{10}) = -\,P^2 + 2i\omega\not{P}\Gamma_{10} = 2i\omega(\not{P}\Gamma_{10}) \;. \tag{2.6}$$

Using (2.5), we find

$$-i\not{P}\Gamma_{10} = \omega\,(\mathbf{1} - i\Gamma_9\,\Gamma_{10}) \;. \tag{2.7}$$

Furthermore

$$\begin{aligned} i\Gamma_9\Gamma_{10} &= (\Gamma_1\Gamma_2\ldots\Gamma_8)\;\Gamma^{(11)} \\ &= \Gamma^{(9)}\Gamma^{(11)} = \Gamma^{(11)}\Gamma^{(9)} \;, \end{aligned} \tag{2.8}$$

where $\Gamma^{(9)}$ is the analogue of γ_5 for the d=8 Clifford algebra associated with the transverse momenta,

$$\Gamma^{(9)} \equiv \Gamma_1\Gamma_2\ldots\Gamma_8 \;. \tag{2.9}$$

Combining (2.1), (2.3), (2.7) and (2.8) we derive the anticommutation relation

$$\{Q_\alpha, Q^\dagger_\beta\} = -2i\not{p}\Gamma_{10} = 2\omega(\mathbf{1} + \Gamma^{(9)})_{\alpha\beta} , \qquad (2.10)$$

which is thus proportional to a chiral projector in d=8 spinor space.

We will restrict ourselves to representations of the supersymmetry algebra that contain only states of positive norm. Therefore, the right-hand side of (2.10) implies that charges Q with negative d=8 chirality ($\Gamma^{(9)}Q = -Q$) should vanish. Hence, the Q's are restricted by $\Gamma^{(9)}Q = Q$ and $\Gamma^{(11)}Q = -Q$, and in this subspace we may write

$$\{Q_\alpha, Q^\dagger_\beta\} = 4\omega\, \delta_{\alpha\beta} . \qquad (2.11)$$

Because Q is now restricted to 8 components (and transforms as an SO(8) Majorana-Weyl spinor) (2.11) defines an 8-dimensional Clifford algebra, which has a unique 16-dimensional representation. Consequently massless supersymmetry representations must decompose into 16-dimensional representations, which in turn consist of two 8-dimensional SO(8) representations. As is well-known SO(8) representations appear in a three-fold variety (triality). With the exception of certain representations, such as the adjoint 28-dimensional one, the three types of representations are inequivalent, and are distinguished by labels s, v and c (see, for instance [1]). We shall denote the representation according to which Q transforms (the positive chirality representation) as $\mathbf{8}_v$; the 16-dimensional representation of (2.11) then decomposes into the $\mathbf{8}_s$ and $\mathbf{8}_c$ representations.

Table 1: Massless N=1 supermultiplets in ten space-time dimensions containing 8+8 or 64+64 bosonic and fermionic degrees of freedom.

supermultiplet	bosons	fermions
Yang-Mills multiplet	$\mathbf{8}_s$	$\mathbf{8}_c$
graviton multiplet	$\mathbf{1} + \mathbf{28} + \mathbf{35}_s$	$\mathbf{8}_v + \mathbf{56}_v$
gravitino multiplet	$\mathbf{1} + \mathbf{28} + \mathbf{35}_c$	$\mathbf{8}_v + \mathbf{56}_v$
gravitino multiplet	$\mathbf{8}_s + \mathbf{56}_s$	$\mathbf{8}_c + \mathbf{56}_c$

The smallest massless supermultiplet has now been constructed, and consists of 8 fermionic and 8 bosonic states, which we assign to the $\mathbf{8}_c$

and $\mathbf{8}_s$ representation, respectively. This is just the supersymmetric Yang-Mills multiplet in 10 dimensions [2]. In order to obtain the supersymmetry representations relevant for supergravity we consider tensor products of the Yang-Mills multiplet with each of the 8-dimensional representations. Multiplying $\mathbf{8}_s$ with $\mathbf{8}_s + \mathbf{8}_c$ yields $\mathbf{8}_s \times \mathbf{8}_s$ bosonic and $\mathbf{8}_s \times \mathbf{8}_c$ fermionic states, and, among others, leads to a $\mathbf{35}_s$ representation which can be associated with the states of the graviton in d=10 dimensions. Consequently this multiplet will be called the graviton multiplet. Multiplication with $\mathbf{8}_c$ or $\mathbf{8}_v$ goes in the same fashion, except that we will associate the $\mathbf{8}_c$ and $\mathbf{8}_v$ representations with fermionic quantities (note that these are the representations to which the fermion states of the Yang-Mills multiplet and the supersymmetry charges are assigned). Consequently, we interchange the boson and fermion assignments in these products. Multiplication with $\mathbf{8}_c$ then leads to $\mathbf{8}_c \times \mathbf{8}_c$ bosonic and $\mathbf{8}_c \times \mathbf{8}_s$ fermionic states, whereas multiplication with $\mathbf{8}_v$ gives $\mathbf{8}_v \times \mathbf{8}_c$ bosonic and $\mathbf{8}_v \times \mathbf{8}_s$ fermionic states. These supermultiplets contain fermions transforming according to the $\mathbf{56}_v$ and $\mathbf{56}_c$ representations, respectively, which can be associated with gravitino states, but no graviton states as those transform in the $\mathbf{35}_s$ representation. Therefore these two supermultiplets are called gravitino multiplets, and we have thus established the existence of two inequivalent gravitino multiplets. The explicit SO(8) decompositions of the Yang-Mills, graviton and gravitino supermultiplets are summarized in table 1.

By combining a graviton and a gravitino multiplet it is possible to construct an N=2 supermultiplet of 128 + 128 bosonic and fermionic

Table 2: The chiral N=2 supergravity multiplet in ten space-time dimensions.

$(\mathbf{8}_s + \mathbf{8}_c) \times (\mathbf{8}_s + \mathbf{8}_c)$	bosons:	$\mathbf{1} + \mathbf{1} + \mathbf{28} + \mathbf{28} + \mathbf{35}_s + \mathbf{35}_c$
	fermions:	$\mathbf{8}_v + \mathbf{8}_v + \mathbf{56}_v + \mathbf{56}_v$

states. However, since there are two inequivalent gravitino multiplets, there will also be two inequivalent N=2 supermultiplets containing the states corresponding to a graviton and 2 gravitini. One N=2 supermultiplet may be viewed as the tensor product of two identical supermultiplets (namely $\mathbf{8}_s+\mathbf{8}_c$). Such a multiplet follows if one starts from a supersymmetry algebra based on two Majorana-Weyl spinor charges Q with the <u>same</u> chirality. One can perform rotations between these spinor charges which leave the supersymmetry algebra unaffected, and this feature should result in a certain degeneracy of some of the states of the supermultiplet. Indeed, the explicit SO(8) decomposition in table 2 shows such a degeneracy for the states assigned to the **1**, **28**, $\mathbf{8}_v$ and $\mathbf{56}_v$ representations. The theory based on this multiplet is chiral N=2 supergravity [3] (sometimes called type 2b supergravity).

The second N=2 supermultiplet may be viewed as the tensor product of a $(\mathbf{8}_s+\mathbf{8}_c)$ supermultiplet with a second supermultiplet $(\mathbf{8}_s+\mathbf{8}_v)$. The fermionic states then appear with both chiralities. Such a multiplet can be derived from a supersymmetry algebra based on <u>two</u> Majorana-Weyl spinor charges Q, but now with <u>opposite</u> chirality. The theory based on this multiplet is nonchiral N=2 supergravity (sometimes called type 2a supergravity), which can be obtained by a straightforward reduction of d=11 supergravity [4]. The latter follows from the fact that two d=10 Majorana-Weyl spinors with opposite chirality can be combined into a single d=11 Majorana spinor. Table 3 presents the field content of nonchiral N=2 supergravity, and indicates how these fields combine into the fields of d=11 supergravity: the gravitino field Ψ_M, the elfbein field $E_M{}^A$ and a 3-rank antisymmetric tensor gauge field A_{MNP}. This theory can be truncated to N=1 supergravity [5,6], which has the interesting feature is that it can be coupled to supersymmetric Yang-Mills theory. The field representation then consists of a zehnbein field $e_\mu{}^a$, a Majorana-Weyl Rarita-Schwinger field ψ_μ, an antisymmetric tensor gauge field $A_{\mu\nu}$, a Majorana-Weyl spinor field λ and a scalar field ϕ. However, there is also an alternative formulation of this theory, in which the rank-2 tensor field is replaced by a rank-6 tensor gauge field [6-9]. An intriguing feature of this second version becomes apparent when counting the off-shell degrees of freedom. With the rank-2 tensor field we have 45+1+36=82 bosonic and 144+16=160 fermionic field components (after subtracting gauge degrees of freedom), as is shown in table 3, while with a rank-6 tensor field we count 45+1+84=130 bosonic and, again, 160 fermionic field components. Let us compare these numbers to the number of physical states of the smallest <u>massive</u> N=1 supermultiplet. That multiplet consists of 128 bosonic and 128 fermionic states, decomposing into the **44**, **84** and **128** representations of SO(9). Therefore the theory with the rank-6 tensor field contains more field components than the number of states of a massive supermultiplet. It should therefore be possible to reduce this number somewhat and find an off-shell formulation of N=1 supergravity based on 128 + 128 degrees of freedom. Indeed, this

Table 3: Fields of nonchiral N=2 supergravity. The first column shows the fields decomposed into the two N=1 submultiplets. The superscript ± on the fermion fields indicates their chirality. The second column gives the corresponding numbers of field components (after subtraction of gauge degrees of freedom), and the third column the SO(8) representations of the corresponding massless states. The last column exhibits how the fields combine into those of d=11 supergravity.

fermions	$\psi_\mu^+ + \psi_\mu^-$	144 + 144	$\mathbf{56}_v + \mathbf{56}_c$	Ψ_M
	$\lambda^- + \lambda^+$	16 + 16	$\mathbf{8}_v + \mathbf{8}_c$	
bosons	$e_\mu^a + A_\mu$	45 + 9	$\mathbf{35}_s + \mathbf{8}_s$	$E_M{}^A$
	ϕ	1	$\mathbf{1}$	
	$A_{\mu\nu} + A_{\mu\nu\rho}$	36 + 84	$\mathbf{28} + \mathbf{56}_s$	A_{MNP}

turns out to be the case, and the resulting theory is conformal supergravity which was constructed in [8]. We emphasize that, sofar, this situation where an on-shell formulation of Poincaré supergravity is based on more field components than the number of states of a massive supermultiplet, has only been encountered in 10 space-time dimensions. In all other cases one needs extra field components (auxiliary fields) to define an off-shell formulation of the theory.

The supersymmetry transformations in the version with a rank-6 tensor field are (modulo fermionic bilinears in $\delta\psi_\mu$ and $\delta\lambda$)

$$\delta e_\mu{}^a = \tfrac{1}{2}\,\bar\varepsilon\Gamma^a\psi_\mu \ ,$$

$$\delta\psi_\mu = D_\mu\varepsilon + \frac{1}{8!}\,e^{-\phi}\big(\Gamma_\mu\Gamma^{(7)} - 2\Gamma^{(7)}\Gamma_\mu\big)\,R(A)_{(7)}\varepsilon \ ,$$

$$\delta A_{\mu_1\ldots\mu_6} = \frac{3}{2}\,e^{\phi}\,\bar\varepsilon\big(\Gamma_{[\mu_1\ldots\mu_5}\psi_{\mu_6]} + \tfrac{1}{6}\,\Gamma_{\mu_1\ldots\mu_6}\lambda\big) \ ,$$

$$\delta\lambda = \tfrac{1}{2}\,\not\partial\phi\,\varepsilon - \frac{1}{4.7!}\,e^{-\phi}\,\Gamma^{(7)}R(A)_{(7)}\,\varepsilon \ ,$$

$$\delta\phi = \tfrac{1}{2}\,\bar\varepsilon\lambda \ , \qquad (2.12)$$

where we use the definition

$$R(A)_{\mu_1\ldots\mu_7} = 7\,\partial_{[\mu_1}A_{\mu_2\ldots\mu_7]}\ . \tag{2.13}$$

These and subsequent results have been taken from [8], where also the higher-order fermion terms are given, after making the substitutions

$$A_{\mu_1\ldots\mu_6} \rightarrow \frac{1}{2.6!}\,A_{\mu_1\ldots\mu_6}\ ,$$

$$\lambda \rightarrow \frac{1}{6}\,\lambda\ ,$$

$$\phi \rightarrow \exp\left(\frac{w}{6}\,\phi\right)\ . \tag{2.14}$$

The Poincaré supergravity lagrangian that is invariant under (2.12) takes the form (we set $\kappa=1$)

$$e^{-1}L_P = -\tfrac{1}{2}R - \tfrac{1}{2}\,\bar{\psi}_\mu\Gamma^{\mu\rho\sigma}D_\rho\psi_\sigma - \bar{\lambda}\not{D}\lambda - \frac{1}{7!}\,e^{-2\phi}\left(R(A)_{(7)}\right)^2 - (\partial_\mu\phi)^2$$

$$+\ \bar{\psi}_\mu\not{\partial}\phi\Gamma^\mu\lambda$$

$$+\ \frac{1}{4.7!}\,e^{-\phi}R(A)_{(7)}\left(\bar{\psi}_\mu\Gamma^{[\mu}\Gamma^{(7)}\Gamma^{\nu]}\psi_\nu + 2\bar{\psi}_\mu\Gamma^{(7)}\Gamma^\mu\lambda\right)$$

$$+\ \text{quartic fermion terms.} \tag{2.15}$$

However, we emphasize that the transformations (2.12), which henceforth will be called Q-supersymmetry transformations, can be derived entirely from off-shell arguments without the need to refer to a specific invariant action. This will be explained shortly.

As outlined above, we are dealing with 130 bosonic and 160 fermionic field components. As a first step in reducing these numbers we now introduce <u>local</u> dilatations (denoted by D) and S-supersymmetry (sometimes called conformal supersymmetry) transformations on the fields:

$$\delta e_\mu{}^a = -\Lambda_D e_\mu{}^a\ ,\qquad \delta\psi_\mu = -\tfrac{1}{2}\,\Lambda_D\psi_\mu - \Gamma_\mu\eta\ ,$$

$$\delta\lambda = \tfrac{1}{2}\,\Lambda_D\lambda + 6\eta\ ,\qquad \delta\phi = 6\Lambda_D\ , \tag{2.16}$$

where Λ_D and η are the infinitesimal parameters of these transformations. Note that the tensor gauge field $A_{\mu\nu\rho\sigma\lambda\tau}$ is inert under both D and S. Because of the local D and S transformations the field configuration is reduced to 129 bosonic and 144 fermionic degrees of freedom. This still exceeds the number of bosonic and fermionic states of the smallest massive supermultiplet by 1 and 16, respectively. To exhibit the extra conditions needed for a further reduction one may construct the commutator algebra of infinitesimal gauge transformations. The most relevant commutation relations are

$$[\delta_Q(\varepsilon_1),\delta_Q(\varepsilon_2)] = \delta_{gct}(\xi^\mu) + \delta_Q(\varepsilon_3) + \delta_S(\eta')$$

$$+ \delta_M(\varepsilon^{ab}) + \delta_A(\xi_{\mu\nu\rho\sigma\lambda}) \ , \qquad (2.17)$$

$$[\delta_Q(\varepsilon),\delta_S(\eta)] = \delta_D(-\tfrac{1}{2}\bar{\eta}\varepsilon) + \delta_M(\tfrac{1}{2}\bar{\varepsilon}\Gamma^{ab}\eta) + \delta_S(\eta'') \ , \qquad (2.18)$$

$$[\delta_S(\eta_1),\delta_S(\eta_2)] = 0 \ , \qquad (2.19)$$

$$[\delta_S(\eta),\delta_D(\Lambda_D)] = \delta_S(\tfrac{1}{2}\Lambda_D\eta) \ , \qquad (2.20)$$

$$[\delta_Q(\varepsilon),\delta_D(\Lambda_D)] = \delta_Q(-\tfrac{1}{2}\Lambda_D\varepsilon) + \delta_S(-\tfrac{1}{2}\partial\!\!\!/\Lambda_D\varepsilon) \ , \qquad (2.21)$$

where the parameters of general coordinate transformations and tensor gauge transformations that appear in (2.17) are given by

$$\xi^\mu = \tfrac{1}{2}\bar{\varepsilon}_2\Gamma^\mu\varepsilon_1 \ ,$$

$$\xi_{\mu\nu\rho\sigma\lambda} = -\frac{3}{2}\, e^\phi\, \bar{\varepsilon}_2\Gamma_{\mu\nu\rho\sigma\lambda}\varepsilon_1 \ . \qquad (2.22)$$

The Q-, S- and M-transformation parameters on the right-hand sides of (2.17-18) take a rather complicated form. For instance, ε_3 and η'' are equal to

$$\varepsilon_3 = -\xi^\mu(\psi_\mu + \frac{7}{16}\Gamma_\mu\lambda) + \frac{1}{64.5!}\left(\bar{\varepsilon}_2\Gamma^{(5)}\varepsilon_1\right)\Gamma_{(5)}\lambda \ , \qquad (2.23)$$

$$\eta'' = \frac{1}{64}\{(\bar{\eta}\varepsilon)(-34\lambda - \Gamma^\mu\psi_\mu) + (\bar{\eta}\Gamma^{(2)}\varepsilon)(3\Gamma_{(2)}\lambda + \frac{1}{2}\Gamma^\mu\Gamma_{(2)}\psi_\mu)$$

$$+ \frac{1}{24}(\bar{\eta}\Gamma^{(4)}\varepsilon)(2\Gamma_{(4)}\lambda - \Gamma^\mu\Gamma_{(4)}\psi_\mu)\} \ . \qquad (2.24)$$

The above commutation relations, which follow from [8] in a special gauge (namely $b_\mu = 0$), are only valid modulo to the following expression

$$\Psi = -\frac{1}{3}e^{-2\phi/3}\,\partial\!\!\!/(e^{2\phi/3}\lambda) + \frac{1}{18.7!}\Gamma^{(7)}\lambda\, e^{-\phi}\, R(A)_{(7)} \ . \qquad (2.25)$$

Under the superconformal transformations Ψ and a second quantity C, defined by

$$C = e^{-2\phi/3}\{\frac{1}{4} D^2 e^{2\phi/3} + \frac{1}{9}\bar{\lambda}\not{D}(e^{2\phi/3}\lambda)\} + \frac{1}{9.7!} e^{-2\phi}(R(A)_{(7)})^2 , \tag{2.26}$$

transform into each other. For instance, under Q, S and D we have

$$\delta\Psi = - C\varepsilon - \frac{17}{32}\varepsilon(\bar{\lambda}\Psi) - \frac{3}{64}\Gamma^{(2)}\varepsilon(\bar{\lambda}\Gamma_{(2)}\Psi) + \frac{1}{768}\Gamma^{(4)}\varepsilon(\bar{\lambda}\Gamma_{(4)}\Psi) + \frac{3}{2}\Lambda_D\Psi , \tag{2.27a}$$

$$\delta C = - \frac{1}{4}\bar{\varepsilon}\not{D}\Psi - \frac{1}{2.8!}\bar{\varepsilon}\Gamma^{(7)}\Psi\, e^{-\phi}R(A)_{(7)} + \frac{1}{512}\bar{\varepsilon}\Gamma^{(3)}\Psi(\bar{\lambda}\Gamma_{(3)}\lambda) - \frac{3}{2}\bar{\eta}\Psi + 2\Lambda_D C , \tag{2.27b}$$

The derivatives D_μ in (2.25- 27) are fully covariant with respect to all superconformal symmetries. Observe that the covariant Dirac operator in (2.25) acts on $\exp(2\phi/3)\lambda$, which is a spinor with Weyl weight (i.e. scale dimension) $w=9/2$, whereas the D'Alembertian D^2 in (2.26) acts on $\exp(2\phi/3)$, which is a scalar with weight $w=4$. This is necessary for Ψ and C to transform covariantly under dilatations. It is important to realize that, because of S covariance, $\not{D}[\exp(2\phi/3)\lambda]$ contains a term $1/3 \exp(2\phi/3)\,\Gamma^{\mu\nu}\psi_{\mu\nu}$, where $\psi_{\mu\nu}$ is the Rarita-Schwinger field strength $\psi_{\mu\nu} = \partial_\mu\psi_\nu - \partial_\nu\psi_\mu$, while, because of D covariance, $D^2\exp(2\phi/3)$ contains a term $2/9\, R \exp(2\phi/3)$, where R is the Ricci scalar.

According to (2.27) we can consistently put C and Ψ to zero, which reduces the degrees of freedom to precisely 128 + 128. As these numbers coincide with the numbers of bosonic and fermionic states of a massive supermultiplet, it is plausible that the resulting field configuration defines an off-shell multiplet. It was shown in [8] that this is indeed the case: all superconformal transformations close if we impose the condition $\Psi = 0$. Furthermore, after imposing certain gauge choices and linearizing the constraint equations $C = \Psi = 0$, one can show that their solutions are not restricted to be massless! Conformal supergravity in 10 dimensions is thus defined in terms of the fields $e_\mu{}^a$, ψ_μ, $A_{\mu_1\ldots\mu_6}$, λ and ϕ, subject to the constraints $C = \Psi = 0$. Its gauge symmetries are general coordinate transformations, local Lorentz transformations, Q and S supersymmetry, dilations and tensor gauge transformations. Modulo field redefinitions, the full transformation rules as well as the explicit expressions for Ψ and C follow uniquely from requiring the closure of the superconformal transformations. To prove this, one starts from the linearized Q transformations, and proceeds by iteration. Therefore, conformal supergravity is defined in terms of a consistent field representation which is meaningful outside the context of a specific invariant action.

Because the super-Poincaré action is invariant under Q supersymmetry, the commutator of two Q transformations should decompose into the various symmetries of that theory, modulo terms that are proportional to the super-Poincaré field equations. On the other hand, from the above results we know that the $\{Q,Q\}$ commutator decomposes also into superconformal transformations, which do not all leave the super-Poincaré action invariant, up to terms proportional to Ψ (it turns out that C does not appear). This proves that Ψ and C must correspond to linear combinations of super-Poincaré field equations. This is indeed confirmed by considering D and S variations of the Poincaré supergravity action corresponding to (2.15)

$$\delta S_P = \int d^{10}x \; e\{72\Lambda_D(C - \tfrac{1}{4}\bar{\psi}_\mu \Gamma^\mu \Psi) + 36\bar{\eta}\Psi\} \;, \tag{2.28}$$

which also demonstrates that Poincaré supergravity is not invariant under D and S symmetry. Furthermore the coefficient η' of the S transformation in the $\{Q,Q\}$ commutator (2.17) must be proportional to the super-Poincaré field equations. Explicit calculations have confirmed that this is the case [8].

It is also possible to implement the superconformal gauge algebra on the fields of supersymmetric Yang-Mills theory. If one first assumes the obvious the Q-supersymmetry transformation rule

$$\delta A_\mu = \tfrac{1}{2}\bar{\varepsilon}\Gamma_\mu \chi, \tag{2.29}$$

for the gauge potential, the realization of the full superconformal algebra on A_μ requires the following Q and D variations of χ:

$$\delta\chi = -\tfrac{1}{4}\Gamma^{\mu\nu}\varepsilon\big(\hat{F}_{\mu\nu} + \tfrac{7}{32}\bar{\lambda}\Gamma_{\mu\nu}\chi\big) + \tfrac{7}{64}\,\varepsilon\,(\bar{\lambda}\chi) - \tfrac{1}{1536}\,\Gamma^{\mu\nu\rho\sigma}\varepsilon\,\big(\bar{\lambda}\Gamma_{\mu\nu\rho\sigma}\chi\big) + \tfrac{3}{2}\Lambda_D\chi, \tag{2.30}$$

while $\delta_S\chi = 0$. In (2.30) $\hat{F}_{\mu\nu}$ is the supercovariant field strength defined by

$$\hat{F}_{\mu\nu} = \partial_\mu A_\nu - \partial_\nu A_\mu - [A_\mu, A_\nu] - \bar{\psi}_{[\mu}\Gamma_{\nu]}\chi \;, \tag{2.31}$$

and we use a Lie-algebra valued notation for χ and A_μ. However, on χ the superconformal transformations close modulo an expression which turns out to correspond to a generalization of the Dirac equation for χ. Therefore the super-Yang-Mills fields A_μ and χ define only an on-shell representation of the superconformal algebra. In view of this result it comes as no surprise that there exists an action which leads to the same equation for χ and is invariant under all superconformal transformations, provided one imposes the constraint $\Psi=0$ on the supergravity fields. This action follows from

$$L_{YM} = g^{-2}\, e\, e^{\phi}\, \{\mathrm{Tr}(\tfrac{1}{4}F_{\mu\nu}F^{\mu\nu} + \tfrac{1}{2}\bar{\chi}\Gamma^{\mu}D_{\mu}\chi) \;+\; \bar{J}_{\mu}(\psi_{\mu} + \tfrac{1}{6}\Gamma_{\mu}\lambda)\}$$

$$+ \tfrac{i}{6!}\, g^{-2}\, \varepsilon^{\mu_1\cdots\mu_{10}}\, \Omega_{\mu_1\cdots\mu_4}\, A_{\mu_5\cdots\mu_{10}}$$

$$+ \tfrac{1}{4.6!}\, g^{-2} e\, \mathrm{Tr}(\bar{\chi}\, \Gamma^{\mu_1\cdots\mu_7}\, \chi)\, \partial_{\mu_1}A_{\mu_2\cdots\mu_7}$$

$$+ \text{ quartic fermion terms }, \qquad (2.32)$$

where

$$J_{\mu} = \tfrac{1}{4}\, \Gamma^{\rho\sigma}\Gamma_{\mu}\, \mathrm{Tr}(\chi F_{\rho\sigma})\ , \qquad (2.33)$$

is the Yang-Mills supercurrent, and $\Omega_{\mu\nu\rho\sigma}$ is defined by

$$\Omega_{\mu\nu\rho\sigma} = \tfrac{1}{4}\, \mathrm{Tr}\big(F_{[\mu\nu}\, F_{\rho\sigma]}\big). \qquad (2.34)$$

Note that Ω, which, in four dimensions, corresponds to the Pontryagin density, is closed (but not exact),

$$\partial_{[\mu}\Omega_{\nu\rho\sigma\lambda]} = 0\ , \qquad (2.35)$$

as a consequence of the Bianchi identity on F. This ensures the invariance of (2.32) under tensor gauge transformations.

Hence, supersymmetric Yang-Mills theory couples only to the subset of supergravitational fields corresponding to conformal supergravity. If we refrain from imposing the constraint $\Psi=0$, then the action corresponding to (2.32) is no longer invariant under Q supersymmetry, and changes according to

$$\delta_Q S_{YM} = \tfrac{3}{64}\, g^{-2} \int d^{10}x\ e\, e^{\phi}\, \mathrm{Tr}(\bar{\chi}\Gamma^{\mu\nu\rho}\chi)\, \bar{\varepsilon}\Gamma_{\mu\nu\rho}\Psi\ . \qquad (2.36)$$

Comparing (2.36) to (2.28) it is now easy to derive the supersymmetry transformations of Einstein-Yang-Mills supergravity, decomposed into superconformal transformations. Requiring that

$$\delta(\varepsilon)\left(S_P + S_{YM}\right) = 0 \quad , \tag{2.37}$$

one finds

$$\delta(\varepsilon) = \delta_Q(\varepsilon) + \delta_S(\eta) \quad , \tag{2.38}$$

where

$$\eta = -\frac{1}{768} g^{-2} e^{\phi} \bar{\chi}\Gamma^{\mu\nu\rho}\chi \, \Gamma_{\mu\nu\rho}\varepsilon \; . \tag{2.39}$$

In order to convert the rank-6 tensor field to a rank-2 field, one first observes that, locally, it is possible to write $\Omega_{\mu\nu\rho\sigma}$ as the curl of a rank-3 antisymmetric tensor,

$$\Omega_{\mu\nu\rho\sigma} = \partial_{[\mu}\omega_{\nu\rho\sigma]} \; . \tag{2.40}$$

The tensor $\omega_{\mu\nu\rho}$ is known as the Chern-Simons term and is equal to

$$\omega_{\mu\nu\rho} = \mathrm{Tr}\left(A_{[\mu}\, \partial_\nu A_{\rho]} - \frac{2}{3} A_{[\mu}\, A_\nu\, A_{\rho]}\right) \; . \tag{2.41}$$

Using (2.40) the Ω-A coupling in (2.32) can be rewritten as

$$L' = \frac{i}{7!} g^{-2} \; \varepsilon^{\mu_1 \ldots \mu_{10}} \; \omega_{\mu_1\mu_2\mu_3} \; R(A)_{\mu_4 \ldots \mu_{10}} \; , \tag{2.42}$$

modulo a total divergence. Subsequently a duality transformation leads to the uniform replacement

$$\tilde{R}(A)_{\mu\nu\rho} = \frac{3}{\sqrt{2}} e^{2\phi}\left\{\partial_{[\mu}A_{\nu\rho]} + \sqrt{2}\, g^{-2}\left(\omega_{\mu\nu\rho} + \frac{1}{24}\mathrm{Tr}(\bar{\chi}\Gamma_{\mu\nu\rho}\chi)\right) + \bar{\psi}\psi,\ \bar{\psi}\lambda\text{-terms}\right\}. \tag{2.43}$$

where $\tilde{R}(A)^{\mu\nu\rho}$ is the dual field strength

$$\tilde{R}(A)^{\mu\nu\rho} \equiv \frac{i\, e^{-1}\, \varepsilon^{\mu\nu\rho\sigma_1 \ldots \sigma_7}}{7!} \; R(A)_{\sigma_1 \ldots \sigma_7} \; . \tag{2.44}$$

Both sides of (2.43) are invariant under tensor gauge transformations. However, invariance under Yang-Mills gauge transformations requires that A_μ and $A_{\mu\nu}$ transform according to

$$\delta A_\mu = \partial_\mu \Lambda - [A_\mu,\Lambda] \ ,$$

$$\delta A_{\mu\nu} = - \sqrt{2}\, g^{-2}\, \mathrm{Tr}\big(\Lambda\, \partial_{[\mu} A_{\nu]}\big) \ , \tag{2.45}$$

where Λ is the Lie-algebra valued Yang-Mills parameter. For a more detailed discussion of the duality transformation leading to (2.43) with further references we refer to [10]. The Chern-Simons term was first found in [6] for Maxwell-Einstein supergravity. Subsequently the nonabelian extension was given in [11]. As is well-known Chern-Simons terms play an important role in the cancellation of anomalies [12,9,13].

In view of the differential nature of constraints $C = \Psi = 0$ the superconformal framework may be regarded as a partially off-shell formulation of Poincaré supergravity. As an intermediate step towards a fully off-shell formulation one may write the super-Poincaré lagrangian in a superconformally invariant form by introducing compensating fields A and ξ. The super-Poincaré lagrangian (2.15) then takes the form

$$e^{-1} L_P = -9e\big\{ A(C - \tfrac{1}{4}\bar{\psi}_\mu \Gamma^\mu \Psi) + \bar{\xi}\Psi \big\} \ , \tag{2.46}$$

which coincides with (2.15) after imposing the gauge conditions

$$A = 1 \quad , \ \xi = 0 \ . \tag{2.47}$$

The constraints $C = \Psi = 0$ can now be understood as the equations of motion corresponding to A and ξ, which act as Lagrange multipliers in (2.46). The form of (2.46) suggests an immediate generalization, in which A and ξ are extended to a full scalar multiplet S of Lagrange multipliers, whereas C and Ψ are regarded as the highest-dimensional components of another scalar multiplet Φ. Together with the superconformal field representation Φ combines into an unconstrained off-shell multiplet. The lagrangian is then based on a product of the two corresponding scalar superfields, and, in the superconformal gauge (2.47), acquires the same structure as the linearized superspace lagrangian presented in [14]. The superconformal field configuration of 128 + 128 components, has now been extended with two scalar multiplets, each consisting of $2^{15} + 2^{15}$ components. For further details and references, we refer to [10].

3. COMPACTIFICATION OF EXTRA DIMENSIONS AND CONSISTENT TRUNCATIONS

For realistic applications of supergravity in higher-dimensional space-times it is necessary that the extra (spatial) dimensions are (spontaneously) compactified to a manifold whose size is sufficiently small to remain unobservable at present energies. In four dimensions this theory will describe an infinite number of states. The mass spectrum of these states follows from analyzing small fluctuations about the field configuration that characterizes the ground state. Gauge symmetries associated with massless states are often related to the isometry group of this ground state.

In practice one often ignores the massive states, and concentrates on an effective low-energy theory that describes the dynamics of the massless states. In the initial lagrangian this amounts to dropping all massive modes in harmonic expansions of the fields. However, a truncation of the original theory must be consistent in the sense that discarded states do not reappear through the relevant symmetry transformations associated with the ground state (i.e. for all $\phi=0$ one should have $\delta\phi=0$), or through the interactions. By the latter we mean that the original higher-dimensional theory should not contain interactions that are linearly proportional to discarded modes, so that solutions of the truncated field equations are automatically solutions of the full field equations. It is possible to obtain the same result as in a consistent truncation by integrating out, in tree approximation, the modes to be suppressed in the truncation. In other words, in the context of a consistent truncation the exchange of discarded modes in tree diagrams does not induce interactions among the modes that are kept in the truncation (as emphasized in [15] these interactions, if present, are not always suppressed by large mass factors).

Some time ago, the question of consistency became relevant in the context of Kaluza-Klein supergravity in attemps to understand the S^7 compactification of d=11 supergravity. There it was noted that the supersymmetry transformations were not automatically consistent upon truncation to the massless modes [16] (although, on the other hand, the d=4 Einstein equation was shown to be consistent upon truncation to the graviton-Yang-Mills sector, in contradistinction with generic Kaluza-Klein theory [17]). For superstring theory we should point out that a truncation of the superstring to supergravity is not consistent according to the above definition, as there are interactions that involve a single massive string mode with several massless modes. However, for these theories the effective lagrangian is obtained directly from the string amplitudes, which already contain the effect of the massive string modes [18]. Therefore, the massive modes have been taken into account and one is not dealing with a truncation.

Nevertheless, in those cases where one does make a truncation one has to find means to ensure its consistency. One way to do this is to impose the restriction that all the fields are invariant under a certain subgroup of the isometry group. An obvious example of this situation arises in the torus truncation, where one only retains the fields that are independent of the extra coordinates y^m (see for instance [19]). The gauge parameters should satisfy the same restriction, so that the transformation rules are consistent upon the truncation (obviously, both ϕ and $\delta\phi$ are now y-independent), and so are the field equations. This way of achieving consistency has also been exploited in [20].

Another approach emphasizes <u>covariance</u> under the full isometry group. One expects that all modes are contained in an infinite set of irreducible representations of the isometry group, so that the truncation can be effected by restricting this infinite set to a finite number of representations. However, representations cannot be discarded arbitrarily. Viewed as a four-dimensional theory we are dealing with a gauge theory coupled to an infinite number of matter multiplets, and while it is often possible in such theories to discard all or some of the matter multiplets, it is essential to retain at least the massless gauge multiplet in order to have a consistent truncation. In fact it turns out that the naive expectation that the modes transform as an infinite set of irreducible representations of the isometry group is not immediately realized [16], as inconsistencies may already appear at the linearized level. To see how this is possible, let us emphasize that the modes are usually identified by examining small fluctuations of the fields about the background, and that the y-dependence of these modes is determined in the context of certain gauge conditions. Consequently there is an inherent ambiguity in determining the y-dependence of these modes. When calculating the linearized symmetry variation of a particular mode of a certain mass one may find contributions from modes with different masses, suggesting that the anticipated decomposition in terms of irreducible multiplets has not been realized. However, this interpretation is not correct: the transitions between different multiplets are due to a mismatch in the y-dependence of the various modes, which is a reflection of the aforementioned ambiguity in the y-dependence. The remedy, at least in linearized approximation, is clear: the transformation parameters, whose y-dependence has been identified by requiring that they leave the ground state invariant (so that they are characterized in terms of Killing spinors or vectors), should be modified by field-dependent terms to correct for this mismatch. This nonlinear modification contributes at the linearized level, because gauge transformations always contain inhomogeneous terms.

In principle, it is thus clear how to regain consistency at the linearized level, so that all multiplets transform among themselves (for the massless modes of the S^7 compactification of d=11 supergravity this was demonstrated in [21]). However, to determine all the nonlinear modifications is a much more difficult task, as it turns out that also the fields are in general subject to nonlinear redefinitions. These redefinitions are important for determining the interactions and the non-linearly realized symmetries of the truncated theory. Contrary to what is sometimes stated in the literature the nonlinear modifications, although originating from modes with Planck-size masses, do contribute to the renormalizable sector of the truncated theory in the dimensionless four-point couplings (see, e.g. [22]; as emphasized above, the same conclusion follows from integrating over the massive modes).

It is evident that invariance of the truncated action under the truncated transformation rules is only guaranteed if the transformation rules are consistent upon the truncation. Consequently the truncated transformation rules must close under commutation, possibly modulo the field equations of the truncated action. In general consistency of the transformation rules does not guarantee that the truncation will be consistent with regard to the field equations. However, for supersymmetry transformations, which close only modulo the <u>full</u> higher-dimensional field equations, the closure of the truncated supersymmetry transformations implies the validity of the full higher-dimensional field equations (of course these arguments can only be applied for a supersymmetric ground state).

The above observations have been used to prove the consistency of the truncation to the massless sector of the S^7 compactification of d=11 supergravity to all orders [23]. The results of this work exhibit all aspects that we have discussed above, and have led to the complete embedding of gauged N=8 supergravity into d=11 supergravity. To appreciate their complexity we present the complete ansatz for the d=11 metric g_{MN}, truncated to gauged N=8 supergravity:

$$g_{MN}(x,y) = \begin{pmatrix} \Delta^{-1}(x,y)g_{\mu\nu}(x) + B^p_\mu(x,y)B^q_\nu(x,y)g_{pq}(x,y) & g_{np}(x,y)B^p_\mu(x,y) \\ g_{mp}(x,y)B^p_\nu(x,y) & g_{mn}(x,y) \end{pmatrix} , \tag{3.1}$$

with

$$\Delta(x,y) = \left(\frac{g(x,y)}{\mathring{g}(y)}\right)^{\frac{1}{2}} , \quad (g \equiv \det g_{mn}) \tag{3.2}$$

$$\Delta^{-1}(x,y)g^{mn}(x,y) = \frac{1}{8} K^{mIJ}(y)\ K^{nKL}(y) \times \left(u_{ij}{}^{IJ}(x) + v_{ijIJ}(x)\right)\left(u^{ij}{}_{KL}(x) + v^{ijKL}(x)\right) , \tag{3.3}$$

$$B^m_\mu(x,y) = -\frac{\sqrt{2}}{4} A^{IJ}_\mu(x)\ K^{mIJ}(y) . \tag{3.4}$$

Here $\mathring{g}_{mn}(y)$ is the S^7 metric and $K^{mIJ}(y)$ are the 28 normalized Killing vectors on S^7 labelled by antisymmetric index pairs [IJ] (I,J = 1,...,8); $g_{\mu\nu}(x)$, $A^{IJ}_\mu(x)$, $u_{ij}{}^{IJ}(x)$ and $v_{ijIJ}(x)$ are the fields of N=8 supergravity associated with the graviton, the 28 spin-1 fields and the 70 spin-0 fields. We recall that the latter parametrize the $E_7/SU(8)$ coset space, and that $u_{ij}{}^{IJ}$, v_{ijIJ} and their complex conjugates form a 56×56 E_7 matrix, sometimes called the 56-bein (cf.[19,24]).

The expression for $g^{mn}(x,y)$ shows that the form of the truncation to N=8 supergravity is quite complicated, and deviates substantially from the result for linearized fluctuations about the S^7 background [25]. To recover the linearized result one substitutes $u_{ij}{}^{IJ} \simeq \delta^I_{[i}\ \delta^J_{j]}$ and $v_{ijIJ} \propto \phi_{ijIJ}$, where ϕ_{IJKL}, which is antisymmetric in IJKL and complex selfdual, describes the 70 scalar and pseudoscalar fields; this yields for (3.3)

$$\left(\frac{\mathring{g}(y)}{g(x,y)}\right)^{\frac{1}{2}} g^{mn}(x,y) \simeq \mathring{g}^{mn}(y) + \tfrac{1}{4}\mathrm{Re}\ \phi_{IJKL}(x)\ K^{mIJ}(y)\ K^{nKL}(y) , \tag{3.5}$$

where we have used

$$K^{mIJ}(y)\ K^{nIJ}(y) = 8\ \mathring{g}^{mn}(y) . \tag{3.6}$$

Hence linearized fluctuations of the metric describe only the scalar N=8 modes, corresponding to the real part of ϕ_{IJKL}, in contrast with finite deviations of the metric which contain both scalar and pseudoscalar modes. As emphasized before the nonlinear aspects do not affect the mass spectrum, which follows from an analysis of small fluctuations, but they are crucial for determining the interactions (and possibly nonlinearly realized symmetries) of the truncated theory, even of its renormalizable sector. For more complete results we refer the reader to [23].

Another interesting feature of the above results pertains the so-called warp factor Δ, shown in the decompostion (3.1) of the higher-dimensional metric. This warp factor must in general be included when investigating Kaluza-Klein solutions for which the higher-dimensional space has the topology of a product space,

$$M^d \to M^4 \times M^{d-4} \,. \tag{3.7}$$

For the ground state one assumes that M^4 is a maximally symmetric space-time (i.e. Minkowski or (anti-)de Sitter space), with metric $g_{\mu\nu}(x)$. The d-dimensional metric then decomposes as

$$g_{MN}(x,y) = \begin{pmatrix} \Delta^{-1}(y)\; g_{\mu\nu}(x) & 0 \\ 0 & g_{mn}(y) \end{pmatrix} . \tag{3.8}$$

The Ricci tensor associated with (3.8), which we denote by $\hat{R}_{MN}$, has the following components

$$\begin{aligned} \hat{R}_{\mu\nu} &= R_{\mu\nu} - \tfrac{1}{2}\, g_{\mu\nu}\; \Delta\; D^m(\Delta^{-3}\partial_m \Delta) \,, \\ \hat{R}_{mn} &= R_{mn} - 2\, \Delta^{\frac{1}{2}}\; D_m(\Delta^{-3/2}\partial_n \Delta) \,, \end{aligned} \tag{3.9}$$

where $R_{\mu\nu}$ and R_{mn} are the Ricci tensors corresponding to $g_{\mu\nu}$ and g_{mn}, respectively. Using a similar decomposition for the higher-dimensional energy-momentum tensor $\hat{T}_{MN}$,

$$\begin{aligned} \hat{T}_{\mu\nu} &= \Delta^{-1}\; g_{\mu\nu}\; t \,, \\ \hat{T}_{mn} &= T_{mn} \quad , \quad T \equiv g^{mn}\; T_{mn} \,, \end{aligned} \tag{3.10}$$

the Einstein equation implies

$$3m_4^2 - \tfrac{1}{2}\,\Delta\, D^m\big(\Delta^{-3}\partial_m\Delta\big) + \frac{(d-6)t-T}{(d-2)\Delta} = 0\ ,$$

$$R_{mn} - 2\,\Delta^{\frac{1}{2}}\, D_m\big(\Delta^{-3/2}\partial_n\Delta\big) + T_{mn} - \frac{4t+T}{d-2}\, g_{mn} = 0\ . \tag{3.11}$$

For a large class of energy-momentum tensors, including those of d=10, 11 supergravity, it can be shown that $(d-6)t < T$. In that case it follows that there are no de Sitter solutions ($m_4^2 < 0$) when the extra dimensions parametrize a compact space. Furthermore, for Minkowski solutions ($m_4^2 = 0$) the warp factor must be constant and only scalar fields may take nonzero values, whereas for anti-de Sitter solutions ($m_4^2 > 0$) the warp factor may be nontrivial and there are no obvious restrictions on the values of the various fields other than those implied by the maximal symmetry of the 4-dimensional subspace [26]. Indeed, the solutions of d=11 supergravity where the warp factor is relevant describe an anti-de Sitter space. Finally we note that the warp factor will also appear in the supersymmetry transformations laws, and must be taken into account when examining the possible residual supersymmetry of the field configuration [25-27].

4. NEW TOPOLOGICAL TERMS IN N=2 SUPERGRAVITY

Some time ago a study of general N=2 supergravity models in the more familiar 4-dimensional setting has revealed the existence of an interesting class of topological terms [28]. These terms involve noninvariant constant tensors belonging to nontrivial Lie-algebra cohomology classes [29]. In this section we shall briefly describe how these terms arise in N=2 supergravity and present the relevant mathematical ingredients for their construction in general even-dimensional space-times. For more extensive treatments the reader is referred to [29] and [30].

Gauge theories with N=2 supersymmetry can be described in terms of reduced chiral superfields $W^I(x,\theta)$ [31], where the index I labels the vector multiplets. In the nonabelian case, these multiplets are assigned to the adjoint representation of the gauge group. In the θ-expansion of $W^I(x,\theta)$ one finds a complex scalar ϕ^I, a (Yang-Mills or Maxwell) field strength $F^I_{\mu\nu}$, a doublet of Majorana spinors and auxiliary fields; in particular ϕ^I and $F^I_{\mu\nu}$ appear according to

$$W^I(x,\theta) = \phi^I(x) + \varepsilon_{ij}\,\bar\theta^i\sigma^{\mu\nu}\theta^j\, F^I_{\mu\nu}(x) + \ldots \tag{4.1}$$

The simplest action for vector multiplets is obtained by squaring $W^I(x,\theta)$ and integrating over chiral superspace coordinates

$$\int d^4x\, d^4\theta\, \left(W^I(x,\theta)\right)^2$$

$$\propto \int d^4x\, \left\{-\tfrac{1}{4}F^{I\mu\nu}\left(F^I_{\mu\nu}+i\tilde{F}^I_{\mu\nu}\right) + \tfrac{1}{2}\phi^I\, \square\, \phi^{*I} + \ldots\right\} , \qquad (4.2)$$

where we have only indicated the terms depending on $F^I_{\mu\nu}$ and ϕ^I ($\tilde{F}^I$ is the dual of F^I and $\square$ denotes the gauge covariant D'Alembertian). The real part of (4.2) is the action of supersymmetric Yang-Mills theory, while the imaginary part containing $F\tilde{F}$ and its supersymmetric completion is the integral over a total divergence.

However, one may also choose a more complicated function $S(W(x,\theta))$ and integrate over the chiral superspace. This leads to an action

$$\mathrm{Re} \int d^4x\, d^4\theta\, S(W(x,\theta))$$

$$\propto \int d^4x\, \left\{\tfrac{1}{4}\mathrm{Re}\left(S_{,IJ}(\phi)\right) F^{I\mu\nu}\, F^J_{\mu\nu} - \tfrac{1}{4}\, \mathrm{Im}\left(S_{,IJ}(\phi)\right) F^{I\mu\nu}\, \tilde{F}^J_{\mu\nu} + \ldots\right\} , \qquad (4.3)$$

where $S_{,IJ}$ denotes the second derivative of S. The term proportional to $F\tilde{F}$ is now real, so that the lagrangian is a total divergence if $S_{,IJ}$ is purely imaginary and constant; the latter is the case if the function S is a quadratic polynomial with imaginary coefficients.

In deriving the above result we have tacitly assumed that the function S is gauge invariant, a requirement that turns out to be too strong: under an infinitesimal gauge transformation S may transform into a quadratic polynomial with imaginary coefficients, viz.

$$\delta S(W(x,\theta)) = -\tfrac{1}{2}i\; C_{K,IJ}\; \Lambda^K(x)\; W^I(x,\theta)\; W^J(x,\theta) , \qquad (4.4)$$

so that the variation of the lagrangian corresponding to (4.3) is equal to (there are no fermionic contributions)

$$\delta L = C_{K,IJ}\; \Lambda^K\; F^{I\mu\nu}\; \tilde{F}^J_{\mu\nu} . \qquad (4.5)$$

From the requirement that the gauge group is consistently realized on the function S it follows that the coefficients $C_{K,IJ}$ should satisfy the condition

$$\tfrac{1}{2}\, f^{M}_{LK}\, C_{M,IJ} + f^{M}_{I[K}\, C_{L],MJ} + f^{M}_{J[K}\, C_{L],MI} = 0 \; . \tag{4.6}$$

A second condition is

$$C_{(I,JK)} = 0 \; , \tag{4.7}$$

which can be derived from the requirement that S is a homogeneous function of second degree. The homogeneity requirement follows from local supersymmetry, i.e. the coupling to supergravity.

It is now possible to show that the addition of

$$L^{add} = -\tfrac{2}{3}\, i\varepsilon^{\mu\nu\rho\sigma}\, C_{K,IJ}\, A^{K}_{\mu}\, A^{I}_{\nu}\, \left(\partial_{\rho} A^{J}_{\sigma} - \tfrac{3}{8}\, f^{J}_{MN}\, A^{M}_{\sigma} A^{N}_{\sigma}\right) \tag{4.8}$$

to the action restores gauge invariance, as well as (local) supersymmetry. The structure of this term is reminiscent of a Chern-Simons term, but the obvious difference is that C is not necessarily an invariant tensor according to (4.6). Its underlying mathematical structure turns out to be related to Lie algebra cohomology, as we shall now explain. To keep the formulae compact we make use of differential forms.

To define Lie algebra cohomology, consider

$$C(\theta,\xi) = C_{I_1 \ldots I_n,\alpha}\, \theta^{I_1} \ldots \theta^{I_n}\, \xi^{\alpha} \; , \tag{4.9}$$

where $C_{I_1 \ldots I_n,\alpha}$ is a constant tensor, antisymmetric in Lie-algebra indices $I_1 \ldots I_n$, θ^I is anticommuting and transforms under the gauge group as

$$\delta\theta^{I} = f^{I}_{JK}\, \Lambda^{J} \theta^{K} \; , \tag{4.10}$$

and ξ^{α} can be either commuting or anticommuting and transform according to some arbitrary representation of the gauge group with generators T_I,

$$\delta\xi^{\alpha} = (T_I)^{\alpha}{}_{\beta}\, \Lambda^{I}\, \xi^{\beta} \; . \tag{4.11}$$

The operator ∂^* defined by

$$\partial^*\theta^{I} - \tfrac{1}{2} f^{I}_{JK}\, \theta^{J}\theta^{K} = 0 \; , \quad \partial^*(\theta^{I}\theta^{J}) = (\partial^*\theta^{I})\theta^{J} - \theta^{I}\partial^*\theta^{J} \; ,$$

$$\partial^*\xi^{\alpha} - (T_I)^{\alpha}{}_{\beta}\, \theta^{I}\xi^{\beta} = 0 \; , \tag{4.12}$$

satisfies $(\partial^*)^2 = 0$, as follows by explicit computation. Application of ∂^* on (4.9) induces an algebraic operation $\mathfrak{d}$ on the coefficients $C_{I_1 \ldots I_n, \alpha}$:

$$\begin{aligned} \partial^* C(\theta,\xi) &\equiv C_{I_1\ldots I_n,\alpha}\, \partial^*(\theta^{I_1}\ldots\theta^{I_n}\xi^\alpha) \\ &\equiv (\mathfrak{d}C)_{I_o\ldots I_n,\alpha}\, \theta^{I_o}\ldots\theta^{I_n}\xi^\alpha , \end{aligned} \tag{4.13}$$

with

$$(\mathfrak{d}C)_{I_o\ldots I_n,\alpha} = \frac{n}{2}\, C_{K[I_2\ldots I_n,\alpha}\, f^K_{I_o I_1]} + (T_{[I_o})^\beta{}_\alpha\, C_{I_1\ldots I_n],\beta} \;. \tag{4.14}$$

As $(\partial^*)^2 = 0$ this implies that $\mathfrak{d}^2 = 0$, which can also be verified directly from the definition (4.14). The algebraic operator $\mathfrak{d}$ leads to the notion of Lie-algebra cohomology and we can define cohomology classes for $\mathfrak{d}$: if C is $\mathfrak{d}$-closed, i.e. if $\mathfrak{d}C = 0$, then the equivalence class of all C' that are of the form $C' = C + \mathfrak{d}\Lambda$ for some Λ defines the cohomology class of C. The cohomology class is trivial if it contains zero, i.e. if C is $\mathfrak{d}$-exact: $C = \mathfrak{d}\Omega$. For semisimple groups, one can construct the Casimir invariant $C_2(T)^\beta{}_\alpha = g^{IJ}\, T_I{}^\beta{}_\gamma\, T_J{}^\gamma{}_\alpha$, where gIJ is the inverse Cartan-Killing metric. If C_2 has an inverse, one can show that a $\mathfrak{d}$-closed C must be $\mathfrak{d}$-exact. Therefore in nontrivial representations one can only have nontrivial Lie-algebra cohomology classes if the group is nonsemisimple (a proof of this result is presented in [29]).

Consider now the case that ξ^α transforms as the symmetric tensor product of m copies of the Lie algebra itself. Hence ξ^α can be replaced by

$$\xi^\alpha \to u^{J_1}\ldots u^{J_m} , \tag{4.15}$$

where the u^J are commuting variables that obey

$$\partial^* u^J - f^J_{KL}\, \theta^K\, u^L = 0 , \tag{4.16}$$

and the coefficients $C_{I_1\ldots I_n,\alpha}$ are replaced by

$$C_{I_1\ldots I_n,\alpha} \to C_{I_1\ldots I_n,J_1\ldots J_m} , \tag{4.17}$$

Consequently C is antisymmetric in indices $I_1 - I_n$ and symmetric in indices $J_1 - J_m$. The operator $\mathcal{D}$ now takes the form

$$(\mathcal{D}C)_{I_o\ldots I_n, J_1\ldots J_m} = \frac{n}{2}\, C_{K[I_2\ldots I_n, J_1\ldots J_m}\, f^K_{I_o I_1]}$$
$$- m\, f^K_{(J_1[I_o}\, C_{I_1\ldots I_n], KJ_2\ldots J_m)} \quad , \qquad (4.18)$$

where the antisymmetrization applies to $I_o - I_n$ and symmetrization to $J_1 - J_m$. For the restricted class of coefficients C satisfying

$$C_{I_1\ldots(I_n, J_1\ldots J_m)} = 0 \quad , \qquad (4.19)$$

one can show that $(\mathcal{D}C)_{I_o\ldots I_n, J_1\ldots J_m}$ also satisfies (4.19) and is thus in the same class. Introducing gauge potentials A^I and field strengths $F^I \equiv dA^I - \frac{1}{2} f^I_{JK}\, A^J A^K$, it is possible to construct n+2m forms

$$C(A,F) \equiv C_{I_1\ldots I_n, J_1\ldots J_m}\, A^{I_1}\ldots A^{I_n}\, F^{J_1}\ldots F^{J_m} \; . \qquad (4.20)$$

For coefficients C that satisfy (4.19) one can then establish a relation between de Rham and Lie-algebra cohomology, namely

$$dC(A,F) = \mathcal{D}C(A,F) \quad , \qquad (4.21)$$

so that the exterior derivative induces the algebraic $\mathcal{D}$ operator.

Let us now return to gauge invariant actions such as (4.3) and (4.8), involving scalar fields ϕ^i and gauge fields A^I. We start from the action (in Euclidean space)

$$I_o = \int_{M^{2m}} S_{J_1\ldots J_m}(\phi)\, F^{J_1}\ldots F^{J_m} \quad , \qquad (4.22)$$

which is not invariant because $S_{J_1\ldots J_m}$ transforms under gauge transformations as

$$\delta S_{J_1\ldots J_m}(\phi) \equiv S_{J_1\ldots J_m, i}\, k^i_I(\phi)\, \Lambda^I$$
$$= \{m\, S_{K(J_2\ldots J_m}\, f^K_{J_1)I} + C_{I, J_1\ldots J_m}\}\Lambda^I \quad , \qquad (4.23)$$

where ,i denotes differentiation with respect to ϕ^i and $k^i_I(\phi)$ characterize the infinitesimal transformations of the scalar fields

$$\delta\phi^i = k^i_I(\phi)\Lambda^I \,, \tag{4.24}$$

which are usually the Killing vectors associated with some sigma model manifold. The first term in the second equation (4.23) proportional to the structure constants represents the covariant variation of $S_{J_1\ldots J_m}$, whereas the constant tensor $C_{I,J_1\ldots J_m}$ represents a noncovariant piece, whose presence is responsible for the lack of gauge invariance of (4.22).

An important observation is that the gauge transformations should be consistently realized on $S_{J_1\ldots J_m}$ (i.e. $[\delta_1\ \delta_2]S = \delta_3 S$, where δ_3 follows from the group property). This leads to a generalization of (4.6) which is just the condition that C is $\mathfrak{d}$-closed, or, explicitly,

$$\tfrac{1}{2} f^I_{KL}\, C_{I,J_1\ldots J_m} + \frac{m}{2}\left(C_{K,M(J_2\ldots J_m}\, f^M_{J_1)L} - K \leftrightarrow L\right) = 0 \,. \tag{4.25}$$

A second observation is that tensors C that are $\mathfrak{d}$-exact, i.e. $C = \mathfrak{d}s$, where $s_{J_1\ldots J_m}$ is a constant symmetric m-rank tensor, can be absorbed into the definition of $S_{J_1\ldots J_m}$ (cf. (4.23)). Therefore we are only interested in those cases where C belongs to a nontrivial Lie-algebra cohomology class, which implies that the subsequent derivation is only relevant for nonsemisimple gauge groups.

Multiplying the second equation (4.23) by $F^{J_1}\ldots F^{J_2}$, replacing Λ^I by A^I and using the Bianchi identity $dF^I = f^I_{JK}\, A^J\, F^K$, one derives

$$S_{J_1\ldots J_m,i}\, \nabla\phi\, F^{J_1}\ldots F^{J_m} = d\left(S_{J_1\ldots J_m}\, F^{J_1}\ldots F^{J_m}\right) - C(A,F) \,, \tag{4.26}$$

where

$$\nabla\phi^i \equiv d\phi^i - k^i_I(\phi)A^I \tag{4.27}$$

is the covariant derivative of ϕ^i. Because the derivative $S_{J_1\ldots J_m,i}$ is a covariant tensor in view of the fact that $C_{I,J_1\ldots J_m}$ in (4.23) is ϕ-independent, we conclude that both sides of (4.26) are gauge invariant. Furthermore we have already established that $\mathfrak{d}C = 0$ and that the results depend only on the cohomology class of C. Finally, assuming that C also satisfies the condition (4.19) we know from (4.21) that both sides of (4.26) are closed. Therefore locally we can write (4.26) as the exterior derivative of a 2m-form, which is gauge invariant modulo a closed 2m-form. By standard techniques [32], it is possible to find an explicit expression for this 2m-form. The result is

$$S_{J_1\ldots J_m,i}\,\nabla\phi^i\,F^{J_1}\ldots F^{J_m}$$

$$= d\{S_{J_1\ldots J_m}\,F^{J_1}\ldots F^{J_m} + \int_0^1 dt\; mt\; C_{I,J_1\ldots J_m}\,A^I\,A^{J_1}\,\hat{F}^{J_2}(t)\ldots\hat{F}^{J_m}(t)\}\,, \qquad (4.27)$$

where $\hat{F}^I(t)$ is defined by

$$\hat{F}^I(t) = t\, dA^I - \tfrac{1}{2}t^2\, f^I_{JK}\, A^J A^K\,. \qquad (4.28)$$

There is an obvious analogy [30] between (4.27) and the well-known relation between the gauge invariant Pontryagin 2m-form $\mathcal{P}_{2m}(F)$, which can be written as

$$\mathcal{P}_{2m}(F) = C_{J_1\ldots J_m}\,F^{J_1}\ldots F^{J_m}\,, \qquad (4.29)$$

with C the invariant tensor

$$C_{J_1\ldots J_m} \propto \mathrm{Tr}\left(T_{(J_1}\cdots T_{J_m)}\right)\,, \qquad (4.30)$$

and the Chern-Simons (2m-1)-form ω_{2m-1}. This relation is expressed by

$$\mathcal{P}_{2m}(F) = d\omega_{2m-1}(A,F)\,. \qquad (4.31)$$

An explicit representation of ω_{2m-1} is [32]

$$\omega_{2m-1}(A,F) = \int_0^1 dt\; m\; C_{J_1\ldots J_m}\,A^{J_1}\,\hat{F}^{J_2}(t)\ldots\hat{F}^{J_m}(t)\,, \qquad (4.32)$$

which has a similar structure as the second term on the right-hand side of (4.27).

Just as one can construct covariant actions in odd dimensions from the Chern-Simons term, which in three dimensions yields a topological mass term for gauge fields, it is possible to have invariant actions by integrating the 2m-form on the right hand side of (4.27). This action then contains the modification that is necessary to make (4.22) gauge invariant. For m = 2 this is precisely the term (4.8) found in N = 2 supergravity. The lagrangian corresponding to the 2m-form is only gauge invariant modulo a total divergence. Consequently, the action is invariant under infinitesimal transformations that approach the identity at the boundary. For finite gauge transformations, however, the variation of the lagrangian leads to a form that is closed but not exact (as emphasized in [30] this form defines a generalized Wess-Zumino term [33]). Locally the variation of the lagrangian therefore takes the form of a total divergence, but for transformations that are not continuously connected to the identity, this relationship does not hold globally. Therefore the action will not be invariant for homotopically nontrivial transformations. This phenomenon, which is well-known for actions constructed from the Chern-Simons term [34], may also play a role in the present case.

REFERENCES

1. R. Slansky, Phys. Rep. 79 (1981) 1.
2. L. Brink, J. Scherk and J.H. Schwarz, Nucl. Phys. B121 (1977) 77.
   F. Gliozzi, J. Scherk and D. Olive, Nucl. Phys. B122 (1977) 253.
3. M.B. Green and J.H. Schwarz, Phys. Lett. 122B (1983) 143;
   J.H. Schwarz and P.C. West, Phys. Lett. 126B (1983) 301;
   J.H. Schwarz, Nucl. Phys. B226 (1983) 269;
   P.S. Howe and P.C. West, Nucl. Phys. B238 (1984) 181.
4. E. Cremmer, B. Julia and J. Scherk, Phys. Lett. 76B (1978) 409.
5. A.H. Chamseddine, Nucl. Phys. B185 (1981) 403.
6. E. Bergshoeff, M. de Roo, B. de Wit and P. van Nieuwenhuizen, Nucl. Phys. B195 (1982) 97.
7. A.H. Chamseddine, Phys. Rev. 24D (1981) 3065.
8. E. Bergshoeff, M. de Roo and B. de Wit, Nucl. Phys. B217 (1983) 489.
9. S.J. Gates and H. Nishino, Phys. Lett. 157B (1985) 157.
10. B. de Wit, in "Supersymmetry, Supergravity and Superstrings '86", eds. B. de Wit, P. Fayet, M.T. Grisaru (World Scient., to appear).
11. G.F. Chapline and N.S. Manton, Phys. Lett. 120B (1983) 109.
12. M.B. Green and J.H. Schwarz, Phys. Lett. 149B (1984) 117.
13. A. Salam and E. Sezgin, Phys. Scripta 32 (1985) 283.
14. P. Howe, H. Nicolai and A. Van Proeyen, Phys. Lett. 112B (1982) 446.
15. A. Salam and J. Strathdee, Ann. Phys. (N.Y.) 141 (1982) 316.
16. B. de Wit and H. Nicolai. Nucl. Phys. B231 (1984) 506.
17. M.J. Duff, B.E.W. Nilsson, C.N. Pope and N.P. Warner, Phys. Lett. 149B (1984) 90.
18. D.J. Gross, J. Harvey, E. Martinec and R. Rohm, Nucl. Phys. B267 (1986) 75.

M.D. Freeman and C.N. Pope, Phys. Lett. 174B (1986) 48.

M.D. Freeman, C.N. Pope, M.F. Sohnius and K.S. Stelle, Phys. Lett. 178B (1986) 199.

D.J. Gross and E. Witten, Nucl. Phys. 277B (1986) 1.

P. Candelas, M.D. Freeman, C.N. Pope, M.J. Sohnius and K.S. Stelle, Phys. Lett. 177B (1986) 341.

D. Chang and H. Nishino, Maryland preprint PPN 86-178.

19. E. Cremmer and B. Julia, Nucl. Phys. B159 (1979) 141.

20. M.J. Duff and C.N. Pope, Nucl. Phys. B255 (1985) 355.
See also, M.J. Duff, in "Supersymmetry", proc. of the NATO Advanced Study Institute, Bonn, 1984, eds. K. Dietz, R. Flume, G. von Gehlen and V. Rittenberg (Plenum, 1985).

21. M.A. Awada, B.E.W. Nilsson and C.N. Pope, Phys. Rev. D29 (1984) 334.

22. H. Nicolai, in proc. of the 1985 Les Houches Summer School, (North-Holland, to appear).

23. B. de Wit and H. Nicolai, Nucl. Phys. B281 (1987) 211.

24. B. de Wit and H. Nicolai, Nucl. Phys. B208 (1982) 323.

25. B. de Wit, H. Nicolai and N.P. Warner, Nucl. Phys.B255 (1985) 29.

26. B. de Wit, D.J. Smit, N.D. Hari Dass, Nucl. Phys. B283 (1987) 165.

27. A. Strominger, Nucl. Phys. B274 (1986) 253.

28. B. de Wit, P.G. Lauwers and A. Van Proeyen, Nucl. Phys. B255 (1985) 569.

29. B. de Wit, C.M. Hull and M. Roček, M., Utrecht preprint, Phys. Lett. B, to appear.

30. B. de Wit, Physica Scripta 34 (1986)...

31. R. Grimm, M.F. Sohnius and J. Wess, Nucl. Phys. B133 (1978) 275.

32. B. Zumino, Y-S. Wu and A. Zee, Nucl. Phys. B239 (1984) 477.
B. Zumino, in: Relativity, Groups and Topology II (notes by K. Sibold), B.S. DeWitt and R. Stora, eds. (North-Holland, 1984).
Stora, R., in: Recent Progress in Gauge Theories, H. Lehmann et al., eds. NATO ASJ Series B Physics Vol. 115 (Plenum, 1984).
W.A. Bardeen, and B. Zumino, Nucl. Phys. B244 (1984) 421.
L. Alvarez-Gaumé and P. Ginsparg, Ann. Phys. (N.Y.) 161 (1985) 423.

33. J. Wess and B. Zumino, Phys. Lett. 37B (1971) 95.
E. Witten, Nucl. Phys. B223 (1983) 422.

34. S. Deser, R. Jackiw and S. Templeton, Ann. Phys. (N.Y.) 140 (1982) 372.

VACUUM STABILITY IN KALUZA-KLEIN THEORIES

J. Strathdee

International Centre for Theoretical Physics
Trieste, Italy

INTRODUCTION

When it was discovered that string theories are inconsistent except when they are formulated in spacetimes of either 26 or 10 dimensions, a new reason for investigating dynamical geometry was born[1]. Up to this point the general theory of relativity had stood in isolation. The gravitational force, based as it was in the geometry of spacetime, seemed different from the other "fundamental" forces. It is true that, among the proposals for unification, there was an early suggestion by Kaluza[2] that spacetime should be viewed as a 5-dimensional cylinder in order to put electromagnetism on the same geometrical footing as gravitation. Interesting as it was, this idea was somewhat premature in that many fundamental forces had yet to be discovered. This may yet be the case. If electromagnetism is to be explained by such a mechanism then the distance scales (e.g.the cylinder's radius) must be smaller by many orders than what is experimentally resolvable. This means that the most novel and distinctive qualities of the 5-dimensional geometry, the most characteristic predictions of the theory, cannot be tested. For this reason, the Kaluza-Klein theories have remained for many years a curiosity, and Einstein's theory of gravitation maintains in isolation.

In its original form as a theory of hadrons, the string theory's scale parameter, α', was assigned the appropriate size, $\sim(10^{-13}\text{cm})^2$. When the problem with dimensions emerged, this scale was boldly reduced to the gravitational scale $\sim(10^{-33}\text{cm})^2$. The reasons for this drastic change are both elegant and convincing[3]. Firstly, it was known that among the states of the free, closed bosonic string there is a massless tensor. Insofar as they were known the couplings of this tensor to other string states seemed to be governed by a gauge principle, they mimicked the couplings of Einstein's graviton. It therefore seemed possible that general relativity - at least in 26 dimensions - could be incorporated in string theory. The price for this, reducing the scale by 20 orders and foregoing the hadronic interpretation, was compensated by the possibility of "spontaneous compactification"[4]. This phenomenon is central to Kaluza-Klein theory and is the main concern in what follows.

In general relativity theory the geometry of spacetime is dynamical in the sense that its metrical properties are determined by solving equations of motion with appropriate boundary conditions. In particular, the geometry of empty space is determined in this way. The simplest vacuum solution of Einstein's equations is the flat Minkowskian geometry. This solution, whose symmetry is described by the Poincaré group, is certainly stable. It is not the only vacuum solution however. Perhaps the simplest alternative is Kaluza's model which is also flat but with a modified topology. Although classically stable this vacuum may decay spontaneously by a kind of tunneling process. More elaborate vacuum geometries are easy to envisage. Some can be shown to be stable.

One of the main weaknesses in the Kaluza-Klein picture has been the objectionably large degree of arbitrariness in its dynamics. In the early days the basic theory was taken to be pure gravity, described by the Einstein-Hilbert action, perhaps with the addition of a cosmological term, in more than four dimensions. The aim was to set up a vacuum solution with the structure $M^4 \times B^K$ where M^4 is 4-dimensional Minkowski spacetime and B^K is a K-dimensional compact manifold. The size of B^K was to be on the order of the Planck length, 10^{-33}cm. It would thus be too small to be resolved by probes of energy $<10^{19}$GeV, and its only influence in the low energy world of observed phenomena would be through its symmetries. The geometrical symmetries of B^K would govern the gauge interactions of particle physics. The prototype of this was Kaluza's derivation of Maxwell theory by taking the compact space to be a circle. More recently, in order to generate compact manifolds with non-vanishing Ricci curvature[5] and in order to generate chiral symmetries[6], it has been necessary to depart from the purity of the original idea by including non-gravitational components such as vectors and scalars in the supposedly fundamental theory. The programme then becomes very arbitrary.

It was hoped that by requiring supersymmetry, most of the arbitrariness could be lifted. The candidate models were supergravity theories in 10 and 11 dimensions. Of these, the $D = 11$, $N = 1$ theory is yet unable to generate chiral gauge interactions in 4-dimensional spacetime; the $D = 10$, $N = 1$ theory though chiral suffers from anomalies and the $D = 10$, $N = 2$ (chiral) theory has yet to yield any realistic vacuum solutions.

Another fundamental weakness of Kaluza-Klein theories is that, like general relativity in 4 dimensions, they are not renormalizable. They make sense at the classical level but their radiative corrections are out of control.

It may well turn out that all these difficulties are resolved by the 10-dimensional superstring theories. If one of these turns out to be a consistent and finite theory then the vacuum problem will become well defined. If the compactification scale of the 10-dimensional vacuum geometry is significantly larger than the string size parameter then it should be possible to determine this geometry by first extracting a 10-dimensional effective field theory of the massless string states and then treating it as a Kaluza-Klein problem. On the other hand, if the scales are comparable then the problem is basically quantum mechanical and the Kaluza-Klein mechanism would not be suitable.

VACUUM CONFIGURATIONS

Until recently Kaluza-Klein theory has been understood to mean general relativity in spacetimes of $D > 4$ dimensions. The equations of motion are to be obtained from an Einstein-Hilbert action, perhaps supplemented by a cosmological term and various matter terms

$$S = -\int d^D x \sqrt{-g} \left[\lambda + \frac{1}{\kappa^2} R + \frac{1}{4} F^2 + \ldots\right] . \tag{1}$$

This action would be fixed <u>a priori</u> and the resulting dynamics explored. Now we expect the action functional to be determined by the underlying string theory: it is to represent the field theory limit of the string. The metric tensor and Yang-Mills potentials, etc., are associated with massless string states. The terms shown in (1) are of course only the leading terms in an infinite series (terms up to fourth order in the Riemann tensor are now available[7]). Truncation of the series is justified if the dimensionless quantities $\alpha' R$, $\alpha' F$, ... are small.

The vacuum configuration is determined by solving the effective equations of motion,

$$0 = \frac{\delta S}{\delta g^{MN}} = -\sqrt{-g}\left(-\frac{\lambda}{2} g_{MN} + \frac{1}{\kappa^2}\left(R_{MN} - \frac{1}{2} g_{MN} R\right) + \ldots\right)$$

$$0 = \frac{\delta S}{\delta A_M} = \nabla_N F^{NM} + \ldots \tag{2}$$

subject to suitable boundary conditions. In particular, invariance with respect to the 4-dimensional Poincaré group, perhaps extended to $N = 1$ supersymmetry, is usually required.

There can be more than one candidate vacuum solution. For example, D-dimensional Minkowski space, M^D, would be a solution if $\lambda = 0$. This solution is indeed a stable one as demonstrated by Witten[8]. Another solution would be $M^4 \times B^{D-4}$ where B^{D-4} is a compact Ricci flat space (e.g. K_3 in the case $D = 8$ or one of the Calabi-Yau manifolds for $D = 10$)[9].

If the vector fields take up a non-vanishing vacuum configuration then they will contribute to the vacuum stress. The components of T_{MN} will be non-vanishing in the internal directions (tangent space of B^{D-4}) and will therefore generate corresponding components in the Ricci tensor. For internal spaces like S^n or CP^n, which are not Ricci flat, it is essential to have some participation from non-gravitational fields. The simplest example of this occurs in the 6-dimensional Einstein-Maxwell system[5] where the vacuum configuration is the product $M^4 \times S^2$ with the Maxwell field taking the form of a magnetic monopole on S^2. The vector potential of the magnetic monopole happens to be proportional to the spin-connection on S^2. This solution is the first and simplest example of what is now referred to as "embedding the spin connection in the gauge group".

Another possibility is for scalar fields to take up a non-trivial vacuum configuration on the internal manifold. Examples have been given where an internal S^2 maps with non-trivial homotopy into the set of scalars. Such configurations can be shown to be stable.

To illustrate these mechanisms consider the 6-dimensional system comprising in addition to the gravitational field, g^{MN} a $U(1)$ gauge field, A_M and a pair of real scalars ϕ^μ. The action is[10]

$$S = -\int d^6 z \sqrt{-g}\left[\frac{1}{\kappa^2} R + \frac{1}{4} F_{MN}F^{MN} + \lambda + \frac{1}{2t} g^{MN}\partial_M\phi^\mu\partial_N\phi^\nu h_{\mu\nu}(\phi)\right] \tag{3}$$

The scalar fields ϕ^μ are regarded as coordinates on the unit 2-sphere with metric $h_{\mu\nu}(\phi)$. The equations of motion are

$$R_{MN} - \frac{1}{2} g_{MN} R = -\frac{\kappa^2}{2}(T_{MN} - \lambda g_{MN})$$

$$\frac{1}{\sqrt{-g}}\partial_M(\sqrt{-g}\, F^{MN}) = 0$$

$$\frac{1}{\sqrt{-g}}\partial_M(\sqrt{-g}\, g^{MN}\partial_N\phi^\mu) = -\Gamma^\mu_{\nu\lambda}\,\partial_M\phi^\nu\partial_N\phi^\lambda g^{MN}\quad, \tag{4}$$

where the energy momentum tensor is given by

$$T_{MN} = F_{ML}F_N{}^L - \frac{1}{4} g_{MN}F^2 + \frac{1}{t} h_{\mu\nu}(\partial_M\phi^\mu\partial_N\phi^\nu - \frac{1}{2} g_{MN}g^{KL}\partial_K\phi^\mu\partial_L\phi^\nu) \; . \tag{5}$$

The vacuum configuration is described by partitioning the coordinates $z^M = (x^m, y^\mu)$, $m = 0, 1, 2, 3$ and $\mu = 4, 5$ where x^m parametrizes Minkowski spacetime and y^μ the internal 2-sphere,

$$\langle g_{MN}\rangle dz^M dz^N = \eta_{mn}dx^m dx^n + a^2 h_{\mu\nu}(y)dy^\mu dy^\nu \; , \tag{6}$$

where $h_{\mu\nu}$ is the metric contained in (3). In effect we are identifying the internal S^2 with the scalar field configuration,

$$\langle\phi^\mu(y)\rangle = y^\mu \; . \tag{7}$$

In polar coordinates this would read

$$h_{\mu\nu}(y)dy^\mu dy^\nu = d\theta^2 + \sin^2\theta d\varphi^2 \; . \tag{8}$$

In the same coordinates the gauge field configuration is given by

$$\begin{aligned}\langle A_M\rangle dz^M &= \langle A_\mu(y)\rangle dy^\mu \\ &= \frac{n}{2e}(\cos\theta \mp 1)d\varphi \; ,\end{aligned} \tag{9}$$

where n is an integer and e is the $U(1)$ coupling parameter. (The $U(1)$ coupling does not appear explicitly in (3) merely because we have introduced only $U(1)$ neutral fields.)

On substituting this ansatz into the equations of motion (4) one finds a pair of algebraic conditions,

$$\frac{1}{a^2} = \frac{\kappa^2}{2}\left(\frac{n^2}{8e^2a^4} + \frac{1}{t} + \lambda\right) \tag{10a}$$

$$0 = \frac{n^2}{8e^2a^4} - \lambda \tag{10b}$$

which are to be solved for a^2 and λ. The radius a^2 of the internal 2-sphere is thereby fixed, along with the 6-dimensional "cosmological" parameter, λ, in terms of n, e^2, κ^2 and t. The fixing of λ is a kind of fine tuning which adjusts the 4-dimensional vacuum curvature to zero. It represents one of the unrealistic features of these models*. There are two solutions, if $n^2 \lesssim 4e^2t/\kappa^4$,

$$\frac{1}{a^2} = \frac{4e^2}{n^2\kappa^2}\left(1 \pm \sqrt{1 - \frac{n^2}{4e^2}\frac{\kappa^4}{t}}\right) \tag{11a}$$

$$\lambda = \frac{4e^2}{n^2\kappa^4}\left(1 - \frac{n^2}{8e^2}\frac{\kappa^4}{t} \pm \sqrt{1 - \frac{n^2}{4e^2}\frac{\kappa^4}{t}}\right) \; . \tag{11b}$$

Both are stable[10].

---

* This fine tuning of the D-dimensional cosmological parameter in order to flatten the 4-dimensional vacuum can work at the classical level. When quantum effects are allowed for, however, it can be troublesome and may give rise to semiclassical instability. It has been argued by Frieman and Kolb[11] that these configurations can tunnel to a de Sitter geometry where all D - 1 spatial dimensions expand exponentially.

There are of course two distinct mechanisms operating here. It is possible to consider the purely spin-connection induced compactification ($t = \infty$) or the scalar induced ($n = 0$). In the former case the vacuum symmetry includes a local SO(3) corresponding to x-dependent rotations of the internal 2-sphere. In the latter case this symmetry is only global (x-independent).

MODE EXPANSIONS AND STABILITY

We do not yet have general criteria for determining stable ground states in general relativity. We have some partial results concerning particular cases. In particular, the positive energy theorem has been used by Witten to argue that Minkowski spacetime is stable in any number of dimensions[8]. In the case of supergravity the ground state is necessarily stable if it is supersymmetric. Examples of ground states which are stable against weak perturbations have been constructed[12].

Stability against weak perturbations is not sufficient to guarantee quantum stability. Indeed, the 5-dimensional Kaluza cylinder has been tested by Witten and found to be liable to spontaneously form holes[13]. This result is qualified by the observation that the laws of nature may not permit changes of topology. It seems to be true at the classical level that topologically non-trivial ground states are stable, but we do not yet know whether this offers protection against tunneling instabilities.

To test a vacuum solution for classical stability it would be sufficient to construct an energy functional, bilinear in the fluctuation fields, and demonstrate its positivity. This functional generally will contain mass operators, expressed as second order differential operators on the internal manifold, whose spectra must be examined. This can be a difficult problem because the mass operator may couple several distinct fluctuation fields and hence not be manifestly positive. However, if the internal manifold is compact then we know that the spectrum is discrete. If it is a homogeneous space as well then we can use algebraic methods to obtain the spectrum. This has been carried out for S^n and CP^n.

With the vacuum configuration $M^4 \times B^K$ it is natural to choose coordinates (x^m, y^μ) with x^m and y^μ labelling the points of M^4 and B^K, respectively. The fluctuation fields $\phi(x,y)$ can then be expanded in a set of modes,

$$\phi(x,y) = \sum_n \phi_n(x)\, Y_n(y) \quad , \tag{12}$$

where the functions $Y_n(y)$ are orthonormal and complete. The coefficients $\phi_n(x)$, of which there are an infinite number, are dynamical fields which describe the fluctuations from a 4-dimensional viewpoint. If the internal space has some symmetry then the Y_n and corresponding ϕ_n will be classified in representations of this symmetry. In particular if B^K is a quotient space, G/H, then the sum in (12) will run over a selection of the unitary irreducible representations of the symmetry group G and the Y_n will be drawn from the corresponding unitary matrices. To go further with the quotient spaces we must introduce some more specific notation[14].

Let G be a compact Lie group with generators, $Q_{\hat\alpha}$, $\hat\alpha = 1,2\ldots \dim G$. From among the $Q_{\hat\alpha}$ let there be singled out a subset $Q_{\bar\alpha}$ which spans the Lie algebra of H, a subgroup of G. The remainder Q_α, $\alpha = 1,2,\ldots K$ will span the K-dimensional tangent space of G/H. We shall assume that K is reductive,

$$[Q_\alpha, Q_{\bar\beta}] = c_{\alpha\bar\beta}{}^{\gamma} Q_\gamma \ , \tag{13a}$$

but not necessarily symmetric,

$$[Q_\alpha, Q_\beta] = c_{\alpha\beta}{}^{\bar\gamma} Q_{\bar\gamma} + c_{\alpha\beta}{}^{\gamma} Q_\gamma \quad . \tag{13b}$$

(For a symmetric space we would have $c_{\alpha\beta}{}^{\gamma} = 0$.) To complete the Lie algebra we have

$$[Q_{\bar\alpha}, Q_{\bar\beta}] = c_{\bar\alpha\bar\beta}{}^{\bar\gamma} Q_{\bar\gamma} \quad . \tag{13c}$$

To parametrize the space of left cosets, G/H, it is helpful to define the "boosts" L_y which are elements of G. For example one might choose $L_y = \exp(y^\alpha Q_\alpha)$. The action of G on the space G/H is then defined by

$$g\, L_y = L_{gy} h(y,g) \;, \quad g \in G \;\; \text{and} \;\; h \in H. \tag{14}$$

This equation can be solved unambiguously for the transformed coordinates, $(gy)^\alpha$, and h as functions of y and g. Of particular interest is the 1-form $L_y^{-1}\, d\, L_y$ which necessarily belongs to the algebra of G,

$$L_y^{-1}\, dL_y = e^\alpha(y) Q_\alpha + e^{\bar\alpha}(y) Q_{\bar\alpha} \quad . \tag{15}$$

The forms $e^\alpha(y)$ define a basis for the tangent space of G/H. The forms $e^{\bar\alpha}(y)$ provide a spin-connection. The corresponding curvature and torsion 2-forms are given by

$$R = -\frac{1}{2} e^\alpha \wedge e^\beta\, c_{\alpha\beta}{}^{\bar\gamma} Q_{\bar\gamma}$$

$$T^\gamma = \frac{1}{2} e^\alpha \wedge e^\beta\, c_{\alpha\beta}{}^{\gamma} \quad . \tag{16}$$

In particular, if G/H is symmetric then $e^{\bar\alpha}$ is torsionless. If G/H is not symmetric it is still possible to define a torsionless connection,

$$\omega_{\alpha\beta} = e^{\bar\gamma}\, c_{\alpha\bar\gamma\beta} + e^{\gamma}\, \frac{1}{2}\, c_{\alpha\gamma\beta} \quad , \tag{17}$$

which takes its values in the algebra of SO(K), $\omega_{\alpha\beta} = -\omega_{\beta\alpha}$. It contains a piece which lies outside the image of H in SO(K). The embedding of H in the tangent space group, SO(K), is described by the structure constants $c_{\alpha\bar\gamma\beta}$. (As can be seen from (3.2a) these constants characterize the H-content of the tangent space K-vector, Q_α.)

In many cases of interest the 4+K-dimensional gravitational field is coupled to a gauge field. The gauge fields, associated with a group U, may acquire non-vanishing but G-invariant values on G/H. This can happen whenever it is possible to map H into U. Let $q_{\bar\gamma}$ denote the image of $Q_{\bar\gamma}$ in the Lie algebra of U. Then the gauge potential 1-form

$$A = \frac{1}{f} e^{\bar\gamma} q_{\bar\gamma} \tag{18a}$$

and its associated field strength 2-form

$$F = \frac{1}{2f} e^\alpha \wedge e^\beta\, c_{\alpha\beta}{}^{\bar\gamma} q_{\bar\gamma} \tag{18b}$$

can be shown to satisfy the Yang-Mills equations[15]. Moreover, the 2-form F is invariant under left translations of G/H associated with gauge transformations corresponding to the embedding of H in U.

For mode expansions on a quotient space G/H, it is natural to use the matrices of the unitary representations of G. These provide a complete set for representing functions on G. Appropriate subsets are complete on G/H.

To represent a set of functions $\phi_i(x,y)$ belonging to an irreducible representation, $\mathbb{D}$, of the stability group H, the complete set consists of matrix elements of all those irreducible representations of G which contain $\mathbb{D}$ when reduced under H. One can write

$$\phi_i(x,y) = \sum_n \sum_q \sqrt{\frac{d_n}{d_{\mathbb{D}}}}\, D^n_{i,q}(L_y^{-1})\, \phi^n_q(x) \; , \tag{19}$$

where $D^n(g)$ is a unitary matrix of dimension d_n representing $g \in G$. These matrices are needed for the boost elements $g = L_y$. The rows and columns of D^n are labelled by i and q, respectively. All values of the column label must be included in the summation but the row label is not summed, it corresponds to the representation $\mathbb{D}_{ij}(h)$ (of dimension $d_{\mathbb{D}}$) to which ϕ_i belongs. The expansion (19) can be inverted to give

$$\phi^n_q(x) = \frac{1}{V_K}\sqrt{\frac{d_n}{d_{\mathbb{D}}}} \int_{G/H} d\mu \sum_i D^n_{q,i}(L_y)\, \phi_i(x,y) \; , \tag{20}$$

where V_K denotes the volume of G/H and the invariant measure is $d\mu = d^K y\,|\det e_\mu{}^\alpha(y)|$.

A useful property of the functions $D^n(L_y^{-1})$ is their simple response to covariant differentiation. With the connection $e^{\bar{\alpha}}(y)$, the covariant derivative of L_y^{-1} is defined by

$$\nabla L_y^{-1} = dL_y^{-1} + e^{\bar{\alpha}} Q_{\bar{\alpha}} L_y^{-1}$$

$$= -e^{\alpha} Q_\alpha L_y^{-1}$$

on using (15). It follows that the covariant derivative of $D^n(L_y^{-1})$ is given by the matrix elements of Q_α in the representation D^n. Differential operations on G/H are reduced to algebraic manipulations.

In cases where the vacuum configuration is a quotient space G/H, perhaps with G-invariant vectors and scalars, the invariance can be exploited to reduce the fluctuation bilinears into distinct sectors where the stability is easily tested.

STABLE AND UNSTABLE COMPACTIFICATIONS

We consider the compactification of Einstein-Yang-Mills systems in D = 4+K dimensions. It can be shown that if K = 4 and an SU(2) instanton is embedded in the gauge group, U, we can obtain a stable $M^4 \times S^4$ vacuum geometry[12]. It will also be shown that the geometry $M^4 \times B^K$, obtained when the spin connection of B^K is embedded in U, fails the stability test[16].

The Einstein-Yang-Mills action is given by

$$S = -\int d^{4+K}z\, \sqrt{-g}\left(\frac{1}{\kappa^2} R + \frac{1}{4}\vec{F}^2 + \lambda\right) \tag{21}$$

where κ^2 and λ are constants. The field equations derived from (21) reduce for $M^4 \times B^K$ to

$$R_{AB} = -\frac{\kappa^2}{2}\vec{F}_{AC}\cdot\vec{F}_{BC}$$

$$0 = \frac{1}{\kappa^2}R + \frac{1}{4}\vec{F}^2 + \lambda$$

$$\nabla_A F_{AB} = 0 \tag{22}$$

where the indices A, B,... refer to an orthonormal basis on $M^4 \times B^K$. To test for stability introduce the fluctuation h_{MN} and V_M by

$$g_{MN} \to g_{MN} + \kappa h_{MN}\,, \quad \vec{A}_M \to \vec{A}_M + \vec{V}_M\;. \tag{23}$$

We substitute (23) into (21) and retain only the bilinears. To simplify the resulting expression it is advantageous to use the light cone gauge conditions,

$$h_{M-} = 0\,, \quad \vec{V}_- = 0\;. \tag{24}$$

In this gauge the components h_{M+} and V_+ are auxiliary fields and can be eliminated. The remaining components are governed by

$$\begin{aligned}\mathcal{L}_2 = {} & \frac{1}{4}h_{ij}(\partial^2+\nabla^2)h_{ij} + \frac{1}{2}\vec{V}_i(\partial^2+\nabla^2)\vec{V}_i + \frac{1}{2}h_{i\alpha}(\partial^2+\nabla^2)h_{i\alpha} \\ & + \frac{1}{2}R_{\alpha\beta}h_{\alpha i}h_{\alpha i} - \kappa\vec{F}_{\alpha\beta}\cdot\nabla_\beta\vec{V}_i h_{\alpha i} + \frac{1}{2}\vec{V}_\alpha(\partial^2+\nabla^2)\vec{V}_\alpha \\ & + \frac{1}{2}R_{\alpha\beta}\vec{V}_\alpha\cdot\vec{V}_\beta - g\,\vec{F}_{\alpha\beta}\cdot\vec{V}_\alpha\times\vec{V}_\beta - \frac{\kappa^2}{2}(\vec{F}_{\alpha\beta}\cdot\vec{V}_\beta)^2 \\ & + \frac{1}{4}h_{\alpha\beta}(\partial^2+\nabla^2)h_{\alpha\beta} + \frac{2+D}{8D}h\left(\partial^2+\nabla^2+\frac{4R}{D(2+D)}\right)h \\ & + \frac{1}{2}R_{\alpha\beta}h_{\alpha\gamma}h_{\beta\gamma} - \frac{1}{D}R_{\alpha\beta}h_{\alpha\beta}h + \frac{1}{2}\left(R_{\alpha\gamma\beta\delta} - \frac{\kappa^2}{2}\vec{F}_{\alpha\gamma}\cdot\vec{F}_{\beta\delta}\right)h_{\alpha\beta}h_{\gamma\delta} \\ & - \kappa\nabla_\gamma\vec{F}_{\alpha\beta}\cdot\vec{V}_\beta h_{\alpha\gamma} - \kappa\vec{F}_{\alpha\beta}\cdot\nabla_\beta\vec{V}_\gamma h_{\alpha\gamma} + \frac{\kappa}{D}\vec{F}_{\alpha\beta}\cdot\nabla_\beta V_\alpha h\,,\end{aligned} \tag{25}$$

where ∇_α contains both the Yang-Mills and the Riemannian background connections on B^K. The components V_i, $i = 1,2$ and V_α, $\alpha = 3,4,\ldots,D$ refer to transverse directions in M^4 and B^K, respectively. Likewise for h_{ij}, $h_{i\alpha}$ and $h_{\alpha\beta}$. Both h_{ij} and $h_{\alpha\beta}$ are traceless, $h_{ii} = h_{\alpha\alpha} = 0$. Thus, h_{ij} is a field of helicity ± 2 and contains the graviton ($\nabla^2 = 0$) and its massive relatives ($\nabla^2 < 0$). The fields $h_{i\alpha}$ and V_i have helicity ± 1. They include Yang-Mills components corresponding both to unbroken parts of U and to symmetries of the vacuum configuration. The remaining fields, h, $h_{\alpha\beta}$ and V_α have helicity 0. The mixing of these components, governed by the vacuum fields $R_{\alpha\beta\gamma\delta}$ and $\vec{F}_{\alpha\beta}$, can be quite complicated and it is here that instabilities usually arise. (The mixing problem is to a certain extent reduced by observing that each massive helicity ± 2 state must match helicity ± 1 and 0 states, and each remaining helicity ± 1 massive state must have a helicity 0 partner. This is guaranteed by the Poincaré invariance of the vacuum configuration.)

By embedding the spin connection of B^K in the gauge group, $SO(K) \to U$, we make the identification

$$F^i_{\alpha\beta} = \frac{1}{2g}R_{\alpha\beta[\gamma\delta]}c^i_{\gamma\delta} \tag{26}$$

and the equations (22) reduce to

$$\nabla_\alpha R_{\alpha\beta\gamma\delta} = 0$$

$$R_{\alpha\beta} = -\frac{\kappa^2}{2g^2} R_{\alpha\gamma\delta\varepsilon} R_{\beta\gamma\delta\varepsilon}$$

$$\lambda = -\frac{1}{2\kappa^2} R$$

They can be solved by Einstein spaces, $R_{\alpha\beta} = -\Lambda\delta_{\alpha\beta}$ with positive Λ. It is then a matter of detailed computation[15] to show that if the adjoint representation of U contains a vector when decomposed relative to SO(K), then that vector, which must be included among the fluctuations, $\vec{V}_\alpha$, will contain a tachyon. This is because the components V_α^β will contain a singlet $V_\alpha^\beta \sim \delta_\alpha^\beta V_1$ relative to the combined gauge and tangent space action of SO(K), and the equation of motion for V_1 is

$$(\partial^2 + \nabla^2 + \Lambda) V_1 = 0 \ ,$$

where ∇^2 is the Laplace-Beltrami operator on B^K. Since B^K is assumed compact, there is always a zero mode of ∇^2. With $\Lambda > 0$ this yields a tachyon. (It should be remarked that the Ricci flat manifolds like K_3 or Calabi-Yau[9] are not available in the models (21) unless $\vec{F} = 0$.)

Although it has been possible to demonstrate the classical stability (or lack of it) of particular vacuum configurations by exploiting the isometries, none of these examples has phenomenological relevance. They are artificial constructions. However, now that we have a potentially realistic fundamental theory of heterotic strings, the arbitrary and artificial features of previously considered Kaluza-Klein theories may be got rid of. It should now be feasible to set up an effective local field theory in 10-dimensional spacetime to represent the low energy physics. This effective theory should be generally covariant, gauge invariant, supersymmetric, chiral and anomaly free, etc. The stable vacuum configurations of this theory should be interesting.

REFERENCES

1. C. Lovelace, Phys. Lett. 34B(1971)500.
2. Th. Kaluza, Sber. Preuss. Akad. Wiss., K1(1921)966.
3. T. Yoneya, Prog. Theor. Phys., 51(1974)1907.
   J. Scherk & J.H. Schwarz, Nucl. Phys. B81(1974)118.
4. J. Scherk & J.H. Schwarz, Phys. Lett. 57B(1975)463.
5. Z. Horvath, L. Palla, E. Cremmer & J. Scherk, Nucl. Phys. B127(1977)57.
6. E. Witten, "Fermion Quantum Numbers in Kaluza-Klein Theories", in Proc. 1983 Shelter Islan II Conf., MIT Press, Cambridge, MA (1984).
7. D. Gross & E. Witten, Princeton University preprint (1986); M.D. Frieman & C.N. Pope, Imperial College preprint TP/85-86/17 (1986); M.T. Grisaru, A.E.M. van de Ven & D. Zanon, Harvard University preprints HUTP-86/A020, 86/A026 (1986); Y. Kikuchi, C. Marzban & Y.J. Ng, North Carolina preprint IFP-272-UNC (1986).
8. E. Witten, "Kaluza-Klein Theory and the Positive Energy Theorem", Banff Summer Inst. on Particles & Fields (2) (1981).
9. P. Candelas, G. Horowitz, A. Strominger & E. Witten, Nucl. Phys., B258(1985)46.
10. H.J. Shin, Phys. Rev. D34(1986)3102.
11. J.A. Frieman & E.W. Kolb, E. Fermi Inst. preprint 85-29 (1985).
12. S. Randjbar-Daemi, Abdus Salam & J. Strathdee, Nucl. Phys. B242(1984) 447; A.N. Schellekens, Nucl. Phys. B248(1984)706.
13. E. Witten, Nucl. Phys. B195(1982)481.
14. J. Strathdee, Int. J. of Mod. Phys. A1(1986)1.
15. S. Randjbar-Daemi and R. Percacci, Phys. Lett. 117B(1982)41.
16. S. Randjbar-Daemi, Abdus Salam and J. Strathdee, ICTP, Trieste, Internal Report, IC/84/62 (1984).

LECTURES ON NON-LINEAR SIGMA-MODELS AND STRINGS

C.M. Hull

DAMTP, University of Cambridge
Silver St., Cambridge CB3 9EW, England

INTRODUCTION

A non-linear sigma-model is a scalar field theory in which the scalar field takes values in some non-trivial manifold **M**, the <u>target space</u>. The most studied case is that in which **M** is a symmetric space such as a sphere or complex projective space, but here we shall consider general target manifolds. In four space-time dimensions, the sigma-model is non-renormalizable, but has been useful as an effective theory describing the low energy behaviour of scalar mesons and occurs naturally in supergravity theories. In two space-time dimensions, however, the sigma-model is renormalizable [1] and has historically been of interest as a non-trivial "toy" field theory in which calculations are easier than in higher dimensional models such as gauge theories. The supersymmetric sigma-model [2] has a rich geometrical structure [3,4] and has led to interesting mathematical results, such as the construction of new complex geometries [4-7]. Some of these supersymmetric models have been shown to be completely free of ultra-violet divergences [8]. The supersymmetric sigma-model in 0+1 dimensions, i.e. the quantum mechanics of a (super-) particle confined to **M**, has been used to give elegant new proofs of index theorems [9] and Morse inequalities [10].

However, the current interest in sigma-models is largely due to the key role they play in string theory. The action governing the propagation of a (super-) string in a fixed space-time **M** is precisely that of a two-dimensional (supersymmetric) non-linear sigma-model with target space **M** [6,11-16]. The background space-time will be a solution of the string effective field equations if the corresponding sigma-model is (super-) conformally invariant. Thus possible vacuum space-times for (super-) strings can be found by seeking target spaces M that define conformally invariant sigma-models.

In Section I we shall discuss the geometric structure of non-linear sigma-models, their coupling to fermions and the topological quantization condition that must be imposed on adding a Wess-Zumino term. Sigma-model anomalies will be reviewed and it will be shown how to cancel them if there is a Wess-Zumino term present. The extra geometric structure that must be imposed to obtain supersymmetric non-linear sigma-models will be examined in the special case of two space-time dimensions. In Section II the quantization of these models will be discussed, paying particular attention to conformal anomalies. Finally, applications to string theory will be briefly discussed.

I. THE GEOMETRIC STRUCTURE OF THE NON-LINEAR SIGMA-MODEL

1. The Non-Linear Sigma-Model

Consider an n-dimensional space time $(N,\gamma_{\mu\nu})$ with co-ordinates x^{μ} $(\mu,\nu,\ldots=0,1,\ldots,n-1)$ and metric $\gamma_{\mu\nu}(x)$. Then a theory of d free scalar fields $\phi^i(x)$ $(i=1,\ldots,d)$ has the action $(\gamma=|det(\gamma_{\mu\nu})|)$

$$S = \frac{1}{2}\int d^n x\sqrt{\gamma}\gamma^{\mu\nu}\,\delta_{ij}\,\partial_\mu\phi^i\partial_\nu\phi^j \tag{1.1}$$

The ϕ^i can be thought of as co-ordinates for $\mathbb{R}^d$ and (1.1) as a "linear sigma-model" with target space $\mathbb{R}^d$. The natural generalization to a curved target space (M,g_{ij}) with co-ordinates ϕ^i and metric $g_{ij}(\phi)$ is

$$S\Big[\phi,g_{ij}(\phi)\Big] = \frac{1}{2}\mu^{n-2}\int d^n x\sqrt{\gamma}\,\gamma^{\mu\nu}(x)g_{ij}(\phi(x))\,\partial_\mu\phi^i\partial_\nu\phi^j \tag{1.2}$$

Here ϕ^i is chosen to be dimensionless, so that a mass scale μ is in general needed to make the action dimensionless. The presence of this dimensional coupling constant for $n \geqslant 3$ makes the theory non-renormalizable, while its absence in two dimensions leads to renormalizability. We shall often choose units in which $\mu=1$.

It is important that the action (1.2) is independent of the choice of co-ordinates on M as well as on N. Indeed, the action is invariant under the general co-ordinate transformation

$$\phi^i \to \phi'^i(\phi) \qquad g_{ij}(\phi) \to g'_{ij}(\phi') \tag{1.3}$$

where

$$g'_{ij}(\phi') = g_{kl}(\phi) \frac{\partial\phi^k}{\partial\phi'^i} \frac{\delta\phi^l}{\partial\phi'^j} \tag{1.4}$$

or, for infinitesimal transformations with

$$\phi'^i = \phi^i + v^i(\phi) \tag{1.5}$$

for some infinitesimal vector field v^i,

$$g'_{ij}(\phi) = g_{ij}(\phi) - 2\nabla_{(i}v_{j)}(\phi) \tag{1.6}$$

Two metrics g_{ij}, g'_{ij} related by a diffeomorphism as in (1.4) or (1.6) are equivalent and give equivalent sigma-models. Indeed, the action (1.2) with the metric (1.6) satisfies

$$S\left[\phi, g'_{ij}(\phi)\right] = S\left[\tilde{\phi}, g_{ij}(\tilde{\phi})\right] \tag{1.7}$$

where $\tilde{\phi}^i \equiv \phi^i - v^i(\phi)$, (note the change of sign from (1.5)). Thus $S[\phi, g'(\phi)]$ is related to (1.2) by the field redefinition $\phi^i \to \tilde{\phi}^i$ and so (1.2), (1.7) describe the same physics. The sigma-model is then determined by an equivalence class of metrics related by diffeomorphisms [1].

The co-ordinate independence is crucial for ensuring that the action (1.2) is well-defined. A configuration is determined by a map $\hat{\phi}: N \to M$ which, if M and N are contractible, can be specified by the co-ordinate map $x^\mu \to \phi^i(x)$. For most manifolds, however, there will not be a system of co-ordinates that covers the whole of M (or N). Instead it is necessary to introduce a set of co-ordinate patches U_α (labelled by α) with co-ordinates $\phi^i_{(\alpha)}$, such that $M = \bigcup_\alpha U_\alpha$. For any two patches U, U' with co-ordinates ϕ^i, ϕ'^i with non-trivial overlap $U \cap U'$, the metrics $g_{ij}(\phi)$, $g'_{ij}(\phi')$ in U, U' must be related by (1.4) in $U \cap U'$. If x_0^μ is any point that is mapped into this overlap, $\hat{\phi}(x_0) \in U \cap U'$, we can use either ϕ^i and g_{ij} or ϕ'^i and g'_{ij} to define the Lagrangian $L(x_0)$, but we get the same result as, from (1.4), at x_0

$$g_{ij}(\phi)\, \partial_\mu\phi^i \partial^\mu\phi^j = g'_{ij}(\phi')\partial_\mu\phi'^i\partial^\mu\phi'^j \tag{1.8}$$

i.e. the Lagrangian is well-defined. If there were not this co-ordinate independence, it would not in general be possible to continue the

Lagrangian consistently beyond a single co-ordinate patch. Similarly, it is necessary for the consistency of the quantum theory that the effective action be co-ordinate independent at least on-shell. In certain circumstances, as we shall see, co-ordinate independence can be lost owing to sigma-model anomalies.

The sigma-model can be thought of as a field theory with an infinite set of coupling constants $g^{(n)}$, given by the Taylor expansion of the metric

$$g_{ij}(\phi) = g^{(0)}_{ij} + g^{(1)}_{ij,k}\phi^k + \tfrac{1}{2}g^{(2)}_{ij,kl}\phi^k\phi^l + \ldots$$

The general co-ordinate transformations $\phi^i \to \phi'^i$, $g_{ij} \to g'_{ij}$ are not a standard symmetry (there is no Noether current) as they involve a transformation of the coupling constant, as well as the fields. Under the field transformation, $\phi^i \to \phi'^i = \phi^i + k^i(\phi)$, $g_{ij}(\phi) \to g_{ij}(\phi')$, the action changes to

$$S[\phi', g_{ij}(\phi')] = S[\phi, g_{ij}(\phi)] + \int d^n x \sqrt{\gamma}\, \nabla_{(i}k_{j)}\partial_\mu \phi^i \partial^\mu \phi^j \tag{1.9}$$

to linear order in k^i, so that this will be a symmetry only if

$$\nabla_{(i}k_{j)} = 0 \tag{1.10}$$

If this holds, then k^i is a Killing Vector and the symmetry $\delta\phi^i = k^i(\phi)$ is an isometry. The set of all such transformations

$$\delta\phi^i = \Lambda^I k_I{}^i(\phi)$$

forms the isometry group $\mathcal{G}$ parameterized by the constant parameters $\Lambda^I (I = 1, \ldots, \dim(\mathcal{G}))$ and there is a Noether current corresponding to each Killing vector. The isometries can be promoted to a local symmetry by minimal coupling

$$\partial_\mu \phi^i \to D_\mu \phi^i = \partial_\mu \phi^i + a^I_\mu k^i_I. \tag{1.11}$$

where, under an infinitesimal local isometry with parameter $\Lambda^I(x)$, the gauge potential a^I_μ transforms as

$$a^I_\mu \to a^I_\mu + \partial_\mu \Lambda^I + a^J_\mu \Lambda^K c_{JK}{}^I \tag{1.12}$$

where $c_{JK}{}^I$ are the structure constants for $\mathcal{G}$.

Further self-interactions can be added to (1.2), provided they are co-ordinate independent. For example,

$$S_{int} = \int d^n x\sqrt{\gamma}\left[V(\phi) + \Phi(\phi)R^{(\gamma)}(x)\right] \tag{1.13}$$

will be co-ordinate independent provided $V(\phi)$ and $\Phi(\phi)$ are scalars on M. Here $V(\phi)$ is a potential while the term involving the scalar curvature $R^{(\gamma)}(x)$ of N is often referred to as a Fradkin-Tseytlin term [12], which does not affect the dynamics of $\gamma_{\mu\nu}$ if n=2, but leads to a propagating n-metric in n>3. In the context of string theory, g_{ij} is the space-time metric, V is the vacuum expectation value of the tachyon field and Φ is that of the dilaton field [12].

2. The Wess-Zumino Term

In addition to the kinetic term (1.2) constructed with the metric $\gamma_{\mu\nu}$, it is possible (for orientable N) to construct a space-time reparameterization invariant term using the alternating tensor on N $\epsilon_{\mu\nu\ldots\rho}$, (providing $n\leqslant d$)

$$S_{WZ} = \frac{q}{n!}\int d^n x\sqrt{\gamma}\,\epsilon^{\mu_1\ldots\mu_n} b_{i_1\ldots i_n}(\phi)\partial_{\mu_1}\phi^{i_1}\ldots\partial_{\mu_n}\phi^{i_n} \tag{2.1}$$

(Here $\epsilon_{\mu\nu\ldots\rho}$ is a tensor, not a tensor density and q is a constant.) This will be M-co-ordinate invariant if $b_{i_1\ldots i_n}(\phi) = b_{[i_1\ldots i_n]}$ is an antisymmetric tensor on M, corresponding to the n-form on M

$$b = \frac{1}{n!}b_{i_1\ldots i_n}d\phi^{i_1}\wedge\ldots\wedge d\phi^{i_n} \tag{2.2}$$

(See e.g. [17] for a review of differential forms, pull-backs, integration, etc.) Note that (2.1) is only linear in the time derivative of ϕ^i, $\partial_o\phi^i$. The term (2.1) changes only by a surface term under shifts $b\to b+d\lambda$ for any (n-1)-form λ , i.e. under

$$\delta b_{i_1\ldots i_n} = \partial_{[i_1}\lambda_{i_2\ldots i_n]} \tag{2.3}$$

The field equation for ϕ^i obtained by varying $S+S_{WZ}$ is

$$- g_{ij}\left[\frac{1}{\sqrt{(\gamma)}}\partial_\mu\gamma^{\mu\nu}\sqrt{\gamma}\,\partial_\nu\phi^j + \begin{Bmatrix} j \\ k\ l\end{Bmatrix}\partial_\mu\phi^k\partial^\mu\phi^l\right]$$

$$+ q\frac{(n+1)}{n!}H_{ii_1\ldots i_n}\epsilon^{\mu_1\ldots\mu_n}\partial_{\mu_1}\phi^{i_1}\ldots\partial_{\mu_n}\phi^{i_n} = 0 \tag{2.4}$$

and only involves the "field strength"

$$H = db \quad ; \quad H_{i_1 \ldots i_{n+1}} = \partial_{[i_1} b_{i_2 \ldots i_{n+1}]} \tag{2.5}$$

that is invariant under (2.3).

The form b on M and the map $\hat{\phi}$:N→M induces an n-form $\hat{b}$ on N, the <u>pull-back</u> of b, [17]

$$\hat{b} = \frac{1}{n!}\hat{b}_{\mu_1 \ldots \mu_n} dx^{\mu_1} \ldots dx^{\mu_n} \quad , \quad \hat{b}_{\mu_1 \ldots \mu_n} \equiv b_{i_1 \ldots i_n} \partial_{\mu_1} \phi^{i_1} \ldots \partial_{\mu_n} \phi^{i_n} \tag{2.6}$$

The action (2.1) can then be written as an integral of the n-form $\hat{b}$, and so as an integral of b over $\hat{\phi}$(N), the sub-manifold of M given by the image of N under the map $\hat{\phi}$:N→M:

$$S_{WZ} = q\int_N \hat{b} = q\int_{\hat{\phi}(N)} b \tag{2.7}$$

Under (2.3),

$$\delta S_{WZ} = \int_{\hat{\phi}(N)} d\lambda = \int_N d\hat{\lambda} = \int_{\partial N} \hat{\lambda} \tag{2.8}$$

giving an integral over the boundary ∂N of the space-time N which will vanish if ϕ satisfies suitable boundary conditions on ∂N. It is convenient to consider boundary conditions such that $\hat{\phi}$(N) is a compact sub-manifold of M. For example, for flat Euclidean N, if $\phi^i(x) \to \phi_o^i$ for some constant point $\phi_o^i \in M$ as $|x| \to \infty$, then $\hat{\phi}$(N) is topologically an n-sphere.

Suppose B is some n+1 dimensional sub-manifold of M whose boundary is $\hat{\phi}$(N), $\partial B = \hat{\phi}(N)$. Then (2.7) can be rewritten as

$$S_{WZ} = q\int_B H \tag{2.9}$$

as $\int_B H = \int_{\partial B} b$ by (2.5) and Stokes' theorem , and is independent of the choice of the sub-manifold B.

The simplest case is n=1 in which $S+S_{WZ}$ becomes (with $\dot{\phi}^i = \partial_o \phi^i$, $\gamma = \gamma_{oo}$)

$$S = \int dx^1 \sqrt{\gamma}\gamma^{-1}\left[\tfrac{1}{2}\dot{\phi}^i\dot{\phi}^j\, g_{ij}(\phi) + q\dot{\phi}^i b_i(\phi)\right]$$

This describes a particle of charge q constrained to lie on M and coupled to an electromagnetic field with potential b_i and field strength $H_{ij}=\partial_i b_j - \partial_j b_i$, so that (2.3) is the usual Maxwell gauge transformation, $\delta b_i = \partial_i \lambda$. The formalism can be used to describe configurations in which b_i is not globally defined but H_{ij} is, corresponding to the presence of magnetic monopoles. In such cases, H is closed (dH=0) but not exact, i.e. there is no globally defined 1-form b such that H=db. Then in any patch U, a 1-form b can be found such that H=db in U, but any attempt to continue b to a 1-form defined on all of M will lead to a Dirac string singularity in b, although H is well-defined. Potentials b , b′ can be constructed in any patches U, U′, but in the overlap region $U \cap U'$, they will differ by a gauge transformation, $b' = b + d\lambda$ for some λ. As is well known, if H is closed but not exact, i.e. if it is cohomologically non-trivial, the quantum theory can only be consistent if the charge q is quantized (see e.g. [17]). Of course, if b is globally defined, there is no quantization condition.

Although to obtain a well-defined theory it is sufficient that b be a globally defined n-form, it is not necessary; we now extend the general case in analogy with the magnetic monopole n=1 case above. We therefore consider well-defined field strengths H that are closed, dH=0, but not exact so that there are no globally defined potentials b, although they can be defined in patches, differing from patch to patch by "gauge transformations" $b'=b+d\lambda$. When H is not exact (2.1) is known as a Wess-Zumino term but it has become common to extend the use of this name to the case where H is exact and we shall do so here.

The field equation (2.4) is well-defined as it only involves H, so that there is no problem with the classical physics. However, the action (2.1) or (2.7) is not well-defined as it involves b, so that there is a potential obstacle to quantization. In the path integral approach, it is necessary that $\exp(iS_{WZ})$ be well-defined, so that S_{WZ} must be defined up to an integer multiplied by 2π. (In the Euclidean path integral we consider Euclidean N and use $\exp(-S_E)$ where S_E is the Euclidean action; the Euclidean version of (2.1) gains a factor of -i from the Wick rotation of $\epsilon_{\mu\ldots\nu}$, giving $\exp(-S_E) \sim \exp(iS_{WZ})$. One approach is to use the action (2.1) but with different potentials b in different patches and consider the conditions for the Lagrangian to be well-defined in the overlaps of co-ordinate patches in analogy to the discussion in the previous section of the action (1.2) which was also only defined "patch-wise". This has been done by Alvarez [18] who showed that the Lagrangian can, for certain choices of H, be well-defined in multiple overlaps $U \cap U' \cap U'' \ldots$ provided the co-efficient q satisfies a quantization condition.

An alternative approach, which we follow here, is to use the action (2.9) involving the well-defined (n + 1)-form H rather than (2.7), which is no longer well-defined as there is no globally defined b. Note that when H is not exact, Stokes' theorem does not apply and (2.9) is no longer equivalent to (2.7). However, (2.9) is no longer independent of the choice of (n + 1)-manifold B if H is not exact. Consider any two, (n + 1)-dimensional submanifolds of M, B, B', both of which have the boundary $\hat{\phi}(N')$, $\partial B = \partial B' = \hat{\phi}(N)$. Then joining B and B' together at their common boundary, $\hat{\phi}(N)$ gives a closed space without boundary, C. For example, C might be an (n + 1)-sphere with B and B' being upper and lower hemispheres with common boundary an n-sphere ∂B. Then

$$q\int_B H - q\int_{B'} H = q\int_C H \qquad (2.10)$$

(The minus sign in (2.10) comes from considering orientation; as ∂B and $\partial B'$ have the same orientation, to get the boundaries to "cancel", we must take $C = B - B'$.) Thus choosing B or B' gives two different actions (2.9) differing by the right hand side of (2.10) which is not zero in general if H is not exact. For example, in the case n = 1, (2.10) gives the magnetic charge contained in the 2-surface C. For certain choices of H, $\int_C H$ is an integer for all closed C so that choosing $q = (2\pi)$x *integer* would give a well-defined quantum theory.

Let us now investigate the restrictions on H that will ensure that $\int_C H$ is an integer following [20]. Any closed H can be expressed as

$$H = d\tilde{b} + \sum_{m=1}^{B_{n+1}} a_m \omega_m \qquad (2.11)$$

where $\tilde{b}$ is a (globally defined) n-form, a_m are real co-efficients, B_{n+1} is the (n + 1)-th Betti number and $\omega_1, \ldots, \omega_{n+1}$ are a linearly independent set of (n + 1)-forms that are closed but not exact [17]. If M is compact, the ω_m are the harmonic (n + 1)-forms. By de Rham's theorem, [17], the forms ω_m can be chosen such that $\int_C \omega_m$ is an integer for all closed n + 1-

surfaces C; we choose such a basis. Then (2.10) gives

$$\int_C H = \sum_{m=1}^{B_{n+1}} a_m N_m[C] \tag{2.12}$$

where $N_m[C]$ are integers depending only on the topology of C ; they will not change under smooth deformations of C. Then whether or not a consistent quantum theory can be defined will depend on the co-efficients a_m. If the ratios a_m/a_n are rational for all m,n, then there is some real k such that $a_m = n_m k$ for some integers n_m and (2.12) implies

$$q\int_C H = \sum_{m=1}^{B_{n+1}} kq\, n_m N_m[C] \tag{2.13}$$

so that if for some integer N we choose q to be

$$q = \frac{2\pi N}{k} \tag{2.14}$$

(2.13) will be an integer for <u>all</u> choices of C and so $\exp(iS_{wz})$ with $S_{wz} = q\int_B H$ will be independent of the choice of B, i.e. we will have a well-defined quantum theory. However, if any of the ratios a_m/a_n, is irrational, there is no way to ensure that the ambiguity in S_{wz} doesn't affect $\exp(iS_{wz})$ and there is no consistent quantum theory.

In what follows, we shall restrict ourselves to those H that give a consistent quantum theory and absorb the co-efficient q into the definition of b. Not that it is crucial that H be closed ; if $dH \neq 0$, then the action (2.9) changes even under infinitesimal changes in B. The quantization condition (2.14) has been shown to be sufficient for consistency with certain choices of H ; the question of the necessity of (2.14) is briefly considered in [21]. Finally, in the context of string theory, n = 2 and b_{ij} can be interpreted as the expectation value of the antisymmetric tensor gauge field [12].

3. <u>Fermions</u>

Space-time (complex) fermions $\lambda^A(x)$ labelled by an internal index $A = 1,\ldots,r$ can be coupled to the sigma-model through an action

$$S_F = i\int d^n x\sqrt{\gamma}\, G_{AB}(\phi(x))\,\bar{\lambda}^A \rho^\mu (D_\mu \lambda)^B \tag{3.1}$$

where ρ^μ is a space-time γ-matrix satisfying

$$\rho^\mu \rho^\nu + \rho^\nu \rho^\mu = 2\gamma^{\mu\nu} \tag{3.2}$$

and the covariant derivative is

$$(D_\mu \lambda)^A = \nabla_\mu \lambda^A + A_i{}^A{}_B(\phi(x))\partial_\mu \phi^i \lambda^B \tag{3.3}$$

where ∇_μ is a derivative covariant with respect to space-time Lorentz transformations. At this stage, G_{AB} and $A_i{}^A{}_B$ correspond to an infinite number of coupling constants introduced to couple the fermions to the sigma-model. However, there is a symmetry, analogous to (1.3), that suggests the underlying geometry of (3.1) - the symmetry transformations can appear as transition transformations relating the quantities G_{AB}, $A_i{}^A{}_B$ in one "co-ordinate patch" to those in another. Indeed, the action is invariant under the "gauge transformations"

$$\delta\lambda^A = \Lambda^A{}_B(\phi)\lambda^B \qquad \delta G_{AB} = -G_{AC}\Lambda^C{}_B - G_{CB}\Lambda^C{}_A \tag{3.4}$$

$$\delta A_i{}^A{}_B = -\left[\partial\Lambda^A{}_B + [A_i,\Lambda]^A{}_B\right] = -(D_i\Lambda)^A{}_B$$

for any matrices $\Lambda^A{}_B(\phi)$ in the Lie algebra of $GL(r,\mathbb{R})$. The metric G_{AB} is invariant under the $O(r)$ subgroup of $GL(r,\mathbb{R})$ for which

$$\Lambda_{AB} \equiv G_{AC}\Lambda^C{}_B = -\Lambda_{BA} \tag{3.5}$$

The transformations in $GL(r,\mathbb{R})/O(r)$ can be used to bring G_{AB} to the standard form

$$G_{AB}(\phi) = \delta_{AB} \tag{3.6}$$

leaving an $O(r)$ residual invariance. (If G_{AB} is not positive definite, $O(r)$ must be replaced by $O(s,r-s)$ for some s and the R.H.S. of (3.6) becomes the indefinite $O(s,r-s)$ invariant metric.)

The fermions λ^A are a (spinor-valued) vector in $\mathbb{R}^r$. Then (ϕ^i,λ^A) can be thought of as co-ordinates in a vector bundle over M [17] with fibre $\mathbb{R}^r$. That means that the bundle space V can be split into patches W_α, $V = \bigcup_\alpha W_\alpha$, with each patch of the form $W_\alpha = U_\alpha \times \mathbb{R}^r$, where $M = \bigcup_\alpha U_\alpha$. The simplest case is the trivial bundle $V = M \times \mathbb{R}^d$. Then $A_i{}^A{}_B(\phi)$ is a connection on V and $G_{AB}(\phi)$ is a fibre metric.

If two patches W, W' have fibre co-ordinates λ^A, λ'^A, then there is some matrix $U^A{}_B(\phi)$ in GL(r, ℝ) such that $\lambda' = U\lambda$ in $W \cap W'$ and the connections and fibre metrics are related by $G' = -U^T.GU$, $A'_i = U\partial_i U^{-1} + UA_i U^{-1}$. Then just as in section 1, the invariance under (3.4) implies that the Lagrangian can be well-defined over the whole of V, not just locally.

A frequently occurring example is the one in which V = T(M), the tangent bundle of M, so that r = d. Introducing a set of frames $e^a_i(\phi)$ on M, with inverse e^i_a (a = 1,...,d) and

$$G_{ab}(\phi) = g_{ij}e^i{}_a e^j_b \quad , \quad e^a_i e^b_j G_{ab} = g_{ij} \tag{3.7}$$

then locally a basis can be chosen by making a GL(d, ℝ) transformation in which $G_{ab}(\phi) = \delta_{ab}$, i.e. such that the e^a_i are an orthonormal frame. Let ω_i^{ab} be the torsion-free spin-connection, so that

$$\nabla_i e^a_j \equiv \partial_i e^a_j - \begin{Bmatrix} k \\ i\ j \end{Bmatrix} e^a_k + \omega_i^{ab} e^b_j = 0 \tag{3.8}$$

Then the action (3.1) can be rewritten in terms of $\lambda^i(\phi) \equiv e^i{}_a \lambda^a$

$$i\int d^n x\sqrt{\gamma}\, \delta_{ab}\bar{\lambda}^a \gamma^\mu \left[\partial_\mu \lambda^b + \omega_i^{bc}(\partial_\mu \phi^i)\lambda^c\right] \tag{3.9}$$

$$= i\int d^n x\sqrt{\gamma}\, g_{ij}\bar{\lambda}^i \gamma^\mu \left[\partial_\mu \lambda^j + \begin{Bmatrix} j \\ k\ l \end{Bmatrix}(\partial_\mu \phi^k)\lambda^l\right]$$

Finally, the connections $A_i{}^A{}_B$, ω_i^{ab} induce the "pull-back connections" that occur in the action

$$\hat{A}_\mu{}^A{}_B = A_i{}^A{}_B \partial_\mu \phi^i \quad , \quad \tilde{\omega}_\mu^{ab} = \omega_i^{ab} \partial_\mu \phi^i \tag{3.10}$$

transforming as

$$-\delta \hat{A}_\mu{}^A{}_B = \partial_\mu \Lambda^A{}_B + \left[A_\mu, \Lambda\right]^A{}_B \tag{3.11}$$

where

$$\partial_\mu \Lambda = (\partial_i \Lambda)\partial_\mu \phi^i . \tag{3.12}$$

4. Sigma-Model Anomalies

Consider an even-dimensional space-time with n = 2m. It is then possible to define right and left handed complex Weyl fermions

$$\lambda_R = \rho^{n+1}\lambda_R \qquad \chi_L = -\rho^{n+1}\chi_L \tag{4.1}$$

of definite chirality with respect to the "γ_5" on N, $\rho^{n+1} = \rho^0\rho^1\ldots\rho^{n-1}$, $\left[\rho^{n+1}\right]^2 = +1$. Consider a sigma-model coupled to r right-handed fermions λ_R^A (A = 1,...,r) and s left-handed ones χ_L^M (M = 1,...,s) through connections A_{iAB} and B_{iMN} whose pull-backs are $\hat{A}_{\mu AB}$, $\hat{B}_{\mu MN}$. The action is $S + S_{WZ} + S_{F,L} + S_{F,R}$ with

$$S_{F,R} = i\int d^n x\sqrt{\gamma}\,\delta_{AB}\bar{\lambda}_R^A\rho^\mu\left[\nabla_\mu\lambda_R^B + \hat{A}_\mu^{BC}\lambda_R^C\right] \tag{4.2}$$

$$S_{F,L} = i\int d^n x\sqrt{\gamma}\,\delta_{MN}\,\bar{\chi}_L^M\rho^\mu\left[\nabla_\mu\chi_L^N + \hat{B}_\mu^{NP}\chi_L^P\right]$$

With the fibre metrics chosen to be δ_{AB}, δ_{MN} there is no need to distinguish between upper and lower A,B... (or M,N...) indices. There is a local O(r) × O(s) symmetry

$$\delta\lambda^A = \Lambda^{AB}\lambda^B \qquad \delta\hat{A}_\mu^{AB} = -\left[\partial_\mu\Lambda^{AB} + 2\hat{A}_\mu^{C[A}\Lambda^{B]C}\right] \tag{4.3}$$

$$\delta\chi^M = \Theta^{MN}\chi^N \qquad \delta\hat{B}_\mu^{MN} = -\left[\partial_\mu\Theta^{MN} + 2\hat{B}_\mu^{P[M}\Theta^{N]P}\right]$$

where $\Lambda_{AB} = -\Lambda_{BA}$, $\Theta_{MN} = -\Theta_{NM}$, $\Lambda = \Lambda(\phi)$, $\Theta = \Theta(\phi)$, with $\partial_\mu\Theta = \partial_i\Theta(\partial_\mu\phi^i)$ and (3.12).

The actions (4.7) can be thought of as describing chiral fermions coupled to background Yang-Mills fields $\hat{A}_\mu$, $\hat{B}_\mu$ with classical invariance under the gauge transformations (4.3). The path integral over the chiral fermions

$$\exp\left[\frac{i}{\hbar}S_{eff}[\hat{A},\hat{B}]\right] = \int[d\lambda_R][d\chi_L]\exp\left[\frac{i}{\hbar}\left[S_{F,R} + S_{F,L}\right]\right] \tag{4.4}$$

is formally identical to that in a chiral gauge theory. The effective action S_{eff} defined by (4.4) is not gauge invariant; under (4.3), we have the chiral anomaly

$$\delta S_{eff} = \hbar\int_N\left\{\Omega[\hat{A},\Lambda] - \Omega[\hat{B},\Theta]\right\} \tag{4.5}$$

for some n-form on N, Ω. In the present context, this is known as a sigma-model anomaly, first considered in [22]. If the anomalous variation (4.5) could not be cancelled, the breakdown of the gauge invariance (4.3) would have serious consequences. In the last section it was seen that the gauge invariance (4.3) is crucial for a global definition of the theory; if it broke down, it would be possible to define an effective action locally in one co-ordinate patch (of the vector bundle) but it would not be possible to find a well-defined continuation of this to the whole vector bundle. Indeed, the sigma-model anomaly can be understood as a topological obstruction preventing any sensible global definition of an effective action [22].

There is an inherent ambiguity in the definition of S_{eff} - we are free to add any finite local counterterm $\hbar\Delta$ to S_{eff}, $S_{eff} \to S_{eff} + \Delta$. For example, calculating S_{eff} using two different regularization and subtraction schemes would in general give results for S_{eff} differing by some such finite local counterterm. Consequently, the anomaly (4.5) can be changed by the variation of a finite local counterterm, $\hbar\delta\Delta$, for any Δ we choose. The anomaly is said to be trivial if it can be cancelled by adding some such counterterm, i.e. if there is some Δ such that $S'_{eff} = S_{eff} + \hbar\Delta$ is invariant under (4.3).

By considering the commutator $[\delta_\Lambda,\delta_{\Lambda'}]S_{eff}$ and (4.5), we obtain the Wess-Zumino consistency conditions

$$\delta_\Lambda \int \Omega(\hat{A},\Lambda') - \delta_{\Lambda'}\int\Omega(\hat{A},\Lambda) = \int \Omega(\hat{A},[\Lambda,\Lambda']) \tag{4.6}$$

A solution of these consistency conditions is given using the descent equations [23]. First, we define the field strengths

$$\begin{aligned} F_{ij}^{AB} &= 2\partial_{[i}A_{j]}^{AB} + 2A_{[i}^{AC}A_{j]}^{CB}; \qquad G_{ij}^{MN} = 2\partial_{[i}B_{j]}^{MN} + 2B_{[i}^{MP}B_{j]}^{PN} \\ \hat{F}_{\mu\nu}^{AB} &= F_{ij}^{AB}\partial_\mu\phi^i\partial_\nu\phi^j = 2\partial_{[\mu}\hat{A}_{\nu]}^{AB} + 2\hat{A}_{[\mu}^{AC}\hat{A}_{\nu]}^{CB} \\ \hat{G}_{\mu\nu}^{MN} &= G_{ij}^{MN}\partial_\mu\phi^i\partial_\nu\phi^j \end{aligned} \tag{4.7}$$

and the corresponding Lie algebra-valued 2-forms

$$\begin{aligned} F &= \tfrac{1}{2}F_{ij}d\phi^i{}_\wedge d\phi^j = dA + A_\wedge A \quad \hat{F} = \tfrac{1}{2}\hat{F}_{\mu\nu}dx^\mu{}_\wedge dx^\nu = d\hat{A} + \hat{A}_\wedge\hat{A}. \\ G &= \tfrac{1}{2}G_{ij}d\phi^i{}_\wedge d\phi^j \quad \hat{G} = \tfrac{1}{2}\hat{G}_{\mu\nu}dx^\mu{}_\wedge dx^\nu \end{aligned} \tag{4.8}$$

The (m + 1)-th Chern form $C_{m+1}(F)$ is the gauge invariant (2m + 2)-form

$(2m = n)$ formed from the 2-form F

$$C_{m+1}(F) = F^{A_1 A_2}{}_\wedge F^{A_2 A_3}{}_\wedge \ldots {}_\wedge F^{A_m A_{m+1}}{}_\wedge F^{A_{m+1} A_1} = \mathrm{tr}\left[F^{(m+1)}\right] \tag{4.9}$$

The Bianchi identity implies that $dC_{m+1}(F) = 0$, so that at least locally there is some $(2m + 1)$-form $\omega_{2m+1}(A,F)$, such that

$$C_{m+1}(F) = d\omega_{2m+1}(A,F) \tag{4.10}$$

The form ω_{2m+1} is known as the Chern-Simons form. For the case $m = 1$

$$C_2(F) = \mathrm{tr}(F_\wedge F) = d\omega_3 \ , \ \omega_3 = \mathrm{tr}\left[A_\wedge F - \tfrac{1}{3}A_\wedge A_\wedge A\right] \tag{4.11}$$

The Chern-Simons form is not gauge-invariant under (4.3), $\delta_\Lambda \omega_{m+1} \neq 0$, but C_{m+1} is, so that $d\delta_\Lambda \omega_{2m+1} = 0$, and at least locally there is a 2m-form on the space M, $\overset{1}{\omega}_{2m}(A,F,\Lambda)$, such that

$$\delta_\Lambda \omega_{2m+1} = d\overset{1}{\omega}_{2m} \tag{4.12}$$

and its pull-back $\overset{1}{\omega}_{2m}(\hat{A},\hat{F},\Lambda)$ defines a 2m-form on N. For $m = 1$ with (4.11),

$$\overset{1}{\omega}_{2m}(A,F,\Lambda) = \mathrm{tr}\left[\Lambda_\wedge dA\right] \tag{4.13}$$

$$= \mathrm{tr}\left[\Lambda A_{[i,j]}\right] d\phi^i{}_\wedge d\phi^j$$

$$\overset{1}{\omega}_{2m}(\hat{A},\hat{F},\Lambda) = \mathrm{tr}(\Lambda_\wedge d\hat{A}) = \mathrm{tr}\left[\Lambda \hat{A}_{[\mu,\nu]}\right] dx^\mu{}_\wedge dx^\nu$$

$$= \mathrm{tr}\left[\Lambda A_{[i,j]}\right] \partial_\mu \phi^i \partial_\nu \phi^j dx^\mu{}_\wedge dx^\nu .$$

Proceeding similarly, we can define $\omega_{2m+1}(G,B)$ and $\overset{1}{\omega}_{2m}(G,B,\Theta)$.

It can be shown that the anomaly is given (up to the addition of trivial terms) by the form $\overset{1}{\omega}_{2m}$:

$$\Omega[\hat{A},\Lambda] = a_m \overset{1}{\omega}_{2m}[\hat{A},\Lambda] \tag{4.14}$$

$$a_m = \frac{(i)^{m+1}}{(m+1)!(2\pi)^m} \tag{4.15}$$

$$\delta S_{eff} = a_m \hbar \int_N \left[\overset{1}{\omega}_{2m}[\hat{A},\Lambda] - \overset{1}{\omega}_{2m}[\hat{B},\Theta]\right] \tag{4.16}$$

$$= a_m \hbar \int_{\hat{\phi}(N)} \left[\omega^1_{2m}[A,\Lambda] - \omega^1_{2m}(B,\Theta) \right]$$

(It will in general be necessary to add a finite local counterterm to S_{eff} to bring the anomaly to this form.) It follows from the above construction that (4.14) satisfies the consistency conditions (4.6) [23]. It can also be shown that in general (4.16) is non-trivial, i.e. that the anomaly cannot be cancelled by adding a finite local counterterm [24]. For the case $m = 1$, using (4.13), (4.16) can be written as

$$\delta S_{eff} = -\frac{\hbar}{4\pi} \int_N d^2x\sqrt{\gamma}\, \epsilon^{\mu\nu} \partial_\mu \phi^i \partial_\nu \phi^j \left[\Lambda^{AB} \partial_{[i} A^{AB}_{j]} - \Theta^{MN} \partial_{[i} B^{MN}_{j]} \right] \tag{4.17}$$

The anomaly is trivial, however, in the special case in which

$$c_{m+1}(F) = c_{m+1}(G) \tag{4.18}$$

This would hold if $r = s$ and $A_i = B_i$, so that the theory was left-right symmetric, but can hold more generally. Then from (4.10), (4.18)

$$X = \omega_{2m+1}(A,F) - \omega_{2m+1}(B,G) \tag{4.19}$$

is closed, $dX = 0$, while its variation is

$$\delta X = \omega^1_{2m}(A,F,\Lambda) - \omega^1_{2m}(B,G,\Theta) \tag{4.20}$$

Then, as $dX = 0$, if we add to the action the term (2.9) with

$$H = -\left[\hbar a_m / q\right] X \tag{4.21}$$

the variation of this term from (4.20) will precisely cancel the anomaly (4.17). It can be shown that this anomaly cancellation is consistent with the quantization condition for the coefficient, q, of the Wess-Zumino term [25].

Suppose that the anomaly can't be cancelled by modifying S_{eff}. If there is a Wess-Zumino term (2.7) in the classical action, then the anomaly can be cancelled if we demand that the "coupling constants" contained in

$b(\phi)$ transform under $O(r) \times O(s)$ at order $\hbar$ [6]. Under $O(r) \times O(s)$, if in addition to (4.3), b transforms as

$$\delta b = -a_m(\hbar/q)\left[\omega^1_{2m}(A,F,\Lambda) - \omega^1_{2m}(B,G,\Theta)\right] \tag{4.22}$$

then the variation of the Wess-Zumino term is

$$\delta S_{WZ} = \delta\left[q\int_{\hat{\phi}(N)} b\right] = -a_m\hbar\int_{\hat{\phi}(N)}\left[\omega^1_{2m}(A,F,\Lambda) - \omega^1_{2m}(B,G,\Theta)\right] \tag{4.23}$$

and this precisely cancels the anomalous variation (4.16). Then with the transformation (4.22), S_{WZ} is not gauge invariant, but neither are the fermi-determinants contributing to S_{eff}, and the combination $S_{eff} + S_{WZ}$ is gauge invariant.

The $(n + 1)$-form $H = db$ is not invariant under (4.22). However, if we define locally

$$H' = db + a_m(\hbar/q)\left[\omega_{2m+1}(A,F) - \omega_{2m+1}(B,G)\right] \tag{4.24}$$

then $\delta H' = 0$, i.e. H' is invariant under (4.3), (4.22). However, H' cannot always be continued to give a globally defined $(n + 1)$-form on M. Taking the exterior derivative of (4.24) gives

$$dH' = a_m(\hbar/q)\left[c_{m+1}(F) - c_{m+1}(G)\right] \tag{4.25}$$

If the cohomology class of $c_{m+1}(F)$, the $(m + 1)$-th Chern class of F [17], does not equal the $(m + 1)$-the Chern class of G, then there is no globally defined $(2m + 1)$-form H' on M satisfying (4.25) - there is a topological obstruction to the existence of such a form. Equivalently, let S be any $(2m + 1)$-dimensional submanifold of M without boundary, $\partial S = 0$. Then

$$\int_S dH' = a_m(\hbar/q)\int_S\left[c_{m+1}(F) - c_{m+1}(G)\right] \tag{4.26}$$

If H' is a globally defined form on M, then by Stokes' theorem the LHS of (4.26) will vanish for all S. However, the RHS of the (4.26) will only vanish for all S if

$$c_{m+1}(F) - c_{m+1}(G) = d\alpha \tag{4.27}$$

for some globally defined $(m + 1)$-form α, i.e. only if $c_{m+1}(F)$ and $c_{m+1}(G)$

are in the same cohomology class. Otherwise, (4.24) gives only a locally defined gauge-invariant form.

By adding an appropriate finite local counterterm to S_{eff}, it is possible to replace the connections A_i, B_i by

$$A'^{AB}_i = A_i^{AB} + S_i^{AB} \qquad B'^{MN}_i = B_i^{MN} + T_i^{MN} \tag{4.28}$$

for any covariant tensors $S_i^{AB}(\phi)$, $T_i^{MN}(\phi)$ we choose in the anomaly (4.5) and hence in all the subsequent formulae [26]. For example, if m = 1, then on adding the non-gauge invariant finite local counterterm

$$\hbar\Delta = a_1\hbar\int d^2x\sqrt{\gamma}\left[A^{AB}_{[i}S^{AB}_{j]} - B^{MN}_{[i}T^{MN}_{j]}\right]\epsilon^{\mu\nu}\partial_\mu\phi^i\partial_\nu\phi^j \tag{4.29}$$

to S_{eff}, the anomalous variation (4.17) changes, to a similar expression but with A, B replaced by A′, B′ [26].

5. Supersymmetric Non-Linear Sigma-Models

In this section we will investigate supersymmetry in the non-linear sigma-model in two space-time dimensions (n = 2). In two dimensions, spinors have two anti-commuting components. We choose conventions such that a Majorana spinor has real components, $\lambda = \lambda^*$. In a basis in which

$$\rho^3 = \rho^0\rho^1 = \begin{bmatrix}1 & 0\\ 0 & -1\end{bmatrix}$$

a right or left handed Majorana-Weyl (MW) spinor λ_R or χ_L has one real component

$$\lambda_R = \begin{bmatrix}\lambda+\\ 0\end{bmatrix} \qquad \chi_L = \begin{bmatrix}0\\ \chi-\end{bmatrix} \tag{5.1}$$

transforming under a Lorentz transformation with parameter $\alpha_{\mu\nu} = \alpha\epsilon_{\mu\nu}$ as

$$\lambda_+ \to e^{\alpha/2}\lambda_+ \qquad \chi_- \to e^{-\alpha/2}\chi_- \tag{5.2}$$

It is convenient to work with the one component spinors λ_+, χ_-, as they are in irreducible representations of the Lorentz group SO(1,1). In this section we shall take N to be Minkowski space, $\gamma_{\mu\nu} = \eta_{\mu\nu}$, and split 2-vectors V_μ into two pieces, $V^{\pm\pm} = V^0 \pm V^1$, each transforming irreducibly under SO(1,1)

$$V^{++} \to e^{\alpha}V^{++} \qquad V^{--} \to e^{-\alpha}V^{--} \tag{5.3}$$

For example, the chiral fermion action (4.2) becomes in this notation

$$S_{F,R} = \frac{i}{2}\int d^2x\, \delta_{AB}\, \lambda_+^A \left[\partial_{--}\lambda_+^B + \hat{A}_{--}^{BC}\lambda_+^C\right] \tag{5.4}$$

(with factor of ½ for Majorana fermions). Note that the condition for SO(1,1) invariance is that, when all indices are lowered, the number of "plus" indices equals the number of "minus" indices, while $V^{++} = V_{\overline{++}}$, $\lambda^+ = \lambda_-$, $\lambda_+ = -\lambda^-$ etc. with our metric conventions.

The most general sigma-model in flat 2-space with no dimensional parameters has the action

$$S = \frac{1}{2}\int d^2x \Big[g_{ij}(\phi)\partial_{++}\phi^{(i}\partial_{--}\phi^{j)} + b_{ij}(\phi)\partial_{++}\phi^{[i}\partial_{--}\phi^{j]} \tag{5.5}$$

$$+ iG_{AB}(\phi)\lambda_+^A\left[\partial_{--}\lambda_+^B + \hat{A}^B_{--C}\lambda_+^C\right] +$$

$$i\tilde{G}_{MN}(\phi)\chi_-^m\left[\partial_{++}\chi_-^N + \hat{B}^N_{++P}\chi_-^P\right]$$

$$+ U_{ABMN}(\phi)\chi_-^M\chi_-^N\lambda_+^A\lambda_+^B\Big]$$

which is invariant under the chiral fermion $GL(r,\mathbb{R}) \times GL(s,\mathbb{R})$ rotations considered previously, provided that we attribute to $G,\tilde{G},A,B$ the transformation properties considered in previous sections and that U_{ABMN} transforms covariantly. For simplicity, we shall work in the gauge in which

$$G_{AB}(\phi) = \delta_{AB}\ , \qquad \tilde{G}_{MN}(\phi) = \delta_{MN} \tag{5.6}$$

so that the chiral symmetry is reduced to $O(r) \times O(s)$ and so we need not distinguish between upper and lower indices, so that $\lambda^A = \lambda_A$ etc.

The (p,q) superalgebra [6] is generated by the 2-momentum P^μ, p right-handed MW supercharges Q_{+I} ($I = 1,\dots,P$), q left-handed ones $Q_{-I'}$ ($I' = 1,\dots,q$) and the Lorentz charge (acting as in (5.2,5.3))

$$\left\{Q_+^I\ ,\ Q_+^J\right\} = 2P_{++}\delta^{IJ} \qquad \left\{Q_+^I\ ,\ Q_-^{J'}\right\} = 0 \tag{5.7}$$

$$\left\{Q_-^{I'}\ ,\ Q_-^{J'}\right\} = 2P_{--}\delta^{I'J'} \qquad [P_\mu,P_\nu] = 0 \qquad [P_\mu,Q] = 0$$

This algebra can be extended to include central charges and, in the sigma-model context, the generators of target space symmetries (e.g. isometries or fermion rotations) although we shall not consider such extensions here; see [7,27,28].

The conditions for a 2-dimensional sigma-model to have (1,0) or (2,0) supersymmetry were considered in [6]; for (1,1) in [5,29]; for (2,2) and (4,4) in [5] and for (2,1) in [30]. The conditions for the remaining cases follow straightforwardly from these and have also been considered in [7,31,32]. See also [33]. The simpler cases in which $b_{ij} = 0$ had been considered previously in [3]. We give here a more general analysis.

We consider the conditions for (5.5), with (5.6), to be invariant under (p,q) supersymmetries with anti-commuting constant parameters ϵ_-^I $(I = 1,\ldots p)$ $\epsilon_+^{I'}$ $(I' = 1,\ldots,q)$. The most general form for the transformation of ϕ^i we can write without any dimensionful parameters is, for some matrices $f_I(\phi), f_{I'}(\phi)$,

$$\delta\phi^i = f^i_{(I')M}(\phi)\,\epsilon_+^{I'}\chi_-^M + f^i_{(I)A}(\phi)\,\epsilon_-^I\lambda_+^A \tag{5.8}$$

with implicit summation over I and I'. (The ϕ^i are dimensionless, the fermions have dimension +½, the parameters ϵ have dimension −½ and ∂_μ has dimension 1.) The $f^i_{(I')M}$, $f^i_{(I)A}$ must transform as contravariant vectors under diffeomorphisms of M (as the variation $\delta\phi^i$ does). Under $O(r) \times O(s)$, the f's must transform covariantly

$$\delta f^i_{(I)A} = \Lambda^{AB} f^i_{(I)B}, \qquad \delta f^i_{(I')M} = \Theta^{MN} f^i_{(I')N} \tag{5.9}$$

The supersymmetry variation of (5.5) contains terms linear, cubic and quintic in the fermions λ, χ. The terms linear in fermions in δS will cancel only if the f-tensors are covariantly constant ($H_{ijk} \equiv (3/2)\partial_{[i}b_{jk]}$)

$$\nabla_i^{(-)} f^j_{(I)A} \equiv \partial_i f^j_{(I)A} + \Gamma^{(-)j}{}_{ik} f^k_{(I)A} + A_i^{AB} f^j_B = 0 \tag{5.10a}$$

$$\nabla_i^{(+)} f^j_{(I')M} \equiv \partial_i f_{(I')M}{}^j + \Gamma^{(+)}{}^j{}_{ik} f^k_{(I')M} \quad B_i^{MN} f_{(I')N}{}^j = 0 \tag{5.10b}$$

$$\Gamma^{(\pm)}{}^i{}_{jk} = \left\{ {i \atop j\,k} \right\} \pm H_{jk}{}^i \tag{5.11}$$

and if the fermion transformations take the form

$$\delta\chi_-^M = i\epsilon_+^{(I')} f^M_{(I')i}(\partial_{--}\phi^i) + O(\lambda^2, \chi^2, \lambda\chi) \tag{5.12}$$

$$\delta\lambda_+^A = i\epsilon_-^I f^A_{(I)i}(\partial_{++}\phi^i) + O(\lambda^2, \chi^2, \lambda\chi)$$

up to quadratic fermi terms, where $f_i^{\ M} = g_{ij}\delta^{MN}f^j_{\ N}$ etc. The most general Lorentz covariant form (5.12) can take with no dimensionful parameters is

$$\delta\lambda_+^A + \left[\delta\phi^i\right]A_i^{AB}\lambda_+^B = \epsilon_-^I C^{ABC}_{(I)}\lambda_+^B\lambda_+^C + i\epsilon_-^I f^{\ A}_{(I)i}(\partial_{++}\phi^i) \qquad (5.13)$$

$$+ \epsilon_+^{I'} D^{ABM}_{(I')}\lambda_+^B\chi_-^M$$

$$\delta\chi_-^M + (\delta\phi^i)B_i^{MN}\chi_-^N = \epsilon_+^{I'} C^{MNP}_{(I')}\chi_-^N\chi_-^P + i\epsilon_+^{I'} f^{\ M}_{(I')i}(\partial_{--}\phi^i)$$

$$+ \epsilon_-^I D^{MNA}_{(I)}\chi_-^N\lambda_+^A$$

for some tensors $C(\phi), D(\phi)$. The terms involving explicit connections A_i, B_i, could have been absorbed into the definitions of C,D but are displayed explicitly to make both sides of (5.13) $O(r) \times O(s)$ covariant.

The 3-fermion terms in δS cancel if

$$D^{ABM}_{(I')} = 0 \qquad D_I^{MNA} = 0 \qquad (5.14)$$

$$\frac{3}{2}f^{i[A}_{(I)} F^{BC]}_{ij} = \nabla_j C^{ABC}_{(I)} \equiv \partial_j C^{ABC}_{(I)} - 3A_j^{D[A}C^{BC]D}_{(I)} \qquad (5.15a)$$

$$\frac{3}{2}f^{i[M}_{(I')}G^{NP]}_{ij} = \nabla_j C^{MNP}_{(I')} \qquad (5.15b)$$

$$f^{\ i}_{(I)A}G^{MN}_{ij} = 2U_{ABMN}f^{\ B}_{(I)\ j} \qquad (5.16a)$$

$$f^{\ i}_{(I')M}F^{AB}_{ij} = 2U_{ABMN}f^{\ N}_{(I')j} \qquad (5.16b)$$

with the field strengths F,G as defined in (4.6). Finally, cancellation of the 5-fermi terms in δS requires

$$\left[\nabla_i U_{AB[MN}\right] f^{(I')i}_{\ \ \ P]} = -2U_{ABQ[P}C^{(I')}_{\ \ MN]Q} \qquad (5.17a)$$

$$f_{(I)i[C}\left[\nabla^i U_{AB]MN}\right] = -2C^{(I)}_{D[AB}U_{C]DMN} \qquad (5.17b)$$

The commutator of two transformations acting on ϕ^i gives

$$\left[\delta_\eta, \delta_\epsilon\right]\phi^i = i\epsilon_-^I\eta_-^J\left[f^{\ i}_{(I)A}\,f^{\ A}_{(J)j} + (I \leftrightarrow J)\right]\partial_{++}\phi^j \qquad (5.18)$$

$$+ i\epsilon_+^{I'}\eta_-^{J'}\left[f^{\ i}_{(I')A}\,f^{\ A}_{(J')\ j} + (I' \leftrightarrow J')\right]\partial_{--}\phi^j$$

$$+ 2\epsilon_-^{(I}\eta_-^{J)}\lambda_+^A\lambda_+^B\left[-H^i_{\ jk}\,f^{\ j}_{(I)A}\,f^{\ k}_{(J)B} + f^{\ i}_{(I)C}\,C^{ABC}_{(J)}\right]$$

$$+\ \epsilon_+^{(I'}\eta_+^{J')}\chi_-^M\chi_-^N\left[H^i_{jk}\, f^j_{(I')M}\, f^k_{(J')N} + f^i_{(I')P}C^{MNP}_{(J')}\right]$$

where (5.10) has been used. Then when (5.10,5.17) are satisfied, the action (5.5) is invariant under p right-handed supersymmetries and q left-handed ones, but as can be seen from (5.18) the algebra will not be (5.7) but a generalization of it. We shall only discuss the cases in which the algebra <u>is</u> (5.7) and for which d = s and d ⩽ r, here; see [7,28,37] for generalizations.

For (5.18) to correspond to the (p,q) algebra (5.7), we must impose

$$f_{(I)}{}^i{}_A\, f_{(J)}{}^A{}_j + f_{(J)}{}^i{}_A\, f_{(I)}{}^A{}_j = 2\delta_{IJ}\delta^i{}_j \tag{5.19a}$$

$$f_{(I')}{}^i{}_M\, f_{(J')}{}^M{}_j + f_{(J')}{}^i{}_M\, f_{(J')}{}^m{}_j = 2\delta_{I'J'}\delta^i{}_j \tag{5.19b}$$

and

$$H^i_{jk}\left[f^j_{(I)}{}_A f^k_{(J)B} + f^j_{(J)A} f^k_{(I)B}\right] = f^i_{(I)C}\, C^{ABC}_{(J)} + f^i_{(J)C}\, C^{ABC}_{(I)} \tag{5.20a}$$

$$H^i_{jk}\left[f^j_{(I')M} f^k_{(J')N} + f^j_{(J')M} f^k_{(I')N}\right] = -f^i_{(I')P}\, C^{MNP}_{(J')} + f^i_{(J')}\, C^{MNP}_{(I')} \tag{5.20b}$$

If s = d, $\left[\delta_\epsilon,\delta_\eta\right]\chi$ will agree with (5.7) on-shell if

$$f^M_{(I')i}\, f_{(J')}{}^i{}_N + f_{(J')}{}^M{}_i\, f^i_{(I')N} = 2\delta_{I'J'}\delta^M{}_N \tag{5.21}$$

so that from (5.20), for each I′

$$C_{(I')}{}^{MNP} = -H_{ijk}\, f_{(I')}{}^{iM}\, f_{(I')}{}^{iN}\, f_{(I')}{}^{kP} \tag{5.22}$$

with no sum over I′. Defining the curvature with torsion

$$R^{(\pm)k}{}_{lij} = \partial_i\Gamma^{(\pm)k}_{jl} + \Gamma^{(\pm)k}_{im}\Gamma^{(\pm)m}_{jl} - (i \longleftrightarrow j) \tag{5.23}$$

the integrability condition for (5.10b) implies

$$G^{MN}_{ij} = R^{(+)}{}_{klij}\, f_{(I')}{}^{kM}\, f_{(I')}{}^{lN} \tag{5.24}$$

for each value of I′. Then (5.16b) follows from (5.22,5.24) and the identity

$$R^{(\pm)}_{[ijk]l} = \pm\tfrac{2}{3}\nabla^{(\pm)}_l H_{ijk} \tag{5.25}$$

which follows from (5.23) if $\partial_{[l}H_{ijk]} = 0$. From (5.16b)

$$U_{ABMN} = \frac{1}{2} F_{ijAB} f^i_{(I')M} f^j_{(I')N} \qquad (5.26)$$

for each I', and (5.17b) then follows from (5.26) and the Bianchi identity

$$\nabla^{(+)}_{[i} F^{AB}_{jk]} = -2H_{[ij}{}^{l} F^{AB}_{k]l} \qquad (5.27)$$

There will then be at least (0,q) supersymmetry if (5.10b,5.19b,5.16b,5.26,5.20b,5.21) hold with the definitions as above.

The simplest solutions are those in which (ϕ^i, χ^M) parameterize the tangent bundle. It will be useful to introduce orthonormal frames $e^a_i(\phi)$ for M (a = 1,...,d) satisfying

$$e^a_i e^b_j \delta_{ab} = g_{ij} \qquad g^{ij} e^a_i e^b_j = \delta^{ab} \qquad (5.28)$$

and spin-connections $\omega^{(\pm)ab}_i$

$$\omega^{(\pm)ab}_i = \omega^{ab}_i \mp H_{ijk} e^{ja} e^{kb} \, , \quad \nabla^{(\pm)}_i e^a_j = 0. \qquad (5.29)$$

Then there is always at least (0,1) supersymmetry, with one of the $f_{(I')\,i}{}^{M}$ taken to be e^a_i, (so that the indices M,N,.. are replaced by a,b,..). Then the transformations (5.8), (5.13) become, on defining $\chi^i_- = e^i_a(\phi)\chi^a_-$,

$$\delta\phi^i = \epsilon_+\chi^i_- \, , \quad \delta\chi^i_- = i\epsilon_+\left[\partial_{--}\phi^i\right], \quad \delta\lambda^A_+ + \left[\delta\phi^i\right] A^{AB}_i \lambda^B_+ = 0 \qquad (5.30)$$

while the action (5.5) becomes that of [6]

$$S = ½\int d^2x \Big[(g_{ij} + b_{ij})\partial_{++}\phi^i \partial_{--}\phi^j + i\lambda^A_+ \Big[\partial_{--}\lambda^A_+ \qquad (5.31)$$

$$+ A^{AB}_i(\partial_{--}\phi^i)\lambda^B \Big] + i\chi^a_- \Big[\partial_{++}\chi^a_- + \omega^{(+)ab}_i (\partial_{++}\phi^i)\chi^b_- + ¼F^{AB}_{ab}\lambda^A_+\lambda^B_+\chi^a_-\chi^b_- \Big]$$

where from (5.24) B^{MN}_i has become $\omega^{(+)ab}_i$ while from (5.26) U_{ABMN} has become $½F^{AB}_{ab}$.

For (0,q) supersymmetry of (5.31), if one of the $f^i_{(I')M}$ is e^i_a (I' = q, say) then the remaining q - 1 quantities $f_{(I')}$, I' = 1,...,q - 1 will be some tensors $J^{(y)i}{}_a$ (y = 1,...,q - 1), so that

$$\left\{ f^i_{(I')M} : I' = 1,...q \right\} = \left\{ J^{(y)i}{}_a, e^i_a; \; y = 1,...,q - 1 \right\}$$

Defining the tensor $J^{(y)i}{}_j = J^{(y)i}{}_a e^a_j$, then (5.10), (5.26) give

$$\nabla_i^{(+)} J^{(y)j}{}_k = 0 \tag{5.32}$$

$$F^{AB}_{kl} J^{(y)k}{}_i J^{(y)l}{}_j = F^{AB}_{ij} \tag{5.33}$$

for each y. From (5.19b), (5.21) with $I' = q$, $J' = y$,

$$J^{(y)}_{ij} = -J^{(y)}_{ji} \tag{5.34}$$

while the same equations with $I' = y, J' = z$ give

$$J^{(y)i}{}_j J^{(z)j}{}_k + J^{(z)i}{}_j J^{(y)j}{}_k = -2\delta^{yz}\delta^i{}_k \tag{5.35}$$

From (5.22) and (5.20b) with $(I',J') = (q,y)$

$$J^{(y)i}{}_j J^{(z)j}{}_k + J^{(z)i}{}_j J^{(y)j}{}_k = -2\delta^{yz}\delta^i{}_k \tag{5.36}$$

with no sum on y, while the same equations with $(I',J') = (y,z)$ give

$$H_{lm[i} J^{(y)l}{}_j J^{(z)m}{}_{k]} + (y \leftrightarrow z) = -\frac{2}{3}\delta_{yz} H_{ijk} \tag{5.37}$$

Then with (5.32-5.37) the action (5.31) is invariant under (0,q) supersymmetries of the form (5.18, 5.13) with, for each y,

$$C_{(y)}{}^{abc} = -H^{ijk} J^{(y)a}{}_i J^{(y)b}{}_j J^{(y)c}_k \tag{5.38}$$

The significance of these conditions is as follows. For each y, there is a tensor $J^i{}_j$, satisfying $J^i{}_j J^j{}_k = -\delta^i{}_k$ and an integrability condition (the vanishing of the Nijenhuis tensor) [5] following from (5.32,5.36), so that each $J^i{}_j$ is a complex structure and M is a complex manifold (see [5,17,40] for details of complex geometry). From (5.34) the metric g_{ij} is hermitian with respect to each complex structure while (5.33) implies that A is a holomorphic connection. Equation (5.32) can be used to solve for H_{ijk} (5.40)

$$H_{ijk} = J_{(y)i}{}^l J_{(y)j}{}^m J_{(y)k}{}^n J_{(y)[lm,n]} \tag{5.39}$$

for each y.

The existence of a complex structure implies that the dimension d must be even and that we can choose a system of complex co-ordinates $z^\alpha, z^{\bar\beta} = (z^\beta)^*$, $\alpha, \bar\beta = 1,\ldots,d/2$, such that the complex structure is constant and diagonal

$$J^i{}_j = \begin{bmatrix} J^\alpha{}_\beta & J^\alpha{}_{\bar\beta} \\ J^{\bar\alpha}{}_\beta & J^{\bar\alpha}{}_{\bar\beta} \end{bmatrix} = \begin{bmatrix} i\delta^\alpha{}_\beta & 0 \\ 0 & -i\delta^{\bar\alpha}{}_{\bar\beta} \end{bmatrix} \tag{5.40}$$

Then the hermiticity condition (5.34) implies $g_{\alpha\beta} = 0$, $ds^2 = 2g_{\alpha\bar\beta}\, dz^\alpha{}_\wedge dz^{\bar\beta}$ while (5.33) gives $F^{AB}_{\alpha\beta} = 0$, $F^{AB}_{\alpha\bar\beta} \neq 0$. From (5.38)

$$H_{\alpha\beta\gamma} = 0 \qquad H_{\alpha\beta\bar\gamma} = 2\,\partial_{[\alpha} g_{\beta]\gamma} \tag{5.41}$$

so that if $dH = 0$ the metric must satisfy

$$g_{\alpha[\bar\beta,\bar\gamma]\delta} - g_{\delta[\bar\beta,\bar\gamma]\alpha} = 0 \tag{5.42}$$

It is useful to introduce holomorphic exterior derivatives [17], $d = \partial + \bar\partial$, $\partial = dz^\alpha{}_\wedge\, \partial/\partial z^\alpha$, $\bar\partial = (\partial)^*$ and the fundamental 2-form

$$J = \tfrac{1}{2} J_{ij}\, d\phi^i{}_\wedge d\phi^j = i g_{\alpha\bar\beta}\, dz^\alpha{}_\wedge dz^{\bar\beta} \tag{5.43}$$

Then (5.40,5.41) become

$$H = i(\partial J - \bar\partial J) = db, \quad i\partial\bar\partial J = 0 \tag{5.44}$$

which can be solved locally in terms of some 1-form $k = k_\alpha dz^\alpha$, $(\bar k_{\bar\alpha} = (k_\alpha)^*$ [6,30]

$$J = i(\partial\bar k - \bar\partial k) \qquad g_{\alpha\bar\beta} = \partial_\alpha \bar k_{\bar\beta} + \partial_{\bar\beta} k_\alpha \tag{5.45}$$

$$b = \partial\bar k + \bar\partial k \qquad b_{\alpha\bar\beta} = \partial_\alpha \bar k_{\bar\beta} - \partial_{\bar\beta} k_\alpha$$

In the special case $H = 0$, M is Kahler with

$$dJ = 0 \ , \ k = \partial K \ , \ J = i\partial\bar{\partial}K \ , \ g_{\alpha\bar{\beta}} = 2\partial_\alpha \partial_{\bar{\beta}} K \tag{5.46}$$

for some locally defined real function K, the Kahler potential. If $H \neq 0$, this is the generalization of Kahler geometry introduced in [6].

For $(0,q)$ supersymmetry there are then $q - 1$ complex structures each satisfying (5.32–5.37). Note that in the complex co-ordinates in which $J^{(1)}$, say, is of the form (5.39), the remaining complex structures will not in general be constant. From (5.35) the complex structures will satisfy a Clifford algebra.

Unless M is parallelizable, q can only take the value 0,1,2 or 4. To see this, it is useful to introduce the holonomy group. The connection A on a vector bundle is a Lie-algebra valued one-form. The smallest Lie algebra in which it takes values is the algebra of the holonomy group for A, $h(A)$. If there is a covariantly constant tensor, then that tensor must be $h(A)$-invariant. For the connection $\omega^{(\pm)}$ in (5.29) on the tangent bundle, $h(\omega^{\pm}) \subseteq O(d)$ as the connection preserves the metric δ^{ab}. If there is a covariantly constant complex structure $\nabla_{i(+)} J_{jk} = 0$, then $h(\omega^{(+)}) \subseteq U(d/2)$, the sub-algebra of $O(d)$ commuting with the anti-symmetric tensor J_{ab}. If there are two complex structures $J^{(1)}, J^{(2)}$ satisfying (5.35) then the product

$$J^{(3)i}{}_j = J^{(1)i}{}_k J^{(2)k}{}_j$$

is a third and $J^{(1)}, J^{(2)}, J^{(3)}$ satisfy the algebra of the (imaginary) quaternions. Then d must be a multiple of four and $h(\omega^{(+)}) \subseteq Sp(d/4)$, (where $Sp(1) \simeq SU(2)$, $Sp(n) = USp(2n)$), the sub-algebra of $O(d)$ preserving such a quaternionic structure. If the torsion vanishes, an $Sp(d/4)$-holonomy manifold is known as hyper-Kahler [3]. Note that in general $h(\omega^{(+)}) \neq h(\omega^{(-)})$ unless $H = 0$.

For general q the holonomy group must be the subgroup of $O(d)$ commuting with $(q - 1)$ complex structures satisfying the Clifford algebra (5.35). For $q > 4$, this implies $h(\omega^{+})$ is trivial [3], so that M must be parallelizable with H the parellelizing torsion. For $dH = 0$ to hold, M must be a group manifold and indeed, using suitable group manifolds, $(0,q)$-supersymmetric models can be constructed for arbitrary q [7].

The "tangent bundle" model (5.31) is by no means the only solution to the (0,q) supersymmetry conditions. Before proceeding to the (p,q) case, we consider the important case in which (ϕ,χ) parameterize the spin-bundle S(M), i.e. in which χ transforms as a spinor under O(d) tangent space rotations of M. In the case d = 8 the indices a,b,... are vector (8_v) indices under SO(8). We choose χ^M (M = 1,...,8) to transform as a right-handed Weyl spinor (8_s) of SO(8) and let M′,N′ label the left-handed (8_c) 8-spinor representation of SO(8). The 16 × 16 gamma-matrices Γ_i of SO(8) are chosen to take the block form (t denotes transpose)

$$\Gamma^i = \begin{bmatrix} 0 & \gamma^i_{MN'} \\ \tilde{\gamma}^i_{P'Q} & 0 \end{bmatrix} \qquad \tilde{\gamma}^i_{P'Q} = \left[\gamma^i\right]^t_{P'Q} \tag{5.47}$$

with

$$\Gamma^i\Gamma^j + \Gamma^j\Gamma^i = 2g^{ij}, \quad \gamma^{MN}_{ij} = \gamma^{MP'}_{[i}\gamma^{P'N}_{j]}, \quad \gamma^{iM}_{M'N'P'} = \gamma^{iN}_{[M'}\gamma^{Nj}_{N'}\gamma^M_{P']j} \tag{5.48}$$

etc. Then the existence of q covariantly constant tensors f^{iM} implies the existence of q covariantly constant commuting spinors $\eta_{M'}$ and anti-symmetric tensors $L_{M'N'P'}$, defined by

$$\eta_{M'} = \gamma^i_{M'N}f^N_{\ i} \; ; \; L_{M'N'P'} = \gamma^{iM}_{M'N'P'}f_{iM} \tag{5.49}$$

(group theoretically, $8_v \times 8_s = 8_c + 56_c$) satisfying

$$\nabla^{(+)}_i\eta_{M'} \equiv \partial_i\eta_{M'} + \tfrac{1}{4}\omega^{(+)ab}_i\gamma^{M'N'}_{ab}\eta_{N'} = 0 \tag{5.50}$$

and $\nabla^{(+)}_i\eta_{M'N'P'} = 0$. We will consider here only f^i_M satisfying

$$f_{iM}\gamma^{iM}_{M'N'P'} = 0$$

so that $L_{M'N'P'} = 0$ and

$$f^i_{(I')M} = \gamma^i_{MN'}\eta^{N'}_{(I')} \tag{5.51}$$

where we have normalized $\eta^{N'}_{(I')}$ such that

$$\eta^{N'}_{(I')}\eta^{N'}_{(J')} = \delta_{I'J'} \tag{5.52}$$

(which is possible as (5.49) implies $\eta^{N'}_{(I')}\eta^{(J')}_{N'}$ is a constant). Then

(5.49) implies $\left[\nabla_i^{(+)},\nabla_j^{(+)}\right]\eta = 0$, i.e. for each $\eta_{(I')}$, suppressing the I′ label,

$$R^{(+)ab}{}_{ij}\gamma_{ab}^{M'N'}\eta_{N'} = 0 \tag{5.53}$$

Multiplying by $\eta_{P'}\gamma_{cd}^{P'M'}$ gives, using (5.51),

$$R^{(+)}{}_{klij} = \tfrac{1}{2}X_{kl}{}^{mn}R^{(+)}_{mnij} \tag{5.54}$$

where $X_{ijkl} \equiv \eta_{M'}\gamma_{ijkl}^{M'N'}\eta_{N'}$. Then choosing

$$B_i^{MN} = \frac{1}{4}\omega_i^{(+)ab}\gamma_{ab}^{MN} \tag{5.55}$$

(5.24) is satisfied as a result of (4.6), (5.53) and the identity

$$f_{[i}^{M}f_{j]}^{N} = \frac{1}{8}\left[\gamma_{ij}^{MN} + \frac{1}{16}X_{ij}{}^{kl}\gamma_{kl}^{MN}\right] \tag{5.56}$$

Then the action (5.5), with (5.6) and

$$U_{ABMN} = \frac{1}{16}F_{ij}^{AB}\left[\gamma_{MN}^{ij} + \tfrac{1}{2}X^{ijkl}\gamma_{kl}^{MN}\right] \tag{5.57}$$

from (5.26,5.55) gives a supersymmetric action. If we further demand that

$$F_{ij}^{AB}\gamma_{M'N'}^{ij}\eta^{N'} = 0 \tag{5.58}$$

then it follows that

$$F_{ij}^{AB} = \tfrac{1}{2}X_{ij}{}^{kl}F_{kl}^{AB}\ ,\quad U_{ABMN} = \frac{1}{8}F_{ij}^{AB}\gamma_{mn}^{ij} \tag{5.59}$$

and the action (5.5) no longer depends on the spinor η or the tensor X_{ijkl}, becoming

$$\begin{aligned} S = \tfrac{1}{2}\int d^2x\Big[&\left[g_{ij} + b_{ij}\right]\partial_{++}\phi^i\partial_{--}\phi^j + i\lambda_+^A\left[\partial_{--}\lambda_+^A + A_i^{AB}\partial_{--}\phi^i\lambda_+^B\right] \\ &+ i\chi_-^m\left[\partial_{++}\chi_-^m + \frac{1}{4}\omega_i^{(+)ab}\gamma_{ab}^{MN}(\partial_{++}\phi^i)\chi_-^N\right] \\ &+ \frac{1}{8}F_{ij_{AB}}\gamma_{M'N'}^{ij}\lambda_+^A\lambda_+^B\chi_-^{M'}\chi_-^{N'}\Big] \end{aligned} \tag{5.60}$$

This action was first considered (with $b_{ij} = 0$) in [34] and later discussed in [28,35,36,37]. Defining (for anti-commuting ϵ_+)

$$\epsilon_+^{M'} = \epsilon_+ \eta^{M'} \tag{5.61}$$

the supersymmetry transformations become [28]

$$\delta\phi^i = \epsilon_+^{M'}\gamma^i_{M'N}\chi_-^N \qquad \delta\lambda_+^A + (\delta\phi^i)A_i^{AB}\lambda_+^B = 0 \tag{5.62}$$

$$\delta\chi_-^M + \frac{1}{4}(\delta\phi^i)\omega_i^{(+)ab}\gamma_{ab}^{MN}\chi_-^N = i\epsilon_+^{M'}\gamma_i^{M'N}(\partial_{--}\phi^i)$$

$$+ \frac{1}{8}\epsilon_+^{M'}\gamma^i_{M'M}\left[H_{ijk} + H_{imn}X^{mn}{}_{jk}\right]\gamma^{jk}_{NP}\chi_-^N\chi_-^P$$

If we further demand that η satisfy

$$\left[H^{ijk}\gamma_{ijk}^{MN'} - 3V^i\gamma_i^{MN'}\right]\eta_{N'} = 0 \tag{5.63}$$

for some vector V^i, then (5.61) can be rewritten without any dependence on X_{ijkl} [28]. The equations (5.49), (5.57) (5.62) have integrability conditions which are closely related to the effective field equations of a field theory limit of the heterotic string [28,37].

For (0,q) supersymmetry of (5.59) there must be q covariantly constant spinors $\eta_{(I')}(I' = 1,\ldots,q)$ and hence $q(q-1)/2$ covariantly constant tensors

$$J_{ij}^{[I'J']} = \eta^{I'}_{M'}\gamma_{ij}^{M'N'}\eta^{J'}_{N'}, \tag{5.64}$$

satisfying (5.32–5.37) (with the index pair $[I'J']$ replaced by y) and the important extra conditions [28,36] (following from (5.49,5.57))

$$C_{ij} \equiv J^{kl}R^{(+)}_{klij} = 0 \tag{5.65}$$

and (5.58) implies $J^{ij}F_{ij} = 0$. If (5.62) holds, then

$$V^i = J^{ij}J^{kl}H_{jkl} \tag{5.66}$$

Again, if the holonomy group $h(\omega^{(+)})$ is non-trivial, the number of complex structures satisfying these conditions can only be 0,1 or 3, so that q = 0,1,2 or 4. Note that because of the extra conditions not every M that can support a (0,2) supersymmetric action of the form (5.31) can also support one of the form (5.59). For example, the vanishing of the U(1) part of the curvature C_{ij} implies $h(\omega^{(+)}) \subseteq SU(d/2)$, a necessary condition for which is the vanishing of the first Chern class of M [28].

We now consider the conditions for a (0,q) supersymmetric model to in fact be invariant under (p,q) supersymmetry. There must then be p tensors $f^{i}_{(I)\ A}$ satisfying the conditions given above. From requiring $\left[\delta_{\epsilon},\delta_{\eta}\right]\lambda$ to be consistent with the (p,q) algebra we obtain

$$f_{(I)i}{}^{A} f_{(J)}{}^{i}{}_{B} + f_{(J)}{}^{A}{}_{i} f_{(I)}{}^{i}{}_{B} = 2\delta_{IJ}\Pi^{A}{}_{B} \qquad (5.67)$$

where Π_{AB} is some symmetric covariantly constant tensor, $\nabla^{(-)}_{i}\Pi_{AB} = 0$ (from 5.10a). The typical case [7] is that in which $\Pi^{A}{}_{B}$ is a projector, projecting the λ^{A} ($A = 1,\ldots,r$; $r \geqslant d$) onto a d-dimensional subspace, constituting the d super-partners of the ϕ^{i}.

We focus first on just one of the $f^{A}_{(I)i}$, $f^{A}{}_{i}$ say. From (5.16) we obtain the relation between the two field strengths F,G

$$\Pi_{A}{}^{C}\Pi_{B}{}^{D}F_{ij}{}^{CD}f^{i}{}_{M}f^{j}{}_{N} = G_{ijMN}f^{i}_{A}f^{j}_{B} \qquad (5.68)$$

and from (5.10a) we obtain the integrability condition

$$R^{(-)}{}_{klij} = F_{ij}{}^{AB}f_{Ak}f_{Bl} \qquad (5.69)$$

Using $\Pi^{AB} f^{i}_{B} = f^{i}_{A}$ and (5.16), we find

$$\Pi^{A}{}_{B}F^{BC}_{ij} = F^{AC}_{ij}\ ,\quad \Pi^{AB}U_{ACMN} = U_{BCMN} \qquad (5.70)$$

Let us take the tangent bundle case, with action (5.31). Then with $f^{i}{}_{M} \to e^{i}{}_{a}$, $G_{ijMN} \to R^{(+)}_{abij}$, (5.68) (5.69) are seen to be consistent on using the identity

$$R^{(+)}{}_{ijkl} = R^{(-)}_{klij} \qquad (5.71)$$

which holds if $dH = 0$. The tensor $f^{A}_{a}(\phi) = e^{i}_{a}f^{A}_{i}$ transforms under $O(d) \times O(r)$ and, for each point ϕ, it gives an embedding of $O(d)$ into $O(r)$, so that $f^{A}_{a}(\phi)$ gives an embedding of the tangent bundle T(M) into the vector bundle parameterized by (ϕ^{i},λ^{A}). The existence of the covariantly constant projector Π_{AB} implies $h(A) \subseteq O(d) \times O(r - d)$ and (5.70) implies that in fact $h(A) \subseteq O(d)$. (5.69) implies that an $O(d) \times O(r)$ gauge can be chosen such that

$$\omega^{(-)ab}_{i} = f^{a}{}_{A}f^{b}{}_{B}A^{AB}_{i} \qquad (5.72)$$

relating the embedded spin-connection ω to the gauge connection A. This leaves a residual diagonal $O(d) \subseteq O(d) \times O(r)$ invariance.

Writing

$$\lambda^A = \left[\Pi + (1-\Pi)\right]^{AB} \lambda^B = f^{Aa} \lambda^a + (1 - \Pi)^{AB} \lambda^B \tag{5.73}$$

$$\lambda^a \equiv f^{aA} \lambda^A$$

and using (5.10,5.69) the action (5.31) becomes $S = S_1 + S_2$ where S_1 is the usual (1,1)-σ-model of [5.29]

$$S = \tfrac{1}{2}\int d^2x \left[\left(g_{ij} + b_{ij}\right) \partial_{++}\phi^i \partial_{--}\phi^i + i\chi_-^a \left[\nabla_{++}^{(+)}\chi\right]^b \right. \tag{5.74}$$

$$\left. + i\lambda_+^a \left[\partial_{--}\lambda_+^a + \omega_i^{(-)ab}(\partial_{--}\phi^i)\lambda_+^b\right] + \tfrac{1}{2}R_{abcd}^{(-)}\lambda_+^a\lambda_+^b\chi_-^c\chi_-^d \right]$$

while S_2 is a free action for the $(r - d)$ fermions $(1 - \Pi)\lambda$ which do not transform under supersymmetry.

The action (5.74) is invariant under (1,q) supersymmetry if there are (q − 1) complex structures satisfying (5.32,5.34,5.35,5.36,5.37) as (5.33) then follows using (5.69), (5.71). For (p,q) supersymmetry then, arguing as before, there must be p − 1 more complex structures $J_{(Y)j}^{i}$ $(Y = 1,\ldots,P - 1)$ satisfying

$$\nabla_i^{(-)} J_{jk}^{(Y)} = 0 \; , \quad J_{ij}^{(Y)} = -J_{ji}^{(Y)} \tag{5.75}$$

$$J^{(Y)i}{}_j J^{(Z)j}{}_k + J^{(Z)i}{}_j J^{(Y)j}{}_k = -\delta^{YZ}\delta^i{}_j \tag{5.76}$$

$$J^{(Y)i}{}_l J^{(Y)j}{}_m J^{(Y)k}{}_n H^{lmn} = +3J^{(Y)[i}{}_l H^{jk]l} \tag{5.77}$$

$$H_{lm[i} J^{(Y)l}{}_j J^{(Z)m}{}_{k]} + (Y \leftrightarrow Z) = +\tfrac{2}{3}\delta_{YZ} H_{ijk} \tag{5.78}$$

with no sum over Y in (5.77). These are very similar to the equations satisfied by the $J^{(Y)}$, apart from some sign changes.

As before, for p = 2, we must have $h(\omega^{(-)}) \subseteq U(d/2)$, for p = 4, $h(\omega^{(-)}) \subseteq Sp(d/4)$ while for p > 4 M must be parallelizable. Note that in general $h(\omega^{(-)}) \neq h(\omega^{(+)})$ and so $p \neq q$, but if $H_{ijk} = 0$, then $\omega^{(+)} = \omega^{(-)}$ and p = q, i.e. there are equal numbers of left and right supersymmetries.

With these conditions the (p,q) supersymmetry algebra closes on-shell. For off-shell closure it is necessary to demand that the $J_{(Y)}$ commute with the $J_{(y)}$ if $p \geqslant 2, q \geqslant 2$,

$$J_{(Y)}{}^{i}{}_{j} J_{(y)}{}^{j}{}_{k} = J_{(y)}{}^{i}{}_{j} J_{(Y)}{}^{j}{}_{k} \tag{5.79}$$

so that there are then $(p - 1) \times (q - 1)$ local product structures

$$\Pi^{yY}_{ij} = J_{i}^{(y)k} J^{(Y)}{}_{kj} \tag{5.80}$$

each satisfying

$$\Pi_{ij} = \Pi_{ji} \;, \; \Pi^{i}{}_{j}\Pi^{j}{}_{k} = +\delta^{i}{}_{k} \tag{5.81}$$

These lead to the so-called twisted chiral models; see [5] for details.

Finally, cancellation of the $O(d) \times O(r)$ anomaly for (5.31) or (5.59) leads to the transformation (see sections 4 and 12)

$$\delta b_{ij} = \frac{\hbar}{4\pi}\left[\Lambda^{AB}\partial_{[i}A^{AB}_{j]} - \theta^{ab}\partial_{[i}\tilde{\omega}^{ab}_{j]}\right] \tag{5.82}$$

with

$$\tilde{\omega}_{i}{}^{ab} = \omega_{i}{}^{ab} + T_{i}{}^{ab} \tag{5.83}$$

and $T_{i}{}^{ab}$ can be chosen arbitrarily by adding the counterterm (4.29) to the effective action, with $B^{MN}_{[i}T^{MN}_{j]}$ replaced by $\omega^{ab}_{[i}T^{ab}_{j]}$ [26]. Then

$$H' = db - \frac{\hbar}{4\pi}\left[\omega_{3}(A,F) - \omega_{3}(\tilde{\omega},\tilde{R})\right] \tag{5.84}$$

is gauge invariant where the curvature 2-form for (5.83) is $\tilde{R} = d\tilde{\omega} + \tilde{\omega}_{\wedge}\tilde{\omega}$ and

$$dH' = -\frac{\hbar}{4\pi}\left[tr(F_{\wedge}F) - tr(\tilde{R}_{\wedge}\tilde{R})\right] \tag{5.85}$$

In many of the equations in this section, it is necessary to replace H by H′ to obtain full gauge invariance; see [36,39,41] for details. Note that choosing $\tilde{\omega}^{ab}_{i} = \omega^{(-)ab}_{i}$, $T_{i}{}^{ab} = -H'^{ab}_{i}$ in (5.83) is consistent with all world-sheet and spacetime supersymmetry and, indeed, appears to be the only consistent choice in string theories [28].

6. Path Integral Quantization

We now turn to the quantization of the n-dimensional non-linear sigma-model with action S, which may include a Wess-Zumino term, a potential term, etc., but for the moment we will exclude fermion couplings. As a configuration is given by a map $\hat{\phi}$ from the space-time N to the target space M, the quantum partition function Z can be given formally by an integral over the space of functions $\hat{\phi}: N \to M$ with some measure $\int[d\hat{\phi}]$,

$$Z = \int[d\hat{\phi}] \exp\left[-\hbar^{-1} S_J\right] \tag{6.1}$$

where S_J is a Euclidean form of S, plus some source term to be specified later. If N is flat, we "Euclideanize" by Wick rotation. If N is not flat, it is not always possible to analytically continue a given Lorentzian space-time metric to a positive definite one; here we shall not dwell on this point, but simply consider N to be a "Euclidean space-time" with positive definite metric $\gamma_{\mu\nu}$.

The functional measure in (6.1) will be chosen to respect the symmetries of the sigma-model action. If fermions are coupled to the sigma-model, there will in addition be an integration over the fermi fields. For chiral fermions, the sigma-model anomaly discussed in Section 4 can be interpreted as due to the difficulty in defining a functional measure consistent with the sigma-model symmetry involving local chiral rotations of the fermi fields.

In general the space of functions $\hat{\phi}$ will consist of a number of disconnected components, corresponding to different instanton sectors. To proceed, we need some convenient parameterization of the function space. The simplest case is that in which N is topologically equivalent to $\mathbb{R}^n$ and M to $\mathbb{R}^d$ so that M,N can both be covered by a single co-ordinate chart. Then any configuration is described by a single function $\phi^i(x^\mu)$ and (at least for flat N) we can choose the co-ordinate independent measure given formally by [1]

$$[d\phi] = \prod_x \sqrt{g(\phi(x))}\, d\phi^1(x) d\phi^2(x) \ldots d\phi^d(x) \tag{6.2}$$

i.e. the product over x of volume elements on M. Then (6.1) becomes a functional integration over d scalar fields. We will return to the case of general M,N at the end of Section 8.

Next we must specify the source terms in S_J. The simplest choice is

$$S_J = S + \int d^n x\, J_i(x)\,\phi^i(x) \equiv S + \phi.J \tag{6.3}$$

for some source field $J_i(x)$. Note that the notation $\phi.J$ denotes both an index summation and an integration over space-time. Then (6.1) becomes

$$Z_J \equiv \exp\left[-\hbar^{-1}W\right] = \int[d\phi]\exp\left[-\hbar^{-1}(S[\phi] + \phi.J)\right] \tag{6.4}$$

Defining the "mean field"

$$\bar{\phi}^i(x) = \frac{\delta W}{\delta J_i} \tag{6.5}$$

and using (6.5) to express J in terms of $\bar{\phi}$, $J_i = J_i(\bar{\phi}(x))$, we can define an effective action by taking the Legendre transform of W,

$$\Gamma[\bar{\phi}] = W[J(\bar{\phi})] - \bar{\phi}.J(\bar{\phi}) \tag{6.6}$$

As is well-known, W generates connected Feynman diagrams while Γ generates one-particle irreducible (1PI) ones. Using (6.4) and the identity $\delta\Gamma/\delta\bar{\phi}^i = -J_i$ we find that Γ has the functional integral representation

$$\exp\left[-\hbar^{-1}\Gamma[\bar{\phi}]\right] = \int[d\phi]\exp\left\{-\hbar^{-1}(S[\phi] - (\phi - \bar{\phi}).\delta\Gamma/\delta\bar{\phi})\right\} \tag{6.7}$$

and defining the fluctuation

$$\pi^i(x) = \phi^i - \bar{\phi}^i \tag{6.8}$$

this can be rewritten as

$$\exp\left[-\hbar^{-1}\Gamma[\bar{\phi}]\right] = \int[d\pi]\exp\left\{-\hbar^{-1}(S[\bar{\phi} + \pi] - \pi.\delta\Gamma/\delta\bar{\phi})\right\} \tag{6.9}$$

Equation (6.9) gives the background field representation of Γ. Although Γ occurs on both sides of (6.9), it can be calculated perturbatively, with

$\Gamma[\bar{\phi}] = S[\bar{\phi}] + O(\hbar)$. The background field method [42] amounts to calculating Γ using the diagrammatic expansion of (6.9). That is, we integrate over fluctuations $\pi^i(x)$ about some fixed background field configuration $\bar{\phi}^i(x)$, so that $\pi^i(x)$ is regarded as a quantum field (operator) while $\bar{\phi}^i$ is treated as a classical c-number field. The subtraction of $\pi.\delta\Gamma/\delta\bar{\phi}$ removes vertices linear in the quantum field π and graphs that are not 1PI. Note that as π is the difference between two co-ordinates (6.8), it does not transform covariantly, so that expanding $S[\bar{\phi}+\pi]$ in terms of π will give non-covariant vertices and the linear term subtracted in (6.9), and hence Γ, will be co-ordinate dependent off-shell. As we shall see, this co-ordinate dependence will be a drawback and in later sections we shall develop alternative techniques to avoid this.

Fig. 1. Divergent 1-loop graphs. Single lines denote quantum fields, double lines denote background fields.

To illustrate this, we consider the two-dimensional non-linear sigma-model given by (1.1) with $n = 2$ and $\gamma_{\mu\nu} = \delta_{\mu\nu}$ and calculate the divergences given by the graphs in Figure 1. The Feynman rules are obtained from the Taylor expansion of $S(\bar{\phi} + \pi)$, which gives

$$S[\bar{\phi} + \pi] - \pi.\frac{\delta S}{\delta\bar{\phi}}[\bar{\phi}] = S[\bar{\phi}] + \int d^2x\Big\{\tfrac{1}{2}g_{ij}\bar{\phi}\partial_\mu\pi^i\partial^\mu\pi^j \tag{6.10}$$

$$+ A^\mu_{ij}(\bar{\phi})\pi^i\partial_\mu\pi^j + \tfrac{1}{2}B_{ij}(\bar{\phi})\pi^i\pi^j\Big\} + O(\pi^3)$$

where the vertices are given by the non-covariant quantities

$$A^\mu_{ij}(\bar{\phi}) = g_{jk,i}\partial^\mu\bar{\phi}^k \tag{6.11}$$

$$B_{ij}(\bar{\phi}) = \tfrac{1}{2}g_{kl,ij}\partial_\mu\bar{\phi}^k\partial^\mu\bar{\phi}^l$$

Adding a mass-term $\int d^2x(m^2/2)\delta_{ij}\pi^i\pi^j$ to (6.10) to regulate infra-red divergences, we obtain the momentum-space propagator for π^i

$$\pi^i \pi^j = g^{ij}(\bar{\phi}) \frac{1}{p^2+m^2} \tag{6.12}$$

Then at one loop there is a divergent contribution to Γ from the two graphs in Figure 1.

Using dimensional regularization, i.e. continuing the dimension of space-time to $n = 2 + \epsilon$, the graphs in Fig. 1 give the one-loop contribution Γ_1 to Γ

$$\Gamma_1 = I\left[g^{ij} B_{ij} - A_{\mu ij} A^{\mu ij}\right] \tag{6.13}$$

$$= \frac{I}{2} \partial_\mu \bar{\phi}^i \partial^\mu \bar{\phi}^j \left[\frac{1}{2} g_{ij,kl} g^{kl} - g_{ij,m} g_{jl,n} g^{kl} g^{mn}\right]$$

where

$$I = \int \frac{d^n k}{(2\pi)^2} \frac{1}{(k^2+m^2)} = \frac{1}{4\pi} m^\epsilon \Gamma\left[-\frac{\epsilon}{2}\right] \tag{6.14}$$

$$= m^\epsilon \left[-\frac{1}{2\pi\epsilon} + \text{finite}\right]$$

After considerable effort, we find that

$$\frac{1}{2} g_{ij,kl} g^{kl} - g_{ik,m} g_{il,n} g^{kl} g^{mn} = -R_{ij} + \nabla_{(i} X_{j)} \tag{6.15}$$

$$X^i \equiv g^{jk} \left\{ {i \atop j\ k} \right\} \tag{6.16}$$

These one-loop divergences can be cancelled if the "bare" metric g^0_{ij} occuring in the action (1.1) is expressed in terms of some new "renormalized" metric g_{ij} by

$$g^0_{ij} = \mu^\epsilon \left[g_{ij} - \frac{\hbar}{2\pi\epsilon}\left[R_{ij}(g) - \nabla_{(i} X_{j)}(g)\right]\right] \tag{6.17}$$

for some mass scale μ. Then $\Gamma(\bar{\phi}, g_{ij})$ is finite at one-loop when regarded as a function of the renormalized metric, g_{ij}. This metric renormalization, first introduced by Friedan [1], is a generalization of standard coupling constant renormalization. As we have seen, the metric can be thought of as an infinite set of coupling constants, so that this corresponds to a renormalization of the infinite number of coupling constants, which mix with one another in a complicated way under renormalization.

The presence of the non-covariant quantity (6.16) in the divergence is at first disturbing. However, note the identity

$$\nabla_{(i}X_{j)}\,\partial_\mu\bar\phi^i\partial^\mu\bar\phi^j = \nabla_\mu(X_i\partial^\mu\bar\phi^i) - X_i\nabla_\mu\partial^\mu\bar\phi^i \qquad (6.18)$$

where the second term is proportional to the classical field equation and so can be written as $X_i\,\delta S/\delta\bar\phi^i$. Then the X_i-contribution to Γ is a surface-term plus a term proportional to the field-equation so that it vanishes on-shell to this order (as $\delta\Gamma/\delta\bar\phi^i = 0$ implies $\delta S/\delta\bar\phi^i = 0 + O(\hbar)$). Alternatively, the X_i dependence can be absorbed into a renormalization of the background field, $\bar\phi^i \to \bar\phi^i - \frac{\hbar}{2\pi\epsilon}X^i(\bar\phi)$. In any case, we see that the non-covariant part of the divergence proportional to X^i is trivial in that it does not influence physical on-shell quantities, such as the S-matrix.

However, this example illustrates a number of drawbacks of using conventional techniques for the sigma-model. First, the vertices corresponding to terms such as (6.10), (6.11) are non-covariant and so the corresponding perturbation theory is not manifestly covariant. Tedious calculations such as that leading to (6.15) are then needed to re-express results in terms of covariant quantities. Secondly, the effective action (6.9) is non-covariant off-shell (i.e. if $\delta\Gamma/\delta\bar\phi^i \neq 0$) as π^i does not transform covariantly. Note that Z in (6.4) is also non-covariant as the source $\phi.J$ in (6.3) is co-ordinate dependent.

To avoid these problems, it is advantageous to adopt a different definition of effective action that agrees with (6.9) on-shell but differs off-shell and is fully covariant. This will give the same physical predictions as the previous definition since the S-matrix only depends on the on-shell effective action. However, the covariance is very useful both for formal manipulations and actual calculations. An essential ingredient is the use of a non-linear split of ϕ into background and quantum fields so as to obtain covariant Feynman rules. We discuss this in the next section and then define covariant effective actions in Section 8.

7. The Background/Quantum Split

We saw in the last section that the non-covariance of the effective action was largely due to the fact that the quantum field $\pi^i(x)$ was not a vector. We therefore seek a new quantum variable that does transform covariantly. We express $\pi^i(x)$ in terms of some new field $\xi^i(x)$ by

$$\pi^i(x) = \xi^i(x) + \sum_{n=2}^{\infty} \frac{1}{n!} M^i_{j_1..j_n}(\bar{\phi})\xi^{j_1}\xi^{j_2}...\xi^{j_n} \qquad (7.1)$$

for some coefficients $M^i_{j...k}(\bar{\phi})$. (We shall assume that (7.1) can be inverted to give ξ in terms of π and $\bar{\phi}$.) We then use (7.1) to split the total field $\phi(x)$ into a background field $\bar{\phi}^i(x)$ and the new field $\xi^i(x)$ so that $\phi(x) = \bar{\phi}(x) + \pi(\xi(x))$. We will now regard ξ rather than π as the fundamental quantum field, over which we shall integrate, expanding $S[\phi] = S[\bar{\phi} + \pi(\xi)]$ in terms of ξ to obtain ξ-vertices. This will lead to a new effective action as, for example, diagrams that are 1PI with respect to ξ will not be with respect to π and vice versa, since ξ and π are non-linearly related. If $M^i_{j...k} \equiv 0$, $\xi = \pi$ and we have the linear split used in the previous section. However, for certain choices of the coefficients $M^i_{j...k}$, the $\xi^i(x)$ will transform covariantly and we will obtain covariant Feynman rules.

The simplest choice giving a covariant perturbation theory is that in which the ξ^i are <u>normal co-ordinates</u>. These are the co-ordinates most frequently used in this context [1,20,43,44,45]. We consider a curve in M, $\phi^i(s)(0 \leqslant s \leqslant 1)$, joining the background field and total field , i.e. $\phi^i(0) = \bar{\phi}^i$, $\phi^i(1) = \bar{\phi}^i + \pi^i$, with tangent vector $\xi^i(s) = \dot{\phi}^i(s) = d\phi^i/ds$. We choose the tangent vector at $\bar{\phi}^i$ to be the quantum variable $\xi^i = \xi(0)$, which clearly transforms covariantly. Finally, we demand that the curve be a geodesic, i.e. that it satisfy the differential equation

$$\ddot{\phi}^i + \Gamma^i_{jk}(\phi)\dot{\phi}^j\dot{\phi}^k = 0 \qquad (7.2)$$

with Γ^i_{jk} the Christoffel connection. For small enough perturbations, the geodesic will be unique and the ξ will provide a good choice of variable. As

$$\phi^i(1) = \sum_{n=0}^{\infty} \frac{1}{n!} \frac{d^n}{ds^n} \phi^i(s)\Big|_{s=1}$$

repeated differentiation of (7.2) to obtain $d^n\phi/ds^n$ gives (7.1) with the M's inductively defined by

$$M^i_{jk} = -\Gamma^i_{jk} \qquad (7.3)$$

$$M^i_{j_1...j_{n+1}} = \partial_{(j_{n+1}} M^i_{j_1...j_n)} - n\Gamma^k_{(j_1 j_2} M^i_{j_3...j_{n+1})k}$$

It is then straightforward to expand $S[\phi]$ in terms of the normal coordinate ξ^i, [20,44,45] using

$$S[\bar\phi + \pi(\xi)] = \int d^n x \sqrt{\gamma}\, L[\bar\phi, \xi], \qquad L = L(\bar\phi) + \sum_{n=1}^{\infty} \frac{1}{n!} L^{(n)}, \tag{7.4}$$

$$L^{(n)} = \frac{d^n}{ds^n} L(\phi(s))\Big|_{s=0} = D^n L(\phi)(s)\Big|_{s=0}$$

where the derivative D is the covariantization of d/ds with respect to Γ^i_{jk}, so that (for any tensor T and vector V)

$$D\left[T_{i\ldots j}(\phi(s))\right] = \left[\nabla_k T_{i\ldots j}\right]\xi^k(s), \qquad D\xi = 0$$

$$D\left[\partial_\mu \phi^i\right] = \nabla_\mu \xi^i + \Gamma^i_{jk}(\partial_\mu \phi^j)\xi^k \tag{7.5}$$

$$[D, \nabla_\mu]V^i = R^i{}_{jkl} V^j \xi^k \partial_\mu \phi^l$$

For example, if

$$L(\phi) = \frac{1}{2} g_{ij} \partial_\mu \phi^i \partial^\mu \phi^j + \frac{1}{n!} b_{i_1 \ldots i_n} \epsilon^{\mu_1 \ldots \mu_n} \partial_{\mu_1} \phi^{i_1} \ldots \partial_{\mu_n} \phi^{i_n}$$

$$+ \Phi R^{(\gamma)} + V \tag{7.6}$$

then dropping surface terms and defining

$$H_{i_1 \ldots i_{n+1}} = \frac{n+1}{n!} \partial_{[i_1} b_{i_2 \ldots i_{n+1}]} \tag{7.7}$$

we obtain

$$L^{(1)} = -\xi^i \frac{\delta S}{\delta \bar\phi^i}(\bar\phi) \tag{7.8}$$

$$L^{(2)} = \frac{1}{2} g_{ij} \nabla_\mu \xi^i \nabla^\mu \xi^j + \frac{1}{2}\xi^i \xi^j \Big[R_{kijl} \partial_\mu \bar\phi^k \partial^\mu \bar\phi^l$$

$$+ \nabla_i \nabla_j V + \left[\nabla_i \nabla_j \Phi\right] R^{(\gamma)} + \nabla_i H_{jk_1 \ldots k_n} \epsilon^{\mu_1 \ldots \mu_n} \partial_{\mu_1} \bar\phi^{i_1} \ldots \partial_{\mu_n} \bar\phi^{i_n} \Big]$$

$$+ \frac{n}{2} \xi^i \nabla_\nu \xi^j H_{ijk_1 \ldots k_{n-1}} \epsilon^{\nu\mu_1 \ldots \mu_{n-1}} \partial_{\mu_1} \bar\phi^{k_1} \ldots \partial_{\mu_{n-1}} \bar\phi^{k_{n-1}}$$

$$L^{(3)} = \frac{2}{3}\xi^i\xi^j\nabla_\mu\xi^k\partial^\mu\bar{\phi}^l R_{kijl} + \frac{1}{6}\xi^i\xi^j\xi^k\Big[\nabla_i\nabla_j\nabla_k V$$

$$+ \left[\nabla_i\nabla_j\nabla_k\Phi\right]R^{(\gamma)} + \nabla_i R_{jmnk}\partial_\mu\bar{\phi}^m\partial^\mu\bar{\phi}^n\Big] + O(H)$$

For higher order terms, see e.g. [20,44,46]. Note that $b_{i\ldots j}$ always occurs in $L^{(n)}$ through its curl (7.7) while Φ and V always occur through their derivatives.

This is by no means the only choice that leads to a covariant perturbation theory. We could, for example, repeat the normal co-ordinate construction but take the curve joining $\bar{\phi}^i$ and $\bar{\phi}^i + \pi^i$ to satisfy (7.2) with some non-minimal connection

$$\Gamma^i_{jk} = \left\{ {i \atop j\ k} \right\} + S^i_{jk} \tag{7.9}$$

for some tensor S^i_{jk}, so that (7.3) still holds, but with the connection as defined in (7.9) [47]. The tangent vector to this curve will again (obviously) be a vector and by choosing D to be the covariant derivative with the connection (7.9) and making corresponding modifications to (7.5), $S[\phi + \pi(\xi)]$ can be expanded as before. These might be called "abnormal" co-ordinates.

In general, a co-ordinate ξ^i defined by (7.1) with some choice of M can be expressed in terms of the normal co-ordinate $\tilde{\xi}^i$, say, using

$$\xi^i = \tilde{\xi}^i + \sum_{n=2}^{\infty}\frac{1}{n!}N^i_{j_1\ldots j_n}(\bar{\phi})\tilde{\xi}^{j_1}\ldots\tilde{\xi}^{j_n} \tag{7.10}$$

for some $N^i_{j\ldots k}$ constructed from the $M^i_{j\ldots k}$. The ξ^i will then be a vector if and only if all the $N^i_{j\ldots k}$ are tensors on M. For the abnormal co-ordinates defined with the connection (7.9), $N^i_{jk} = S^i_{jk}$, $N^i_{jkl} = \nabla_{(j}S^i_{kl)} + S^i_{m(j}S^m_{kl)}$, etc.. Expanding the action $S[\bar{\phi} + \pi(\xi)]$ in terms of these general co-ordinates ξ gives extra N-dependent terms to (7.8) involving the N co-efficients. For the lowest order terms, these changes are

$$L_1 \to L_1, \quad L_2 \to L_2 - \frac{1}{2}N^i_{jk}(\bar{\phi})\xi^j\xi^k\delta S/\delta\bar{\phi}^i \tag{7.11}$$

If the N's are tensors, then ξ and ξ are both covariant and equally good choices for the quantum fields.

8. Covariant Effective Actions

First, we choose some background/quantum split as in (7.1) so that ξ^i is a vector. We then seek a definition of effective action given by an integration over ξ^i or, more conveniently, over $\xi^a(x) \equiv \xi^i e_i^a(\bar{\phi})$ for some vielbein e_i^a, using the measure

$$\int[d\xi] = \int \prod_x \prod_{a=1}^{d} d\xi^a(x) \qquad (8.1)$$

(The Jacobian obtained in changing from the measure (6.2) is unity as $\delta^n(0) = 0$ for non-integer n in dimensional regularization, which we shall use here.) Vilkovisky has suggested replacing (6.9) by [48]

$$\exp\left[-\hbar^{-1}\Gamma_V[\bar{\phi}]\right] = \int[d\xi]\exp\left\{-\hbar^{-1}\left[S[\bar{\phi} + \pi(\xi)] - \xi.\delta\Gamma_V/\delta\bar{\phi}\right]\right\} \qquad (8.2)$$

As ξ is a vector, the linear term in the exponent on the right-hand-side of (8.2) is now covariant and $\Gamma_V[\bar{\phi}]$ is a co-ordinate independent functional. However, (8.2), as we shall see, does not seem to correspond to 1PI diagrams at higher loops off-shell and the renormalization of Γ_V may be problematic.

As we shall see, it seems preferable to start by modifying (6.3) or (6.4). Friedan modifies the source term (6.3) to become [1]

$$S_J = S[\phi] + \int d^n x\, h(x,\phi(x)) \qquad (8.3)$$

for some scalar function h. Then (6.4) can be evaluated by a change of variables to ξ, giving a covariant perturbation theory. We can then take a Legendre transform of W[h] with respect to the non-linear source h, obtaining a quantity perhaps more like a "free-energy" than a conventional effective action. See [1] for details of this approach.

Here, we prefer simply to add a covariant linear source term $J.\xi$ for the quantum field ξ, to obtain [49]

$$\exp\left[-\hbar^{-1}W[\bar{\phi},J]\right] = \int[d\xi]\exp\left\{-\hbar^{-1}\left[S[\bar{\phi} + \pi(\xi)] + \int d^n x\, J_a(x)\xi^a(x)\right]\right\} \qquad (8.4)$$

With $J_a(x)$ chosen to transform as a vector under diffeomorphisms of M and as a scalar density on the space-time N, W is co-ordinate independent. Defining a "mean quantum field"

$$\bar{\xi}^a(x) = \frac{\delta W}{\delta J_a(x)} \tag{8.5}$$

we can take the Legendre transform of W

$$\Gamma[\bar{\phi},\bar{\xi}] = W[\bar{\phi},J(\bar{\xi})] - \bar{\xi}.J = S[\bar{\phi} + \pi(\bar{\xi})] + O(\hbar) \tag{8.6}$$

to obtain a covariant effective action Γ, corresponding to the set of all diagrams that are 1PI with respect to the quantum field $\xi^a(x)$. Using $J_a = -\delta\Gamma/\delta\bar{\xi}^a$ and (8.4), Γ can be written as

$$\exp\left[-\hbar^{-1}\Gamma[\bar{\phi},\bar{\xi}]\right] = \int[d\xi]\exp\left\{-\hbar^{-1}\left[S[\bar{\phi} + \pi(\xi)] - (\xi - \bar{\xi}).\delta\Gamma/\delta\bar{\xi}\right]\right\}$$

$$= \int[d\xi']\exp\left\{-\hbar^{-1}\left[S[\bar{\phi} + \pi(\xi' + \bar{\xi})] - \xi'.\delta\Gamma/\delta\bar{\xi}\right]\right\} \tag{8.7}$$

where $\xi'^a = \xi^a - \bar{\xi}^a$. This can be interpreted as a kind of "double background field expansion", in which ϕ is split into a classical background $\bar{\phi}$, plus a field ξ which is in turn split into a classical field $\bar{\xi}$, plus a quantum field (operator) ξ', over which we integrate.

The expectation value of some quantity $A(\bar{\phi} + \pi(\xi))$ or, more generally, of $A(\bar{\phi},\xi)$, is defined by

$$\langle A\rangle(\bar{\phi},\bar{\xi}) = e^{\Gamma/\hbar}\int[d\xi']\exp\left\{-\hbar^{-1}\left[S[\bar{\phi} + \pi(\xi' + \bar{\xi})] - \xi'.\delta\Gamma/\delta\bar{\xi}\right]\right\}$$

$$\times A(\bar{\phi},\xi' + \bar{\xi}). \tag{8.8}$$

so that $\langle 1\rangle = 1$, $\langle\delta S/\delta\bar{\phi}\rangle = \delta\Gamma/\delta\bar{\phi}$ etc.

The classical action $S[\bar{\phi} + \pi(\xi)]$ clearly satisfies

$$\frac{\delta S}{\delta\bar{\phi}^i} = F^j{}_i\frac{\delta S}{\delta\xi^j}, \tag{8.9}$$

$$F^j{}_i[\bar{\phi},\xi] \equiv \frac{\partial\xi^j}{\partial\pi^i} = \delta^j{}_i - \tfrac{1}{2}M^j{}_{ik}(\bar{\phi})\xi^k + O(\xi^2) \tag{8.10}$$

Then taking the expectation value of (8.9) and discarding terms proportional to $\delta^n(0)$ gives the Ward identity [49]

$$\frac{\delta\Gamma}{\delta\bar{\phi}^i}[\bar{\phi},\bar{\xi}] = \langle F^j{}_i\rangle(\bar{\phi},\bar{\xi})\frac{\delta\Gamma}{\delta\bar{\xi}^j}(\bar{\phi},\bar{\xi}) \tag{8.11}$$

For example, if ξ is a normal co-ordinate, the appearance of $\langle F\rangle$ rather than F in (8.11) implies that $\Gamma[\bar{\phi},\bar{\xi}]$ is <u>not</u> the normal co-ordinate expansion of $\Gamma[\bar{\phi}] \equiv \Gamma[\bar{\phi},\bar{\xi} = 0]$. This has the important consequence that divergences in $\Gamma[\bar{\phi},\bar{\xi}]$ cannot be removed by simply adding the counterterms given by the normal co-ordinate expansion of the counterterms used to remove the divergences from $\Gamma[\bar{\phi}]$ [49], contrary to claims sometimes made in the literature.

We are now in a position to compare the effective action Γ (8.7) corresponding to 1PI diagrams with that of Vilkovisky, (8.2). Using (8.11) in (8.7) and taking $\bar{\xi} = 0$, we obtain, with $\langle F\rangle \equiv \langle F\rangle(\bar{\phi},\bar{\xi} = 0)$,

$$\exp\left\{-\hbar^{-1}\Gamma[\bar{\phi}]\right\} = \int[d\xi']\exp\left\{-\hbar^{-1}(S[\bar{\phi} + \pi(\xi')] - \langle F\rangle^{-1}.\delta\Gamma/\delta\bar{\phi}\right\} \tag{8.12}$$

which would only agree with (8.2) if $\langle F\rangle = 1$. However, $\langle F\rangle = 1 + O(\hbar)$ so that $\Gamma[\bar{\phi}]$ and $\Gamma_V[\bar{\phi}]$ disagree at the 2-loop order, $\Gamma[\bar{\phi}] - \Gamma_V[\bar{\phi}] = O(\hbar^2)$. For example, there is a divergent contribution at one loop to $\langle F\rangle^{-1}$

$$(\langle F\rangle^{-1})^i{}_j\Big|_{\bar{\xi}=0} = \delta^i{}_j - \frac{\hbar}{2\pi\epsilon}m^{\epsilon}M^i_{jkl}g^{kl} + \ldots \tag{8.13}$$

as is easily verified. Then the Vilkovisky effective action does not correspond only to diagrams that are 1PI with respect to ξ at more than one loop, while Γ does. However, on-shell ($\delta\Gamma/\delta\bar{\phi} = 0$) the effective action (8.12) has formally the same functional integral representation as Γ_V does when $\delta\Gamma_V/\delta\bar{\phi} = 0$, although the field equations $\delta\Gamma/\delta\bar{\phi} = 0$ and $\delta\Gamma_V/\delta\bar{\phi} = 0$ have a different form since $\Gamma \neq \Gamma_V$. In the following sections, we will use Γ rather than Γ_V, as its properties are closer to those of the conventional effective action (6.7) and its renormalization is better understood.

Since the beginning of Section 6, we have been assuming that M and N were topologically equivalent to $\mathbb{R}^d$ and $\mathbb{R}^n$, respectively, so that configurations corresponded to functions $\mathbb{R}^n \to \mathbb{R}^d$, $x^\mu \to \phi^i(x)$. In Section 7, we also tacitly assumed that the co-ordinates ξ were "good" co-ordinates globally on M. For such choices of M and N, the effective action constructed above correctly encodes all the quantum physics of the model. If, however, M cannot be covered by a single co-ordinate chart then the space of maps from N to M is not correctly parameterized by a single function $\phi^i(x)$. For such M, (6.9) doesn't correctly take into account quantum fluctuations that are large enough to take one out of the original co-ordinate patch containing the background $\overline{\phi}$; also, changing variables to, say, normal co-ordinates doesn't help. However, small fluctuations <u>are</u> correctly represented and so the effective action (8.7) should give the correct short distance physics (ultra-violet divergences, conformal anomaly, etc.) but will not in general describe boundary conditions, infra-red divergences, topological effects etc correctly.

A simple example is that in which the target space is a circle, $M = S^1$. Then M needs at least two patches to cover it and normal co-ordinates do not give a good description - there are infinitely many geodesics joining any two points, winding round the circle different numbers of times. The normal co-ordinate prescription will give the quantum mechanics of a free particle on the covering space of the circle, the real line (parameterized by ξ), with a continuous spectrum. This is rather different from the physics of a particle on a circle, which leads to a discrete spectrum, etc.

In the general case when more than one patch is needed, no convenient global parameterization of the theory is known, except in special cases such as group manifolds or homogeneous spaces [50]. We will use the effective action (8.7) in the general case to obtain the short distance structure of the theory, which is our chief interest here, and ignore questions in which large scale effects play a role.

9. Regularization and Renormalization

The functional integrals that have been manipulated formally in the previous sections are finite and must be regulated. It is desirable to have a covariant regularization scheme that preserves as many of the sigma-model symmetries as possible. We regulate the infra-red divergences by introducing a mass m into the theory. For actual calculations, it is convenient to add the covariant quantum field mass term

$$S_m = \frac{m^2}{2}\int d^n x\sqrt{\gamma}\, \delta_{ab}\xi^a\xi^b \qquad (9.1a)$$

to $S(\bar{\phi}+\pi(\xi))$ so that $(p^2)^{-1} \to (p^2 + m^2)^{-1}$ in the ξ-propagator, while for general arguments it is often more useful instead to introduce a scalar potential $m^2P(\phi)$ so that

$$V(\phi) \to V' = V(\phi) + m^2P(\phi) \qquad (9.1b)$$

in (1.13), where P is chosen so that the mass matrix $\nabla_i\nabla_j V'$ is positive definite for large enough m. We shall usually omit the prime on V.

If there is no Wess-Zumino term, we regulate the ultra-violet divergences using dimensional regularization, continuing the space-time dimension n to non-integer values $n' = n + \epsilon$. If there is a Wess-Zumino term, however, we need some prescription for dealing with the alternating tensor $\epsilon_{\mu_1\ldots\mu_n}$. We continue the metric $\gamma_{\mu\nu}$ to some n'-dimensional metric $\hat{\gamma}_{\mu\nu}$, which satisfes $\hat{\gamma}_{\mu\nu}\hat{\gamma}^{\mu\nu} = n + \epsilon$, and continue $\epsilon_{\mu_1\ldots\mu_n}$ to some rank-n anti-symmetric tensor in n' dimensions, $\hat{\epsilon}_{\mu_1\ldots\mu_n}$. While for all integer dimensions n', an m-th rank anti-symmetric tensor with $m > n'$ must vanish, $T_{[\mu_1\ldots\mu_m]} = 0$, we do not assume this to be true for non-integer dimensions n', so that in particular $\hat{\epsilon}_{\mu_1\ldots\mu_n}$ need not vanish in $n-|\epsilon|$ dimensions. Note that many of the special properties (e.g. invariance) enjoyed by $\epsilon_{\mu_1\ldots\mu_n}$ in n-dimensions are not satisfied by $\hat{\epsilon}_{\mu_1\ldots\mu_n}$ in n'-dimensions. For the general arguments and one-loop calculations presented in these lectures, we will not need to specify the properties of $\hat{\epsilon}_{\mu_1\ldots\mu_n}$ in $n + \epsilon$ dimensions, but will need only to know that there is some consistent prescription.

However, it is of interest to see what properties we would need to impose on $\hat{\epsilon}_{\mu_1\ldots\mu_n}$. For concreteness, consider the case n = 2 for which the sigma-model is renormalizable, and which is the only case we shall consider in any detail in what follows. We need a prescription for simplifying products of $\hat{\epsilon}$-tensors. While in 2-dimensions the identity $\epsilon_{\mu\nu}\epsilon^{\rho\sigma} = -2\delta^{[\rho}_{[\mu}\delta^{\sigma]}_{\nu]}$ holds, in $(2+\epsilon)$ dimensions it would be inconsistent to demand that $\hat{\epsilon}_{\mu\nu}\hat{\epsilon}^{\rho\sigma} = -2f(\epsilon)\delta^{[\rho}_{[\mu}\delta^{\sigma]}_{\nu]}$ for any function $f(\epsilon)$, as it is easily shown that this would lead to the equation $\epsilon f(\epsilon) = 0$. Fortunately, an ansatz for products of $\hat{\epsilon}$-tensors without any contractions is not needed. For local divergences, we need only demand that

$$\hat{\epsilon}_{\mu\nu}\hat{\epsilon}_{\rho\sigma}\hat{\gamma}^{\nu\rho} = g(\epsilon)\hat{\gamma}^{\nu\rho} \tag{9.2}$$

for some function $g(\epsilon)$ satisfying $g(0) = 1$. This is because local divergences involving $\hat{\epsilon}$-tensors are proportional either to $\hat{\epsilon}_{\mu\nu}\hat{\epsilon}_{\rho\sigma}\ldots\hat{\epsilon}_{\alpha\beta}\partial_\lambda\phi\partial_\tau\phi$ or to $(V \text{ or } m^2 \text{ or } R^{(\gamma)})\hat{\epsilon}_{\mu\nu}\hat{\epsilon}_{\rho\sigma}\ldots\hat{\epsilon}_{\alpha\beta}$, with all indices contracted using $(2+\epsilon)$-dimensional metrics. In both types of term all the space-time indices must be contracted with each other, so that the $\partial\phi\partial\phi$ term will completely reduce, on using (9.2), to something proportional to either $\hat{\gamma}^{\mu\nu}\partial_\mu\phi\partial_\nu\phi$ or $\hat{\epsilon}^{\mu\nu}\partial_\mu\phi\partial_\nu\phi$, while the other term becomes simply a scalar function. The divergences will then be proportional to terms of the form found in the classical Lagrangian and can, as we shall see, be removed by renormalization. Note that this would not be true for any tensor other than the metric on the right hand side of (9.2). Although uncontracted products of $\hat{\epsilon}$-tensors might occur in non-local divergences, these are cancelled by counterterm graphs without needing to make any assumptions about $\hat{\epsilon}$. Once the theory has been renormalized, we can take the $\epsilon = 0$ limit and then use 2-dimensional identities.

Thus we only need assume that $\hat{\epsilon}$ satisfies (9.2) for some function g, which we must now specify - for example, different functions g give different results for the 2-loop β-functions [46]. Using a dimensional reduction regularization of the form used in [51] would correspond to using (9.2) with $g(\epsilon) = 1 + \epsilon$; however, it is well known that dimensional reduction regularization has inconsistencies. We choose $g(\epsilon) = 1$, for reasons explained in more detail elsewhere [46,52]. There are two main reasons for this choice; first, it is only with this choice that certain consistency conditions are satisfed by the β-functions. Second, it is only this choice that preserves the natural splitting of modes into left-movers and right-movers. We now adopt this scheme and henceforth drop the "hats" from $\hat{\gamma}_{\mu\nu}$ and $\hat{\epsilon}_{\mu\nu}$. We use the convention that $\epsilon_{\mu\nu}$ is real in Minkowski space and imaginary in Euclidean space. (One alternative to this prescription would be to use higher derivative regularization, but this is harder to calculate with and requires a separate prescription for regulating one-loop divergences.)

The effective action can now be evaluated perturbatively using Feynman diagram techniques. If the space-time dimension $n > 2$, then there will be L-loop divergent contributions to the Lagrangian of canonical dimension $D = n + (L - 1)(n - 2)$, (e.g. terms with D space-time derivatives, $(\partial\phi)^D$) and so the theory is not renormalizable. For $n = 2$, all divergences are of dimension 2 and the theory is renormalizable [1]. We will therefore restrict ourselves to the two-dimensional case with bare action in $n = 2+\epsilon$ dimensions

$$S\left[\phi_0, \Psi^0\right] = \int d^n x \sqrt{\gamma}\left[\frac{1}{2}(g^0_{ij} + b^0_{ij})(\gamma^{\mu\nu} + \epsilon^{\mu\nu})\partial_\mu \phi^i_0 \partial_\nu \phi^j_0 + \Phi^0 R^{(\gamma)} + V^0\right] \tag{9.3}$$

where the subscripts 0 denote bare quantities and the background/quantum split is now $\phi^i = \bar{\phi}^i_0 + \pi^i_0(\xi)$. It is useful to introduce a "vector" notation for the metric, potential, etc. $\Psi^0 = \left\{V^0; g^0_{ij}, b^0_{ij}, \Phi^0\right\}$. We shall refer to the bare functionals (8.4),(8.7) defined using (9.3) as W_0, Γ_0, respectively. Note that if we scale $\Psi_0 \to \Omega^{-1}\Psi_0$ for some constant Ω, then an L-loop 1PI graph scales as $\Omega^{(L-1)}$, so that the larger Ψ_0 is, the more convergent the perturbation theory. Then the "size" of M is an effective coupling constant, (scaling g^0_{ij} scales the volume of M).

We will not prove renormalizability here; see [1,49]. Divergences will be removed by expressing $\Psi_0, \bar{\phi}_0, \xi_0$ in terms of renormalized quantities $\Psi \equiv (V, g_{ij}, b_{ij}, \Phi), \bar{\phi}$ and ξ, together with a renormalization mass scale μ, by

$$\Psi_0 = \mu^\epsilon\left[\Psi + \sum_{\nu=1}^{\infty} \epsilon^{-\nu} T^{(\nu)}(\Psi)\right]$$

$$\bar{\phi}^i_0 = \bar{\phi}^i + \sum_{\nu=1}^{\infty} \epsilon^{-\nu} Y^i_{(\nu)}(\bar{\phi}, \Psi) \tag{9.4}$$

$$\xi^a_0 = \xi^a + \sum_{\nu=1}^{\infty} \epsilon^{-\nu} X^a_{(\nu)}(\xi, \bar{\phi}, \Psi) \equiv \xi^a + X^a$$

with $T^{(\nu)}$ representing the "vector" $T^{(\nu)} = (T^{(\nu,V)}, T^{(\nu,g)}_{ij}, T^{(\nu,b)}_{ij}, T^{(\nu,\Phi)})$. We define the renormalized functional $W[\bar{\phi}, J, \Psi]$ as in [49]

$$\exp(\hbar^{-1} W) = \int [d\xi] \exp\left\{-\hbar^{-1}\left[S[\bar{\phi}_0, \xi_0, \Psi_0] + J.\xi\right]\right\} \tag{9.5}$$

It is important that the source couples to the renormalized field ξ, not ξ_0, in (9.5). Taking the Legendre transform with $\bar{\xi} = \delta W/\delta J$ gives

$$\exp(-\hbar^{-1}\Gamma) = \int [d\xi] \exp\left\{-\hbar^{-1}\left[S[\bar{\phi}_0, \xi_0, \Psi_0] - (\xi - \bar{\xi}).\delta\Gamma/\delta\bar{\xi}\right]\right\} \tag{9.6}$$

which can be rewritten using (9.4) as

$$\exp(-\hbar^{-1}\Gamma) = \int[d\xi']\exp\left\{-\hbar^{-1}\left[S[\bar\phi_0,\xi' + \bar\xi + X(\xi'+\bar\xi)] - \xi'.\delta\Gamma/\delta\bar\xi\right]\right\} \tag{9.7}$$

with $\xi' = \xi - \bar\xi$. The residues $T^{(\nu)}, X^{(\nu)}, Y^{(\nu)}$ in (9.4) are chosen so as to render W and Γ finite when expressed in terms of the renormalized quantities $\Psi, \bar\phi, \bar\xi$ using (9.4). As the Feynman rules are covariant (for appropriate choices of coefficients in (7.1)) the counterterms must be covariant, that is $T^{(\nu,g)}_{ij}$ and $T^{(\nu,b)}_{ij}$ are tensors, $T^{(\nu,\Phi)}$ and $T^{(\nu,V)}$ are scalars and $X^i_{(\nu)}, Y^i_{(\nu)}$ are vectors. Moreover, they are universal in that they have the same functional form whatever the choice we make for $\Psi = (V, g_{ij}, \ldots)$.

As the bare quantities on the left-hand-side of (9.4) are independent of μ, the renormalized fields must be μ-dependent. We define

$$\gamma^i_\phi = -\mu\frac{\partial}{\partial_\mu}\bar\phi^i, \quad \gamma^a_\xi = -\mu\frac{\partial}{\partial_\mu}\xi^a, \quad \gamma^a_{\bar\xi} = -\mu\frac{\partial}{\partial_\mu}\bar\xi^a \tag{9.8}$$

The β and γ functions are given in terms of the $1/\epsilon$ poles by

$$\hat\beta^\Psi = \epsilon\Psi + \beta^\Psi \tag{9.9}$$

$$\beta^\Psi = \left[\left[1 + \Lambda\frac{\partial}{\partial\Lambda}\right]T^{(1)}(\Lambda^{-1}\Psi)\right]\Bigg|_{\Lambda=1}$$

$$\gamma_\phi = \left[\frac{\partial}{\partial\Lambda}Y^{(1)}(\bar\phi,\Lambda^{-1}\Psi)\right]\Bigg|_{\Lambda=1}$$

$$\gamma_\xi = \left[\frac{\partial}{\partial\Lambda}X^{(1)}(\bar\phi,\xi,\Lambda^{-1}\Psi)\right]\Bigg|_{\Lambda=1}$$

while the residues of the higher order poles are constrained by renormalization group pole equations [43]. We shall calculate the β-functions to lowest order in section 11.

Consider now the μ-dependence of the renormalized effective action, $\Gamma[\bar\phi,\Psi,\bar\xi]$

$$\mu\frac{d\Gamma}{d\mu} \equiv \mu\frac{\partial\Gamma}{\partial\mu} - \beta^\Psi.\frac{\delta\Gamma}{\delta\Psi} - \gamma^i_\phi.\frac{\delta\Gamma}{\delta\bar\phi^i} - \gamma^a_{\bar\xi}.\frac{\delta\Gamma}{\delta\bar\xi^a} \tag{9.10}$$

with the notation

$$\beta^\Psi.\frac{\delta\Gamma}{\delta\Psi} = \beta^V.\frac{\delta\Gamma}{\delta V} + \beta^g_{ij}.\frac{\delta\Gamma}{\delta g_{ij}} + \beta^b_{ij}.\frac{\delta\Gamma}{\delta b_{ij}} + \beta^\Phi.\frac{\delta\Gamma}{\delta\Phi} \tag{9.11}$$

$$\beta^g_{ij}.\frac{\delta\Gamma}{\delta g_{ij}} = \lim_{\eta\to 0}\left\{\frac{1}{\eta}\left[\Gamma[g_{ij} + \eta\beta^g_{ij};\ldots] - \Gamma[g_{ij};\ldots]\right]\right\}$$

etc. The left-hand-side of (9.10) does not vanish; careful evaluation of the μ-dependence of (9.6) gives

$$\mu\frac{d\Gamma}{d\mu} = -(\gamma^a_{\bar{\xi}} - \langle\gamma^a_{\xi}\rangle).\frac{\delta\Gamma}{\delta\bar{\xi}^a} \tag{9.12}$$

so that (9.10), (9.12) give the renormalization group equation [52]

$$\mu\frac{\partial\Gamma}{\partial\mu} = \beta^{\Psi}.\frac{\delta\Gamma}{\delta\Psi} + \gamma^i_{\phi}.\frac{\delta\Gamma}{\delta\bar{\phi}^i} + \langle\gamma^a_{\xi}\rangle.\frac{\delta\Gamma}{\delta\bar{\xi}^a} \tag{9.13}$$

We now use $\langle A\rangle_0$ for the bare expectation value (8.8), while the corresponding "renormalized" average used in (9.13) is (c.f. (8.8,9.7))

$$\langle A\rangle(\bar{\phi},\bar{\xi}) = e^{\Gamma/\hbar}\int[d\xi']\exp\left\{-\hbar^{-1}S[\bar{\phi}_0,\ \xi' + \bar{\xi} + X(\xi'+\bar{\xi})] - \xi'.\delta\Gamma/\delta\bar{\xi}\right\}$$

$$\times\ A(\bar{\phi}_0,\ \xi' + \bar{\xi}) \tag{9.14}$$

There are two chief kinds of ambiguity in the above scheme, as will be seen explicitly in the calculations of section 11. First, instead of the minimal subtraction (9.4), we could allow finite renormalizations, so that in the summations in (9.4) ν runs from 0 to ∞ instead of from 1 to ∞. This amounts simply to a finite field redefinition of the renormalized variables,

$$\Psi \to \Psi + T^{(0)}(\Psi) \qquad \xi \to \xi + X^{(0)}(\xi,\bar{\phi},\Psi) \tag{9.15}$$

$$\bar{\phi} \to \bar{\phi} + Y^{(0)}(\bar{\phi},\Psi)$$

and doesn't affect physical quantities [13,52,53,54].

Second, even with minimal subtraction, the counterterms in (9.4) are not uniquely determined. We can, for example, choose not to renormalize $\bar{\phi}^i$, so that $Y^{(\nu)} = 0$ for all ν. Then the $X^{(\nu)}$ and $T^{(\nu)}$ are uniquely determined and the renormalized effective action satisfies the Ward identity (8.11) [49]. In particular, the $X^{(\nu)}$ are non-zero; the quantum field wave function renormalization is needed for the cancellation of sub-divergences even if we take $\bar{\xi} = 0$ [49]. However, it is not necessary to use this scheme in which $Y^{(\nu)} = 0$ to cancel the divergences. To see the extent of the ambiguity, we start with any consistent renormalization (9.4) for some counterterm tensors T,X,Y and then make an infinitesimal change in the $\epsilon^{-\nu}$ counterterm $Y^{(\nu)}$ for some ν, $Y^{(\nu)} \to Y^{(\nu)} + \delta Y^{(\nu)}$, changing Γ by

$$\delta\Gamma = -\epsilon^{-\nu}(\delta Y^{(\nu)}).\delta\Gamma/\delta\bar{\phi} + O(\epsilon^{-(\nu+1)})$$

so that $\Gamma + \delta\Gamma$ is no longer finite. We then attempt to cancel this new $\epsilon^{-\nu}$ divergence by modifying the $T^{(\nu)}$ counterterm, $T^{(\nu)} \to T^{(\nu)} + \delta T^{(\nu)}$, so that $\delta\Gamma$ becomes

$$\delta\Gamma = -\epsilon^{-\nu}\left[\delta Y^{(\nu)}.\delta\Gamma/\delta\bar{\phi} + \delta T^{(\nu)}.\delta\Gamma/\delta\Psi\right] + O\left[\epsilon^{-(\nu+1)}\right] \tag{9.16}$$

To find some $\delta T^{(\nu)}$ such that (9.16) vanishes (to order $\epsilon^{-\nu}$), we need a relation between $\delta\Gamma/\delta\bar{\phi}$ and $\delta\Gamma/\delta\Psi$. Classically, the co-ordinate independence of $S[\phi,\Psi]$ gives the identity

$$(\underset{u}{L}\Psi).\frac{\delta S}{\delta\Psi} = u.\frac{\delta S}{\delta\phi} \tag{9.17}$$

for any vector $u^i(\phi)$, where $\underset{u}{L}$ denotes Lie differentiation with respect to u^i, so that

$$\underset{u}{L}\Psi = (\underset{u}{L}V,\ \underset{u}{L}g_{ij},\ \underset{u}{L}b_{ij},\ \underset{u}{L}\Phi)$$

$$\underset{u}{L}V = u^i V_{,i} \qquad \underset{u}{L}g_{ij} = \nabla_i u_j + \nabla_j u_i \tag{9.18}$$

$$\underset{u}{L}\Phi = u^i\Phi_{,i} \qquad \underset{u}{L}b_{ij} = 2H_{ij}{}^k u_k + 2\partial_{[i}\left\{b_{j]k}u^k\right\}$$

Then (9.17) leads to the Ward identity [52]

$$(\underset{u}{L}\Psi).\frac{\delta\Gamma}{\delta\Psi} = u.\frac{\delta\Gamma}{\delta\bar{\phi}} \tag{9.19}$$

for any $u^i(\bar{\phi})$. Then we see that the $\epsilon^{-\nu}$ pole in (9.16) will cancel if we choose $\delta T^{(\nu)}$ to be

$$\delta T^{(\nu)} = -\underset{k}{L}\Psi, \qquad k^i = \delta Y^i_{(\nu)} \tag{9.20}$$

Further, the poles of order $\nu' > \nu$ introduced by (9.16) can also be cancelled by modifying $T^{(\nu')}, Y^{(\nu')}$ giving a new finite effective action. If $\nu = 1$, the $1/\epsilon$ poles and hence the β- and γ-functions are modified:

$$\beta^{\Psi} \to \beta^{\Psi} - \underset{w}{L}\Psi \qquad \gamma_{\phi}^{i} \to \gamma_{\phi}^{i} + w^{i}$$

$$w^{i} \equiv \left[\frac{\partial}{\partial\Lambda}\delta Y^{i}_{(1)}(\bar{\phi},\Lambda^{-1}\Psi)\right|_{\Lambda=1} \tag{9.21}$$

Explicitly, the change in the β-functions is

$$\beta^{g}_{ij} \to \beta^{g}_{ij} - 2\nabla_{(i}w_{j)}, \qquad \beta^{\Phi} \to \beta^{\Phi} - w^{i}\Phi_{,i} \tag{9.22}$$

$$\beta^{b}_{ij} \to \beta^{b}_{ij} - 2H_{ij}{}^{k}w_{k} - 2\partial_{[i}\left\{b_{j]k}w^{k}\right\} \qquad \beta^{V} \to \beta^{V} - w^{i}V_{,i}$$

Thus we have seen two main kinds of ambiguity in the β- and γ-functions, corresponding to dependence on the renormalization scheme. The first is due to the possibility of adding finite local counterterms to Γ, i.e. to making a non-minimal subtraction, and is analogous to the ambiguity in the Yang-Mills β-function at more than two loops. Its effect is to allow local redefinitions of V, g_{ij}, b_{ij}, Φ. The second kind is due to the existence of a symmetry (the diffeomorphisms of M) acting on the infinite set of coupling constants represented by the metric etc., allowing shifts in the β- and γ-functions as in (9.21), (9.22).

There are a number of other ambiguities, however. First the Ward identity (8.11) allows us to compensate for a change in the wave function renormalization of $\bar{\phi}^{i}$ by a change in that of ξ^{a}, so that both γ_{ϕ} and γ_{ξ} change. Secondly, the classical symmetry (2.3) implies that changing the b_{ij}-counterterms by

$$T^{(\nu,b)}_{ij} \to T^{(\nu,b)}_{ij} + \partial_{[i}\rho^{(\nu)}_{j]} \tag{9.23}$$

for some $\rho^{(\nu)}_{j}$ changes Γ by a divergent surface term. We assume that the boundary conditions are such that surface terms vanish (e.g. N compact) so that this change in Γ is zero. However, there is then a change in the β^{b}-function

$$\beta^{b}_{ij} \to \beta^{b}_{ij} + \partial_{[i}\lambda_{j]} \tag{9.24}$$

$$\lambda_{i} = \left[\left[1 + \Lambda\frac{\partial}{\partial\Lambda}\right]\rho^{(1)}_{i}(\bar{\phi},\Lambda^{-1}\Psi,\xi)\right]\Big|_{\Lambda=1}$$

Finally, changing the background/quantum split, i.e. changing the M-coefficients in (7.1), leads to a change in the off-shell part of Γ. More generally, all sensible definitions of the effective action should agree on-shell but may differ off-shell. Any infinitesimal change in Γ that doesn't change the on-shell part of Γ must be proportional to the effective field equation $\delta\Gamma/\delta\overline{\phi}$ i.e. $\delta\Gamma = u.\delta\Gamma/\delta\overline{\phi}$ for some infinitesimal u^i. Using (9.19), this can also be written as $\delta\Gamma = (\mathcal{L}_u\Psi).\delta\Gamma/\delta\Psi$. Then if $u^i = \epsilon^{-\nu}v^i(\overline{\phi},\Psi)$, for some infinitesimal vector field v^i, the divergences in the new effective action $\Gamma + \delta\Gamma$ can be removed by either changing $T^{(\nu)} \to T^{(\nu)} - \mathcal{L}_v\Psi$ or by changing $Y_i^{(\nu)} \to Y_i^{(\nu)} - v_i$. In particular, for $\nu = 1$, with $w^i = \partial/\partial\Lambda[v^i(\Lambda^{-1}\Psi)]|_{\Lambda=1}$, changing $T^{(1)}$ gives $\beta^\Psi \to \beta^\Psi - \mathcal{L}_w\Psi$, while changing $Y^{(1)}$ yields $\gamma^i_\phi \to \gamma^i_\phi - w^i$. The two possibilities correspond, of course, to different renormalization schemes, related as in the discussion leading to (9.21), (9.22). Given the ambiguity (9.21), (9.22), we can regard this as solely an ambiguity in the γ-functions. For example, the effective action (6.9) with non-covariant one-loop contributions (6.13) differs from the covariant effective action (8.7), with ξ a normal co-ordinate, by precisely such a term with $u^i = \hbar X^i\epsilon^{-1}+\ldots$, where X^i is as defined in (6.16). As all choices of definition of Γ that give the same on-shell results are equally good and describe the same physics, we find again an inherent ambiguity in the β- and γ-functions. See [13,52,54] for further discussion.

10. Conformal Invariance

In n space-time dimensions, consider the group of general co-ordinate transformations and local scale (Weyl) transformations. Under an infinitesimal transformation, the space-time metric changes by

$$\delta\gamma_{\mu\nu} = \nabla_\mu k_\nu + \nabla_\nu k_\mu + 2\Lambda\gamma_{\mu\nu} \tag{10.1}$$

where $k^\mu(x)$ parameterizes the diffeomorphism and $\Lambda(x)$ is the local scale parameter. If we choose some fixed metric, $\overline{\gamma}_{\mu\nu}$ say, we can seek the transformations leaving $\overline{\gamma}_{\mu\nu}$ invariant, i.e. the parameters (k_μ,Λ) such that $\delta\overline{\gamma}_{\mu\nu} = 0$. The vectors k^μ for which some Λ can be found so that $\delta\overline{\gamma}_{\mu\nu} = 0$ are the conformal Killing vectors and the set of combined diffeomorphisms and scalings preserving $\bar{\gamma}_{\mu\nu}$ form the conformal group (of $\bar{\gamma}_{\mu\nu}$), G_{conf}. If $\bar{\gamma}_{\mu\nu}$ is the n-dimensional Minkowski metric, $\bar{\gamma}_{\mu\nu} = \eta_{\mu\nu}$, then for $n > 2$, $G_{conf} = SO(n,2)$, while for $n = 2$ it is an infinite dimensional group with $SO(2,2) \approx SO(2,1) \times SO(2,1)$ as a finite dimensional subgroup and with Lie algebra $V \times V$, where V is the Virasoro algebra. For all dimensions, there is an $ISO(n-1,1) \times SO(1,1) = H_W$ subgroup of G_{conf} consisting of the Poincaré group and dilatations. A generally covariant field theory that is also local scale invariant will, in Minkowski spacetime, have conserved currents corresponding to each of the generators of the conformal group [55], while a covariant theory that is only rigid scale invariant, i.e. invariant under (10.1) only for constant Λ, will have conserved currents corresponding only to the generators of H_W. These symmetries can break down in the quantum theory, due to conformal anomalies. We shall be interested in $n = 2$ sigma-models that give local scale invariant quantum field theories or, in the context of string theories, in models whose local scale anomaly is cancelled by ghost contributions (see sections 14–16). As we shall see, in some cases quantum corrections can break the local scale symmetry down to rigid scale invariance, so that in Minkowski space there are only conserved currents corresponding to the generators of H_W instead of those of the infinite dimensional conformal groups G_{conf} [47].

Under local scale transformations in $n = 2 + \epsilon$ dimensions,

$$\gamma_{\mu\nu} \to e^{2\Lambda(x)}\gamma_{\mu\nu}, \qquad \epsilon_{\mu\nu} \to e^{2\Lambda(x)}\epsilon_{\mu\nu} \tag{10.2}$$

so that

$$\sqrt{\gamma} \to \epsilon^{n\Lambda}\sqrt{\gamma}, \qquad \sqrt{\gamma}(\gamma^{\mu\nu} + \epsilon^{\mu\nu}) \to e^{\epsilon\Lambda}\sqrt{\gamma}(\gamma^{\mu\nu} + \epsilon^{\mu\nu}) \tag{10.3}$$

$$\sqrt{\gamma}R^{(\gamma)} \to e^{\epsilon\Lambda}\sqrt{\gamma}\left[R^{(\gamma)} - 2(1+\epsilon)\Box\Lambda - (1+\epsilon)\epsilon(\partial_\mu\Lambda)^2\right]$$

For infinitesimal Λ, the variation of an action S gives

$$\delta S = \int d^n x\sqrt{\gamma}\Lambda(x)T \tag{10.4}$$

$$\sqrt{\gamma}T \equiv 2\left[\gamma^{\mu\nu}\frac{\delta S}{\delta\gamma^{\mu\nu}} + \epsilon^{\mu\nu}\frac{\delta S}{\delta\epsilon^{\mu\nu}}\right]$$

so that T consists of the trace of the energy momentum tensor $\sqrt{\gamma}T_{\mu\nu} \equiv \delta S/\delta\gamma^{\mu\nu}$, plus a term coming from the variation of $\epsilon^{\mu\nu}$. With

$S = \int d^n x \sqrt{\gamma} L$ as given in (9.3), we have, using (10.3),

$$T = \epsilon L(\Psi^0) - 2(1 + \epsilon)\Box\Phi^0 + 2V^0 \tag{10.5}$$

with

$$\Box\Phi = \gamma^{-1/2}\partial_\mu\left[\gamma^{1/2}\partial^\mu\Phi\right] = \left[\nabla_i\nabla_j\Phi\right]\partial_\mu\phi^i\partial^\mu\phi^j + \left[\nabla_i\Phi\right]\nabla_\mu\partial^\mu\phi^i \tag{10.6}$$

Then the theory is classically local scale invariant (i.e. $T = 0 + O(\epsilon)$) only if $V^0 = 0$ and $\Phi^0 =$ constant. In the quantum theory, under an infinitesimal local scale transformation

$$\delta\Gamma = \int d^n x \sqrt{\gamma}\Lambda(x)\langle T\rangle \tag{10.7}$$

with

$$\langle T\rangle = 2\gamma^{-1/2}\left[\gamma^{\mu\nu}\frac{\delta}{\delta\gamma^{\mu\nu}} + \epsilon^{\mu\nu}\frac{\delta}{\delta\epsilon^{\mu\nu}}\right]\Gamma \tag{10.8}$$

$$= \langle \epsilon L - (1 + \epsilon)\Box\Phi^0 + 2V^2\rangle$$

and the renormalized expectation value of (9.14). We implicitly include an infra-red regulator term (9.1b) in V in this section.

The quantum theory is local scale invariant if $\langle T\rangle$ vanishes on-shell, while rigid scale invariance, i.e. invariance under (10.2) for constant Λ, requires only that the <u>integrated</u> trace anomaly $\int d^n x\sqrt{\gamma}\langle T\rangle$ vanish on-shell in the $\epsilon \to 0$ limit. For this to happen for all $\gamma_{\mu\nu}$ requires $\langle T\rangle$ to be a total derivative on-shell. As $\langle T\rangle = T + O(\hbar)$, local scale invariance, to lowest order in $\hbar$, requires $V = O(\hbar)$, $\Phi =$ constant $+ O(\hbar)$, while for rigid scale invariance to this order, V must still vanish but Φ is unrestricted as its "classical" contribution to $\langle T\rangle$ is a total derivative.

The integrated trace anomaly is (dropping all surface terms)

$$\int d^n x\sqrt{\gamma}\langle T\rangle = \langle \epsilon S\rangle + \int d^n x\sqrt{\gamma}\langle V^0\rangle \tag{10.9}$$

We now obtain a simple formula for this. As $S[\phi_0,\Psi_0]$ is linear in Ψ_0 (9.3), we have (with the notation of (9.11))

$$\epsilon S = (\epsilon\Psi^0).\frac{\delta S}{\delta\Psi^0} \tag{10.10}$$

Differentiating (9.4) with respect to μ gives

$$\epsilon\Psi^0 = \hat{\beta}^{\Psi}.\frac{\delta\Psi_0}{\delta\Psi} \tag{10.11}$$

so that

$$\langle\epsilon S\rangle = \langle\hat{\beta}^{\Psi}.\frac{\delta\Psi_0}{\delta\Psi}.\frac{\delta S}{\delta\Psi_0}\rangle = \hat{\beta}^{\Psi}.\frac{\delta\Gamma}{\delta\Psi} \tag{10.12}$$

Writing $V = M^2\bar{V}$ for some mass scale M, we have

$$\int d^n x\sqrt{\gamma}\langle V_0\rangle = M^2\frac{\partial}{\partial M^2}\langle S\rangle = M^2\frac{\partial}{\partial M^2}\Gamma = V.\frac{\delta\Gamma}{\delta V} \tag{10.13}$$

as all counterterms T,X,Y are independent of V except $T^{(\nu,V)}$, which is linear in V. Putting this together gives the following expression for the integrated trace anomaly [13,52,54]

$$\int d^n x\sqrt{\gamma}\langle T\rangle = \hat{\beta}^{\Psi}.\frac{\delta\Gamma}{\delta\Psi} + V.\frac{\delta\Gamma}{\delta V} \tag{10.14}$$

$$= \hat{\beta}^g_{ij}.\frac{\delta\Gamma}{\delta g_{ij}} + \hat{\beta}^b_{ij}.\frac{\delta\Gamma}{\delta b_{ij}} + \hat{\beta}^{\Phi}.\frac{\delta\Gamma}{\delta\Phi} + (\hat{\beta}^V + V).\frac{\delta\Gamma}{\delta V}$$

However, we have seen that the $\hat{\beta}$ functions are ambiguous. By changing the renormalization scheme as explained in Section 9, for example, we are free to shift the β-functions as in (9.21, 9.22, 9.24), for some w^i,λ^i. However, using the Ward identities (9.13) and

$$(\partial_{[i}\lambda_{j]}).\frac{\delta\Gamma}{\delta b_{ij}} = 0 \tag{10.15}$$

(the latter Ward identity corresponding to the classical invariance (2.3)) the change in (10.14) is proportional to $\delta\Gamma/\delta\bar{\phi}$ and so vanishes on-shell. Then although the β-functions are ambiguous, (10.14) is not.

The integrated anomaly determines ⟨T⟩ up to surface terms. Defining an effective Lagrangian L_{eff} by $\Gamma = \int d^n x\sqrt{\gamma}L_{eff}$, it can be shown that [13,52,54]

$$\langle T\rangle = B^{\Psi}.\frac{\delta}{\delta\Psi}L_{eff} + C^i.\frac{\delta}{\delta\bar{\phi}^i}L_{eff} \tag{10.16}$$

for certain coefficients $B^{\Psi} = (B^V, B^g_{ij}, B^b_{ij}, B^V)$ given by

$$B^g_{ij} = \hat{\beta}^g_{ij} + 2\nabla_{(i}v_{j)}$$

$$B^b_{ij} = \hat{\beta}^b_{ij} + 2v^k H_{kij} + \partial_{[i}\kappa_{j]} \qquad (10.17)$$

$$B^{\Phi} = \hat{\beta}^{\Phi} + v^i\partial_i\Phi$$

$$B^V = \hat{\beta}^V + V + v^i\partial_i V$$

for some definite vectors $v^i(\Psi)$, $\kappa^i(\Psi)$, $c^i(\Psi)$. These vectors are scheme-dependent, as are the β-functions, but the combinations appearing in (10.16) can be shown to be unambiguous on-shell [52]. For example, as we shall see in the next section,

$$v^i = \gamma^i{}_{\phi} - 4\,\nabla^i\Phi + O(\hbar^2) \qquad (10.18)$$

$$\kappa^i = 0 + O(\hbar^2), \qquad c_i = 2\nabla_i\Phi + O(\hbar^2)$$

to lowest order.

The effective action will be local scale invariant if $\langle T\rangle = 0$, which requires

$$B^{\Psi}(\Psi) = 0 \qquad (10.19)$$

However, the effective action has, as we have seen, certain ambiguities corresponding to renormalization scheme dependence, so that the physical criterion for local scale invariance is whether there is some Γ', equivalent to Γ, that is scale invariant, i.e. whether any lack of scale invariance can be absorbed into the ambiguities in Γ. The ambiguity (9.21) in the β-functions does not effect $\langle T\rangle$ (on-shell) but the freedom to add finite local counterterms does. If a finite local counterterm $\hbar^L\Delta$ to be added to the action S is not to spoil renormalizability, Δ must be of the same form as the classical action (9.3)

$$\Delta = \int d^2x\sqrt{\gamma}\Big[\tfrac{1}{2}(h_{ij} + k_{ij})(\gamma^{\mu\nu} + \epsilon^{\mu\nu})\partial_\mu\phi^i\partial_\nu\phi^j + W + \Omega R^{(\gamma)}\Big] \qquad (10.20)$$

corresponding to a redefinition $\Psi \to \Psi' + \hbar^L(\delta\Psi)$, $\delta\Psi \equiv (W, h_{ij}, k_{ij}, \Omega)$. Further, it is natural to consider only counterterms constructed from the metric and other original renormalized "coupling constants" in the theory, $\delta\Psi = \delta\Psi(\Psi)$, so that the "universality" of the perturbation theory is maintained, i.e. all quantities depend only on the original variables Ψ, not on any new ones introduced by hand. Then the β-functions (and hence the scale anomaly) changes by

$$B^{\Psi}(\Psi) \to \tilde{B}^{\Psi}(\Psi) = B^{\Psi}(\Psi + \hbar^{L}\delta\Psi) \tag{10.21}$$

If some $\tilde{\Psi}$ defines a conformally invariant sigma-model in the original scheme, $B^{\Psi}(\tilde{\Psi}) = 0$, then in the new non-minimal scheme $\tilde{\Psi}$ will no longer define a locally scale invariant theory, $\tilde{B}^{\Psi}(\tilde{\Psi}) \neq 0$, but the shifted "coupling constants" $\tilde{\Psi} - \delta\Psi(\tilde{\Psi})$ will define a local scale invariant sigma-model, $\tilde{B}^{\Psi}(\tilde{\Psi} - \hbar^{L}\delta\Psi(\tilde{\Psi})) = 0$. Then for every local scale invariant sigma-model in a given scheme, there is a corresponding one in every other scheme, related to the original one by a redefinition of the "coupling constants" Ψ. Thus in seeking conformally invariant sigma-models, it is sufficient to seek solutions Ψ of (10.19) in some definite renormalization scheme. However, if a theory is local scale invariant order by order in perturbation theory in some definite scheme, it will not be so in most other subtraction schemes and it is often advantageous to seek to perturbatively define a subtraction scheme in which $\langle T\rangle = 0$ order by order.

While local scale invariance requires that (10.16) vanish, rigid scale invariance requires only that $\langle T\rangle$ be a total derivative. If

$$B^{g}_{ij} = 2\nabla_{(i}w_{j)} \qquad B^{\Phi} = w^{i}\Phi_{,i} \tag{10.22}$$

$$B^{b}_{ij} = 2H_{ij}{}^{k}w_{k} + \partial_{[i}\lambda_{j]} \qquad B^{V} = w^{i}V_{,i}$$

for <u>any</u> vectors w^{i},λ^{i} then the Ward identities (9.13, 10.15) imply that on-shell (10.16) is a total derivative and (10.14) is a surface term which vanishes by our assumed boundary conditions.

Then the theory is <u>locally</u> <u>scale</u> invariant if (10.19) holds, while <u>rigid</u> <u>scale</u> invariance requires only that (10.22) hold for some vectors w^{i},λ^{i}. The β-functions will be trivial if (10.22) is satisfied with the B-functions replaced by β-functions. If the β-functions are trivial, then from (9.13), $\mu\partial\Gamma/\partial\mu = 0$ (on-shell and up to surface terms) so that Γ is independent of the renormalization mass μ. From (10.17), the condition for rigid scale independence (up to terms proportional to the infra-red regulator mass m^{2}) is that the potential vanish up to m-dependent terms,

– i.e. $V = 0 + O(m^2)$. This is because the potential term V breaks scale invariance classically, but doesn't introduce any μ-dependence. Finally, if the β-functions vanish order by order in perturbation theory, then so do the $1/\epsilon$ pole residues $T^{(1)}$ and (by the renormalization group pole equations) the $T^{(\nu)}$. However, if β^{Ψ} has a non-trivial fixed point, then (9.9) does not imply that the $1/\epsilon$ poles cancel, i.e. non-perturbative triviality of the β-functions is not the same thing as finiteness [56].

11. One-loop Calculations

We now illustrate the previous sections with some one-loop calculations. For a general background/quantum split (7.1), expanding $S[\bar{\phi} + \pi(\xi)]$ using (7.8), (7.11) gives the terms quadratic in ξ^a (with $n = 2 + \epsilon$)

$$S^{(2)} = \int d^n x \sqrt{\gamma} \left\{ \frac{1}{2}(\tilde{\nabla}_\mu \xi)^a (\tilde{\nabla}^\mu \xi)^a + \frac{1}{2}\xi^a \xi^b \left[R^{(+)}_{i(ab)j}(\gamma^{\mu\nu} + \epsilon^{\mu\nu})\partial_\mu \bar{\phi}^i \partial_\nu \bar{\phi}^j \right.\right.$$

$$\left.\left. + R^{(\gamma)}\nabla_a \nabla_b \Phi + \nabla_a \nabla_b V - N^i_{ab}(\delta S/\delta \bar{\phi}^i)\right]\right\} \qquad (11.1)$$

where

$$(\tilde{\nabla}_\mu \xi)^a = \partial_\mu \xi^a + \tilde{\omega}_\mu^{\ ab} \xi^b \ ; \qquad \tilde{\omega}_\mu^{\ ab} = \omega^{ab}_i \partial_\mu \bar{\phi}^i - H_{iab}\epsilon_\mu^{\ \nu} \partial_\nu \bar{\phi}^i \qquad (11.2)$$

Adding the infra-red regulating mass term (9.1a), expanding the covariant derivatives (11.2) and writing $\gamma_{\mu\nu} = \delta_{\mu\nu} + h_{\mu\nu}$, we obtain

$$S^{(2)} = \int d^n x \left\{ \frac{1}{2}(\partial_\mu \xi^a)(\partial^\mu \xi^a) + A^\mu_{ab}\xi^b \partial_\mu \xi^a + \frac{1}{2}B_{ab}\xi^a \xi^b \right.$$

$$\left. - \bar{h}^{\mu\nu}\partial_\mu \xi^a \partial_\nu \xi^b + \ldots \right\} \qquad (11.3)$$

where $\delta_{\mu\nu}$ and $\epsilon_{\mu\nu}$ are n-dimensional (see section 9) and

$$A_\mu^{\ ab} = \tilde{\omega}_\mu^{\ ab}$$

$$B_{ab} = R^{(+)}_{i(ab)j}\partial_\mu \bar{\phi}^i \partial_\nu \bar{\phi}^j (\delta^{\mu\nu} + \epsilon^{\mu\nu}) + R^{(\gamma)}\nabla_a \nabla_b \Phi + \nabla_a \nabla_b V$$

$$+ \frac{1}{2}m^2 h \delta_{ab} - N^i_{ab}\delta S/\delta \bar{\phi}^i + \tilde{\omega}_{\mu ac}\tilde{\omega}^{\mu bc} \qquad (11.4)$$

$$\bar{h}_{\mu\nu} = h_{\mu\nu} - \tfrac{1}{2}\delta_{\mu\nu}h \quad , \quad h = h_{\mu\nu}\delta^{\mu\nu}$$

and terms in (11.3) of higher order in $h_{\mu\nu}$ have been omitted. We shall calculate to low orders in $h_{\mu\nu}$ and use general covariance to fix higher order terms, following [47]. The momentum space ξ-propagator is

$$\xi^a \xi^b = \delta^{ab}\frac{1}{p^2 + m^2} \tag{11.5}$$

As in Section 6, the diagrams in fig.1 give a divergent contribution to the effective action

$$\frac{\hbar}{2}\int d^n x \; I\,(B_{aa} - A_{\mu ab}A_{\mu ab}) \tag{11.6}$$

with I the divergent integral (6.14). Note that, on using (11.4), the connections $\tilde{\omega}$ cancel in (11.6) to give a result that is covariant with respect to target space diffeomorphisms. Relevant diagrams involving the $\bar{h}\partial\xi\partial\xi$ vertex are shown in fig.2.

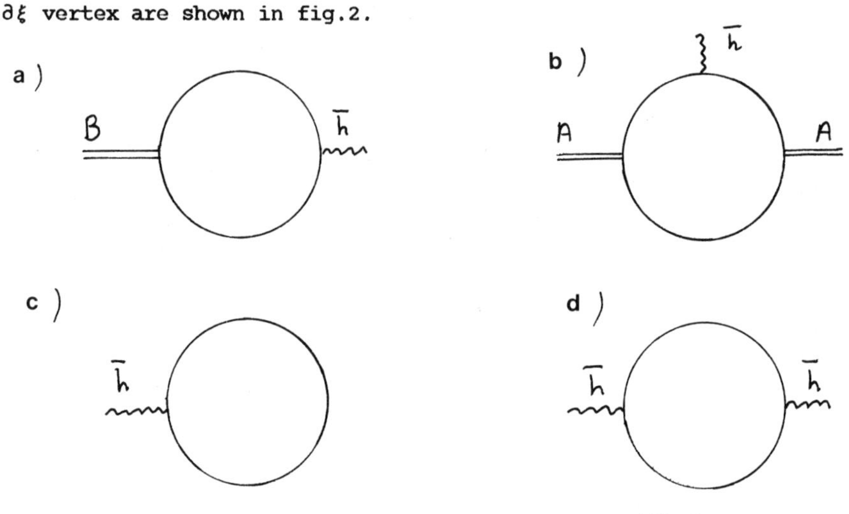

Fig.2. Feynman diagrams involving the $\bar{h}\partial\xi\partial\xi$ vertex.

Figures (2a,2b) give a term whose space-time covariantization is the non-local term

$$\frac{-\hbar}{8\pi}\int d^n x\sqrt{\gamma}\left[B_{aa} - A_{\mu ab}A^{\mu ab}\right]\Box^{-1}R(\gamma) \tag{11.7}$$

where $\Box f = \gamma^{-1/2}\partial_\mu(\gamma^{1/2}\partial^\mu f)$ and we use the shorthand

$$\int d^n x\sqrt{\gamma}M\Box^{-1}N \equiv \int d^n x d^n y\sqrt{\gamma(x)}M(x)K(x,y)N(y) \tag{11.8}$$

for the Green function K satisfying

$$\Box_x K(x,y) = \delta^n(x,y) \tag{11.9}$$

Fig (2c) gives a finite n-dimensional cosmological constant proportional to m^2 (in addition to the divergent cosmological constant contribution contained in (11.6)). Fig (2d) gives a term whose covariantization gives (up to finite local terms)

$$\hbar\int d^n x \sqrt{\gamma}\; d \left[\frac{1}{12}R^{(\gamma)} + \frac{1}{96\pi}R^{(\gamma)}\Box^{-1}R^{(\gamma)}\right] \tag{11.10}$$

where the dimension d of M comes from a trace of δ_{ab}. Adding these terms gives a contribution to the bare effective action (8.7)

$$\Gamma_0 = S + \hbar\int d^n x\sqrt{\gamma}\left\{ \frac{1}{2}\left[-R^{(+)}_{ij}\partial_\mu\bar{\phi}^i\partial_\nu\bar{\phi}^j(\gamma^{\mu\nu} + \epsilon^{\mu\nu}) \right.\right. \tag{11.11}$$

$$\left. + \nabla^2 V + \frac{d}{2}m^2 + (d/6 + \nabla^2\Phi)R^{(\gamma)} - N^i\delta S/\delta\bar{\phi}^i\right]$$

$$+ \frac{1}{8\pi}\left[R^{(+)}_{ij}(\gamma^{\mu\nu} + \epsilon^{\mu\nu})\partial_\mu\bar{\phi}^i\partial_\nu\bar{\phi}^j - \nabla^2 V - \frac{dm^2}{2}\right.$$

$$\left.\left. + \left[\frac{d}{16} - \nabla^2\Phi\right]R^{(\gamma)} + N^i(\delta S/\delta\bar{\phi}^i)\right]\Box^{-1}R^{(\gamma)} + \frac{1}{4\pi}m^2 d + \ldots\right\}$$

with $N^i \equiv \delta^{ab}N^i_{ab}$ and $R^{(+)}_{ij} = R^{(+)k}{}_{ikj}$. All other one-loop contributions are ultra-violet finite and local scale invariant or are proportional to $\bar{\xi}^a$. Note that all explicit $\tilde{\omega}$ connections have cancelled, leaving a covariant result.

We drop all surface terms but work off-shell, i.e. we do not assume that $\bar{\phi}^i(x)$ satisfies any field equation. Then, for example, the term $I.N^i\delta S/\delta\bar{\phi}^i$ in (11.11) is not zero but can be rewritten as $I(L_N\Psi).\delta S/\delta\Psi$ using (9.17). The divergences can be removed if we make the renormalization (9.4) with the counterterm tensors

$$Y^i_{(1)} = \frac{1}{4\pi}V^i(\bar{\phi})$$

$$T^{(1,g)}_{ij} = -\frac{1}{2\pi}\left[R^{(+)}_{(ij)} + \nabla_{(i}V_{j)} + \nabla_{(i}N_{j)}\right]$$

$$T^{(1,b)}_{ij} = -\frac{1}{2\pi}\left[R^{(+)}_{[ij]} + (V^k + N^k)H_{ijk} + \partial_{[i}\lambda_{j]}\right] \tag{11.12}$$

$$T^{(1,\Phi)} = \frac{d}{24\pi} + \frac{1}{4\pi}\left[\nabla^2\Phi - (v^i + N^i)\partial_i\Phi\right]$$

$$T^{(1,V)} = \frac{1}{8\pi} m^2 d + \frac{1}{4\pi}\left[\nabla^2 V - (v^i + N^i)\partial_i V\right]$$

for any vectors $v^i(\Psi,\bar{\phi})$, $\lambda^i(\Psi,\bar{\phi})$ we choose. In Γ, the v-dependent divergences introduced by the $\bar{\phi}$ renormalization are cancelled by the v-dependent terms in the Ψ-renormalization, while the $\partial_{[i}\lambda_{j]}$ term in $T^{(1,b)}_{ij}$ changes Γ by a divergent surface term, which vanishes with our boundary conditions. There will also be divergences in Γ_0 involving $\bar{\xi}^a$ and to remove these a further renormalization of ξ^a as in (9.4) is needed [49]. The β- and γ-functions are then, using (9.9),

$$\beta^g_{ij} = -\frac{\hbar}{2\pi}\left[R^{(+)}_{(ij)} + \nabla_{(i}\tilde{v}_{j)} + \nabla_{(i}\tilde{N}_{j)}\right] + O(\hbar^2)$$

$$\beta^b_{ij} = -\frac{\hbar}{2\pi}\left[R^{(+)}_{[ij]} + (\tilde{v}^k + \tilde{N}^k)H_{ijk} + \partial_{[i}\tilde{\lambda}_{j]}\right] + O(\hbar^2)$$

$$\beta^\Phi = \frac{d\hbar}{24\pi} + \frac{\hbar}{4\pi}\left[\nabla^2\Phi - (\tilde{v}^i + \tilde{N}^i)\partial_i\Phi\right] + O(\hbar^2)$$

$$\beta^V = \frac{\hbar}{4\pi}\left[\nabla^2 V + \frac{m^2 d}{2} - (\tilde{v}^i + \tilde{N}^i)\partial_i V\right] + O(\hbar^2)$$

$$\gamma^i_\phi = \frac{\hbar}{4\pi}\tilde{v}^i \tag{11.13}$$

where

$$\tilde{v}^i = \left[\frac{\partial}{\partial\Lambda}v^i(\Lambda^{-1}\Psi,\bar{\phi})\right]\Big|_{\Lambda=1} \qquad \tilde{N}^i = \left[\frac{\partial}{\partial\Lambda}N^i(\Lambda^{-1}\Psi,\bar{\phi})\right]\Big|_{\Lambda=1}$$

$$\tilde{\lambda}_i = \left[\left(1 + \Lambda\frac{\partial}{\partial\Lambda}\right)\lambda_i(\Lambda^{-1}\Psi,\bar{\phi})\right]\Big|_{\Lambda=1} \tag{11.14}$$

We see explicitly the ambiguity (9.21,9.22) in the β- and γ-functions parameterized by $\tilde{v}^i, \tilde{\lambda}^i$ and also the dependence on the choice of the background/quantum split through the N^i_{jk} coefficient. For example, if we take $v^i = 0$ and use normal co-ordinates, we find $N^i_{jk} = 0$ and $T^{(1,g)}_{ij} = -(2\pi)^{-1}R^{(+)}_{(ij)}$, while for the linear split described in Section 6, $N^i_{jk} = -\left\{{i \atop j\,k}\right\}$ and $N^i = \tilde{N}^i = -X^i$ (with X^i as defined in (6.16)) and we obtain $T^{(1,g)}_{ij} = -(2\pi)^{-1}(R^{(+)}_{(ij)} - \nabla_{(i}X_{j)})$, so that the result (6.17) is recovered. By choosing v^i so that $\tilde{v}^i = -\tilde{N}^i$, all dependence on the N-tensors can be removed from the β-functions at the expense of introducing it into γ_ϕ. Thus for the linear split of Section 6, choosing $v^i = x^i$ removes the non-covariant term from the 1-loop β-function.

The renormalization (9.4), (11.12) leaves the following contribution to the $\bar{\xi}$-independent part of the one-loop renormalized effective action (9.7) (taking $\epsilon \to 0$ and using $m^\epsilon = 1 + \epsilon\log(m) + O(\epsilon^2)$)

$$
\begin{aligned}
\Gamma = S + \hbar\int d^2x\sqrt{\gamma}\Bigg[\frac{1}{2\pi}\log(m/\mu)&\Bigg\{\frac{1}{2}R^{(+)}_{ij} + \nabla_{(i}\left[\tilde{v}_{j)} + \tilde{N}_{j)}\right] \\
&+ H_{ijk}(\tilde{v}^k + \tilde{N}^k) + \partial_{[i}\tilde{\lambda}_{j]}\Bigg\}(\gamma^{\mu\nu} + \epsilon^{\mu\nu})\partial_\mu\bar{\phi}^i\partial_\nu\bar{\phi}^j \\
&- \frac{1}{2}\left[\frac{d}{6} + \nabla^2\Phi + (\tilde{v}^i + \tilde{N}^i)\partial_i\Phi\right]R^{(\gamma)} - \frac{1}{2}\Bigg\{\nabla^2 V + \frac{m^2 d}{2} \\
&\qquad - (\tilde{v}^i + \tilde{N}^i)\partial_i V\Bigg] - \frac{1}{2}\tilde{v}^i(\delta S/\delta\bar{\phi}^i)\Bigg\} \\
&+ \frac{1}{8\pi}\Bigg\{R^{(+)}_{ij}(\gamma^{\mu\nu} + \epsilon^{\mu\nu})\partial_\mu\bar{\phi}^i\partial_\nu\bar{\phi}^j - \nabla^2 V - \frac{dm^2}{2} \\
&+ \left[\frac{d}{16} - \nabla^2\Phi\right]R^{(\gamma)} - N^i(\delta S/\delta\bar{\phi}^i)\Bigg]\Box^{-1}R^{(\gamma)} + \frac{1}{4\pi}m^2 d\ldots\Bigg]
\end{aligned}
\tag{11.15}
$$

which satisfies the renormalization group equation (9.13), up to terms proportional to $\bar{\xi}$ which are omitted in (11.15). Note that the terms in (11.15) involving $\tilde{v}^i$ cancel leaving no $\tilde{v}$-dependence, as was to be expected from Section 9.

Under an infinitesimal local scale transformation, (10.2), Γ changes by (10.7) where, using (10.3),

$$
\begin{aligned}
\langle T\rangle = \frac{1}{2}\partial_\mu\bar{\phi}^i\partial_\nu\bar{\phi}^j&(\gamma^{\mu\nu} + \epsilon^{\mu\nu})\left[-\frac{\hbar}{2\pi}R^{(+)}_{ij}\right] \\
+ 2\left[V + \frac{\hbar}{8\pi}(\nabla^2 V + 3dm^2/2)\right] &+ \left[-2\Box\Phi + \hbar\left[-\frac{d}{24\pi} + \frac{1}{4\pi}\nabla^2\Phi\right]\right]R^{(\gamma)} \\
+ \frac{\hbar}{4\pi}N^i(\delta S/\delta\bar{\phi}^i) &- \hbar(\nabla^2 V + dm^2/2)\frac{1}{4\pi}\left[\Box^{-1}R^{(\gamma)} + 2\log(m/\mu)\right] \\
&+ \frac{\hbar}{4\pi}\Box\left[\nabla^2\Phi\left[\Box^{-1}R^{(\gamma)} + 2\log(m/\mu)\right]\right]
\end{aligned}
\tag{11.16}
$$

$$
+\frac{\hbar}{4\pi}(\partial_i V)\left[N^i\Box^{-1}R^{(\gamma)} + 2\tilde{N}^i\log(m/\mu)\right] - \frac{\hbar}{4\pi}\Box\left[(\partial_i\Phi)(N^i\Box^{-1}R^{(\gamma)} + 2\tilde{N}^i\log(m/\mu))\right] + O(\hbar^2)
$$

At first sight, the presence of non-local terms and terms proportional to $\log(m/\mu)$ in ‹T› seems surprising. However, if the theory is locally scale invariant classically, i.e. if $V = 0 + O(\hbar)$ and $\nabla_i\Phi = 0 + O(\hbar)$, and we take the limit $m^2 \to 0$, then the non-local terms (and the $\log(m/\mu)$) terms in (11.16) vanish. Indeed, if local scale invariance breaks down at L loops, then ‹T› will be local at L-loop order but non-local and μ-dependent at higher loop orders.

The rather complicated expression for ‹T› (11.16) can, using (10.16), (11.15), be recast in the comparatively simple form (10.16), with the B-functions given by

$$B^g_{ij} = -\frac{\hbar}{2\pi}R^{(+)}_{(ij)} - 4\nabla_i\nabla_j\Phi + O(\hbar^2)$$

$$B^b_{ij} = -\frac{\hbar}{2\pi}R^{(+)}_{[ij]} - 4H_{ij}{}^k\nabla_k\Phi + O(\hbar^2)$$

$$B^\Phi = d\frac{\hbar}{24\pi} + \frac{\hbar}{4\pi}\nabla^2\Phi - 2(\partial_i\Phi)(\partial^i\Phi) + O(\hbar^2)$$

$$B^V = \frac{\hbar}{4\pi}\left[\nabla^2 V + \frac{3m^2 d}{2}\right] + 2V - 2(\partial_i\Phi)(\partial^i V) + O(\hbar^2)$$

$$C^i = -\frac{\hbar}{4\pi}N^i + 2\nabla^i\Phi + O(\hbar^2) \qquad (11.17)$$

These are in agreement with (10.17, 10.18) and the general arguments of section 9 and 10 are borne out by these calculations. There are classical terms in (11.17) involving Φ and V, corresponding to the lack of classical scale invariance of (1.13). Note that while the β-functions depend on v^i, N^i_{jk} etc, the B-functions do not and so are unambiguous. The only N-dependence in ‹T› is proportional to the effective field equation and so vanishes on-shell. Comparing (11.13) and (11.17), we see explicitly the difference between the β-functions and the B-functions in a general scheme.

In contrast, consider the local scale variation of the bare effective action (11.11). Remarkably, if the theory is classically scale invariant ($V = 0$ and $\partial_i\Phi = 0$), then the variation of the non-local terms in (11.11) comletely cancel the variation of the divergent terms, up to terms that are proportional to ϵ or m, i.e. $\delta\Gamma_0 = 0 + O(\epsilon) + O(m^2)$, expressed in terms of the bare variables [47]. For example, there is a "classical" term in the variation proportional to $\epsilon g^0_{ij}\partial_\mu\bar{\phi}^i\partial^\mu\bar{\phi}^j$, which although vanishing in the $\epsilon\to 0$ limit if g^0_{ij} is kept fixed, gives a finite result on re-expressing in terms of renormalized quantities. Indeed, using $g^0_{ij} = \mu^\epsilon(g_{ij} -$

$-\ \hbar(2\pi\epsilon)^{-1}R_{ij} + \ldots)$, the variation becomes $\delta\Gamma_0 = -\hbar(2\pi)^{-1}R_{ij} + \ldots$ which is finite on taking the limit $\epsilon\to 0$ while keeping the renormalized metric fixed, in agreement with (11.16). From another viewpoint, the lack of scale invariance of (11.15) can be understood as due to the lack of scale invariance of the counterterms added to cancel the divergences, although the bare effective action is scale invariant to this order. At higher loops, the bare effective action is no longer scale invariant, so that there are contributions from both the bare action and from the counterterms to $\langle T\rangle$.

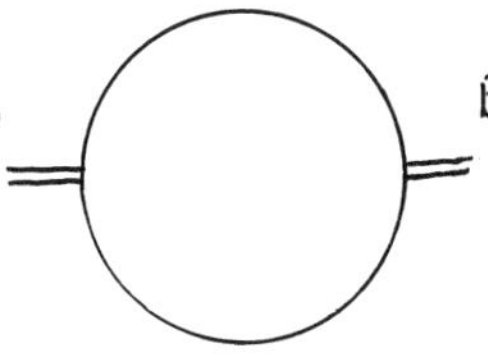

Fig.3. Graph giving one-loop divergence in four-dimensional non-linear sigma model.

Finally, we calculate the one-loop divergence for the four-dimensional sigma model. For simplicity, taking $H_{ijklm} = 0$, $V = 0$, the vertex $\frac{1}{2}B_{ab}\xi^a\xi^b$ in (7.8) with $B_{ab} = R_{i(ab)j}\partial_\mu\bar{\phi}^i\partial^\mu\bar{\phi}^j$ gives, from the graph in fig.3, a result proportional to

$$B_{ab}B^{ab} = R_{i(ab)j}R_{k(ab)l}\partial_\mu\bar{\phi}^i\partial^\mu\bar{\phi}^j\partial_\mu\bar{\phi}^k\partial^\mu\bar{\phi}^l \tag{11.18}$$

times a logarithmically divergent momentum integral. This divergence with four derivatives cannot be removed by any renormalization and the higher the loop order, the more derivatives there will be, so that the sigma model in non-renormalizable in four space-time dimensions.

12. Quantization with Fermions

Consider adding to the bare action (9.3) the fermion action described in sections 5 :

$$S_F = \frac{i}{2}\int d^n x\Big[\lambda_+^A(\partial_{--}\lambda_+^A + A_i^{AB}\partial_{--}\phi^i\lambda_+^B) \tag{12.1}$$

$$+ \chi_-^M(\partial_{++}\chi_-^N + B_i^{MN}\partial_{--}\phi^i\chi_+^N) + U_{ABMN}\chi_-^M\chi_-^N\lambda_+^A\lambda_+^B\Big]$$

We can generalize the methods described in the preceding sections to quantize the theory described by $S + S_F$ using path integral techniques. First, we would like to introduce a background/quantum split for the fermions so as to obtain Feynman rules that are covariant with respect to the "gauge" symmetry (4.3). As the λ^A, χ^M already transform covariantly, it is tempting to just use a linear split of the total field λ^A into a background field $\bar{\lambda}^A$ (which would usually be taken to be zero) and a quantum fluctuation, $\lambda^A = \bar{\lambda}^A + \psi^A$, and similarly for χ^M. (Note that here $\bar{\lambda}$ denotes a mean field, not a Dirac conjugate!) However, if this is done simultaneously with a normal co-ordinate expansion as described in section 7, the expansion of $S_F[\phi,\lambda] = S_F[\bar{\phi} + \pi(\xi),\bar{\lambda} + \psi]$ (where for the moment we set $\chi^A = 0$) does not lead to covariant Feynman rules, as is easily checked. One way of seeing why this is the case is to note that the derivative D of section 7 used to give the normal co-ordinate expansion is not covariant with respect to the symmetry (4.3).

A convenient way of generating covariant Feynman rules is to generalize the normal co-ordinates on M to some "normal co-ordinates" on the vector bundle parameterized by (ϕ^i,λ^A). We will reinstate χ-dependence later. We then introduce a curve $(\phi^i(s),\lambda^A(s))$, $0 \leqslant s \leqslant 1$, in the vector bundle joining the background field to the total field, so that $(\phi^i(0),\lambda^A(0)) = (\bar{\phi}^i,\bar{\lambda}^A)$ and $(\phi^i(1),\lambda^A(1)) = (\bar{\phi}^i + \pi^i,\bar{\lambda}^A + \psi^A)$. We define D to be a generalization of d/ds that is both general co-ordinate covariant and covariant with respect to the "gauge" symmetry (4.3), satisfying (7.5) and

$$D\lambda^A(s) = \frac{d\lambda^A}{ds} + \frac{d\phi^i}{ds}A_i^{AB}\lambda^B \tag{12.2}$$

Then

$$\xi^i(s) = d\phi^i/ds \ , \qquad \rho_+^A(s) = D\lambda_+^A \tag{12.3}$$

defines covariant "tangent vectors" $\xi^i(s),\rho_+^A(s)$ and we choose our quantum fields ξ,ρ to be the tangents at $s = 0$, i.e. $\xi^i = \xi^i(0)$, $\rho_+^A = \rho_+^A(0)$. Finally, we fix the curve by demanding that it satisfy the generalization of (7.2)

$$D(d\phi^i/ds) = 0 \ , \qquad D(D\lambda^A(s)) = 0 \tag{12.4}$$

As in section 7, there are a number of ways of modifying this background field expansion without spoiling covariance. We could alter the expansion by changing the connection in (7.2) from the Chrisoffel connection to that of (7.9) and by changing the connection appearing in the derivative (12.2) from A_i^{AB} to $A_i^{AB} + T_i^{AB}$ for some covariant T_i^{AB}. In some cases it is useful to modify the second equation in (12.4) to become

$$D(D\lambda^A(s)) = T_{ij}^{AB}(\phi(s))\ \lambda^B(s)\xi^i(s)\xi^j(s) \tag{12.5}$$

for some covariant tensor T_{ij}^{AB}. For example, in models with (0,1) supersymmetry, under which $\delta\phi^i = f^i{}_A(\phi)\epsilon_i\lambda_+^A$ etc. as described in section 5, the choice

$$T_{ij}^{AB} = R_{k(ij)l}f^{kA}f^{lB} \tag{12.6}$$

in (12.4) gives a perturbation theory which respects supersymmetry, whereas (12.4) does not. (The conditions (12.5,12.6) can be obtained by taking a normal co-ordinate expansion of the theory written in (0,1) superfields, as described in [39], and making a covariant component expansion using covariant super-derivatives.) Dependence of the theory on these modifications of the background quantum split should be trivial, just as the dependence on the $M^i_{j...k}$ in (7.1) was seen to be, and we shall set $T_{ij}^{AB} = 0$ and use (12.4) in what follows.

The normal co-ordinate expansion of $S[\bar\phi+\pi(\xi), \bar\lambda+\psi(\rho)]$ can be obtained as in (7.4) using (7.5, 12.2, 12.2 12.4) and the identity

$$\left[D,\nabla_{--}\right]V^A = F_{ij}^{AB}\xi^i\partial_{--}\phi^jV^B \tag{12.7}$$

giving the following additional fermi-terms $L_f^{(n)}$ to the $L_{(n)}$ of (7.8)

$$L_f^{(1)} = i\rho_+^A\left[\nabla_{--}\bar\lambda_+\right]^A + \frac{i}{2}F_{ij}^{AB}\xi^i\partial_{--}\bar\phi^j\bar\lambda_+^A\bar\lambda_+^B$$

$$L_f^{(2)} = \frac{i}{2}\rho_+^A\left[\nabla_{--}\rho_+\right]^A - \frac{i}{2}F_{ij}^{AB}\xi^i\partial_{--}\bar\phi^j\rho_+^A\bar\lambda_+^B \tag{12.8}$$

$$+ \frac{i}{4}F_{ij}^{AB}\lambda_+^A\lambda_+^B\xi^i\nabla_{--}^{(+)}\xi^j + \frac{i}{4}\left[\nabla_k^{(+)}F_{ij}^{AB}\right]\xi^i\xi^k(\partial_{--}\bar\phi^j)\bar\lambda_+^A\bar\lambda_+^B$$

$$L_f^{(3)} = \frac{i}{2}F_{ij}^{AB}\xi^i(\partial_{--}\bar\phi^j)\rho_+^A\rho_+^B + O(\bar\lambda)$$

$$L_f^{(4)} = \frac{i}{4}\xi^i\rho_+^A\rho_+^B\left[F_{ij}^{AB}\nabla_{--}\xi^j + \left[\nabla_kF_{ij}^{AB}\right]\xi^k\partial_{--}\bar\phi^j\right] + O(\bar\lambda)$$

where ∇ includes both Γ^i_{jk} and A_i^{AB} connections.

The bare effective action (8.7) is now generalized to include a fermion integration

$$\exp\left[-\hbar^{-1}\Gamma[\bar{\phi},\bar{\xi},\bar{\lambda},\bar{\rho}]\right] = \int[d\xi][d\rho]\exp\left\{-\hbar^{-1}\left[S[\bar{\phi} + \pi(\xi + \bar{\xi}),\bar{\lambda} + \psi(\rho + \bar{\rho})] - \xi.\delta\Gamma/\delta\bar{\xi} - \rho.\delta\Gamma/\delta\bar{\rho}\right]\right\} \qquad (12.9)$$

We now consider the renormalization of (12.9). It is useful to include A_{iAB} in the "vector" Ψ, so that now $\Psi = (V,g_{ij},b_{ij},\Phi,A_{iAB})$. Then we renormalize the bare quantities $\Psi_0,\bar{\lambda}_0,\rho_0,\xi_0,\bar{\phi}_0$ occurring in (12.9) by expressing then in terms of renormalized quantities by

$$\bar{\lambda}^A_{0+} = \bar{\lambda}^A_+ + \sum_{\nu=1}^{\infty} \epsilon^{-\nu} U^{(\nu)A}_+(\Psi,\bar{\lambda},\bar{\phi}) \qquad (12.10)$$

$$\rho^A_{0+} = \rho^A_+ + \sum_{\nu=1}^{\infty} \epsilon^{-\nu} V^{(\nu)A}_+(\Psi,\bar{\lambda},\bar{\phi},\rho,\xi)$$

$$\equiv \rho^A_+ + V^A_+$$

together with (9.4), which now includes a renormalization of A_{iAB}

$$A^0_{iAB} = \mu^{\epsilon}\left[A_{iAB} + \sum_{\nu=1}^{\infty} \epsilon^{-\nu} T^{(A,\nu)}_{iAB}(\Psi)\right] \qquad (12.11)$$

Then β^{Ψ} as defined in (9.9) now includes a β-function corresponding to (12.11), β^A_{iAB}, and there are also γ-functions for ρ and $\bar{\lambda}$,

$$\gamma^{\bar{A}}_{\lambda} = \frac{\partial}{\partial\Lambda} U^{(1)A}_+(\Lambda^{-1}\Psi,\bar{\lambda},\bar{\phi})\Big|_{\Lambda=1} ; \qquad (12.12)$$

$$\gamma^{\bar{A}}_{\rho} = \frac{\partial}{\partial\Lambda} V^{(1)A}_+(\Lambda^{-1}\Psi,\bar{\lambda},\bar{\phi},\bar{\xi},\bar{\rho})\Big|_{\Lambda=1}.$$

The renormalized effective action generalizing (9.7) and generating graphs that are 1PI with respect to both of the renormalized quantum fields ξ^a,ρ^A_+, is

$$\exp\left[-\hbar^{-1}\Gamma\right]=\int[d\xi][d\rho]\exp\left\{-\hbar^{-1}\left[S(\bar{\phi}_0,\xi + \bar{\xi} + X(\xi + \bar{\xi}),\bar{\lambda}_0,\rho + \bar{\rho} + V(\rho + \bar{\xi})) - \xi.\delta\Gamma/\delta\bar{\xi} - \rho.\delta\Gamma/\delta\bar{\rho}\right]\right\} \qquad (12.13)$$

At one loop, the fermi vertices (12.8) do not contribute to the β-functions calculated in section 11 but the diagram in fig.4 does contribute to β^A.

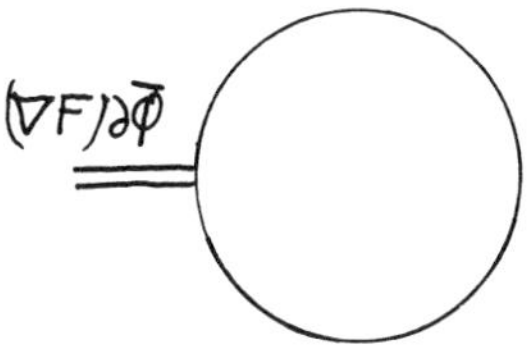

Fig.4. One-loop diagram contributing to β^A.

The diagram corresponds to contracting the $\xi\xi$ term in (12.8), i.e. $\frac{i}{4}(\nabla^{(+)}_a F^{AB}_{bj})\partial_{--}\bar{\phi}^j\bar{\lambda}^A\bar{\lambda}^B\xi^a\xi^b$, and gives a $1/\epsilon$ pole that can be cancelled by taking

$$T^{(A,1)}_{iAB} = \frac{1}{4\pi}\left[\nabla^{(+)j}F_{ji}\right]^{AB}, \tag{12.14}$$

so that (in this scheme)

$$\beta^A_{iAB} = \frac{\hbar}{4\pi}\left[\nabla^{(+)j}F_{ji}\right]^{AB} \tag{12.15}$$

where $\nabla^{(+)}_j$ includes $\Gamma^{(+)i}_{\ \ ik}$ and A^{AB}_i connections.

We can now calculate directly the anomalies in the symmetry (4.3) discussed in section 4. The anomalous contribution to Γ is given by the diagram in fig.5.

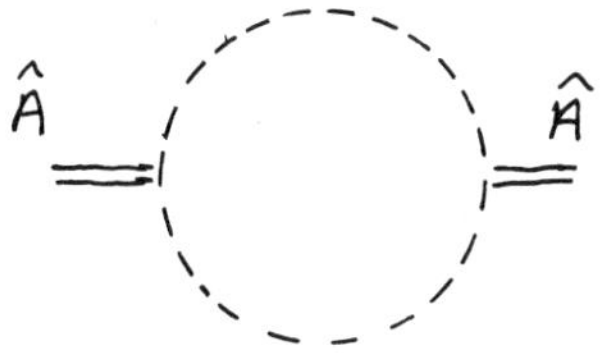

Fig.5. Diagram giving sigma-model anomaly. Dotted line indicates a fermion loop

Writing the kinetic term in (12.8) as

$$\frac{i}{2}\rho^A_+(\nabla_{--}\rho_+)^A = \frac{i}{2}\rho^A\partial_{--}\rho^A + \frac{i}{2}\hat{A}^{AB}_{--}\rho^A_+\rho^B_+ \tag{12.16}$$

with the notation $\hat{A}^{AB}_{--} = A^{AB}_i(\partial_{--}\bar{\phi}^i)$, the first term gives the momentum-space propagator $\overline{\hat{A}\hat{B}}\rho_+\rho_+ = -i\delta^{AB}p_+p_+/p^2$, while the second term gives a non-covariant vertex and it is this vertex that appears in fig.5. The diagram gives (in momentum space) the finite contribution to Γ [39]

$$\frac{-\hbar}{16\pi}\hat{A}^{AB}_{--}(-p)\hat{A}^{AB}_{--}(p)\frac{p_{++}p_{++}}{p^2} \tag{12.17}$$

using the prescription of section 9. (Using other prescriptions would give an extra finite local term proportional to $\epsilon I A_{\mu AB}A^{\mu AB}$ which is trivial in that it could be cancelled by adding a finite local counterterm to the action. For further discussion of the regularization of this graph, see [73].) In 2-dimensional x-space (10.17) gives an anomalous contribution to Γ given by [39] (with $\hat{A}_\mu = A_i\partial_\mu\bar{\phi}^i$)

$$\Gamma_{an} = \frac{-\hbar}{8\pi}\int d^2x\left[(\delta^{\mu\nu}+\epsilon^{\mu\nu})\partial_\mu\hat{A}^{AB}_\nu\Box^{-1}\partial^\rho\hat{A}^{AB}_\rho - \frac{1}{2}\hat{A}^{AB}_\mu\hat{A}^{\mu AB}\right] \tag{12.18}$$

whose variation under (4.3) is

$$\delta\Gamma_{an} = \frac{\hbar}{8\pi}\int d^2x\ \Lambda^{AB}\left[\delta^{\mu\nu}+\epsilon^{\mu\nu}\right]\partial_\mu\hat{A}^{AB}_\nu \tag{12.19}$$

and can be cancelled as in section 4. First, we add the finite local counterterm to S

$$\Delta = \frac{\hbar}{16\pi}\int d^2x\ \hat{A}^{AB}_\mu\hat{A}^{\mu AB} \tag{12.20}$$

corresponding to redefining the metric, $g_{ij} \to g_{ij}+(\hbar/2\pi)A^{AB}_i A^{AB}_j$, so that part of (12.19) is cancelled, leaving

$$\delta\left[\Gamma_{an}+\Delta\right] = \frac{\hbar}{8\pi}\int d^2x\ \Lambda^{AB}\epsilon^{\mu\nu}\partial_i A^{AB}_j\partial_\mu\bar{\phi}^i\partial_\nu\bar{\phi}^j \tag{12.21}$$

which, taking into account the Wick rotation, is precisely half the result (4.17), which is correct as (4.17) is the anomaly for Dirac fermions and should be double the result (12.21) for Majorana fermions. Then, as we saw in section 4, demanding that b_{ij} transforms as

$$\delta b_{ij} = -\frac{\hbar}{4\pi}\Lambda^{AB}\partial_{[i}A^{AB}_{j]} \tag{12.22}$$

under the symmetry (4.3) leads to a complete cancellation of the anomaly. Alternatively, the anomalous variation (12.19) can be cancelled <u>without</u>

adding a finite local counterterm if, in addition to (12.22), we demand that the metric g_{ij} also transform under (4.3) [39]

$$\delta g_{ij} = -\frac{\hbar}{4\pi}\Lambda^{AB}\partial_{(i}A_{j)}^{AB} \tag{12.23}$$

Note the similarity between (12.22) and (12.23).

As the anomaly (12.19) is equivalent to the standard chiral anomaly, it is only a one-loop effect, i.e. once the anomaly is cancelled at one-loop, it is cancelled to all orders (possibly after adding suitable finite local counterterms to Γ). Then we have the Ward identity (setting $\bar{\xi} = \bar{\rho} = 0$)

$$\Lambda^{BA}\bar{\lambda}^{A}.\frac{\delta\Gamma}{\delta\bar{\lambda}^{B}} - (\nabla_i\Lambda^{AB}).\frac{\delta\Gamma}{\delta A_{iAB}} = \frac{\hbar}{4\pi}\Lambda^{AB}(\partial_i A_j^{AB}).\left[\frac{\delta\Gamma}{\delta g_{ij}} + \frac{\delta\Gamma}{\delta b_{ij}}\right] \tag{12.24}$$

The diffeomorphism Ward identity (9.17) now includes an extra term involving the Lie derivative of A_i^{AB}

$$\underset{u}{L}A_i^{AB} = u^j F_{ji}^{AB} + \left[\nabla_i(u^j A_j)\right]^{AB} \tag{12.25}$$

and so under (9.21) $\beta_{iAB}^{A} \rightarrow \beta_{iAB}^{A} - \underset{w}{L}A_{iAB}$. Just as the diffeomorphism invariance (9.17) led to the possibility of changing the counterterms as in (9.16, 9.20) and the β-functions as in (9.21, 9.22), the Ward identity (12.24) leads to an additional ambiguity in the renormalization scheme. If we change $U_{+}^{(\nu)A}$ in (12.10) by

$$\delta U_{+}^{(\nu)A} = L^{A}{}_{B}(\bar{\phi},\Psi)\bar{\lambda}_{+}^{B} \tag{12.26}$$

for some $L^{A}{}_{B}$, the divergences are still cancelled provided we also make the modifications

$$\begin{aligned}
\delta T_{iAB}^{(\nu,A)} &= -\nabla_i L_{AB} \\
\delta T_{ij}^{(\nu,g)} &= -\frac{\hbar}{4\pi}L^{AB}\partial_{(i}A_{j)}^{AB} \\
\delta T_{ij}^{(\nu,b)} &= -\frac{\hbar}{4\pi}L^{AB}\partial_{[i}A_{j]}^{AB}
\end{aligned} \tag{12.27}$$

as can be checked using (12.24). Then, defining

$$\Theta_{AB} = \left[\frac{\partial}{\partial\Lambda}(L_{AB}(\bar{\phi},\Lambda^{-1}\bar{\Psi}))\right]\Bigg|_{\Lambda=1} \qquad (12.28)$$

the β- and γ-functions are modified by

$$\gamma^{A}_{\lambda+} \to \gamma^{A}_{\lambda+} + \Theta^{AB}\bar{\lambda}^{B}_{+}$$

$$\beta^{A}_{iAB} \to \beta^{A}_{iAB} - \nabla_i\Theta_{AB} \qquad (12.29)$$

$$\beta^{g}_{ij} \to \beta^{g}_{ij} - \frac{\hbar}{4\pi}\Theta^{AB}\partial_{(i}A^{AB}_{j)}$$

$$\beta_{ij} \to \beta^{b}_{ij} - \frac{\hbar}{4\pi}\Theta^{AB}\partial_{[i}A^{AB}_{j]}$$

Then there are ambiguities in the β-functions (9.22, 9.24, 12.29) corresponding to the classical diffeomorphism invariance (1.3), the "antisymmetric tensor gauge invariance" (2.3) and the gauge invariance (4.3), respectively. This relation between β-function ambiguities and classical invariances is a general feature, associated with the fact that the infinite "sets of coupling constants" $g_{ij}(\phi), b_{ij}(\phi), A_{iAB}(\phi)$, are only defined up to "gauge transformations" which lead to physically equivalent couplings.

This quantization can be generalized to a curved space-time $(N,\gamma_{\mu\nu})$ and the analysis of section 10 repeated. The trace anomaly is again given by (10.16), which now contains a term $B^{A}_{iAB}.(\delta L_{eff}/\delta A_{iAB})$ where (cf. (10.17))

$$B^{A}_{iAB} = \beta^{A}_{iAB} + v^{j}F^{AB}_{ji} + (\nabla_i\Lambda)_{AB} \qquad (12.30)$$

for some Λ_{AB}, with v^{i} as in (10.18). The B-functions do not suffer from the ambiguity (12.29), just as they did not from (9.22).

If we now reinstate χ^{M} in (12.1) the quantization is rather similar, but renormalizations of the coefficients B_{iMN} and U_{ABMN} are now required, with corresponding β-functions given to lowest order by

$$\beta^{B}_{iMN} = \frac{\hbar}{4\pi}\left\{\nabla^{(-)j}G_{ji}\right\}^{MN} \qquad (12.31)$$

$$\beta^{U}_{ABMN} = \frac{\hbar}{4\pi}\left\{\nabla^{i}\nabla_{i}U\right\}_{ABMN} \qquad (12.32)$$

13. Quantization of the Supersymmetric Non-Linear Sigma Model

We have considered the quantization of the general non-linear sigma-model with fermions (5.5, 5.6). It is of interest to investigate the special cases in which the classical action is supersymmetric. The classical theory is supersymmetric only if the coupling coefficients $\Psi = (g_{ij}, b_{ij}, A_{iAB}, B_{iMN}, U_{ABMN})$ satisfy certain constraints. If supersymmetry is to be maintained in the quantum theory, the renormalized geometric couplings Ψ should satisfy similar constraints, although there may be some quantum modifications to these restrictions. These place restrictions on the possible counterterms and tell us much about the quantum theory. It is also of interest to know whether supersymmetry can be maintained simultaneously with the other symmetries of the theory. It is unfortunate that there is no known regularization scheme that is consistent with supersymmetry and is fully satisfactory. However, in the absence of such a scheme, we might proceed along the lines discussed in [57].

Consider first the (1,1) supersymmetric model with action (5.74). Here g_{ij} and b_{ij} are the only independent couplings, the co-efficients of the Yukawa interactions A_i^{AB}, B_i^{MN} being essentially the spin-connections $\omega_i^{(-)ab}$, $\omega_i^{(+)ab}$ respectively, while U_{ABMN} is the curvature $R^{(-)}_{abij}$ (see Section 5). The symmetries of the theory are target space co-ordinate invariance, (1,1) supersymmetry and the anti-symmetric tensor gauge invariance (2.3). A perturbation theory manifestly maintaining all these symmetries can be constructed using normal co-ordinates in superspace [43, 51]. In (covariant) components, this corresponds to the prescription described in previous sections in which ϕ is expanded using normal co-ordinates and the background/quantum expansion of the fermi fields uses equations of the type (12.5), (12.6). This ensures that the relations between coupling constants are maintained so that, for example, the renormalized co-efficient of the $\lambda^a\lambda^b\partial\phi^i$ interaction is the spin-connection $\omega_i^{(-)ab}$ constructed from the renormalised g_{ij}, b_{ij}, at least up to the ambiguities described in Section 9 corresponding to dependence on the renormalization scheme. This can be checked explicitly using the results of section 12. Thus the only independent renormalizations are those of g_{ij} and b_{ij}, the other renormalizations being determined in terms of these by supersymmetry.

The situation is more subtle for (1,2) supersymmetry [30]. The model (5.74) has (1,2) supersymmetry if the target space is a complex manifold in which the g_{ij} and b_{ij} satisfy (5.41, 5.42). The theory can be formulated in (1,2) superspace and, using a superspace background field method with a linear background/quantum split of the type discussed in Section 6, a manifestly (1,2) supersymmetric perturbation theory can be developed. However, as we have seen, using a linear split gives non-covariant Feynman rules, so that target space co-ordinate invariance is not manifest. There is, however, no way of choosing a non-linear split in superspace (e.g. using normal co-ordinates) in such a way as to obtain a covariant perturbation theory. (This is because the "chirality" constraints on the (1,2) superfields are inconsistent with geodesic-type equations. One way round this problem may be to solve for the constrained superfields in terms of pre-potentials, as in [58].) On the other hand, we could use a (1,1)-superspace normal co-ordinate quantization which would maintain manifest general co-variance, but only (1,1) supersymmetry invariance would be manifest. The extra supersymmetry would be no longer manifest in this approach. However, the results from these two prescriptions should be physically equivalent, so that the counterterms and β-functions calculated in the two approaches should be equivalent, up to the ambiguities discussed in Section 9.

Let us illustrate this at one-loop. Using normal co-ordinates, we have seen that the one-loop divergences can be minimally subtracted using the counterterms

$$T_{ij}^{(1,g)} = -\frac{\hbar}{2\pi} R^{(+)}_{(ij)}, \qquad T_{ij}^{(1,b)} = -\frac{\hbar}{2\pi} R^{(+)}_{[ij]} \tag{13.1}$$

and (1,1) superspace normal co-ordinates give the same results. However, these counterterms are not (1,2) supersymmetric [30]. For example, (1,2) supersymmetry requires that the renormalized metric be hermitian, i.e. that in complex co-ordinates (see Section 5) it satisfy $g_{\alpha\beta} = 0$, $g_{\overline{\alpha\beta}} = 0$. However, a calculation shows that this is not the case here, as the one-loop corrections to the metric are not hermitian, $R^{(+)}_{(\alpha\beta)} \neq 0$, $R^{(+)}_{(\overline{\alpha\beta})} \neq 0$ [30]. Using a linear split (either in components or in (1,2) superspace), on the other hand, we obtain to lowest order

$$T_{ij}^{(1,g)} = -\frac{\hbar}{2\pi}\left[R^{(+)}_{(ij)} - \nabla_i X_j\right] \tag{13.2}$$

$$T_{ij}^{(1,b)} = -\frac{\hbar}{2\pi}\left[R^{(+)}_{[ij]} - X^k H_{ijk} - \partial_{[i} X_{j]}\right]$$

where X^i is defined in (6.16). From (5.45), this will be consistent with (1,2) supersymmetry if there is some k_i such that, in a complex co-ordinate system,

$$T^{(1,g)}_{\alpha\beta} = 0 \qquad T^{(1,g)}_{\alpha\bar\beta} = -\frac{\hbar}{4\pi}\left[\partial_\alpha k_{\bar\beta} + \partial_{\bar\beta} k_\alpha\right] \tag{13.3}$$

$$T^{(1,b)}_{\alpha\beta} = 0 \qquad T^{(1,b)}_{\alpha\bar\beta} = -\frac{\hbar}{4\pi}\left[\partial_\alpha k_{\bar\beta} - \partial_{\bar\beta} k_\alpha\right]$$

together with the complex conjugate equations. For geometries that define a sigma-model with (1,2) supersymmetry classically, (13.2) indeed satisfies (13.3) with [30]

$$k_i = J_i{}^k \Gamma^{(+)m}_k{}_n J^n{}_m \equiv J_i{}^k \Gamma_k \tag{13.4}$$

$$k_\alpha = \left[\Gamma^{(+)\beta}_{\alpha}{}_{\beta} - \Gamma^{(+)\bar\beta}_{\alpha}{}_{\bar\beta}\right]$$

Thus the counterterms (13.2) maintain (1,2) supersymmetry off-shell. On the other hand, as X^i is in general non-covariant, it might appear that general covariance is lost. However, for hermitian geometries satisfying (5.45), X^i can be rewritten in terms of covariant quantities

$$X^i \equiv g^{jk}\left\{{i \atop j\ k}\right\} = J^{ij}J^{kl}H_{jkl} \tag{13.5}$$

so that it is a vector for these geometries; in fact it is the same vector as that which occurred in (5.66). Then the counterterms (13.2) are consistent with both (1,2) supersymmetry and general covariance.

Note that the U(1) part of the curvature C_{ij} (5.65) satisfies

$$C_{ij} \equiv J^l{}_k R^{(+)k}{}_{lij} = 2\partial_{[i}\Gamma_{j]} \tag{13.6}$$

so that the one-loop β-functions corresponding to (13.2) can be written as

$$\beta^g_{ij} = \frac{\hbar}{4\pi} J_{(i}{}^k C_{j)k} + O(\hbar^2) \tag{13.7}$$

$$\beta^b_{ij} = \frac{-\hbar}{4\pi}(J_{[i}{}^k C_{j]k} + 2\partial_{[i}k_{j]}) + O(\hbar^2)$$

We see in particular that the β-functions will be trivial at one-loop if $C_{ij} = 0$, which is precisely the condition that the holonomy group for the connection $\omega^{(+)ab}_i$ be SU(d/2) (or a subgroup thereof), $h(\omega^+) \subseteq SU(d/2)$. In fact, there are no further non-trivial contributions to the β-functions at two loops [51].

If we had used normal co-ordinates, the counterterms (13.1) would give the β-functions (11.13) related to (13.7) by the transformations (9.22). In general, using the linear split and (1,2) superspace, we would obtain an effective action that was fully (1,2) supersymmetric off-shell but which may only be covariant on-shell, while using normal co-ordinates in (1,1) superspace we would obtain an effective action that was covariant and (1,1) supersymmetric off-shell, but only fully (1,2) supersymmetric on-shell. However, the two effective actions agree on-shell.

We now restrict ourselves to the case in which $H_{ijk} = 0$, so that the space is Kahler and there is (2,2) supersymmetry. The metric is given in terms of a Kahler potential K by $g_{\alpha\bar{\beta}} = 2\partial_\alpha\partial_{\bar{\beta}}K$ in complex co-ordinates. The superspace background field expansion of [58] ensures that (2,2) supersymmetry and reparameterization invariance are maintained simultaneously, so that the counterterm tensors $T^{(\nu,g)}_{ij}$ can be chosen to be "Kahler tensors", satisfying

$$T^{(\nu,g)}_{\alpha\beta} = 0 \qquad T^{(\nu,g)}_{\alpha\bar{\beta}} = \partial_\alpha\partial_{\bar{\beta}} \sum_{L=1}^{\infty} \hbar^L S^{(L,\nu)} \tag{13.8}$$

for some $S^{(L,\nu)}$ [8,58,59]. It can be shown that for $L > 1$, the $S^{(L,\nu)}$ are scalars on M [58], while at one-loop $S^{(1,1)} = -(2\pi)^{-1} 2\log\det\left[g_{\alpha\bar{\beta}}\right]$, as on a Kahler manifold the Ricci tensor satisfies

$$R_{\alpha\bar{\beta}} = 2\partial_\alpha\partial_{\bar{\beta}} \log\det\left[g_{\gamma\bar{\delta}}\right] . \tag{13.9}$$

Then in a complex co-ordinate system the β^g-function takes the form $\beta^g_{\alpha\beta} = 0$,

$$\beta^g_{\alpha\bar{\beta}} = -\frac{\hbar}{2\pi} R_{\alpha\bar{\beta}} + \sum_{L=2}^{\infty} \hbar^L W^{(L)}_{\alpha\bar{\beta}} \tag{13.10}$$

$$W^{(L)}_{\alpha\bar{\beta}} = \partial_\alpha\partial_{\bar{\beta}} S^{(L)} \qquad S^{(L)} \equiv L S^{(1,L)}$$

The two and three loop corrections $S^{(2)}$, $S^{(3)}$ vanish so that for Ricci-flat Kahler metrics the β-function vanishes to three loop order and it was conjectured that $\beta^g = 0$ to all orders for such spaces [59].

Ricci-flat Kahler spaces have a number of interesting properties. For a Kahler space

$$C_{ij} = -2J_{[i}{}^{k}R_{j]k}\,, \qquad C_{\alpha\bar{\beta}} = -2iR_{\alpha\bar{\beta}} \tag{13.11}$$

so that Ricci-flatness is equivalent to the condition $C_{ij} = 0$ for the U(1) part of the holonomy group to be absent, i.e. for $h(\omega) \subseteq SU(d/2)$. The cohomology class containing the 2-form $\frac{1}{2}C_{ij}d\phi^i{}_\wedge d\phi^j$ is known as the first Chern class and depends only on the topology and complex structure of a complex space, not on its metric or connection. Thus for a complex space to admit a Kahler-Ricci-flat metric it is necessary that the first Chern class be trivial. For compact complex spaces, it was conjectured by Calabi [60] and proved by Yau [61] that the vanishing of the first Chern class was a sufficient condition for the existence of a Kahler-Ricci-flat metric on the space. Calabi showed that this metric was unique (for a given complex structure and cohomology class of the fundamental form J (5.43)) and compact Kahler-Ricci-flat spaces have become known as Calabi-Yau spaces [16].

Unfortunately, for a Kahler-Ricci-flat space $S^{(4)}$ is not zero [62] so that the β-function does not vanish. However, it is possible to perturbatively construct a Kahler metric g_{ij} such that $\beta^g_{ij}(g) = 0$ [63]. Consider a Kahler metric g_{ij} defined formally as a perturbation series about the Calabi-Yau metric $\bar{g}_{ij}$

$$g_{ij} = \bar{g}_{ij} + \sum_{L=3}^{\infty} \hbar^L g^{(L)}_{ij}\,, \qquad g^{(L)}_{\alpha\bar{\beta}} = \partial_\alpha \partial_{\bar{\beta}} U^{(L)} \tag{13.12}$$

for some corrections $U^{(L)}$ to the Kahler potential. We now attempt to find functions $U^{(L)}$ such that the β-function vanishes for the metric g_{ij}, following [63]. To lowest order, using (13.9) we find

$$R_{\alpha\bar{\beta}}(g) = R_{\alpha\bar{\beta}}(\bar{g}) + 2\hbar^3 \partial_\alpha \partial_{\bar{\beta}} \nabla^2 U^{(3)} + O(\hbar^4) \tag{13.13}$$

so that, as $R_{\alpha\bar{\beta}}(\bar{g}) = 0$,

$$\beta^g_{\alpha\bar{\beta}} = \hbar^4 \partial_\alpha \partial_{\bar{\beta}} \left\{ \frac{-1}{2\pi} \nabla^2 U^{(3)} + S^{(4)} \right\} + O(\hbar^5) \tag{13.14}$$

Then if $U^{(3)}$ can be chosen such that for some constant C

$$\nabla^2 U^{(3)} = 2\pi (S^{(4)} + C) \tag{13.15}$$

$\beta^g_{\alpha\bar{\beta}}(g)$ will vanish to four-loop order. Integrating (13.15) over M, which we assume to be compact with volume V, we find that the constant C must be

$$C = -\frac{1}{V}\int d^d\phi\sqrt{g}S^{(4)} \tag{13.16}$$

Then it is necessary for the constant to be chosen as in (13.16) for there to be some $U^{(3)}$ satisfying (13.15). In fact this is also sufficient, choosing C as in (13.16) ensures that there exists a function $U^{(3)}$ satisfying (13.15) and using this function in (13.12) gives a metric g_{ij} such that $\beta^g_{ij}(g) = 0 + O(\hbar^5)$, i.e. we have cancelled the four-loop contribution to the β-function. We then proceed inductively. If $U^{(L)}$ have been found for all $L < N$ for some integer N, so that $\beta^g(g)$ vanishes to order $\hbar^N$, then

$$\beta^g_{\alpha\bar{\beta}} = 2\hbar^{N+1}\partial_\alpha\partial_{\bar{\beta}}\left[-\frac{1}{2\pi}\nabla^2 U^{(N)} + 2S^{(N+1)} + f^{(N)}\right] + O(\hbar^{N+2}) \tag{13.17}$$

where $f^{(N)}$ is a function of the $U^{(L)}$, $S^{(L+1)}$ for $L < N$, $f = f(U^{(L)}, S^{(L+1)})$. Then choosing $U^{(N)}$ to be the solution of

$$\frac{1}{2\pi}\nabla^2 U^{(N)} = S^{(N+1)} + f - \frac{1}{V}\int d^d\phi\sqrt{g}(S^{(N+1)} + f^{(N)}) \tag{13.18}$$

will ensure that $\beta^g(g)$ vanishes to order $\hbar^{N+1}$. By induction the metric defined by the formal power series (13.12) with the U^L satisfying (13.18) will then give a sigma-model with vanishing β-function. However, it is not known whether (13.12) converges i.e. whether, for this choice of $U^{(L)}$, the metric g_{ij} is well-defined, within some circle of convergence.

Next we specialise further to the torsion-free case with (4,4) supersymmetry, so that g_{ij} is hyper-Kahler i.e. a Kahler space with real dimension $d = 4m$, for some m, and with holonomy Sp(m). As $Sp(m) \subseteq SU(2m)$, hyper-Kahler spaces are special cases of Kahler Ricci-flat manifolds. If the (4,4) supersymmetry is to be maintained in the quantum theory, the renormalized metric must also be hyper-Kahler. We also know that the renormalized metric $g_{\alpha\bar{\beta}}$ and the bare metric $g^0_{\alpha\bar{\beta}}$ are related by $g^0_{\alpha\bar{\beta}} = \mu^\epsilon(g_{\alpha\bar{\beta}} + \partial_\alpha\partial_{\bar{\beta}}S)$ for some globally defined scalar S. Then $g_{\alpha\bar{\beta}}$ and $g_{\alpha\bar{\beta}} + \partial_\alpha\partial_{\bar{\beta}}S$ are two hyper-Kahler metrics for which the fundamental 2-forms $J = ig_{\alpha\bar{\beta}}dz^\alpha \wedge d\bar{z}^{\beta}$ and $J + i\partial\bar{\partial}S$ are cohomologically equivalent. If the manifold is compact, then Calabi's theorem applies and we conclude that the scalar S must be a constant so that the theory is finite and $\beta^g_{ij} = 0$. It can also be shown that the β-function vanishes to all orders for (4,4) sigma-models defined on hyper-Kahler manifolds that are non-compact [8]. This is subject to the assumption that the (4,4) supersymmetry is maintained in the quantum theory. Harmonic superspace techniques give a manifestly (4,4) supersymmetric perturbation theory (see fourth reference in [8]).

The quantization of the (0,q) supersymmetric sigma-models is very similar to that of the (p,q) models with $p \geq 1$ we have just considered. For example, the (0,1) model (5.31) can be quantized using superspace normal co-ordinates so as to obtain covariant Feynman rules that respect the supersymmetry. The sigma-model anomalies that arise at one loop can be cancelled by the mechanism described in Section 4 and it was shown in [39] that this can be done in a way that is consistent with the supersymmetry. As the (1,1) sigma model is a special case of the (0,1) model in which $\omega_i^{(-)ab}$ and A_i^{AB} are identified as in (5.72), the β-functions and effective action for the (0,1) model must reduce to those of the (1,1) model on making the identification (5.72), up to the ambiguities described in Section 9. For example, the metric β-function β^g_{ij} for the (0,1) model is known [14] to contain at two loops terms proportional to $g^{kl}F^{AB}_{ik}F^{AB}_{jl}$ and to $R_{iabc}R_{jabc} + O(\hbar)$. As these must reduce to the (1,1) β-function on identifying $\omega_i^{(-)}$ and A_i, the β-function must, in some scheme, be

$$\beta^g_{ij} = -\frac{\hbar}{2\pi}R^{(+)}_{(ij)} + a\hbar^2\left[F^{AB}_{ik}F^{AB}_{jl} - R^{(-)}_{abik}R^{(-)}_{abjl}\right]g^{kl} + O(\hbar^3) \qquad (13.19)$$

for some constant a [28, 30], where the torsion in the curvatures in this formula contains Chern-Simons forms, as in (5.84).

The (0,2) case is similar to the (1,2) case, but with the torsion H replaced by the torsion H′ including Chern-Simons forms (5.85). For example, the renormalized g_{ij}, b_{ij} still satisfy the first equation in (5.44), $H' = i(\partial J - \bar{\partial} J)$, but the second equation becomes, using (5.85)

$$2i\partial\bar{\partial}J = -\frac{\hbar}{8\pi}\mathrm{tr}[(F_\wedge F) - (\tilde{R}_\wedge\tilde{R})] \qquad (13.20)$$

Note that the left-hand side is proportional to $g_{\alpha\bar{\beta},\gamma\bar{\delta}}\,dz^\alpha{}_\wedge dz^\gamma{}_\wedge d\bar{z}^{\bar{\beta}}{}_\wedge d\bar{z}^{\bar{\delta}}$, the right-hand-side must also have two dz factors and two $d\bar{z}$ factors, i.e. must be a (2,2)-form. Then $\tilde{\omega}_i{}^{ab}$, as defined in (5.83), must be chosen such that $\mathrm{tr}(\tilde{R}{}_\wedge\tilde{R})$ is a (2,2)-form. One choice that does this is that described at the end of Section 5, in which $\tilde{\omega}_i{}^{ab} = \omega_i^{(-)ab}$ [28]; an alternative choice meeting this requirement was proposed in [36].

14. Strings in Flat Space

In this section we briefly review those aspects of string theory we wish to use here; for a complete account, see [64] and references therein. Let (M,η_{ij}) be d-dimensional Minkowski space-time with co-ordinates ϕ^i, $i = 0,\ldots,d-1$. Then a (bosonic) string moving in M sweeps out a two-dimensional surface, N, known as the world-sheet. If x^μ $(\mu = 0,1)$ are co-ordinates for the world-sheet, then the position of a string in space-time is given by $\phi^i(x)$. An action for the dynamics of a string is given by

$$S = \frac{1}{4\pi\alpha'}\int d^2x\sqrt{\gamma}\,\gamma^{\mu\nu}\eta_{ij}\partial_\mu\phi^i\partial_\nu\phi^j \tag{14.1}$$

where $\gamma_{\mu\nu}$ is a metric on the world-sheet. Varying $\gamma_{\mu\nu}$ in (14.1) gives a constraint equation which can be used to eliminate $\gamma_{\mu\nu}$ from the classical theory. The space-time co-ordinates ϕ^i are taken to have the dimensions of length, so that a dimensional parameter α' known as the Regge slope or inverse string tension is introduced. If we were to rescale the co-ordinates $\phi^i \to \sqrt{2\pi\alpha'}\phi^i$ so that they become dimensionless, we would regain the free action (1.1) (but with δ_{ij} replaced by the indefinite metric η_{ij}). One quantum effect of the constraints obtained by eliminating $\gamma_{\mu\nu}$ is to eliminate negative norm states from the theory [64].

Remarkably, the quantum theory of the string can be interpreted in terms of the physics of an infinite number of particles in space-time. The quantum theory for the closed string is described by a Euclidean functional integral [65, 66]

$$Z = \sum_g \lambda^g \int [d\gamma]_g [d\phi]\exp(-S') \tag{14.2}$$

where $\hbar = 1$, the action (14.1) is continued to Euclidean space (so that $\eta_{ij} \to \delta_{ij}$ and $\gamma_{\mu\nu}$ is now Euclidean) and we include a source term and a cosmological constant term (for renormalizability)

$$S' = S + \frac{1}{2\pi\alpha'}\int d^2x(\sqrt{\gamma}\Lambda + \phi^i(x)J_i(x)) \tag{14.3}$$

The path integral over $\phi^i(x)$ can be evaluated perturbatively in α', using the methods of previous sections. The integral $\int[d\gamma]_g$ is over all <u>inequivalent</u> Euclidean metrics $\gamma_{\mu\nu}$ of compact spaces of genus g i.e. of spaces topologically equivalent to spheres with g handles. Finally, there is a sum over g weighted with a coupling constant λ, so that Z can be evaluated as a double perturbation series in terms of the form $\lambda^g(\alpha')^{L-1}$ where L counts the number of "sigma-model loops" and g the number of "string loops" or handles. Amplitudes for the scattering of space-time particles are obtained by inserting certain vertex functions in the path integral (14.2) [64].

The ϕ^i-integration gives $\exp(-W[J,\gamma_{\mu\nu}])$ as in (6.4). For J = 0, this is given to lowest order in α' by (11.11) on setting $g_{ij} = \delta_{ij}$, $b_{ij} = \Phi = 0$, $V = \Lambda$, $\bar{\phi}$ = constant, $\hbar \to 2\pi\alpha'$,

$$W_0 = \int d^nx\sqrt{\gamma}\Big\{-\frac{d}{24\pi\epsilon}R^{(\gamma)} + \frac{d}{96\pi}R^{(\gamma)}\Box^{-1}R^{(\gamma)} \tag{14.4}$$

$$-\frac{m^2d}{8\pi\epsilon} - \frac{dm^2}{16\pi}\Box^{-1}R^{(\gamma)} + \frac{1}{4\pi}m^2d + \frac{1}{2\pi\alpha'}\Lambda\Big\}$$

so that minimally subtracting the poles leaves the renormalized functional

$$W = \int d^nx\sqrt{\gamma}\left[\frac{d}{96\pi}R^{(\gamma)}\Box^{-1}R^{(\gamma)} - \frac{dm^2}{16\pi}\Box^{-1}R^{(\gamma)} + \frac{1}{4\pi}m^2d + \frac{1}{2\pi\alpha'}\Lambda\right] \tag{14.5}$$

and we can take the limit $\epsilon \to 0$, $n \to 2$. Next, we must integrate e^{-W} over inequivalent metrics $\gamma_{\mu\nu}$, where we regard two metrics as equivalent if they are related by a diffeomorphism. Any metric on a compact orientable 2-surface N can be brought to the form

$$\gamma_{\mu\nu} = e^{2\sigma(x)}\tilde{\gamma}_{\mu\nu}(x;\tau_1,\ldots,\tau_n) \tag{14.6}$$

for some conformal factor $\sigma(x)$, with $\tilde{\gamma}$ a metric of constant curvature, by a diffeomorphism. Here $\tau_1,\ldots,\tau_n$ are a finite set of parameters - moduli - labelling the inequivalent metrics of constant curvature. The number n of moduli depends on the genus. It is convenient to fix the world-sheet diffeomorphism invariance by choosing the gauge in which the metric takes the form (14.6), taking care of the Jacobian factor by introducing Faddeev-Popov ghosts in the usual way; see [64]. Then the integral over metrics reduces to an integration over ghosts, over moduli and over conformal factors $\sigma(x)$.

Here we will consider only genus zero surfaces i.e. tree level in the string loop expansion, for which n = 0 so that there are no moduli and $\tilde{\gamma}_{\mu\nu}$ is the metric on the sphere, which can be brought to the form $\tilde{\gamma}_{\mu\nu} = \delta_{\mu\nu}$

by a conformal transformation (note that we now include the "point of infinity" in the x-plane, so that we are dealing with the Riemann sphere). The ghost integral gives the correction Δ to W [65]

$$\Delta = -\int d^2x \left[\frac{26}{24\pi} \sigma \Box \sigma + a \frac{m^2}{8\pi} e^{2\sigma} \right] \tag{14.7}$$

for some number a, so that substituting $\gamma_{\mu\nu} = \delta_{\mu\nu} e^{2\sigma}$ in (14.5) gives

$$W + \Delta = \int d^2x \left\{ \left[\frac{d-26}{24\pi} \right] \sigma \Box \sigma + e^{2\sigma} \left[\frac{\Lambda}{2\pi\alpha'} + \frac{(3d-a)}{8\pi} m^2 \right] \right\} \tag{14.8}$$

and we are left with the functional integral $\int [d\sigma] \exp[-(W+\Delta)]$. Note that (14.8) includes a kinetic term and Liouville mass term for σ, so that in general σ will become an extra dynamical field. The standard string theory is defined as that in which there are no extra dynamical fields, i.e. in which σ drops out. This requires that the local scale anomaly of W is cancelled by the ghost contribution (14.7). This happens only in the critical dimension d = 26 and only if the cosmological constant Λ is chosen as

$$\Lambda = -\frac{\alpha'}{8}(3d-a)m^2 \tag{14.9}$$

in which case the integration over σ is trivial, yielding an (infinite) constant factor. Then only in 26 dimensions can we define a string theory without an extra propagating mode and we will only consider this case here. Note that in the above discussion, it was important to take the renormalized functional W rather than W_0. If instead we evaluated the bare functional W_0 in the conformal gauge $\gamma_{\mu\nu} = e^{2\sigma}\delta_{\mu\nu}$ using (10.3), we would find that the σ-dependence of the divergences cancels that of the non-local terms up to $O(\epsilon)$, so that in this gauge (after choosing Λ so that the Liouville mass term vanishes) $W_0 = 0 + O(\epsilon)$ [47]. (This is a reflection of the fact that at one loop W_0 is scale invariant, as was seen in section 11.) This suggests that going to conformal gauge before renormalizing W_0 can be problematic, to say the least.

15. Strings in Background Fields

The natural generalization of the flat-space action (14.1) to a curved space-time M with metric $g_{ij}(\phi)$ is [11-15]

$$S = \frac{1}{4\pi\alpha'} \int d^2x \sqrt{\gamma} \gamma^{\mu\nu}(x) g_{ij}(\phi(x)) \partial_\mu \phi^i \partial_\nu \phi^j \tag{15.1}$$

On rescaling $\phi^i \to \sqrt{2\pi\alpha'}\phi^i$, this becomes just the non-linear sigma-model action (1.2), but with M regarded as space-time rather than as the target space and with N regarded as the world-sheet rather than the space-time. As a check on the choice of action (15.1), note that if $g_{ij} = \eta_{ij} + h_{ij}$, it can be shown that (15.1) gives the correct interactions of a string with a linearized graviton h_{ij} [15]. The graviton is one of the massless modes of the oriented closed string, the others being an anti-symmetric tensor gauge field and a scalar field, the dilaton. There is one tachyonic scalar field, all other particles "contained" in the string being massive. If the anti-symmetric tensor gauge field, dilaton and tachyon have expectation values $b_{ij}(\phi)$, $\Phi(\phi)$, $V(\phi)$ respectively, then (15.1) is modified to become [12, 14]

$$S = \frac{1}{4\pi\alpha'}\int d^2x\sqrt{\gamma}\left[\left(\gamma^{\mu\nu}g_{ij} + \epsilon^{\mu\nu}b_{ij}\right)\partial_\mu\phi^i\partial_\nu\phi^j - \frac{\alpha'}{2}\Phi R^{(\gamma)} + 4\alpha' V\right] \tag{15.2}$$

corresponding to the most general renormalizable sigma-model (9.3) but with Φ and V scaled by factors of α' for later convenience, $\Phi \to (-\alpha'/4)\Phi$, $V \to 2\alpha' V$. With these rescalings, $2\pi\alpha' S$ is local scale invariant at zero'th order in α'. The cosmological constant term in (14.3) is absorbed into the definition of V. As before, we integrate over configurations $\phi^i(x)$ and ghosts and then demand that dependence on the conformal factor $\sigma(x)$ drops out, so that the final integration over σ is trivial. Here something rather interesting happens; whereas before we obtained the constraint d = 26, in curved space we obtain constraints on the curvature etc that can be interpreted as field equations for the fields g_{ij}, b_{ij}, Φ, V [14]. Thus a string theory can be defined in a background only if that background satisfies these effective field equations.

Instead of introducing a source term and integrating out ϕ completely, it is convenient to use the background field method to obtain a renormalized effective action Γ, as defined in (9.7). Having calculated and renormalized Γ, we take $\epsilon \to 0$ and go to the conformal gauge, in which

$$\gamma_{\mu\nu} = e^{2\sigma}\delta_{\mu\nu}, \qquad \epsilon_{\mu\nu} = e^{2\sigma}\bar{\epsilon}_{\mu\nu}, \tag{15.3}$$

$$\sqrt{\gamma}R^{(\gamma)} = -2\Box\sigma$$

with $\bar{\epsilon}_{\mu\nu}$ the constant alternating tensor of flat space. Then $\delta\Gamma/\delta\sigma = \langle T\rangle$, the local scale anomaly of section 10, which can be expressed in terms of B-functions as in (10.16). The ghosts do not couple to $\phi^i(x)$, so that the ghost contribution is still given by (14.7). Writing $\Gamma = \int d^2x\sqrt{\gamma}L_{eff}$, we find that

$$\frac{\delta}{\delta\sigma}(\Gamma+\Delta) = \hat{B}^g_{ij}.\frac{\delta L_{eff}}{\delta g_{ij}} + \hat{B}^b_{ij}.\frac{\delta L_{eff}}{\delta b_{ij}} + \hat{B}^\Phi.\frac{\delta L_{eff}}{\delta \Phi} + \hat{B}^V.\frac{\delta L_{eff}}{\delta V} \qquad (15.4)$$

with the notation of section 10, where the $\hat{B}$-functions differ from the B-functions of section 10 only by some α'-dependent rescalings of the fields and some ghost contributions. In (15.4) we have neglected a term proportional to $\delta\Gamma/\delta\bar{\phi}$, which vanishes on-shell. Using the one-loop results of section 11 and the two-loop results of [14] we find

$$\hat{B}^g_{ij} = -\alpha'\left[R_{ij} - H_{ikl}H_j{}^{kl} - \nabla_i\nabla_j\Phi\right] + O(\alpha'^2) \qquad (15.5)$$

$$\hat{B}^b_{ij} = \alpha'\left[\nabla^k H_{kij} + H^k{}_{ij}\nabla_k\Phi\right] + O(\alpha'^2)$$

$$\hat{B}^\Phi = -8\alpha'\left[\frac{d-26}{24\pi}\right] - \alpha'^2\left[R - \frac{1}{3}H^2 + 2\nabla^2\Phi - (\nabla\Phi)^2\right] + O(\alpha'^3)$$

$$\hat{B}^V = -\frac{\alpha'^2}{2}\left[\nabla^2 V + \frac{4}{\alpha'}V + (\nabla_i\Phi)(\nabla^i V)\right] + O(m^2) + O(\alpha'^3)$$

The α' dependence in (15.2) is such that the $\hat{B}$-functions all vanish in the $\alpha' \to 0$ limit in the critical dimension d = 26, so that the V and Φ terms in (15.2) can be thought of as one-loop counterterms, while the remaining terms are classically scale invariant. The constant part of the $O(m^2)$ contribution to $\hat{B}^V$ can be cancelled by shifting V by a constant.

Then the condition for the σ-dependence to drop out is that the $\hat{B}$-functions (15.5) vanish to all orders. These equations $\hat{B} = 0$ can be interpreted as field equations for g_{ij}, b_{ij}, Φ and V. To lowest order in α', they can be derived from the d-dimensional action [14]

$$I = \int d^d\phi\sqrt{g}e^{\Phi}\left[-R + \frac{1}{3}H^2 - 3(\nabla\Phi)^2 + \alpha'^2(\nabla V)^2 - 4\alpha' V^2 + O(\alpha')\right] \qquad (15.6)$$

with the usual kinetic terms, a tachyonic mass term for V, and an overall factor of e^{Φ}. Note that as was seen in sections 9 and 10, these field equations are independent of the renormalization scheme, unlike the conditions that the β-functions vanish. There will presumably be corrections to (15.5) from string loops as well as from sigma-model loops. There are some subtleties involved in obtaining higher order V contributions [67].

Then quantum string theories can be consistently defined (without extra propagating fields) only in background field configurations satisfying the effective field equations $\hat{B} = 0$. Such solutions are then candidate ground states for the theory. For the bosonic string the only known exact solutions of the form $M = M_n \times K$ (with M_n n-dimensional Minkowski-space and K some (d-n)-dimensional space) are either flat, with K a torus, or are such that K is a group manifold [11]. For K a group manifold with g_{ij} the group-invariant metric, H_{ijk} the parallelizing torsion and $\Phi = V = 0$, $\hat{B}^g_{ij} = \hat{B}^b_{ij} = 0$ order by order in α', while, at least for simply-laced groups [11], B^Φ has a non-trivial fixed point. Indeed, $\hat{B}^\Phi$ is proportional to $[(d - 26) + f(\alpha'/r)]$ where f is some function of the dimensionless quantity α'/r and r is the radius of K. Then the function f is such that $\hat{B}^\Phi$ vanishes for some definite radius r_0, in units of α'.

16. Superstrings in Background Fields

The classical action for a superstring in a non-trivial background is given by a (p,q) supersymmetric extension of (15.2). In the Ramond-Neveu-Schwarz formalism [64], this corresponds to a (p,q) sigma-model of the form (5.31) or (5.74) coupled to the two-dimensional (p,q) supergravity multiplet so as to have (p,q) local world-sheet supersymmetry. (See [68] for a discussion of coupling (p,q) sigma-models to supergravity.)

In addition to the two-dimensional "graviton" $\gamma_{\mu\nu}$, there are now its superpartners, the gravitini. In the path integral, as well as integrating over $\bar{\phi}$ and $\gamma_{\mu\nu}$, we must also integrate over the gravitini and the sigma-model fermions, λ, χ. As well as fixing the world-sheet reparameterization invariance, we must also fix the (p,q) local supersymmetry, so that there are now more ghost fields than in the bosonic case and the ghost contribution to the local scale anomaly changes. The critical dimension in which, for flat space-time M, the sigma-model scale anomaly cancels against the ghost contribution is now 10 for (1,1) (1,0) or (0,1) supersymmetry, 2 for (2,r) or (r,2) supersymmetry ($r \leqslant 2$) and is negative if p or q is 4 [64, 69].

The (1,1) case corresponds to the type II superstring [64] while the (0,1) or (1,0) case corresponds to the heterotic string [70]. Strictly speaking the superstring is obtained by projecting onto a subspace of the modes of the Ramond-Neveu-Schwarz model, which in particular removes the tachyon mode. Then there should be no potential $V(\phi)$ in the corresponding supersymmetric sigma-model, as this would correspond to a tachyon expectation value. This is in fact the case, as the (p,q)-supersymmetric extension of the potential term is inconsistent with the projection onto the superstring.

While g_{ij} is interpreted as the space-time metric and b_{ij} as the vacuum expectation value of the anti-symmetric tensor gauge field, A_{iAB} in (5.31) is interpreted as a vacuum expectation value for the Yang-Mills gauge potential that occurs as a massless mode in the heterotic string. The symmetries (1.3), (2.3) then become general co-ordinate invariance and the anti-symmetric tensor gauge invariance, while (3.4) represents Yang-Mills invariance. For the model (5.31), the anomalies of section 4 are Yang-Mills and local Lorentz anomalies, which are cancelled to all orders in α' but only at tree level in string loops by the mechanism described in section 4. In the string loop expansion, there are further anomalies that would render the theory inconsistent, but which cancel if the number of fermions λ^A in (5.31) is 32 ($A = 1,\ldots,32$) and the Yang-Mills group is $Spin(32)/\mathbb{Z}_2$ or $E_8 \times E_8$ [70]. In the $E_8 \times E_8$ case, the field A_{iAB} in (5.31) is the Yang-Mills field for the $O(16) \times O(16)$ subgroup of $E_8 \times E_8$. (The action of the generators of $E_8 \times E_8$ not in the $O(16) \times O(16)$ subgroup is through spin operators that change periodic fermions into anti-periodic ones and vice-versa [70].)

Consider the case of the heterotic string, corresponding to the sigma-model (5.31) coupled to (0,1) supergravity, including a supersymmetrization of the Fradkin-Tseytlin term with dilaton field Φ. The only independent couplings are then $\Psi = (g_{ij}, b_{ij}, A_{iAB}, \Phi)$ and the condition for the local scale anomaly of the sigma-model to be cancelled by ghost contributions is that certain $\hat{B}$-functions (generalizing those of the last section) vanish. To lowest order in α' and at tree level in string loops, these are [14, 28]

$$\hat{B}^g_{ij} = -\alpha'\left[R^{(+)}_{(ij)} - \nabla_i\nabla_j\Phi\right] - \frac{(\alpha')^2}{4}\left[R^{(-)}_{abci}R^{(-)abc}{}_j - F^{AB}_{ik}F^{ABk}_j\right] + O(\alpha'^3)$$

$$\hat{B}^b_{ij} = \alpha'\left[R^{(+)}_{[ij]} - H_{ij}{}^k\nabla_k\Phi\right] + O(\alpha'^3) \qquad (16.1)$$

$$\hat{B}^\Phi = -8\alpha'\left[\frac{d-10}{16\pi}\right] - \alpha'^2\left[R - \frac{1}{3}H^2 - 2\nabla^2\Phi - (\nabla\Phi)^2\right]$$

$$\hat{B}^A_{iAB} = \frac{\alpha'}{2}\left[\nabla^{(+)j}F^{AB}_{ji} + (\nabla^j\Phi)F^{AB}_{ji}\right] + O(\alpha'^2)$$

up to the field redefinition ambiguities discussed in earlier sections.

In order to be able to consistently formulate a heterotic string theory in a given background field configuration, it is necessary, then, that the background fields $g_{ij}, b_{ij}\Phi, A_{iAB}$ satisfy the effective field equations $\hat{B} = 0$, together with the Bianchi identity (5.85) needed for anomaly cancellation. A background satisfying these conditions would be a candidate vacuum for the theory and it is of considerable importance to find a compactifying vacuum in the critical dimension d = 10 in which $M = M_4 \times K$ with M_4 four-dimensional Minkowski space and K a compact 6-space, so as to yield effective four-dimensional physics in M_4. We now briefly consider some "vacua" satisfying these field equations, at least to lowest order in α'.

Consider a background $M_4 \times K$ such that, on setting $\gamma_{\mu\nu} = \eta_{\mu\nu}$ and the two dimensional gravitino to zero, the corresponding (0,1) sigma-model (5.31) in fact has (0,2) world-sheet supersymmetry, i.e. K is a complex manifold satisfying the constraints described in section 5, with the holonomy group of $\omega_i^{(+)ab}$ satisfying $h(\omega^{(+)}) \subseteq U(3)$. Such extra world-sheet supersymmetries restrict the form of the counterterms and hence the form of the scale anomaly, so that one might expect this to make it easier to find solutions, and this is indeed the case. For such a space, the U(1) part of the curvature C_{ij} defined in (5.65) satisfies [28,30]

$$-\frac{1}{2}J_i{}^k C_{jk} = R_{ij}^{(+)} - \nabla_{(i}V_{j)} - H_{ij}{}^k V_k - \frac{1}{2}\partial_{[i}V_{j]}$$
$$+ \frac{1}{2}J_i{}^k J_j{}^l \partial_{[k}V_{l]} \tag{16.2}$$

where $V_i = J_{ij}J_{kl}H^{jkl}$, as in (5.66). Then if $C_{ij} = 0 + O(\alpha')$, so that in the $\alpha' \to 0$ limit the space has SU(3) holonomy, i.e. $h(\omega^{(+)}) \subseteq SU(3)$, comparing with (16.1) we have, to lowest order,

$$\hat{B}^g_{ij} = \nabla_{(i}\bar{V}_{j)} \qquad \hat{B}^b_{ij} = H_{ij}{}^k \bar{V}_k + \partial_{[i}\bar{V}_{j]} \tag{16.3}$$

and $\bar{V}_i = V_i - \partial_i\Phi$. This is sufficient for the $\hat{B}^g, \hat{B}^b$ contributions to the trace anomaly to be a total derivative, so that there is rigid scale invariance to $O(\alpha')$ (ignoring for the moment the $\hat{B}^A$ contribution) and β^g_{ij} and β^b_{ij} are trivial at one loop, as we saw in section 13. However, for string consistency we need <u>local</u> scale invariance, so that we must further demand that

$$V_i = \partial_i\Phi \tag{16.4}$$

for the $\hat{B}^g$ and $\hat{B}^b$ to vanish to this order. Remarkably, this is also sufficient for $\hat{B}^\Phi$ to vanish up to $O(\alpha'^3)$. It will only be possible to choose Φ to satisfy (16.4) if the V_i defined by (5.66) satisfies the integrability condition $\partial_{[i}V_{j]} = 0$. This condition will not be satisfied for general (0,2) models, but is in fact automatically satisfied by the geometries that admit (2,2) twisted chiral sigma-models. This condition is sufficient if the first cohomology group $H^1(K)$ is trivial. (H^1 is trivial for any compact six-manifold that admits an SU(3) holonomy metric [64].) Finally, if the Yang-Mills field satisfies

$$J_{[i}{}^{k} F_{j]k}{}^{AB} = 0 \qquad J^{ij} F_{ij}{}^{AB} = 0 \tag{16.5}$$

the $\hat{B}^A$ will also vanish to lowest order and $\hat{B}^g$ and $\hat{B}^b$ will in fact vanish up to $O(\alpha'^3)$ provided $\tilde{\omega}_i{}^{ab} = \omega_i^{(-)ab}$ in the Bianchi identity (5.85) [28]. There will usually be solutions to (16.5) provided certain topological integrability conditions are satisfied [72]. We have, then, solutions to the effective field equations $\hat{B} = 0$ to lowest order in α'.

An ansatz for which (16.5) is automatically satisfied is the one which leads to the underlying sigma-model (5.31) having (1,2) supersymmetry, namely the identification of the spin-connection $\omega_i^{(-)ab}$ with the Yang-Mills potential A_{iAB} under some embedding of the holonomy group of K (and M) into the Yang-Mills group, $A \sim \omega^{(-)}$, corresponding to the solution proposed in [30]. The simplest case is that in which $b_{ij} = 0$ and $A \sim \omega$, corresponding to a (2,2) sigma-model on a Kahler-Ricci-flat space K, i.e. in which K is a Calabi-Yau space [16]. One might wonder whether there are any spaces satisfying all of these conditions with non-zero torsion. In fact, a number of examples are known. Suppose that $A \sim \omega^{(-)}$, so that the underlying sigma-model has at least (1,2) supersymmetry and that $K = T^2 \times K_4$ with K_4 a four-dimensional manifold, and T^2 the two-torus with flat metric. Then choosing (K_4, g_{ij}, b_{ij}) so as to give a (4,4) twisted chiral model (see section 5), $C_{ij} = 0$ and it can be shown that $\partial_{[i}V_{j]} = 0$, so that there is a Φ satisfying (6.4), at least locally [28,30].

Explicit forms are known locally for the g_{ij} and b_{ij} of the (4,4) twisted chiral model [5], so that we then have an explicit solution to the effective field equations $\hat{B} = 0 + O(\alpha')$, at least in some co-ordinate patch, so that the field equations we have obtained are at least not inconsistent. However, it is not known whether these local solutions can be continued to give a non-singular complete globally defined solution. Other examples in which A is not identified with $\omega^{(+)}$ were discussed in [36, 72].

It is important to know whether these spaces satisfy the effective field equations at higher orders. For the equations $\hat{B} = 0$ to be satisfied, it is necessary but <u>not</u> sufficient that the corresponding β-

functions be trivial. For the Calabi-Yau spaces, we saw in section 13 that $\beta^g_{ij} \neq 0$, but a new metric (13.12) could be constructed perturbatively about the Calabi-Yau metric for which the β^g-function vanished. This implies rigid scale invariance but is necessary, not sufficient, for local scale invariance. The triviality of the β-function implies only that $\hat{B}^g_{ij} = \nabla_{(i}w_{j)}$ for some w, while we need here the stronger condition $\hat{B}^g_{ij} = 0$ together with $\hat{B}^\Phi = 0$. It is not known whether these are satisfied to all orders. Similarly, the (4,4) twisted chiral models are known to be finite (see third reference in [8]), a necessary condition for the example of the previous paragraph to be a solution to all orders, but by no means a sufficient condition.

Giving the Yang-Mills potential a vacuum expectation value spontaneously breaks the gauge symmetry [16]. If the background Yang-Mills potential A is equal (under some embedding of the tangent space group into the gauge group) to the spin-connection $\omega^{(-)}$, (so that there is (1,2) world-sheet supersymmetry), then the gauge group G breaks down to the subgroup that commutes with the holonomy group $h(\omega^{(-)})$. For the Calabi-Yau compactifications with $A \approx \omega$, $H = 0$, $h(\omega) = SU(3)$ so that $E_8 \times E_8$ is broken to $E_8 \times E_6$ [16]. For the metric (13.12) defined perturbatively about the Calabi-Yau metric, $h(\omega) = U(3)$ so that if A is still identified with ω, the $E_8 \times E_8$ symmetry would break to $E_8 \times SO(10) \times U(1)$. However, in [71] it is argued that the "embedding condition" $A \approx \omega$ must be modified at order α'^4 in such a way that the unbroken gauge group remains $E_8 \times E_6$.

For the (1,2) models with torsion, in the $\alpha' \to 0$ limit $h(\omega^{(+)}) = SU(3)$ and at higher orders it is to be expected that this becomes U(3), as in the torsion-free case. However, for (1,2) supersymmetry, the gauge connection is identified with $\omega^{(-)}$, not $\omega^{(+)}$, and the holonomy group of $\omega^{(-)}$ is a priori unrelated to that of $\omega^{(+)}$. Thus even in the classical ($\alpha' \to 0$) limit, the gauge group could break to $E_8 \times SO(10)$, for example, if the holonomy group for $\omega^{(-)}$ is SO(6) and clearly there is a lot more freedom in choosing a symmetry breaking scenario.

Instead of using the Ramond-Neveu-Schwarz formalism with world-sheet supersymmetry, we could have used the Green-Schwarz formalism with manifest space-time supersymmetry. For $M = M_4 \times K$, this would correspond to using the sigma-model (5.60) defined on K, with i,j = 1,...,8, the ϕ^i being coordinates on the "transverse space" $\mathbb{R}^2 \times K$. For backgrounds admitting covariantly constant spinors $\eta_{M'}$, satisfying (5.50, 5.58, 5.63), the action (5.60) is invariant under the supersymmetry transformation (5.62). Such a supersymmetry is interpreted as a space-time supersymmetry invariance of the background field configuration and the number of these symmetries corresponds to the number of space-time supersymmetries left unbroken by the vacuum. For example, for the effective field theory in M_4 the N = 4 local supersymmetry will be unbroken if there are 8 linearly independent "Killing spinors" satisfying (5.50, 5.58, 5.63) (which will be the case if K is flat), while the N = 4 local supersymmetry will be spontaneously broken down to N = 1 by backgrounds with only 2 Killing spinors [16,34,28,36]. Remarkably, the conditions for there to be two such Killing spinors led us in section 5 to precisely the spaces for which (5.65) and (16.5) hold, i.e. to precisely the spaces that we have seen give trivial one-loop β-functions (16.3) and which will be solutions of $\hat{B}=0$ to lowest order if the extra condition (16.4) is imposed. Thus, to lowest order in α', the spaces we have been considering as solutions are precisely the ones that lead to N = 1 supersymmetry in four dimensions. Indeed, the Calabi-Yau spaces were originally found by seeking those compactifications leading to N = 1 supersymmetry in four dimensions [16].

This relation between the world-sheet supersymmetric approach and the space-time supersymmetric approach can be understood as follows; for details, see [28,37,71]. If there is a Killing spinor $\eta_{M'}$ satisfying (5.50,5.58,5.63), then (classically and ignoring boundary conditions) the Ramond-Neveu-Schwarz action (5.31), which is invariant under the world-sheet supersymmetry (5.30), becomes precisely the Green-Schwarz action (5.60), which is invariant under the "space-time supersymmetry" (5.62) on making the substitution $\chi^a = \eta^{M'} \gamma^a_{M'N} \chi^N$ (and (5.61)). Conversely, setting $\chi^N = \eta_{M'} \gamma_a^{M'N} \chi^a$ converts (5.60) to (5.31). Thus, in the presence of Killing spinors, the space-time supersymmetric action and the world-sheet supersymmetric one are equivalent, being related by a "triality" rotation. However, the four-loop corrections of [62] render the relation between the world-sheet supersymmetry approach and the space-time supersymmetry approach rather subtle; see e.g. [71].

<u>References</u>

[1] D Friedan, Phys. Rev. Lett. 45 (1980) 1057 and Ann. Phys. 163 (1985) 318.

[2] D Z Freedman and P K Townsend, Nucl. Phys. B177 (1981) 282; B Zumino, Phys. Lett. 87B (1979) 203.

[3] L Alvarez-Gaumé and D Z Freedman, Commun. Math. Phys. 80 (1981) 443.

[4] U Lindstrom and M Roček, Nucl. Phys. B222 (1983) 285; N Hitchin, A Karlhede, U Lindstrom and M Rocek, ITP Stockholm preprint (1986).

[5] S J Gates Jr, C M Hull and M Roček, Nucl. Phys. B248 (1984) 157.

[6] C M Hull and E Witten, Phys. Lett. 160B (1985) 398.

[7] C M Hull, in preparation.

[8] C M Hull, Nucl. Phys. B260 (1985) 182;
L Alvarez-Gaume and P Ginsparg, Commun. Math. Phys. 102 (1985) 311;
J Grundberg, A Karlhede, U Lindstrom and G Theodoridis, Nucl. Phys. B282 (1987) 142;
A Galperin, E Ivanov, V Ogievetsky and E Sokatchev, Class. Quantum Grav. 2 (1985) 617;
K S Stelle and E Sokatchev, Imperial College preprint (1986).

[9] L Alvarez-Gaumé, Commun. Math. Phys. 90 (1983) 161;
D Friedan and P Windey, Nucl. Phys. B235 (1984) 395;
E Witten, unpublished.

[10] E Witten, J. Diff. Geom. 17 (1982) 661.

[11] D Friedan and S Shenker, unpublished lectures at the Aspen Summer Institute (1984);
C Lovelace, Phys. Lett. 35B (1984) 75;
D Nemeschansky and S Yankielowicz, Phys. Rev. Lett. 54 (1985) 620.
E Bergshoeff, S Randjbar-Daemi, A Salam, H Sarmadi and E Sezgin, Nucl Phys B269 (1986) 77.

[12] E S Fradkin and A A Tseytlin, Phys. Lett. 158B (1985) 316 and Nucl. Phys. B261 (1985) 1;
A A Tseytlin, Lebedev Inst. preprints (1985,1986).

[13] A A Tseytlin, Phys. Lett. 178B (1986) 34.

[14] C G Callan, D Friedan, E Martinec and M Perry, Nucl. Phys. B262 (1985) 593.

[15] A Sen, Phys. Rev. 32D (1985) 2162, Phys. Ref. Lett. 55 (1985) 1846.

[16] P Candelas, G Horowitz, A Strominger and E Witten, Nucl. Phys. B258 (1985) 46.

[17] T Eguchi, P B Gilkey and A J Hanson, Phys. Rep. 66 (1980) 213.

[18] O Alvarez, Commun. Math. Phys. 100 (1985) 279.

[19] E Witten, Nucl. Phys. B223 (1983) 422.

[20] E Braaten, T Curtright and C Zachos, Nucl. Phys. B260 (1985) 630.

[21] R Rohm and E Witten, Ann. Phys. 170 (1986) 454.

[22] G Moore and P Nelson, Phys. Rev. Lett. 53 (1984) 1519; Commun. Math. Phys. 100 (1985) 83.

[23] B Zumino, Y S Wu and A Zee, Nucl. Phys. B239 (1984) 477;
B Zumino, in "Relativity Groups and Topology II", B S DeWitt and R Stora, eds., North Holland, Amsterdam (1984);
W Bardeen and B Zumino, Nucl. Phys. B244 (1984) 421.

[24] B Zumino, in "Supersymmetry and its Applications", G W Gibbons, S W Hawking and P K Townsend, eds., C.U.P. Cambridge (1986).

[25] J Bagger, D Nemeschansky and S Yankielowicz, Nucl. Phys. B262 (1985) 478.

[26] C M Hull, Phys. Lett. 167B (1986) 51.

[27] L Alvarez-Gaumé and D Z Freedman, Commun. Math. Phys. 91 (1983) 87.
S J Gates Jr., Nucl. Phys. B238 (1984) 349.

[28] C M Hull, Phys. Lett. 178B (1986) 357.

[29] T Curtright and C Zachos, Phys. Rev. Lett. 53 (1984) 1799;
P Howe and H Sierra, Phys. Lett. 148B (1984) 451;
R Rohm, Princeton preprint (1984).

[30] C M Hull, Nucl. Phys. B267 (1986) 266.

[31] P Howe and G Papadopoulos, King's College, London preprint (1986).

[32] E Bergshoeff and E Sezgin, ICTP Trieste preprint (1986).

[33] H W Braden and P H Frampton, Phys. Rev. Lett. 57 (1986) 2112.

[37] P Candelas, G Horowitz, A Strominger and E Witten, in "Proceedings of the Argonne Symposium on Anomalies, Geometry and Topology", (1985).

[35] I Bars, D Nemeschansky and S Yankielowicz, Nucl. Phys. B278 (1986) 632.

[36] A Strominger, Nucl. Phys. B274 (1986) 253.

[37] C M Hull in "Superunification and Extra Dimensions", R D'Auria and P Fre, eds., World Scientific, Singapore (1986).

[38] C M Hull in "Supersymmetry and its Applications", G W Gibbons, S W Hawking and P K Townsend, eds., C.U.P. Cambridge (1986).

[39] C M Hull and P K Townsend, Phys. Lett. 178B (1986) 187.

[40] K Yano, "Differential Geometry on Complex and Almost Complex Spaces", Pergamon, Oxford (1965).

[41] A Sen, Nucl. Phys. B278 (1986) 289; Phys. Lett. 166B (1986) 300; 174B (1986) 277.

[42] B S DeWitt, Phys. Rev. 162 (1967) 1195, 1239;
J Honerkamp, Nucl. Phys. B26 (1972) 130;
J Honerkamp, F Krause and M Schennert, Nucl. Phys. B69 (1974) 618;
L F Abbott, Nucl. Phys. B185 (1981) 189.

[43] L Alvarez-Gaumé, D Z Freedman and S Mukhi, Ann. Phys. 134 (1981) 85.

[44] S Mukhi, Nucl. Phys. B264 (1986) 640.

[45] D G Boulware and L S Brown, Ann. Phys. 138 (1982) 392.

[46] C M Hull and P K Townsend, DAMTP preprint (1987).

[47] C M Hull and P K Townsend, Nucl. Phys. B274 (1986) 349.

[48] G A Vilkovisky, Nucl. Phys. B234 (1984) 125.

[49] P Howe, G Papadopoulos and K S Stelle, Princeton IAS preprint (1986).

[50] C J Isham, in "Relativity, Groups and Topology II", B S DeWitt and R Stora, eds., North Holland, Amsterdam (1984).

[51] B Fridling and A E M van de Ven, Nucl. Phys. B268 (1986) 719.

[52] C M Hull and P K Townsend, in preparation.

[53] D J Gross and E Witten, Nucl. Phys. B277 (1986) 1.

[54] G Shore, Berne pre-print (1986).

[55] B Zumino, in "Lectures in Elementary Particle Physics and Quantum Field Theory", Proc. 1970 Brandeis Summer Institute, vol.2 (MIT Press, 1970).

[56] C G Callan, I R Klebanov and M J Perry, Nucl. Phys. B272 (1986) 111.

[57] P Breitenlohner and D Maison, in "Supersymmetry and its Applications", G W Gibbons, S W Hawking and P K Townsend, eds., C.U.P. Cambridge (1986).

[58] P Howe, G Papadopoulos and K S Stelle, Phys Lett 174 (1986) 405.

[59] L Alvarez-Gaumé and D Z Freedman, Phys Rev D15 (1980) 846.

[60] E Calabi in "Algebraic Geometry, A Symposium in Honour of S Lefschitz", Princeton University Press, (1957).

[61] S T Yau, Proc Natl Acad Sci, 74 (1977) 1798.

[62] M T Grisaru, A E M van de Ven and D Zanon, Phys. Lett. 173B (1986)423; Nucl. Phys. B277 (1986) 388,409.

[63] D Nemeschansky and A Sen, Phys. Lett. 178B (1986) 365.

[64] M B Green, J H Schwarz and E Witten, "Superstrings", C.U.P. Cambridge, (1987).

[65] A M Polyakov, Phys.lett. 103B, (1981) 207.

[66] D Friedan, in Proceedings of "Les Houches Summer School, 1982" J B Zuber and R Stora, eds. North Holland, Amsterdam, (1984).

[67] S R Das and B Sathiapalan, Phys. Rev. Lett. 56 (1986) 2664.

[68] R Brooks, F Muhammed and S J Gates Jr, Nucl. Phys. B268 (1986) 599;
P Nelson, G Moore and J Polchinski, Phys. Lett. 169B (1986) 599;
B A Ovrut and M Evans, Phys. Lett. 171B (1986) 177;
E Bergshoeff and E Sezgin, ICTP Trieste pre-prints (1985,1986).

[69] E S Fradkin and A A Tseytlin, Lebedev Inst. pre-print (1985).

[70] D J Gross, J H Harvey, E Martinec and R Rohm, Phys. Rev. Lett. 54 (1985) 502; Nucl. Phys.B256 (1985) 253; Nucl. Phys. B267 (1986) 75.

[71] M D Freeman, C N Pope, C M Hull and K S Stelle, CERN pre-print TH 4563/86.

[72] E Witten, Nucl. Phys. B268 (1986) 79.

[73] M T Grisaru, L Mezincescu and P K Townsend, Phys. Lett. 257B (1986) 247;
S J Gates Jr., M T Grisaru, L Mezincescu and P K Townsend, Brandeis pre-print (1986).

TORUS COMPACTIFICATION OF THE BOSONIC STRING AND THEIR SUPERSTRING CONTEXT

F. Englert

Université Libre de Bruxelles, Bruxelles
Campus Plaine, C.P. 225
Boulevard du Triomphe, 1050-Bruxelles, Belgium

1. INTRODUCTION

The string theory approach to the unification of gravity and matter is the outcome of the relativistic string introduced by Nambu[1], Nielsen[2] and Susskind[3] to interpret the Veneziano dual model[4]. Indeed, as explained below, the bosonic string theory requires for consistency a massless spin 2 excitation in the closed string sector. Thus the theory acquires in Hilbert space invariance under general coordinate transformations and reduces, in the low energy limit, to general relativity coupled to matter.

To understand the origin of the graviton, one may describe the string by the action of two dimensional general relativity coupled to 26 bosonic fields x^μ ($\mu=1,\cdots 26$). The x^μ,s, which may be interpreted as the string coordinates in a 26-dimensional flat space-time, render the 2-dimensional field theory conformally invariant at the quantum level[5]. Namely these fields cancel the conformal anomaly of the Fadeev-Popov ghost in the metric $g_{\alpha\beta} = \rho(\sigma,\tau)\eta_{\alpha\beta}$; $\eta_{\tau\tau} = -\eta_{\sigma\sigma} = 1$; $\eta_{\sigma\tau} = 0$. The equation of motion for the field operators may then be written in the Weyl gauge $g_{\alpha\beta} = \eta_{\alpha\beta}$. These equations are still invariant under the conformal diffeomorphisms generated by the Virasoro algebra[6]. One may use this additional gauge freedom to select a light-cone gauge in which the light-cone frame evolution coordinate-field x^+ coincides with τ. The physical fields are then the transverse coordinates x^I ($I = 1,\cdots 24$) and the states

$$\alpha^{(I}_{-1,L}\,\alpha^{S)}_{-1,R}\,|0_L;0_R\rangle \tag{1}$$

describe transverse spin 2 modes. Here $(1/\sqrt{|n|})\alpha^I_{-n,L(R)}$ are for $n > 0$ the creation operators for the left-(right-) moving n^{th} harmonic mode of the closed string and $|0_L\rangle$ ($|0_R\rangle$) the corresponding vacuum. The transverse spin 2 modes cannot be completed by vector helicity states because the states $\alpha^I_{-2,L}|0_L\rangle$ and $\alpha^S_{-2,R}|0_R\rangle$ having different left and right masses violate the Virasoro algebra for closed strings. Hence the states (1) describe a massless spin 2 particle. The resulting invariance under general coordinate transformation in 26-dimensions is thus rooted in the 2-dimensional conformal invariance described by the Virasoro algebra. This connexion becomes explicit in the second quantized version of the string theory[7].

Unfortunately, the emergence of general relativity from the bosonic string appears inevitably linked to the existence of tachyonic vacuum states because the massless graviton state described by (1) is an excited state. To remove the tachyon one may try to compensate the vacuum fluctuations of the bosonic excitations by introducing strings with fermionic degrees of freedom. One constructs in this way supersymmetric strings. Such "superstrings" are free of tachyons but nevertheless still reduce at low energy to (super)gravity coupled to (supersymmetric) matter. Again one may try to take for the superstring action conformal (super)gravity in two dimensions: the conformal anomalies of both gravitational and supersymmetry Fadeev-Popov ghosts are now cancelled by those of ten bosonic and ten fermionic Majorana fields[10]. Thus superstrings are expected to live in ten dimensions.

In the old formalism[8] the connextion between 2-dimensional supergravity and 10-dimensional supergravity appears however only after projecting out excitations which violate 10-dimensional supersymmetry[11]. The projection is automatically included in the new formalism[9] but the explicit Lorentz invariance of the theory is then lost. Thus, superstrings do not share the simple 2-dimensional characterization of the bosonic string. It is therefore of interest to inquire whether superstrings are not contained in a larger and more fundamental string theory.

Ten dimensional superstrings indeed emerge as solutions of the 26-dimensional closed bosonic string theory[12]. Namely we shall show that the states and the interactions defining a superstring form a subset of states and interactions of the bosonic string; this subset is fully determined by choosing for the vacuum an appropriate soliton state. The emergence of spacetime fermions and of supersymmetry, anticipated by Freund[13], is an impressive property of the bosonic string, whose apparent unphysical nature merely reflected the inadequacy of the tachyonic vacuum. The 26-dimensional bosonic string thus appears at a fundamental level as the only relevant string theory.

II. TORUS COMPACTIFICATION

Before presenting this result, we briefly summarize the torus compactification scheme[14] which will play an essential role in its derivation. The mass formula for an open string with r dimensions compactified on a maximal torus of a simply laced group* G is (choosing the slope parameter α' equal to 1/2)

$$\frac{m^2}{2} = \frac{\vec{p}_\gamma^{\,2}}{2} + N - 1; \qquad N = \sum_{n=1}^{\infty} \sum_{I=1}^{24} \alpha_{-n}^{I} \, \alpha_{n}^{+I} \tag{2}$$

where $\vec{p}_\gamma \in \Lambda_R$, the root lattice of G. In (2), the length squared of the roots $\vec{\beta}$ has been normed to 2.** Then all states of the compactified bosonic string fall into representations of the global symmetry group. An example is provided by the massless scalars

$$\alpha_{-1}^{a} |0,0\rangle \qquad a = 1,\cdots r \tag{3}$$

$$|0;\, \vec{p}_\gamma = \vec{\beta}\rangle \tag{4}$$

---

* A compact group G is called simply laced if all its roots have equal length. The simple simply laced groups are A_n, D_n, E6, E7, E8.

** One may extend the results of section II to the case $\vec{p}_\gamma \in \Lambda$ if Λ is an even lattice containing Λ_R as a sublattice.

where the vacuum is a direct product of the oscillator and of the compactified momentum vacua. Note that, for simplicity the space time momentum k ($k^2 = 0$) has been omitted in (3) and (4). The r states (3) and the (dim G-r) states (4) combine to form the adjoint representation of G. The global invariance of the free string spectrum can be generalized to the interacting theory by multiplying the usual tachyonic emission veritces by phase factors depending on the compactified momenta. The twisted amplitude in the tree approximation are then deducible from a given amplitude by a modified twist operator which commutes with the group generators, namely[15]

$$\Omega = (-1)^{N-1-\vec{p}_\gamma^{\,2}/2} \tag{5}$$

The twist operator (5), when applied to non-planar loops of open strings, yield a spectrum for closed string invariant under G. Moreover the G-invariance appears then as a local symmetry because the closed strong spectrum admits massless vector states belonging to the adjoint representation of G[15]. The mass formula for closed string is indeed

$$\frac{m^2}{4} = \left(\frac{\vec{p}_\gamma^{\,2}}{4} + \vec{L}^2\right) + N_L + N_R - 2 , \tag{6}$$

where $N_{L(R)}$ is given in (2) with α_{-n} replaced by $\alpha_{-n,L(R)}$. Here $\vec{L}$ is a winding vector which satisfies the closed string Virasoro constraint,

$$\vec{p}_\gamma \cdot \vec{L} = N_R - N_L \tag{7}$$

The solutions of Eq. (7) consistent with the twist (5) are[15]

$$\vec{L} = \vec{W} + \frac{\vec{p}_\gamma}{2} \tag{8}$$

where $\vec{W}$ belongs to the weight lattice $\Lambda_W = \Lambda_R^+$. It is now easily verified that the massless vector counterpart of the graviton states (1), namely

$$\alpha^i_{-1,L}\, \alpha^a_{-1,R} | 0_L;\ \ 0_R;\ \ \vec{p}_\gamma = 0,\ \vec{W} = 0 \rangle ,$$

$$[i = 1, \cdots 24-r],\ [a = 1, \cdots r] . \tag{9}$$

$$\alpha^i_{-1,L} | 0_L;\ \ 0_R;\ \ \vec{p}_\gamma = \vec{\beta},\ \vec{W} = 0 \rangle \tag{10}$$

belong to the adjoint representation of G. In fact the gauge group acting on the closed string sector is G × G for non-oriented strings.

From (8), we define left and right weights $\vec{W}_L$, $\vec{W}_R$ satisfying

$$\vec{p}_\gamma = \vec{W}_R - \vec{W}_L \tag{11}$$

$$\vec{L} = (\vec{W}_R + \vec{W}_L)/2 \tag{12}$$

It then follows from (6) and (7) that

$$\frac{1}{8} m^2_{L(R)} = \frac{1}{2} \vec{W}^2_{L(R)} + N_{L(R)} - 1 \tag{13}$$

with

$$m_L^2 = m_R^2 = m^2 \tag{14}$$

We then rewrite (9) and (10) in the form

$$\alpha^i_{-1,L}\ \alpha^a_{-1,R}\left|\Omega_L;\ \Omega_R\right> \tag{15}$$

$$\alpha^i_{-1,L}\left|\Omega_L;\ \vec{W}_R=\vec{\beta}\right> \tag{16}$$

where $\left|\Omega_{L(R)}\right>$ are vacua both with respect to oscillators and compactified momenta. Eq. (13) shows that closed strings behave like two open strings with momenta quantized in the weight lattice, subject to the mass constraint (14) and the group constraint (11). Using (13), (14) and (11), it is straightforward to verify that loop amplitudes of closed strings are modular invariant; this invariance is essential for the consistency of a theory of closed strings.

The group constraint (11) which is essential for the proof of modular invariance simply means that the left-moving and right-moving components of a physical state must belong to the same conjugation class of G. One can formulate this more elegantly by considering $(\vec{W}_L, \vec{W}_R)$ as a vector on the lattice $\Lambda_{left} \times \Lambda_{right}$ with a Lorentzian metric with signature (+•••+, -•••-). One then considers a sublattice, obtained by removing all points for which $\vec{W}_L$ and $\vec{W}_R$ do not belong to the same conjugation class. As was pointed out by Narain[16], this sublattice is even and self-duel with respect to the Lorentzian metric.

Thus the closed string sector can be compactified on the local gauge group G × G when the open strings are globally compactified on G. For theories with only closed strings we may contemplate more general possibilities, such as compactification on G × G' (G ≠ G'). A minimal requirement is modular invariance at the level of the closed string one-loop amplitudes. Then in addition to G × G with the constraint (11) one can choose a Lorentzian lattice $\Lambda_{left} + \Lambda'_{right}$ where Λ and Λ' are even, self-dual Euclidean lattices. The constraint (11) has no meaning in this case, since it was only derived for compactification of left and right sectors on the weight lattice of the same group G. Instead, the modular invariance is in this case an immediate consequence of the one-loop calculations in Refs. 15, 12 and 16. Even, self-dual Euclidean lattices exist only in dimension 8n, the lowest-dimensional examples being E_8 for r = 8 and $E_8 \times E_8$ or $Spin(32)/Z_2$ for r = 16. This yields two new closed string compactifications, namely,

$$(Spin(32)/Z_2)_L \times (Spin(32)/Z_2)_R,\ (Spin(32)/Z_2)_L \times (E_8 \times E_8)_R \tag{17}$$

[Since the root lattice of E_8 is identical to its weight lattice, the cases $(E_8)_L \times (E_8)_R$ in r = 8 and $(E_8 \times E_8)_L \times (E_8 \times E_8)_R$, in r = 16 are of the general G × G type. Notice that we have now two lattices which lead to an SO(32) × SO(32) Lie-algebra in 16 dimensions.] For r > 16 it is still possible to choose G'≠G when $(\vec{W}_R, \vec{W}_L)$ form a more general self-dual Lorentzian lattice.[16]

To conclude this section we point out that it is possible to generalize the above construction to accommodate closed string compactification on subgroups of the simply laced group G. The subgroup $\bar{G}$ G must be chosen in such a way that the Fock space of the interacting string, generated by the

$\bar{G}$ covariant vertices only, form an irreducible representation of the original Virazoro algebra. This leads to the condition

$$r = j \dim \bar{G}/(j+g) \tag{18}$$

where j is the Dynkin index of the imbedding of $\bar{G}$ into G and of its (suitability normed) Casimir invariant. As $\bar{G}$ need not be simply laced, corresponding solutions of Eq. (18) for $r < 24$ provides a way to compactify the interacting bosonic string on non-simply laced groups.[17]

III. SPACE-TIME FERMIONS AND SUPERSYMMETRY

In addition to the chiral N = 1 open and closed superstrings, which require the SO(32) Chan-Paton factors for anomaly cancellation[18] there exist four consistent closed superstring theories in 10 dimensions. These describe non-chiral and chiral anomaly-free[19] N = 2 superstrings and N = 1 heterotic strings with internal symmetry group $Spin(32)/Z_2$ or $E_8 \times E_8$.[20] Heterotic closed superstrings in less than 10 dimension have been obtained by Narain using the torus compactification on more general (Lorentzian) self-dual lattices.[16] To construct all these superstrings from the bosonic string we shall impose three requirements.

a) To generate space-time fermionic states in d+2 dimensions we shall compactify r = 24-d transverse dimensions. In this way, the continuum of bosonic zero modes is removed.

b) The internal group G resulting from the compactification will contain as subgroup the group Spin(d) isomorphic to the covering of the transverse group SO(d) of the non-compactified dimensions. If the diagonal subgroup $\widetilde{SO}(d)$ of Spin(d) × SO(d) can be identified with a new transverse group, spinor representations of $\widetilde{SO}(d)$ will describe fermionic states because a rotation in space will induce a half angle rotation on these states. The consistency of the above procedure depends critically on the possibility of extending the algebra $\widetilde{so}(d)$ to the full Lorentz algebra $\widetilde{so}(d+1,1)$. This leads to a third requirement.

c) To break the old Lorentz group in 26 dimensions in favour of the new one a consistent truncation must be performed on the spectrum of the bosonic string. This truncation is in fact also crucial for generating the supersymmetric vacuum, thereby removing the tachyon. Indeed supersymmetry in D = d+2 dimensions is apparently linked to a super Virasoro algebra rooted in a local supersymmetry of the world sheet. Hence, as will be discussed later on, some bosonic degrees of freedom are expected to decouple from the physical transverse states to account for the conformal anomaly of the unphysical Fadeev-Popov superghost. More precisely in a supersymmetric sector of the closed string, the states involving p bosonic operators pertaining to the r compactified dimensions must decouple (except possibly for some zero modes) where p is determined by the cancellation of two-dimensional conformal anomalies. In units such that the conformal anomaly is -1 for a scalar and -1/2 for a Majorana fermion, the supersymmetry ghost has an anomaly -11.[20] Matching the anomaly of the superghost and of the unphysical longitudinal and time-like Majorana fermions to the anomaly of the decoupled bosonic states yields the prediction $p(-1) = -11 + 2(-1/2)$, hence p = 12. To ensure a consistent decoupling, we require Spin(d) to belong to a regular embedding of G so that the root lattice of Spin(d) is contained in the root lattice of g. Hence d is even and the d/2 bosonic dimensions generating Spin(d) belong to a subset of their dimensions generating G. As $p + (d/2) \leq r = 24-d$, we get when p = 12,

$$d \leq 8 \tag{19}$$

The highest available value of d will correspond to ten-dimensional superstring and the lower values to Narain strings.

We first show that the zero mass states for the ten-dimensional heterotic strings are contained in the spectrum of the compactified bosonic string.

To remove the group constraint (11) we restrict the compactification of the closed bosonic string to self-dual groups. Taking the left sector to be the supersymmetric one we can then choose, according to requirement a) with $r = 24-8 = 16$, the group G to be $E_8 \times E_8$ or $Spin(32)/Z_2$. However, requirement b) select uniquely $E_8 \times E_8$. Indeed the regular imbeddings of the algebra in these two groups are given by

$$Spin(32)/Z_2 : (496) \to (28) + 24\,(8_i) + \text{singlets} \tag{20}$$

$$E_8 : (248) \to (28) + 8(8_v + 8_s + 8_c) + \text{singlets} \tag{21}$$

Thus only E_8 contains the covering of SO(8) as a regular imbedding.

To perform the decoupling imposed by requirement c) we consider the branching rule

$$E_8 \to Spin(16) \to Spin^1(8) \times Spin^2(8) \tag{22}$$

and keep only the $Spin^1(8)$ subgroup of one E_8, thereby decoupling (except for the zero-modes discussed below) the other E_8 and $Spin^2(8)$. This indeed yields $p = 8+4 = 12$. To understand the zero mode problem, we consider the decomposition of (248) according to (22). We get

$$[248] \xrightarrow[Spin(16)]{} [120] + [128] \xrightarrow[Spin^1(8)\times Spin^2(8)]{}$$

$$[(28,1) + (1,28) + (8^1_c,8^2_c)] + [(8^1_v,8^2_v) + (8^1_s,8^2_s)] \tag{23}$$

To obtain 8^1_c, 8^1_v and 8^1_s as roots of E_8 we must fix unit vectors $\vec{\eta}_c$, $\vec{\eta}_v$, $\vec{\eta}_s$ in 8^2_c, 8^2_v, 8^2_s. This removes the eight-fold degeneracy of (21) to the root vectors of E_8 of square length two. Actually one may choose the "hypercharge vectors $\vec{\eta}$ to be in a plane (and hence to form an equilateral triangle) because the [128] must reproduce itself under the action of the Spin(16) generators belonging to the [120]. We thus define E_8-roots by

$$\vec{\beta}^{\pm}_{c,(v),(s)} = \vec{\alpha}_{c,(v),(s)} \pm \vec{\eta}_{c,(v),(s)} \tag{24}$$

One then easily verifies that the following zero mass states

$$\alpha^i_{-1,R}|\vec{\beta}^+_v \, ; \, 0_R> \tag{25}$$

$$\alpha^i_{-1,R}|\vec{\beta}^+_s \, ; \, 0_R> \tag{26}$$

form a supergravity multiplet in ten dimensions while

$$\alpha^a_{-1,R}|\vec{\beta}^+_v \, ; \, 0_R> \quad ; \quad |\vec{\beta}^+_v \, ; \, \vec{\beta}> \tag{27}$$

$$\alpha^a_{-1,R}|\vec{\beta}^+_s \, ; \, 0_R> \quad ; \quad |\vec{\beta}^+_s \, ; \, \vec{\beta}> \tag{28}$$

form the matter supermultiplet of gauge groups given by the roots $\vec{\beta}$. The latter can be chosen to be $E_8 \times E_8$ or $Spin(32)/Z_2$. However this identification is valid if and only if the states (26) and (28) describe space-time fermions. This will be the case if the diagonal SO(8) algebra closes to SO(9,1). This is indeed the case[12].

To get the full spectrum of excitation one can use the vertex opeators of the bosonic string factorized into states of fixed hypercharge $\pm\vec{\eta}_c$. A simpler procedure is to consider the E_8 partition function. If we ignore the contributions of the transverse oscillators α^i_{-n} and the second E_8 group (which are trivial to take into account), the partition function for the E_8 excitations is[21]

$$P_{E_8}(q) = \frac{1}{2q}\left[\prod_{n=1}^{\infty}(1+q^{n-1/2})^{16} + \prod_{n=1}^{\infty}(1-q^{n-1/2})^{16}\right] + 128\prod_{n=1}^{\infty}(1+q^n)^{16} \tag{29}$$

This may be recognized as the partition function of a 16-dimensional NSR model with a Gliozzi-Scherk-Olive projection[11] removing positive G-parity states. Indeed, this function has a simple interpretation in terms of the SO(16) subgroup of E_8; the first term gives the number of "bosonic" SO(16) states, while the second term gives the number of fermionic ones at each level. We can factorize this expression so that it has an $SO(8)^1 \times SO(8)^2$ interpretation:

$$P_{E_8}(q) = \left\{\frac{1}{2\sqrt{q}}\left[\prod_{n=1}^{\infty}(1+q^{n-1/2})^8 + \prod_{n=1}^{\infty}(1-q^{n-1/2})^8\right]\right\}^2$$

$$+ \left\{\frac{1}{2\sqrt{q}}\left[\prod_{n=1}^{\infty}(1+q^{n-1/2})^8 - \prod_{n=1}^{\infty}(1-q^{n-1/2})^8\right]\right\}^2 + 2\left[8\prod_{n=1}^{\infty}(1+q^n)^8\right]^2 \tag{30}$$

The first term corresponds to the product of two eight-dimensional Neveu-Scharz models with only odd G-parity states. In each factor these states consist of the tachyon and everything created from it by an <u>even</u> number of NS operators. None of the states in $SO(8)^2$ obtained this way is an SO(8) vector, so that all these states are eliminated by our truncation. The second term has no poles at $q = 0$, and is therefore free of tachyons. Here we have the even G-parity states of two NS models. Our truncation prescription is to keep only states with $SO(8)^2$ weight $\vec{\eta}$ and not to allow Cartan subalgebra excitations in $SO(8)^2$. This implies that only the massless ground state of the second factor is kept, so that we should take the limit $q \to 0$ in this factor. Furthermore we must divide by 8, because we keep only one of the eight components of the ground state. The third term corresponds to two Ramond models. Again we take $q = 0$ in the second factor, divide by eight, and then by two because we keep only one of the two hypercharge vectors $\vec{\eta}_s$ and $\vec{\eta}_c$. After these operations the partition function of the truncated model is

$$P^{trunc}_{E_8}(q) = \frac{1}{2\sqrt{q}}\left[\prod_{n-1}^{\infty}(1+q^{n-1/2})^8 - \prod_{n=1}^{\infty}(1-q^{n-1/2})^8\right] + 8\prod_{n=1}^{\infty}(1+q^n)^8 \tag{31}$$

This is indeed the partition function of the spinning string or the superstring.

The spectrum of all closed ten-dimensional superstrings can now be obtained from the 26-dimensional bosonic string in a straightforward way. To get the heterotic strings, it suffices to combine the left supersymmetric sector contained in the $E_8 \times E_8$ compactification after projecting out the irrelevant states with a right sector compactified either on $E_8 \times E_8$ or on $Spin(32)/Z_2$. To get the N = 2 superstrings one compactifies both sectors on $E_8 \times E_8$ and performs the projection on both sides. In this way, one gets indeed both the chiral and the vector-like N = 2 theories as the identification of the fermionic ground states to $(8_s, \vec{\eta}_s)$ or to $(8_c, \vec{\eta}_c)$ can be done independently in the right and in the left sectors.*

One can also obtain heterotic superstrings in less than ten dimensions[16] by compactifying the ten-dimensional one on the torus of a simply laced group G of rank m. To this effect one uses m of the eight transverse <u>bosonic</u> operators in both sectors but one can get of course massless gauge vectors only in the right sector. In this case the gauge group becomes $E_8 \times E_8 \times G$ or $Spin(32)/Z_2 \times G$. The self-dual Lorentzian lattice is $E_8 \times E_8 \times G \times G$ with the constraint (11) on $G \times G$. More general self-dual Lorentzian lattices can be used and can generate new gauge groups in $d < 8$ dimensions[16]. Such superstrings, which are not necessarily compactifications of the ten-dimensional heterotic string, are always contained in the original 26-dimensional bosonic string: one may simply extend any of the lattices used in ref. 16 with $E_8 \times E_8$ lattice. The new lattice (which is obviously even, Lorentzian self-dual) is then used for compactification of the bosonic string, and one obtains a superstring sector from the additional $E_8 \times E_8$.

Finally, one can show that the interaction of the superstrings are also contained in the interacting 26-dimensional bosonic string theory[12].

To conclude this section we first discuss the possible dynamical mechanism by which the decoupling of the erstwhile physical states and their conversion into unphysical states could actually be realized. We then comment upon the possible relevance of the "unified" description of superstrings and the bosonic string.

It is clear that the light-cone gauge is not a suitable framework for handling the problem since it describes only physical states. Rather one must resort to a covariant formulation of the theory in which all unphysical degrees of freedom are kept. In such a formulation, the physical subspace is defined by the condition[22]

$$Q_{BRS}|\text{phys}\rangle = 0 \qquad (32)$$

where Q_{BRS} is the BRS-charge operator which obeys

$$Q^2_{BRS} = 0 \qquad (33)$$

The formalism is the same for ordinary strings and superstrings [see, e.g., Ref. 23 and references therein] with the only difference being that the nilpotency of Q_{BRS} requires D = 26 and D = 10, respectively, with the corresponding values of the intercepts; these conditions are again equivalent to the vanishing of the conformal anomaly on the world-sheet. By imposing (32) one also defines and eliminates the unphysical states. We now conjecture that the BRS-charge of the D = 26 theory is actually background dependent and will therefore change as one moves away from the tachyonic vacuum (by "background" we here mean possible vacuum expectation values for all the higher excitations for the strings and not just the D = 26 gravitational background). To see how this is possible we recall that the full action in covariant string field theory is[7]

$$S = \phi\, Q_{BRS}\phi + \cdots \qquad (34)$$

---

* One would also obtain in this way open (and closed) type-I superstring if one allowed for suitable Chan-Paton factors.

where ϕ is the fundamental string field (containing infinitely many ordinary fields of arbitrarily high spin) and the dots stand for possible interaction terms. Observe that, in (34), Q_{BRS} plays the role of the kinetic operator. At a non-trivial soliton-like solution of the equations of motion, which follow from (34), ϕ will acquire a vacuum expectation value (possibly corresponding to vacuum expectation values for infinitely many ordinary fields). Shifting to the new vacuum, we see that Q_{BRS} is also modified through the interaction terms such that (34) becomes

$$S = \phi' \tilde{Q}_{BRS} \phi' + \cdots \tag{35}$$

with a new "kinetic operator" $\tilde{Q}_{BRS}$. The nilpotency condition (33) for this new operator, which is necessary for consistency, puts non-trivial restrictions on the possible soliton-like solutions of the D = 26 theory.* While the actual solution that corresponds to the superstring remains to be constructed, these arguments strongly point towards its existence.

Assuming this to be the case, what would be the interest of such a result? Clearly if the correct ultimate unified theory would be a superstring theory, the possibility of defining a larger theory of which it would only be a sector would be physically irrelevant, even if in this larger theory superstring could be dynamically reached. However, it is hard to believe that under unusual circumstances, as, for instance the very high temperatures of the early world, the supersymmetric sector would not communicate with other possible sectors of the larger theory. More important perhaps, there is no convincing evidence that supersymmetric string and more generally string theory do solve the quantum gravity puzzles revealed by quantum cosmology, at least in the present formulation of the theory. To make further theoretical progress, it may therefore be useful to use as a laboratory, instead of superstrings, the possibly more fundamental bosonic strings.

References

1. Y. Nambu, Proc. Int. Conf. on Symm. and Quark Models, Wayne State University (1969), (Gordon & Breach, 1970).
2. H.B. Nielsen, 15th Int. Conf. on High Energy Physics, Kiev (1970).
3. L. Susskin, Nuovo Cimento 69A(1970)457.
4. G. Veneziano, Il Nuovo Cimento 57A(1968)190.
5. A.M. Polyakov, Phys. Lett. 103B(1981)207.
6. M.A. Virasoro, Phys. Rev. D1(1970)2933.
7. W. Siegel, Phys. Lett. 149B(1985)162; 151B(1985)391,396; A. Neveu, H. Nicolai and P.C. West, CERN preprint TH 4233/85 (1985); T. Banks and M.E. Peskin, SLAC-PUB-3740 (1985); E. Witten, Nucl. Phys. B268(1986)253.
8. P. Ramond, Phys. Rev. D3(1971)2415; A. Neveu and J.H. Schwarz, Nucl. Phys. B31(1971)86.
9. M.B. Green and J.Y. Schwarz, Phys. Lett. 109B(1982)444.
10. A.M. Polyakov, Phys. Lett. 103B(1981)211.
11. F. Gliozzi, J. Scherk and D. Olive, Phys. Lett. 65B(1976)282; Nucl. Phys. B122(1977)253.
12. A. Casher, F. Englert, H. Nicolai and A. Taormina, Phys. Lett. 162B (1985)121; F. Englert, H. Nicolai and A. Schellekens, CERN preprint TH-4360/86 (1986).
13. P.G.O. Freund, Phys. Lett. 151B(1985)387.
14. J.B. Frenkel and V.G. Kac, Inv. Math. 62(1980)23; P. Goddard and D. Olive, DAMPT preprint 83/22 (1983); T. Eguchi and K. Higashijima, "Recent Developments in Quantum Field Theory", Eds. J. Ambjörn, B.J. Durhuus and J.L. Petersen, (North Holland, 1985) pp.57-65.

* It is perhaps instructive to note similarities with 't Hooft's anomaly matching conditions[24].

15. F. Englert and A. Neveu, Phys. Let. 163B(1985)349.
16. K.S. Narain, Rutherford preprint, RAL-85-097 (1985).
17. F.A. Bais, F. Englert, A. Taormina and P. Zizzi, ULB preprint TH86/05 (1986).
18. M.B. Green and J.H. Schwarz, Phys. Lett. 149B(1984)117; 151B(1985)21.
19. L. Alvarez-Gaumé and E. Witten, Nucl. Phys. B234(1983)269.
20. D.J. Gross, J.A. Harvey, E. Martinec and R. Rohm, Phys. Rev. Lett. 54(1985)502; Princeton University preprint (1985).
21. V.G. Kac, Proc. Nat. Acad. Sc. USA 77(1980)5048.
22. M. Kato and K. Ogawa, Nucl. Phys. B212(1983)443.
23. J.H. Schwarz, preprint CALT-68-1304 (1985).
24. G. 't Hooft, in "Recent Developments in Gauge Theories", Cargèse 1979, Eds. G. 't Hooft et al., (Plenum, New York and London).

HIDDEN SYMMETRY OF TWO DIMENSIONAL σ-MODEL DEFINED ON SYMMETRIC COSET SPACE

K.C. Chou

Institute of Theoretical Physics
Academia Sinica
Beijing, China

ABSTRACT

The hidden symmetry of two dimensional σ-model defined on symmetric coset space and its integrability properties such as the existence of infinite numbers of conservation laws are reviewed. Chiral models with Wess-Zumino-Witten terms are shown to have similar behaviors.

0. INTRODUCTION

Nonlinear σ-models and their effective actions are of particular interest in recent years.[1-10] They are useful not only in the context of chiral dynamics but also in the geometrical interpretation and possibly compactification of some superstring theories.

It was pointed out by Witten that Wess-Zumino term arising from the quantum anomalies is of great importance in the theory of two dimensional chiral models.[11] A zero mass free fermion theory is shown to be equivalent to a chiral model with special Wess-Zumino term in two dimensions.

The integrability of some of the σ-models in two dimensions is related to the existence of an Abelian hidden symmetry depending on a continuous parameter ξ. It leads immediately to infinite numbers of conservation laws, linear equation of Lax pair and Bäcklund transformation. Most of the nonlinear σ-models having these properties are defined on symmetric coset space. Chiral models with or without Wess-Zumino anomaly term belong to this category.

The first part of this talk is devoted to the hidden symmetry and the related integrability properties for models without the anomaly term. It will be generalized to chiral models with the Wess-Zumino term in the second part. Application to some supersymmetric theories will be briefly discussed in the third part.

I. HIDDEN SYMMETRY OF A TWO DIMENSIONAL σ-MODEL DEFINED ON A SYMMETRIC COSET SPACE

Consider a group G with a subgroup H. A nonlinear σ-model can be defined on the coset manifold G/H with its local coordinates as the field variables. Any group element $g \in G$ can be decomposed into a product of two elements $\phi \in G/H$ and $h \in H$,

$$g = \phi h \,. \tag{1.1}$$

The field of the nonlinear σ-model is then described by a function $\phi(x)$ G/H. Under a group transformation the element $g \cdot \phi(x)$ can again be decomposed in the form of Eq. (1.1).

$$g\phi(x) = \phi(g,\phi(x))h(g,\phi(x)) \tag{1.2}$$

where $\phi(g,\phi(x)) \in G/H$ and $h(g,\phi(x)) \in H$ satisfying the group multiplication law

$$\phi(g',\phi(g,\phi(x))) = \phi(g'g,\phi(x)) \tag{1.3}$$

$$h(g',\phi(g,\phi(x)))h(g,\phi(x)) = h(g'g,\phi(x)) \tag{1.4}$$

They form nonlinear representations of the group G. The field $\phi(x)$ then transforms as

$$\phi(x) \to \phi'(x) = g\phi(x)h^{-1}(g,\phi(x)) = \phi(g,\phi(x)) \tag{1.5}$$

The 1-form $\phi^{-1}(x)d\phi(x)$ is valued in the Lie algebra of the group G and can be decomposed into two parts

$$\phi^{-1}d\phi = H(x) + K(x) \tag{1.6}$$

where $H(x)$ is a 1-form valued in the Lie algebra of the subgroup H and $K(x)$ that of the coset G/H. They transform under Eq. (1.5) in the following way

$$H \to H' = h(g,\phi(x))\left[d + H + (\phi^{-1}(g^{-1}dg)\phi)_H\right]h^{-1}(g,\phi(x)) \tag{1.7}$$

and

$$K \to K' = h(g,\phi(x))\left[K + (\phi^{-1}(g^{-1}dg)\phi)_K\right]h^{-1}(g,\phi(x)) \tag{1.8}$$

For a global transformation $g^{-1}dg = 0$, $H(x)$ behaves as a gauge field in the subgroup and $K(x)$ as a vector field.

The action of the nonlinear σ-model is usually chosen to be

$$S = \frac{1}{2\lambda}\int tr(K{*}K) \tag{1.9}$$

where $*K$ is the Hodge star of K. In two dimensions

$$*K = *K_\mu dx^\mu = \varepsilon_{\mu\nu}K^\nu dx^\mu \tag{1.10}$$

From Eq. (1.8) it is easily seen that the action is invariant under the global transformation of G. The corresponding current can be easily found to be

$$J(x) = \frac{-i}{\lambda}\,\phi(x)K(x)\phi^{-1}(x) \tag{1.11}$$

The equation of motion is

$$D{*}K = d{*}K + H{*}K + {*}KH = 0 \tag{1.12}$$

from which the conservation of the current follows

$$d{*}J = 0\ . \tag{1.13}$$

The current J(x) can also be written in the form

$$J(x) = \frac{i}{\lambda} \phi(x)(d + H(x))\phi^{-1}(x) \tag{1.14}$$

Under a local transformation the current becomes

$$J(x) \to J'(x) = g(x)\big(d + J(x)\big)g^{-1}(x)$$

$$+ \frac{i}{\lambda} g(x)\phi(x)\left[\phi^{-1}(x)(g^{-1}dg)\phi(x)\right]_H \phi^{-1}(x)g^{-1}(x) \tag{1.15}$$

It transforms like a vector in the adjoint representation under a global g.

Write

$$J(x) = \frac{i}{2\lambda} A(x) \tag{1.16}$$

one easily finds from Eqs. (1.13), (1.11) and (1.6) that

$$dA = \frac{1}{2} A^2 - 2\phi F_H \phi^{-1} \tag{1.17}$$

where

$$F_H = dH + H^2$$

is the field strength of the gauge field H.

From now on we shall restrict ourselves to the two dimensional σ-model defined on a symmetric coset space, where the commutator of two generators in the coset lies in the Lie algebra of the subgroup. In this case we find from the pure gauge nature of $\phi^{-1}d\phi$ the following

$$F_H = dH + H^2 = -K^2 \tag{1.18}$$

$$dK + KH + HK = 0 \tag{1.19}$$

Substituting Eq. (1.18) into Eq. (1.17) we obtain

$$dA = A^2 \tag{1.20}$$

The potential A(x) is also a conserved quantity

$$d{*A} = 0 \tag{1.21}$$

In two dimensions we always have

$$B{*C} + {*B}C = 0$$

$$*({*B}) = B \tag{1.22}$$

for arbitrary 1-forms B and C. Using this fact we write together with the equation of motion and the compatibility conditions

$$d{*K} + {*K}H + H{*K} = 0 \tag{1.23}$$

$$dK + KH + HK = 0 \tag{1.24}$$

$$dH + H^2 = -K^2 = -\frac{1}{2}(K^2 - {*K}^2) \tag{1.25}$$

These equations show us an explicit dual symmetry between K and *K. Therefore any linear transformation that makes $K^2 - {*K}^2$ invariant is a good symmetry of Eqs. (1.23) - (1.25).

This dual symmetry first discovered in this way is a hidden symmetry of the field involved. It has the form

$$K \to K' = \text{ch}\xi K + \text{sh}\xi {*K} \tag{1.26}$$

It is an Abelian symmetry with the group parameter ξ added under successive transforations. The action Eq. (1.9) is also invariant under this transformation.

In order to get the transformation law for the field $\phi(x)$ we construct a linear equation

$$\begin{aligned} g^{-1}dg &= H + K' \\ &= H + \text{ch}\xi K + \text{sh}\xi {*K} \end{aligned} \tag{1.27}$$

The integrability condition of the above equation is guaranteed by the hidden symmetry that makes H + K' a pure gauge potential.

Write $g = \psi\cdot\phi$ in Eq. (1.27) we obtain from Eqs. (1.6), (1.11) and (1.16) that

$$\begin{aligned} \psi^{-1}d\psi &= \phi[(\text{ch}\xi-1)K + \text{sh}\xi{*K}]\phi^{-1} \\ &= \frac{1}{2}(\text{ch}\xi-1)A + \frac{1}{2}\text{sh}\xi{*A} \end{aligned} \tag{1.28}$$

Integrability of Eq. (1.28) can be verified directly from the conservation equation and the compatibility condition (1.20)

$$d{*A} = 0, \qquad dA = A^2 \tag{1.29}$$

It is possible to define a conserved current

$$j(\xi,A) = {*d}\psi(\xi,A) \tag{1.30}$$

so that

$$d{*j}(\xi,A) = 0$$

Expanding $j(\xi,A)$ as a series in ξ each term will be a separate conserved current. Hence we get an infinite number of conserved currents. This set of conserved currents was first obtained by Berezin et al.[2] in a slightly different way. However they are not the Noether currents of the σ-model. When Eq. (1.28) has been solved the transformed field $\phi'(x)$ is found to be

$$\phi(x) \to \phi'(x) = \psi(\xi,A)\phi(x)h^{-1}(\psi,\phi) = \phi(\psi(\xi,A)\phi) \tag{1.31}$$

The Noether current corresponding to this Abelian transformation is

$$J_\xi(x) = \frac{-i}{\lambda}\,\text{tr}(A(x)\int^x {*A}(y)) \tag{1.32}$$

which is conserved.

After a ξ-transformation the Noether current J(x) becomes

$$J(x) \to J(\xi,x) = \psi(\xi,A)[\text{ch}\xi J(x) + \text{sh}\xi{*J}(x)]\psi^{-1}(\xi,A) = \frac{-i}{\lambda}A(\xi,x) \tag{1.33}$$

$J(\xi,x)$ is also a conserved current depending on a continuous parameter ξ. Infinite numbers of conserved currents then follow by expansion in ξ. It is easily proved that

$$dA(\xi,x) = A^2(\xi,x), \quad d{*}A(\xi,x) = 0 \tag{1.34}$$

By successive transformations we have

$$\psi(\eta,A(\xi,x))\psi(\xi,A(x)) = \psi(\xi+\eta,A(x)) \tag{1.35}$$

Therefore $\psi(\xi,A)$ forms a nonlinear representation of the Abelian group.

Let the operator in the Hilbert space that generates the ξ-transformation be

$$U(\xi) = \exp\{\xi Q_\xi\} \tag{1.36}$$

where Q_ξ is the charge of the current (1.32). We have

$$U(\xi)\phi(x)U^{-1}(\xi) = \phi(\psi(\xi,A),\phi(x)) \tag{1.37}$$

Write the operator generating transforamtion of the group G as

$$U(\lambda_a,Q_a) = \exp\{i\lambda_a Q_a\} \tag{1.38}$$

with λ_a and Q_a the group parameters and the corresponding charges respectively. It is evident that

$$U(\lambda_a,Q_a(\xi))U(\xi) = U(\xi)U(\lambda_a,Q_a) \equiv U(\xi,\lambda_a) \tag{1.39}$$

where

$$Q_a(\xi) = U(\xi)Q_a U^{-1}(\xi) \tag{1.40}$$

is the ξ-transformed charges.

Consider now an infinitesimal transformation in G

$$U(\xi,\lambda_a) = U(\xi) + i\lambda_a M_a(\xi) \tag{1.41}$$

Then $M_a(\xi)$ will satisfy the commutation relations

$$[M_a(\xi),\ M_b(\eta)] = if_{abc}M_c(\xi+\eta) \tag{1.42}$$

where f_{abc} are the structure constants of the group G. Expanding $M_a(\xi)$ in power series in ξ

$$M_a(\xi) = \sum_h \frac{1}{n!}\,\xi^n M_{na} \tag{1.43}$$

one easily deduces from Eq. (1.42) that

$$[M_{ma},M_{mb}] = if_{abc}M_{n+m,c} \tag{1.44}$$

This is a Kac-Moody algebra. It may have center terms when quantum anomalies are present. We shall not discuss this point further.

It is noted that the charges from the Noether currents $J(\xi)$ satisfy an algebra at equal ξ

$$[Q_a(\xi),Q_b(\xi)] = if_{abc}Q_c(\xi) \tag{1.45}$$

which is different from that of $M_a(\xi)$. These Taylor expansion coefficients do not satisfy a Kac-Moody algebra.

When there are no anomalies coming from the quantum loops it is possible to solve the linear equation (1.28) by the inverse scattering method and find the S-matrix of the system. Since it is a long story we shall stop here giving only the symmetry properties. The interested reader can consult the original literature.[5,16,17]

II. CHIRAL MODELS WITH WESS-ZUMINO TERMS[11-15]

A chiral model is defined in the coset space $G_L \times G_R/G_{L+R}$ for some general group G. The group element can be represented by a matrix

$$\begin{pmatrix} g_L & 0 \\ 0 & g_R \end{pmatrix} \tag{2.1}$$

with the coset element

$$\begin{pmatrix} \phi & 0 \\ 0 & \phi^{-1} \end{pmatrix} \tag{2.2}$$

and the subgroup element

$$\begin{pmatrix} h & 0 \\ 0 & h \end{pmatrix} \tag{2.3}$$

Therefore

$$g_L = \phi h, \quad g_R = \phi^{-1}h \tag{2.4}$$

The chiral field is defined to be

$$g(x) = \phi^2(x) \tag{2.5}$$

that transforms under $G_L \times G_R$ as

$$g(x) \to g'(x) = g_L g(x) g_R^{-1} \tag{2.6}$$

It was pointed out by Wess and Zumino[16] long ago that additional terms for the action may arise due to the anomalies appearing on field manifolds with nontrivial topological structures. Witten and others stressed more recently the importance of such terms in the construction of many interesting theories.

It is well known that a gauge field might break the current conservation by anomalies, an effect that can be described by adding to the action a term proportional to the Chern-Simons characteristic class in one dimension higher. In the chiral model under consideration the gauge field is

$$A = \begin{pmatrix} A_L & 0 \\ 0 & A_R \end{pmatrix} = \begin{pmatrix} dg\, g^{-1} & 0 \\ 0 & dg^{-1} g \end{pmatrix} \tag{2.7}$$

The action has the form

$$S = S_0 + S_{W-Z} \tag{2.8}$$

where

$$S_0 = \frac{1}{4\lambda} \int tr(A*A) \tag{2.9}$$

and

$$S_{W-Z} = \frac{n}{24\pi} \int_D tr(\gamma_5 A^3) \tag{2.10}$$

Here γ_5 is a matrix

$$\gamma_5 = \begin{pmatrix} 1 & 0 \\ 0 & -1 \end{pmatrix} \tag{2.11}$$

that differentiates left from right; D is a three dimensional disk with the two dimensional spacetime as its only boundary and n is an integer.

The action is invariant under global $G_L \times G_R$ transformation. The corresponding conserved currents are

$$J(x) = \frac{-i}{\lambda} \left[A(x) - K\gamma_5 {*}A(x) \right] \tag{2.12}$$

where

$$K = \frac{\lambda n}{4\pi} \tag{2.13}$$

is a constant. The conservation equation and the compatibility condition are

$$d{*}A = \gamma_5 K dA = \gamma_5 K A^2 \tag{2.14}$$

and

$$dA = A^2 \ . \tag{2.15}$$

Before going on to find the hidden Abelian symmetry we shall first establish an isomorphism of the classical solutions among models with different values of K. For this purpose let

$$B = A - \gamma_5 K{*}A \tag{2.16}$$

We obtain from Eqs. (2.14), (2.15) and (1.22) that

$$d{*}B = 0 \tag{2.17}$$

$$dB = (1-K^2)A^2 = B^2 \ . \tag{2.18}$$

Hence B is the current of a chiral model without the Wess-Zumino term, i.e. K = 0. Eq. (2.16) establishes an isomorphism of classical solutions for these two models. However, it should be noted that the energy-momentum tensor and other properties of these isomorphic solutions are in general different for different K.

In light-cone coordinates

$$x^{\pm} = \frac{1}{\sqrt{2}}(t \pm x) \tag{2.19}$$

the energy-momentum tensor has two independent components

$$\Theta^{A}_{\pm\pm} = \frac{1}{2}(H \pm P) = \frac{1}{4\lambda}\int \mathrm{tr}(A^{2}_{\pm L})d^{2}x \tag{2.20}$$

The Wess-Zumino term containing only one time or space derivative is of magnetic nature and has no contribution to the energy-momentum tensor when written in terms of the generalized coordinates only. From the Eq. (2.16) we get

$$\Theta^{B}_{\pm\pm} = (1 \mp K)^{2}\, \Theta^{A}_{\pm\pm} \tag{2.21}$$

The points $K = \pm 1$ are singular implying the existence of zero mass states moving on the light-cone in the B-model. Since B-models have hidden symmetry as discussed in the previous section, it will be no surprise that A-models also have this symmetry.

We now turn to the study of hidden symmetry. The experience gained in the previous section suggests the existence of a linear equation

$$\psi_{K}^{-1}d\psi_{K} = \frac{1}{2}\{(\mathrm{ch}\xi - 1 - \gamma_{5}K\mathrm{sh}\xi)A + (\mathrm{sh}\xi - \gamma_{5}K(\mathrm{ch}\xi - 1)*A\} \tag{2.22}$$

The integrability condition for Eq. (2.22) is provided by Eqs. (2.14) and (2.15). Write ψ in the matrix form

$$\psi_{K} = \begin{pmatrix} \psi_{KL} & 0 \\ 0 & \psi_{KR} \end{pmatrix} \tag{2.23}$$

the chiral field $g(x)$ can be shown to transform as

$$g(x) \to g_{K}(\xi,A) = \psi_{KL}(\xi,A)g(x)\psi_{KR}^{-1}(\xi,A) \tag{2.24}$$

The Noether current of the chiral group after the ξ-transformation becomes

$$\begin{aligned} J(x) \to J_{K}(\xi,A) &= \psi_{K}(\xi,A)[\mathrm{ch}\xi J(x) + \mathrm{sh}\xi * J(x)]\psi_{K}^{-1}(\xi,A) \\ &= \frac{-i}{\lambda}A_{K}(\xi,A) \end{aligned} \tag{2.25}$$

It can be verified directly that

$$A_{K}(\xi,A) = \begin{pmatrix} A_{KL} & 0 \\ 0 & A_{KR} \end{pmatrix} \tag{2.26}$$

where

$$A_{KL}(\xi,A) = dg_{K}(\xi,A)g_{K}^{-1}(\xi,A) \tag{2.27}$$

and

$$A_{KR}(\xi,A) = dg_{K}^{-1}(\xi,A)g_{K}(\xi,A) \tag{2.28}$$

They satisfy the equations

$$d{*}A_K(\xi,A) = 0, \qquad dA_K(\xi,A) = A_K^2(\xi,A) \tag{2.29}$$

Therefore as in the previous section an infinite number of conserved currents follow.

Next we shall discuss the current algebra or the equal time commutators of the currents in the chiral model.[17-18] The current 1-form can be expressed as

$$J_{L,R}(x) = J^a_{\mu L,R}\, I^a dx^\mu \tag{2.30}$$

where I^a, $a = 1, \cdots n_G$ are generators of the group G. The charges

$$Q^a_{L,R} = \int dx\, J^a_{0L,R} \tag{2.31}$$

are generators of the global transformation as proved in general in quantum mechanics. They satisfy the commutation relations

$$\begin{aligned} [Q^a_L, Q^b_L] &= if_{abc} Q^c_L \\ [Q^a_L, Q^b_R] &= 0 \\ [Q^a_R, Q^b_R] &= i\, f_{abc} Q^c_R \end{aligned} \tag{2.32}$$

For any operator O(x) that transforms under infinitesimal global transformations described by group parameters ε_a

$$O(x) \to O'(x) = O(x) + \varepsilon_a \delta^a\, O(x) \tag{2.33}$$

we always have

$$[Q^a_L, O(x)] = -\,\delta^a_L\, O(x) \tag{2.34}$$

$$[Q^a_R, O(x)] = -\,\delta^a_R\, O(x) \tag{2.35}$$

The usual derivation of the current algebra requires quantization with constraints which are not easy to comply with. A simple way to obtain the correct result for bosonic models is to use the fact that the charge density $J^a_{0L,R}(x,t)$ can be considered as generators for local transformations at different space points but at the same time. The local transformations of the gauge fields are

$$\begin{aligned} A_L(x) \to A'_L(x) &= g_L(x)[A_L(x)-d]g_L^{-1}(x) \\ &\quad + g_L(x)g(x)[dg_R^{-1}(x)g_R(x)]g^{-1}(x)g_L^{-1}(x) \end{aligned} \tag{2.36}$$

and

$$\begin{aligned} A_R(x) \to A'_R(x) &= g_R(x)[A_R(x)-d]g_R^{-1}(x) \\ &\quad + g_R(x)g^{-1}(x)[dg_L^{-1}(x)g_L(x)]g(x)g_R^{-1}(x) \end{aligned} \tag{2.37}$$

From Eq. (2.36) and (2.37) it is easily shown that

$$[J^a_{0L}(x,t),\ A^b_{0L}(y,t)] = i\delta(x-y)f_{abc}A^c_{0L}(y,t) \tag{2.38}$$

$$[J^a_{0L}(x,t),\ A^b_{1L}(y,t)] = i\delta(x-y)f_{abc}A^c_{1L}(y,t) + i\delta_{ab}\delta'(x-y) \tag{2.39}$$

$$[J^a_{0L}(x,t),\ A^b_{0R}(y,t)] = 0 \tag{2.40}$$

$$[J^a_{0L}(x,t),\ A^b_{1R}(y,t)] = -i\delta'(x-y)tr(I^b g^{-1}(x)I^a g(x)) \tag{2.41}$$

Similar commutation relations for J^a_{0R} can be found. Using Eqs. (2.38)-(2.41) we find the current algebra

$$[J^a_{0L}(x,t),\ J^b_{0L}(y,t)] = if_{abc}J^c_{0L}(y,t)\delta(x-y) - \frac{K}{\lambda}\delta_{ab}\delta'(x-y) \tag{2.42}$$

$$[J^a_{0L}(x,t),\ J^b_{1L}(y,t)] = if_{abc}J^c_{1L}(y,t)\delta(x-y) + \frac{1}{\lambda}\delta_{ab}\delta'(x-y) \tag{2.43}$$

Since A^a (x,t) does not contain time derivatives, it is a function of the generalized coordinates only. Therefore the equal time commutator of two A_{1L} vanishes and we have

$$[J^a_{1L}(x,t) + KJ^a_{0L}(x,t),\ J^b_{1L}(y,t) + KJ^b_{0L}(y,t)] = 0 \tag{2.44}$$

Eqs. (2.42)-(2.44) provide the current algebra of the left-handed currents of the model. A similar algebra holds for the right-handed currents. These two algebras are not independent. They are related through Eq. (2.41),

$$[J^a_{0L}(x,t),\ J^b_{0R}(y,t)] = \frac{K}{\lambda}\delta'(x-y)tr(g^{-1}I^a gI^b)$$

$$[J^a_{0L}(x,t),\ J^b_{1R}(y,t)] = -\frac{1}{\lambda}\delta'(x-y)tr(g^{-1}I^a gI^b)$$

$$[J^a_{0R}(x,t),\ J^a_{1L}(y,t)] + -\frac{1}{\lambda}\delta'(x-y)tr(gI^a g^{-1}I^b) \tag{2.45}$$

and

$$[J^a_{0L}(x,t) + KJ^a_{1L}(x,t),\ J^b_{0R}(y,t) + KJ^b_{1R}(y,t)] = 0 \tag{2.46}$$

In the special case of K = ±1 we have

$$J^a_{0L} = \mp J^a_{1L},\qquad J^a_{0R} = \pm J^a_{1R}$$

In terms of light-cone components only one component $J^a_{\mp L}$ and $J^a_{\pm R}$ for each current survives in the K = ±1 case. Furthermore conservation equation implies that $J_\mp$ are functions of $x^\mp$ only. These special cases have been studied by Witten in the light-cone quantization and are shown to be equivalent to free massless fermions.[11]

III. SUPERSYMMETRIC SIGMA-MODELS

The results of the previous sections can be extended to a class of supersymmetry theories in two dimensions. It is more convenient in this case to work in the light-cone coordinates where massless particles with definite charges move to the left or to the right. We shall introduce two supercharges. They are real Majorana-Weyl spinors satisfying

$$Q_\pm^2 = P_\pm = i\partial/\partial x^\pm \tag{3.1}$$

$$Q_+Q_- + Q_-Q_+ = 0 \ . \tag{3.2}$$

Eq. (3.1) tells us that a left mover with $P_+ = 0$ has zero Q_+ and a right mover with $P_- = 0$ has zero Q_-. A theory with two supercharges with opposite chiralities has (1,1) supersymmetry. We use a superspace with two bosonic coordinates $x^\pm$ and two fermionic coordinates $\theta^\pm$ with opposite chirality. The generators of the (1,1) supersymmetry are

$$Q_\pm = i\left(\frac{\partial}{\partial\theta^\pm} - i\theta^\pm \frac{\partial}{\partial x^\pm}\right) \tag{3.3}$$

The chiral field now is a function of the superspace coordinates. The super-covariant derivatives are

$$D_\pm = \frac{\partial}{\partial\theta^\pm} + i\theta^\pm \frac{\partial}{\partial x^\pm} \tag{3.4}$$

They satisfy the relations

$$\begin{aligned} &D_\pm^2 = P_\pm \\ &D_+D_- + D_-D_+ = 0 \end{aligned} \tag{3.5}$$

The components of the gauge potential for the chiral field are

$$A_\pm = \begin{pmatrix} A_\pm^L & 0 \\ 0 & A_\pm^R \end{pmatrix} \tag{3.6}$$

where

$$A_\pm^L = (D_\pm g)g^{-1} \tag{3.7}$$

and

$$A_\pm^R = (D_\pm g^{-1})g \tag{3.8}$$

The action is

$$S_0 = \frac{1}{4\lambda} \int d^2x d^2\theta \ \mathrm{tr}(A_+A_- - A_-A_+) \tag{3.9}$$

from which one can deduce the equation of motion

$$D_+A_- - D_-A_+ = 0 \tag{3.10}$$

and the compatibility condition

$$D_+A_- + D_-A_+ = A_+A_- + A_-A_+ \tag{3.11}$$

To find the hidden symmetry we construct a linear equation

$$\phi^{-1}D_\pm\phi = \frac{1}{2}\left[\mathrm{ch}\xi - 1 \pm \mathrm{sh}\xi\right]A_\pm \tag{3.12}$$

The integrability condition for Eq. (3.12) can be directly verified by imposing Eqs. (3.10) and (3.11).

The transformed chiral field and the gauge potential can be easily found to be

$$g(x,\theta) \rightarrow g_\xi(x,\theta) = \phi_L(\xi,A)g(x,\theta)\phi_R^{-1}(\xi,A) \tag{3.13}$$

and

$$A_\pm \rightarrow A'_\pm(\xi) = (\mathrm{ch}\xi \pm \mathrm{sh}\xi)\phi(\xi,A)A_\pm\phi^{-1}(\xi,A) \tag{3.14}$$

It is easily verified that $A_\pm(\xi)$ satisfy

$$\begin{aligned} &D_+A_-(\xi) - D_-A_+(\xi) = 0 \\ &D_+A_-(\xi) + D_-A_+(\xi) = A_+(\xi)A_-(\xi) + A_-(\xi)A_+(\xi) \end{aligned} \tag{3.15}$$

The current

$$J_\pm(\xi) = \frac{-i}{\lambda} A_\pm(\xi)$$

is conserved, which provides an infinite number of conservation laws by expansion in terms of ξ.

Models with Wess-Zumino terms can be treated in a similar way. We shall leave this as an exercise for the interested reader.

References

1. Lüscher, M. & Phlmeyer, K. Nucl. Phys. B137,(1978)46.
2. Brézin, E., Itzykson, C., Zimm-Justin, J. & Zuber, J.B. Phys. Lett. 82B,(1979)442.
3. Eichenherr, H., & Forger, M. Nucl. Phys. B155,(1979)381.
4. de Vega, H.J., Phys. Lett. 87B,(1979)233.
5. Zakharov, V.E. & Mikhailov, A.V. JETP 47,(1978)1017.
6. Ogielsk, A.T., Pracad, M.K., Sinha, A. & Chan, L.L. Phys. Lett. 91B,(1980)387.
7. Dolan, L. Phys. Rev. Lett. 47,(1981)1371.
8. Veno, K. & Nakmura, Y. Phys. Lett. 117,(1982)208.
9. Hou, B.Y. Commun. Theor. Phys. 1,(1981)333.
10. Ge, M.L. & Wu, Y.S. Phys. Lett. 108B,(1982)411.
11. Witten, E. Comm. Math. Phys., 92,(1984)455.
12. de Vega, H.J. Phys. Lett. 87B,(1979)233.
13. Chou, K.C. & Dai, Y.B. Comm. Theor. Phys. 3,(1984)767.
14. Aballa, M.C.B. Phys. Lett., 215 (1985).
15. Bhattacharya, G. & Rajeev, S. Nucl. Phys. B246,(1984)157.
16. Wess, J. and Zumino, B., Phys. Lett. 37B(1971)95.
17. Faddeev, L.D. les Houches lectures, North-Holland (1983).
18. de Vega, H.J., Eichenherr, H. & Maillet, J.M. Comm. Math. Phys. 92,(1984)507.

REPRESENTATIONS OF THE TWO-DIMENSIONAL CONFORMAL GROUP

Roman Jackiw

Center for Theoretical Physics
Department of Physics
Massachusetts Institute of Technology
Cambridge, Massachusetts 02139 U.S.A.

CONTENTS

I. INTRODUCTION TO THE CONFORMAL GROUP

The dynamics of a field A in D-dimensional Minkowski space-time can be invariant against a group of coordinate transformations, given infinitesimally by

$$\delta_f x^\mu = -f^\mu \tag{1.1}$$

which is larger than the Poincaré group of translations and Lorentz rotations. When the energy-momentum tensor $\theta^{\mu\nu}$ is not only conserved [as a consequence of translation invariance] and symmetric [as a consequence of Lorentz invariance] but is also traceless

$$\theta^\mu_\mu = 0 \tag{1.2}$$

then the following are conserved currents

$$j^\mu_f = \theta^{\mu\nu} f_\nu \tag{1.3}$$

provided f_μ is a conformal Killing vector.

$$\partial_\mu f_\nu + \partial_\nu f_\mu - \frac{2}{D} g_{\mu\nu} \partial_\alpha f^\alpha = 0 \tag{1.4}$$

For $D > 2$ the only solutions to this equation are the $(D+2)(D+1)/2$ vectors, at most quadratic in x^μ, which generate the $SO(D,2)$ special conformal group.

A linear transformation law for the multicomponent field A is

$$\delta_f A = f^\mu \partial_\mu A + \partial_\mu f_\nu \left(g^{\mu\nu} \frac{d}{D} + \frac{1}{2} \Sigma^{\mu\nu} \right) A \tag{1.5}$$

Here, d is the scale dimension of A and $\Sigma^{\mu\nu}$ is its spin matrix. The conserved charges

$$Q_f = \int d\mathbf{r}\, \theta^{0\nu} f_\nu \tag{1.6}$$

generate this transformation by commutation.

$$i\,[Q_f,\, A] = \delta_f A \tag{1.7}$$

[Classically, the commutator is replaced by the Poisson bracket, divided by i.] Moreover, the charge algebra follows the Lie algebra of the group.

$$[\delta_f, \delta_g] = -\delta_{(f,g)} \qquad (f,g)^\mu \equiv f^\alpha \partial_\alpha g^\mu - g^\alpha \partial_\alpha f^\mu \tag{1.8}$$

$$[Q_f, Q_g] = iQ_{(f,g)} \tag{1.9}$$

For further discussion, see Refs. [1] and [2].

At $D = 2$, the above continues to hold and describes an $SO(2,2) = SO(2,1) \otimes SO(2,1)$ invariance group. However, there are additional conformal Killing vectors: *any* f^μ with $+$ component depending on x^+, $f^+(x^+)$, and $-$ component depending on x^-, $f^-(x^-)$, satisfies (1.4). [Light-cone components are defined by $\pm \equiv \frac{1}{\sqrt{2}}(0 \pm 1)$.] These give rise to an infinite dimensional Lie group, whose action on the coordinates of two-dimensional Minkowski space-time [$x^\mu \equiv (t, x)$, metric $g_{\mu\nu} = \mathrm{diag}(1, -1)$] takes light-cone components into arbitrary functions of themselves. The Lie algebra of these coordinate transformations is as in (1.8); the transformation of the fields is as in (1.5), except that in two dimensions $\Sigma^{\mu\nu}$ may be represented by $\epsilon^{\mu\nu}\Sigma$, hence (1.5) becomes

$$\delta_f A = f^\mu \partial_\mu A + \frac{1}{2} \partial_+ f^+ (d + \Sigma) A + \frac{1}{2} \partial_- f^- (d - \Sigma) A \tag{1.10}$$

Eq. (1.6) now gives rise to an infinite set of conserved charges. The energy-momentum tensor has only two independent components owing to its symmetry and tracelessness: $\theta^{01} = \theta^{10}$, $\theta^{00} = \theta^{11}$. Equivalently, the two components may be taken as θ_{++} and θ_{--}, since θ_{+-} vanishes. Moreover, conservation insures that θ_{++} is only a function of x^+ and θ_{--} only of x^-. Hence, (1.6) may be presented as

$$\begin{aligned} Q_f &= \frac{1}{\sqrt{2}} \int dx \left[\theta_{++}(x^+) f^+(x^+) + \theta_{--}(x^-) f^-(x^-) \right] \\ &= \int dz \left[\theta_{++}(z) f^+(z) + \theta_{--}(z) f^-(z) \right] \end{aligned} \tag{1.11}$$

where the t-independence is explicit in the second formula. These charges generate the transformation (1.7), (1.10); however, upon quantization their algebra does not reproduce the group's Lie algebra (1.8), *i.e.* while (1.9) is a valid classical formula

involving Poisson brackets, it is no longer true in the quantum theory when f^μ and g^μ range beyond the quadratic expressions appropriate to $SO(2,2)$.

To prove this, observe that (1.9) for arbitrary $f^\pm(x^\pm)$, $g^\pm(x^\pm)$ implies the following equal-time commutation relations.

$$[\theta_{\pm\pm}(t,x),\ \theta_{\pm\pm}(t,y)] = \pm 2i\,(\theta_{\pm\pm}(t,x) + \theta_{\pm\pm}(t,y))\,\delta'(x-y) \qquad (1.12)$$

$$[\theta_{\pm\pm}(t,x),\ \theta_{\mp\mp}(t,y)] = 0 \qquad (1.13)$$

[In fact, $\theta_{\pm\pm}$ depends only the combination $\frac{1}{\sqrt{2}}(t \pm x)$.] By Lorentz invariance, a traceless and symmetric tensor has vanishing expected value in the vacuum $|0\rangle$, so the above requires $\langle 0\,|[\theta_{\pm\pm}(t,x),\ \theta_{\pm\pm}(t,y)]|\,0\rangle$ to be zero. But one may show that in a theory with a non-negative energy spectrum, a unique ground [vacuum] state and positive norm in the Hilbert space, the commutator's vacuum expectation value cannot vanish. Consider the positive expectation of the Hamiltonian H in the states $|\alpha_\pm\rangle \equiv \int dx\,\alpha(x)\theta_{\pm\pm}(t,x)|0\rangle$, where α is an arbitrary real function with bounded support: $0 < \langle\alpha_\pm|H|\alpha_\pm\rangle = \int dxdy\,\alpha(x)\alpha(y)\,\langle 0\,|\theta_{\pm\pm}(t,x)\,H\,\theta_{\pm\pm}(t,y)|\,0\rangle$. Two equivalent formulas for this expression are $\int dxdy\,\alpha(x)\alpha(y)\,\langle 0\,|[\theta_{\pm\pm}(t,x),H]\,\theta_{\pm\pm}(t,y)|\,0\rangle$ and $\int dxdy\,\alpha(x)\alpha(y)\,\langle 0\,|\theta_{\pm\pm}(t,x)\,[H,\theta_{\pm\pm}(t,y)]|\,0\rangle$. Since $i\,[H,\theta_{\pm\pm}] = \dot\theta_{\pm\pm} = \pm\theta'_{\pm\pm}$, where the dot [dash] indicates differentiation with respect to time [space], we find

$$\begin{aligned} 0 &< 2\,\langle\alpha_\pm\,|H|\,\alpha_\pm\rangle \\ &= \int dxdy\alpha(x)\alpha(y)\{\pm i\,\langle 0\,|\theta'_{\pm\pm}(t,x)\theta_{\pm\pm}(t,y)|\,0\rangle \mp i\,\langle 0\,|\theta_{\pm\pm}(t,x)\theta'_{\pm\pm}(t,y)|\,0\rangle\} \\ &= \mp i\int dxdy\ \alpha'(x)\alpha(y)\,\langle 0\,|[\theta_{\pm\pm}(t,x),\ \theta_{\pm\pm}(t,y)]|\,0\rangle \end{aligned} \qquad (1.14)$$

Thus, the vacuum expectation value of the commutator cannot vanish. Eq. (1.12), which is a valid classical formula when Poisson brackets are used, must acquire a quantum mechanical correction. Dimensional analysis suggests that the simplest modification of (1.12) is

$$[\theta_{\pm\pm}(t,x),\theta_{\pm\pm}(t,y)] = \pm 2i\,(\theta_{\pm\pm}(t,x) + \theta_{\pm\pm}(t,y))\,\delta'(x-y) \mp \frac{ic_\pm}{6\pi}\delta'''(x-y) \quad (1.15)$$

where according to (1.14), $c_\pm > 0$. In a parity invariant theory, which we shall henceforth consider, $c_+ = c_- \equiv c$. Of course, more complicated modifications are also possible.

The addition to the commutator is frequently called a *Schwinger* term because the argument for its occurrence is an adaptation to the present context of Schwinger's proof that current commutators possess non-canonical contributions involving derivatives of delta functions. Also, it is a non-canonical commutator *anomaly*, giving rise to an *extension* or a *center* to the Lie algebra of generators.

$$\begin{aligned} [Q_f,\ Q_g] &= iQ_{(f,g)} \\ &\quad - \frac{ic}{24\pi}\int dx\Big\{\Big(f^+\frac{d^3}{dx^3}g^+ - g^+\frac{d^3}{dx^3}f^+\Big) - \Big(f^-\frac{d^3}{dx^3}g^- - g^-\frac{d^3}{dx^3}f^-\Big)\Big\} \end{aligned} \qquad (1.16)$$

Notice that the extension vanishes for the $SO(2,2)$ subgroup when the conformal Killing vectors are at most quadratic. Also it vanishes for the subgroup of transformations for which $f^+(z) = -f^-(-z)$. We call these the ***Rindler*** transformations; they arise when $x+t$ and $x-t$ are transformed by the same function, which means that $t=0$ is left unchanged. The transformation to an accelerated coordinate system, frequently studied as an aid to understanding black holes, belongs to this class.[3]

The technical reason for the difference between the classical (1.12) and quantal (1.15) commutators is that the energy-momentum tensor is not well-defined in the quantum theory since it involves products of non-commuting operators. In the conventional resolution of this problem, one selects a Fock vacuum [not necessarily the true vacuum of the theory] and normal orders $\theta_{\mu\nu}$ with respect to this state. The commutator anomaly then emerges when the right-hand side of (1.12) is normal ordered. Alternatively, one may compute the true vacuum expectation value of the commutator in perturbation theory, using the Bjorken, Johnson, Low technique.

Here is presented another approach, which makes no reference to the vacuum state. Rather, we first explore, in Section II, the implication of a commutator anomaly for the representation theory of the conformal group: the representations are necessarily projective. Then, in Section III we construct the representation, determine the projective phase, and extract from it the extension (1.16) with $c=1$. In Section IV, it is shown how to construct representations which lead to arbitrary $c>1$. Finally, in Section V we discuss how the states of the theory transform under conformal transformations, with special attention to particle production in accelerated coordinates. References [4] and [5] comprise the original research papers on which these lectures are based.

II. COCYCLES AND EXTENSIONS IN REPRESENTATION THEORY

When one is dealing with a group of transformations g, which obey the composition law

$$g_1 g_2 = g_{1\circ 2} \tag{2.1}$$

and transform variables q according to a definite rule,

$$q \underset{g}{\longrightarrow} q^g \tag{2.2}$$

then a representation of this on functions of q, $\Psi(q)$, is gotten by associating with g an operator $U(g)$, which realizes (2.2) according to

$$U(g)qU^{-1}(g) = q^g \tag{2.3}$$

In the simplest case, $U(g)$ is defined to act on $\Psi(q)$ according to

$$U(g)\Psi(q) = \Psi(q^g) \tag{2.4}$$

and to satisfy a composition law which parallels (2.1).

$$U(g_1)U(g_2) = U(g_{1\circ 2}) \tag{2.5}$$

The multiplication in (2.5) is associative.

$$\Big(U(g_1)U(g_2)\Big)U(g_3) = U(g_1)\Big(U(g_2)U(g_3)\Big) \tag{2.6}$$

However, it is possible to elaborate this simplest realization by introducing phases into formulas (2.4), (2.5) and (2.6). Such phases are called cocycles and the various conditions that they must satisfy have been elaborated in mathematics.[6]

It has been known for some time that cocycles arise in quantum mechanics of point particles. Moreover, recently it was appreciated that the anomaly phenomena of field theory makes use of them as well.

1 – Cocycle

In the simplest generalization, a phase factor is inserted in the action of the representation. Rather than (2.4) we have

$$U(g)\Psi(q) = e^{i2\pi\omega_1(q,g)}\Psi(q^g) \tag{2.7}$$

Consistency with (2.5) requires that ω_1 satisfy

$$\omega_1(q^{g_1};g_2) - \omega_1(q;g_{1\circ 2}) + \omega_1(q;g_1) = 0 \text{ (mod integer)} \tag{2.8}$$

A quantity $\omega_1(q,g)$ depending on one member of the transformation group, g, and possibly on the variable acted upon, q, is called a *1-cocycle*, if it obeys (2.8).

Representations of Galilean boosts in quantum mechanics make use of a 1-cocycle. The Abelian group of Galileo transformations on the position vector $\mathbf{r}$ [q in the general discussion] is defined to act as $\mathbf{r} \to \mathbf{r}+\mathbf{v}t$. Also, we have $\mathbf{p} = m\dot{\mathbf{r}} \to \mathbf{p}+m\mathbf{v}$ [m is the particle mass]. The transformation is labelled by $\mathbf{v}$ [corresponding to g in the general discussion] and is implemented on wave functions $\Psi(\mathbf{r})$ by the operator

$$\begin{aligned} U(\mathbf{v}) &= e^{i\mathbf{v}\cdot(\mathbf{p}t-m\mathbf{r})} \\ U(\mathbf{v})\mathbf{r}U^{-1}(\mathbf{v}) &= \mathbf{r}+\mathbf{v}t \\ U(\mathbf{v})\mathbf{p}U^{-1}(\mathbf{v}) &= \mathbf{p}+m\mathbf{v} \end{aligned} \tag{2.9}$$

as is established with the help of Heisenberg's algebra.

$$[r^i,r^j] = 0 \quad , \quad [r^i,p^i] = i\delta^{ij} \quad , \quad [p^i,p^j] = 0 \tag{2.10}$$

One verifies that (2.5) is satisfied, but $U(\mathbf{v})$ acts on wavefunctions $\Psi(\mathbf{r})$ as in (2.7) with

$$2\pi\omega_1(\mathbf{r};\mathbf{v}) = -\left(m\mathbf{v}\cdot\mathbf{r} + \frac{1}{2}mv^2t\right) \tag{2.11}$$

which satisfies (2.8).

2 – Cocycle

In the next generalization, a phase is introduced into the composition law (2.5).

$$U(g_1)U(g_2) = e^{i2\pi\omega_2(q;g_1,g_2)}U(g_{1\circ 2}) \tag{2.12}$$

A consistency condition on ω_2 follows from associativity (2.6).

$$\omega_2(q^{g_1};g_2,g_3) - \omega_2(q;g_{1\circ 2},g_3) + \omega_2(q;g_1,g_{2\circ 3}) - \omega_2(q;g_1,g_2) = 0 \text{ (mod integer)} \tag{2.13}$$

When a quantity depends on two group elements and possibly on q, and also satisfies (2.13), it is called a *2-cocycle*. Representations that make use of 2-cocycles are called *projective* or *ray* representations and they too occur in quantum mechanics.

Translations on phase space: $\mathbf{r} \to \mathbf{r} + \mathbf{a}$, $\mathbf{p} \to \mathbf{p} + \mathbf{b}$, which arise when Galileo boosts are supplemented by spatial translations, are represented by the operator

$$\begin{aligned} U(\mathbf{a}, \mathbf{b}) &= e^{i(\mathbf{a}\cdot\mathbf{p} - \mathbf{b}\cdot\mathbf{r})} \\ U(\mathbf{a}, \mathbf{b})\mathbf{r}U^{-1}(\mathbf{a}, \mathbf{b}) &= \mathbf{r} + \mathbf{a} \\ U(\mathbf{a}, \mathbf{b})\mathbf{p}U^{-1}(\mathbf{a}, \mathbf{b}) &= \mathbf{p} + \mathbf{b} \end{aligned} \tag{2.14}$$

which composes according to (2.12) with

$$2\pi\omega_2(\mathbf{r}; \mathbf{a}_1\mathbf{b}_1, \mathbf{a}_2\mathbf{b}_2) = \frac{1}{2}(\mathbf{a}_2 \cdot \mathbf{b}_1 - \mathbf{a}_1 \cdot \mathbf{b}_2) \tag{2.15}$$

The above obeys the consistency condition (2.13).

When dealing with a continuous Lie group of transformations, the discussion may be carried out in infinitesimal terms. Corresponding to the finite group element g, there is an infinitesimal quantity θ, $g = e^{\theta}$, and the composition law (2.1) is reflected in a Lie algebra.

$$[\theta_1, \theta_2] = \theta_{(1,2)} \tag{2.16}$$

The infinitesimal action of the transformation on q is given by

$$q \to q + \delta_\theta q \tag{2.17}$$

and is represented by a generator G_θ.

$$i\,[G_\theta, q] = \delta_\theta q \tag{2.18}$$

The operator representing the finite transformation is the exponentiated generator, $U(g) = e^{iG_\theta}$, and the composition law without a cocycle (2.5) insures that G_θ satisfies the Lie algebra of the group.

$$i\,[G_{\theta_1}, G_{\theta_2}] = G_{\theta_{(1,2)}} \tag{2.19}$$

The presence of a 2-cocycle in the composition law, as in (2.12), means that the commutator algebra of the generators acquires an extension,

$$i\,[G_{\theta_1}, G_{\theta_2}] = G_{\theta_{(1,2)}} + \delta\omega_2 \tag{2.20}$$

where $\delta\omega_2$ is [proportional to] the infinitesimal portion of the 2-cocycle, whose consistency condition (2.13) insures that (2.20) does not contradict the Jacobi identity, which is the infinitesimal version of (2.6).

Thus, the Heisenberg algebra (2.10) is an extension of the Abelian algebra of translations, and we appreciate that the essence of quantum mechanics is a 2-cocycle. Also, we see that the extension in (1.16) signals that quantum mechanically the conformal transformation group is realized projectively.

Higher Cocycles

I shall not pursue the subject of cocycles further, beyond mentioning that, not unexpectedly, a *3-cocycle* involves abandoning associativity: Eq. (2.6) acquires a phase, while infinitesimally the Jacobi identity fails. This happens for gauge invariant velocity operators of a charged particle in the presence of an external magnetic field which is not source-free. The 3-cocycle then measures the magnetic flux enclosed by an arbitrary surface. The only physically conceivable magnetic sources are magnetic monopoles, and the magnetic flux they produce through closed surfaces is quantized in integer units, as a consequence of the Dirac quantization condition. Hence, the non-associativity is invisible since $e^{i2\pi n} = 1$. Indeed, one may view the demand of associativity as the origin of the Dirac quantization condition.

Finally, it must be stressed that cocycles are of interest only when they are non-trivial; *i.e.* when they cannot be removed by redefining phases. For example, a 2-cocycle of the form

$$\omega_2(q; g_1, g_2) = \alpha(q; g_1) + \alpha(q^{g_1}; g_2) - \alpha(q; g_{1\circ 2}) \tag{2.21}$$

satisfies (2.13) with arbitrary α, but may be removed by redefining the phase of $U(g)$ by $2\pi\alpha(q; g)$. Hence it is trivial.

III. FUNCTIONAL REPRESENTATION OF CONFORMAL TRANSFORMATIONS, $c = 1$

We now construct a representation of the conformal group for the simplest case: a two-dimensional spinless, dimensionless and massless free field Φ. The transformation law (1.10) reduces to the Lie derivative of Φ.

$$\delta_f \Phi = f^\alpha \partial_\alpha \Phi \tag{3.1}$$

We further restrict the discussion to one component of the group, transforming x^+ but not x^-. Since transformations belonging to different components act independently, we can always construct the full transformation sequentially. Thus we set $f^- = 0$ and define $f(x) \equiv \sqrt{2} f^+(x/\sqrt{2})$.

We adopt a canonical, fixed-time $[t = 0]$ description, where one deals with a pair of canonical variables Φ and $\Pi \equiv \dot{\Phi}$, which depend on x [but not on t] and satisfy the usual commutation relations. In terms of these, the transformation (3.1) is

$$\begin{aligned} \delta_f \Phi &= \frac{1}{2} f(\Pi + \Phi') \\ \delta_f \Pi &= \frac{1}{2} f(\Pi + \Phi')' + \frac{1}{2} f'(\Pi + \Phi') \end{aligned} \tag{3.2}$$

It is advantageous to introduce a self-dual field $\mathcal{X}$ [it depends only on x^+]

$$\begin{aligned} \mathcal{X} &\equiv \frac{1}{\sqrt{2}}(\Pi + \Phi') = \partial_+ \Phi \\ \dot{\mathcal{X}} &= \mathcal{X}' \\ [\mathcal{X}(x), \ \mathcal{X}(y)] &= i\delta'(x - y) \equiv k(x, y) \end{aligned} \tag{3.3}$$

Then, (3.2) reads

$$\delta_f \mathcal{X} = (f\mathcal{X})' \tag{3.4}$$

The traceless energy momentum tensor of the theory

$$\theta_{\mu\nu} = \partial_\mu \Phi \partial_\nu \Phi - \frac{1}{2} g_{\mu\nu} \partial^\alpha \Phi \partial_\alpha \Phi \tag{3.5a}$$

may also be written in the Sugawara-Sommerfield form in terms of a conserved current $J_\mu = \frac{1}{\sqrt{\pi}} \epsilon_{\mu\nu} \partial^\nu \Phi$.

$$\theta_{\mu\nu} = \pi \left(J_\mu J_\nu - \frac{1}{2} g_{\mu\nu} J^\alpha J_\alpha \right) \tag{3.5b}$$

[Similar formulas hold for $U(N)$ multiplets of free massless fermions, provided products of $U(N)$ currents are properly defined, as well as for bosons and fermions interacting with background gauge or gravitational fields.[7]] The generator of (3.4) involves θ_{++},

$$Q_f = \frac{1}{2} \int dx\, f \mathcal{X}^2 \tag{3.6}$$

$$i[Q_f, \mathcal{X}] = \delta_f \mathcal{X} \tag{3.7}$$

and the finite transformation is given by

$$e^{i\tau Q_f} \mathcal{X}(x) e^{-i\tau Q_f} = \frac{f(X)}{f(x)} \mathcal{X}(X) = \frac{\partial X(x)}{\partial x} \mathcal{X}(X)$$

$$X(x) = \mathcal{F}^{-1}(\tau + \mathcal{F}(x)) \quad , \qquad \mathcal{F}(x) \equiv \int^x \frac{dz}{f(z)} \tag{3.8}$$

[We assume that $f(x)$ has all the properties needed to justify the manipulations that we perform with it.]

Of course, Eqs. (3.6) – (3.8) are formal, since powers of the non-commuting operator $\mathcal{X}(x)$ at the same point which occur in Q_f need to be defined. The difficulty with working with (3.6) is not apparent in (3.7) and (3.8), but when the commutator of two charges is evaluated formally, the necessary extension is not found. Rather than handling the problem conventionally by normal ordering $\mathcal{X}^2$, we proceed by constructing a representation of the group of transformations (3.4). The Lie algebra may then be determined.

In order to build a well-defined generator, $f(x)$ is promoted to a real, symmetric bilocal function $F(x,y)$ which in the local limit becomes $f(x)\delta(x-y)$. The regulated generator

$$Q_F = \frac{1}{2} \int dx dy\, \mathcal{X}(x) F(x,y) \mathcal{X}(y) \tag{3.9}$$

is not beset by singularities; it satisfies the algebra without center.

$$[Q_F, Q_g] = i Q_{(F,G)} \quad , \qquad i(F,G) \equiv FkG - GkF \tag{3.10}$$

but Q_F does not possess a local limit. [A functional matrix notation is being used: *e.g.* $(FkG)(x,y) \equiv \int dz dz'\; F(x,z) k(z,z') G(z',y) = i \int dz\, F(x,z) \frac{\partial}{\partial z} G(z,y)$.] Of course, Q_F no longer generates a symmetry, but it is a generator of a linear canonical transformation.[8]

As we shall see, a local limit can be taken if we add a c-number to Q_F, which is linear in F. So we define

$$:Q_F: \equiv Q_F - \frac{1}{4}\,\mathrm{tr}\left\{F^{1/2}\Omega F^{1/2}\right\} \tag{3.11}$$

and the formal definition (3.6) is replaced by $Q_f = \lim_{F\to f} :Q_F:$. $\Omega(x,y)$ will be determined by the requirement that the representation of $e^{-i\tau :Q_F:}$ possesses a local limit. We emphasize that the colons do not signify normal ordering with respect to the Fock vacuum; they are defined by (3.11). Since the modification is a c-number, it does not affect the infinitesimal (3.7) nor finite (3.8) transformation law. However, non-linear relations like (3.10) are changed

$$[:Q_F:,\ :Q_G:] = i:Q_{(F,G)}: + \frac{i}{4}\,\mathrm{tr}\left\{(F,G)\Omega\right\} \tag{3.12}$$

To construct the representation, we consider eigenstates $|\varphi\rangle$ of Φ.

$$\Phi(x)|\varphi\rangle = \varphi(x)|\varphi\rangle \ , \qquad \langle\varphi_1|\varphi_2\rangle = \delta\left(\varphi_1(x) - \varphi_2(x)\right) \tag{3.13}$$

The scalar product involves a functional δ-function and functional integration is used to sum over φ. In this basis, $\mathcal{X}$ is represented by

$$\langle\varphi_1|\mathcal{X}(x)|\varphi_2\rangle = \left(\frac{1}{i}\,\frac{\delta}{\delta\varphi_1(x)} + \varphi_1'(x)\right)\delta(\varphi_1 - \varphi_2)\ . \tag{3.14}$$

We seek the form of

$$U(\varphi_1,\varphi_2;\tau F) \equiv \langle\varphi_1|e^{-i\tau Q_F}|\varphi_2\rangle = U^*(\varphi_2,\varphi_1;-\tau F)\ , \tag{3.15}$$

which must be a Gaussian in φ_1 and φ_2 since Q_F is quadratic in $\mathcal{X}$. U may be determined by solving the Schrödinger-like equation obtained by differentiating (3.15) with respect to τ, evaluating the matrix element of Q_F from (3.14) and using the boundary condition (3.13). The result is

$$\begin{aligned} U(\varphi_1,\varphi_2;\tau F) &= e^{\Gamma(\tau F)}\, e^{i\int dx\, \varphi_1'(x)\varphi_2(x)} \\ &\times \exp\frac{i}{2}\int dx dy\, [\varphi_1(x) - \varphi_2(x)] K_{\tau F}(x,y)\, [\varphi_1(y) - \varphi_2(y)] \end{aligned} \tag{3.16}$$

where

$$K_{\tau F} = F^{-1/2}\left\{k_F\,\mathrm{ctn}\left(\frac{\tau}{2}k_F\right)\right\}F^{-1/2} \tag{3.17a}$$

$$\Gamma(\tau F) = -\frac{1}{2}\,\mathrm{tr}\ln\left\{F^{1/2}\frac{2\pi i\sin\left(\frac{\tau}{2}k_F\right)}{k_F}F^{1/2}\right\} \tag{3.17b}$$

$$k_F \equiv F^{1/2}kF^{1/2}\ . \tag{3.17c}$$

The functions of the kernels are defined either by power series, or by diagonalizing k_F. The composition law can be verified.

$$\int \mathcal{D}\varphi_2 U(\varphi_1,\varphi_2;\tau F)U(\varphi_2,\varphi_3;\tau G) = U(\varphi_1,\varphi_3;\tau F\circ G) \tag{3.18}$$

Let us now examine the local limit $F \to f$. The spectrum of k_f is the real line and the normalized and complete eigenfunctions are

$$\int dy\, k_f(x,y)\psi_\lambda^f(y) = \lambda\psi_\lambda^f(x)$$
$$\psi_\lambda^f(x) = \frac{1}{\sqrt{2\pi f(x)}}\exp\left\{-i\lambda\int^x \frac{dz}{f(z)}\right\} . \tag{3.19}$$

Consequently,

$$K_{\tau f}(x,y) = \frac{1}{f(x)}\int \frac{d\lambda}{2\pi}\lambda\,\mathrm{ctn}\left(\frac{\tau}{2}\lambda\right)\exp\left\{-i\lambda\int_y^x \frac{dz}{f(z)}\right\}\frac{1}{f(y)} . \tag{3.20}$$

The λ integration requires a prescription for handling the singularities: continue to the imaginary axis $\tau \to -i\tau'$, evaluate the integral and continue back $\tau' \to i\tau + 0^+$. This gives [for $\tau > 0$]

$$K_{\tau f}(x,y) = -i\pi\frac{1}{\tau f(x)}P\csc^2\left\{\pi\int_y^x \frac{dz}{\tau f(z)}\right\}\frac{1}{\tau f(y)} \tag{3.21}$$

(P means principal value.) For the prefactor, Γ, we proceed as follows. After continuation, we have

$$\begin{aligned}\Gamma(-i\tau' F) &= -\frac{1}{2}\operatorname{tr}\ln\left\{F^{1/2}\frac{2\pi\sinh\left(\frac{\tau'}{2}k_F\right)}{k_F}F^{1/2}\right\}\\ &= -\frac{1}{2}\operatorname{tr}\ln\left\{F^{1/2}\frac{2\pi\sinh\left(\frac{\tau'}{2}\omega_F\right)}{\omega_F}F^{1/2}\right\}\\ &= \frac{1}{2}\operatorname{tr}\ln\left(\frac{1}{\pi}\omega\right) - \frac{1}{2}\operatorname{tr}\ln\left\{1 - e^{-\tau'\omega_F}\right\} - \frac{\tau'}{4}\operatorname{tr}\omega_F\end{aligned} \tag{3.22}$$

where $\omega_F(x,y) \equiv |k_F|(x,y)$ and $\omega(x,y) \equiv |k|(x,y) = \int \frac{d\lambda}{2\pi}e^{-i\lambda(x-y)}|\lambda| = -P\frac{1}{\pi(x-y)^2}$. The first term is an infinite τ'-independent constant, $\ln Z \equiv \frac{1}{2}\operatorname{tr}\ln\left(\frac{1}{\pi}\omega\right) \sim \frac{1}{2}\int dx\int\frac{d\lambda}{2\pi}\ln\frac{1}{\pi}|\lambda|$, which may be removed by redefining the functional integration measure $D\varphi \to Z^{-1}D\varphi$. The second attains a finite local limit.

$$\gamma(-i\tau' f) = -\frac{1}{2}\int\frac{dx}{f(x)}\int\frac{d\lambda}{2\pi}\ln(1 - e^{-\tau'|\lambda|}) = \frac{\pi}{12}\int\frac{dx}{\tau' f(x)} \tag{3.23}$$

The third has no local limit. However, we observe that

$$\begin{aligned}\omega_f(x,y) &= \frac{1}{f^{1/2}(x)}\int\frac{d\lambda}{2\pi}|\lambda|\exp\left\{-i\lambda\int_y^x\frac{dz}{f(z)}\right\}\frac{1}{f^{1/2}(y)}\\ &= -\frac{1}{f^{1/2}(x)}P\frac{1}{\pi\left(\int_y^x\frac{dz}{f(z)}\right)^2}\frac{1}{f^{1/2}(y)}\\ &= \frac{1}{f^{1/2}(x)}\frac{(x-y)^2}{\left(\int_y^x\frac{dz}{f(z)}\right)^2}\frac{1}{f^{1/2}(y)}\omega(x,y)\end{aligned} \tag{3.24}$$

Since the quantity multiplying ω in the last formula of (3.25) becomes $f(x)$ for $x = y$ we see that

$$\operatorname{tr}\omega_F = \operatorname{tr}(F^{1/2}\omega F^{1/2}) + \left\{\begin{array}{c}\text{terms that vanish}\\ \text{in the local limit}\end{array}\right\} . \tag{3.25}$$

Thus, if we chose Ω in (3.11) to coincide with ω, $e^{-i\tau:Q_F:}$ has the representation

$$Z^{-1/2}\left\langle\varphi_1\left|e^{-i\tau:Q_F:}\right|\varphi_2\right\rangle Z^{-1/2} = Z^{-1}\, e^{\frac{i\tau}{4}\operatorname{tr}(F\omega)}U(\varphi_1,\varphi_2;\tau F) \tag{3.26}$$

which tends in the local limit to

$$:U:(\varphi_1,\varphi_2;\tau f) = \exp -i\left\{\frac{\pi}{12}\int\frac{dx}{\tau f(x)} - \int dx\,\varphi_1'(x)\varphi_2(x)\right\}\times$$
$$\exp\frac{\pi}{2}\int dxdy[\varphi_1(x)-\varphi_2(x)]\frac{1}{\tau f(x)}P\csc^2\left(\pi\int_y^x\frac{dz}{\tau f(z)}\right)\frac{1}{\tau f(y)}[\varphi_1(y)-\varphi_2(y)] \tag{3.27}$$

But the composition law changes. From (3.18) it follows that instead of that formula a projective rule with a 2-cocycle is satisfied.

$$\int\mathcal{D}\varphi_2 :U:(\varphi_1,\varphi_2;\tau F):U:(\varphi_2,\varphi_3;\tau G) = e^{i\omega_2(F,G)}:U:(\varphi_1,\varphi_3;\tau F\circ G) \tag{3.28}$$

The 2-cocycle in the local limit is given by

$$\omega_2(f,g) = \lim_{\substack{F\to f\\ G\to g}}\frac{\tau}{4}\operatorname{tr}\{(F+G-F\circ G)\omega\} . \tag{3.29}$$

Before the limit is taken, the cocycle is trivial in the sense (2.21); it arises because the phase of U has been redefined, and ω_2 can similarly be removed. However, after the local limit, a non-trivial expression remains which to $0(\tau^2)$ produces the extension in (1.16). This is evaluated as follows. The $0(\tau^2)$ contribution to (3.29) is $-\frac{\tau^2}{8}\operatorname{tr}\{(F,G)\omega\}$. In terms of a Fourier representation $F(x,y) = \int\frac{dp}{2\pi}\frac{dq}{2\pi}e^{-i\frac{p}{2}(x+y)}e^{-iq(x-y)}\tilde F(p,q)$, $\tilde F(p,q) = \tilde F(p,-q) \xrightarrow[F\to f]{} \tilde f(p) = \int dx\; e^{ipx}f(x)$, and similarly for G, this reads

$$\begin{aligned}-\frac{\tau^2}{8}\operatorname{tr}\{(F,G)\omega\} &= \frac{i\tau^2}{8}\operatorname{tr}\{(FkG-GkF)\omega\}\\
&= \frac{i\tau^2}{8}\left\{\int\frac{dp}{2\pi}\frac{dq}{2\pi}\epsilon\left(q+\frac{p}{2}\right)\left(q^2-\frac{p^2}{4}\right)\tilde F(p,q)\tilde G(-p,q)-(F\leftrightarrow G)\right\}\\
&= \frac{i\tau^2}{8}\left\{\int\frac{dp}{2\pi}\int_{-p/2}^{p/2}\frac{dq}{2\pi}\left(q^2-\frac{p^2}{4}\right)\tilde F(p,q)\tilde G(-p,q)-(F\leftrightarrow G)\right\}\end{aligned}$$

where the last expression follows from the previous owing to the the symmetry of $\tilde F$ and $\tilde G$ in q. Next, we pass to the local limit, perform the q integral and are left with

$$-\frac{i\tau^2}{96\pi}\int\frac{dp}{2\pi}p^3\left\{\tilde f(p)\tilde g(-p)-\tilde g(p)\tilde f(-p)\right\} = -\frac{\tau^2}{2}\frac{1}{48\pi}\int dx\,(f'''g-g'''f)$$

This corresponds to the charge algebra

$$[Q_f,Q_g] = iQ_{(f,g)} + \frac{i}{48\pi}\int dx\,(f'''g-g'''f) \tag{3.30}$$

i.e. in (1.16), $c=1$.

Note the following properties of the representation functional (3.27). $: U :$ is invariant against the replacement of $\varphi_i(x)$ by $\varphi_i(X(x))$. This is a consequence of the identity $\frac{dX}{f(X)} = \frac{dx}{f(x)}$ and the coordinate invariance of the symplectic structure $\int dx\, \varphi_1' \varphi_2 = -\int \varphi_1 d\varphi_2$. The occurrence of this symplectic form is natural for self-dual scalar fields. A Lagrangian for a self-dual field is

$$L_{\mathcal{X}} = \frac{1}{4} \int dx dy\, \mathcal{X}(t,x) \epsilon(x-y) \dot{\mathcal{X}}(t,y) - \frac{1}{2} \int dx\, \mathcal{X}^2(t,x) \tag{3.31}$$

which leads to the Hamiltonian

$$H_{\mathcal{X}} = \frac{1}{2} \int dx\, \mathcal{X}^2 = Q_1 \tag{3.32}$$

and canonical quantization produces precisely the commutator (3.3). Moreover, $L_{\mathcal{X}}$ is invariant under (3.4), and Q_f of (3.6) is the associated Noether constant of motion. So we recognize that $: U : (\varphi_1, \varphi_2; t)$ is the propagation kernel for time evolution of a self-dual field. Moreover, Q_f and $: U : (\varphi_1, \varphi_2; tf)$ may be viewed as the Hamiltonian and propagation kernel for a self-dual field $f\mathcal{X}$ whose propagation is described with time intervals that vary in space as $\Delta t = \frac{1}{f(x)}$.

The self-dual field $\mathcal{X}$, with its non-vanishing commutation relation (3.3), is somewhat analogous to the velocity operator of a charged particle in an external magnetic field. That operator also does not commute with itself; the commutator gives rise to the external magnetic field. So we may view a self-dual field as defined on a [functional] configuration space with a [functional] $U(1)$ connection that gives rise to a constant [functional] curvature given by (3.3).[9] Indeed, the propagator for a charged particle in a constant magnetic field bears close analogy to our functional propagators.[5]

Finally, let us observe that the kernel which connects an eigenstate of Π with that of Φ is given by an elegant expression. We record the formula for the unregulated bilocal transformation function.

$$\begin{aligned}
\tilde{U}(\pi_1, \varphi_2; \tau F) &\equiv \langle \pi_1 | e^{-i\tau Q_F} | \varphi_2 \rangle / \langle \pi_1 | \varphi_2 \rangle \\
\int \mathcal{D}\varphi_1 e^{-i \int dx \pi_1 (\varphi_1 - \varphi_2)} U(\varphi_1, \varphi_2; \tau F) &= e^{\tilde{\Gamma}(\tau F)} \exp -\frac{i}{2} \int dx\, \chi(x) K^{-1}_{\tau F}(x,y) \chi(y) \\
\chi &\equiv \pi_1 + \varphi_2' \\
K^{-1}_{\tau F} &= F^{1/2} \left\{ \frac{\tan\left(\frac{\tau}{2} k_F\right)}{k_F} \right\} F^{1/2} \\
\tilde{\Gamma}(\tau F) &= -\frac{1}{2} \operatorname{tr} \ln \left\{ \cos \frac{\tau}{2} k_F \right\}
\end{aligned} \tag{3.33}$$

As mentioned earlier, the representation functional for a full transformation acting both on x^+ and x^- may be obtained by composing the individual ones. In general, only formal expressions emerge which involve separate kernels K corresponding to the two separate transformations. A simple formula arises in the Rindler case, $-f^-(-x) = f^+(x) = \frac{1}{\sqrt{2}} f(\sqrt{2}x)$,

$$: U : (\varphi_1, \varphi_2; \tau f) = \delta\left(\varphi_1\left(X^{-1}(x)\right) - \varphi_2(x)\right) \tag{3.34}$$

where $X(x)$ is given in (3.8). As discussed in the Introduction, no center is present in the Lie algebra of the Rindler subgroup of transformations. Correspondingly, the representation (3.34) is faithful; it composes without a cocycle.

A closed formula for the representation functional may also be obtained when $f^+ = f^-$.[5]

IV. FUNCTIONAL REPRESENTATION OF CONFORMAL TRANSFORMATIONS $c > 1$

It is also possible to find representations which lead to an extension in (1.16) with $c > 1$. We consider again the scalar field Φ but postulate a formula for the energy-momentum tensor which modifies and "improves" upon (3.5).

$$\begin{aligned} \theta^{\lambda}_{\mu\nu} &= \theta_{\mu\nu} + 2\lambda\left(g_{\mu\nu}\Box - \partial_\mu\partial_\nu\right)\Phi \\ &= \theta_{\mu\nu} - 2\lambda\partial_\mu\partial_\nu\Phi \end{aligned} \tag{4.1}$$

$\theta^{\lambda}_{\mu\nu}$ remains conserved, symmetric and traceless, and we may construct the conformal charges from $\theta^{\lambda}_{\mu\nu}$, rather than from $\theta_{\mu\nu}$.

$$Q^{\lambda}_f = Q_f - 2\lambda \int dx\, f^\mu \partial_\mu \dot{\Phi} \tag{4.2}$$

The modified transformation generated by Q^{λ}_f is inhomogeneous.

$$i[Q^{\lambda}_f, \Phi] = f^\mu\partial_\mu\Phi + \lambda\partial_\mu f^\mu \tag{4.3}$$

This is a symmetry operation because $\lambda\partial_\mu f^\mu$ shifts Φ by a harmonic function $[\Box\partial_\mu f^\mu = 0$, when f^μ is a conformal Killing vector$]$ which is an invariance of a free, massless field.

In terms of the previously defined canonical variables we have

$$Q^{\lambda}_f = Q_f + \sqrt{2}\lambda \int dx\, f' \chi \tag{4.4}$$

It is now easy to compute the modified Lie algebra from (3.7), (3.30). Eq. (3.30) is again true for Q^{λ}_f except the center acquires the factor $1 + 48\pi\lambda^2$; *i.e.* $c = 1 + 48\pi\lambda^2$. Also, the representation functional for the finite transformation acquires a further phase.

$$:U^{\lambda}:(\varphi_1,\varphi_2;\tau f) =: U:(\varphi_1,\varphi_2;\tau f) \times \exp -i2\lambda \int dx\,(\ln\tau f)'\left(\varphi_1 - \varphi_2 - \frac{\tau\lambda}{2} f'\right) \tag{4.5}$$

The extension is already present classically: evaluation of the Lie algebra with Poisson brackets produces $c = 48\pi\lambda^2$, while the quantum correction adds the additional 1. It should be recalled that while the free theory allows any value of λ, there exists a 2-dimensional interacting theory, the Liouville model, with conformal invariance realized inhomogeneously as in (4.3), where λ is identified with the Liouville coupling constant.[10]

Are there models with $c < 1$? It is known that unitarity restricts values of c below 1.[11] The first permitted is $c = \frac{1}{2}$ and this is realized for free Majorana fermions. [Dirac fermions, even when interacting as in the Thirring model,[2] possess $c = 1$.] For $\frac{1}{2} < c < 1$, there are known statistical models at discrete values of c, but apparently there are no field theories. However, I know of no proof that for quantum field theory $c = 1/2$ or $c \geq 1$. Perhaps a spectral sum rule of the Källen-Lehmann sort is called for.

V. CONFORMALLY TRANSFORMED STATES AND PARTICLE PRODUCTION IN ACCELERATED FRAMES

I shall now discuss how the states of our massless free field theory transform under conformal transformations. Of course, the general result is clear: for any state, $\Psi(\varphi)$, which in our representation is given by a functional of φ, the transformed state $\tilde{\Psi}(\varphi)$ is

$$\tilde{\Psi}(\varphi) = \int \mathcal{D}\tilde{\varphi} : U : (\varphi, \tilde{\varphi}; \tau f)\Psi(\tilde{\varphi}) \tag{5.1a}$$

In order to deal with well-behaved expressions, it is better to view (5.1a) as the local limit of

$$\Psi_{\tau F}(\varphi) = \int \mathcal{D}\tilde{\varphi} : U : (\varphi, \tilde{\varphi}; \tau F)\Psi(\tilde{\varphi}) \tag{5.1b}$$

We present some explicit formulas.

An arbitrary Gaussian

$$\Psi_\Omega(\varphi) = e^{-\frac{1}{2}\int dx dy\, \varphi(x)\Omega(x,y)\varphi(y)} \tag{5.2}$$

transforms according to (5.1) into

$$\Psi_{\Omega_{\tau F}}(\varphi) = e^{\sigma(\tau F)}\; e^{-\frac{1}{2}\int dx dy\, \varphi(x)\Omega_{\tau F}(x,y)\varphi(y)} \tag{5.3}$$

where [see (3.17) and (3.23)].

$$\Omega_{\tau F} = \Omega - (\Omega + k)(\Omega - iK_{\tau F})^{-1}(\Omega - k) \tag{5.4a}$$

$$\sigma(\tau F) = \gamma(\tau F) - \frac{1}{2}\,\mathrm{tr}\ln\frac{1}{2\pi}(\Omega - iK_{\tau F}) \tag{5.4b}$$

Infinitesimally, these formulas read

$$\delta_F\Omega = -\frac{i\tau}{2}(\Omega + k)F(\Omega - k) \tag{5.5a}$$

$$\delta_F\sigma = -\frac{i\tau}{4}\,\mathrm{tr}\,F(\Omega - \omega) \tag{5.5b}$$

and one verifies that (5.5a) gives rise to the Lie algebra without center, which resides in (5.5b).

Of particular interest, of course, is the vacuum state for the Hamiltonian of our theory.

$$H = \frac{1}{2}\int dx\, \left(\Pi^2 + \Phi'^2\right) \tag{5.6}$$

The ground state functional is Gaussian

$$\begin{aligned}\Psi_\omega(\varphi) &= e^{-\frac{1}{2}\int dx dy\, \varphi(x)\omega(x,y)\varphi(y)} \\ \omega(x,y) &= -\frac{1}{\pi} P \frac{1}{(x-y)^2}\end{aligned} \tag{5.7}$$

The occurrence of precisely the kernel ω in the ground state shows that our group theoretically motivated subtraction in Q_F coincides with the conventional subtraction. From (3.24) we see that ω_f may also be written as $\omega_f(x,y) = f^{1/2}(x)\left\{\frac{1}{f(x)}\omega\left(\int^x \frac{dz}{f(z)}, \int^y \frac{dz}{f(z)}\right)\frac{1}{f(y)}\right\} f^{1/2}(y)$. A Gaussian with the kernel as in the curly brackets is an eigenstate of the unsubtracted Q_f, Eq. (3.6), with eigenvalue $\frac{1}{4}\operatorname{tr}\omega_f$. Hence, we may also view our regularization of Q_f as normal ordering with respect to the "vacuum" of Q_f.

The conformally transformed vacuum is given by Eqs. (5.3) and (5.4) with $\Omega = \omega$. By iteration, one can obtain the relevant expressions for the case when both x^+ and x^- are transformed.

Evidently, the ground state is not invariant under arbitrary conformal transformations. This, of course, is no surprise: when a transformation is represented projectively there can be no invariant states; equivalently, a central extension in the Lie algebra of generators prevents the generators from annihilating states. On the other hand, since the $SO(2,2)$ subgroup possess no center, one expects invariance of the ground state. Indeed, one can confirm this from (5.4): with $\Omega = \omega$ and $F \to f = 1, x, x^2$, the second member of the right-hand side of (5.4a) vanishes.[12]

The Rindler subgroup also has no center in the Lie algebra and is represented faithfully by (3.34). But the ground state is not invariant, as follows from (3.34) and (5.1). The transformed vacuum is a Gaussian with kernel $\tilde{\omega}$.

$$\begin{aligned}\tilde{\omega}(x,y) &= \frac{\partial X(x)}{\partial x}\omega\left(X(x), Y(y)\right)\frac{\partial Y(y)}{\partial y} \\ \frac{\partial X(x)}{\partial x} &= \frac{f(X)}{f(x)}\ , \qquad \frac{\partial Y(y)}{\partial y} = \frac{f(Y)}{f(y)}\end{aligned} \tag{5.8}$$

Thus for the conformal group, with the exception of the $SO(2,2)$ subgroup, we are facing a kind of "spontaneous symmetry breaking", which is triggered by the central extension in general, but remains mysterious for the Rindler subgroup.[12] Moreover, this breaking can be related to known physical effects in an accelerated reference frame, which is what a conformally transformed coordinate system describes. We define the number operator

$$N(p,p') = a^\dagger(p)a(p') \tag{5.9}$$

where

$$a(p) = \int \frac{dx}{\sqrt{4\pi|p|}} e^{-ipx}\left[|p|\Phi(x) + i\Pi(x)\right] \tag{5.10}$$

and compute its expectation in the transformed vacuum $\Psi_{\tilde{\omega}}$. An alternative and equivalent computation is the expectation of the transformed number operator in the standard vacuum (5.2). An explicit formula can be given,[5] which for the Rindler subgroup is

$$\langle N(p,p')\rangle = \frac{1}{4\pi\sqrt{|pp'|}} \int dx dy \; e^{+ipx} \; e^{-ip'y} \times$$
$$\times \left(\frac{\partial X^{-1}(x)}{\partial x} \omega \left(X^{-1}(x), Y^{-1}(y)\right) \frac{\partial Y^{-1}(y)}{\partial y} \mp k(x,y) \right) \tag{5.11}$$

where the upper sign holds for p and p' positive, the lower for p and p' negative, and $\langle N\rangle$ vanishes otherwise. The nature of the spectrum is determined by the analytic properties of the new coordinates in terms of the old.[13] If in particular,

$$X^{-1}(x) = \frac{1}{\mu} e^{\mu x} \tag{5.12}$$

i.e. the transformation is to a system with uniform acceleration μ, Eq. (5.11) reduces to the well-known thermal spectrum.[14]

$$\langle N(p,p')\rangle = \delta(p-p') \frac{1}{e^{\frac{2\pi}{\mu}|p|} - 1} \tag{5.13}$$

[It should be emphasized that (5.12) is not properly a member of a group, rather it belongs to a semi-group.] We thus arrive at an interpretation for (5.11), (5.13) and for particle production in an accelerated frame as spontaneous breaking of conformal symmetry.

REFERENCES

1. S. Treiman, R. Jackiw, B. Zumino and E. Witten, *Current Algebra and Anomalies*, World Scientific, Singapore (1985), p. 168.

2. S. Fubini, A. Hanson and R. Jackiw, *Phys. Rev. D* **7**, 1732 (1973).

3. For a review, see N. Birrell and P. Davies, *Quantum Fields in Curved Space*, Cambridge University Press, Cambridge (1982).

4. R. Floreanini and R. Jackiw, *Phys. Lett. B* (in press) (MIT preprint CTP# 1351).

5. R. Floreanini, to appear (MIT preprint CTP# 1370).

6. For a review, see Ref. [1], p. 311.

7. M. Bos, *Phys. Lett. B* (in press) (MIT preprint CTP# 1348) and to appear (MIT preprint CTP# 1376).

8. Representing linear canonical transformations in field theory without choosing a ground state has been discussed by I. Segal, *Mathematical Problems of Relativistic Physics*, American Mathematical Society, Providence, RI (1963).

9. Functional $U(1)$ connections and curvatures have also arisen in gauge theories with topological effects. For a review, see Ref. [1], p. 309.

10. E. D'Hoker and R. Jackiw, *Phys. Rev. D* **26**, 3517 (1982), *Phys. Rev. Lett.* **50**, 1719 (1983); E. D'Hoker, D. Freedman and R. Jackiw, *Phys. Rev. D* **28**, 2583 (1983). For a review, see R. Jackiw in *Progress in Quantum Field Theory*, H. Ezawa and S. Kamefuchi, eds., North-Holland, Amsterdam (1986).

11. For a review, see D. Friedan and S. Shenker in *Unified String Theories*, M. Green and D. Gross, eds., World Scientific, Singapore (1986).

12. However, it has been argued that in the Liouville theory the ground state is not invariant under the full $SO(2,2) = S)(2,1) \otimes SO(2,1)$, rather under the diagonal $SO(2,1)$ subgroup; see Ref. [10].

13. N. Sanchez, *Phys. Rev. D.* **24**, 2100 (1981).

14. For a recent discussion, see T. D. Lee, *Nucl. Phys.* **B264**, 537 (1986).

RECENT DEVELOPMENTS IN THE PATH INTEGRAL APPROACH TO ANOMALIES

Kazuo Fujikawa

Research Institute for
Theoretical Physics
Hiroshima University
Takehara, Hiroshima 725

Abstract

After a brief summary of the path integral approach to anomalous identities, some of the recent developments in this approach are discussed. The topics discussed include

(i) Construction of the effective action by means of the covariant current,
(ii) Gauss law constraint in anomalous gauge theories,
(iii) Path integral approach to anomalies in superconformal transformations,
(iv) Conformal and ghost number anomalies in string theory in analogy with the instanton calculation,
(v) Covariant local Lorentz anomaly and its connection with the mathematical construction of the consistent anomaly.

I. Introduction

The symmetry principle plays fundamental roles in particle physics. The symmetries and the related currents are well specified in the classical level as Noether's theorem. When one deals with the quantum theory with an infinite number of degrees of freedom, however, some of the fundamental classical symmetries are spoiled by the quantization procedure. This phenomenon is generally referred to as the anomaly.[1] Recent studies in field theory revealed that the anomaly is a general and fundamental property of quantum field theory.

In the present lecture, I would like to discuss one particular approach to the anomaly on the basis of the Feynman path integral.[2,3] Each approach to field theory has its own advantages. The path integral approach allows us to understand all the known anomalies as the Jacobian factors arising from the symmetry transformations of path integral variables. This universality and the physically intuitive understanding of the anomaly may be the main features of the path integral method. The path integral method also provides a convenient bridge between the conventional field theoretical formulation and the differential geometrical formulation of the anomaly.[4] The connection between the anomaly and the topological properties of the background gauge and gravitational fields thus become transparent.

In the following I would like to discuss the basic idea of the path integral approach to anomalies and some of the recent developments. The topics covered here are rather limited; for example, the application of the path integral method to two-dimensional field theories is not discussed except for the string path integral. For this reason, I would like to apologize those authors whose contributions to the recent developments in this field are not adequately treated.

II. Path Integral Approach to Chiral Gauge Anomalies

2a, Chiral U(1) Anomaly

We illustrate the basic idea of the path integral approach to anomalies by considering the chiral U(1) anomaly in the QCD Lagrangian

$$\mathcal{L} = \bar{\psi}(i\not{D} - m)\psi - \frac{1}{2g^2}\mathrm{Tr}F^{\mu\nu}F_{\mu\nu} \tag{2.1}$$

$$\not{D} = \gamma^\mu(\partial_\mu - iA_\mu)$$

$$A_\mu \equiv A^a_\mu(x)T^a$$

$$[T^a, T^b] = if^{abc}T^c, \qquad \mathrm{Tr}T^aT^b = \frac{1}{2}\delta^{ab}.$$

We note that the γ-matrices in our convention become anti-hermitean after Wick-rotation to the Enclidean theory

$$\gamma^{\mu\dagger} = -\gamma^\mu \tag{2.2}$$

The path integral is then defined by

$$\int \prod_x \mathcal{D}\bar{\psi}(x)\mathcal{D}\psi(x)[\mathcal{D}A_\mu(x)]\exp[\int \mathcal{L}(x)dx] \tag{2.3}$$

where $[\mathcal{D}A_\mu(x)]$ includes the gauge fixing and compensating terms. For the moment, we forget $[\mathcal{D}A_\mu]$ and define

$$d\mu(\psi) = \prod_x \mathcal{D}\bar{\psi}(x)\mathcal{D}\psi(x) \tag{2.4}$$

Under the infinitesimal chiral U(1) transformation specified by a localized parameter $\alpha(x)$

$$\begin{aligned}\psi(x) &\to \psi'(x) = e^{i\alpha(x)\gamma_5}\psi(x)\\ \bar{\psi}(x) &\to \bar{\psi}(x)' = \bar{\psi}(x)e^{i\alpha(x)\gamma_5}\end{aligned} \tag{2.5}$$

the Lagrangian changes as

$$\mathcal{L} \to \mathcal{L} - \partial_\mu\alpha(x)\bar{\psi}\gamma^\mu\gamma_5\psi(x) - \alpha(x)2mi\bar{\psi}\gamma_5\psi(x) \tag{2.6}$$

where the coefficient of $\partial_\mu\alpha(x)$ defines the symmetry current. If the measure (2.4) should remain invariant under (2.5), we would have

$$\int d\mu(\psi)e^{S(\psi)} \equiv \int d\mu(\psi')e^{S(\psi')} = \int d\mu(\psi)e^{S(\psi')} \tag{2.7}$$

with $S(\psi) = \int \mathcal{L}(x)dx$. By expanding (2.7) in powers in $\alpha(x)$, one obtains the naive identity

$$\partial_\mu \langle\bar{\psi}\gamma^\mu\gamma_5\psi(x)\rangle - 2im\langle\bar{\psi}\gamma_5\psi(x)\rangle = 0 \tag{2.8}$$

where $\langle O(x)\rangle$ stands for the average of $O(x)$ in the path integral formula.

A careful estimate of the Jacobian factor for the transformation (2.5) in fact gives

$$d\mu(\psi') = d\mu(\psi)\exp[-i\int \alpha(x)A(x)dx] \tag{2.9}$$

and the relation (2.8) is modified to

$$\partial_\mu \langle\bar{\psi}\gamma^\mu\gamma_5\psi(x)\rangle - 2im\langle\bar{\psi}\gamma_5\psi(x)\rangle = iA(x) \tag{2.10}$$

the well-known anomalous chiral U(1) identity.

To evaluate the Jacobian factor in (2.9), it becomes more transparent if one expands $\psi(x)$ and $\bar{\psi}(x)$ as

$$\begin{aligned}\psi(x) &= \sum_n a_n \varphi_n(x) = \sum_n a_n \langle x|\varphi_n\rangle \\ \bar{\psi}(x) &= \sum_n \bar{b}_n \varphi_n(x)^\dagger = \sum_n \bar{b}_n \langle\varphi_n|x\rangle\end{aligned} \tag{2.11}$$

with

$$\not{D}\varphi_n(x) = \lambda_n \varphi_n(x) \ , \qquad \int \varphi_m(x)^\dagger \varphi_n(x)dx = \delta_{m,n} \ . \tag{2.12}$$

The coefficients a_n and $\bar{b}_n$ in (2.11) are the elements of the Grassmann algebra. The expansion such as (2.11) is well-known in the path integral formalism. The path integral measure is then given by

$$d\mu(\psi) = \frac{1}{\det[\langle x|\varphi_n\rangle]\det[\langle\varphi_n|x\rangle]} \prod_n da_n d\bar{b}_n = \prod_n da_n d\bar{b}_n \ . \tag{2.13}$$

In the present case, where the basic operator $\not{D}$ in (2.12) is hermitean, the Jacobian for the transformation from $\psi(x)$ and $\bar{\psi}(x)$ to a_n and $\bar{b}_n$ in (2.13) becomes unity. In terms of (2.11), the fermionic action in (2.1) becomes

$$\int \bar{\psi}(i\not{D} - m)\psi dx = \sum_n (i\lambda_n - m)\bar{b}_n a_n \ . \tag{2.14}$$

Namely, the fermionic action is formally diagonalized, which may justify the use of the particular basis set in (2.12).

Under the chiral transformation (2.5), the coefficient a_n, for example, is transformed as

$$\psi'(x) = \sum_n a_n' \varphi_n(x) = \sum_m e^{i\alpha(x)\gamma_5} a_m \varphi_m(x) \ . \tag{2.15}$$

Namely,

$$a_m' = \sum_n \int \varphi_m(x)^\dagger e^{i\alpha(x)\gamma_5} \varphi_n(x) dx \, a_n \tag{2.16}$$

and

$$\prod_m da_m' = \det[\int \varphi_m(x)^\dagger e^{i\alpha(x)\gamma_5} \varphi_n(x) dx]^{-1} \prod_n da_n \tag{2.17}$$
$$= \exp[-i\sum_n \int \varphi_n(x)^\dagger \alpha(x)\gamma_5 \varphi_n(x) dx] \prod_n da_n \ .$$

One may sum the series in (2.17) starting from small eigenvalues ($|\lambda_n| \leq M$ and $M \to \infty$ later)

$$\sum_n \varphi_n(x)^\dagger \gamma_5 \varphi_n(x) \equiv \lim_{M\to\infty} \sum \varphi_n(x)^\dagger \gamma_5 e^{-\lambda_n^2/M^2} \varphi_n(x)$$
$$\equiv \lim_{M\to\infty} \sum \varphi_n(x) \gamma_5 e^{-\not{D}^2/M^2} \varphi_n(x) \ . \tag{2.18}$$

In this regularized form, one may change the basis set to plane waves (interaction picture), and one obtains

$$\lim_{M\to\infty} \mathrm{Tr} \int \frac{d^4k}{(2\pi)^4} e^{-ikx} \gamma_5 e^{-\not{D}^2/M^2} e^{ikx} \tag{2.19}$$

$$= \lim_{M\to\infty} \mathrm{Tr} \int \frac{d^4k}{(2\pi)^4} \gamma_5 \exp[-\frac{(ik_\mu + D_\mu)(ik^\mu + D^\mu)}{M^2} + \frac{i}{4M^2}[\gamma^\mu, \gamma^\nu]F_{\mu\nu}]$$

since

$$\not{D}^2 = D_\mu D^\mu - \frac{i}{4}[\gamma^\mu, \gamma^\nu]F_{\mu\nu} \ . \tag{2.20}$$

The trace in (2.19) runs over the Dirac and internal symmetry indices.

After re-scaling $k_\mu \to Mk_\mu$, one may perform the $1/M$ expansion in (2.19). If one takes the trace with γ_5 into account, (2.19) can be rewritten as

$$\lim_{M\to\infty} \mathrm{Tr} M^4 \int \frac{d^4k}{(2\pi)^4} \gamma_5 \exp[-k_\mu k_\mu + \frac{i}{4M^2}[\gamma^\mu, \gamma^\nu]F_{\mu\nu}]$$

$$= \lim_{M\to\infty} \mathrm{Tr}\gamma_5 \frac{1}{2!}(\frac{i}{4}[\gamma^\mu, \gamma^\nu]F_{\mu\nu})^2 \int \frac{d^4k}{(2\pi)^4} e^{-k_\mu k_\mu}$$

$$= \frac{1}{32\pi^2} \mathrm{Tr}\, \varepsilon^{\mu\nu\alpha\beta} F_{\mu\nu} F_{\alpha\beta} \ . \tag{2.21}$$

In 2n-dimensions, the calculation proceeds in an identical manner, and one obtains

$$\mathrm{Tr} \frac{1}{n!} \gamma_5 (\frac{i}{4}[\gamma^\mu, \gamma^\nu]F_{\mu\nu})^n \int \frac{d^{2n}k}{(2\pi)^{2n}} e^{-k_\mu k_\mu} = \frac{1}{(2\pi)^n} \frac{i^n}{n!} \mathrm{Tr} F^n \tag{2.22}$$

with the form notation

$$F \equiv \frac{1}{2} F_{\mu\nu} dx^\mu dx^\nu \ . \tag{2.23}$$

The γ-matrices in (2.22) play the role of dx^μ in (2.23) in the precence of $\mathrm{Tr}\gamma_5$; the precise sign factor in (2.23) depends on the definition of γ_5. We note that the final result (2.21) or (2.23) is independent of the particular form of the regulator $\exp[-\lambda_n^2/M^2]$ in (2.18).

Coming back to 4-dimensions, we have equal contributions from $\mathcal{D}\psi$ and $\mathcal{D}\bar{\psi}$, and

$$d\mu \to d\mu \exp[-2i\mathrm{Tr}\int dx\alpha(x)\frac{\varepsilon^{\mu\nu\alpha\beta}}{32\pi^2}F_{\mu\nu}F_{\alpha\beta}] \tag{2.24}$$

and the anomalous chiral U(1) identity

$$\partial_\mu\langle\bar{\psi}\gamma^\mu\gamma_5\psi(x)\rangle - 2im\langle\bar{\psi}\gamma_5\psi(x)\rangle = \langle\frac{i}{16\pi^2}\mathrm{Tr}\,\varepsilon^{\mu\nu\alpha\beta}F_{\mu\nu}F_{\alpha\beta}\rangle\,. \tag{2.25}$$

We note that the calculation (2.18) corresponds to a local version of the Atiyah-Singer index theorem.[5] If one assumes that a global limit $\alpha(x) \to$ constant can be taken in (2.24), one obtains

$$d\mu \to \sum_\nu d\mu_{(\nu)}\exp[-2i\alpha\nu] \tag{2.26}$$

with ν the Pontryagin index, and the θ-vacuum structure.[6,7]

2b, Non-Abelian Anomaly

We next discuss the non-Abelian anomaly,[8] namely the anomaly in non-Abelian gauge transformations. We illustrate the basic features of the non-Abelian anomaly by considering the model

$$\mathcal{L} = \bar{\psi}i\not{D}\psi$$

$$\not{D} = \gamma^\mu(\partial_\mu - iV_\mu^aT^a - iA_\mu^aT^a\gamma_5) \tag{2.27}$$

$$[T^a, T^b] = if^{abc}T^c, \qquad \mathrm{Tr}T^aT^b = \frac{1}{2}\delta^{ab}\,.$$

A salient feature of (2.27) is that the basic operator $\not{D}$ is <u>not</u> hermitean in the Euclidean sense

$$(\Phi, \not{D}\Psi) \equiv (\not{D}^\dagger\Phi, \Psi) \tag{2.27}$$

for

$$(\Phi, \Psi) = \int\Phi(x)^\dagger\Psi(x)dx\,. \tag{2.28}$$

To be precise

$$\not{D}^\dagger = \gamma^\mu(\partial_\mu - iV_\mu + iA_\mu\gamma_5) \neq \not{D}\,. \tag{2.29}$$

There are basically two distinct ways to handle this situation:

(i) "Analytic" continuation in A_μ, $A_\mu \to iA_\mu$, and[9]

$$\not{D} \equiv \gamma^{\mu}(\partial_{\mu} - iV_{\mu} + A_{\mu}\gamma_5) = \not{D}^{\dagger} . \tag{2.30}$$

This gives rise to the integrable (or consistent) anomaly,[10] if one uses the basis set

$$\not{D}\varphi_n(x) = \lambda_n \varphi_n(x) \tag{2.31}$$

and the corresponding regulator with a hermitean $\not{D}$ in (2.30) as

$$e^{-\not{D}^2/M^2} . \tag{2.32}$$

For example, one obtains the Jacobian factor

$$-2i\mathrm{Tr}\int \frac{d^4k}{(2\pi)^4}\, e^{-ikx}\gamma_5 T^a e^{-\not{D}^2/M^2} e^{ikx} \tag{2.33}$$

for the chiral transformation

$$\begin{aligned} \psi(x) &\rightarrow \psi'(x) = e^{i\alpha^a(x)T^a\gamma_5}\psi(x), \\ \bar{\psi}(x) &\rightarrow \bar{\psi}(x)' = \bar{\psi}(x)e^{i\alpha^a(x)T^a\gamma_5} . \end{aligned} \tag{2.34}$$

The calculation (2.33) is straightforward but tedious, and one obtains the result

$$\frac{1}{24\pi^2}\mathrm{Tr}T^a\varepsilon^{\mu\nu\alpha\beta}\partial_{\mu}(W_{\nu}\partial_{\alpha}W_{\beta} - \frac{i}{2}W_{\nu}W_{\alpha}W_{\beta}) \tag{2.35}$$

if one sets $\frac{1}{2}W_{\mu} = V_{\mu} = -A_{\mu}$ after the calculation of (2.33).

A notable feature of the prescription (2.30) is that the vector transformation

$$\begin{aligned} \psi(x) &\rightarrow \psi'(x) = e^{i\alpha^a(x)T^a}\psi(x) \\ \bar{\psi}(x) &\rightarrow \bar{\psi}'(x) = \bar{\psi}(x)\, e^{-i\alpha^a(x)T^a} \end{aligned} \tag{2.36}$$

is always anomaly-free.

(ii) The second method is to use the polar decomposition of the original $\not{D}$ in (2.27) by means of the eigenvalue equations,[11]

$$\begin{aligned} \not{D}^{\dagger}\not{D}\varphi_n(x) &= \lambda_n^2\varphi_n(x) \\ \not{D}\not{D}^{\dagger}\phi_n(x) &= \lambda_n^2\phi_n(x) \\ \psi(x) &= \sum_n a_n\phi_n(x) \equiv \sum_n a_n\langle x|\varphi_n\rangle \\ \bar{\psi}(x) &= \sum_n \bar{b}_n\phi_n(x)^{\dagger} = \sum_n \bar{b}_n\langle\phi_n|x\rangle \end{aligned} \tag{2.37}$$

and

$$d\mu = \Pi \mathcal{D}\bar{\psi}\mathcal{D}\psi = \det[\langle x|\varphi_n\rangle]^{-1}\det[\langle\phi_n|x\rangle]^{-1}\prod_n d\bar{b}_n da_n$$

$$= \det[\langle x|\phi_n\rangle]\det[\langle\varphi_n|x\rangle]\prod_n d\bar{b}_n da_n \ .$$

One thus obtains

$$\begin{aligned} \det \not{D} &\equiv \int d\mu \, e^{\int \mathcal{L} dx} \\ &= \int d\mu \, \exp[\sum_n i\lambda_n \bar{b}_n a_n] \\ &= \det[\langle x|\phi_n\rangle]\prod_n \lambda_n \det[\langle\varphi_n|x\rangle] \end{aligned} \tag{2.38}$$

if one notes

$$\not{D}\varphi_n(x) = \lambda_n \phi_n(x) \tag{2.39}$$

for a suitable choice of the phase factor. Eq.(2.38) corresponds to the polar decomposition of a finite dimensional matrix M and det M = $\det[U\Lambda V^\dagger]$ with two unitary matrices U and V and a diagonal Λ. [The phase factors in (2.38) are not well-regularized even if one cuts off the large eigenvalues λ_n. The anomaly and the associated current are however well specified, as will be explained later].

The Jacobian factor for the transformation (2.34) is thus given by

$$\begin{aligned} &(-i)\sum_n [\varphi_n(x)^\dagger\gamma_5 T^a \varphi_n(x) + \phi_n(x)^\dagger\gamma_5 T^a\phi_n(x)] \\ &\equiv (-i)\lim_{M\to\infty}\sum_n [\varphi_n(x)^\dagger\gamma_5 T^a e^{-\not{D}^\dagger\not{D}/M^2}\varphi_n(x) \\ &\qquad + \phi_n(x)^\dagger\gamma_5 T^a e^{-\not{D}\not{D}^\dagger/M^2}\phi_n(x)] \\ &= (-i)\lim_{M\to\infty} \mathrm{Tr}\int\frac{d^4k}{(2\pi)^4} e^{-ikx}\gamma_5 T^a [e^{-\not{D}^\dagger\not{D}/M^2} + e^{-\not{D}\not{D}^\dagger/M^2}]e^{ikx} \ . \end{aligned} \tag{2.40}$$

The calculation of (2.40) becomes transparent if one writes

$$\not{D} = \gamma^\mu(\partial_\mu - iL_\mu)L + \gamma^\mu(\partial_\mu - iR_\mu)R \tag{2.41}$$

with

$$\begin{aligned} L_\mu &= V_\mu - A_\mu \\ R_\mu &= V_\mu + A_\mu \\ L = (1-\gamma_5)/2 \ , &\quad R = (1+\gamma_5)/2 \ . \end{aligned} \tag{2.42}$$

One can then confirm

$$\begin{aligned} \not{D}^\dagger\not{D} &= \not{D}(L)^2 L + \not{D}(R)^2 R \\ \not{D}\not{D}^\dagger &= \not{D}(L)^2 R + \not{D}(R)^2 L \end{aligned} \tag{2.43}$$

and

$$e^{-\not{D}^{\dagger}\not{D}/M^2} \pm e^{-\not{D}\not{D}^{\dagger}/M^2}$$
$$= (L \pm R)e^{-\not{D}(L)^2/M^2} + (R \pm L)e^{-\not{D}(R)^2/M^2} . \tag{2.44}$$

The evaluation of (2.40) is then essentially reduced to that of the chiral U(1) anomaly in (2.19), and we obtain

$$(-i)\frac{1}{32\pi^2} \mathrm{Tr}T^a \varepsilon^{\mu\nu\alpha\beta}[F_{\mu\nu}(R)F_{\alpha\beta}(R) + F_{\mu\nu}(L)F_{\alpha\beta}(L)] . \tag{2.45}$$

The calculation of the "covariant" anomaly (2.45) in arbitrary 2n-dimensions proceeds just as in (2.22).

A salient feature of the present prescription is that the Jacobian for the vector transformation (2.36) contains the anomaly

$$(-i)\frac{1}{32\pi^2} \mathrm{Tr}T^a \varepsilon^{\mu\nu\alpha\beta}[F_{\mu\nu}(R)F_{\alpha\beta}(R) - F_{\mu\nu}(L)F_{\alpha\beta}(L)] . \tag{2.46}$$

In particular, if one sets $R_\mu = 0$ and $L_\mu = W_\mu$ in (2.46) one obtains

$$i\frac{1}{32\pi^2} \mathrm{Tr}T^a \varepsilon^{\mu\nu\alpha\beta}F_{\mu\nu}(W)F_{\alpha\beta}(W) \tag{2.47}$$

which leads to the baryon (and lepton) number non-conservation (if one sets $T^a = 1$) in the Weinberg-Salam theory in the presence of instantons.[6]

In summary, the non-Abelian anomalies can be characterized by two different forms of anomalies corresponding to the two different definitions of composite current operators involved. The anomaly cancellation condition, for example, becomes identical for those two different forms of the anomaly. The integrable form of the non-Abelian anomaly (2.35) allows the integration of the anomalous identities in the form of the Wess-Zumino term.

We also note that the "covariant" form of the anomaly in d = 2n dimensions is obtained from the U(1)-type anomaly such as (2.22) in d = 2n + 2 dimensions by the replacement

$$F \rightarrow F + \omega(x) \tag{2.48}$$

and picking up the term linear in the transformation parameter $\omega(x) = \omega^a(x)T^a$, which is the zero-form. Starting from (2.22) in d = 2n + 2 dimensions, one obtains

$$\frac{i^{n+1}}{(2\pi)^{n+1}}\frac{1}{(n+1)!}\mathrm{Tr}(F + \omega)^{n+1} \rightarrow (\frac{i}{2\pi})\frac{i^n}{(2\pi)^n}\frac{1}{n!}\mathrm{Tr}\,\omega F^n \tag{2.49}$$

which gives the generalization of (2.45) with $L_\mu = 0$, for example. A similar property for the consistent anomaly is well-known.[4] Those properties show that the anomaly cancellation in the level of d = 2n + 2 U(1)-type anomaly ensures the non-Abelian anomaly cancellation in d = 2n.

III. Construction of the Effective Action by means of the Covariant Current

3a, Regularized Covariant Current

The non-Abelian anomaly (2.45) is produced if one considers the axial current[11]

$$
\begin{aligned}
J_5^{\mu a}(x) &= \langle \bar{\psi}\gamma^\mu \gamma_5 T^a \psi(x)\rangle \\
&\equiv \sum_n \phi_n(x)^\dagger \gamma^\mu \gamma_5 T^a \varphi_n(x) \frac{1}{i\lambda_n} e^{-\lambda_n^2/M^2}
\end{aligned}
\tag{3.1}
$$

where we assume that $\lambda_n \neq 0$ by suitably adjusting the global behavior of the background gauge field. We then obtain

$$
\begin{aligned}
D_\mu J_5^{\mu a}(x) &= \sum_n [(\not{D}^\dagger \phi_n(x))^\dagger \gamma_5 T^a \varphi_n(x) \\
&\quad + \phi_n(x)^\dagger \gamma_5 T^a \not{D} \varphi_n(x)] \frac{-1}{i\lambda_n} e^{-\lambda_n^2/M^2} \\
&= i \sum_n [\varphi_n(x)^\dagger \gamma_5 T^a \varphi_n(x) + \phi_n(x)^\dagger \gamma_5 T^a \phi_n(x)] e^{-\lambda_n^2/M^2}
\end{aligned}
\tag{3.2}
$$

which gives rise to the anomaly factor (2.40). One can also confirm that the anomaly factor remains unchanged even if one replaces[11]

$$
e^{-\lambda_n^2/M^2} \to f(\lambda_n^2/M^2)
\tag{3.3}
$$

for any smooth function with $f(0) = 1$, $f(\infty) = f'(\infty) = f''(\infty) = \ldots = 0$. For the perturbative calculation in the Minkowski metric, it is convenient to choose

$$
f(x) = \left(\frac{1}{1+x}\right)^n
\tag{3.4}
$$

with n a positive integer. The current (3.1) is then rewritten as

$$
\begin{aligned}
J_5^{\mu a}(x) &= \mathrm{Tr}[\gamma^\mu \gamma_5 T^a \frac{1}{i\not{D}} \left(\frac{M^2}{M^2+\not{D}\not{D}^\dagger}\right)^n \delta(x-y)]_{y\to x} \\
&= \mathrm{Tr}\int \frac{d^4k}{(2\pi)^4} e^{-ikx} [\gamma^\mu \gamma_5 T^a \frac{1}{i\not{D}} \left(\frac{M^2}{M^2+\not{D}\not{D}^\dagger}\right)^n] e^{ikx}
\end{aligned}
\tag{3.5}
$$

In any case, the currents such as (3.1) and (3.5) are well-regularized, and one may expand (3.5) in powers of gauge fields A_μ and V_μ. By this way one can define (3.5) as a well-defined functional of A_μ and V_μ.

3b, Definition of the Vacuum Functional in Terms of Covariant Current

Recently, H. Banerjee, R. Banerjee and P. Mitra[12] showed that one can define the vacuum functional

$$
\exp\{W(L_\mu)\} \equiv \int \Pi \mathcal{D}\bar{\psi}\mathcal{D}\psi\, e^{\int [\bar{\psi} i\not{D}(L)\left(\frac{1-\gamma_5}{2}\right)\psi + \bar{\psi} i\not{\partial}\left(\frac{1+\gamma_5}{2}\right)\psi] dx}
\tag{3.6}
$$

by

$$
W(L_\mu) \equiv \int_0^1 \int dt L_\mu^a(x) J^{\mu a}(tL_\mu)(x) dx
\tag{3.7}
$$

in terms of the covariant current (3.5) with the replacement $\gamma_5 \to (\frac{1-\gamma_5}{2})$. We also set $V_\mu = -A_\mu = \frac{1}{2}L_\mu$ in (3.5). The formal derivation of (3.7) proceeds as

$$\frac{\partial}{\partial t} W(tL_\mu) = \int L^a_\mu(x) \frac{\delta}{\delta(tL^a_\mu(x))} W(tL_\mu)dx$$
$$= \int L^a_\mu(x) J^{\mu a}(tL_\mu)(x)dx \qquad (3.8)$$

The integration of (3.8) gives (3.7) after discarding L_μ independent terms.

The manipulation (3.7) is formal, but one can take (3.7) as the definition of $W(L_\mu)$. This definition of W satisfies the following properties:

(i) Those parts of $W(L_\mu)$ which remain finite for $M \to \infty$ agree with the conventional definition, since they are independent of the regularization in the limit $M \to \infty$.

(ii) The anomaly generated from $W(L_\mu)$ satisfies the Wess-Zumino integrability condition, since it is generated from the well regularized finite functional $W(L_\mu)$. In particular, the "consistent" current is defined by

$$j^{\mu a}(x) \equiv \frac{\delta}{\delta L^a_\mu(x)} W(L_\mu) \qquad (3.9)$$
$$= \int_0^1 dt\ [J^{\mu a}(tL_\mu)(x) + \int dy L^b_\alpha(y) \frac{\delta}{\delta L^a_\mu(x)} J^{\alpha b}(tL_\mu)(y)]$$

The property (i) above may be obvious. As for the property (ii), it is sufficient to show that the leading term in the anomaly generated from $W(L_\mu)$ agrees with the conventional definition, since the higher terms in the gauge fields are fixed by the Wess-Zumino condition. The leading term of the non-Abelian anomaly has the same gauge property as in the chiral Abelian theory, and one may study the chiral Abelian model. In this case one obtains

$$W(L_\mu + \partial_\mu\omega) = \int_0^1 dt \int (L_\mu + \partial_\mu\omega) J^\mu(t(L_\mu + \partial_\mu\omega))dx$$
$$= \int_0^1 dt \int (L_\mu + \partial_\mu\omega) J^\mu(tL_\mu)dx \qquad (3.10)$$

namely,

$$\delta W = -\int_0^1 dt \int \omega(x) \partial_\mu J^\mu(tL_\mu)dx$$
$$= -\int_0^1 dt\ t^2 (\frac{-i}{32\pi^2}) \varepsilon^{\mu\nu\alpha\beta} F_{\mu\nu}(L) F_{\alpha\beta}(L)$$
$$= (\frac{1}{3}) \frac{i}{32\pi^2} \varepsilon^{\mu\nu\alpha\beta} F_{\mu\nu}(L) F_{\alpha\beta}(L) \qquad (3.11)$$

We thus recover the bose symmetrization factor $1/(n+1)$ for $d = 2n$ dimensions. In (3.10), we used the fact that the covariant current is gauge invariant for the Abelian theory. [For non-Abelian theory, the factor t in front of L_μ in $J^{\mu a}(tL_\mu)$ spoils the gauge covariance of $J^{\mu a}$.]

One can thus see that $W(L_\mu)$ defines the acceptable vacuum functional and $j^{\mu a}$ in (3.9) defines the consistent current. This definition of $W(L_\mu)$ in terms of the covariant current is quite attractive, since the covariant regularization of the current (3.1) and (3.5) can be readily defined. For further discussions of this approach, interested readers are referred to Ref. 12.

IV. Quantization of Anomalous Gauge Theories

4a, Gauss Law Constraint

The quantization of anomalous gauge theories received much attention recently, mainly because of the influential work by Faddeev.[13] On the basis of differential geometrical formulation of non-Abelian anomalies, Faddeev presented an elegant derivation of anomalous commutators among Gauss law constraint operators. Those commutators have also been studied in detail in the regularized perturbation theory.[14,15]

I would like to comment on this interesting problem from the view point of the path integral. In the path integral approach, it is relatively easy to understand why it is difficult to quantize the anomalous gauge theory in the conventional procedure. For this purpose, we start with the chiral Lagrangian

$$\mathcal{L} = \bar{\psi} i\gamma^\mu(\partial_\mu - igA_\mu)\left(\frac{1-\gamma_5}{2}\right)\psi + \bar{\psi} i\partial\!\!\!/\left(\frac{1+\gamma_5}{2}\right)\psi - \frac{1}{4}F^a_{\mu\nu}F^{a\mu\nu}$$
$$A_\mu = A^a_\mu T^a, \qquad [T^a, T^b] = if^{abc}T^c, \qquad \mathrm{Tr}T^aT^b = \frac{1}{2}\delta^{ab}. \tag{4.1}$$

We consider the gauge group SU(n) with the n-dimensional representation of the fermion field ψ. The theory (4.1) is anomalous for $n \geq 3$. The anomalous U(1) theory is also included in (4.1) by setting $T^a = 1$ there.

We first study the strict time-like axial gauge

$$A^a_0(x) = 0 \tag{4.2}$$

The path integral is then defined by

$$Z = \int \Pi\mathcal{D}\bar{\psi}\mathcal{D}\psi\mathcal{D}A^a_k \exp[i\int\mathcal{L}d^4x + i\int\mathcal{L}_J d^4x] \tag{4.3}$$

where A^a_0 is set to zero in $\mathcal{L}$. The source term is often discarded in the following expressions for notational simplicity. One may consider the change of path integral variables in (4.3)

$$\psi_L \to e^{i\omega^a(\vec{x},t)T^a}\psi_L$$
$$\psi_R \to \psi_R \tag{4.4}$$
$$A^a_k \to A^a_k + \partial_k\omega^a + gf^{abc}A^b_k\omega^c .$$

One then obtains the local Ward-Takahashi identity by collecting the variations of the action and the path integral measure[16]

$$\partial_0 G^a(\vec{x},t) = -gH^a(x) \tag{4.5}$$

with the Gauss law operator

$$G^a(\vec{x},t) \equiv D_k F^{a0k} - g\bar{\psi}\gamma^0 T^a (\frac{1-\gamma_5}{2})\psi \tag{4.6}$$

and the (consistent) anomaly factor

$$H^a \equiv \mathrm{Tr}T^a \frac{1}{48\pi^2} \varepsilon^{\mu\nu\alpha\beta}\partial_\mu (A_\nu \partial_\alpha A_\beta + \partial_\nu A_\alpha A_\beta + A_\nu A_\alpha A_\beta) \tag{4.7}$$

with $A_\mu \equiv -igA_\mu^a T^a$: The identity is the anomalous identity associated with the invariance of the action in (4.3) under the time independent gauge transformation.

The relation (4.5) clearly shows that the Gauss law operator becomes time dependent, i.e., does not commute with the Hamiltonian. The Gauss law operator ceases to be the constraint of the system, as the right-hand side of (4.5) cannot be set to zero: This fact can be understood by recalling the $\pi^0 \to \gamma\gamma$ decay where the anomaly factor has a non-vanishing matrix element between physical states. One may thus conclude that the conventional Gauss law constraint cannot be imposed on the states as the physical state condition.

We also note that the relation (4.5) can be understood in the operator language by noting[16]

$$\begin{aligned}
\partial_0 G^a &= \partial_0 D_k F^{a0k} - g\partial_0(\bar{\psi}_L \gamma^0 T^a \psi_L) \\
&= D_k \partial_0 F^{a0k} - g\partial_0(\bar{\psi}_L \gamma^0 T^a \psi_L) \\
&= -gD_k(\bar{\psi}_L \gamma^k T^a \psi_L) - g\partial_0(\bar{\psi}_L \gamma^0 T^a \psi_L) \\
&= -gD_\mu(\bar{\psi}_L \gamma^\mu T^a \psi_L) = -gH^a
\end{aligned} \tag{4.8}$$

where we used the equation of motion for A_k^a

$$\partial_0 F^{a0k} = -g\bar{\psi}_L \gamma^k T^a \psi_L \tag{4.9}$$

and the fact that $[\partial_0, D_k]F^{a0k} = 0$ in the $A_0 = 0$ gauge.

4b, BRS Symmetry, Slavnov-Taylor Identity and Anomalous Commutators

A more general consideration of anomalous theory is possible by considering the general Faddeev-Popov path integral defined by

$$Z = \int \Pi \mathcal{D}\bar{\psi}\mathcal{D}\psi\mathcal{D}A_\mu \mathcal{D}B\mathcal{D}\xi\mathcal{D}\eta \exp[i\int \mathcal{L}_{eff} dx] \tag{4.10}$$

with

$$\mathcal{L}_{eff} = \mathcal{L} + B^a A_0^a - i\xi^a(\delta^{ab}\partial_0 + gf^{acb}A_0^c)\eta^b \tag{4.11}$$

where ξ and η stand for the Faddeev-Popov anti-ghost and ghost, respectively. The action in (4.10) is invariant under the so-called BRS transformation

$$\delta A^a_\mu(x) = i\lambda(D_\mu\eta)^a(x) = i\lambda(\partial_\mu\eta^a + gf^{abc}A^b_\mu\eta^c)$$

$$\delta\eta^a(x) = -\lambda\,\frac{g}{2}\,f^{abc}\eta^b(x)\eta^c(x)$$

$$\delta\xi^a(x) = \lambda B^a(x)\ , \qquad \delta B^a(x) = 0 \tag{4.12}$$

$$\delta\psi_L(x) = -\lambda g\eta^a(x)\psi_L(x)\ , \qquad \delta\psi_R(x) = 0$$

where λ is an element of the Grassmann algebra.

One can then write the Slavnov-Taylor identity starting with the matrix element

$$<\xi^a(x)> \tag{4.13}$$

and applying the transformation (4.12). By collecting the terms linear in λ, one obtains

$$<B^a(x)> = <T^*\xi^a(x)\int\eta^b(z)gH^b(z)dz> \tag{4.14}$$

where the right-hand side of (4.14) arises from the anomalous Jacobian. If one combines (4.14) with the equation of motion for $A^a_0(x)$,

$$<G^a(x) - B^a(x) + ig\xi^b(x)f^{bac}\eta^c(x)> = 0 \tag{4.15}$$

one obtains

$$<G^a(x)> = g<T^*\xi^a(x)\int\eta^b(z)H^b(z)dz> + ig<\xi^b(x)f^{bac}\eta^c(x)>\ . \tag{4.16}$$

After integrating over $B^a(x)$ in (4.10), A^a_0 is set to zero and the Faddeev-Popov ghosts completely decouple from other fields. The connected component of (4.16) is then given by[16]

$$<G^a(x)> = g<\int G(x-z)H^a(z)dz> \tag{4.17}$$

with

$$<T^*\xi^a(x)\eta^b(x)> \equiv \delta^{ab}G(x-z)$$

$$= i\delta^{ab}\int\frac{d^4k}{(2\pi)^4}\,e^{ik(x-z)}\,\frac{1}{k^0} \tag{4.18}$$

There are some freedom in the definition of (4.18) (e.g., the principal part or other prescriptions), but (4.17) definitely shows that the matrix element of G^a is not zero in general in anomalous gauge theory with suitable physical sources inserted.

By the natural use of the BRS transformation, the Slavnov-Taylor identities and the Bjorken-Johnson-Low limit, one can also derive the anomalous commutation relation[16]

$$[G^a(x), G^b(y)]\delta(x^0-y^0) + igf^{abc}G^c(x)\delta(x-y)$$

$$= \frac{g^2}{48\pi^2}\varepsilon^{0\mu\nu\lambda}\{\mathrm{Tr}[T^a, T^b](\partial_\mu A_\nu A_\lambda + A_\mu\partial_\nu A_\lambda + A_\mu A_\nu A_\lambda)$$

$$+ \mathrm{Tr}T^a\partial_\mu(A_\nu T^b A_\lambda)\}\delta(x-y) \qquad (4.19)$$

which agrees with the perturbative calculation.[14,15] It is known that (4.19) is essentially equivalent to the result of Faddeev.[13] A salient feature of (4.19) is that the anomalous term vanishes for the Abelian theory. Our consideration in (4.5) and (4.17) is more restrictive in the sense that it holds for the Abelian theory as well; this arises from the fact that the Hamiltonian is flavor (or color) singlet.

4c, Modified Path Integral Prescription

Faddeev and Shatashvili[17] recently suggested a modified path integral prescription for anomalous gauge theories. They suggest adding the Wess-Zumino term with a chiral scalar field g to the action

$$S_{eff} = S + \Gamma_{WZ}(A, g) \qquad (4.20)$$

in such a way that the anomalous Jacobian is cancelled by the variation of Γ_{WZ}. One may then apply the conventional gauge fixing procedure to the effective action (4.20).

We here briefly comment on an interesting interpretation of the above prescription[17]. [See also papers in Ref.(18)]. The Faddeev-Shatashvili prescription may be regarded as an integration over the entire gauge orbit instead of the conventional Faddeev-Popov prescription. Namely,

$$Z = \int \Pi\mathcal{D}\bar{\psi}\mathcal{D}\psi[\mathcal{D}A_\mu]_{rep}\mathcal{D}g\ e^{S(\psi,A_\mu)+\Gamma_{WZ}(A,g)} \qquad (4.21)$$

where the field A_μ in $S(\psi, A_\mu)$ is regarded as a representative field configuration specified by the gauge fixing and $\mathcal{D}g$ the gauge orbit volume element. This is based on the decomposition

$$[\mathcal{D}A_\mu]_{gen} = [\mathcal{D}A_\mu]_{rep}\mathcal{D}g \qquad (4.22)$$

and the generic gauge field in the original action $S(\psi, (A_\mu)_{gen})$ is brought to the representative configuration $(A_\mu)_{rep}$ by gauge transforming the fermion field, thus giving rise to Γ_{WZ} as the Jacobian factor. The prescription (4.21) thus formally ensures the independence of Z of the choice of $(A_\mu)_{rep}$, since one is integrating over the entire gauge field configuration. This interpretation is interesting, since the Polyakov string theory for D < 26 can be regarded as an example of the integration over the entire gauge orbit in the anomalous Weyl-gauge theory,[19] as is noted in Ref.(17).

V. Anomaly in Superconformal Transformation

One of the recent developments in the path integral approach to anomalous identities is that the anomaly associated with supersymmetry has been clearly formulated. In the first place, P. van Nieuwenhuizen and his collaborators[20] clarified the path integral measure for supersymmetry and supergravity in general. K. Shizuya[21] presented an elegant treatment of the anomaly in the Wess-Zumino model and supersymmetric QED. H. Suzuki discussed this problem from a slightly different view point.[22] The supersymmetry generally contains some notational complications. Nevertheless, I would like to briefly comment on those elegant developments on the basis of the simplest case, namely the superconformal anomaly in the Wess-Zumino model.[21]

The Wess-Zumino model is described by[23]

$$S = \int d^8z\bar{\Phi}\Phi + \{\int d^6z(\frac{1}{2}\,m\Phi^2 - \frac{1}{6}\,g\Phi^3) + \text{h.c.}\} \tag{5.1}$$

with the chiral and anti-chiral superfields $\Phi(z)$ and $\bar{\Phi}(z)$, respectively, and $d^8z = d^4xd^2\theta d^2\bar{\theta}$, $d^6z = d^4xd^2\theta$, and $d^6\bar{z} = d^4xd^2\bar{\theta}$. The action (5.1) is invariant under the so-called (global) R-transformation ($n = 2/3$)

$$\Phi(x,\ \theta,\ \bar{\theta}) \rightarrow e^{-in\beta}\Phi(x,\ e^{i\beta}\theta,\ e^{-i\beta}\bar{\theta}) \tag{5.2}$$

except for the mass term. The supercurrent multiplet contains the R-generator as the first component.[24] A local generalization of (5.2) is given by[21]

$$\begin{aligned}\delta\Phi(z) &= -\frac{1}{4}\,\bar{D}^2[\Omega^\alpha D_\alpha + \frac{1}{3}(D^\alpha\Omega_\alpha)]\Phi(z)\\ \delta\bar{\Phi}(z) &= -\frac{1}{4}\,D^2[\bar{\Omega}_{\dot{\alpha}}\bar{D}^{\dot{\alpha}} + \frac{1}{3}(\bar{D}_{\dot{\alpha}}\bar{\Omega}^{\dot{\alpha}})]\bar{\Phi}(z)\end{aligned} \tag{5.3}$$

where $\Omega^\alpha(z)$ is an <u>arbitrary</u> (non-chiral) spinor superfield and $\bar{\Omega}^{\dot{\alpha}}$ its hermitean conjugate. Under this transformation, the interaction term in (5.1) is invariant and the action changes as

$$\delta S = \int d^8z[\frac{1}{2}(D^\alpha\bar{\Omega}^{\dot{\alpha}} - \bar{D}^{\dot{\alpha}}\Omega^\alpha)R_{\alpha\dot{\alpha}} + \frac{m}{6}(\Omega^\alpha D_\alpha\Phi^2 + \bar{\Omega}_{\dot{\alpha}}\bar{D}^{\dot{\alpha}}\bar{\Phi}^2)] \tag{5.4}$$

where the supercurrent is defined by

$$R_{\alpha\dot{\alpha}} = D_\alpha\Phi\bar{D}_{\dot{\alpha}}\bar{\Phi} - \frac{1}{3}[D_\alpha,\ \bar{D}_{\dot{\alpha}}]\bar{\Phi}\Phi\ . \tag{5.5}$$

The classical conservation law arising from (5.4) is given by

$$\begin{aligned}\bar{D}^{\dot{\alpha}}R_{\alpha\dot{\alpha}} - \frac{m}{3}\,D_\alpha\Phi^2 &= -2\,\frac{\delta}{\delta\Omega^\alpha}\int d^6z\,\delta\Phi\,\frac{\delta}{\delta\Phi}\,S\\ D^\alpha R_{\alpha\dot{\alpha}} - \frac{m}{3}\,\bar{D}_{\dot{\alpha}}\bar{\Phi}^2 &= 2\,\frac{\delta}{\delta\bar{\Omega}^{\dot{\alpha}}}\int d^6\bar{z}\,\delta\bar{\Phi}\,\frac{\delta}{\delta\bar{\Phi}}\,S\end{aligned} \tag{5.6}$$

The right-hand sides in (5.6) naively vanish if the equations of motion are used, but they generally survive as the anomaly in quantum theory; in fact, the evaluation of the right-hand sides is equivalent to the evaluation of the Jacobian factor in the path integral.[2]

To calculate the anomaly, we split $\Phi(z)$ into the classical part $\phi(z)$ and the quantum part $\eta(z)$ as

$$\Phi(z) = \phi(z) + \eta(z) \tag{5.7}$$

and the one-loop quantum property is specified by the quadratic part

$$S_2(\phi,\eta) = \frac{1}{2}(\eta,\bar{\eta})\cdot\begin{pmatrix} (m-g\phi)1_- & 1_-1_+ \\ 1_+1_- & (m-g\bar{\phi})1_+ \end{pmatrix}\cdot\begin{pmatrix}\eta \\ \bar{\eta}\end{pmatrix} \qquad (5.8)$$

$$\equiv \frac{1}{2} X^t\cdot\Gamma(\phi,\bar{\phi})\cdot X$$

where $X = (\eta, \bar{\eta})^t$. The dot $\cdot$ implies a summation over superspace coordinate labels of appropriate chirality using d^6z or $d^{\bar{6}}\bar{z}$. The operators $1_- = -\frac{1}{4}D^2$ or $1_+ = -\frac{1}{4}\bar{D}^2$ combine with the chiral measure to produce d^8z.

The variation (5.3) induces the transformation of X-variables $\delta X = B\cdot X$ with

$$B = \text{diag}[1_-\{\Omega D + \frac{1}{3}(D\Omega)\}1_- \,, \quad 1_+\{\bar{\Omega}\bar{D} + \frac{1}{3}(\bar{D}\bar{\Omega})\}1_+] \,. \qquad (5.9)$$

The Jacobian (anomaly) for this transformation in the superfield path integral is given by

$$\langle \mathcal{A} \rangle = i\langle \text{Tr}B\rangle$$

$$\equiv \lim_{\tau\to 0} i\text{Tr}(B\cdot e^{\tau\Gamma(\phi,\bar{\phi})^2}) \qquad (5.10)$$

Shizuya shows that a careful calculation of (5.10) gives[21]

$$\frac{\delta}{\delta\Omega^\alpha}\langle \mathcal{A} \rangle = \frac{1}{768\pi^2} D_\alpha \bar{D}^2[2m - g(\phi+\bar{\phi})]^2 \,. \qquad (5.11)$$

After a suitable redefinition of the operator

$$\hat{R}_{\alpha\dot{\alpha}} \equiv R_{\alpha\dot{\alpha}} - \frac{i}{96\pi^2}(\sigma^\mu)_{\alpha\dot{\alpha}}\partial_\mu[N(\phi) - N(\bar{\phi})] \qquad (5.12)$$

with $N(x) \equiv gx^2 - 4mgx$, one finally obtains

$$\bar{D}^{\dot{\alpha}}\hat{R}_{\alpha\dot{\alpha}} = \frac{1}{3} mD_\alpha\Phi^2 - \frac{g^2}{192\pi^2} D_\alpha\bar{D}^2(\bar{\Phi}\Phi) \qquad (5.13)$$

and the current $\hat{R}^\mu = \frac{1}{2}(\sigma^\mu)^{\dot{\alpha}\alpha}\hat{R}_{\alpha\dot{\alpha}}$ is anomaly-free. The formula (5.13) reproduces the result in the different method[25] of calculation (and in different notational conventions).

VI. Conformal and Ghost Number Anomalies in String Theory

String theories have received much attention recently. Here I would like to briefly discuss the anomalies appearing in the string path integral. Although the calculations of those anomalies themselves are not new, their discussion nicely illustrates the general path integral treatment of gravitational interactions. We discuss only the bosonic string theory originally discussed by Polyakov. The extension of the present path integral method to superstrings both in flat and curved space-time has been discussed in detail by P. Bouwknegt and P. van Nieuwenhuizen,[26] and by A. Eastaugh, L. Mezincescu, E. Sezgin and P. van Nieuwenhuizen.[27] See also Ref.(20). The interested readers are referred to those more recent developments of the path integral approach.

6a, Gravitational Path Integral Measure

The basic ingredient of the path integral I am going to use is the observation that the weight 1/2 variable for the (real) world scalar quantity S(x)

$$\tilde{S}(x) \equiv \sqrt[4]{g(x)}\, S(x) \tag{6.1}$$

defines the general coordinate invariant measure.[28,29] This is intuitively understood by noting that

$$\int \prod_x \mathcal{D}\tilde{S}(x)\, e^{-\int \sqrt{g} S(x)^2 dx} = \int \prod_x \mathcal{D}\tilde{S}(x)\, e^{-\int \tilde{S}(x)^2 dx} = \text{constant} \tag{6.2}$$

and the fact that the action in the integrand is general coordinate invariant. For the world vector S_μ and the second rank symmetric tensor $S_{\mu\nu}$, we define the world scalar quantities

$$\begin{aligned} S_a &\equiv e_a{}^\mu S_\mu \\ S_{ab} &\equiv e_a{}^\mu e_b{}^\nu S_{\mu\nu} \end{aligned} \tag{6.3}$$

and the invariant measure

$$\begin{aligned} \Pi \mathcal{D}\tilde{S}_a(x) &= \Pi \mathcal{D}[\sqrt[4]{g(x)}\, S_a(x)] \\ \Pi \mathcal{D}\tilde{S}_{ab}(x) &= \Pi \mathcal{D}[\sqrt[4]{g(x)}\, S_{ab}(x)] \end{aligned} \tag{6.4}$$

To treat the viel-bein $e_\mu{}^a$ and the metric $g_{\mu\nu}$, we rewrite (6.4) as

$$\begin{aligned} \Pi \mathcal{D}\tilde{S}_a(x) &= \Pi \mathcal{D}[g(x)^{(n-2)/4n}\, S_\mu(x)] \\ \Pi \mathcal{D}\tilde{S}_{ab}(x) &= \Pi \mathcal{D}[g(x)^{(n-4)/4n}\, S_{\mu\nu}(x)] \end{aligned} \tag{6.5}$$

in n-dimensional space-time. We thus define the invariant gravitational measure[29]

$$\begin{aligned} &\Pi \mathcal{D}[g^{(n-2)/4n} e_\mu{}^a(x)] \\ &\Pi \mathcal{D}[g^{(n-4)/4n} g_{\mu\nu}(x)] \; . \end{aligned} \tag{6.6}$$

Finally, the Faddeev-Popov ghost field η^μ associated with the general coordinate transformation is assigned the weight as the contravariant vector

$$\Pi \mathcal{D}\tilde{\eta}^\mu(x) = \prod_x \mathcal{D}[g^{(n+2)/4n} \eta^\mu(x)] \tag{6.7}$$

in n dimensions.[29]

6b, Conformal and Ghost Number Anomaly

We start with the bosonic action

$$\int \mathcal{L}\, d^2\sigma = - \int \frac{1}{2}\sqrt{g}\, g^{ab}\partial_a X^\mu \partial_b X^\mu d^2\sigma \tag{6.8}$$

which is regarded as the action for the two-dimensional gravitational interactions with the world scalar X^μ, $\mu = 1 \sim D$. The conformal gauge[19] $g_{ab} = \rho\delta_{ab}$, which is well-defined at least locally, is specified by

$$\tilde{g}_{12} = 0 = \frac{1}{2}(\tilde{g}_{11} - \tilde{g}_{22}) = 0\ . \tag{6.9}$$

We use the integration variables in (6.1), (6.6) and (6.7) with n = 2. The conventional Faddeev-Popov prescription then gives rise to[30]

$$Z = \int \Pi\, \mathcal{D}\sqrt{\rho}\,\mathcal{D}\tilde{X}^\mu \mathcal{D}\xi\,\mathcal{D}\tilde{\eta}\, \exp\{\int [-\frac{1}{2}\partial_a(\frac{\tilde{X}^\mu}{\sqrt{\rho}})\partial_a(\frac{\tilde{X}^\mu}{\sqrt{\rho}}) + \xi\sqrt{\rho}\,\not{\partial}\,\frac{1}{\rho}\,\tilde{\eta}]d\sigma\} \tag{6.10}$$

with the Faddeev-Popov ghost η and the anti-ghost ξ

$$\tilde{\eta} = \begin{pmatrix} \rho\eta^1 \\ \rho\eta^2 \end{pmatrix}, \qquad \xi = \begin{pmatrix} \xi_1 \\ \xi_2 \end{pmatrix} \tag{6.11}$$

and

$$\not{\partial} \equiv \sigma^1\partial_1 + \sigma^3\partial_2\ . \tag{6.12}$$

The conformal anomalies are defined for the variations

$$\begin{aligned} \tilde{X}^\mu &\to e^{\frac{1}{2}\alpha(x)}\tilde{X}^\mu \\ \xi &\to e^{-\frac{1}{2}\alpha(x)}\xi \\ \tilde{\eta} &\to e^{\alpha(x)}\tilde{\eta}\ . \end{aligned} \tag{6.13}$$

The basis vectors to define the Jacobian factors are defined by

$$H\varphi_m = \lambda_m^2 \varphi_m \tag{6.14}$$

with

$$H = -\rho^{-(n+1)/2}\not{\partial}\rho^n\not{\partial}\rho^{-(n+1)/2} \tag{6.15}$$

We choose n = 0, -2, and 1 for $\tilde{X}^\mu$, ξ and η, respectively. The calculation of the Jacobian factor is then reduced to the calculation[30]

$$\begin{aligned} \lim_{\beta\to 0} \mathrm{Tr}(e^{-\beta H}) &= \lim_{\beta\to 0}\int \frac{d^2k}{(2\pi)^2} e^{-ikx}e^{-\beta H}e^{ikx} \\ &= \lim_{\beta\to 0} 2\{\frac{3n+1}{24\pi}[-\partial^2 \ln\rho] + \frac{\rho}{4\pi}\frac{1}{\beta}\}\ . \end{aligned} \tag{6.16}$$

The anomaly factor for the transformation (6.13) is then given by

$$G(\rho) = -(\frac{D-26}{24})[-\partial^2 \ln\rho + 2\mu^2\rho] \ . \tag{6.17}$$

The "integration" of (6.17) gives the Liouville action.[19]

Another interesting anomaly is the Faddeev-Popov ghost "number" anomaly defined for

$$\tilde{\eta} \to e^{\alpha(x)}\tilde{\eta} \ , \qquad \xi \to e^{-\alpha(x)}\xi \tag{6.18}$$

and it is expressed as[30]

$$\partial_\mu[\xi\sqrt{\rho}\sigma^\mu \frac{1}{\rho}\tilde{\eta}] = \frac{3}{4\pi}\partial^2 \ln\rho = -\frac{3}{4\pi}\sqrt{g}\,R \ . \tag{6.19}$$

This ghost number anomaly corresponds to a local version of the Riemann-Roch theorem, which is described in detail in O. Alvarez.[31]

6c, Ghost Number Anomaly and the "θ-vacuum"

To treat the ghost number anomaly globally for the general Riemannian surface, one may define the background gauge[31,32] (at D=26, where the naive manipulations are justified)

$$g_{ab}(\sigma) \equiv \hat{g}_{ab}(\tau,\sigma) + h_{ab}(\sigma) \tag{6.20}$$

with

$$h_{ab} = h_{ba} \ , \qquad g^{ab}h_{ab} = 0 \tag{6.21}$$

The background metric $\hat{g}_{ab}(\tau,\sigma)$ with the Teichmuller parameter τ stands for the constant curvature metric consistent with the Gauss-Bonnet theorem. By imposing $h_{ab} = 0$, one obtains

$$Z = \int \Pi \mathcal{D}\hat{g}_{ab}\mathcal{D}X^\mu \mathcal{D}\xi\mathcal{D}\eta \ \exp\{-\int\sqrt{\hat{g}}\,\hat{g}^{ab}\partial_a X^\mu \partial_b X^\mu d^2\sigma + \int \xi P(\hat{g})\eta d^2\sigma\} \ . \tag{6.22}$$

We note that this formula (6.22) can be regarded as the general "instanton" calculation if one makes the identifications:

$\hat{g}_{ab}(\tau,\sigma) \longleftrightarrow$ Instanton $\hat{A}_\mu(\tau,x)$

Teichmuller parameter $\longleftrightarrow$ Collective coordinate of $\hat{A}_\mu(\tau,x)$

ghost fields $\xi,\eta \longleftrightarrow$ chiral fermion $\bar{\psi}_L$, ψ_L

ghost number anomaly $\longleftrightarrow$ chiral anomaly.

Inside the topologically non-trivial string configuration, therefore, the "vacuum" becomes something like the θ-vacuum in the instanton theory.[6] For example, the tree amplitude for the open-string may be visualized as

$$n_i\text{-particles} \to \text{hemi-sphere} \to n_f\text{-particles} \tag{6.23}$$

where the hemi-sphere or disc stands for the string configuration, which absorbs n_i particles and emits n_f particles. The hemi-sphere in this process appears and then disappears only inside the vacuum. The string configuration can thus be regarded as the "instanton".

Since the Euler number $\chi = 1$ for the hemi-sphere, (6.19) suggests

$$\Delta Q_{ghost} = -3 \tag{6.24}$$

and the vacuum amplitude vanishes

$$\langle -3|0\rangle = 0 \tag{6.25}$$

or one may equally write (6.25) as

$$\langle -\tfrac{3}{2}|-\tfrac{3}{2}\rangle = 0 \tag{6.26}$$

since the ghost number is the scale transformation and not the phase transformation as in the case of instanton. Just as in the case of the instanton calculation, one can see

$$\langle -\tfrac{3}{2}|\eta(\sigma_1)\eta(\sigma_2)\eta(\sigma_3)|-\tfrac{3}{2}\rangle \neq 0 \tag{6.27}$$

which corresponds to the fact that we have three conformal Killing vectors[31] for the hemi-sphere. In fact, if one recalls that the conformal Killing vectors are spanned by

$$1, \quad z, \quad z^2 \tag{6.28}$$

associated with the SL(2,R) transformation when one projects the unit hemi-sphere to the upper half-plane, (6.27) gives

$$\begin{aligned} &\langle -\tfrac{3}{2}|\eta_1^{(0)}\eta_2^{(0)}\eta_3^{(0)}|-\tfrac{3}{2}\rangle \det\begin{pmatrix} 1 & z_1 & z_1^2 \\ 1 & z_2 & z_2^2 \\ 1 & z_3 & z_3^2 \end{pmatrix} \\ &= \langle -\tfrac{3}{2}|\eta_1^{(0)}\eta_2^{(0)}\eta_3^{(0)}|-\tfrac{3}{2}\rangle (z_1-z_2)(z_2-z_3)(z_3-z_1) \ . \end{aligned} \tag{6.29}$$

We note that our path integral measure in (6.22) (in contrast to the prescription in Ref.(31)) contains the ghost zero-modes

$$\int d\eta_1^{(0)} d\eta_2^{(0)} d\eta_3^{(0)} . \tag{6.30}$$

The factor in (6.29) naturally agrees with the Jacobian factor from $dz_1 dz_2 dz_3$ to $d\varepsilon_1 d\varepsilon_2 d\varepsilon_3$ for the infinitesimal SL(2,R) transformation

$$\delta z = \varepsilon_1 + \varepsilon_2 z + \varepsilon_3 z^2 \ . \tag{6.31}$$

The instanton-like structure in the string path integral in connection with the ghost number anomaly was first suggested in Ref.(30). The implications of the ghost number anomaly have been analysed in greater detail by Friedan, Martinec and Shenker[33], and by Witten[34] in connection with the covariant string field theory.

VII. Gravitational Anomaly

The gravitational anomaly in d=4k+2 dimensions has been analysed in detail by Alvarez-Gaumé and Witten.[35] The path integral approach, which is closely related to the discussions in the present note, has also been discussed.[36] The relation between the covariant and consistent forms of gravitational anomalies has been elegantly analysed by Bardeen and Zumino.[4] I here briefly comment on two technical developments in the evaluation of gravitational anomalies.

The first technical development is that the Feynman gauge for the Rarita-Schwinger field

$$\mathcal{L} = \bar{\psi}^{\mu}(x) i \not{D} \psi_{\mu}(x) + \mathcal{L}_{F-P} \tag{7.1}$$

in the space-time with d > 4 has been formulated by Endo and Takao.[37] The gauge (7.1), which was originally introduced in Ref.(35), simplifies the various calculations. However, the derivation of the formula (7.1) from the view point of the conventional gauge fixing procedure has not been clear in the past. The authors of Ref.(37) derived the formula (7.1) on the basis of the conventional gauge fixing

$$\mathcal{L}_{GF} = \frac{i}{\alpha} \sqrt{g}\, (\bar{\psi}\gamma) \not{D} (\gamma\psi) \tag{7.2}$$

with $\alpha = 4/(d-2)$ combined with the point transformation

$$\phi_{\mu} = \psi_{\mu} - \frac{1}{2} \gamma_{\mu} (\gamma\psi) \ . \tag{7.3}$$

Another interesting analysis is related to the explicit evaluation of the local Lorentz anomaly. The local Lorentz anomaly is based on the Jacobian factor of the form[38]

$$\frac{i}{2} \sum_{n} \phi_{n}(x)^{\dagger} \sigma^{ab} \gamma_{5} \phi_{n}(x) e^{-\beta\lambda_{n}^{2}} \tag{7.4}$$

The explicit evaluation of (7.4) in d=2 and 6 is known.[38,36] The mixed local Lorentz anomaly in d=4 in the presence of U(1) gauge field has also been calculated to be[39,40]

$$\langle T_{ab} - T_{ba} \rangle = \frac{i}{12(4\pi)^{2}} \varepsilon_{ab\mu\nu} \{ R F^{\mu\nu} - R^{\mu\nu\lambda\rho} F_{\lambda\rho} + 2F^{\mu\nu}{}_{|\lambda}{}^{\lambda} \} \tag{7.5}$$

The general feature of the evaluation (7.4) is that the indices a and b of σ_{ab} appear in the ε-tensor, as is seen in (7.5). On the other hand, the mathematical construction of the (mixed) local Lorentz anomaly gives[40] (when rewritten in the covariant form)

$$\varepsilon_{\mu\nu\rho\sigma} R_{ab}{}^{\rho\sigma} F^{\mu\nu} \tag{7.6}$$

which contains the indices a and b in the Riemann tensor. The equivalence between the explicit calculation (7.5) and the mathematical construction (7.6) is not obvious.

Yajima and Kimura[40] recently showed that (7.5) and (7.6) are identical up to the local counter term

$$\mathcal{L}_{c} = \varepsilon_{ab\mu\nu} \omega^{ab\mu} F^{\rho\nu}{}_{|\rho} \tag{7.7}$$

with ω^{ab}_{μ} the spin connection. In d=2, the equivalence of the local Lorentz anomaly evaluated by the path integral and mathematical methods is well understood.[36] The present illustration of the equivalence in the case of the mixed local Lorentz anomaly in d=4 suggests that the explicit evaluation of (7.4) and the mathematical construction are generally equivalent up to a suitable local counter term.

VIII. Conclusion

The path integral approach to anomalous Ward-Takahashi identities has been extended to all the known anomalies in field theory, thanks to the efforts of many authors. As for the future problems of this approach, a further technical refinement related to the supersymmetry and non-linear interactions in general has to be performed. The path integral treatment of the anomaly in string field theory, for example, is also an interesting subject. As for the general implications of anomalies, the anomaly-related phenomena in various branches in physics remain to be studied.[41]

References

1. S. Adler, Lectures on elementary particles and quantum field theory, Eds. S. Deser et al. (MIT Press, Cambridge, MA, 1970). R. Jackiw, Lectures on current algebra and its applications, Eds. S. Treiman et al. (Princeton U.P., Princeton, N.J. 1972) and references therein.
2. K. Fujikawa, Phys. Rev. Lett. 42(1979)1195; Phys. Rev. Lett. 44(1980) 1733; Phys. Rev. D21(1980)2848.
3. The path integral approach is reviewed in K. Fujikawa, Proc. Kyoto Summer Institute, May 1985; Ed. T. Inami (World Scientific, Singapore, 1986).
4. W.A. Bardeen & B. Zumino, Nucl. Phys. B244(1984)421 and refs. therein.
5. M.F. Atiyah & I.M. Singer, Ann. Math. 87(1968)484. M.F. Atiyah, R. Bott & V. Patodi, Invent. Math. 19(1973)279.
6. G. 't Hooft, Phys. Rev. Lett. 37(1976)8; Phys. Rev. D14(1976)3432.
7. R. Jackiw & C. Rebbi, Phys. Rev. Lett. 37(1976)172. C. Callan, R. Dashen & D. Gross, Phys. Lett. 63B(1976)334.
8. W.A. Bardeen, Phys. Rev. 184(1969)1848. R.W. Brown, C.C. Shih & B.L. Young, Phys. Rev. 186(1969)1491.
9. A.P. Balachandran, G. Marmo, V.P. Nair & C.G. Trahern, Phys. Rev. D25(1982)2718. M.B. Einhorn & D.R.T. Jones, Phys. Rev. D29(1984)331. S.K. Hu, B.L. Young & D.W. Mckay, Phys. Rev. D30(1984)836. A. Andrianov & L. Bonora, Nucl. Phys. B233(1984)232. L. Alvarez-Guamé & P. Ginsparg, Nucl. Phys. B243(1984)449. R.E. Gamboa-Saravi, M.A. Muschietti, F.A. Schaposnik & J.E. Solomin, Phys. Lett. 138B(1984)145.
10. J. Wess & B. Zumino, Phys. Lett. B37(1971)95.
11. K. Fujikawa, Phys. Rev. D29(1984)285.
12. H. Banerjee, R. Banerjee & P. Mitra, Saha Inst. of Nuclear Research report, March 1986 (to be published in Z. Phys. C).
13. L.D. Faddeev, Phys. Lett. B145(1984)81.
14. S.G. Jo, Nucl. Phys. B259(1985)616; Phys. Lett. B163(1985)353.
15. M. Kobayashi, K. Seo & A. Sugamoto, KEK report, KEK-TH-112 (1985).
16. K. Fujikawa, Phys. Lett. B171(1986)424. See also M. Testa & K. Yoshida, Phys. Lett. 167B(1986)83.
17. L.D. Faddeev & S.L. Shatashvili, Phys. Lett. 167B(1986)225.
18. K. Harada & I. Tsutsui, Tokyo Inst. of Technology report, HEP-94, 1986; O. Babelon, F.A. Schaposnik & C.M. Viallet, Paris report, LPTHE 86/31.
19. A.M. Polyakov, Phys. Lett. 103B(1981)207 & 211.
20. M. Rocek, P. van Nieuwenhuizen & S.C. Zhang, SUNY report, ITP-SB-86-18.

21. K. Shizuya, Tohoku report, TU-86-299; see also K. Shizuya & H. Tsukahara, TU-86-296 (to appear in Z. Phys. C).
22. H. Suzuki, Phys. Rev. Lett. 56(1986)1534.
23. J. Wess & B. Zumino, Nucl. Phys. B70(1974)39. As for the superfield notation, J. Wess & J. Bagger, Supersymmetry & Supergravity, (Princeton U.P., Princeton, N.J., 1983).
24. S. Ferrara & B. Zumino, Nucl. Phys. B87(1975)207.
25. O. Piguet & M. Schweda, Nucl. Phys. B92(1975)334.
T.E. Clark, O. Piguet & K. Sibold, Nucl. Phys. B143(1978)445.
M.T. Grisaru & P.C. West, Nucl. Phys. B254(1985)249.
26. P. Bouwknegt & P. van Nieuwenhuizen, Class. & Quantum Gravity, 3(1986)207.
27. A. Eastaugh, L. Mezincescu, E. Sezgin & P. van Nieuwenhuizen, SUNY report, ITP-SB-86-11.
28. S. Hawking, Comm. Math. Phys. 55(1977)133.
29. K. Fujikawa, Nucl. Phys. B226(1983)437; K. Fujikawa & O. Yasuda, Nucl. Phys. B245(1984)436.
30. K. Fujikawa, Phys. Rev. D25(1982)2584.
31. O. Alvarez, Nucl. Phys. B216(1983)125.
32. D. Friedan, in Recent Advances in Field Theory & Statistical Mechanics, J.B. Zuber & R. Stora eds. (Elsevier, 1984). See also J.L. Gervais & B. Sakita, Phys. Rev. Lett. 30(1973)716.
33. D. Friedan, E. Martinec & S.H. Shenker, Phys. Lett. 160B(1985)55; Chicago report, 85-89 (1985).
34. E. Witten, Nucl. Phys. B268(1986)253; Princeton report, March, 1986.
35. L. Alvarez-Gaumé & E. Witten, Nucl. Phys. B234(1984)269.
36. K. Fujikawa, M. Tomiya & O. Yasuda, Z. Phys. C28(1985)289.
37. R. Endo & M. Takao, Phys. Lett. 161B(1985)155.
38. L.N. Chang & H.T. Nieh, Phys. Rev. Lett. 53(1984)21.
39. H.T. Nieh, Phys. Rev. Lett. 53(1984)2219.
40. S. Yajima & T. Kumura, Phys. Lett. 173B(1986)154.
41. See, e.g., Zhao-bin Su & B. Sakita, Phys. Rev. Lett. 56(1986)780.

Note added

(i) The possible quantization of anomalous gauge theories has been discussed by G. Semenoff and by C.M. Viallet at this Workshop, where more detailed related references are found.

(ii) The gravitational path integral measure has been recently analysed by D. Toms (Univ. of New Castle preprint, NCL-86 TP6, 1986). I believe that his prescription is essentially the same as the one used in this note, although the phrasing is somewhat different.

(iii) Various aspects of anomalous identities have also been discussed by R.B. Mann, V. Elias, G. McKeon and H.C. Lee at this Workshop.

REPRESENTATIONS OF KAC-MOODY AND VIRASORO ALGEBRAS

Peter Goddard†

Institute for Theoretical Physics
University of California
Santa Barbara, CA 93106

1. INTRODUCTION

In these lectures we shall discuss aspects of the representation theory of certain infinite dimensional Lie algebras, the Virasoro algebra, $\hat{v}$, and the untwisted affine Kac-Moody algebras, $\hat{g}$. (Much of what we shall say extends easily to the case of twisted affine Kac-Moody algebras but we shall stick to the untwisted case for simplicity.)

Let us first define the algebras we are to discuss. The Virasoro algebra, $\hat{v}$, has a basis labelled by the integers L_n, $n \in \mathbf{Z}$, together with a central element, c, that is an element which commutes with all the other generators. The algebra is specified by the commutation relations,

$$[L_m, L_n] = (m-n)L_{m+n} + \frac{c}{12}m(m^2-1)\delta_{m,-n} \tag{1.1}$$

where $m, n \in \mathbf{Z}$, together with $[L_n, c] = 0$. We shall be concerned with representations of this algebra which are unitary in the sense that they satisfy the hermiticity condition,

$$L_n^\dagger = L_{-n}. \tag{1.2}$$

In an irreducible unitary representation, c may be ascribed a fixed numerical value.

To define an untwisted affine Kac-Moody algebra, $\hat{g}$, we start from a compact finite dimensional Lie algebra g, with basis t^a, $1 \leq a \leq \dim g$, specified by commutation relations

$$[t^a, t^b] = i f_{abc} t^c \tag{1.3}$$

where the structure constants f_{abc} are totally antisymmetric. A basis for the associated untwisted affine Kac-Moody algebra $\hat{g}$ consists of generators T^a_m, $1 \leq a \leq \dim g$,

† Permanent address: Department of Applied Mathematics and Theoretical Physics, University of Cambridge, Silver Street, Cambridge CB3 9EW, United Kingdom.

$m \in \mathbf{Z}$, again together with a central element k, and the commutation relations are

$$[T_m^a, T_n^b] = i f_{abc} T_{m+n}^c + k m \delta^{ab} \delta_{m,-n}, \tag{1.4}$$

where $m, n \in \mathbf{Z}$, together with $[T_m^a, k] = 0$. In this case, we shall be interested in representations which are unitary in the sense that

$$T_n^{a\dagger} = T_{-n}^a \tag{1.5}$$

and in an irreducible representation of this sort k will take a single numerical value. Note that $\hat{g}$ has a subalgebra, with basis T_0^a, $1 \leq a \leq \dim g$, which is isomorphic to g and which we will identify with it.

In most of what follows, to avoid complication, we shall assume that g is simple unless we specify to the contrary. If it is not we can clearly introduce an independent central term k for each simple or $U(1)$ factor. At certain places in our discussion, we shall need to refer to particular non-simple algebras and we shall then explain how the results we establish are generalized straightforwardly to the non-simple case.

These algebras have applications in a wide variety of different areas of mathematics and physics, including sporadic finite simple groups, the theory of modular forms, "completely integrable" dynamical systems, string theories of fundamental particles and their interactions, conformally invariant two-dimensional field theories, and critical phenomena in two-dimensional statistical systems. The occurrence of these algebras and, associated with their representation theory, modular forms, can be seen as a unifying feature underlying connections between such disparate, or tenuously connected, branches of mathematics and physics.

For a longer and more detailed review of the material covered in these lectures see ref. [1]. For other reviews of Kac-Moody algebras and their physical applications consult refs. [2] and [3]. For a synopsis of the mathematical development of the subject see the article of Lepowsky[4], or other contributions in the same volume, the review by MacDonald[5] and the book by Kac.[6]

We can easily get a more geometric picture of the algebras we have introduced by defining infinite-dimensional groups of which they are the Lie algebras, provided that we omit their centers, c and k. Denote the corresponding algebras by $\hat{g}_0$ and $\hat{v}_0$, respectively.

Suppose g is the algebra of a compact finite-dimensional Lie group, G. The infinite-dimensional Lie group $\mathcal{G}$ with Lie algebra $\hat{g}_0$ is just the loop group of G, that is the group of (suitably smooth) maps from the circle S^1 to G, with multiplication defined pointwise. To be more explicit, realize S^1 as the unit circle in the complex plane

$$S^1 = \{z \in \mathbf{C} : |z| = 1\}. \tag{1.6}$$

Then the group product of two elements, $\gamma_1, \gamma_2 \in \mathcal{G}$ is defined by

$$\gamma_1\gamma_2(z) = \gamma_1(z)\gamma_2(z). \tag{1.7}$$

To calculate the Lie algebra of $\mathcal{G}$, consider an element near the identity,

$$\begin{aligned}\gamma(z) &= exp\{-it^a\theta_a(z)\} \\ &\cong 1 - it^a\theta_a(z)\end{aligned} \tag{1.8}$$

where $\theta_a(z)$ is small and has a Laurent expansion

$$\theta_a(z) = \sum_{n=-\infty}^{\infty} \theta_a^{-n} z^n. \tag{1.9}$$

So we see that, for $T_n^a = t^a z^n$

$$\gamma(z) \cong 1 - i\sum \theta_a^{-n} T_n^a \tag{1.10}$$

showing that $\hat{g}_0$ has a basis of generators T_n^a satisfying (1.4) with $k = 0$.

To construct an infinite-dimensional group with algebra $\hat{v}_0$ consider the group $\mathcal{V}$ of smooth one-to-one maps $S^1 \to S^1$ with the group law in this case defined by composition; that is we define the product of two such maps $\xi_1, \xi_2 : S^1 \to S^1$ by

$$\xi_1\xi_2(z) = \xi_1(\xi_2(z)). \tag{1.11}$$

The Lie algebra of $\mathcal{V}$ can be calculated by considering its action on functions $f : S^1 \to V$, where V is any vector space, defined by

$$\xi f(z) = f(\xi^{-1}(z)) \tag{1.12}$$

for $\xi \in \mathcal{V}$. An element close to the identity can be written

$$\begin{aligned}\xi(z) &= ze^{-i\epsilon(z)} \\ &\cong z(1 - i\epsilon(z))\end{aligned} \tag{1.13}$$

so that

$$\xi f(z) \cong f(z) + i\epsilon(z)\frac{d}{dz}f(z). \tag{1.14}$$

Making a Laurent expansion of $\epsilon(z)$ we are led to introduce generators

$$L_n = -z^{n+1}\frac{d}{dz} \tag{1.15}$$

which satisfy (1.1) with $c = 0$.

There is a natural interconnection between the groups $\mathcal{V}$ and $\mathcal{G}$, since $\mathcal{V}$ can be made to act on $\mathcal{G}$ by way of reparameterization. Thus they go together to form a semi-direct-product, and with the consequent interaction of the algebras:

$$[L_m, T_n^a] = -mT_{m+n}^a. \tag{1.16}$$

We shall often find it profitable to think of the algebra defined by (1.1), (1.4) and (1.16), that is

$$[T^a_m, T^b_n] = if_{abc}T^c_{m+n} + km\delta^{ab}\delta_{m,-n}, \tag{1.17a}$$

$$[L_m, T^a_n] = -nT^a_{m+n}, \tag{1.17b}$$

$$[L_m, L_n] = (m-n)L_{m+n} + \frac{c}{12}m(m^2-1)\delta_{m,-n}. \tag{1.17c}$$

In physical applications L_0 normally has an interpretation which requires it to have a spectrum which is bounded below (e.g. it corresponds to something like an energy or dilatation). Then it follows that the states for which the eigenvalue of L_0 is least must be annihilated by the generators T^a_n, L_n with $n > 0$, and that an invariant and irreducible subspace of the representation is built up by applying the T^a_n, $n \leq 0$, or L_n, $n < 0$, or both sets [depending on whether we are considering $\hat{g}$ of $\hat{v}$ or the whole algebra (1.17)] to a "vacuum state" Ψ satisfying

$$T^a_n\Psi = L_n\Psi = 0, \qquad n > 0. \tag{1.18}$$

A representation built up in this way from vacuum states satisfying (1.18) is called a highest weight representation. In Sect. 2 we shall discuss for which values of c and k unitary highest weight representation exist and how are they labelled. (By the term unitary here we need that the hermicity conditions (1.2) and (1.5) should hold with respect to some positive definite scalar product.) In Sect. 3 we shall describe the construction of a wide class of representations of untwisted affine Kac-Moody algebras using fermion fields. To construct the representations of the Virasoro algebra in Sect. 4 we exploit the Sugawara construction. To obtain the remaining representations of the Kac-Moody algebras we are studying, we develop the vertex operator construction in Sects. 5 to 8.

2. HIGHEST WEIGHT REPRESENTATIONS

First let us consider how to label the highest weight representations of the Virasoro and Kac-Moody algebras. Considering $\hat{v}$ alone, if h is the lowest eigenvalue of L_0, the states of an irreducible unitary highest weight representation can be built up from a single vacuum state $|h\rangle$ satisfying

$$L_0|h\rangle = |h\rangle, \tag{2.1}$$

$$L_n|h\rangle = 0, \qquad n > 0, \tag{2.2}$$

by applying the algebra. Thus we can label such representations by the pair of real numbers (c, h). We shall return to the question of for which values of (c, h) such unitary representations exist.

For an untwisted affine Kac-Moody algebra $\hat{g}$, we in general have a space of vacuum states Ψ satisfying

$$T^a_n \Psi = 0, \qquad n > 0, \tag{2.3}$$

from which $\hat{g}$, or in fact just T^a_{-m}, $m > 0$, generates the whole of a unitary highest weight representation. The vacuum space (2.3) must form an irreducible representation of $g \cong \{T^a_0\}$; for an irreducible representation of $\hat{g}$ this vacuum representation of g must be itself irreducible as it is clear that $\hat{g}$ would generate orthogonal invariant subspaces by its action on different irreducible vacuum representations. The irreducible unitary highest weight representations of $\hat{g}$ are thus labelled by a vacuum representation, or equivalently its highest weight λ, and the value of k.

It is much easier to establish necessary and sufficient conditions for the existence of unitary highest representations in the Kac-Moody case than in the Virasoro case. Given that g is simple, the conditions for $\hat{g}$ take the form that

$$x = \frac{2k}{\psi^2} \tag{2.4a}$$

be a non-negative integer, where ψ is a long root of g, and that the highest weight of the vacuum representation satisfy

$$|\alpha \cdot \lambda| \leq k \tag{2.4b}$$

for each root α of g. The integer x is called the level of the representation. (If $x = 0$ the representation is necessarily trivial, i.e., $T^a_n = 0$.)

To prove the conditions (2.4) are necessary we only have to consider appropriate $su(2)$ subalgebras of $\hat{g}$. We first write g in a basis consisting of a Cartan subalgebra H^i, $1 \leq i \leq rank\ g$, and step operators, E^α, $\alpha \in \Phi$, where Φ denotes the set of roots of g. This basis the algebra g takes the form

$$[H^i, H^j] = 0 \tag{2.5}$$

$$[H^i, E^\alpha] = \alpha^i E^\alpha \tag{2.6}$$

$$[E^\alpha, E^\beta] = \epsilon(\alpha, \beta) E^{\alpha+\beta}, \quad \text{if} \quad \alpha + \beta \in \Phi, \tag{2.7a}$$

$$= 2\alpha \cdot H/\alpha^2, \quad \text{if} \quad \alpha = -\beta \tag{2.7b}$$

$$= 0, \quad \text{otherwise.} \tag{2.7c}$$

Using such a basis, $\hat{g}$ can be written

$$[H^i_m, H^j_m] = km\delta^{ij}\delta_{m,-n} \tag{2.8}$$

$$[H^i_m, E^\alpha_m] = \alpha^i E^\alpha_{m+n} \tag{2.9}$$

$$[E^\alpha_m, E^\beta_n] = \epsilon(\alpha, \beta) E^{\alpha+\beta}_{m+n}, \quad \text{if} \quad \alpha + \beta \in \Phi, \tag{2.10a}$$

$$= \frac{2}{\alpha^2}\left\{\alpha \cdot H_{m+n} + km\delta_{m,-n}\right\}, \quad \text{if} \quad \alpha = -\beta, \tag{2.10b}$$

$$= 0, \quad \text{otherwise.} \tag{2.10c}$$

From these commutation relations we see that $\hat{g}$ has an $su(2)$ subalgebra consisting of $I_+ = E_1^{-\alpha}, I_- = E_{-1}^{\alpha}$, $2I_3 = \frac{2}{\alpha^2}(k - \alpha \cdot H_0)$; that is

$$[I_+, I_-] = 2I_3, \quad [I_3, I_\pm] = \pm I_\pm. \tag{2.11}$$

Since any unitary representation of $\hat{g}$ must provide unitary representations of this $su(2)$, it follows that the spectrum of $2I_3$ must be integral. Consider any simultaneous eigenvector $|\mu\rangle$ of H_0^i,

$$H_0^i|\mu\rangle = \mu^i|\mu\rangle, \tag{2.12}$$

so that μ is a weight of g; on $|\mu\rangle$, I_3 takes the value

$$\frac{2k}{\alpha^2} - \frac{2\alpha \cdot \mu}{\alpha^2} \in \mathbf{Z}. \tag{2.13}$$

But, by the usual theory of roots and weights (for a summary see ref. 1), $2\alpha \cdot \mu / \alpha^2 \in \mathbf{Z}$, implying that

$$\frac{2k}{\alpha^2} \in \mathbf{Z} \quad \text{if} \quad \alpha \in \Phi. \tag{2.14}$$

Since g is assumed simple, it has at most two distinct squared root lengths and if there are two they are in the ratio 2:1 or 3:1. In any case we see that if (2.14) holds for a long root it holds for all roots, amounting to the condition that (2.4a) be integral. To obtain the remaining part of (2.4) take $|\mu\rangle$ to be a vacuum state, so that

$$E_1^{-\alpha}|\mu\rangle = 0 \tag{2.15}$$

and consider the norm

$$\begin{aligned} ||E_{-1}^{\alpha}|\mu\rangle||^2 &= \langle\mu|E_1^{-\alpha}E_{-1}^{\alpha}|\mu\rangle \\ &= \langle\mu|\left[E_1^{-\alpha}, E_{-1}^{\alpha}\right]|\mu\rangle \\ &= \frac{2}{\alpha^2}(k - \alpha \cdot \mu)||\,|\mu\rangle\,||^2. \end{aligned} \tag{2.16}$$

Thus

$$k \geq \alpha \cdot \mu \tag{2.17}$$

for all roots α and all vacuum weights μ. In particular, by changing the sign of α if necessary, we see that $k \geq 0$ and also (2.4b) follows. This shows that the conditions (2.4) are necessary. We can show that they are sufficient by actually constructing them. We shall discuss this further in Sect. 3, and finally describe how to produce all of them in Sect. 8.

The discussion of unitary height weight representations of the Virasoro algebra is more difficult. It is easy to see, by considering

$$\begin{aligned} ||L_{-n}|h\rangle||^2 &= \langle h|L_n L_{-n}|h\rangle \\ &= \langle h|[L_n, L_{-n}]|h\rangle \\ &= \left\{2nh + n(n^2 - 1)\frac{c}{12}\right\}||\,|h\rangle||^2, \end{aligned} \tag{2.18}$$

and taking $n = 1$ and then n large, that unitarity requires that

$$h \geq 0 \quad \text{and} \quad c \geq 0. \tag{2.19}$$

To progress further we need to consider the matrix $M_N(c,h)$ of inner products of states of the form

$$L_{-1}^{n_1} L_{-2}^{n_2} \dots L_{-m}^{n_m} |h\rangle \tag{2.20}$$

with a given eigenvalue, N, of L_0, i.e., for a given value of $N = \sum_{j=1}^{m} j n_j$. The representation will be unitary, in the sense of the hermicity condition $L_n^\dagger = L_{-n}$ holding with respect to some positive definite scalar product, provided that the matrix $M_N(c,h)$ is positive semi-definite for each value of N. (If $M_N(c,h)$ has some zero eigenvalues, but its remaining eigenvalues are positive, we can consistently set to zero the linear combinations of vectors (2.20) corresponding to the zero eigenvalues.)

This problem can be explored case by case for low values of N, but since $\dim M_N(c,h)$ grows like $\pi(N)$ the number of partitions of N, defined by the formula

$$\prod_{n=1}^{\infty} (1 - q^n)^{-1} = \sum_{N=0}^{\infty} \pi(N) q^N, \tag{2.21}$$

it quickly becomes impractible. However a remarkable formula for det $M_N(c,h)$ was given by Kac[7] and proved by Feigin and Fuchs.[8] From this formula it follows quite easily that the unitarity condition is satisfied if $h \geq 0$ and $c \geq 0$. Analyzing the formula in detail, Friedan, Qiu and Shenker[9] were able to show that unitarity requires, if $0 \leq c < 1$, that c be a number in the series

$$c = 1 - \frac{6}{m(m+1)}, \qquad m = 2, 3 \dots, \tag{2.22}$$

i.e., c is one of the number $0, \frac{1}{2}, \frac{7}{10}, \frac{4}{5}, \frac{6}{7}, \frac{25}{28}, \dots$. For each value of c in this series (the "discrete series" of unitary representations of $\hat{v}$) there are $\frac{1}{2}m(m-1)$ possible values of h given by

$$h = h_{p,q}^{(c)} = \frac{[(m+1)p - mq]^2 - 1}{4m(m+1)} \tag{2.23}$$

where $1 \leq p \leq m-1$, $1 \leq q \leq p$. [In terms of the definitions provided by eqs. (2.22) and (2.23) the Kac formula can be stated as

$$\det \; M_N(c,h) = \prod_{k=1}^{N} \eta_k(c,h)^{\pi(N-k)}, \tag{2.24}$$

where

$$\eta_k(c,h) = \prod_{pq=k} [h - h_{p,q}(c)] \tag{2.25}$$

where p and q now range over the positive integers and m in (2.23) is defined as a function of c by (2.22).] The first few possible values of (c,h) are thus

$$c = 0; \quad h = 0; \tag{2.26a}$$
$$c = \frac{1}{2}; \quad h = 0, \ \frac{1}{16} \text{ or } \frac{1}{2}; \tag{2.26b}$$
$$c = \frac{7}{10}; \quad h = 0, \ \frac{3}{80}, \frac{1}{10}, \frac{7}{16}, \frac{3}{5}, \text{ or } \frac{3}{2}; \tag{2.26c}$$
$$c = \frac{6}{7}; \quad h = 0, \ \frac{1}{40}, \frac{1}{15}, \frac{2}{5}, \frac{21}{40}, \frac{2}{3}, \frac{7}{5}, \frac{13}{8}, \text{ or } 3. \tag{2.26d}$$

The first of these possibilities $c = h = 0$, corresponds to the trivial representation $L_n = 0$. The result, that if $c = 0$, for a highest weight unitary representation of $\hat{v}$, then $L_n = 0$, is a very powerful one. It follows from the general analysis of Friedan, Qiu and Shenker,[9] but a simple direct proof was provided by Gomes[10] by considering the matrix of inner products of the two states

$$L_{-2N}|h\rangle \quad \text{and} \quad L_{-N}^2|h\rangle. \tag{2.27}$$

This has determinant

$$4N^3h^2(4h - 5N) \tag{2.28}$$

if $c = 0$, from which it follows, by taking $N > h$, that $h = 0$. It is then easy to see that all $L_n = 0$ because the inner product of any two states of the form (2.20) vanishes.

The representations with $c = \frac{1}{2}$ have been known in the context of string theories since 1971 where they occurred implicitly in the fermionic theory of Neveu and Schwarz[11] and Ramond.[12] We shall discuss this case in the next section and the more elaborate construction necessary to obtain the other values of c given by (2.22) in the following section.

3. THE FERMION CONSTRUCTION

We shall first describe the fermion construction of representations of the Virasoro and untwisted affine Kac-Moody algebras. We consider a d-component real (i.e. Majorana) fermi field $H^\alpha(z)$, $1 \leq \alpha \leq d$, defined on the unit circle, $|z| = 1$. Since physical quantites involve even products of fermi fields, $H^\alpha(z)$ is either periodic (Ramond[12] case) or periodic (Neveu-Schwarz[11] case). Thus

$$H^\alpha(z) = \sum_r b_r^\alpha z^{-r} \tag{3.1}$$

where the sum is *either* over $r \in \mathbf{Z}$ (Ramond case) *or* over $r \in \mathbf{Z} + \frac{1}{2}$ (Neveu-Schwarz case). The oscillators b_r^α satisfy the hermicity condition

$$b_r^{\alpha\dagger} = b_{-r}^\alpha, \tag{3.2}$$

and the canonical anti-commutation relations

$$\{b_r^\alpha, b_s^\beta\} = \delta^{\alpha\beta}\delta_{r,-s}. \tag{3.3}$$

The field $H_r^\alpha(z)$ acts in a space generated by the b_r^α, $r < 0$, from a vacuum $|0\rangle$ satisfying

$$b_r^\alpha|0\rangle = 0, \ r > 0. \tag{3.4}$$

In the NS case we may assume that we have a non-degenerate vacuum but in the R case it has to be of dimension $2^{d/2}$, d even, or $2^{(d-1)/2}$, d odd, in order to provide a representation of the Dirac algebra of the b_0^α.

In this space we can write down a representation of the Virasoro algebra using what is essentially the energy-momentum tensor for the fermion field $H^\alpha(z)$,

$$L(z) \equiv \sum_n L_n z^{-n} = \frac{1}{2} z : \frac{dH}{dz} H : + \epsilon d. \tag{3.5}$$

Here $\epsilon = 0$ in the NS case and $\frac{1}{16}$ in the R case, and the colon denotes normal ordering with respect to the fermion oscillators b_r^α, to be precise

$$\begin{aligned} : b_r^\alpha b_s^\beta : &= -b_s^\beta b_r^\alpha \quad \text{if} \quad r > 0, \\ &= \frac{1}{2}[b_r^\alpha, b_s^\beta] \quad \text{if} \quad r = 0, \\ &= b_r^\alpha b_s^\beta \quad \text{if} \quad r < 0. \end{aligned} \tag{3.6}$$

From the definition (3.5) it is straightforward to show that

$$[L_m, L_n] = (m-n)L_{m+n} + \frac{d}{24} m(m^2-1)\delta_{m,-n}, \tag{3.7}$$

i.e., L_n satisfies the algebra (1.1) with $c = \frac{d}{2}$. This construction provides just one value of $c < 1$, namely $c = \frac{1}{2}$ from $d = 1$. We obtain all the corresponding values of h given by (2.26b): $h = 0$ and $h = \frac{1}{2}$ are provided by the states $|0\rangle$ and $b_{-\frac{1}{2}}|0\rangle$ in the NS case, and $h = \frac{1}{16}$ is provided by the vacuum $|0\rangle$ in the R case. (These are all the highest weight states for $d = 1$.) We shall discuss later, in Sect. 4, how to construct the other representations in the discrete series.

Using these fermi fields we can also construct representations of $\hat{g}$. Consider any real representation of g in terms of d dimensional antisymmetric matrices M:

$$[M^a, M^b] = f_{abc} M^c. \tag{3.8}$$

(Complex representations can be incorporated within this framework by considering the direct sum of the representation and its complex conjugate, which will be real.) If we define

$$T^a(z) \equiv \sum_n T_n^a z^{-n} = \frac{i}{2} H^\alpha(z) M_{\alpha\beta}^a H^\beta(z) \tag{3.9}$$

where $H^\alpha(z)$ is either R or NS, then

$$[T_m^a, T_n^b] = i f_{abc} T_{m+n}^c + \frac{\kappa}{2} m \delta^{nb} \delta_{m,-n} \tag{3.10}$$

where

$$\mathrm{tr}(M^a M^b) = -\kappa \delta^{ab}. \tag{3.11}$$

Thus we obtain (1.4) with $k = \frac{1}{2}\kappa$ so that the level of the representation of $\hat{g}$ is

$$x = \frac{2k}{\psi^2} = \frac{\kappa}{\psi^2}, \tag{3.12}$$

and this integer is called the Dynkin index of the representation M.

From such representations we can obtain, for the classical algebras $g = so(n)$, $su(n)$ or $sp(n)$, all the representations permitted by (2.4). Consider a specific classical algebra g with ψ^2 normalized to 2, so that $x = k$. For each highest weight λ, the maximum value, ℓ, say, of $|\alpha.\lambda|$, for $\alpha \in \Phi$, is the lowest level at which the vacuum representation can have highest weight λ. Suppose that the fundamental weights of g are $\lambda_1, \lambda_2, \ldots, \lambda_r$. The lowest permitted levels corresponding to taking λ to be one of these fundamental weights, $\ell_1, \ell_2, \ldots, \ell_r$, say, are all either 1 or 2 for g classical. For $g = su(n)$ or $sp(n)$ each $\ell_j = 1$, while for $so(n)$, $\ell_j = 1$ for the vector and fundamental spinor representations and $\ell_j = 2$ for all other fundamental weights.

Given an arbitrary highest weight $\lambda = \sum n_j \lambda_j$, so that $n_j \geq 0$, it follows that $\ell = \sum n_j \ell_j$. To construct the representation of level $k \geq \ell$ in which the vacuum has highest weight λ, we take the tensor product of n_j copies of the representations with highest vacuum weight λ_j and level ℓ_j and $k - \ell$ copies of the representation with level 1 and scalar vacuum, and then take the subspace generated from the irreducible component of the vacuum representation with highest weight λ. Thus it remains to construct the representation with highest weights λ_j and levels ℓ_j and with highest weight 0 and level 1. This can be done using as follows: using (3.8) with M the n-dimensional representation of $so(n)$, we obtain level 1 representations with the vacuum representations being the scalar and the vector in the NS case and the spinor representations in the R case. The level 1 representations of $su(n)$ and $sp(n)$ are given by the inclusions $su(n) \subset so(2n)$ and $sp(n) \subset su(2n) \subset so(4n)$. The level 2 antisymmetric tensor representations of $so(n)$ are given by the inclusion $so(n) \subset su(n) \subset so(2n)$, completing the list.

4. THE SUGAWARA CONSTRUCTION

In order to consturct representations of the Virasoro algebra with values of c in the discrete series (2.22) other than 0 and $\frac{1}{2}$ we need to consider the interrelation of the Virasoro and Kac-Moody algebras in the semi-direct product (1.15). Suppose we start with a given unitary highest weight representation of an untwisted affine Kac-Moody algebra, $\hat{v}$. This can always be extended in a certain "minimal" way to provide a representation of the whole semi-direct product by introducing the Virasoro generators $\mathcal{L}_n$ as quadratic functions of the T^a_n,

$$\mathcal{L}^g_n = \frac{1}{2\beta^g} \sum_{m,a} {}^\circ_\circ T^a_m T^a_{n-m} {}^\circ_\circ, \tag{4.1}$$

where the open does denote normal ordering with respect to the T^a_n,

$$\begin{aligned} {}^\circ_\circ T^a_m T^a_n {}^\circ_\circ &= T^a_n T^a_m, \qquad m \geq 0, & (4.2a) \\ &= T^a_m T^a_n, \qquad m < 0. & (4.2b) \end{aligned}$$

With this definition

$$[\mathcal{L}^g_m, \mathcal{L}^g_n] = (m-n)\mathcal{L}^g_{m+n} + \frac{c^g}{12}m(m^2-1)\delta_{m,-n} \tag{4.3}$$

and

$$[\mathcal{L}^g_m, T^a_n] = -T^a_{m+n}, \qquad 1 \leq a \leq \dim\ g \tag{4.4}$$

provided that

$$\beta = k + \frac{1}{2}Q^g_\psi, \tag{4.5}$$

where Q^g_ψ is the value of the quadratic Casimir operator for g in the adjoint representation, i.e.,

$$f_{abc}f_{abd} = Q^g_\psi \delta_{cd}. \tag{4.6}$$

The value of the central element c^g in (4.3) is given by the formula

$$c^g = \frac{2k \dim g}{Q^g_\psi + 2k}. \tag{4.7}$$

We can rewrite this in terms of the level x of the representation of $\hat{g}$ and the quantity

$$\tilde{h} = Q^g_\psi / \psi^2 \tag{4.8}$$

which is an integer called the dual Coxeter number. In these terms,

$$c^g = \frac{x \dim g}{x + \tilde{h}}. \tag{4.9}$$

Thus c^g is a rational number. It satisfies

$$\text{rank}\ g \leq c^g \leq \dim\ g \tag{4.10}$$

so that this construction can not directly provide us with any of the discrete series with $c < 1$. The second of the inequalities (4.10) is evident from (4.9). To see the first, we use that

$$\tilde{h} = \left\{ n_L + n_S\ S^2/L^2 \right\} / \text{rank}\ g \tag{4.11}$$

where n_L, n_S are the numbers of long and short roots of g, respectively, and S^2/L^2 is the square of the ratio of the length of a short root to the length of a long root. Since $\dim\ g = n_L + n_S +$ rank g,

$$c^g - \text{rank}\ g = \frac{n_L(x-1) + n_S(x - S^2/L^2)}{x + \tilde{h}}, \tag{4.12}$$

showing the first part of (4.10), provided that the representation of $\hat{g}$ is nontrivial.

To get the series values of $c < 1$ we need to consider[13] a subalgebra $h \subset g$ and the corresponding $\hat{h} \subset \hat{g}$. Let us suppose that T^a_m, $1 \leq a \leq \dim\ h$, $m \in \mathbf{Z}$, and k provide a basis for $\hat{h}$, whilst if $1 \leq a \leq \dim\ g$ we have a basis for $\hat{g}$. Then we can apply the construction (4.1) to $\hat{h}$,

$$\mathcal{L}_n^h = \frac{1}{2\beta^h} \sum_{m,a=1}^{\dim h} : T_m^a T_{n-m}^a : \tag{4.13}$$

with

$$\beta^h = k + \frac{1}{2} Q_\psi^h \tag{4.14}$$

and with the corresponding value of c,

$$c^h = \frac{2k \dim\ h}{Q_\psi^h + 2k}. \tag{4.15}$$

Then

$$[\mathcal{L}_m^h, T_n^a] = -nT_{m+n}^a, \qquad 1 \leq a \leq \dim\ h, \tag{4.16}$$

so that if we introduce the difference

$$K_m = \mathcal{L}_m^g - \mathcal{L}_m^h \tag{4.17}$$

it follows that

$$[K_m, T_n^a] = 0 \qquad 1 \leq a \leq \dim\ h. \tag{4.18}$$

Since K_m commutes with the whole of $\hat{h}$,

$$[K_m, \mathcal{L}_n^h] = 0 \tag{4.19}$$

and so

$$[\mathcal{L}_m^g, \mathcal{L}_n^g] = [\mathcal{L}_m^h, \mathcal{L}_n^h] + [K_m, K_n]. \tag{4.20}$$

Hence K_m satisfies the Virasoro algebra

$$[K_m, K_n] = (m-n)K_{m+n} + \frac{c_K}{12} m(m^2-1)\delta_{m,-n} \tag{4.21}$$

with

$$c_K = c^g - c^h. \tag{4.22}$$

This construction yields the entire discrete series of representation of $\hat{v}$ but to describe this we must explain how the construction (4.1) extends to the case where g is not simple. If

$$g = g_1 \oplus g_2 \oplus \cdots \oplus g_M, \tag{4.23}$$

with g_j simple, then we replace (4.1) by

$$\mathcal{L}_n^g = \mathcal{L}_n^{g_1} + \mathcal{L}_n^{g_2} + \cdots + \mathcal{L}_n^{g_M}, \tag{4.24}$$

with each $\mathcal{L}_n^{g_j}$ given by (4.1), and the corresponding c is

$$c^g = c^{g_1} + c^{g_2} + \cdots + c^{g_M} \tag{4.25}$$

with each c^{g_j} given by (4.9).

One way to obtain the discrete series is to take $g = su(2) \oplus su(2)$ and h to be the diagonal $su(2)$ subalgebra.[13] In general for $su(2)$ if we take the root length $\psi = 1$, as usual as in (2.11), the quadratic Casimir operator takes the value $\ell(\ell+1)$ in the spin ℓ representation and $Q_\psi = 1$. So for a level x representation of $su(2)$,

$$c^{su(2)} = \frac{3x}{x+2}. \tag{4.26}$$

If we choose a representation of $\hat{g} = s\hat{u}(2) \oplus s\hat{u}(2)$ which has levels $n-1$ and 1 for the two factors, respectively, it induces a level n representation of the diagonal subalgebra h. Then

$$\begin{aligned} c_K = c^g - c^h &= \frac{3(n-1)}{n+1} + 1 - \frac{3n}{n+2} \\ &= 1 - \frac{6}{(n+1)(n+2)} \end{aligned} \tag{4.27}$$

providing all of the sequence (2.22).

To show this construction yields all the corresponding values of h given by (2.23) we need to consider the representations of $su(2)$ in a little more detail. As we have seen in Sect. 2, these can be labelled by (x, ℓ) where x is the level and ℓ is the spin of the vacuum representation. Since $\psi = 1$ is the only root of $su(2)$, the conditions (2.4) amount to x being a non-negative integer together with

$$x \geq 2\ell, \tag{4.28}$$

so that for $x = 1$ we have the possibilities $\ell = 0, \frac{1}{2}$. Now we shall show that[13] if we take an irreducible representation of $s\hat{u}(2) \oplus s\hat{u}(2)$ which is $(n-1, \ell)$ for the first factor and $(1, \delta)$, where $\delta = 0$ or $\frac{1}{2}$, for the second, it decomposes into a direct sum of irreducible representations of algebra consisting of the diagonal $s\hat{u}(2)$ together with the Virasoro algebra, K_n, with each irreducible subspace transforming under a different representation

$$(n-1, \ell) \times (1, \delta) = \bigoplus_q \left(n, \frac{1}{2}\{q-1\}\right)(c_K, h_{p,q}(c)) \tag{4.29}$$

where c_K is given by (4.27), $h_{p,q}(c)$ is given as in (2.23) by

$$h_{p,q}(c) = \frac{[(n+2)p - (n+1)q]^2 - 1}{4(n+1)(n+2)} \tag{4.30}$$

with $p = 2\ell + 1$, $p - q \in \mathbf{Z} + 2\delta$ and the direct sum in (4.29) being over permitted values of q in the range $1 \leq q \leq n+1$. This provides all the necessary values of h.

The decomposition (4.29) is established by proving the equality of the characters of the representations on the two sides of the equation. For the representation (n, ℓ) of $s\hat{u}(2)$ the character is defined by

$$\chi_{n,\ell}(z, \theta) = \operatorname{tr}\left(z^{\mathcal{L}_0} e^{iT_3^0\theta}\right), \tag{4.31}$$

and for the representation (c, h) of $\hat{v}$, it is defined by

$$\chi^v_{c,h}(z) = \mathrm{tr}(z^{L_0}). \tag{4.32}$$

Equation (4.31) is then equivalent to the identity,

$$\chi_{n-1,\ell}(z,\theta)\chi_{1,\delta}(z,\theta) = \sum_q \chi_{n,\frac{1}{2}(q-1)}(z,\theta)\chi^v_{c,h}(z) \tag{4.33}$$

which can be checked using the known explicit formulae for the characters. For $s\hat{u}(2)$ the Weyl-Kac formula gives

$$\chi_{n,\ell}(z,\theta) = \Delta_{n,\ell}(z,\theta)\prod_{m=1}^{\infty}(1-z^m)^{-1}(1-z^m e^{i\theta})^{-1}(1-z^{m-1}e^{-i\theta})^{-1} \tag{4.34}$$

where

$$\Delta_{n,\ell}(z,\theta) = z^{\ell(\ell+1)/\lambda}\sum_{m\in\mathbf{Z}} z^{\lambda m^2+(2\ell+1)m}\left\{e^{i(\ell+\lambda m)\theta} - e^{-i(\ell+1+\lambda m)\theta}\right\} \tag{4.35}$$

with $\lambda = n+2$. For $n=1$ we have the alternative expressions

$$\chi_{1,\delta}(z,\theta) = \sum_{r\in\mathbf{Z}+\delta} z^{\frac{1}{2}r^2}e^{ir\theta}\prod_{m=1}^{\infty}(1-z^m)^{-1}. \tag{4.36}$$

The required character formulae for $\hat{v}$ have been given by Rocha-Caridi[14]

$$\chi^v_{c,h}(z) = \Delta^n_{p,q}(z)\prod_{m=1}^{\infty}(1-z^m)^{-1} \tag{4.37}$$

where

$$\Delta^n_{p,q}(z) = \sum_{m\in\mathbf{Z}}\left\{z^{\alpha^n_{pq}(m)} - z^{\beta^n_{pq}(m)}\right\} \tag{4.38}$$

with

$$\alpha^n_{pq}(m) = \frac{[2(n+1)(n+2)m - q(n+1) + p(n+2)]^2 - 1}{4(n+1)(n+2)} \tag{4.39a}$$

and

$$\beta^n_{pq}(m) = \frac{[2(n+1)(n+2)m + q(n+1) + p(n+2)]^2 - 1}{4(n+1)(n+2)} \tag{4.39b}$$

With these formulae it is a matter of straightforward algebra to establish (4.33).

By using the Sugawara construction for an appropriate choice of an algebra and a subalgebra, $g \supset h$, we have obtained all of the discrete series representations of the Virasoro algebra. (Other choices are possible, e.g. $sp(n) \supset sp(n-1)\times sp(1)$.[13]) By a slightly different application of essentially the same ideas we can see what are the possible representations of the semi-direct product (1.17). Using (1.17b) and (4.4), we see that if we now define K_n by

$$K_n = L_n - \mathcal{L}^g_n \tag{4.40}$$

we have that K_n satisfies a Virasoro algebra[15] with

$$c_K = c - c^g \tag{4.41}$$

and K_n and $\hat{g}$ commute. Thus the semi-direct product (1.17) has been written as the direct product of two commuting algebras whose representations can be chosen independently. Thus, in particular, either

$$c \geq c^g + 1, \tag{4.42a}$$

or

$$c = c^g + 1 - \frac{6}{m(m+1)} \tag{4.42b}$$

where $m = 2, 3, \ldots$.

We can use these results, together with (4.12), to establish a lower bound on the number of independent free fermion and free bosons fields necessary to construct a representation of an untwisted affine Kac-Moody algebra $\hat{g}$ at a particular level, x. In the Fock space generated by n_B independent free boson fields and n_F independent free fermion fields, we have the standard representation of the Virasoro algebra, L_n (given by the sum of (3.5) for the fermion fields and (4.1), applied in the case of an Abelian algebra, for the boson fields). For this algebra the central term

$$c = n_B + \frac{1}{2} n_F. \tag{4.43}$$

If a representation of $\hat{g}$ satisfies the equation

$$[L_m, T_n^a] = -n T_{n+m}^a \tag{4.44}$$

with respect to this Virasoro algebra, it follows from (4.37) and (4.12) that

$$n_B + \frac{1}{2} n_F \geq \text{rank } g + \frac{n_L(x-1) + n_S(x - S^2/L^2)}{x + \tilde{h}} \tag{4.45}$$

so that a construction might be possible with rank g boson fields if (and only if) $x = 1$ and $S^2 = L^2$, *i.e.*, the algebra is simply-laced. It is indeed possible to give a construction of the level 1 representation of simply-laced algebras in this way. This is the Frenkel-Kac-Segal construction[16–19] based on the vertex operator of dual string theory. In the next section we shall motivate this construction by reviewing at an elementary level relevant aspects of string theory.

5. STRING AND VERTEX OPERATORS

A history of a string is described by a world sheet $x^\mu(\sigma, \tau)$, $0 \leq \sigma \leq \pi$, $-\infty < \tau < \infty$, where μ ranges over d values. The motion of the string is determined by the action

$$S = k \int [(\dot{x}.x')^2 - \dot{x}^2 x'^2]^{\frac{1}{2}} d\sigma d\tau \tag{5.1}$$

where

$$\dot{x} = \frac{\partial x}{\partial \tau}, \quad x' = \frac{\partial x}{\partial \sigma}, \quad \text{and} \quad k = T_0/c \tag{5.2}$$

is a constant, with T_0 having the dimensions of tension. If the reparameterization invariance of S is used to impose the orthnormality conditions,

$$\dot{x}.x' = 0, \quad \dot{x}^2 + x'^2 = 0, \tag{5.3}$$

the equations of motion and boundary conditions take the form

$$\ddot{x} = x'' \tag{5.4}$$

and

$$x' = 0 \quad \text{at} \quad \sigma = 0, \pi. \tag{5.5}$$

These have the general solution

$$\frac{1}{\kappa} x^r(\sigma, \tau) = q^\mu + p^\mu \tau + \sum_{h=1}^{\infty} (q_n^r \cos n\tau + \dot{q}_n^r \sin n\tau) \cos n\sigma \tag{5.6}$$

where the constant κ has been introduced to absorb the dimensions of length. It is convenient to introduce harmonic oscillator variables

$$\alpha_n^\mu = \frac{1}{2}\{\dot{q}_n^\mu - inq_n^\mu\} \tag{5.7}$$

so that

$$\alpha_n^{\mu\dagger} = \alpha_{-n}^\mu \tag{5.8}$$

where, classically, the dagger denotes complex conjugation, to become hermitian conjugation on quantization. We can then rewrite (5.6) as

$$\frac{1}{\kappa} x^\mu(\sigma, \tau) = q^\mu + p^\mu \tau + \frac{i}{2} \sum_{n \neq 0} \left\{ \frac{\alpha_n^\mu}{n} e^{-in(\tau+\sigma)} + \frac{\alpha_n^\mu}{n} e^{-in(\tau-\sigma)} \right\} \tag{5.9}$$

$$\equiv \frac{1}{2} Q^\mu(e^{i(\tau+\sigma)}) + \frac{1}{2} Q^\mu(e^{i(\tau-\sigma)}), \tag{5.10}$$

for

$$Q^\mu(z) = q^\mu - ip^\mu \log z + i \sum_{n \neq 0} \frac{\alpha_n^\mu}{n} z^{-n}. \tag{5.11}$$

In terms of the harmonic oscillator variables (5.7), the canonical commutation relations take the form

$$[\alpha_m^\mu, \alpha_n^v] = m g^{\mu\nu} \delta_{m,-n}, \tag{5.12}$$

$$[q^\mu, p^v] = i g^{\mu\nu}, \tag{5.13}$$

provided that

$$\kappa^2 k \pi = \hbar. \tag{5.14}$$

The creation and annihilation operators α_n^μ act in a Hilbert space which they generate from certain "vacuum" states $|\gamma\rangle$ which carry only momentum,

$$\alpha_n^\mu |\gamma\rangle = 0, \quad n > 0; \quad p^\mu |\gamma\rangle = \gamma^\mu |\gamma\rangle, \tag{5.15}$$

and can be generated from a true vacuum $|0\rangle$ in the usual way,

$$|\gamma\rangle = e^{iq\cdot\gamma} |0\rangle. \tag{5.16}$$

The constraints (5.3) correspond, in terms of α_n^μ, to the vanishing of

$$L_n = \frac{1}{2} \sum_{m=-\infty}^{\infty} : \alpha_m . \alpha_{n-m} : \tag{5.17}$$

where the normal ordering has been introduced to ensure that

$$L_0 = \frac{1}{2} + \sum_{n=1}^{\infty} \alpha_n^{\dagger} . \alpha_n \tag{5.18}$$

is well-defined. So defined, these operators satisfy the Virasoro algebra,

$$[L_m, L_n] = (m-n)L_{m+n} + \frac{d}{12} m(m^2-1)\delta_{m,-n} \tag{5.19}$$

In this covariant treatment of the quantum theory, the classical constraints (5.3) are applied quantum mechanically, as conditions

$$L_n|\psi\rangle = 0, \quad n > 0, \tag{5.20a}$$

$$L_0|\psi\rangle = \lambda|\psi\rangle \tag{5.20b}$$

where λ is an arbitrary constant which can be regarded as an ambiguity associated with normal ordering. It can be shown that the space of physical states, defined by (5.20), is free of ghosts, i.e., negative norm states resulting from the Lorentzian metric in (5.12), if and only if

$$\textit{either} \qquad d = 26 \quad \text{and} \quad \lambda = 1 \tag{5.21}$$

$$\textit{or} \qquad 1 \leq d \leq 25 \quad \text{and} \quad \lambda \leq 1. \tag{5.22}$$

However the spectrum of states for the free open string that we have been discussing is much neater if $d = 26$ and $\lambda = 1$ and it is only in this case that an apparently reasonably consistent theory of interacting strings exists. Even this theory possesses a tachyon, namely the ground state of the string described by unexcited states $|\gamma\rangle$, of the form (5.16) satisfying (5.20) with $\lambda = 1$ and so having

$$\gamma^2 = 2. \tag{5.23}$$

With the metric we are using, such a momentum γ is space-like. To avoid this difficulty one needs to move to the $d = 10$ theory of Neveu, Schwarz and Ramond,[11,12] which also involves fermionic degrees of freedom and has a sectror which is tachyon-free and supersymmetric in ten-dimensional Minkowski space, the "superstring" theory of Green and Schwarz.[23]

Living with the tachyon, the interactions of this $d = 26$ bosonic string are defined perturbatively, starting from tree diagram Born terms. If we introduce the vertex operator

$$U(\gamma, z) =: exp\{i\gamma.Q(z)\} : \tag{5.24}$$

$$\equiv exp\{i\gamma.Q_<(z)\} \exp\{i\gamma.Q_0(z)\} exp\{i\gamma.Q_>(z)\} \tag{5.25}$$

where

$$Q_>^\mu(z) = i\sum_{n>0}\frac{\alpha_n^\mu}{n}z^{-n}, \quad Q_<^\mu(z) = i\sum_{n<0}\frac{\alpha_n^\mu}{n}z^{-n} \tag{5.26}$$

and

$$Q_0^\mu(z) = q^\mu - ip^\mu \log z \tag{5.27}$$

so that

$$e^{i\gamma Q_0(z)} = z^{\frac{1}{2}\gamma^2} e^{i\gamma\cdot q} z^{\gamma\cdot p}, \tag{5.28}$$

the Born term amplitude for the interaction of N ground state strings of momenta $\gamma_1, \ldots, \gamma_N$ (with $\gamma_i^2 = 2, \Sigma\gamma_i = 0$) is

$$\int \langle 0|U_{\gamma_1,\ldots,\gamma_N}(z_1,\ldots,z_N)|0\rangle \prod_{i=1}^N \frac{dz_i}{z_i}/d\gamma \tag{5.29}$$

where

$$U_{\gamma_1,\ldots,\gamma_N}(z_1,\ldots,z_N) = U(\gamma_1,z_1)\ldots U(\gamma_N,\, z_N), \tag{5.30}$$

$$d\gamma = \frac{dz_a dz_b dz_c}{(z_a - z_b)(z_b - z_c)(z_c - z_a)}, \tag{5.31}$$

a, b, c being any fixed three of the indices $1, \ldots, N$, and the integral is taken with the variables $z_1, \ldots, z_N$ varying over the real line, subjcct to the constraints that they maintain their order $z_1 > z_2 > \cdots > z_N$, and the points z_a, z_b, z_c remain fixed (as is implied by dividing by $d\gamma$). Independence of the choice of z_a, z_b, z_c follows from the invariance of the integrand of (5.29) under the group of Möbius transformations

$$z \mapsto (az+b)/(cz+d). \tag{5.32}$$

The way that one proves this Möbius invariance and, more generally, the fact that the only states that couple as intermediate states in such amplitudes are physical states in the sense of (5.20) is to exploit the relationship of the vertex (5.24) to the Virasoro algebra, which has the Möbius algebra as the three-dimensional subalgebra given by $n = 1,\ 0,\ -1$,

$$[L_n, U(\gamma,z)] = z^n\left(z\frac{d}{dz} + \frac{\gamma^2}{2}n\right)U(\gamma,z). \tag{5.33}$$

A particular consequence, if $\gamma^2 = 2$, is that then

$$[L_n, U(\gamma,z)] = z\frac{d}{dz}(z^n U(\gamma,z)). \tag{5.34}$$

It follows that

$$A_\gamma = \frac{1}{2\pi i}\oint \frac{dz}{z}U(\gamma,z), \tag{5.35}$$

where the integration contour encircles the origin once, is a physical state creation operator,

$$[L_n, A_\gamma] = 0, \tag{5.36}$$

i.e., it maps physical states into physical states, *provided that it is well-defined.*

Consider A_γ acting on a state with momentum λ. The only problem is with the factor $exp\{i\gamma.Q_0(z)\}$,

$$exp\{i\gamma.Q_0(z)\}|\lambda\rangle = z^{\frac{1}{2}\gamma^2+\gamma.\lambda}|\lambda\rangle. \tag{5.37}$$

Since $\gamma^2 = 2$, this will be single-valued of $\gamma.\lambda$ is an integer. The easiest way to achieve this is for both γ and λ to be members of the same integral lattice Λ, that is a lattice such that the scalar between any two of its points is an integer. We will have operators A_γ acting within the physical state space if we restrict the momenta, or at least some of their components, to lie on an integral lattice. This is not such an unfamiliar restriction. If we consider a particle moving on a torus, formed by taking a d-dimensional space modulo some lattice Γ, its wave function $exp\{i\gamma x/\hbar\}$ will be single-valued on the torus provided that

$$\frac{\gamma.a}{2\pi\hbar} \in \mathbf{Z}, \tag{5.38}$$

that is provided that $\gamma \in 2\pi\hbar\Gamma^*, = \Lambda$ say, where Γ^* is the lattice dual to Γ in the sense defined in Sect. 6.

At this point we shall rather abruptly leave our discussion of string theory amplitudes because we have enough to motivate the introduction of vertex operators with momenta lying on an integral lattice. This motivation has necessarily been rather sketchy in order to avoid too big a digression.

6. LATTICES

In the next section we shall describe how the vertex operators of string theory associate an algebra with a (suitable) lattice. This will be done with a variant of the Frenkel-Kac-Segal construction,[16–18] following an approach developed by Goddard and Olive[19] (see also the work of Frenkel[24]). The construction provides the algebras with a natural string theory interpretation: they are symmetries of the physical states for a string whose momenta lie on the lattice, i.e., a string moving on the torus dual to the lattice.

In this section we shall review some properties of lattices. (See, e.g., refs. 19 or 25 for more details.) A *lattice* is a set of points in a vector space of the form

$$\Lambda = \{\Sigma n_i e_i \ : \ n_i \in \mathbf{Z}\} \tag{6.1}$$

where $\{e_i\}$ is a basis for the space, but not necessarily an orthonormal one. Such a lattice is said to be *integral* if

$$x.y \in \mathbf{Z} \quad \text{for all} \quad x, y \in \Lambda, \tag{6.2}$$

i.e., $\Lambda \subset \Lambda^*$, the lattice *dual* to Λ,

$$\Lambda^* = \{x : x.y \in \mathbf{Z} \quad \text{for all} \quad y \in \Lambda\}. \tag{6.3}$$

A lattice is said to *unimodular* if it has one point per unit volume. It is easy to see that a lattice is *self-dual* if and only if it is both integral and unimodular. A lattice is called *even* if the squared lengths of all of its points are even integers.

Even self-dual lattices are not only particularly interesting to consider from a mathematical point of view; their study is also motivated by physical considerations. Self-dual lattices occur naturally in monopole theory if one seeks a duality between electricity and magnetism.[26] They also appear to be necessary for the consistency of certain string theories.[27]

An even self-dual Euclidean lattice must have a dimension which is a multiple of 8. Thus the lowest dimensional such lattices are 8-dimensional. Up to isomorphism there is only one, the root lattice of E_8. In dimension 16 there are two inequivalent even self-dual lattices, the root lattice of $E_8 \times E_8$ and the weight lattice of $spin(32)/\mathbf{Z}_2$. (For each of E_8 and $E_8 \times E_8$ the weight and root lattices coincide.) In dimension 24 there precisely 24 inequivalent such lattices, including one for which the minimum squared distance between points is 4, namely the Leech lattice. After this the number grows very rapidly.

In Minkowski space the situation is simpler. A Lorentzian even self-dual lattice must have a dimension of the form $8n + 2$, where n is an integer, and it is essentially unique for each n. It is usually denoted $II^{8n+1,1}$ and consists of points of the form $\underline{m}$ and $\underline{m} + \underline{\frac{1}{2}}$ where

$$\underline{m} = (m_1, m_2, \ldots, m_{8n+1}; m_{8n+2}), m_i \in \mathbf{Z}, \Sigma m_i \in 2\mathbf{Z} \tag{6.4}$$

and

$$\underline{\frac{1}{2}} = \left(\frac{1}{2}, \frac{1}{2}, \ldots, \frac{1}{2}; \frac{1}{2}\right). \tag{6.5}$$

It is easy to check that the lattice consisting of such points is even and self-dual.

There is a relationship betwen Euclidean even self-dual lattices in dimension $8n$ and light-like vectors in $II^{8n+1,1}$. Given a primitive light-like vector k in the Lorentzian lattice, (i.e., k is not a multiple of any other lattice vector except $-k$), we consider the points of this lattice modulo translations by k. In this way we obtain an even self-dual Euclidean lattice. It can be shown that all such lattices can be obtained in this way and that there is a one to one correspondence between their equivalence classes and the primitive light-like vectors k modulo automorphisms of the lattice $II^{8n+1,1}$. This construction is clearly related to the polarization states of a massless vector particle (photon) in the appropriate dimension. For $n = 3$, Conway and Sloane have shown that taking

$$k = (0, 1, 2, \ldots, 24; 70) \tag{6.6}$$

yields the Leech lattice from this construction.

7. THE ALGEBRA ASSOCIATED WITH AN INTEGRAL LATTICE

We shall now explore further the string model formalism of Sect. 5, taking the momenta to lie on a lattice Λ. For the purposes of developing the mathematics of the construction, this could be taken to lie in a space with any signature we wish; it could be Lorentzian, Euclidean or there could even be certain directions which were totally null, i.e., the metric tensor might be singular. (More complicatedly it could be that only the projection of the momenta on a certain subspace lies on a discrete lattice, corresponding to a configuration space which is the product of Minkowski space and a torus; this is indeed the situation in the applications which are considered to be promising from the point of view of physical relevance.)

Specifically we consider the Hilbert space, $\mathcal{H}$, generated by the operators α_n^μ, $n \in \mathbf{Z}$, $1 \leq \mu \leq d$, satisfying the commutation relations

$$[\alpha_m^\mu, \alpha_n^v] = m g^{\mu\nu} \delta_{m,-n} \tag{7.1}$$

and the hermiticity conditions

$$\alpha_m^{\mu\dagger} = \alpha_{-m}^\mu \tag{7.2}$$

from states $|\gamma\rangle$, $\gamma \in \Lambda \subset \mathbf{R}^d$, obeying

$$p^\mu|\gamma\rangle = \gamma^\mu|\gamma\rangle \tag{7.3a}$$

$$\alpha_m^\mu|\gamma\rangle = 0, \quad m > 0, \tag{7.3b}$$

$$\langle\gamma'|\gamma\rangle = \delta_{\gamma'\gamma}, \tag{7.3c}$$

where $p^\mu \equiv \alpha_0^\mu$. In this context we cannot define the position operator q^μ but we can define $exp\{i\gamma.q\}$ for $\gamma \in \Lambda$ by

$$e^{i\gamma q}|\gamma'\rangle = |\gamma + \gamma'\rangle \tag{7.4}$$

and the condition it commutes with α_n^μ, $n \neq 0$. Then the whole of $\mathcal{H}$ is generated from the vacuum vector $|0\rangle$ by α_n^μ and $exp(i\gamma.q)$, $\gamma \in \Lambda$.

We wish to investigate the algebra generated by the operators

$$A_r = \frac{1}{2\pi i}\oint \frac{dz}{z} U(r, z), \tag{7.5}$$

for

$$r \in \Lambda_2 = \{r \in \Lambda \ : \ r^2 = 2\}. \tag{7.6}$$

How does one calculate $[A_r, A_s]$? Consider

$$A_r A_s = \frac{1}{2\pi i}\oint \frac{dz}{z} U(r, z) \frac{1}{2\pi i}\oint \frac{d\varsigma}{\varsigma} U(s, \varsigma). \tag{7.7}$$

To normal order the integrand in (7.7) we use

$$\begin{aligned} & exp\{ir.Q_>(z)\}exp\{is.Q_<(\varsigma)\} \\ &= (1-\varsigma/z)^{r.s} exp\{is.Q_<(\varsigma)\}exp\{ir.Q_>(z)\}, \quad \text{if} \quad |\varsigma| < |z|, \end{aligned} \tag{7.8}$$

which follows from

$$[Q^{\mu}_>(z), Q^{\nu}_<(\varsigma)] = g^{\mu\nu} \log(1-\varsigma/z), \quad \text{if } |\varsigma| < |z|, \tag{7.9}$$

and

$$e^A e^B = e^{[A,B]} e^B e^A, \tag{7.10}$$

which holds if $[A, B]$ is a c-number. It follows that

$$\begin{aligned} U(r,z)U(s,\varsigma) &= exp\{ir.Q_<(z) + is.Q_<(\varsigma)\} \\ &\cdot (z-\varsigma)^{r.s} z\varsigma e^{i(r+s).q} z^{r.p} \varsigma^{s.p} \\ &\cdot exp\{ir.Q_>(z) + is.Q_>(\varsigma)\} \quad \text{for} \quad |\varsigma| < |z| \\ &\equiv U_{r,s}(z,\varsigma) \end{aligned} \tag{7.11}$$

The right-hand side of (7.11) is a regular function of z, ς and except for $z = 0$, $\varsigma = 0$ and $z = \varsigma$. If we can calculate the product in the other order we see that

$$U(s,\varsigma)U(r,z)$$

will be given by the same function of z and ς apart from a factor of $(-1)^{r.s}$, which could be either $+1$ or -1. Thus what we can easily calculate is

$$A_r A_s - (-1)^{r.s} A_s A_r = \frac{1}{(2\pi i)^2} \left\{ \oint_0 \frac{d\varsigma}{\varsigma} \oint_{|z|>|\varsigma|} \frac{dz}{z} - \oint \frac{d\varsigma}{\varsigma} \oint_{|\varsigma|>|z|} \frac{dz}{z} \right\} U_{r,s}(z,\varsigma) \tag{7.12}$$

The z contours in these two integrals, illustrated in Fig. 1, have a difference which is

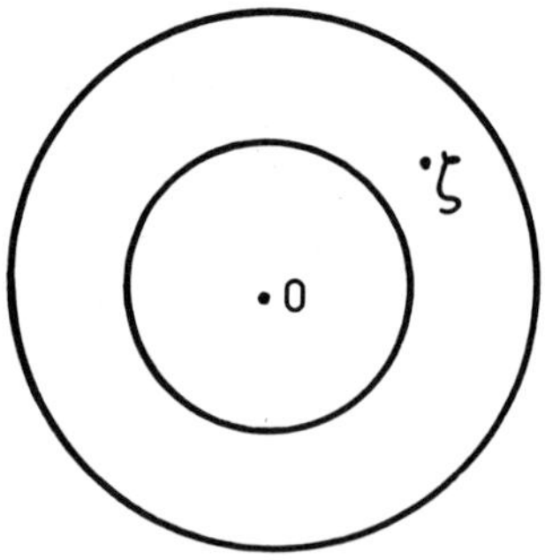

Figure 1

equivalent to a small loop about ς. Thus

$$A_r A_s - (-1)^{r.s} A_s A_r = \frac{1}{(2\pi i)^2} \oint_0 \frac{d\varsigma}{\varsigma} \oint_\varsigma \frac{dz}{z} U_{r,s}(z,\varsigma), \tag{7.13}$$

where the z contour positively encircles ς once and the ς contour positively encircles the origin.

In certain circumstances (7.13) is easy to evaluate. Firstly, it vanishes if $r.s \geq 0$ because in this case $U_{r,s}(z,\varsigma)$ is non-singular at $z = \varsigma$. Secondly if $r.s = -1$, there is a simple pole at $z = \varsigma$ and so

$$\begin{aligned} A_r A_s - (-1)^{r.s} A_s A_r &= \frac{1}{2\pi i} \oint d\varsigma \; e^{i(r+s).q} \varsigma^{(r+s).p} exp\{i(r+s).Q_<(\varsigma)\} exp\{i(r+s).Q_>(\varsigma)\}, \\ &= \frac{1}{2\pi i} \oint \frac{d\varsigma}{\varsigma} U(r+s,\varsigma), \qquad (7.14) \\ &= A_{r+s}. \end{aligned}$$

using (5.24–28) and since $(r+s)^2 = 2$ in this case. Thirdly, if $r,s = -2$ there is a double pole at $z = \varsigma$ whose contribution is particularly easy to evaluate in the special case where $r = -s$. In this case it gives

$$A_r A_s - (-1)^{r.s} A_s A_r = \frac{1}{2\pi i} \oint \frac{d\varsigma}{\varsigma} r.P(\varsigma), \qquad (7.15)$$

where

$$P^\mu(z) = iz \frac{dQ^\mu(z)}{dz}, \qquad (7.16a)$$

$$= \sum_{n=-\infty}^{\infty} \alpha_n^\mu z^{-n}. \qquad (7.16b)$$

Thus we have calculated the results:

$$A_r A_s - (-1)^{r.s} A_s A_r = 0, \qquad \text{if} \quad r.s \geq 0, \qquad (7.17a)$$

$$= A_{r+s}, \qquad \text{if} \quad r.s = -1, \qquad (7.17b)$$

$$A_r A_s - (-1)^{r.s} A_s A_r = rp \qquad \text{if} \quad r = -s \qquad (7.17c)$$

Additionally we have that

$$[p^\mu, A_r] = r^\mu A_r \qquad (7.17d)$$

because A_r carries momentum r. The possibilities listed in eqs. (7.17) are exhaustive if we are dealing with a Euclidean lattice because then, if $r,s \in \Lambda_2$, $|r.s| \leq 2$; and $r.s = -2$ if and only if $r = -s$.

Clearly the factor $(-1)^{r.s}$ is annoying. Without it we would have a Lie algebra. It can be removed by a trick introduced by Frenkel and Kac[17]: we just multiply the vertex operators A_r by certain suitable chosen functions of momenta (which are thus functions of a discrete variable, since the momenta lie on a lattice). Suppose we denote by Λ_R the sublattice of Λ generated by the points $r \in \Lambda_2$. For each $u \in \Lambda_R$, we introduce a function of momentum c_u with properties such that if $\hat{c}_u = exp(iq.u)c_u$

$$\hat{c}_u \hat{c}_v = (-1)^{u.v} \hat{c}_v \hat{c}_u, \qquad (7.18a)$$

$$\hat{c}_u \hat{c}_{-u} = 1, \tag{7.18b}$$

$$\hat{c}_u \hat{c}_v = \epsilon(u,v) \hat{c}_{u+v} \tag{7.18c}$$

where for each pair $u, v \in \Lambda$, $\epsilon(u,v)$ is either $+1$ or -1. If we can find such an object it will solve our problems because we can then set

$$E^r = A_r c_r, \quad r \in \Lambda, \tag{7.19}$$

and we have in consequence

$$\begin{aligned} [E^r, E^s] &= 0, && \text{if} \quad r.s \geq 0, && (7.20a) \\ &= \epsilon(r,s) E^{r+s}, && \text{if} \quad r.s = -1, && (7.20b) \\ &= r.p, && \text{if} \quad r = -s, && (7.20c) \end{aligned}$$

and

$$[p^\mu, E^r] = r^\mu E^r, \tag{7.20d}$$

$$[p^\mu, p^\nu] = 0. \tag{7.20e}$$

Further, provided that c_u is hermitian, p^μ, E^r, satisfy the hermicity conditions

$$E^{r\dagger} = E^{-r}, \ p^{\mu\dagger} = p^{-\mu} \tag{7.21}$$

Elementary explanations of how to construct c_u are given in refs. 1 and 19.

In the Euclidean case, eqs. (7.20) define a closed finite dimensional Lie algebra g_Λ, which we can regard as the Lie algebra of some compact Lie group G_Λ associated with the integral lattice Λ. We have obtained it in a basis just like that of (2.5-7). The momentum operators $p^\mu, 1 \leq \mu \leq d$, form a Cartan subalgebra for g_Λ and E^r are the step operators, so that

$$\text{rank } g_\Lambda = d, \tag{7.22a}$$

$$\dim \ g_\Lambda = d + |\Lambda_2| \tag{7.22b}$$

using $|\Lambda_2|$ for the number of points of length squared 2 on Λ; Λ_2 is the set of roots of g_Λ. (We have not completely established that there are no elements of g_Λ commuting with all the p^μ which are not linear combinations of them. For this see ref. 19.) The algebra g_Λ will be non-Abelian provided that Λ_2 is non-empty, and it will be semi-simple provided that dim Λ_R = dim Λ.

We can generalize what we have said here a little: it is not really necessary for the lattice Λ to be integral. We need only that the operators E^r are well-defined for $r \in \Lambda_2$ and this requires that

$$e^{ir.Q_0(z)}|\lambda\rangle = z^{1+\lambda.r}|\lambda + r\rangle \tag{7.23}$$

be single valued [cf. (5.37)]. This happens if and only if $\lambda.r \in \mathbf{Z}$ whenever $r \in \Lambda_2$ and

$\lambda \in \Lambda$. This condition can be equivalently written

$$\Lambda_R \subset \Lambda^* \tag{7.24}$$

which is a weaker condition than that we previously imposed, that the lattice be integral, i.e., $\Lambda \subset \Lambda^*$.

Let us consider some examples of Euclidean lattices. The simplest possibility is to take $\Lambda = \mathbf{Z}^d$, $d > 1$, the hypercubic lattice in d dimensions. There are $2d(d-1)$ points of squared length 2 on this lattice. These are the points for which only two of the coordinates are non-zero and these are each ± 1. Thus Λ_R consists of points of the form

$$(n_1, n_2, \ldots, n_d), \; n_i \in \mathbf{Z}, \; \Sigma n_i^2 \in 2\mathbf{Z}. \tag{7.25}$$

It is not difficult to check that Λ_2 is the root system of $so(2d)$ and, of course, the dimensions check

$$d + |\Lambda_2| = d(2d-1) = \dim so(2d). \tag{7.26}$$

If we exponentiate the generators we have constructed we get a specific group G_Λ with g_Λ as algebra, that is we get a specific global structure. The torus from which we have started is essentially the maximal torus (the maximal Abelian subgroup) of G_Λ. We can determine this global structure by looking at which eigenvalues of the generators $p^\mu, 1 \leq \mu \leq d$, occur, and these are given exactly by the original lattice Λ. In other words Λ is the set of weights of G_Λ. The condition that λ be a weight for g_Λ is that

$$2r.\lambda/r^2 \in \mathbf{Z} \quad \text{for all roots } r \tag{7.27}$$

and this is precisely (7.24) because in our construction all the roots of g_Λ inevitably have $r^2 = 2$. In the case $\Lambda = \mathbf{Z}^d$, the points of length 1, which are just those with only one non-zero coordinate and that coordinate being ± 1, correspond to the weights of the vector representation of $so(2d)$. Since these weights generate additively all of $\mathbf{Z}^d$, we have

$$G_{\mathbf{Z}^d} = so(2d). \tag{7.28}$$

The largest lattice permitted, if we are to have g_Λ equal to $so(2d)$, is the dual of (7.25), namely the lattice consisting of the points $(n_1, n_2, \ldots, n_d)$ where

$$\textit{either} \quad n_i \in \mathbf{Z}, \; 1 \leq i \leq d, \qquad \textit{or} \quad n_i + \frac{1}{2} \in \mathbf{Z}, \; 1 \leq i \leq d. \tag{7.29}$$

This is the lattice of all weights of $so(2d)$. Taking this lattice instead of $\mathbf{Z}^d$, we still have $g_\Lambda = so(2d)$, provided that $d \neq 8$, but we have $G_\Lambda = spin(2d)$, the simply connected covering group of $so(2d)$. On the other hand, the smallest possible choice is to take Λ to be just Λ_R, in which case we would have

$$G_\Lambda = so(2d)/\mathbf{Z}_2, \quad \text{where} \quad \mathbf{Z}_2 = \{\pm 1\} \subset so(2d).$$

If d is even, there are yet other possible choices for Λ. We could take Λ to be the lattice consisting of Λ_R together with the points

$$\left(n_1 + \frac{1}{2}, n_2 + \frac{1}{2}, \ldots, n_d + \frac{1}{2}\right), \quad n_i \in \mathbf{Z}, \Sigma n_i \in 2\mathbf{Z}. \tag{7.30}$$

This lattice is only integral if d is a multiple of 4 and in that case it is, like $\mathbf{Z}^d$, self-dual, i.e., $\Lambda = \Lambda^*$. But when d is a multiple of 8 it has the additional property, not possessed by $\mathbf{Z}^d$, of being even, like $II^{8n+1,1}$, which it resembles in construction. In general, this choice of Λ produces a group G_Λ, which is $spin(2d)$ divided by a different subgroup, which is not isomorphic to $so(2d)$. The exceptions occur when $d = 4$ or 8. If $d = 8$, the lattice has extra points of squared length 2, namely the 128 permitted points each of whose coordinates is $\pm\frac{1}{2}$. The corresponding operators E^r increase in number from 112 to 240 and in fact $G_\Lambda = E_8$. This is the essentially unique self-dual Euclidean lattice of dimension 8 mentioned in Sect. 2. (Of course, it is open to us not to add these extra points and stick with $so(16)$.) If $d = 4$, the resulting group G_Λ is actually isomorphic to $so(8)$. Actually, the condition in (7.30) that Σn_i be even could be replaced by the condition that Σn_i be odd; for any d this would produce an isomorphic G_Λ and for $d = 4$ it produces a third copy of $so(8)$. These three $so(8)$ algebras are related by other isomrophisms associated with the symmetry of the correesponding Dynkin diagram; see Fig. 2. This is known as the triality property of $so(8)$.

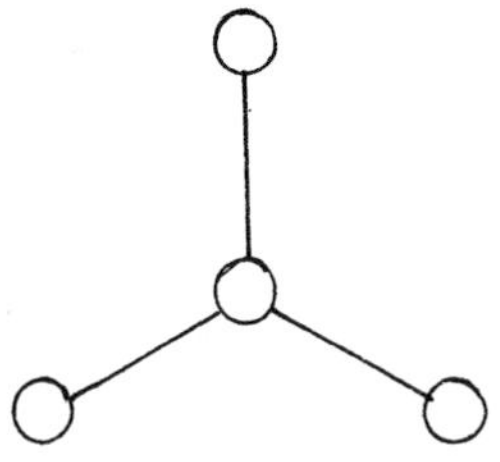

Figure 2

When $d = 16$ the lattice specified by adding the points of (7.30) to Λ_R, is again even and self-dual. The only other possibility for such a lattice, with dimension 16, is the direct sum of two E_8 lattices. These even self dual lattices produce the groups $spin(32)/\mathbf{Z}_2$ and $E_8 \times E_8$, respectively, which have excited a very great deal of interest recently.[27]

Another exercise is to consider the possibilities in one dimension. There if Λ is to satisfy (7.24) and have points r with $r^2 = 2$, it must be either $\sqrt{2}\mathbf{Z}$ or $(1/\sqrt{2})\mathbf{Z}$. In the former case $G_\Lambda = so(3)$, in the latter $su(2)$. In two dimensions taking $\Lambda = \mathbf{Z}^2$ given $G_\Lambda = so(4) = su(2) \times su(2)/\mathbf{Z}_2$ whilst taking $\Lambda = \sqrt{2}\mathbf{Z}^2$ gives $G_\Lambda = su(2) \times su(2)$.

In the general Euclidean case, we see that Λ is the weight lattice of some group with algebra g_Λ, a sublattice of the lattice of all weights of g_Λ or, equivalently of the weight lattice of the simply connected group with algebra g_Λ. This Lie algebra is inevitably simply-laced, that is all the roots have the same squared length, 2. If g_Λ is simple, the only possibilities are that it is one of $A_n = su(n+1)$, $D_n = so(2n)$, E_6, E_7 and E_8; otherwise it is a direct sum of copies of these and $u(1)$ factors. All simply-laced compact Lie groups can be obtained in this way. To obtain analogous constructions of the non-simply-laced compact-Lie groups it is necessary to generalize the construction in a way which is, in most cases, quite subtle.[28,29]

Having obtained a representation of the finite-dimensional Lie algebra G_Λ, we can extend it to a representation of the associated untwisted affine Kac-Moody algebra $\hat{g}_\Lambda$ by considering all the Laurent coefficients of $U(r,z)$. This is the original Frenkel-Kac-Segal construction.[17,18] We generalize (7.19) by introducing

$$E_n^r = \frac{1}{2z_i} \oint \frac{dz}{z} z^n U(r,z) c_r, \tag{7.31}$$

so that

$$U(r,z) c_r = \sum_{n=-\infty}^{\infty} E_n^r z^{-n}, \tag{7.32}$$

and we set

$$H_n^\mu = \alpha_n^\mu. \tag{7.33}$$

Calculations like those of (7.14) then show that this provides a representation of $\hat{g}_\Lambda$ as in (2.8-10), with $k = 1$, i.e., a level one representation.

8. NON-EUCLIDEAN LATTICES

We consider what happens in the construction of the last section if Λ is not Euclidean. Then g_Λ is defined to be the Lie algebra generated by the $p^\mu, 1 \le \mu \le d$; E^r, $r \in \Lambda_2$. Thus it is spanned by elements of the form

$$E' = [E^{r_1}, [E^{r_2}, [\ldots [E^{r_{m-1}}, E^{r_m}] \ldots]]]. \tag{8.1}$$

This is a step operator corresponding to the root

$$v = \sum_{i=1}^{m} r_i. \tag{8.2}$$

Using dual model techniques we can show that

$$E' = 0 \quad \text{unless} \quad v^2 \le 2. \tag{8.3}$$

Further it follows from Eq. (5.36) that

$$[L_n, E'] = 0, \tag{8.4}$$

that is the whole of g_Λ commutes with the Virasoro algebra. In the Lorentzian case we can think of the elements of g_Λ as being associated with states of a string moving on a Minkowski torus $2\pi\hbar\mathbf{R}^d/\Lambda$. Actually an isomorphism can be set up between g_Λ and a subspace of the physical states, which has interesting consequences if Λ is Lorentzian.[19,24]

A new feature occurs when we move away from Λ being Euclidean: there is no longer in general just one step operator for non-zero root v. Roots occur with variable multiplicities. To illustrate this let us consider the calculation of

$$[E^r, E^s], \quad r, s \in \Lambda_2, \tag{8.5}$$

a little further in the non-Euclidean case. This vanishes if $r.s \geq 0$ and the case $r.s = -1$ is as before. The first new feature occurs when $r.s = -2$. This no longer forces $r = -s$ but its does imply that $r + s$ is light-like,

$$(r+s)^2 = 0. \tag{8.6}$$

There is a double pole in (7.13) and evaluation of the appropriate residue gives

$$[A_r, A_s] = \frac{1}{2}(r-s)_\mu A^\mu_{r+s} \tag{8.7}$$

where for light-like vectors $k \in \Lambda_R$ we define

$$A^\mu_k = \frac{1}{2\pi i}\oint \frac{dz}{z} : P^\mu(z) exp\{ikQ(z)\} : . \tag{8.8}$$

This operator satisfies

$$k_\mu A^\mu_k = 0, \tag{8.9}$$

$$[L_n, \epsilon_\mu A^\mu_k] = 0 \quad \text{if} \quad \epsilon.k = 0. \tag{8.10}$$

Note that $(r-s).(r+s) = 0$. The operators $\epsilon.A_k$ describe the coupling of massless states with polarization vectors ϵ. They are the original "photon" vertex operators of Del Giudice, Di Vecchia and Fubini.[30] For each $k \in \Lambda_R$ with $k^2 = 0$, there are $d-2$ independent step operators $\epsilon.A_k$ corresponding to the possible states of a massless vector particle in d dimensions. The root multiplicity of such a light-like vector $k \in \Lambda_R$ is thus $d-2$. For a Lorentzian lattice these root multiplicities grow fast with $-v^2$.

A particularly interesting case to consider is the *affine* case,[19] where we take a d-dimensional Euclidean lattice Λ and enlarge it by the addition of an orthogonal light-like vector k to obtain a new lattice Λ'. Thus

$$k^2 = 0, \; x.k = 0 \quad \text{for all} \quad x \in \Lambda \tag{8.11}$$

and

$$\Lambda' = \{x + nk : x \in \Lambda, \; n \in \mathbf{Z}\}. \tag{8.12}$$

Now one needs to be a little careful because the metric tensor $g^{\mu\nu}$ for Λ' is singular, in that det $g = 0$, and so the inverse $g_{\mu\nu}$ does not exist. In consequence we cannot construct the Virasoro algebra (5.17) for this lattice.

For the affine lattice, the points of squared length 2 are

$$\Lambda_2' = \{r + nk : r \in \Lambda_2, n \in \mathbf{Z}\} \tag{8.13}$$

and using the construction of the last section with oscillators $\alpha_n^\mu, 1 \leq \mu \leq d+1$, g_Λ, has a basis consisting of

$$E_n^r \equiv E^{r+nk}, \quad r \in \Lambda_2,\ n \in \mathbf{Z}, \tag{8.14}$$

$$A_n^\mu = \oint \frac{dz}{z} P^\mu(z)\ exp\{ink.Q(z)\}, \quad 1 \leq \mu \leq d. \tag{8.15}$$

These satisfy the commutation relations of the untwisted affine Kac-Moody algebra $\hat{g}_\Lambda$

$$[A_m^\mu, A_n^\nu] = \delta^{\mu\nu}\delta_{m,-n}k.p \tag{8.16}$$

$$[A_m^\mu, E_n^r] = r^\mu E_{m+n}^r \tag{8.17}$$

$$[E_m^r, E_n^s] = 0 \qquad r.s \geq 0, \tag{8.18a}$$

$$= \epsilon(r,s)E_{m+n}^{r+s} \qquad , r.s = -1, \tag{8.18b}$$

$$= r.A_{m+n} + mk.p\delta_{m,-n}, \quad r = -s \tag{8.18c}$$

The operators $k \cdot \alpha_n$ commute with the whole of $g_{\Lambda'}$.

If we restrict the eigenvalues of p to the lattice Λ', $k.p$ would vanish identically but it is not necessary to do this; p can take values in any lattice $\tilde{\Lambda} \supset \Lambda'$ with the property that $x.y \in \mathbf{Z}$ if $x \in \tilde{\Lambda}$, $y \in \Lambda'$. A convenient construction is to take $\tilde{\Lambda}$ to be the Lorentzian lattice obtained by adjoining to Λ a second light-like vector k' with

$$k.k' = 1, \quad k'^2 = 0 \tag{8.19}$$

Then we have

$$\tilde{\Lambda} = \{x + mk + nk' : x \in \Lambda,\ m, n \in \mathbf{Z}\} \tag{8.20}$$

and

$$\Lambda' = \{x \in \tilde{\Lambda} :\ k.x = 0\}. \tag{8.21}$$

We can now reintroduce the Virasoro operators if we wish and

$$[L_n, \hat{g}] = [k.\alpha_n, \hat{g}] = 0. \tag{8.22}$$

This construction of the Kac-Moody algebra $\hat{g}_\Lambda$ is very analogous to constructing the physical states of dual string theory in the covariant formation, whilst the original Frenkel-Kac-Segal construction[17,18] is the analogue of working directly in the light-cone gauge. These are not really very different ways of doing things. We could deduce the algebra satisfied by E_n^r, as defined by (7.31) and α_m^μ as follows. Since the $k.\alpha_n$ with all the operators involved in the definitions of E_n^r, A_n^μ by (8.14) and (8.15), we will not affect the algebra (8.16-18) if we set

$$k.\alpha_n \to 0, \; n \neq 0, \qquad k.p \to 1 \tag{8.23}$$

Doing this yields the definition (7.21) of E_n^r and sends

$$A_n^\mu \to \alpha_n^\mu. \tag{8.24}$$

However the "covariant" construction of (8.12-21), which is a sort of generalization of the relation between Euclidean and Lorentzian lattices explained in Sect. 6, contains more because we can find within it directly all the representations of simply-laced untwisted affine Kac-Moody algebras. Given a simply-laced finite-dimensional algebra g, any weight lattice, Λ, of g, then we can construct $\tilde{\Lambda}$ as in (8.20). Then the state $|\lambda + xk\rangle$, where x is a positive integer, is a vacuum state for a level representation of $\hat{g}$ provided that is is annihilated by (8.14-15) for $n > 0$. This requires

$$x \geq |r \cdot \lambda| \tag{8.25}$$

for all $r \in \Lambda_2$ which is just condition (2.4). In particular the construction provides us with all the representations of the affine Kac-Moody algebras associated with E_6, E_7 and E_8. Using the inclusions $su(n) \subset so(2n)$, $sp(n) \subset so(4n)$, explained in Sect. 3, together with $so(2n+1) \subset so(2n+2)$, $G_2 \subset so(8)$, $F_4 \subset E_6$, each of which preserves the level of the associated Kac-Moody algebra (cf. ref. 29) we verify that the conditions (2.4) are sufficient as well as necessary.

ACKNOWLEDGEMENT

I am very grateful to Adrian Kent, Werner Nahm and David Olive for discussions on which many aspects of the presentation here are based. This research was supported in part by the National Science Foundation under Grant No. PHY82-17853, supplemented by funds from the National Aeronautics and Space Administration, at the University of California at Santa Barbara.

REFERENCES

1. P. Goddard and D. Olive, *Kac-Moody and Virasoro Algebras in Relation to Quantum Physics,* *International Journal of Modern Physics* **A1**, 303 (1986).
2. L. Dolan, *Phys. Rep.* **109**, 1 (1984).
3. B. Julia, A.M.S. Lectures in Applied Mathematics, **21**, 355 (1985), and in *Vertex Operators in Mathematics and Physics* (ed. J. Lepowsky *et al.*, Springer, 1984) 393.
4. J. Lepowsky, in *Vertex Operators in Mathematics and Physics* (ed. J. Lepowsky *et al.*,Springer, 1984) 1.

5. I. MacDonald, Lecture Notes in Mathematics **901**, 258, (1981).

6. V.G. Kac, Infinite Dimensional Lie Algebras (2nd edition, Cambridge University Press, 1985).

7. V.G. Kac, Proceedings of the International Congress of Mathematicians, Helsinki, 299 (1978).

8. B.L. Feigin and D.B. Fuchs, *Funct. Anal. App.* **16**, 114 (1982).

9. D. Friedan, Z. Qiu and S. Shenker, *Phys. Rev. Lett.* **52**, 1575 (1984); in *Vertex Operators in Mathematics and Physics* (ed. J. Lepowsky *et al.*, Springer, 1984) 491.

10. J.F. Gomes, *Phys. Lett.* **171B**, 75 (1986).

11. A. Neveu and J.H. Schwarz, *Nucl. Phys.* **B31**, 86 (1971); *Phys. Rev.* **D4**, 1109 (1971).

12. P. Ramond, *Phys. Rev.* **D3**, 2415 (1971).

13. P. Goddard, A. Kent and D. Olive, *Phys. Lett.* **152B**, 88 (1985); *Comm. Math. Phys.* **103**, 105 (1986).

14. A. Rocha-Caridi in *Vertex Operators in Mathematics and Physics* (ed. J. Lepowsky *et al.*, Springer, 1984) 451.

15. P. Goddard and D. Olive, *Nucl. Phys.* **257** [FS14], 226 (1985).

16. M. Halpern, *Phys. Rev.* **D12**, 1684 (1975).

17. I.B. Frenkel and V.G. Kac, *Inv. Math.* **62**, 23 (1980).

18. G. Segal, *Commun. Math. Phys.* **80**, 301 (1981).

19. P. Goddard and D. Olive, in *Vertex Operators in Mathematics and Physics*, (ed. J. Lepowsky *et al.*, Springer, 1984) 51.

20. R.C. Brower, *Phys. Rev.* **D6**, 1655 (1972).

21. P.Goddard and C.B. Thorn, *Phys. Lett.* **40B**, 235 (1972).

22. C.B. Thorn, in *Vertex Operators in Mathematics and Physics* (ed. J. Lepowsky *et al.*, Springer, 1984) 411.

23. F. Gliozzi, D. Olive and J. Scherk, *Nucl. Phys.* **B122**, 253 (1977); M. Green and J. Schwarz, *Nucl. Phys.* **B181**, 502 (1981).

24. I.B. Frenkel, *Am. Math. Soc. Lectures in Appl. Math.* **21**, 325 (1985).

25. J.P. Serre, *A Course in Arithmetic* (Springer, 1973).

26. D. Olive, in *Monopoles in Quantum Field Theory* (ed. N. Craigie *et al.*, Springer, 1982). 157.

27. M. Green and J. Schwarz, *Phys. Lett.* **149B**, 117 (1984); D. Gross, J.A. Harvey, E. Martinec and R. Rohm, *Phys. Rev. Lett.* **54**, 502 (1985); *Nucl. Phys.* **B256**, 253 (1985).

28. J. Lepowsky and M. Primc in *Vertex Operators in Mathematics and Physics* (ed. J. Lepowsky *et al.*, Springer, 1984) 143.

29. P. Goddard, W. Nahm, D. Olive and A Schwimmer, DAMTP preprint 86/10, to appear in Commun. Math. Phys.

30. E. Del Giudice, P. Di Vecchia and S. Fubini, *Ann. Phys.* **70**, 378 (1972).

DIMENSIONS, INDICES AND CONGRUENCE CLASSES OF REPRESENTATIONS OF AFFINE KAC-MOODY ALGEBRAS (with examples for affine E_8)

S. N. Kass
Centre de recherches mathématiques, Université de Montréal
Montréal, Québec, Canada
and
Theoretical Division, Los Alamos National Laboratory
Los Alamos, NM 87545, USA

J. Patera
Centre de recherches mathématiques, Université de Montréal
Montréal, Québec, Canada

INTRODUCTION

The purpose of this lecture is to introduce, describe and illustrate affine generalizations of some familiar notions from the representation theory of semisimple Lie algebras/groups. We touch upon the multiplicity of a weight and the dimension, congruence class, and indices of a representation. Our examples of the highest weight representations of affine E_8 can be considered as a preview of far more extensive results of this type to appear (Kass et al., 1987).

THE WEIGHT SYSTEM OF A REPRESENTATION

First we recall some familiar facts. Finite-dimensional irreducible representations of $SU(2)$ can be specified by the "angular momenta" J, $J = 0,\frac{1}{2},1,\frac{3}{2},2,\ldots$ of the representation, and the basis vectors or "angular momentum states" within the representation J can be labelled by the "projection" M of J, where M takes on each value $J,J-1,\ldots,-J$. Convenience and consistency with higher rank algebras leads us to change these conventions and to take J as twice the angular momentum and M as twice its projection, giving

$$J \in 0,1,2,\ldots\ , \qquad M \in \langle J,J-2,\ldots,-J\rangle = \Omega(J),$$

where $\Omega(J)$ is called the *weight system* of the $SU(2)$ representation J. We will sometimes denote the representation simply by J and sometimes by $L(J)$.

A finite-dimensional irreducible $SU(3)$ representation is determined by its *highest weight* $\Lambda = (p,q)$, with integer $p,q \geq 0$. The inherent geometry of the algebra allows us to draw the weight system $\Omega(\Lambda)$ as a triangle (for highest weight

$(p,0)$ or $(0,q)$), a hexagon (if $pq > 0$), or a point (for the one-dimensional representation with highest weight $(0,0)$). For example, one has the following weight systems $\Omega(\Lambda)$ for the representations $L(\Lambda)$ with highest weights $\Lambda = (1,0)$, $(0,1)$ and $(1,1)$:

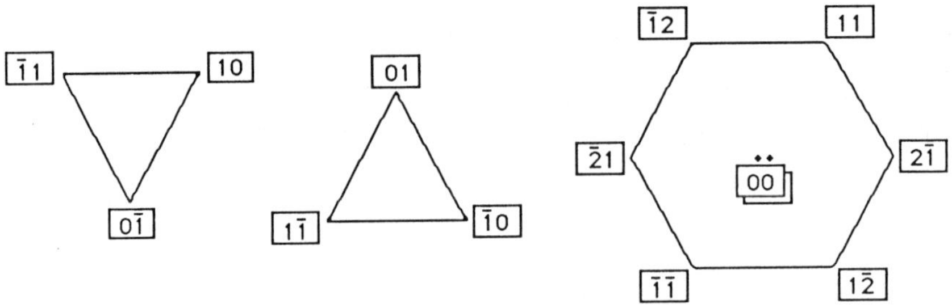

In the last example the symbolism represents the fact that there are two vectors of weight $(0,0)$. In general we suppress the commas between the coordinates of the weight and economize space by using an overbar instead of a minus sign.

Briefly, the weights of a finite-dimensional simple complex Lie algebra g "live" in a Euclidean lattice, and their coordinates are given in terms of the basis of *fundamental weights* $\lambda_1, \lambda_2, \ldots \lambda_\ell$ of that lattice. This basis and its dual basis of *simple roots* $\alpha_1, \alpha_2, \ldots, \alpha_\ell$ are related by the matrix CM, the *Cartan matrix*, which is the change of basis matrix between the two bases, and which completely describes the algebra. The same information contained in CM can be given by a graph, the *Dynkin diagram* of g (DD). The Cartan matrices and Dynkin diagrams for each of the finite-dimensional simple complex Lie algebras can be found in many places (see Bremner et al., 1985). A thorough development of the representation theory of these algebras can be found in (Humphreys, 1972).

For an affine Kac-Moody algebra g an irreducible representation is again given by the tuple Λ, which now has $\ell + 1$ coordinates $\lambda_0, \lambda_1, \ldots \lambda_\ell$. Although these coordinates completely describe a module, they do not completely describe Λ within the weight lattice, as we shall soon see, and we generally call this $\ell + 1$-tuple the *weight label* of Λ.

The distinct weights of $\Omega(\Lambda)$ can be calculated recursively from the highest weight $\Lambda = (p_0, \ldots, p_\ell)$ (ℓ is the *rank* of g, and of CM) with non-negative integer coordinates p_k (if g is finite, there is no p_0) using the algorithm:

(a) Put Λ into $\Omega(\Lambda)$ and let $\mu = \Lambda$.

(b) For $\mu \in \Omega(\Lambda)$, for any positive coordinate p_k of μ, add the p_k weights $\Lambda - \alpha_k, \Lambda - 2\alpha_k, \ldots, \Lambda - p_k\alpha_k$ to $\Omega(\Lambda)$. In the basis of fundamental weights, the vector α_k is the k-th row of CM for g (in some conventions α_k is the k-th column of CM).

(c) Repeat (b), replacing μ by each of the weights found in (b).
The procedure terminates for finite-dimensional Lie algebras and continues forever in the affine case.

The above algorithm computes only the list of distinct weights of a module. The number of vectors of weight μ in $L(\Lambda)$ (the *multiplicity* of μ in $L(\Lambda)$) is difficult to compute in general, and can be found in published tables (Bremner et al., 1985 (finite case); Kass et al., 1987 (affine case)).

We now consider a few examples in the affine case. Each of the finite-dimensional simple Lie algebras has an "affinization." For $g = A_1$, the Lie algebra of $SU(2)$, the affinization, $A_1^{(1)}$ has $CM = \begin{pmatrix} 2 & -2 \\ -2 & 2 \end{pmatrix}$. Consider its representations with highest weights $\Lambda = (10)$ and (11). Arranging $\Omega(\Lambda)$ into horizontal slices according to the number of simple roots which have been subtracted from Λ (we call this the *principal slicing* of the representation), we obtain the following diagrams:

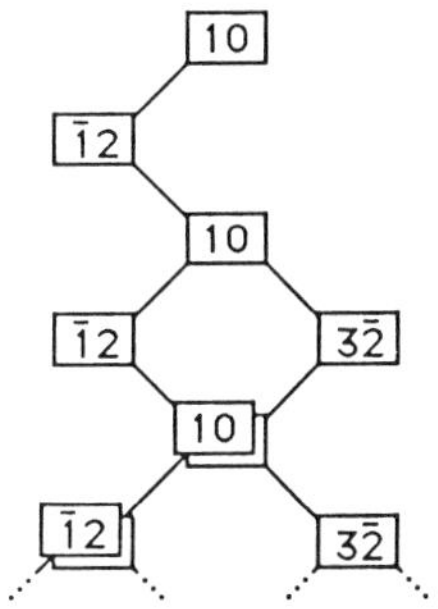

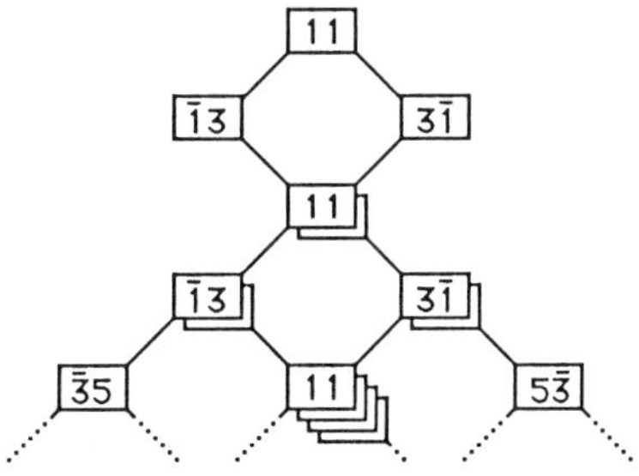

Lines indicate subtraction of some simple root α_1. Note in particular that the weight labels reappear, shifted by units of $\alpha_0 + \alpha_1$. In the affine case, the simple roots are linearly dependent when projected onto the space generated by the fundamental weights, and an additional label would be needed to separate the repeated occurences of each weight label. Already we can see one of the generalizations of a notion in the finite-dimensional case. While we could find the multiplicity of each weight and treat these as separate identities as is generally done in the finite case, it is instructive to write a generating function for the multiplicities of each weight label, using powers of q to separate the occurences. Thus in the first example above, where the weight (10) occurs with multiplicities $1, 1, 2, \ldots$, we could say that its multiplicity is $1 + q + 2q^2 + \ldots.$ Remarkably, this particular power series is exactly the generating function of the classical partition function, where the coefficient of q^n is the number of partitions of the integer n. In general, the generalized multiplicity (with an appropriate power of q on the outside) is a modular form, and many interesting identities and properties are known for these series (see Kac and Peterson, 1984). As the next examples we take two representations of the algebra $E_8^{(1)}$, the affinization of the 248-dimensional algebra E_8. To remain independent of numbering conventions, we write the coordinates of the simple roots as well as the weight coordinates in Dynkin diagram form. One has (in the principal slicing):

$$\alpha_0 = 2\bar{1}000\overset{0}{0}00$$
$$\alpha_1 = \bar{1}2\bar{1}00\overset{0}{0}00$$
$$\alpha_2 = 0\bar{1}2\bar{1}0\overset{0}{0}00$$
$$\alpha_3 = 00\bar{1}2\bar{1}\overset{0}{0}00$$
$$\alpha_4 = 000\bar{1}2\overset{0}{\bar{1}}00$$
$$\alpha_5 = 0000\bar{1}\overset{\bar{1}}{2}\bar{1}0$$
$$\alpha_6 = 00000\overset{0}{\bar{1}}2\bar{1}$$
$$\alpha_7 = 00000\overset{0}{0}\bar{1}2$$
$$\alpha_8 = 00000\overset{2}{\bar{1}}00$$

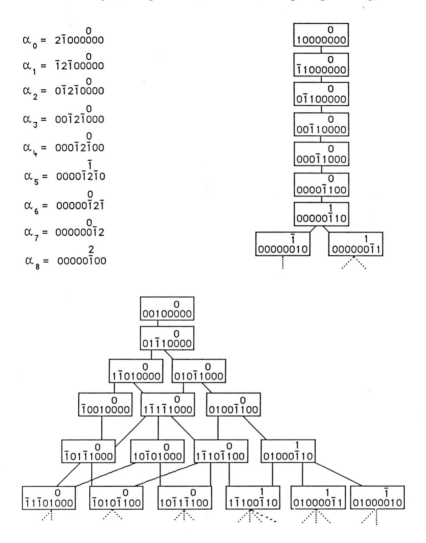

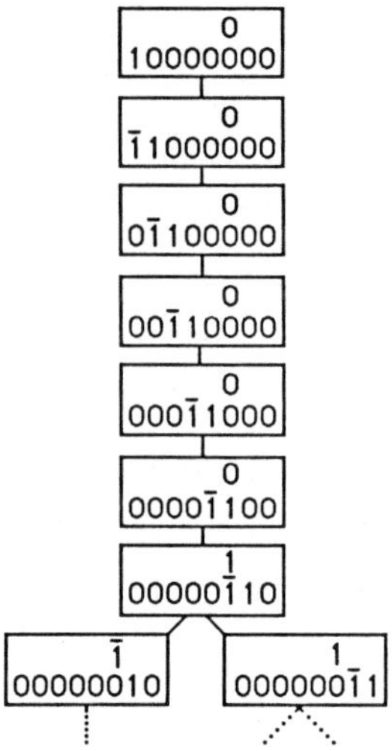

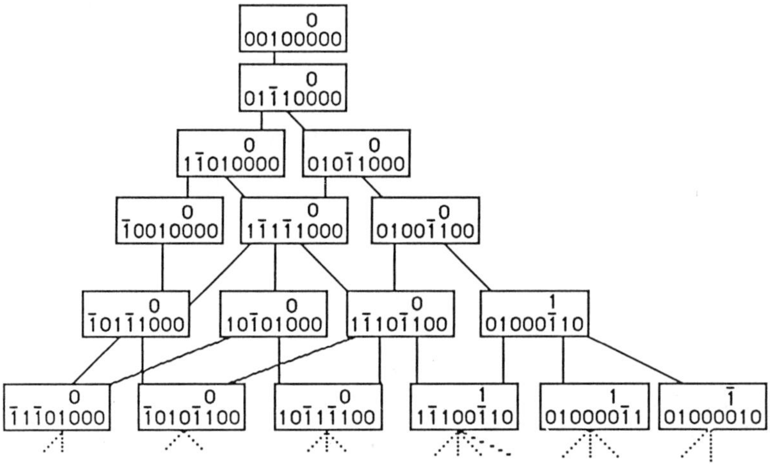

All weights shown here have multiplicity 1. Higher multiplicities occur further down when other weights with non-negative coordinates (dominant weights) appear.

INVARIANT CHARACTERISTICS OF REPRESENTATIONS

There exist several quantities which are easy to determine for a representation Λ of g of any type, and which are often very useful in calculations. Suppose that one has been given g and a representation $L(\Lambda)$ of g with highest weight $\Lambda = (p_0, p_1, \ldots, p_\ell)$. (If g is finite-dimensional, simply let $p_0 = 0$.) Consider the two integers:

Congruence number of Λ $\quad C(\Lambda) = \Sigma_{i=1}^{\ell} b_i p_i \quad \mathrm{mod}(\det CM)$.

Level of Λ $\quad L(\Lambda) = \Sigma_{i=0}^{\ell} c_i p_i$.

Here $\det CM$ is the determinant of the Cartan matrix. The coefficients b and c depend on the algebra only. They are found, for example, in (Bremner et al., 1984) and (Kass et al., 1987). For more details and examples of $C(\Lambda)$ see (Lemire and Patera, 1982). In the case of $SU(2)$, $C(J) = 0\,(1)$ for odd (even) dimensional representations; for $SU(3)$, $C(\Lambda)$ is the familiar triality number.

Two representations Λ and Λ' of g belong to the same *congruence class* provided

$C(\Lambda) = C(\Lambda')$ for finite g;

$C(\Lambda) = C(\Lambda')$ and $L(\Lambda) = L(\Lambda')$ for affine g.

It is always true that $C(\alpha_i) = 0$ and $L(\alpha_i) = 0$ for a simple root α_i, hence $C(\Lambda) = C(\mu)$ and $L(\Lambda) = L(\mu)$ for any $\mu \in \Omega(\Lambda)$, since μ is obtained from Λ by the subtraction of simple roots. It can also be shown that for g finite-dimensional, there are a finite number $(\det CM)$ of congruence classes each containing infinitely many representations. In contrast, an affine g has infinitely many congruence classes each containing a finite number of irreducible representations.

The most common use of $C(\Lambda)$ is in computing tensor products of representations or tensor powers of a single representation (with or without a particular permutational symmetry). Thus if

$$\Lambda\otimes\Lambda' = \Lambda_1 \oplus \ldots \oplus \Lambda_1 \quad \text{and} \quad \Lambda\otimes\ldots\otimes\Lambda_{YT} = \Lambda_1 \oplus \ldots \oplus \Lambda_n,$$

where the subscript YT (=Young tableau with k boxes) denotes a permutation symmmetry of the k-th tensor power of Λ, one has

$$C(\Lambda\otimes\Lambda') = C(\Lambda) + C(\Lambda') = C(\Lambda_1) = .. = C(\Lambda_j) \ \mathrm{mod}(\det CM),$$
$$C(\Lambda\otimes\ldots\otimes\Lambda)_{YT} = kC(\Lambda) = C(\Lambda_1) = .. = C(\Lambda_n) \ \mathrm{mod}(\det CM).$$

The *dimension* $d(\Lambda)$ of a representation Λ of an affine algebra g is of course infinite, but we can proceed as we did with the multiplicity of a weight and write the dimension as a power series in q according to some particular slicing of g, letting the coefficient of q^k be the dimension of the $k+1$-st slice. Such a series will have many of the useful properties of the dimension in the finite case. The dimensions (in the principal slicing) of the above representations of affine A_1 and E_8 are

$$d(10) = 1 + q + q^2 + 2q^3 + 2q^4 + 3q^5 + \ldots ,$$
$$d(11) = 1 + 2q + 2q^2 + 4q^3 + 6q^4 + \ldots ,$$
$$d\left({}^{0}_{10000000}\right) = 1 + q + q^2 + q^3 + q^4 + q^5 + q^6 + 2q^7 + \ldots,$$
$$d\left({}^{0}_{00100000}\right) = 1 + q + 2q^2 + 3q^3 + 4q^4 + 6q^5 + \ldots .$$

Another type of slicing giving useful information is a slicing into representations of a finite-dimensional subalgebra of g. If one removes the k-th node from the

Dynkin diagram of g (equivalently, the k-th row and column from CM), one obtains DD or CM for a maximal finite-dimensional subalgebra of g, which we denote $g(k)$. (The algebras so obtained are essentially all the maximal finite-dimensional semisimple subalgebras of g. (R. V. Moody, private communication, 1986)) We can then define the $g(k)$-slicing of $L(\Lambda)$ so that the weights of the p-th slice are those weights of the form

$$\mu = \Lambda - (p-1)\alpha_k - (\text{other } \alpha_i\text{'s}) \in \Omega(\Lambda).$$

For the affine A_1 modules above, we can arrange the weights to present the $g(0)$ slices as horizontal slices and the $g(1)$ slices as vertical slices. In this case both $g(0)$ and $g(1)$ are isomorphic to $SU(2)$, hence the slices are (generally reducible) $SU(2)$ modules.

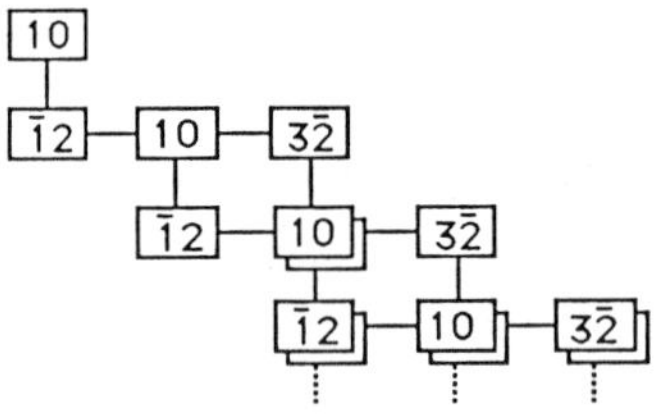

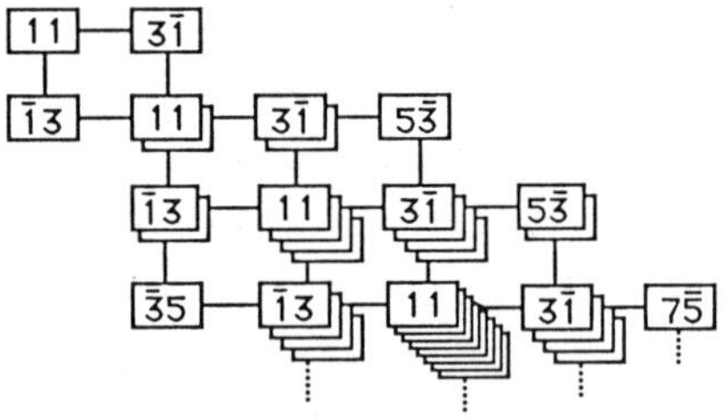

The first few slices of $L(10)$ and $L(11)$ in the $g(0)$ slicing are

$$\Gamma_1=(0),\ \Gamma_2=(2),\ \Gamma_3=(2)\oplus(0),\ \Gamma_4=(2)\oplus(2),\ldots;$$

$$\Gamma_1=(1),\ \Gamma_2=(3)\oplus(1),\ \Gamma_3=(3)\oplus(3)\oplus(1)\oplus(1),\ldots\ ,$$

with corresponding dimension series $\Sigma_i d(\Gamma_i)q^{i-1}$, here giving

$$1 + 3q + 4q^2 + 6q^3 + \ldots;$$

$$2 + 6q + 12q^2 + 18q^3 + \ldots .$$

More interesting are the E_8 examples. Consider $\Lambda = \left(\begin{smallmatrix} & & & & & 0 & & \\ 1&0&0&0&0&0&0&0\end{smallmatrix}\right)$ of $g(0) = E_8 \subset$ affine E_8. One has

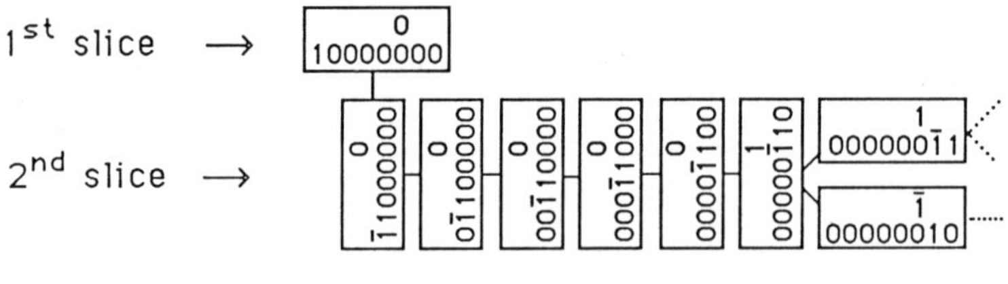

Viewing the slices by ignoring the p_0 coordinate of each weight, one sees immediately that the top slice here is a one-dimensional E_8 (scalar) representation, while the second slice contains only the representation $\Gamma_2 = \left(\begin{smallmatrix} & & & & 0 & & \\ 1&0&0&0&0&0&0\end{smallmatrix}\right)$ of dimension 248 (another path down from the first slice would be necessary for another highest weight to occur). The dimension series for this slicing thus begins with

$$d\begin{pmatrix} 0 \\ 10000000 \end{pmatrix} = 1 + 248q + \ldots .$$

For $\Lambda = \begin{pmatrix} 0 \\ 00100000 \end{pmatrix}$, we have

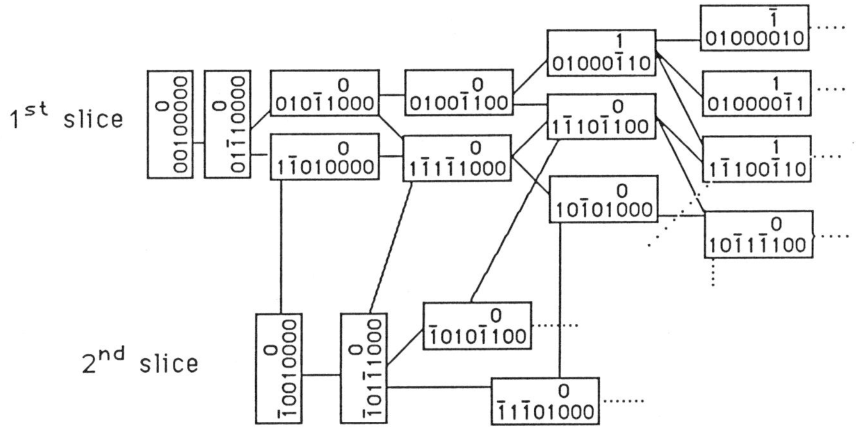

Here the first slice is the E_8 representation $\Gamma_1 = \begin{pmatrix} 0 \\ 0100000 \end{pmatrix}$ with $d(\Gamma_1)$ = 27 000, and the second one contains the representation $\Gamma_2 = \begin{pmatrix} 0 \\ 0010000 \end{pmatrix}$ of dimension 2 450 240.

Next let us take $g(2) = SU(3) + E_6$. The slices now are defined by the number of times α_2 has been subtracted from Λ, and affine E_8 weights are read as the subalgebra weights as follows:

$$\begin{pmatrix} p_8 \\ p_0p_1p_2p_3p_4p_5p_6p_7 \end{pmatrix} = (p_0p_1)\begin{pmatrix} p_8 \\ p_3p_4p_5p_6p_7 \end{pmatrix}$$

In $g(2)$-slicing the top slice of the basic representation $\Lambda = \begin{pmatrix} 0 \\ 10000000 \end{pmatrix}$ of affine E_8 is an $SU(3)$ triplet with a E_6 singlet, $\Gamma_1 = (10)\begin{pmatrix} 0 \\ 00000 \end{pmatrix}$; the second slice contains $\Gamma_2 = (00)\begin{pmatrix} 0 \\ 10000 \end{pmatrix}$, the singlet of $SU(3)$ and the 27-plet of E_6 :

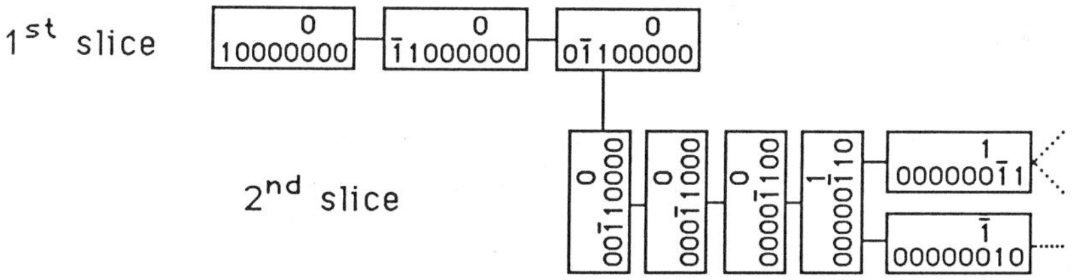

The other E_8 representation slices with respect to $g(2) = SU(3) + E_6$ as follows:

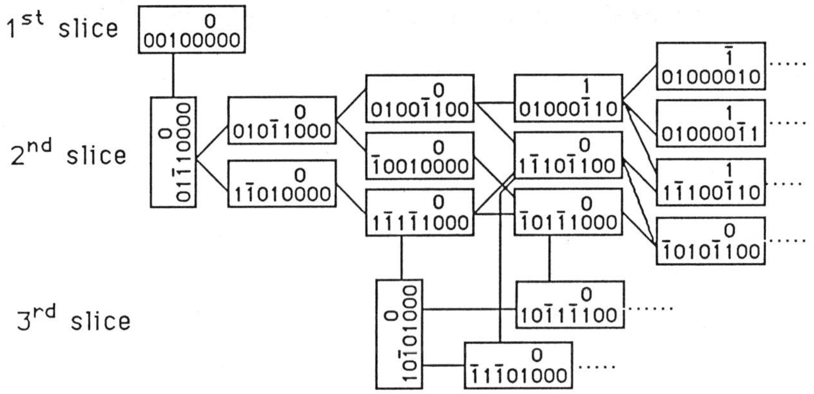

The top is a singlet of $SU(3)+E_6$, $\Gamma_1 = (00)\binom{0}{00000}$. Next is a triplet of $SU(3)$ and an E_6 representation of dimension 27, $\Gamma_2 = (01)\binom{0}{10000}$; the third slice is $\Gamma_3 = (10)\binom{0}{01000} \oplus (02)\binom{0}{00001} \oplus (10)\binom{0}{00001}$. The dimension is then $d\binom{0}{00100000}$ $= d(\Gamma_1) + d(\Gamma_2)q + \ldots = 1 + 81q + 1296q^2 + \ldots$.

Finally let us point out a general property of the representations Γ of the subalgebra $g(k)$ in different slices: all irreducible components of a slice belong to the same congruence class, though the congruence class may vary from slice to slice. It is not difficult to write specific rules for each case.

The *indices of representations* of finite Lie algebras g (Patera et al., 1976) as well as the anomaly numbers (Patera, Sharp, 1981) also generalize into power series (different series in different slicings) retaining all their useful properties. The index of degree k can be defined in the affine case as the power series

$$I^{(k)}(\Lambda) = \sum_{j=1}^{\infty} q^{j-1} I^{(k)}(\Gamma_j) .$$

Here $I^{(k)}(\Gamma_j)$ is the index of degree k (Patera et al., 1976; Patera, Sharp, 1981) in the finite-dimensional case. Using its definition, we have, for example

$$I^{(2k)}(\Lambda) = \sum_{j=1}^{\infty} q^{j-1} I^{(2k)}(\Gamma_j) = \sum_{j=1}^{\infty} q^{j-1} \sum_{\mu\in\Omega(\Gamma_j)} (\mu,\mu)^k .$$

Let $I^{(k)}(\Lambda)$ be the index and $d(\Lambda)$ the dimension of a representation Λ. Then we have all the properties from (Patera, Sharp, 1981; McKay et al., 1981). Examples:

$$I^{(2)}(\Lambda\otimes\Lambda') = I^{(2)}(\Lambda)d(\Lambda') + d(\Lambda)I^{(2)}(\Lambda'),$$

or, denoting by a Young tableau a permutation symmetry component of a tensor power $\Lambda\otimes\ldots\otimes\Lambda$,

$$I^{(2)}(\underbrace{\square\square\cdots\square}_{\leftarrow k \rightarrow}) = \frac{(d(\Lambda) + k)!}{(d(\Lambda) + 1)!(k-1)!}\, I^{(2)}(\Lambda) \quad ,$$

$$I^{(2)}\left(k \left\updownarrow \begin{matrix}\square\\ \vdots \\ \square\\ \square\end{matrix}\right.\right) = \frac{(d(\Lambda) - 2)!}{(d(\Lambda) - k - 1)!(k-1)!}\, I^{(2)}(\Lambda) \quad ,$$

$$I^{(4)}(\square\square) = (d(\Lambda) + 8)\, I^{(4)}(\Lambda) + \frac{\ell + 2}{\ell}\left(I^{(2)}(\Lambda)\right)^2 \quad ,$$

$$I^{(4)}\left(\begin{matrix}\square\\ \square\end{matrix}\right) = (d(\Lambda) - 8)\, I^{(4)}(\Lambda) + \frac{\ell + 2}{\ell}\left(I^{(2)}(\Lambda)\right)^2 .$$

Using the anomaly-index (Patera, Sharp, 1981) as $I^{(k)}(\Gamma_j)$, one has also

$$I^{(3)}(\square\square) = (d(\Lambda) + 4) I^{(3)}(\Lambda) \, ,$$

$$I^{(3)}\left(\begin{matrix}\square\\ \square\end{matrix}\right) = (d(\Lambda) - 4) I^{(3)}(\Lambda) , \quad \text{etc.}$$

ACKNOWLEDGEMENTS

The authors thank the Aspen Center for Physics and Los Alamos National Laboratory for their hospitality as well as R. Slansky and R. V. Moody for useful conversations.

REFERENCES

M. R. Bremner, R.V. Moody, J. Patera, "Tables of dominant weight multiplicities for representations of simple Lie algebras," M. Dekker, New York (1985).

J. E. Humphreys, "Introduction to Lie algebras and representation theory," Springer-Verlag, New York (1972).

V. J. Kac, D. H. Peterson, Infinite-dimensional Lie algebras, theta functions and modular forms, Adv. in Math. 53:125 (1984).

S. N. Kass, R.V. Moody, J. Patera, R. Slansky, "Weight multiplicities and branching rules for affine Kac-Moody algebras," (to be published, 1987).

F.W. Lemire, J. Patera, Congruence number, a generalization of SU(3) triality, J. Math. Phys. 21:2026 (1980).

J. McKay, J. Patera, R.T. Sharp, Second and fourth indices of plethysms, J. Math. Phys. 22:2770 (1981).

J. Patera, R.T. Sharp, On the triangle anomaly number, J. Math. Phys. 22:2352 (1981).

J. Patera, R.T. Sharp, P. Winternitz, Higher indices of group representations, J. Math. Phys. 17:1972 (1976).

FOUR-LOOP SIGMA-MODEL BETA-FUNCTIONS VERSUS α'^3 CORRECTIONS TO SUPERSTRING EFFECTIVE ACTIONS

D. Zanon*

Lyman Laboratory of Physics
Harvard University
Cambridge, Massachusetts 02138, U.S.A.

ABSTRACT

We present results on higher-order calculations for the N=1 supersymmetric σ-model in two dimensions. The condition of vanishing four-loop β-function, interpreted as the classical equations of motion of the associated string theory, can be derived from an effective action. It represents the generalization to an arbitrary Riemannian manifold of the corresponding low-energy action obtained from the four-graviton scattering amplitude for type II superstring. Thus the σ-model approach indeed describes the tree-level dynamics of the superstring.

The low-energy physics of string theories can be described by a local field approximation[1] as an expansion in the inverse string tension α'. In this limit one obtains an effective field theory for the massless fields of the string with all the massive modes integrated out. The corresponding action, which is given by an infinite sum of terms with increasing number of derivatives, generates all the string tree-level amplitudes of the massless particles. The motivations for computing this effective action reside primarily in the attempt to derive consistent string compactifications and viable phenomenology. Moreover one can hope that a knowledge of higher derivative terms in the expansion will give insights on general invariance principles underlying the full string theory. While a **priori** straightforward, the actual procedure for such a construction requires considerable effort.

It has recently become clear that an alternative way to get information about the string low-energy effective theory is provided by the σ-model approach[2-7]. The fact that non-linear σ-models in two dimensions are strictly related to tree string S-matrix elements can be easily understood. Let us consider for simplicity the purely gravitational bosonic σ-model. The action is

$$S = (1/4\pi\alpha') \int d^2x \; g_{ij}(\phi)\partial^\mu\phi^i\partial_\mu\phi^j \qquad (1)$$

---

* On leave of absence from Dipartimento di Fisica, Universita' di Milano and INFN, Sezione di Milano, Italy. Supported in part by NSF grants PHY-82-15249, PHY-83-13243.

where ϕ^i are the spacetime coordinates and g_{ij} is the metric tensor. If one splits the metric around a flat background $g_{ij} = \delta_{ij} + h_{ij}$ and expands the exponential in the functional integral, one obtains

$$Z = \int D\phi \exp S$$

$$= \sum_n \frac{1}{n!} \int D\phi \left[\frac{1}{4\pi\alpha'} \int d^2x\, h_{ij}(\phi)\partial^\mu\phi^i \partial_\mu\phi^j\right]^n \exp S_0 \tag{2}$$

where

$$S_0 = (1/4\pi\alpha') \int d^2x\, \partial^\mu\phi^i\, \partial_\mu\phi_i \tag{3}$$

is the free string action. By taking a momentum Fourier-transform and imposing on-shell conditions, one is led to consider the following expression

$$\sum_n \frac{1}{n!} \prod_r \left(\frac{\tilde{h}_{i_r j_r}(k_r)}{\alpha'}\right) \langle \partial\phi^{i_1}\, \partial\phi^{j_1}\, e^{ik_1\cdot\phi} \cdots \partial\phi^{i_n}\partial\phi^{j_n}\, e^{ik_n\cdot\phi} \rangle_{S_0} \tag{4}$$

which reproduces the n-point scattering amplitudes for the graviton field of the string. More generally, one expands $g_{ij} = g^0_{ij} + h_{ij}$ where g^0_{ij} is a classical configuration and h_{ij} is the on-shell field. Thus the σ-model can be viewed as the generating functional of string S-matrices[5,6]. Indeed, it has been shown that loop graphs of the two-dimensional theory rearrange themselves to give tree graphs of the string[7]. In particular, the σ-model β-function becomes the tadpole amplitude of the string. As in ordinary field theory, the tree-level tadpole amplitude corresponds to the classical Euler-Lagrange equations. Therefore a knowledge of the β-function of the σ-model implies a knowledge of the equations of motion of the massless fields of the string. (An equivalent, complementary way to obtain the condition of vanishing β-function is to demand conformal invariance of the two-dimensional theory[3,4].)

It is well-known that the one-loop gravitational β-function for the non-linear σ-model gives the correct equations of motion of the Einstein-Hilbert action. It corresponds to the zero-order term in the α' expansion of the effective action and as such it does not contain "string" physics. In order to include string effects one has to go beyond the leading approximation.

I will report on recent calculations of higher-order corrections to the β-function for supersymmetric non-linear σ-models[8]. Supersymmetry enters here since our aim is to make contact with superstring theories. More precisely, I will consider the purely metric σ-model with N=1 supersymmetry (the antisymmetric tensor field is set equal to zero and the dilation is treated as a constant field). For this model, the two- and three-loop corrections to the β-function are identically zero[9,10,8]. The next-leading contribution arises at four loops[8]. A superfield formulation and supergraph techniques are essential tools to perform such a high-order perturbative calculation. The four-loop correction to the N=1 gravitational β-function is given by a rather complicated expression involving structures of conformal weight three in the Riemann tensors.

According to the superstring-σ-model correspondence, the equations β=0 should be derivable from an action that, as explained above, should be identified with the low-energy effective theory for the graviton field of the string. Indeed the action whose variation reproduces the equations β=0 to order α'^3 has recently been derived[11,12]. This result can be analyzed in a superstring context. For type II superstring theory the correction to the Einstein-Hilbert action has been obtained through the same order in α' by studying tree-level four-graviton scattering amplitudes[13]. The one predicted by the N=1 σ-model calculation is just its proper generalization to an arbitrary Riemannian manifold. This remarkable agreement proves once more that the σ-model approach indeed describes the low-energy dynamics of string theories.

The superspace action of the N=1 two-dimensional supersymmetric σ-model[14] defined on a general Riemannian manifold is given by (we set $2\pi\alpha'=1$ and use the conventions of Ref. 15)

$$S = (1/4) \int d^2x\, d^2\theta\, g_{ij}(\Phi)\, D_\alpha\Phi^i D^\alpha\Phi^j \tag{5}$$

which is just the direct extension of the purely bosonic σ-model action in (1). $g_{ij}(\Phi)$ is the metric tensor and the real superfields $\Phi^i(x,\theta)$ are the coordinates on the manifold. D_α is the covariant spinor derivative

$$D_\alpha \equiv \partial_\alpha + i\theta^\beta \partial_{\alpha\beta} \tag{6}$$

Non-linear σ-models in two dimensions are renormalizable in the sense that the geometrical structure can be renormalized order by order in the loop expansion[16]. Superspace power counting (the $d^2xd^2\theta$ integral has dimension -1, while $\dim\Phi = 0$, $\dim D_\alpha = 1/2$) shows that the renormalization counterterms are local expressions of dimension one of the form

$$(1/4)\, T_{ij}(\Phi)\, D_\alpha\Phi^i D^\alpha\Phi^j \tag{7}$$

They appear as changes of the metric, whose corresponding renormalization is given by

$$g^B_{ij} = g^R_{ij} + \sum_{n=1}^{\infty} \frac{1}{\varepsilon^n} T^{(n)}_{ij}(g^R) \tag{8}$$

The β-function is determined by the coefficient of the simple pole in (8)

$$\beta_{ij}(g^R) = \varepsilon\, g^R_{ij} + \left(1 + \lambda \frac{\partial}{\partial\lambda}\right) T^{(1)}_{ij}(\lambda^{-1}g^R)\Big|_{\lambda=1} \tag{9}$$

Perturbative calculations are most efficiently performed using the background field method in terms of normal coordinates. One splits the superfields Φ^i in terms of a classical solution and a quantum field

$$\Phi^i = \Phi^i_0 + \Phi^i_Q \tag{10}$$

Since Φ^i_Q does not transform simply under reparametrization, it is convenient to reexpress it as a local power series in a new field ξ^i which transforms as a contravariant vector. In this way one obtains a manifestly covariant expansion. In order to define the quantum field ξ^i, one considers the geodesic $\phi^i(s)$ joining $\Phi^i_0(s=0)$ to $\Phi^i(s=1)$. ξ^i is then defined as the tangent vector to this geodesic at Φ^i_0

$$\xi^i = \frac{d\Phi^i}{ds}\Big|_{s=0} \tag{11}$$

With these conditions, the solution of the geodesic equation

$$\frac{d^2\Phi^i}{ds^2} + \Gamma^i_{jk} \frac{d\Phi^j}{ds}\frac{d\Phi^k}{ds} = 0 \tag{12}$$

is

$$\Phi^i = \Phi^i_0 + \xi^i - \frac{1}{2}\Gamma^i_{jk}\xi^j\xi^k + \cdots \tag{13}$$

Substituting (13) into the action (5), one finds that the coefficients of the expansion in the quantum fields ξ^i are in terms of covariant tensor structures. One obtains the following terms to second, third and fourth order in the quantum fields ($d^4z \equiv d^2x d^2\theta$)

$$\begin{aligned}
S^{(2)} &= \frac{1}{4}\int d^4z\; g_{ij}\nabla_\alpha\xi^i\nabla^\alpha\xi^j + \frac{1}{4}\int d^4z\; R_{ikmj}D_\alpha\Phi^i D^\alpha\Phi^j\xi^k\xi^m \\
S^{(3)} &= \frac{1}{12}\int d^4z\; R_{ijmj;n}D_\alpha\Phi^i D^\alpha\Phi^j\xi^k\xi^m\xi^n + \frac{1}{3}\int d^4z\; R_{ikmj}D_\alpha\Phi^i\nabla^\alpha\xi^j\xi^k\xi^m \\
S^{(4)} &= \frac{1}{48}\int d^4z (R_{ikhj;mn} + 4\,R^p{}_{khi}R_{pmnj})D_\alpha\Phi^i D^\alpha\Phi^j\xi^k\xi^h\xi^m\xi^n \\
&\quad + \frac{1}{8}\int d^4z\; R_{ikhj;m}D_\alpha\Phi^i\nabla^\alpha\xi^j\xi^k\xi^h\xi^m + \frac{1}{2}\int d^4z\; R_{ikhj}\nabla_\alpha\xi^i\nabla^\alpha\xi^j\xi^k\xi^h
\end{aligned} \tag{14}$$

where Φ^i denotes the background superfields and ξ^i are the quantum fields that transform as vectors under coordinates changes. The spinor derivative ∇_α is defined as

$$\nabla_\alpha\xi^i \equiv D_\alpha\xi^i + \Gamma^i_{jk}D_\alpha\Phi^k\xi^j \tag{15}$$

In order to perform diagrammatic calculations one has to derive the superspace Feynman rules of the theory. From the quadratic part of the action in (14) one obtains standard quantum propagators if the quantum fields are rotated to tangent frames on the Riemannian manifold[9]

$$\xi^a = e^a_i(\Phi)\xi^i, \quad \nabla_\alpha\xi^a \equiv (D_\alpha - i\Gamma_\alpha)\xi^a = D_\alpha\xi^a + \omega^{ab}_i D_\alpha\Phi^i\xi^b \tag{16}$$

One then obtains

$$\begin{aligned}
\frac{1}{4}\int d^4z\; g_{ij}\nabla_\alpha\xi^i\nabla^\alpha\xi^j &= \frac{1}{4}\int d^4z\; \nabla_\alpha\xi^a\nabla^\alpha\xi^a \\
&= \frac{1}{2}\int d^4z\; \xi^a\nabla^2\xi^a
\end{aligned} \tag{17}$$

It is simple to prove that Γ_α in (16) transforms as a superfield gauge spinor connection under tangent frame rotations. The lowest dimensional gauge-invariant term one can construct from it has dimension two. Since as noticed before, the ultraviolet counterterms will be local expressions of dimension one, the spinor connection Γ_α can contribute to them only when it covariantizes the derivatives on tensor structures. Therefore for the purpose of computing divergences, one can use flat spinor derivatives and covariantize the result at the end. This implies that the quadratic part of the action in (17) gives a free kinetic term for the quantum fields ξ^a. One obtains usual propagators

$$\langle \xi^a(z)\xi^b(z')\rangle = -\delta^{ab}\Box^{-1} D^2\delta^2(x-x')\delta^2(\theta-\theta') \tag{18}$$

and quantum-background vertices from the expansion in (14), with $\nabla_\alpha\xi^a$ replaced by $D_\alpha\xi^a$.

At this point perturbative calculations are performed using standard supergraph techniques[15]: the background fields always appear as external fields in the diagrams and conventional D-algebra leads to contributions that can be rewritten in momentum space and evaluated in standard fashion. Divergent integrals are computed using supersymmetric dimensional regularization and minimal subtraction.

At one loop the divergence structure is given by the tadpole integral ($n=2+\varepsilon$)

$$I = \int \frac{d^n p}{(2\pi)^n} \frac{1}{(p^2+\mu^2)} \tag{19}$$

where μ^2 is an infrared cutoff. The one-loop divergnce proportional to the Ricci tensor is[9]

$$- (1/8\pi\varepsilon)\, R_{ij}\, D_\alpha\Phi^i D^\alpha\Phi^j \tag{20}$$

At two and three-loop orders all the divergent terms are given by tadpole type integrals. At four-loops one obtains divergent structures that are again proportional to tadpoles plus other contributions whose ultraviolet behavior is given by the integral[17,8]

$$A_4 = \int \frac{d^n k\, d^n q\, d^n r\, d^n t}{(2\pi)^{4n}} \frac{k\cdot(t-k)\; q\cdot(t-q)}{k^2(t-k)^2 q^2(t-q^2)(r^2+\mu^2)[(t-r)^2+\mu^2)]} \tag{21}$$

The structure of the counterterms can be easily computed using the BPHZ-subtraction procedure. Subdivergences corresponding to lower-loop renormalizations are subtracted from the divergent integrals. One then obtains local contributions whose overall divergence determines the counterterms. Tadpole type integrals, after subtraction, lead to contributions proportional to $1/\varepsilon^L$ poles, where $L = 2,3,4$ denotes the loop-order. In addition, at four loops, the integral in (21) gives rise to the following divergent contribution after subtraction of subdivergences

$$A_4 \approx -\frac{8}{(4\pi)^4}\left[\frac{\zeta(3)}{\varepsilon} + \frac{4}{3\varepsilon^4}\right] \tag{22}$$

Since the β-function is determined by the simple-pole divergences (see eq. (9)), we conclude that at two and three loops the β-function is identically zero. The $1/\varepsilon$ pole in (22) is instead a genuine four-loop divergence and the corresponding contribution to the β-function is obtained by computing the complete background dependence of terms proportional to A_4. The final outcome is, up to four loops[8] (we reintroduce the explicit α' dependence for the sake of clarity)

$$\beta_{ij} = R_{ij} + \alpha'^3 \frac{\zeta(3)}{48} T_{(ij)} + O(\alpha'^4) \tag{23}$$

where

$$
\begin{aligned}
T_{ij} = {} & 2\nabla_h \nabla_k R_{nijm} \left(R^{msrk} R^{n}{}_{(sr)}{}^{h} + R^{msrn} R^{k}{}_{sr}{}^{h} \right) \\
& + [\nabla_i, \nabla_h] R_{jkmn} R^{m(rk)s} R^{n}{}_{sr}{}^{h} \\
& + 3\big(\nabla_r R_{ikht} R^{tsrq} \nabla_s R_j{}^{hk}{}_q + \nabla_h R_{irkt} R^{t}{}_{s}{}^{rq} \nabla^k R_j{}^{sh}{}_q \\
& \qquad + 2 \nabla_r R_{jkht} R^{t}{}_{s}{}^{rq} \nabla^h R_j{}^{sk}{}_q \big) \\
& \qquad + (2 \nabla_i R_{rqst} - \nabla_i R_{rsqt}) R^{t}{}_{h}{}^{rk} \nabla_j R^{hqs}{}_k \\
& \qquad - 12 R_{mhki} R_{jrt}{}^{m} \left(R^{k}{}_{qs}{}^{r} R^{tqsh} + R^{k}{}_{qs}{}^{t} R^{hrsq} \right)
\end{aligned} \tag{24}
$$

The four-loop contribution to the β-function in (23) vanishes on hyper-Kahler spaces, a result which is consistent with general arguments of ultraviolet finiteness for σ-models with N=4 supersymmetry[18]. Furthermore if one imposes the constraints required by N=2 extended supersymmetry, i.e. restricts the Riemannian manifold to be Kahler, the four-loop divergence agrees with the corresponding calculation for the N=2 model[17]. In particular it does not vanish on Ricci-flat Kahler manifolds.

In a superstring context, the equations $\beta_{ij} = 0$ should give the classical equations of motion for the graviton modes of the string and as such they should be derivable from an action. Quite generally, given the action S, the equations of motion will be of the form

$$
\frac{\delta S}{\delta g^{ij}} = W_{ij} + \frac{1}{2} g_{ij} A = 0 \tag{25}
$$

where W_{ij} is interpreted as the gravitational β-function and A should be identified with the dilaton β-function[3,19]. In order to determine S, one has to consider the various independent structures of conformal weight three in the Riemann tensors, compute the corresponding variations and choose the coefficients so as to match the expression in (23), (24). At order α'^3 one can set $R_{ij} = 0$ at the equations of motion level, since this condition follows from the leading order $(\alpha')^0$ result. After a lengthy algebraic exercise one finds that the local action whose variation reproduces the σ-model β-function in (23) is[11]

$$
S = \int \sqrt{g} \left\{ -R + \alpha'^3 \frac{\zeta(3)}{8} \left[L_1 - 2L_2 + \frac{2}{9} L_3 \right] + O(\alpha'^4) \right\} \tag{26}
$$

where

$$
\begin{aligned}
L_1 &= R_{hmnk} R_p{}^{mn}{}_q R^{hrsp} R^{q}{}_{rs}{}^{k} + \frac{1}{2} R_{hkmn} R_{pq}{}^{mn} R^{hrsp} R^{q}{}_{rs}{}^{k} \\
L_2 &= R^{hk} \left[\frac{1}{2} R_{htrk} R^{msqt} R_{msq}{}^{r} + \frac{1}{4} R_{htmn} R_k{}^{tqs} R_{qs}{}^{mn} + R_{hmnp} R_{kqs}{}^{p} R^{nqsm} \right] \\
L_3 &= R \left[\frac{1}{4} R_{htmn} R^{htqs} R_{qs}{}^{mn} + R_{hmnp} R^{h}{}_{qs}{}^{p} R^{nqsm} \right]
\end{aligned} \tag{27}
$$

Since we want to interpret this result as an effective action which generates tree-level string amplitudes, the L_2 and L_3 terms in (26) are not interesting for this purpose. A local field redefinition of the metric $g_{ij} \to g_{ij} + \delta g_{ij}$ does not alter S-matrix elements. Therefore only purely

Riemannian structures are relevant to this order, while terms like L_2 and L_3, containing Ricci and scalar curvature tensors, can be eliminated by an appropriate field redefinition of the meteric. We are then left with

$$S = \int \sqrt{g} \{-R + \alpha'^3 \frac{\zeta(3)}{8} [R_{hmnk} R_p{}^{mn}{}_q R^{hrsp} R^q{}_{rs}{}^k + \frac{1}{2} R_{hkmn} R_{pq}{}^{mn} R^{hrsp} R^q{}_{rs}{}^k] + O(\alpha'^4)\} \tag{28}$$

We note that the α'^3 term is the <u>unique</u> combination of Riemann tensors of conformal weight three that vanishes on a Kahler manifold.

Finally we compare the action in (28) with a direct superstring calculation, the α'^3 correction to the Einstein-Hilbert action obtained from an analysis of the tree-level four-graviton superstring amplitude[13]. Since it is derived from an S-matrix element that depends on only three independent momenta, this action is valid up to terms that vanish in less than eight dimensions. The effective Lagrangian can be written as an integral over fermion zero modes

$$Y = \int d\psi_R \, d\psi_L \, \exp(\psi_L \Gamma^{ij} \psi_L \psi_R \Gamma^{kl} \psi_R \, R_{ijkl}) \tag{29}$$

This representation is very convenient in order to analyze the result on a six-dimensional manifold. In this form it has been used to derive the corresponding equations of motion on six-dimensional Kahler manifolds[13,20] and agreement has been found with the condition of vanishing β-function for the N = 2 σ-model[17]. Moreover it has been shown that Y vanishes on such spaces. Since, as emphasized above, the α'^3 term in (28) is the unique combination of Riemann tensors that vanishes to this order on a Kahler manifold, it is clear that the σ-model action is the correct generalization of the superstring result to arbitrary Riemannian manifolds. In fact, when rewritten as a polynomial in the Riemann tensors, Y can be appropriately generalized[12] and recast in a form proportional to the α'^3 term in (28).

The σ-model calculation, which is valid for a general Riemannian manifold of arbitrary dimensions, seems to suggest that the tensorial structure of the α'^3 term in (28) contains all the informations necessary to determine the off-shell correction to the low-energy effective action of the superstring in full generality. A complete understanding of these invariance properties, which clearly go beyond general coordinate invariance, might give further insights on the role of gauge symmetries in string theories.

I will now briefly comment on the implications of these results for superstring phenomenology. Consistent compactifications of the ten dimensional theory must be obtianed as particular solutions of the corresponding equations of motion. Moreover in order to be realistic they must describe a geometry of the form $M_4 \times K_6$, where K_6 is a six-dimensional manifold and M_4 is a four-dimensional Minkowski space with N=1 unbroken supersymmetry. It has been shown that, to zero-order in α', this last requirement constraints the internal six-dimensional space to be Kahler with vanishing first Chern class (Calabi-Yau space), i.e. it admits a Ricci-flat Kahler metric[21]. Since the α'^3 corrections to the equations of motion in (23) modify the Ricci-flat Kahler solution, one might ask if compactification on Calabi-Yau manifolds is still a viable scenario. These issues have been addressed in the recent literature and the following conclusions have been reached:

a) The physical background solution can be viewed as a perturbative deformation of the Ricci-flat metric, obtainable from it through a non-local field redefinition[22].

b) The corrections to the spacetime supersymmetry transformations necessary for consistency with N=1 supersymmetry in the modified vacuum have been analyzed for the case of the heterotic string[23]. Appropriate modifications of the original embedding of the spin-connection in the gauge group lead to the conclusion that the zero-order results are still valid.

Thus satisfactory phenomenology seems to be viable from a perturbative low-energy approach of superstring theories.

Acknowledgements

I am particularly indebed to Marc Grisaru and Anton van de Ven with whom most of the material reviewed in this paper was developed.

References

1. J. Scherk and J.H. Schwarz, Nucl. Phys. B81(1974)118; Phys. Lett. 52B (1974)347.
2. C. Lovelace, Phys. Lett. 135B(1984)75.
3. C.G. Callan, D. Friedan, E.J. Martinec and M.J. Perry, Nucl. Phys. B262(1985)593.
4. A. Sen, Phys. Rev. D32(1985)2102; Phys. Rev. Lett. 55(1985)1846.
5. E.S. Fradkin and A.A. Tseytlin, Phys. Lett. 158B(1985)316; Nucl. Phys. B261(1985)1.
6. B.E. Fridling and A. Jevicki, Brown preprint HET-566 (1985).
7. C. Lovelace, Rutgers preprint Ru-85-51 (1985).
8. M.T. Grisaru, A.E.M. van de Ven and D. Zanon, Harvard preprints HUTP-86/A020, HUTP-86/A027 (1986).
9. L. Alvarez-Gaumé, D.Z. Freedman and S. Mukhi, Ann. Phys. 134(1981)85.
10. L. Alvarez-Gaumé, Nucl. Phys. B184(1981)180.
11. M.T. Grisaru and D. Zanon, Harvard preprint HUTP-86/A046 (1986).
12. M.D. Freeman, C.N. Pope, M.F. Sohnius and K.S. Stelle, preprint Imperial/TP/85-86/27 (1986).
13. D.J. Gross and E. Witten, Princeton preprint (1986).
14. D.Z. Freedman and P.K. Townsend, Nucl. Phys. B177(1981)282.
15. S.J. Gates, M.T. Grisaru, M. Rocek and W. Siegel, "Superspace" (Benjamin Cummings, Reading MA, 1983).
16. D. Friedan, Phys. Rev. Lett. 45(1980)1057; Ann. Phys. 163(1985)318.
17. M.T. Grisaru, A.E.M. van de Ven and D. Zanon, Harvard preprint HUTP-86/A026 (1986).
18. C.M. Hull, Nucl. Phys. B260(1985)182.
L. Alvarez-Gaumé and P. Ginsparg, Comm. Math. Phys. 102(1985)311.
19. C.G. Callan, I.R. Klebanov and M.J. Perry, Princeton preprint (1986).
20. M.D. Freeman and C.N. Pope, preprint Imperial/TP/85-86/17 (1986).
21. P. Candelas, G. Horowitz, A. Strominger and E. Witten, Nucl. Phys. B256(1985)46.
22. D. Nemeschansky and A. Sen, preprint SLAC-PUB-3925 (1986).
23. P. Candelas, M.D. Freeman, C.N. Pope, M.F. Sohnius and K.S. Steele, preprint Imperial/TP/85-86/26 (1986).

MODULAR INVARIANCE AND FINITENESS OF FIVE-POINT CLOSED SUPERSTRINGS

C.S. Lam

Department of Physics, McGill University
3600 University Street
Montreal, Quebec, Canada H3A 2T8

INTRODUCTION

In this talk, I will describe a calculation done by Da-Xi Li and myself[1] concerning the modular invariance and hence the finiteness of five-point heterotic and Type-II superstrings, to the one loop level. A similar calculation has been performed by Fampton, Moxhay and Ng.[2]

It is known that to the one loop level, the M-point function for a superstring vanishes[3] if M<4. For M=4, finiteness of the SO(32) Type I[4], Type II[5], and heterotic[6] strings has been shown in the classical papers of Green & Schwarz, and of Gross, Harvey, Martinec & Rohm. For M=5 Frampton, Moxhay and Ng[7] (FMN) have recently demonstrated the finiteness of the open SO(32) Type I superstring. Our work[1] shows that this is also true for the closed superstrings. All the above calculations rely on the light cone quantization techniques and will be described later. For M>5, nothing rigorous is known using these techniques though it is widely conjectured that all these strings would remain finite. For the SO(32) Type I open string, we heard earlier from Clifford Burgess[8] that finiteness for all M can also be demonstrated using the Polyakov formalism. To the best of my knowledge, nothing similar has been shown for the M-point closed superstrings.

OPEN (TYPE I) SUPERSTRING

The recent revival of the superstring theory began with the observation of Green and Schwarz[9] that the only Type I string without anomaly is the SO(32) string. Even for that string, anomaly occurs separately in the planar diagram and in the diagrams with an odd number of twists but the divergences cancel out in the sum. Later on, they also showed[4] that the same cancellation mechanism works to make the 4-point function of an SO(32) Type I superstring finite. For M = 5, FMN then showed that there are now two potential divergences for each class of diagrams instead of one. The leading ones, diverging logarithmically, are analogous to the ones for the M = 4 case and also to the ones in the (M = 6) anomaly case, and they also got cancelled out in the same way. The new non-leading divergences diverge like log log, but after a very lengthy algebraic calculation, FMN[7] were able to show that the coefficients of them are all zero, thus completing the proof of the finiteness of the M = 5 Type I superstring.

In the following, we will make use of these algebraic calculations of FMN to show that the M = 5 closed superstrings possess modular invariance and finiteness.

CLOSED SUPERSTRINGS

For the closed superstrings (Type II and heterotic), the strategy for demonstrating finiteness is different. There are no longer planar and twisted diagrams: the only one-loop diagram describing the propagation of a closed string is a torus with 'spaghetti'-like external lines attached to it (Fig. 1). Accordingly, finiteness cannot come from cancellation between diagrams and the strategy for its demonstration must also be different.[5] To explain how that goes, let us first consider the variables in Fig. 1.

Kinematical Variables

We describe the 2-dimensional world-sheet in Euclidean space and denote its coordinates by $(\sigma^0,\sigma^1) = (it,\sigma)$, with σ and t real, and $0 \lesssim \sigma \lesssim \pi$. For a closed string, its $\sigma = 0$ end and its $\sigma = \pi$ end are identified so that the world sheet looks like a tube, with 2σ being the variable going around the tube and 2t the one along it. For a 1-loop amplitude, the tube

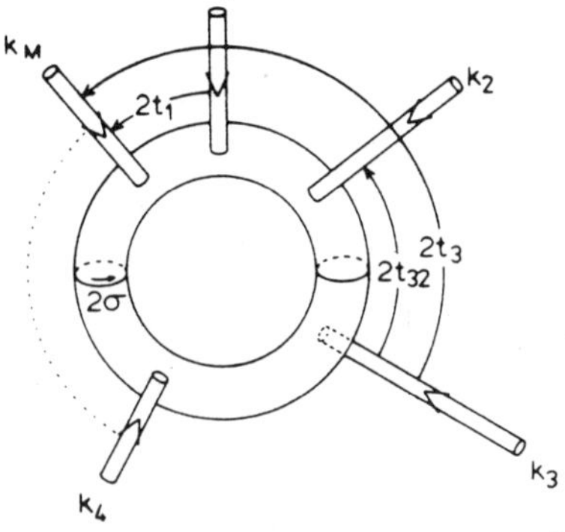

Fig. 1. Variables for a one loop closed-string diagram with M external lines.

bends back to form a torus as shown in Fig. 1. For an M-point amplitude, let k_A by the 10-dimensional momentum, and (it_A, σ_A) be the world-sheet coordinates of the Ath external line $(1 \le A \le M)$. Let $\nu_A = (\sigma_A + it_A)/\pi$. Suppose the Mth line be the one with the largest t: $t_M \geq t_A$ for $A < M$. Then the length of the tube is given by $2t_M = 2\pi \mathrm{Im}\tau$ if $\tau \equiv \nu_M$. The (complex) difference coordinates between the Bth and the Ath lines are then denoted by $\nu_{BA} = \nu_B - \nu_A$. Defined this way, all $t_A \geq 0$ and ν_A lies in a vertical strip of width 1 in the upper-half complex plane. This strip can be mapped into the whole complex plane by the transformation $Z = \exp[2\pi i\nu]$.

Periodicity

In terms of these variables, the 1-loop M-point amplitude is[3]

$$T_M = \int \left(\prod_{A=1}^{M} d^2 Z_A\right) G_M = (2\pi)^{2M} \int d^2\tau \int \prod_{B=1}^{M-1} d^2\nu_B \; G_M(\nu|\tau) \tag{1}$$

The ν_B-integrals are confined to $0 \leq \mathrm{Im}\nu_B \leq \mathrm{Im}\tau$ and $0 \leq \mathrm{Re}\nu_B \leq 1$. In order for $G_M(\nu|\tau)$ to be one-valued on the torus of Fig. 1, one expects the integrand G_M to have to obey the periodicity conditions

$$G_M(\nu|\tau) = G_M(\nu + 1|\tau) = G_M(\nu + \tau|\tau) \tag{2}$$

If this expectation is verified by explicit calculations (see next section), then the ν_B-integral in (1) can be carried out over any unit cell of the lattice which defines the torus of Fig. 1. In particular, we can take the shaded area in Fig. 2.

Modular Invariance

The group SL(2,Z) of modular transformations is defined by

$$\tau \to \tau' = \frac{a\tau + b}{c\tau + d} \tag{3}$$

with a, b, c, d $\in$ Z and $ad - bc = 1$. This group is generated by two elements: $\tau' = \tau + 1$ and $\tau' = -1/\tau$. Every other element of the group can be expressed as the repeated products of these two elements. For reasons explained below, G_M is expected to be modular-invariant, in the sense that

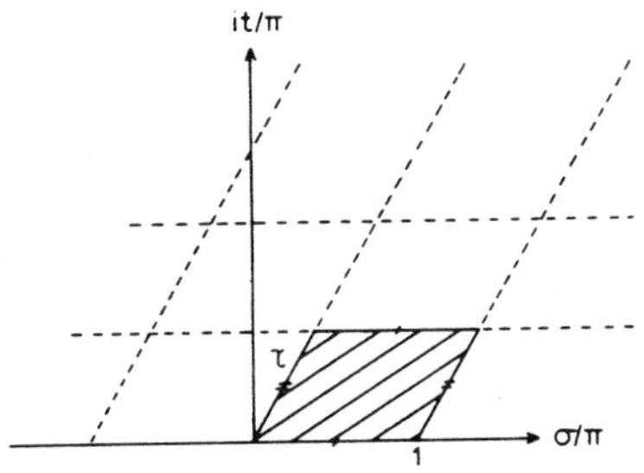

Fig. 2. The lattice defining the torus of Fig. 1.

$$d^2\tau \prod_{B=1}^{M-1} d^2\nu_B \; G_M(\nu|\tau) = d^2\tau' \prod_{B=1}^{M-1} d^2\nu_B' \; G_M(\nu'|\tau') \tag{4}$$

for $(\nu'|\tau') = (\nu|\tau + 1)$ and for $(\nu'|\tau') = (-\nu/\tau|-1/\tau)$. In the next section, this expectation will be verified by explicit calculations for M = 4 and M = 5. Assuming that this is true, then the τ-plane can be divided into regions related to each other by a modular transformation (4). One such region, the 'fundamental region' F, is shown dotted in Fig. 3. Then because of (2), the τ-integration in (1) can be confined to F. To get the amplitude T_M in (1), we merely multiply this by the number of regions in Fig. 3. It is true that this number is infinite, but this infinity is analogous (see next paragraph) to the infinity obtained from the Feynman path integral of a gauge theory if we do not choose a gauge fixing to avoid duplication. Such an infinity is absent when the theory is properly quantized and in any case has nothing to do with the ultraviolet or infrared divergences of the theory. Now return to (1) integrated over F. The function $G_M(\nu|\tau)$ damps at $\tau = \infty$ and has singularities only at $\tau = 0$. Since the region F excludes $\tau = 0$, the integral over F and hence the 1-loop amplitude in this sense is finite. In this way finiteness follows from modular invariance.

We will now explain why modular invariance is expected to be valid. Essentially, this is because a string theory is formulated to be reparametrization-invariant,[3] and modular transformation simply corresponds to a special kind of discrete reparametrization. To see that, consider a torus defined by identifying opposite edges of the parallelogram in Fig. 4(a). Let $2\omega_1$ and $2\omega_2$ be the complex numbers defining the lattice vectors of the parallelogram and let $\tau = \omega_2/\omega_1$. The torus in Fig. 1 is a special case of Fig. 4(a) with $2\omega_1 = 1$. The coordinate axes for $2\omega_1$ and $2\omega_2$ are labelled by 1 and 2 respectively in Figs. 4(b). Now the modular transformation $\tau \to -1/\tau$ corresponds to $\omega_1 \to \omega_2$, $\omega_2 \to -\omega_1$, whence the coordinate axes become those shown in Fig. 4(c). It is clear that this corresponds to a

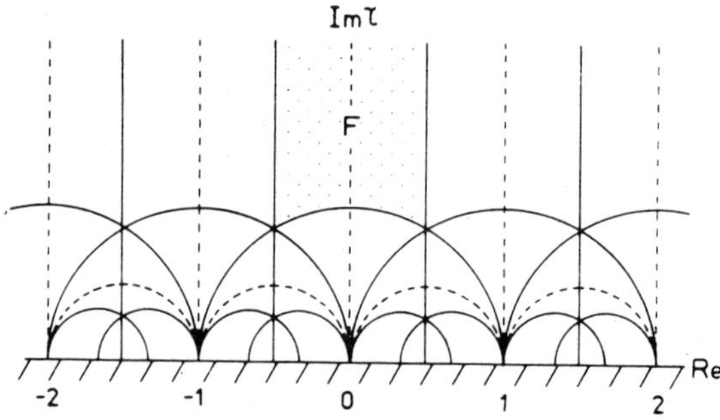

Fig. 3. Regions related to the fundamental region F by modular transformations.

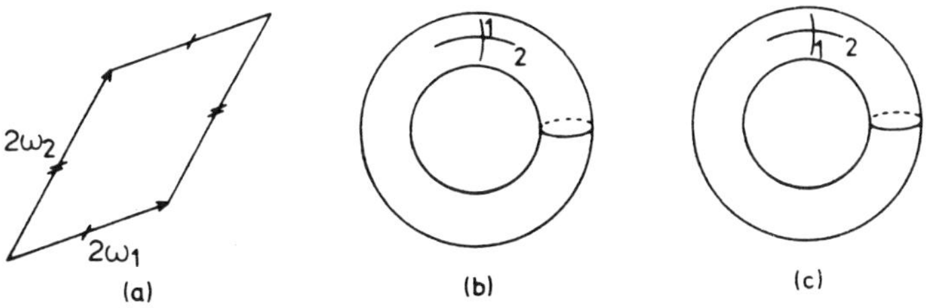

Fig. 4. (a) gives the parallelogram defining the torus in (b) by identifying opposite edges. The coordinate axes $2\omega_1$ and $2\omega_2$ are labelled by 1 and 2 in (b) and (c). The transformation $\tau \to -1/\tau$ causes the two coordinate axes to be interchanged. As a result (b) becomes (c).

discrete transformation of the coordinate system. Similarly, for the other modular transformation $\tau \to \tau + 1$, consider Fig. 5(b). The coordinate axis running along the vector τ in Fig. 5(a) wraps around the length of the tube in Fig. 5(b) once, and is labelled by 2. If τ is changed into $\tau + 1$, then this axis would wrap around the length of the tube (τ) and the breadth of the tube (1) once each, as in Fig. 5(c). This obviously also corresponds to a discrete coordinate transformation (of the τ-axis). Alternatively, the parallelogram in Fig. 4(a) is changed into the shaded parallelogram of Fig. 5(a). The new torus is now obtained by cutting the old torus and twisting along its breadth by 2π before gluing it back together again. In this way we can see also that the τ-axis in Fig. 5(b) also wraps around the breadth as in Fig. 5(c).

CALCULATIONS

It remains now to verify the validity of the periodicity of $G(\nu|\tau)$ (eq. (2)) and its modular invariance (eq. (4)) for M = 4 and 5. This part is simple conceptually, though it is relatively difficult technically.[1,2]

In what follows, we shall sketch the calculation for the heterotic string and indicate at the end where changes have to be made for the Type II string. For details, see Ref. 1.

In the light-cone formalism, the amplitude in (1) is written as (see Fig. 1)

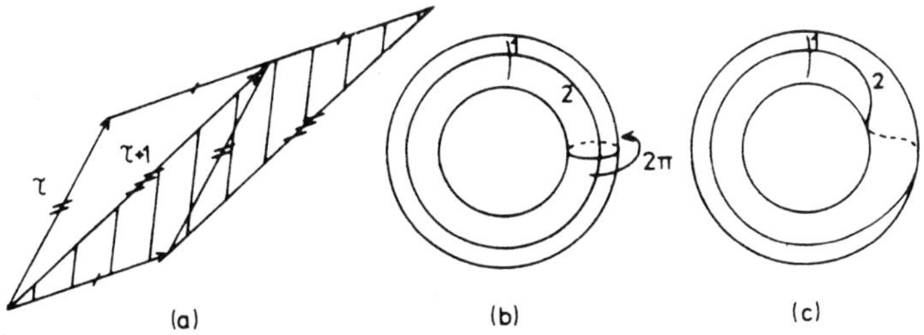

Fig. 5. (a) gives the parallelograms with lattice vectors $(1,\tau)$ and $(1, \tau + 1)$ which correspond to the tori in (b) and (c) respectively. The coordinate axes are again labelled by 1 and 2.

$$T_M = \mathrm{Tr}[\Delta(1)V(1) \; \ldots \; \Delta(M)V(M)] \tag{5}$$

where

$$\Delta(A) = \frac{1}{\pi} \int \frac{d^2 z_A}{|z_A|^2} \; |z_A|^{(p_A^i)^2/4} \; z_A^N \; \bar{z}_A^{\tilde{N}-1+(p_A^I)^2/2} \tag{6}$$

is the Ath propagator, with N and $\tilde{N}$ to be the number operator for the right- and the left-moving oscillators,[3]

$$N = \sum_{r=1}^{\infty} [\alpha_{-r}^i \, \alpha_r^i + \frac{r}{2} \bar{S}_{-r} \gamma^- S_r]$$

$$\tilde{N} = \sum_{r=1}^{\infty} [\tilde{\alpha}_{-r}^i \, \alpha_r^i + \tilde{\alpha}_{-r}^I \, \tilde{\alpha}_r^I] \tag{7}$$

The superscripts i refer to the transverse space-time indices ($1 \leqq i \leqq 8$) and, for the left-moving bosonic string, the superscript I refers to the compactified dimensions ($1 \leqq I \leqq 16$). The momenta p_A refers to the momenta of the Ath propagator. In terms of the loop momenta ℓ^i and L^I, they are

$$p_A^i = \ell^i + Q_A^i$$

$$p_A^I = L^I + Q_A^I \tag{8}$$

$$Q_1 = 0 \; , \; Q_A = \sum_{B=1}^{A-1} k_B$$

The vertex operators are

$$V(A) = g \; W(A) \quad :\exp\{ik_A^i \cdot X^i(0) + 2i \; k_A^I \cdot X^I(0)\}: \quad C(k_A^I) \tag{9}$$

with $C(k_A^I)$ being the cocycle factor and g the coupling constant. Moreover,

$$X^i(0) = \frac{i}{2} \sum_{n \neq 0} (\alpha_n^i + \tilde{\alpha}_n^i)/n$$

$$X^I(0) = \frac{i}{2} \sum_{n \neq 0} \tilde{\alpha}_n^I/n \tag{10}$$

The spin-dependent part W(A), for charged gauge boson external states, is

$$W(A) = \rho^i(k_A) \; B_i(A) \tag{11}$$

with

$$B^i(A) = P_A^i + \frac{1}{2} R^{ij} \; k_A^j$$

$$P_A^i = \frac{1}{2} \, p_A^i + \sum_{n \neq 0} \alpha_{ni} \tag{12}$$

$$R^{ij} = \bar{S} \; \gamma^{ij-} \; S/8 = \frac{1}{8} \sum_{m,n} \bar{S}_m \; \gamma^{ij-} \; S_n$$

and $\rho^i(k_A)$ is the spin 1 wave function with transverse polarizations. Of particular importance in R^{ij} is the term from m = n = 0. This term will be denoted by R_o^{ij}, or R_o for short.

Finally, the trace in (5) is a shorthand for

$$Tr = \int d^{10}\ell^i \sum_{L^I \epsilon \Lambda_{16}} \prod_{a=0}^{4} (tr)_a \tag{13}$$

where Λ_{16} is the root lattice of the group spin $(32)/Z_2$ or $E_8 \times E_8$. The traces $(tr)_a$ are taken over the oscillators S_o, S_n $(n\neq 0)$, α_n^i, $\tilde{\alpha}_n^i$ and $\tilde{\alpha}_n^I$ respectively for a = 0, 1, 2, 3, 4.

In the Introduction it was mentioned that the M-point amplitude T_M vanishes for M < 4. This is because $tr_o[(S_o)^k] = 0$ for k < 8, and hence $tr_o[(R_o)^M] = 0$ for M < 4. Here $(R_o)^M$ is the shorthand for the product of M R_o^{ij}'s, possibly all with different superscript indices ij.

Consider the tr_o operation. For M = 4, the only contribution is proportional to $tr[(R_o)^4]$, which comes from taking the $R_o k$ term for every vertex in (11) and (12). Let us denote the result of this trace by K_4. For M = 5, there are four non-zero terms, denoted now respectively by I_a, I_b, I_c and I_d, and they are proportional to $tr_o[(R_o)^5]$, $tr_o[(R_o)^3|S_n S_o)^2]$ $(n\neq 0)$, $tr_o[(R_o)^4 p]$, and $tr_o[(R_o)^4 P\ (n\neq 0)]$ respectively. The total contribution from this trace is then denoted by $K_5 = I_a + I_b + I_c + I_d$.

Carrying out all the Tr operations in (13), one gets in a rather standard way[3] the following:

$$T_M = \int d^2\tau \int \prod_{B=1}^{M-1} d^2\nu_B\, G_M(\nu|\tau) \tag{14}$$

$$G_M(\nu|\tau) = c\, \left(\frac{2}{Im\tau}\right)^5 \prod_{D<C} [\chi(\nu_{DC}|\tau)]^{\frac{1}{2} k_C^i \cdot k_D^i} J_M \tag{15}$$

where

$$J_M = K_M \cdot \tilde{T}_M \tag{16}$$

$$\chi(\nu|\tau) = 2\pi \exp[-\pi(Im\nu)^2/Im\tau]\,|\theta_1(\nu|\tau)/\theta_1'(0|\tau)| \tag{17}$$

$$\tilde{T}_M = [\frac{1}{\bar{w}} f(\bar{w})^{-24}] \prod_{D<C} [\psi(\bar{\nu}_{DC}|\bar{\tau})]^{k_C^I \cdot k_D^I} \cdot \Lambda \tag{18}$$

$$\psi(\bar{\nu}|\bar{\tau}) = 2\pi i \exp[-\pi i\, \bar{\nu}^2/\bar{\tau}][\theta_1(\bar{\nu}|\bar{\tau})/\theta_1'(0|\bar{\tau})] \tag{19}$$

$$\frac{1}{\bar{w}} f(\bar{w})^{-24} = [\theta_1'(0|\tau)/2\pi]^{-8} \tag{20}$$

and $\theta_i(\nu|\tau)$ are the Jacobi θ-functions.

K_4 is a constant, independent of ν and τ. By definition, $K_5 = I_a + I_b + I_c + I_d$. Calculation[1,2,7] shows that

$$I_a = \text{constant}$$

$$I_b(\nu|\tau) = \sum_{C<D} U_{CD} f_p(\nu_{DC}|\tau)$$

$$I_c(\nu|\tau) = -\sum_{A=1}^{5} \sum_{B=A}^{A-1} W_{AB}\, \mathrm{Im}\nu_{BA}/\mathrm{Im}\tau \tag{21}$$

$$I_d(\nu|\tau) = +\sum_{A=1}^{5} \sum_{B=A}^{A-1} W_{AB}\, f_p(\nu_{BA}|\tau)$$

with

$$f_p(\nu|\tau) = \frac{i}{2\pi}\, \theta_3'(\nu - \frac{\tau}{2} - \frac{1}{2}|\tau)/\theta_3(\nu - \frac{\tau}{2} - \frac{1}{2}|\tau) \tag{22}$$

The sum over B from A to A-1 means the cyclic sum over B = A, A + 1, ..., M, 1, 2, ..., A - 1.

The coefficients U_{CD} and W_{AB} are fairly complicated, but fortunately we do not have to know all the details. Because the polarization vectors are transverse, $W_{AA} = 0$. Momentum conservation implies

$$\sum_{A=1}^{5} W_{AB} = 0 \tag{23}$$

Beyond these relationships, we also need to know two results from Ref. 7, obtained after a lengthy algebraic calculation. In the notation of (21), they are

$$I_a = \frac{1}{2} \sum_{C<D} U_{CD}$$

$$0 = \sum_{C<D} U_{CD}\, \nu_{CD} \tag{24}$$

These are the relations required to show that the non-leading log log divergences of the M = 5 Type 1 open string vanish.[7]

Now using all these relationships, it is straightforward to show[1-3] that under $(\nu|\tau) \to (\nu'|\tau')$, where $(\nu'|\tau')$ is equal to (i) $(\nu + 1|\tau)$, (ii) $(\nu + \tau|\tau)$, or (iii) $(\nu|\tau + 1)$, each of the following quantities is invariant:

$$A = d^2\tau \prod_{B=1}^{M-1} d^2\nu_B$$

$$B = (1/\mathrm{Im}\tau)^5$$

$$C = \prod_{D<C} [\chi]^{\frac{1}{2} k_C^i \cdot k_D^i} \tag{25}$$

$$D = K_4 \cdot \tilde{T}_4$$

$$E = K_5 \cdot \tilde{T}_5$$

When these results are substituted into eqs. (14) and (15), periodicity (2) and half of the modular invariance (4) follow. The remaining task then is to verify the modular invariance under (iv) $(\nu'|\tau') = (-\nu/\tau| - 1/\tau)$. Under this change, calculation shows that the quantities A to E in (25) gains a factor $|\tau|^{-2M-2}$, $|\tau|^{10}$, 1, 1, $|\tau|^2$ respectively. The product ABCD is therefore invariant for M = 4 and the product ABCE is invariant for M = 5, thus proving the modular invariance of the 1-loop amplitude for M = 4 and 5 for the heterotic string.

For Type II superstrings, some changes are needed. The left-going states are states of a 10-dimensional closed superstring instead of a 26-dimensional bosonic string. The external lines in Fig. 1 are no longer charge gauged bosons, because such objects no longer exist. Instead, we take them to be gravitons. Then instead of (11), we have

$$W(A) = \rho^{ij}(k_A)B_i(A)\tilde{B}_j(A) \tag{26}$$

The definition of B_i is given in (12), and $\tilde{B}_i(A)$ is obtained from the $B_i(A)$ by replacing all the right-moving oscillators with the left-moving ones. Correspondingly eq. (13) must be modified. The sum over L^I is no longer present; the product over a is now taken from 0 to 5, with $(tr)_a$ taken to mean the trace over the $\tilde{S}_n$ $(n\neq 0)$ and $\tilde{S}_o$ oscillators for a = 4 and 5 respectively, and no change to its interpretation in (13) for a = 0 to 3. As a result, instead of (16), we have

$$J_M = K_M \cdot \tilde{K}_M + E_M \tag{27}$$

where $\tilde{K}_M$ is produced from the left-moving oscillators the same way that K_M is produced from the right-moving ones. Note that the momenta p_A of the Ath line are common to $B_i(A)$ and $\tilde{B}_i(A)$. As a result, for M = 5, the term $I_c \cdot \tilde{I}_c$, which is proportional to $tr_o[(R_o)^4 p]\, tr_5[(\tilde{R}_o)^4 p]$ (see the paragraph preceeding eq. (14)), involves two p_A's. Upon carrying out the loop integration $d^{10}\ell$, this produces the extra term E_5 in (27). This term is of the form $V/\mathrm{Im}\tau$ where V is a constant. For M = 4, such a phenomenon does not occur and $E_4 = 0$. With these changes, the D and E terms in (25) should be replaced by $D = K_4 \cdot \tilde{K}_4$ and $E = K_5 \cdot \tilde{K}_5 + V/\mathrm{Im}\tau$. Under the transformations (i) - (iii) preceeding (25), D and E are still invariant. Under (iv), D is still invariant and E gains a factor $|\tau|^2$, as was the case for the heterotic string. This shows that the Type II superstring is also modular invariant and hence finite.

REFERENCES

1. C. S. Lam and Da-Xi Li, Phys. Rev. Lett. 56 (1986) 2575.
2. P. H. Frampton, P. Moxhay, and Y. J. Ng, U. of North Carolina preprint IFP-269-UNC.
3. J. H. Schwarz, Phys. Rep. 89 (1982) 223.
4. M. B. Green and J. H. Schwarz, Phys. Lett. 151B (1985) 21.
5. M. B. Green and J. H. Schwarz, Phys. Lett. 109B (1982) 444.
6. D. J. Gross, J. A. Harvey, E. Martinec, and R. Rohm, Nucl. Phys. B256 (1985) 253, and B267 (1986) 75.
7. P. H. Frampton, P. Moxhay, and Y. J. Ng, Phys. Rev. Lett. 55 (1985) 2107; U. of North Carolina preprint IFP-256-UNC.
8. Talk by Clifford Burgess in these proceedings.
9. M. B. Green and J. H. Schwarz, Phys. Lett. 149B (1984) 117; Nucl. Phys. B255 (1985) 93.

SOME TOPICS IN THE LOW-ENERGY PHYSICS FROM SUPERSTRINGS

L.E. Ibáñez

Theory Division, CERN
CH-1211 Geneva 23, Switzerland

INTRODUCTION

The recent revival of string theories[1-5] was mostly motivated by purely phenomenological reasons. The experimental results obtained during the last decades indicate that quarks and leptons come in chiral (complex) representations of a gauge group SU(3) × SU(2) × U(1) (or some extension). The important point of the heterotic $E_8 \times E_8$ superstring[4] is that, since it contains explicit gauge bosons, one may obtain chiral low-energy fermions which transform like the standard quarks and leptons under an SU(3) × SU(2) × U(1) subgroup of $E_8 \times E_8$. Still, the theory has no gauge or gravitational anomalies as shown for the field theory limit by Green and Schwarz[2]. To some extent, the parity-breaking world we observe now is related to the left-right asymmetric construction of the heterotic string. This asymmetric construction is required to have explicit gauge bosons, which in turn are required to get low-energy chiral fermions (quarks and leptons). It is obvious that the possible phenomenological applications of the $E_8 \times E_8$ string are at the root of the present interest for this theory. Other strings (e.g., the anomaly-free chiral type IIb superstring) are simpler and have less contrived constructions, but are certainly less popular, since their phenomenological uses do not seem very bright.

It is reasonable to expect that, in the same way that chirality and observed gauge interactions were important in selecting a superstring candidate, other low-energy phenomenological constraints (an obvious example is the number of generations) could give us further information on the superstring dynamics. This is particularly the case for the compactification process. Understanding the enormous hierarchy of scales between the weak scale and the compactification mass probably requires as a technical ingredient the existence of an unbroken supersymmetry in four dimensions. If, on the other hand, we insist on maintaining an unbroken supersymmetry when compactifying the superstring[5], the possible form of the six-dimensional compact manifold becomes severely restricted [Kähler, SU(3)-holonomy manifolds]. It is in fact interesting that there seem to be two independent reasons to require supersymmetry in our fundamental theory: it seems to be necessary at low energies in order to understand the hierarchy problem, but supersymmetry also gives us the simplest way to obtain tachyon-free string theories.

If one unbroken supersymmetry is left after compactification, the effective low-energy theory will be N = 1, d = 4 supergravity coupled to chiral matter and gauge bosons[6]. In the case of the $E_8 \times E_8$ heterotic string, all the presently-known gauge interactions should be included in one of the E_8's due to the direct product structure. The other E_8 will interact with the known world only through gravitational interactions; it will be a "hidden sector" of the theory. Remarkably enough, this structure for the low-energy limit of the $E_8 \times E_8$ string is very reminiscent of what are called "low-energy supergravity models"[7-9]. These are N = 1, d = 4 supergravity models coupled to quarks, leptons and gauge bosons. The minimal model is just an SU(3) × SU(2) × U(1) theory coupled to three usual families of chiral superfields plus Higgs particles. In these theories, supersymmetry breaking takes place in a "hidden sector" of the theory[8] and it is transmitted to the observable quark-lepton-Higgs world only through gravitational interactions. After SUSY breaking, one is left at low energies with a theory with softly-broken supersymmetry having soft parameters m (universal scalar masses), M (universal gaugino masses) and other soft couplings proportional to the superpotential. A remarkable fact of this class of models is that the $SU(2)_L \times U(1)_Y$ symmetry is broken in a natural way as a radiative effect of supersymmetry breaking[9,10]. It is certainly interesting that this phenomenologically successful structure seems to be embeddable inside the $E_8 \times E_8$ string.

Low-energy physics could give us further restrictions on the superstring (particularly the compactification) dynamics. If, in the next generation of accelerators (Tevatron, SLC, LEP, HERA, LHC, SSC), supersymmetric particles are found, their masses could give us interesting information on the SUSY-breaking dynamics and about the symmetries of the compactifying vacuum. Thus, e.g, if $m \ll M$, it could be an indication of an approximate scale invariance in the original Lagrangian. Also the ratios between the masses of different sparticles (e.g., sleptons and squarks) would give independent tests of the unification idea. It is also important to search for the existence of possible extra Z°'s, whose couplings could tell us a lot about the compactification and gauge-symmetry-breaking dynamics. Of course, quark masses and mixings (Yukawa couplings) also contain potential information about the compactification dynamics, but it will probably be highly non-trivial to extract such information. However, qualitative features such as, for example, the observed hierarchy of fermion masses (or the size of the CP-violating phase) could guide the search for interesting compactifications. Thus a "superstring-inspired" phenomenology could be very fruitful in providing constraints on the superstring dynamics. While this is probably correct, we are far away from a situation (which may never come) in which the superstring gives us any definite prediction able to be tested at low energies. Hence, most probably, "experimental" support for the superstring idea will only come from "circumstantial evidence" based on the possibility of these theories containing all the observed dynamics. I am going to discuss here some topics in "superstring-inspired" phenomenology[11-13] considered in the spirit explained above. It is not the search for experimental consequences of the $E_8 \times E_8$ superstring, but an effort to understand how one could embed the low-energy physics inside it.

SUPERSYMMETRY BREAKING, GAUGINO CONDENSATION, DILATONS, AXIONS, ETC.

The massless sector of the heterotic string consists of a ten-dimensional supergravity[14] sector which includes bosonic (scalar ϕ; antisymmetric B^{MN}; g^{MN}) and fermionic (gravitino ψ^M; spinor λ) fields, and a Yang-Mills sector (gauge bosons A^M and gauginos χ^α) in the adjoint of $E_8 \times E_8$. The bosonic Lagrangian contains the terms

$$E^{-1}\mathcal{L} = -\frac{1}{2}R + \frac{9}{16}\left(\frac{\partial_M \varphi}{\varphi}\right)^2 - \frac{1}{4}\varphi^{-3/4} F_{MN}^{a} F_a^{MN} + \frac{3}{4}\varphi^{-3/2} H_{MNP} H^{MNP} + \ldots \quad (1)$$

plus higher derivative terms. The scalar "dilaton" ϕ plays the rôle of a coupling constant and H_{MNP} is the field strength for B_{MN}

$$H_{MNP} = \partial_{[M} B_{NP]} - (\omega^G_{MNP} - \omega^L_{MNP}) \quad (2)$$

where ω^G and ω^L are the gauge and Lorentz Chern-Simons symbols[2,14]. The tree-level Lagrangian has a classical invariance under the rescalings[15]

$$\left.\begin{array}{l} g_{MN} \to \lambda g_{MN} \\ \varphi \to \lambda^{-4/3}\varphi \\ \Psi_A, \chi^\alpha \to \lambda^{-1/4}\Psi_A, \chi^\alpha \end{array}\right\} \to E^{-1}\mathcal{L} \to \lambda^{-1}E^{-1}\mathcal{L} \qquad M,N,A = 1-10 \tag{3}$$

The whole effective action is not invariant under this symmetry, but all n-loop contributions scale in a definite way since the power of λ measures the number of loops. The existence of this classical scale invariance turns out to be interesting when analyzing the low-energy symmetries.

In order to study the low-energy physics, we have to compactify our ten-dimensional theory down to four dimensions. A first possibility to consider is a compactification at the pointlike field theory level. In this case, if we insist on conserving an unbroken supersymmetry[5], the extra dimensions must curl into a compact Kähler manifold of SU(3) holonomy (Calabi-Yau or some other more or less related type of manifold). Alternatively, one could perhaps compactify the theory already at the string level. This is what is done when compactifying[16] on an "orbifold" [some type of six-torus modded out by some discrete subgroup of SU(3)]. This latter approach is very promising since it keeps much of the simplicity of the original torus manifold. Whatever the compactification procedure may be, up to now no concrete example of compactification completely consistent with phenomenological constraints has been given in the literature (see, however, Ref. 17). Still, there are some general properties of the low-energy states expected to be present in any supersymmetry-preserving compactification, as argued in Ref. 15.

Amongst the d = 4 massless (at least at the tree level) states which one expects, there is a dilaton ϕ (from the original D = 10 dilaton ϕ) and another dilaton σ associated to the size of the compact manifold

$$E = \det|g_{MN}|^{1/2} = e^{-3\sigma(x)} e_4 \quad ; \quad V_6 = e^{3\sigma(x)} \tag{4}$$

Since the low-energy theory is assumed to be supersymmetric, these scalar fields must have a couple of massless pseudoscalar partners. These zero modes $\theta(x)$ and $\eta(x)$ come in fact from the antisymmetric field B_{MN}:

$$H_{\mu\nu\rho} \sim \varepsilon_{\mu\nu\rho\sigma}\,\partial^\sigma\theta(x)\,, \quad \mu,\nu,\rho,\sigma = 1-4 \tag{5a}$$

$$B_{i,\bar{i}} = \varepsilon_{i\bar{i}}\,\eta(x) \tag{5b}$$

i=SU(3)$\in$tangent O(6) index

The precise combination of dilaton fields which are partners of these pseudoscalars $\theta(x)$ and $\eta(x)$ may be shown to be[15]

$$\begin{aligned} S &= \varphi^{-3/4} e^{3\sigma} + i\theta \\ T &= e^{\sigma} \varphi^{3/4} + i\eta \end{aligned} \tag{6}$$

and they have supersymmetric fermionic partners from the D = 10 fields ψ_m and λ. All these fields are gauge singlets which interact with usual matter only gravitationally and hence form part of the "hidden sector" of the theory. The quark, lepton and Higgs fields are expected to arise from zero modes of the Yang-Mills sector, A^m (m = 5-10) and χ^{α}.

Several symmetries are expected to appear in the low-energy theory:

a) Peccei-Quinn Symmetries[18-22]

The pseudoscalars $\theta(x)$ and $\eta(x)$ only have derivative couplings, since the field B_{MN} always appears through its field strength $H_{MNP} = \partial_{[M}B_{NP]} + \ldots$. Thus there are two P-Q invariances under

$$\begin{aligned} \theta &\to \theta + c \\ \eta &\to \eta + c' \end{aligned} \tag{7}$$

This implies that the Kähler potential G, which determines the low-energy Lagrangian, depends on the dilaton superfields S and T only through the combinations $(S+S^*)$ and $(T+T^*)$. These symmetries are, however, broken by both space-time and string-world-sheet non-perturbative effects, as we will discuss below. Let us also remark that there could in fact be extra zero-mode light superfields originating in g_{MN} and B_{MN} apart from those considered above (e.g., moduli).

b) "S-scale Invariance"[15]

The D = 10 scale invariance in Eqs. (3) remains in four dimensions, since we assume that the compactification obeys the equations of motion and no tadpole for ϕ is produced. The form of this four-dimensional invariance is[15,11,23]

$$\left.\begin{array}{l} S \longrightarrow \lambda S \\ T, C_x \longrightarrow T, C_x \\ g_{\mu\nu} \longrightarrow \lambda g_{\mu\nu} \\ \psi_\mu \longrightarrow \lambda^{1/4} \psi_\mu \\ \chi \longrightarrow \lambda^{-3/4} \chi \end{array}\right\} \longrightarrow \quad e^{-1}\mathcal{L} \longrightarrow \lambda^{-1} e^{-1}\mathcal{L} \tag{8}$$

where the fields C_x are the gauge non-singlet scalar fields. Of course, this is a classical scale invariance and any loop effect will break this symmetry. Notice that the T and C_x fields are left untouched by this symmetry, whereas S is not; that is why I call it S-scale invariance here. Notice also that the classical scalar potential of the truncated theory, since it obviously implies no space-time contractions, has necessarily to be of the general form[11]

$$V(S,T,C_x) = \frac{1}{(S+S^*)} V(T+T^*; C_x, C_x^*) \tag{9}$$

with $V(T+T^*;C_x)$ scale invariant, in order to scale like λ^{-1}. Recalling the general N = 1, d = 4 supergravity expression for the scalar potential (proportional to e^G), one obtains that at the classical level the S-dependent part of the Kähler potential G_S should be of the form

$$G_S = -\log(S+S^*) \tag{10}$$

The existence of this classical scale invariance dictates[23] also the form of the gauge kinetic function $f(S,T,C_x)$. This must be an analytic function of its arguments and induces the following bosonic terms in the Lagrangian

$$\mathrm{Re} f_{ab} F^a_{\mu\nu} F^{\mu\nu}_b \quad ; \quad \mathrm{Im} f_{ab} \varepsilon^{\mu\nu\rho\sigma} F^a_{\mu\nu} F^b_{\rho\sigma} \tag{11}$$

In order that these terms scale like λ^{-1}, the function f must necessarily be linear in S. One then has at the classical level[24,23]

$$f_{ab} = \delta_{ab} S \tag{12}$$

One could, in principle, multiply f by an analytic product of scale-invariant fields like T and S, but one can easily see that the analyticity constraint plus the other T-scale invariance we define below forbids that possibility. Notice that the result (12) is the same for both observable and hidden sectors, i.e., for the whole of the unbroken $E_8 \times E_8$ generators.

c) "T-Scale Invariance"[11,23]

The existence of this extra invariance is related to the fact that in the simplified tree-level analysis we are assuming that the compactification scale, which is related to ReT, is not determined. It can be intuitively understood as follows[11]. In the compactification procedure, wherever there is an internal space contraction of indices, one gets a $(T+T^*)^{-1}$ factor from the internal metric g_{mn} (m,n = 5-10). The matter scalar fields C_x come from the A_m's (m = 5-10), and hence whenever you have a field C_x in your Lagrangian, you expect a $(T+T^*)^{-\frac{1}{2}}$ factor from the internal contraction. Thus there should be a scale invariance under[11]

$$\left.\begin{array}{l} T \to (\lambda')^2 T \\ C_x \to \lambda' C_x \\ S \to S \end{array}\right\} \qquad \mathcal{L} \to \mathcal{L} \tag{13}$$

This is also only a classical scale invariance since, as I will show below, there are loop terms which spoil it. This additional symmetry also restricts significantly the form of the tree-level Kähler potential. If the Kähler potential G is to contain the standard piece $\log|W|^2$ (W = superpotential), the only way to make it scale invariant[11,23] is to add a term

$$-3 \operatorname{Log}(T+T^* + \alpha_i |C_i|^2) + \operatorname{Log}|W|^2 \tag{14}$$

where the factor 3 implies that the superpotential is trilinear and the α_i are just constants which may be different for different SU(3) × SU(2) × U(1) representations (or families). Thus, from general symmetry considerations, one expects a tree-level Kähler potential for the low-energy N = 1, d = 4 theory of the general form[15,23,11]

$$G_0 = -\operatorname{Log}(S+S^*) - 3\operatorname{Log}(T+T^* + \alpha_i|C_i|^2) + \operatorname{Log}|W|^2 + F\left((T+T^*)/|C_x|^2\right) \tag{15}$$

where we have added[23] an arbitrary function F of the scale-invariant combination $(T+T^*)/|C_x|^2$. The second term in (15) is very similar to the one appearing in the so-called "no-scale models"[25], but there are a couple of important differences [apart from the other terms in (15)]. First, an SU(n) symmetry (n = number of chiral superfields) is not expected. Second, the term in Eq. (15) is just a tree-level term and it is known that the radiative corrections[26-29] and non-perturbative effects[30] spoil its structure (since it has its origin in a classical invariance).

This is to be contrasted with the very assumptions of the "no-scale" idea[25], in which the second term in Eq. (15) is assumed to represent the exact Kähler potential (including radiative gravitational corrections). Let us comment that the result in Eq. (15) can also be obtained (in a less general form) by performing a supersymmetry truncation of the D = 10 supergravity + Yang-Mills Lagrangian, either from the bosonic[15] or the fermionic[21] sectors.

There are many sources of corrections for the above low-energy interactions. First, there are higher derivatives and/or higher power in the inverse compactification scale terms coming from higher terms in the string world-sheet σ-model. This includes all tree-level string exchanges amongst massless external fields. Amongst these terms are

$$R^2 \; ; \; R^4 \; ; \; (R^2 - F^2)^2 \; ; \; H^4 \; \ldots . \tag{16}$$

couplings which have been shown to appear in string scattering amplitudes[31]. From the four-dimensional point of view, terms of the form[11]

$$\frac{\left| \frac{\partial W}{\partial C_x} \frac{\partial W}{\partial C_y} d_{xyz} \right|^2}{(S+S^*)(T+T^*)^4} \; ; \; \frac{\left| \frac{\partial W}{\partial C_x} \right|^4}{(S+S^*)(T+T^*)^4} \; ; \; \frac{|W|^4}{(S+S^*)(T+T^*)^6} \; ; \; \ldots \tag{17}$$

respecting the symmetries described above will be induced in the scalar potential. Notice also that the Cremmer et al. formalism[6] for N = 1, d = 4 supergravity only includes up to two derivatives and the D = 10 terms in Eq. (16) include more. Thus, a formalism including higher derivative terms in d = 4, N = 1 supergravity is also required for a complete description of the low-energy theory[32].

The terms in Eqs. (16) and (17) are tree level and that explains why the classical invariances described above still apply. Loop effects both in the effective low-energy Lagrangian[26-28] and also including the effect of the string excitations[29] violate those symmetries. The same is true for non-perturbative effects, both in Minkowski space[29] and on the string world-sheet[30]. We will consider these effects below. Let us start with the related problem of residual N = 1, d = 4 supersymmetry breaking.

The most appealing way to break the residual supersymmetry seems to be gaugino condensation[33,24] in the hidden sector of the theory. It is well known that in the presence of non-minimal gauge kinetic terms[34], gaugino condensation $\langle \bar{\chi}\chi \rangle = \Lambda^3 \neq 0$ may break supersymmetry and give a mass to the gravitino $m_{3/2} \sim \Lambda^3/M_P^2$. Also in this scheme one may understand the smallness of the gravitino mass (and hence the weak scale) since $\Lambda \sim e_{M_P}^{-1/(2g^2 b_0)}$ can be exponentially small. It is certainly interesting that this structure is given for free in the $E_8 \times E_8$ heterotic string. Since, in a simple compactification scheme, the vev of S [and hence the value of the coupling constant $g = (\mathrm{Re}\ S)^{-\frac{1}{2}}$] is initially undetermined, instead of the creation of a condensate one can talk of an induced non-perturbative superpotential[24]:

$$W_S \sim (e^{-1/2g^2 b_0} M_P)^3 \sim e^{-3\frac{S}{2b_0}} M_P^3 \tag{18}$$

This term obviously violates the classical S-invariance (it is a non-perturbative effect) and also the Peccei-Quinn symmetry associated with the S field, since (18) induces non-derivative couplings for the pseudoscalar $\theta = \mathrm{Im}\ S$. This implies that one cannot use θ as an axion to solve the strong-CP problem. Furthermore, the existence of the interaction (18) destabilizes the vacuum since it gives a contribution to the scalar potential

$$V_S \sim \frac{1}{(S+S^*)(T+T^*)^3} |W_S|^2 \tag{19}$$

which induces a cosmological constant (except for the unphysical limits S and/or $T \to \infty$). There is, however, an interesting solution to this problem[24]. An alternative to the gauge condensation breaking of supersymmetry is the existence of a vev for the antisymmetric field H_{ijk} in extra dimensions[33,24,35]

$$\langle H \rangle = \langle H_{ijk} \varepsilon^{ijk} \rangle = C \neq 0 \tag{20}$$

which, from the four-dimensional point of view, would look like a constant superpotential. This gives a mass to the gravitino $m_{3/2} \sim C/M_P^2$ and also induces a cosmological constant. The interesting point is that when one considers both SUSY-breaking mechanisms simultaneously, their contribution to the scalar potential comes in the form of a perfect square[24]:

$$V_S \sim \frac{1}{(S+S^*)(T+T^*)^3} \left| \langle H \rangle + W_S + W \right|^2 \tag{21}$$

so that, upon minimization, the S field adjusts itself in such a way that the cosmological constant vanishes. Still, one can check that supersymmetry is broken ($F_T \neq 0$ although $F_S = 0$). This mechanism is very appealing, but one would have to show that it still works when further terms (e.g., H^4 D = 10 terms) are included, and also that it is stable under radiative corrections. One must admit that it would be a miracle if that were the case, but at least this provides us with a possible scenario in which SUSY-breaking and zero cosmological constant (at a certain level) are compatible. It could turn out that the cancellation of the zero-energies between $\langle H\rangle$ and $\langle\bar{\chi}\chi\rangle$ could be more general than the simple arguments given here.

As we stated above, a vev for H induces a gravitino mass $m_{3/2} \sim |C|/M_P^2$. Then, if we want to relate the $m_{3/2}$ mass to the weak scale M_W, one needs to choose $|C| \lll M_P$ by hand and we lose the opportunity offered by the gaugino condensation mechanism of determining $m_{3/2}$ in terms of M_P. Thus we have a nice mechanism for cancelling the cosmological constant, but we can no longer determine the $m_{3/2}$ scale dynamically. I think that a different interpretation of the cancelling cc mechanism is more interesting[11,21,36]. It may well be that the values of $\langle S\rangle$ and $\langle T\rangle$ were determined upon compactification (e.g., by some string effects). In such a case, gaugino condensation will form at a definite mass scale $\Lambda \sim e^{-1/(2g^2 b_0)}_{M_P}$ (e.g., of order 10^{13} GeV). Then the $\langle H\rangle$ vev adjusts itself to cancel the condensate contribution to the vacuum energy. There is nothing wrong with this field having a dynamical rôle, since H_{ijk} contains a piece $\omega^G_{ijk}-\omega^L_{ijk}$ which could take a non-vanishing value $\sim C\,\varepsilon_{ijk}$. This piece vanishes if one identifies gauge and spin connections[5], but this identification should appear as a result of the dynamics of the system, not as an arbitrary input. Thus a small miscancellation[21] $\omega^G_{ijk} - \omega^L_{ijk} \sim C\,\varepsilon_{ijk}$ could appear if one considers <u>simultaneously</u> the dynamics of the compactification <u>and</u> the low-energy dynamics. In this situation the $m_{3/2}$ scale would be dynamically determined by the gaugino condensation scale and not by an arbitrary input parameter C.

The total tree-level (plus condensation effects) scalar potential is essentially given by V_s + the usual <u>globally</u> supersymmetric scalar potential [with appropriate powers of $(S+S^*)$ and $(T+T^*)$ in order to have a scale-invariant result]. However, although SUSY has been broken in the hidden sector ($F_T \neq 0$), the observable world remains supersymmetric at this level

$$M = m = A = 0 \tag{22}$$

where M and m are the soft SUSY-breaking gaugino and scalar masses, and A parametrizes the soft trilinear scalar couplings. The parameters M and A vanish since they are proportional to $(\langle H\rangle + W_s)$ which vanishes upon minimization of the scalar potential Eq. (21). The value of m vanishes because of the scale-invariant structure of Eq. (15). Of course, this is only true at the tree level and one expects SUSY breaking soft terms to appear radiatively. It turns out that the required soft terms are not so easy to generate, at least if one just calculates radiative corrections starting with the low-energy truncated Lagrangian of Refs. 15 and 24. If one starts from A = M = 0, the effective theory has an R-symmetry which forbids non-vanishing A or M to be generated. It is sometimes stated in the literature that the usual observable sector interactions may generate radiatively some modification to the gauge kinetic function f_{ab} involving non-singlet superfields C_x. This is not possible, since radiative corrections are unable to give rise to an analytic contribution to f_{ab}. With a trilinear superpotential, as is the case here, any radiative graph will contain as many outgoing as incoming chiral superfields C_x. The appearance of such a radiative contribution to f_{ab} is also forbidden by the scale invariance in Eq. (13).

Masses for observable scalars are not generated[26-28] at one loop (at least in the truncated Lagrangian), but do have contributions at two loops[27]. However, this source for SUSY breaking would in general be problematic[36] since it gives rise to squark and slepton masses

$$m_{\tilde{q},\tilde{\ell}} \sim h_{q,\ell} \frac{m_{3/2}}{(4\pi)^2} \tag{23}$$

and hence the soft scalar masses will not be universal but proportional to each fermion partner's mass. This would be a disaster for the suppression of flavour-changing neutral currents (FCNC). Thus, although one obtains SUSY-breaking terms from the truncated Lagrangian, the phenomenological prospects do not seem very promising.

To calculate radiative corrections using just the low-energy truncated Lagrangian is probably unreliable since one is neglecting, for example, the effect of heavy string modes. However, quantum corrections involving heavy string modes can be important[29]. Specific examples of such terms are the "Wess-Zumino" one-loop couplings[2,3]

$$\frac{1}{18(2\pi)^5}\left[4\left(\omega^L - \frac{1}{30}\omega^G\right) X_7 - 6 B X_8\right] \tag{24}$$

where $X_8 = dX_7$ and

$$X_8 = \frac{1}{24} Tr F^4 - \frac{1}{7200}(Tr F^2)^2 - \frac{1}{240} Tr F^2 Tr R^2 + \frac{1}{8} Tr R^4 + \frac{1}{32}(Tr R^2)^2 \tag{25}$$

in the notation of Ref. 3. These one-loop terms have to be present in the low-energy D = 10 field theory if one is to understand the fact that anomalies are cancelled at the string level. When one considers a truncation of these couplings down to four dimensions, one obtains, for example, from the second term in Eq. (24) four-dimensional couplings of the form[18]

$$B^{mn} \mathrm{Tr}(F^{\mu\nu} F^{\rho\sigma}) \langle \mathrm{tr}\, R^{pq} R^{rs} \rangle \longrightarrow \sim \frac{1}{(2\pi)^5} \eta F\tilde{F} \tag{26}$$

These are axion-type couplings of the pseudoscalar field η = Im T with the four-dimensional gauge particles. In the $E_8 \times E_8$ case, one can easily check that the coefficients of the observable E_6 and the hidden E_8 (or subgroups) axion couplings are equal and opposite[21,22]

$$-\varepsilon \eta F_6 \tilde{F}_6 \quad ; \; +\varepsilon \eta F_8 \tilde{F}_8 \quad , \qquad \varepsilon \sim (2\pi)^{-5} \tag{27}$$

Since one assumes that the low-energy theory is supersymmetric, one concludes that the gauge kinetic function Eq. (12) gets a one-loop correction from these terms[21,29]

$$f^8_{ab} = \delta_{ab}(S + \varepsilon T) \qquad ; \; f^6_{ab} = \delta_{ab}(S - \varepsilon T) \tag{28}$$

Notice that the couplings in Eq. (27) violate S-scale invariance (since they are one loop) and T-scale invariance (since they are proportional to a large vev $\langle R^2 \rangle \neq 0$). One does not expect Eq. (24) to be the only effective local one-loop terms induced by heavy string mode exchanges. Thus, in order to obtain the supersymmetric counterparts of the axion couplings at low energies, D = 10 couplings of the form[29]

$$\varepsilon \, \epsilon^{MNPQRSTUVW} \varphi^{3/4} (\partial_M B_{NP}) \mathrm{Tr}(\bar{\chi} \Gamma_{QRS} \chi) \mathrm{Tr}(F_{TU} F_{VW}) \tag{29}$$

are also required. By the same token, one also expects analogous gravitino couplings

$$\varepsilon \, \epsilon^{MNPQRSTUVW} \varphi^{3/4} (\partial_M B_{NP}) \mathrm{Tr}(\bar{\Psi}_Q \Gamma_R \Psi_S) \mathrm{Tr}(F_{TU} F_{VW}) \tag{30}$$

which, after truncation, induce four-dimensional terms[29]

$$\frac{i\varepsilon}{(S+S^*)} (\partial_\mu \eta) \, \varepsilon^{\mu\nu\rho\sigma} (\bar{\Psi}_\nu \gamma_\rho \Psi_\sigma) \tag{31}$$

Comparing this result with the corresponding general N = 1, d = 4 supergravity coupling $(G'^T D_\mu T)\varepsilon^{\mu\nu\rho\sigma}(\bar{\phi}_\nu \gamma_\rho \phi_\sigma)$, one concludes that the tree-level Kähler potential G_0 gets a one-loop correction δG[29]:

$$G = G_0 + \delta G \quad ; \quad \delta G = -a\varepsilon \, \frac{(T+T^*)}{(S+S^*)} \tag{32}$$

where a is a number O(1) and G_0 is the tree-level Kähler potential of Eq. (15). As indicated in that equation, one may have extra scale-invariant factors $\sim \left(|C|^2/(T+T^*)\right)$. To obtain the result in Eq. (32), it is in fact enough to consider the transformation properties under the S and T scale invariances. The above modification of G_0 is the only one which leads to terms in the Lagrangian which transform like the axion couplings (i.e., $\eta F\tilde{F} \to \lambda'\lambda^{-2}\eta F\tilde{F}$).

Notice that the results in Eqs. (28) and (32) substantially modify the tree-level situation. First, the fact that the T-scale invariance is broken means that the $\left(-3\log(T+T^*) + \log|W|^2\right)$ structure guaranteeing a positive definite scalar potential disappears. Secondly, if a gaugino condensation is generated in the hidden sector, Eq. (28) implies that a non-perturbative superpotential is also generated for the T-field

$$W_8 \sim \left(e^{-1/2g^2 b_0} M_P\right)^3 \sim e^{-\frac{3S}{2b_0}} \, e^{-\frac{3\varepsilon T}{2b_0}} M_P^3 \tag{33}$$

so that non-derivative interactions appear for η = Im T and the second Peccei-Quinn symmetry is broken, as already happened to the one associated with θ = Im S. In fact, it has recently been argued that this symmetry is also broken by string world-sheet instantons[30], and analogous superpotentials of the form exp(T) may also be generated by that mechanism. Of course, as in the case of the superpotential Eq. (18), the existence of W_8 destabilizes the vacuum unless one finds a mechanism to cancel the cosmological constant.

One interesting observation[29] is that we may define a new field $S' = S + a\varepsilon T$ and reabsorb the δG correction into the $-\log(S'+S'^*)$ term (up to ε^2). Then the gauge kinetic functions may be re-expressed as

$$f^8_{ab} = \delta_{ab}(S' + \varepsilon(1-a)T) \quad ; \quad f^6_{ab} = \delta_{ab}(S' - \varepsilon(1+a)T) \tag{34}$$

where we recall that a is a number O(1) which gives us the coefficient of the one-loop corrections to G. In terms of the redefined field, one then obtains for the scalar potential

$$V = e^{K}\left[\left|W + 3(S'+S'^{*})\frac{W_8}{b_0}\right|^2 + \frac{t_c}{3}\left|\frac{\partial W}{\partial C_x}\right|^2 + D^2_{term}\right] + $$
$$+ e^{K} 3\varepsilon' \frac{t_c}{b_0} W_8^{*}(W_8 + c) + h.c. \tag{35}$$

where $\varepsilon' \equiv \varepsilon(1-a)$, $2tc = (T+T^{*}-2|C_x|^2)$ and $\exp(K) = (S'+S'^{*})^{-1}(t_c^{-3})/8$. The term in brackets is analogous to the tree-level result, but the term in ε' is not positive definite and spoils the mechanism for cancelling the cosmological constant that we explained above. On the other hand, soft SUSY-breaking terms are generated in the observable sector, $M, m \sim \varepsilon\, m_{3/2}$ and $Am \sim \varepsilon\, m_{3/2}$, since the symmetries giving rise to the result in Eq. (22) are no longer present.

An interesting situation[29] would occur if the "a" coefficient of the correction δG were $a = 1$ ($\varepsilon' = 0$). In that situation, the second term in Eq. (35) would vanish and the scalar potential would be completely analogous to the tree-level result, giving rise to SUSY breaking with zero cosmological constant ($F_{S'} = 0$, $F_T \neq 0$). There is, however, an important difference now. Due to the one-loop correction for f^8 and f^6, although the minimization conditions imply vanishing "hidden" gaugino mass $M^8 = 0$, the observable gaugino mass will be in general non-vanishing:

$$M_8 = f^{8\,\prime}_{S'} F_{S'} = 0$$
$$M_6 = f^{6\,\prime}_{S'} F_{S'} + f^{6\,\prime}_{T} F_T = -\varepsilon(1+a) F_T \sim \varepsilon\, m_{3/2} \tag{36}$$

Thus, in the case $a = 1$ ($\varepsilon' = 0$), we would have soft terms[29]

$$M_6 \sim \varepsilon\, m_{3/2} \quad ; \quad m = A = 0 \quad , \quad \varepsilon \sim (2\pi)^{-5} \tag{37}$$

and there would be no problem in transmitting SUSY breaking to the observable sector. Gaugino masses would induce the supersymmetry breaking. Since for phenomenological reasons we want $M \lesssim 1$ TeV one necessarily has

$$m_{3/2} \lesssim 10^{7}\ GeV \tag{38}$$

This is certainly an appealing possibility, but there is no obvious reason why "a" should be equal to one. Furthermore, other (or higher) loop effects may probably modify the mechanism which cancels the cosmological constant.

What is the conclusion of this long excursion through SUSY-breaking-dilaton-axion physics? It seems that both loop and non-perturbative effects substantially alter the results of a first naive tree-level truncation of the theory. The scalar potentials one obtains for the dilatons Re S and Re T lead to unphysical limits ($S,T \to 0$ or ∞). This is probably an indication that the vevs for the dilatons (and axions) are <u>not determined</u> by low-energy physics but by full string dynamics, and it does not make much sense to use S and T as undetermined dynamical variables at low energies. On the optimistic side, one may think of several aspects of the low-energy analysis which may survive. Gaugino condensation in the hidden sector seems a rather appealing mechanism for breaking the residual supersymmetry and obtaining a hierarchically small gravitino mass $m_{3/2} \sim \langle\bar{\chi}\chi\rangle/m_P^2$ which would presumably be related to the weak scale. In any case, there are loop effects which will transmit the supersymmetry breaking to the observable sector so that the result of Eq. (38), $m_{3/2} \lesssim 10^7$ GeV, seems unavoidable if we want to maintain the gauge hierarchy. The cancellation of the gaugino condensation cosmological constant by the vev of H_{ijk} could also be a more general mechanism. If that were the case, one would expect vanishing hidden gaugino mass M_8 and A-parameter, since both are proportional to $(\langle H\rangle - \langle\bar{\chi}_8\chi_8\rangle)$. In this case, the existence of radiative corrections which make $f^8 \neq f^6$ [i.e., Eq. (28)] is important, since it allows for non-vanishing observable gaugino masses M^6 to transmit supersymmetry breaking to the quark-lepton-Higgs world.

SUPERSTRING-INSPIRED LOW-ENERGY MODELS

We have explored in the previous chapter mostly the singlet "hidden" sector of the theory. Since the conclusions were not so positive, one may be led to the conclusion that nothing can be said about the low-energy limit of the $E_8 \times E_8$ string after compactification. This is not correct. There are some general features expected for the low-energy "observable" theory which are more or less independent of the details of the compactification[37-40]. Thus, for example, if we compactify on a "Calabi-Yau" manifold embedding the gauge connection into the spin connection, we know that we obtain an E_6 (or some subgroup) model with several families of <u>27</u>'s (or some subset of states). Also, some pairs of ($\underline{27}+\overline{27}$) could be present to start with[37]. The rank of the group can be lowered further. If the compact manifold has a certain non-Abelian discrete symmetry[37], one can break E_6 down to $SU(3)_C \times SU(2)_L \times U(1)_Y \times U(1)'''$, where $U(1)'''$ is a specific U(1) to be described below. This is not the only possible low-energy group, since one can also obtain other rank = 5 or 4 examples[36-46]. This may be done by using the chiral fields in the $\underline{27+\overline{27}}$ representation which models usually have. A vev for them can lower the rank from 6 down to 5 or 4. This symmetry breaking may occur at an intermediate scale (e.g., $10^{10}-10^{13}$ GeV) by radiative corrections. The final unbroken gauge group can then be just the standard SU(3) × SU(2) × U(1) model (see, e.g., Ref. 17).

There are other possible compactification schemes which may lead to a variety of low-energy models. If one goes beyond the usual recipe of identifying spin- and gauge connections, one may find manifolds[19] which break the original E_8 directly down to an SO(10) or SU(5) subgroup. These may be further broken (e.g., through the Wilson-loop mechanism[37]) down to some rank 5 subgroup or the standard model. Low-energy SO(10) or SU(5) subgroups may also be obtained[16], compactifying on certain classes of "orbifolds". The orbifolds are obtained by modding out some six-torus by a discrete subgroup of SU(3) (so that supersymmetry is preserved). They are in general "singular manifolds" with a discrete holonomy group, which may often be obtained as singular limits of some Calabi-Yau manifolds. If the original torus has some Wilson lines, or if (as in the manifolds mentioned above) one generalizes the usual procedure of identifying spin and gauge connections, one may break E_8 directly down to SU(5). The Wilson-loop mechanism may then give rise again to the standard model.

The opposite possibility exists, i.e., one may get rank > 6 low-energy gauge groups. For example, if the "twisting" group of our orbifold[16] is an Abelian discrete subgroup (as, for example, in the Z-orbifold), one obtains low-energy rank = 8 groups (subgroups from E_8). If, furthermore, one considers the new class of tori compactifications[47] (or some orbifold versions of these) introduced by Narain, the rank of the low-energy gauge theory may be even larger. However, the existence of a large low-energy gauge group is not phenomenologically welcome, so we will restrict ourselves from now on to low-energy groups contained in E_6, which is the simplest possibility.

Since it seems that one may obtain low-energy groups with rank 4, 5 and 6, it is useful to discuss what extra gauge interactions (if any) one may have at low energies. We are going to confine ourselves to models involving only (at most) extra U(1)'s, since non-Abelian generalizations of these lead in general to bad predictions for $\sin^2\theta_W$ and M_x[39,40].

Extra U(1) interactions are conveniently analyzed in terms of the diagonal generators of the $SU(3)_C \times SU(3)_L \times SU(3)_R$ subgroup of E_6. Apart from the Cartan subalgebra of the $SU(3)_C \times SU(2)_L$ group, one has the extra diagonal generators

$$T_L = \begin{pmatrix} 1 & & \\ & 1 & \\ & & -2 \end{pmatrix}_L \; ; \; T_R = \frac{1}{2}\begin{pmatrix} -2 & & \\ & 1 & \\ & & 1 \end{pmatrix}_R \; ; \; T_N = \frac{1}{2}\begin{pmatrix} 0 & & \\ & 1 & \\ & & -1 \end{pmatrix}_R \tag{39}$$

The quantum numbers of the fields in a <u>27</u> of E_6 with respect to these generators are shown in the Table.

This table also shows our notation[36] for the quark, lepton and Higgs superfields. Let us recall that a <u>27</u> of E_6 contains, apart from a standard family, two coloured triplets $D, \bar{D}$, a right-handed neutrino ν_L^C and some other neutral object N. There is a set of Higgses $H+\bar{H}$ per <u>27</u> generation. One can write for the usual hypercharge generator

$$Y = \sqrt{\frac{3}{5}}\left(\frac{2}{3}T_R - \frac{1}{6}T_L\right) \tag{40}$$

	Q_L	U_L^c	e_L^c	d_L^c	L	H	$\bar{D}$	$\bar{H}$	D	ν_L^c	N
$T_L(\times \frac{1}{2\sqrt{3}})$	-1	0	-2	0	1	1	0	1	2	-2	-2
$L_R(\times \frac{1}{\sqrt{3}})$	0	-1	1	1/2	-1/2	-1/2	1/2	1	0	-1/2	-1/2
T_N	0	0	0	1/2	-1/2	-1/2	-1/2	0	0	-1/2	1/2
$Y(\times \sqrt{\frac{3}{5}})$	1/6	-2/3	1	1/3	-1/2	-1/2	1/3	1/2	-1/3	0	0
$Y'(\times \frac{1}{\sqrt{40}})$	-1	-1	-1	-2	-2	3	3	2	2	0	-5
$Y''(\times \frac{1}{\sqrt{40}})$	-1	-1	-1	3	3	-2	-2	2	2	-5	0
$Y'''(\times \frac{1}{\sqrt{15}})$	-1	-1	-1	1/2	1/2	1/2	1/2	2	2	-5/2	-5/2
$Y'_\perp(\times \frac{1}{4\sqrt{6}})$	-2	-2	-2	4	4	-2	-2	4	4	-8	-2
$Y''_\perp(\times \frac{1}{4\sqrt{6}})$	-2	-2	-2	-2	4	4	4	4	4	-2	-8

Any linear combination of T_L, T_R and T_N orthogonal to Y is, in principle, a candidate for extra U(1) and some candidates are shown in the Table.

The superpotential of the low-energy N = 1 supergravity model may, in general, contain any term present in the $(27)^3$ E_6 coupling, although the Yukawa couplings will not usually obey any E_6 relationship, since the

Wilson-loop breaking alters those[37]. Under these conditions, the most general superpotential originating in the $(27)^3$ coupling will be

$$W = \sum_{\text{families}} h_L H L e_L^c + h_d H Q d_L^c + h_u \bar{H} Q u_L^c + \\ + h_\nu \bar{H} L \nu_L^c + h'_\nu D d_L^c \nu_L^c + \\ + \lambda_2 N H \bar{H} + \lambda_3 N D \bar{D} + \\ + \lambda_L L Q \bar{D} + \lambda'_L e_L^c u_L^c D + \\ + \lambda_B Q Q D + \lambda'_B u_L^c d_L^c \bar{D} \tag{41}$$

It is well known that if all the couplings in (41) are present, several phenomenological disasters may occur: (a) the $D,\bar{D}$ fields mediate fast proton decay unless $\lambda_B = \lambda'_B = 0$ or $\lambda_L = \lambda'_L = 0$ (or those fields are heavier than $\sim 10^{10}$ GeV). These couplings cannot all vanish simultaneously since then D and $\bar{D}$ would be absolutely stable, causing cosmological trouble; (b) there are unacceptable Dirac neutrino masses unless $h_\nu = 0$ (or the ν_L^c's have Majorana masses bigger than $\sim$ 10 TeV); (c) unless essentially only one of the three sets of Higgses $H+\bar{H}$ couples to quarks and leptons, tree-level FCNC's will appear (or the extra two Higgses are heavy enough). All these problems become less severe the smaller the rank of the low-energy group is. For a rank = 6 model, the only possibility is to assume that some symmetries exist which forbid the dangerous couplings, since the gauge symmetries do not allow us to give mass to any of the dangerous particles. In rank = 5 models, one may get rid of some of the unwanted particles (D, $\bar{D}$, H, $\bar{H}$ or ν_L^c's), so that one only needs to forbid some of the couplings. Finally, in rank 4 models (i.e., the standard model gauge group) the possibility exists that no dangerous particles at all remain in the low-energy spectrum. Still, one has to be sure that Weinberg-Salam (WS) doublets remain light. We now consider these three cases in turn.

i) Two Extra Z°'s

In this case the relevant gauge group is

$$G_6 = SU(3)_c \times SU(2)_L \times U(1)_Y \times U(1)^2 \tag{42}$$

and all the 27 states in a fundamental of E_6 remain light, since they are all chiral under the G_6 group. Thus one has to assume there is some symmetry forbidding dangerous couplings. Radiative corrections may break

G_6 down to the standard model inducing non-vanishing vevs $\langle H\rangle \neq 0$, $\langle\bar{H}\rangle \neq 0$, $\langle\tilde{\nu}_L^c\rangle \neq 0$ and $\langle N\rangle \neq 0$. However, several diseases will in general appear. A vev for $\tilde{\nu}_L^c$ will give rise to d_L^c-$\bar{D}$ and L-H mixing through the h_ν and h'_ν couplings. This would ruin the GIM mechanism. Also, one gets induced vevs for the left-handed $\tilde{\nu}_L$ giving rise to lepton number violation. All these problems, along with the fact that a good number of symmetries forbidding certain Yukawa couplings are needed, make this model rather contrived.

ii) One Extra Z°

The low-energy gauge group is in this case

$$G_5 = SU(3)_c \times SU(2)_L \times U(1)_Y \times U(1) \tag{43}$$

The extra U(1) may be any of the linear combinations of T_L, T_R and T_N which commute with the hypercharge. The more general form for it will be[36]

$$Y_{extra} = (\text{constant})\,(T_L + T_R + d\,T_N) \tag{44}$$

Thus there is a one-parameter (d) family of possible extra U(1)'s. There are, however, three cases of special physical interest which correspond to the hypercharges Y', Y" and Y''' in the Table.

Y' hypercharge. It is defined as the one under which the ν_L^c field is inert, $Y'(\nu_L^c) = 0$, and is given by[36]

$$Y' = \frac{1}{2\sqrt{10}}\,(T_L + T_R - 5\,T_N) \tag{45}$$

with the same normalization as the hypercharge. The interest of this U(1) is that a large Majorana mass term $\nu_L^c\nu_L^c$ is not forbidden by this symmetry, and hence this leaves the door open to a solution of the neutrino mass problem mentioned above[11,36]. Some of the mechanisms discussed in Ref. 21 may be the source of these ν_L^c masses. A model like this may be obtained from a rank = 6 model if a field N_R with the quantum numbers of a ν_L^c gets a vev at a large scale. Alternatively, it may be obtained after Wilson-loop breaking of an SO(10) model with fermion content $3 \times (\underline{16} + \underline{10})$. However, the embedding of a standard family is unusual in that $\underline{16} = (Q_L, U_L^c, e_L^c;\ H, \bar{D}, N)$, $\underline{10} = (L, d_L^c;\ \bar{H}, D)$. Thus, N does not behave like a right-handed neutrino (it is harmless) and L, d_L^c are inside a $\underline{10}$

and not inside a 16. This extra Z° model has been studied in detail in Ref. 36.

Y" hypercharge. It is defined as the one under which the N field is inert, Y"(N) = 0, and is given by

$$Y'' = \frac{1}{2\sqrt{10}} (T_L + T_R + 5 T_N) \tag{46}$$

In fact, this is obtained from Y' by changing $T_N \rightarrow -T_N$ and just corresponds to U(1)' after redefining $(d_L^c, L, \nu_L^c) \leftrightarrow (\bar{D}, H, N)$. U(1)" is the U(1) inside SO(10) which commutes with SU(5) and the quark-lepton embedding is the standard one. A model with a Y" hypercharge is only physically distinguishable from a Y' model if you destroy the symmetry $T_N \rightarrow -T_N$ in the low-energy spectrum. This is what is done in the models with an intermediate symmetry-breaking scale ($\langle N \rangle \sim 10^{10}-10^{14}$ GeV) of Refs. 21, 39 and 45. In those models, a Higgs field N breaks the U(1) orthogonal to Y and Y" and at the same time gives masses to the unwanted D, $\bar{D}$, H and $\bar{H}$ fields. In this way we may get rid of the fast proton decay problem. The low-energy superpotential is then just

$$W'' = \sum_{\text{families}} h_L H L e_L^c + h_d H Q d_L^c + h_u \bar{H} Q u_L^c + h_\nu \bar{H} L \nu_L^c + \lambda_2 N H \bar{H} \tag{47}$$

Unfortunately[36], with such a superpotential it is hard to obtain the desired pattern of symmetry breaking $SU(2) \times U(1)^2 \rightarrow U(1)_{e.m.}$. In order to forbid Dirac ν-masses, we have to set $h_\nu = 0$, and then it is hard to understand how $\tilde{\nu}_L^c$ may acquire a negative $(\text{mass})^2$ and radiatively break U(1)". Furthermore, one can see that the soft coupling $mA(\tilde{\bar{H}} L \nu_L^c)$ will induce a vev for the left-handed sneutrino $\tilde{\nu}_L$, leading to lepton number violation. All these problems make this U(1) interaction rather unattractive.

Y''' hypercharge. It is just the U(1) generator[37,41,46] orthogonal to Y and T_N

$$Y''' = \frac{1}{\sqrt{15}} (T_L + T_R) \tag{48}$$

The U(1)''' quantum numbers of the different particles are shown in the Table. This model can be obtained directly from E_6 through Wilson-loop breaking. All the particles in a 27 are chiral under this symmetry, so that the complete 27 states must remain light. Thus, one has to rely on the possible existence of symmetries to forbid the dangerous couplings in the superpotential. Particularly, one has to set the couplings $h_\nu = 0$ to avoid Dirac neutrino masses[41,46]. The right-handed neutrinos are then massless, and this could cause a serious cosmological problem, since six massless neutrinos are probably too much for consistent nucleosynthesis[48].

Apart from these three physically interesting U(1)'s, there is a family of other possible U(1)'s as one varies the parameter "d" in Eq. (44).

$$Y_{extra} = \cos\alpha \, Y' + \sin\alpha \, Y'' \tag{49}$$

The U(1)', U(1)" and U(1)''' above correspond to the values $\alpha = 0$, $\pi/2$, $\pi/4$. The new possibilities in (49) have no special advantage over the Y''' case, so we will not consider them any longer here.

From the above discussion, it seems clear that the least contrived extra-U(1) models are those based on the U(1)' and U(1)''' hypercharges. The first one especially, U(1)', has the advantage of being the only one consistent with a heavy right-handed neutrino, and hence may avoid problems with neutrino masses. On the other hand, for the U(1)''' hypercharge, one has to forbid the h_ν neutrino couplings, and then the right-handed neutrinos are massless. This leads to six massless neutrinos which, as we remarked above, could be incompatible with nucleosynthesis bounds. This possibly makes the U(1)' case more interesting. However, one can make a rather general analysis of the process of symmetry breaking

$$SU(2)_L \times U(1)_Y \times U(1) \longrightarrow U(1)_{e.m.} \tag{50}$$

for an arbitrary extra-U(1) model. This is explained in some detail in Ref. 36. The relevant pieces of the low-energy superpotential regarding the process of symmetry breaking are

$$W_r = h_t \bar{H} Q u_L^c + \lambda_2 N H \bar{H} + \lambda_3 N D \bar{D} \tag{51}$$

where the first term corresponds to the top quark. To simplify the computations, we also assume that only one of the three N-fields couples strongly enough to H, $\bar{H}$, D, $\bar{D}$. Then the scalar potential (along the neutral directions) which is relevant for the symmetry breaking is

$$V = \frac{g_1'^2}{2}\left(X_{\bar{H}}|\bar{H}|^2 + X_H|H|^2 + X_N|N|^2\right)^2 + \frac{(g_1^2+g_2^2)}{8}\left(|H|^2-|\bar{H}|^2\right)^2 +$$
$$+ \lambda_2^2\left(|N|^2|H|^2 + |N|^2|\bar{H}|^2 + |H|^2|\bar{H}|^2\right) + m_N^2|N|^2 + m_H^2|H|^2 + m_{\bar{H}}^2|\bar{H}|^2 +$$
$$+ mA_2\lambda_2(NH\bar{H}) + h.c. \qquad (52)$$

where $X_{\bar{H}}$, X_H and X_N are the extra-U(1) charges of the corresponding fields (see the Table). We have included in the potential the usual soft SUSY-breaking terms, i.e., scalar mass terms and trilinear scalar couplings.

All the parameters in the scalar potential are assumed to be taken at the biggest of the two symmetry-breaking scales, i.e., at $Q = M_{Z^\circ}'$. One also expects M_{Z°' not to be very much larger than M_{Z° (e.g., $M_{Z^\circ}' \lesssim$ 500-600 GeV), otherwise we would need to do some unnatural fine tuning in order to avoid radiative corrections giving an unwanted large contribution to M_{Z°. We have then to compute how the parameters m_H^2, $m_{\bar{H}}^2$, m_N^2, λ_2 and A_2 are renormalized in going from the compactification scale down to low energies[36].

Some intuition into how the double symmetry breaking (50) in this type of model occurs can be obtained from a discussion of the form of the scalar potential (52). Since we obviously want to obtain that $M_{Z^\circ}' \gg M_{Z^\circ}$ (because of neutral current constraints), the symmetry breaking should appear as follows. The parameter m_N^2 becomes negative not much above the weak scale, and a vev for N is induced. Upon renormalization one gets $m_{\bar{H}}^2 \ll m_H^2$ (although $m_{\bar{H}}^2$ does not need to be negative) and the extra-U(1) D^2 terms give some (usually negative) contribution to the H and $\bar{H}$ effective masses. This, along with the trilinear term, induces vevs for H and $\bar{H}$. The vev for $\bar{H}$ is usually bigger than that for H because, as we remarked above, renormalization effects yield $m_{\bar{H}}^2 \ll m_H^2$ (the contribution of the D'^2 term is not so important in this respect). It is important to remark that the form of the renormalization group equations for the soft SUSY-breaking terms is such that one may obtain this small hierarchical ($M_{Z^\circ}' > M_{Z^\circ}$) symmetry-breaking for wide ranges of the parameters. This has been checked for specific models in Refs. 36 and 46. In particular, a model based on the U(1)' hypercharge was recently studied numerically[36]. One obtains consistent $SU(2) \times U(1)^2 \to U(1)_{e.m.}$ symmetry breaking for wide ranges of the parameters λ_2, λ_3, A, m_t, m and M. An interesting point to remark is that the suggested boundary conditions in Eq. (37) (i.e., $m/M \ll 1$, $|A| \ll 1$) are consistent with the desired pattern of symmetry breaking, but only for top-quark masses $m_t \lesssim 70$ GeV. The lightest supersymmetric particle (LSP) is usually a Higgsino or a "singlino" $\tilde{N}$ (unless the top quark is very heavy). As in all extra-U(1) models, there are plenty of new particles (D, $\bar{D}$, $\tilde{H}^{\pm}$, $H^{\pm}$, etc.), which could be detected at present accelerators. For a discussion of the typical spectra in these models, see Ref. 36.

Although the models with an extra U(1) are technically viable, they do, however, look rather artificial. To avoid all phenomenological problems, extra-U(1) models require a number of lucky coincides to happen:

i) some Yukawa couplings are absent in order to avoid fast proton decay;

ii) other B-violating couplings are present in order to avoid the absolute stability of the D-fields;

iii) the extra $SU(2)_L$ doublets in the models have negligible couplings to quarks in order to avoid FCNC's;

iv) ν_L-ν_R-$\bar{H}$ Yukawas are absent in order to avoid unobserved neutrino masses [this is, in principle, solved for the U(1)' hypercharge].

There are, in fact, further conditions coming from the fact that the new particles of the first two generations should be heavy enough also to suppress other sources of FCNC. In this situation, one must admit that the models with an extra U(1) are rather contrived. This makes more desirable the obtention of compactification schemes leading directly to the standard model.

iii) No Extra Z°[36]

As we discussed above, one can have compactification schemes which lead to the standard model gauge group SU(3) × SU(2) × U(1) and no additional gauge bosons. This is probably the most interesting possibility, since in this case there is no need to have at low energies any of the extra new particles (D, $\bar{D}$, $H^{\pm}$, ν_R's, etc.) which lead to problems in the models with an extra U(1). The minimal model with gauge group SU(3) × SU(2) × U(1) should contain, apart from the standard quark-lepton superfields, the Higgs doublets H and $\bar{H}$ and, possibly, a singlet N. In the absence of a singlet N°, the resulting model would be precisely the "minimal low-energy supergravity model" studied in the last few years[9,49-51]. A mass coupling of the form $\varepsilon H\bar{H}$ has then to be present in the superpotential to avoid the appearance of an axion and to induce a $\langle H \rangle \neq 0$. Although there are no obvious sources for a term $\varepsilon H\bar{H}$ at the tree level, it could appear through radiative corrections[49]. Then one expects ε to be a small parameter, as would be the corresponding soft term in the scalar potential $\mu_3^2 \sim \varepsilon m$. This leads in turn to a small value for $\langle H \rangle \lll \langle \bar{H} \rangle$ and to a very light "chargino" in the spectrum[9,49] with mass $m_{\tilde{W}} \sim v\bar{v}/M$. If ε is very small [e.g., $\varepsilon \sim (\alpha/\pi)m$], this light chargino may even be incompatible with experiment. As we have seen above, the presence of a singlet $N°$ coupling to $H\bar{H}$ is a very natural feature in superstring-inspired models and it is likely to appear also in the minimal $SU(3) \times SU(2) \times U(1)$ case. In this more general case, $\langle H \rangle$ can be as large as $\langle \bar{H} \rangle$ and there is no danger associated with a possible light chargino. The relevant Higgs superpotential in the general case is

$$W_H = \lambda_2 N H \bar{H} \tag{53}$$

where we define $N \equiv N° + \varepsilon/\lambda_2$. The scalar potential (along the neutral direction) in this model is

$$V = \frac{(g_1^2+g_2^2)}{8}(|H|^2-|\bar{H}|^2)^2 + \lambda_2^2 |H|^2|\bar{H}|^2 + \lambda_2^2 |N|^2 (|H|^2+|\bar{H}|^2) + $$
$$+ m_N^2 |N|^2 + m_H^2 |H|^2 + m_{\bar{H}}^2 |\bar{H}|^2 + m A_2 \lambda_2 (NH\bar{H}) + h.c. \tag{54}$$

Notice that Eq. (54) implicitly assumes that the soft bilinear coupling $B \simeq A_2$. We assume this in order to simplify the computations and also because the term missing in Eq. (54) $[(B-A_2)\varepsilon H\bar{H}]$ is expected to be small for ε small. It also vanishes in the interesting case in which gaugino masses are the only source of supersymmetry breaking ($A = B = 0$ and A_2, B renormalize in a very similar way). Notice, however, that a non-vanishing ε is required in order to give a mass to a would-be Goldstone boson which appears when $N°$ gets a vev.

The scalar potential above looks similar to that of "minimal low-energy supergravity" (MLES), but it has several important differences due to the existence of the $\lambda_2 NH\bar{H}$ coupling. In particular, it has no flat direction ($\langle H \rangle = \langle \bar{H} \rangle$) for $m_H^2 = m_{\bar{H}}^2 = \mu_3^2$ as happened in the MLES case. This was important in that case because one could have radiative $SU(2) \times U(1)$ breaking with a low t-quark mass[49-51], since a large difference between m_H^2 and $m_{\bar{H}}^2$ was not needed. In the present case, one needs a negative $m_{\bar{H}}^2$ to be generated since there is no flat direction, and an extra positive term in the potential $(\lambda_2^2|H|^2|\bar{H}|^2)$ has to be overwhelmed in order to obtain a stable minimum. This has the important consequence that a relatively heavy top quark is required to trigger $SU(2) \times U(1)$ breaking, as happened in the first version of the MLES model[9].

Solving the renormalization group equations for m_N^2, m_H^2, $m_{\bar{H}}^2$, A_2 and λ_2, one can analyze numerically the constraints in the parameters (essentially A, m, M, m_t and λ_2) obtained by imposing the appropriate SU(2) × U(1) breaking. One only finds symmetry breaking for a top-quark mass[36]

$$m_t \gtrsim 70 \text{ GeV}$$

just because of the absence of an approximate flat direction $H = \bar{H}$ in the potential. Radiative corrections generate symmetry breaking through a negative $m^2_{\bar{H}}$ which in turn requires $|\langle\bar{H}\rangle| > |\langle H\rangle|$. Also in this case the suggested boundary conditions in Eq. (37) (i.e., $m/|M| \ll 1$, $|A| \ll 1$) are consistent with the desired minimum of the potential. However, this is only the case for $m_t \lesssim 100$ GeV (particularly for m_t = 70-80 GeV).

The typical SUSY spectra in this "minimal stringy model" is in general lighter[36] than in the extra-U(1) models, since in the latter the overall SUSY-breaking scale is $\sim M'_{Z^\circ}$. For example, the charged sleptons may be relatively light (e.g., 25-60 GeV). The sneutrino $\tilde{\nu}_L$ is often the lightest supersymmetric particle and the decays $W^\pm \to \tilde{e}\tilde{\nu}$ and $Z^\circ \to \tilde{\ell}^+\tilde{\ell}^-$, $\tilde{\nu}\tilde{\bar{\nu}}$ are then possible. Sometimes the LSP is a Higgsino-singlino state ($\tilde{\bar{H}}^\circ$-$\tilde{N}$), but it is rarely the photino. However, unlike the extra-Z° case, in this minimal model there is not a clear signature from the stringy origin of the low-energy Lagrangian (except for, to a small extent, the singlet N). Stringiness could be difficult to reveal.

To conclude, let us remark that the heterotic $E_8 \times E_8$ superstring offers us the possibility of embedding the "low-energy supergravity models" developed in the last few years inside a complete unification scheme which includes gravity. It contains a possible built-in mechanism for residual supersymmetry breaking (gaugino condensation in the "hidden sector"). Although the physics of the "hidden sector" and supersymmetry breaking is not yet clarified, one expects in the "observable" sector some truncated version of an E_6 GUT. The gauge structure of the low-energy Lagrangian may have two, one or no extra Z°'s. Whereas the case of two extra U(1)'s does not seem viable, models with an extra Z°' seem to be consistent with the required symmetry breaking structure. However, they have a variety of phenomenological problems which can only be avoided if one assumes that some Yukawa couplings are absent but others are present. From the low-energy point of view, the absence of any extra Z°' would be rather more simple. This makes specially important the study of compactification schemes leading directly to the standard model gauge structure SU(3) × SU(2) × U(1).

ACKNOWLEDGEMENTS

I acknowledge discussions with F. Del Aguila, J. Mas, C. Muñoz and H.P. Nilles.

REFERENCES

1. J.H. Schwarz, Physics Reports 89:223 (1982);
   M.B. Green, Surveys in High Energy Phys. 3:127 (1982).
2. M.B. Green and J.H. Schwarz, Phys. Lett. 149B:117 (1984); ibid. 151B:21 (1985).
3. M.B. Green, J.H. Schwarz and P. West, Nucl. Phys. B254:327 (1985).
4. D. Gross, J. Harvey, E. Martinec and R. Rohm, Phys. Rev. Lett. 55: 502 (1985); Nucl. Phys. B256:253 (1985).
5. P. Candelas, G. Horowitz, A. Strominger and E. Witten, Nucl. Phys. B258:46 (1985).
6. E. Cremmer, S. Ferrara, L. Girardello and A. Van Proeyen, Nucl. Phys. B212:413 (1983).
7. For reviews, see, e.g.,
   H.P. Nilles, Physics Reports 110:1 (1984);
   L.E. Ibáñez, Proceedings of the 12th International Winter Meeting on Fundamental Physics, J.E.N. Madrid, ed., (1984);
   L. Hall, Proceedings of the Winter School in Theoretical Physics, Mahabaleshwar, India (1984), Springer Verlag Lecture Notes in Physics 208:197 (1984).
8. L.E. Ibáñez, Phys. Lett. 118B:73 (1982);
   H.P. Nilles, Phys. Lett. 115B:193 (1982); Nucl. Phys. B217:366 (1983);
   P. Nath, R. Arnowitt and A.H. Chamseddine, Phys. Rev. Lett. 49:970 (1982);
   R. Barbieri, S. Ferrara and C.A. Savoy, Phys. Lett. 119B:343 (1982).
9. L.E. Ibáñez, Nucl. Phys. B218:54 (1983);
   L.E. Ibáñez and C. Lopez, Phys. Lett. 126B:54 (1983);
   L. Alvarez-Gaumé, J. Polchinski and M. Weise, Nucl. Phys. B221:495 (1983);
   J. Ellis, J. Hagelin, D.V. Nanopoulos and K. Tamvakis, Phys. Lett. 125B:275 (1983).
10. L.E. Ibáñez and G.G. Ross, Phys. Lett. 110B:215 (1982);
    L. Alvarez-Gaumé, M. Claudson and M. Wise, Nucl. Phys. B207:16 (1982);
    B. Ovrut and C. Nappi, Phys. Lett. B113:65 (1982).
11. L.E. Ibáñez, Phenomenology from superstrings, CERN preprint TH.4308/85 (1985), to appear in the proceedings of the 1st Torino Meeting on Superunification and Extra Dimensions, World Scientific, Singapore.
12. H.P. Nilles, CERN preprint TH.4444%86 (1986).
13. G. Segrè, Low-energy physics from superstrings, University of Pennsylvania preprint (1985);
    J. Ellis, CERN preprints TH.4255/85 (1985); TH.4439/86 (1986).
14. A. Chamseddine, Nucl. Phys. B185:403 (1981);
    F. Berghoeff, M. de Roo, B. de Witt and P. van Nieuwenhuizen, Nucl. Phys. B195:97 (1982);
    G. Chapline and N. Manton, Phys. Lett. 120B:105 (1983).
15. E. Witten, Phys. Lett. 155B:151 (1985).
16. L. Dixon, J. Harvey, C. Vafa and E. Witten, Nucl. Phys. B261:651 (1985), and Princeton preprint (1986).
17. R. Greene, K. Kirklin, P. Miron and G. Ross, A three-generation superstring model, parts I and II, Oxford University preprints (1986).
18. E. Witten, Phys. Lett. 153B:243 (1985).
19. E. Witten, New issues in manifolds of SU(3) holonomy, Princeton preprint (1985).
20. X. Wen and E. Witten, Phys. Lett. 166B:397 (1986);
    M. Dine and N. Seiberg, String theory and the strong CP problem, IAS-Princeton preprint (1986).
21. J.-P. Derendinger, L.E. Ibáñez and H.P. Nilles, Nucl. Phys. B267:365 (1986).

22. K. Choi and J. Kim, Phys. Lett. 165B:71 (1985).
23. C. Burgess, A. Font and F. Quevedo, Texas preprint UTTG-31-85 (1985).
24. M. Dine, R. Rohm, N. Seiberg and E. Witten, Phys. Lett. 156B:55 (1985).
25. E. Cremmer, S. Ferrara, C. Kounnas and D.V. Nanopoulos, Phys. Lett. 133B:61 (1983);
J. Ellis, A. Lahanas, D.V. Nanopoulos and K. Tamvakis, Phys. Lett. 134B:429 (1984);
J. Ellis, C. Kounnas and D.V. Nanopoulos, Nucl. Phys. B241:406 (1984); B274:373 (1984).
26. J. Breit, B. Ovrut and G. Segrè, Phys. Lett. 162B:303 (1985);
P. Binétruy and M.K. Gaillard, Phys. Lett. 168B:347 (1986).
27. Y.J. Ahn, J. Breit and G. Segrè, Pennsylvania preprint (1985).
28. M. Quiros, CERN preprint TH.4363/86 (1986).
29. L.E. Ibáñez and H.P. Nilles, Phys. Lett. 169B:354 (1986).
30. M. Dine, N. Seiberg, X. Wen and E. Witten, Non-perturbative effects on the string world sheet, Princeton preprint (1986);
J. Ellis, C. Gomez, D.V. Nanopoulos and M. Quiros, CERN preprint TH.4395/86 (1986).
31. D. Gross and E. Witten, Superstring modifications of Einstein's equations, Princeton preprint (1986);
Y. Kikuchi, C. Marzban and Y. Ng, North Carolina University preprint IFP-272-UNC (1986).
32. S. Cecotti, S. Ferrara, L. Girardello and M. Porrati, CERN preprint TH.4253/85 (1985);
S. Cecotti, S. Ferrara, L. Girardello, A. Pasquinucci and M. Porrati UCLA preprint 85-TEP 24 (1985).
33. J.-P. Derendinger, L.E. Ibañez and H.P. Nilles, Phys. Lett. 155B:65 (1985).
34. S. Ferrara, L. Girardello and H.P. Nilles, Phys. Lett. 125B:457 (1983).
35. R. Nepomechie, Y. Wu and A. Zee, Washington preprint 40048-02 P5 (1985).
36. L.E. Ibáñez and J. Mas, CERN preprint TH.4426/86 (1986).
37. E. Witten, Nucl. Phys. B258:75 (1985).
38. J. Breit, B. Ovrut and G. Segrè, Phys. Lett. 158B:33 (1985).
39. M. Dine, V. Kaplunovsky, M. Mangano, C. Nappi and N. Seiberg, Nucl. Phys. B259:549 (1985).
40. S. Cecotti, J.-P. Derendinger, S. Ferrara, L. Girardello and M. Roncadelli, Phys. Lett. 156B:318 (1985).
41. E. Cohen, J. Ellis, K. Enqvist and D.V. Nanopoulos, Phys. Lett. 165B:76 (1985).
42. S.M. Barr, Phys. Rev. Lett. 55:2778 (1985).
43. M. Drees, N. Falck and M. Glück, Dortmund preprint 85/25 (1985).
44. P. Binétruy, S. Dawson, I. Hinchliffe and M. Sher, LBL preprint 203173 (1985).
45. F. Del Aguila, G. Blair, M. Daniel and G.G. Ross, Nucl. Phys. B272:413 (1986).
46. J. Ellis, K. Enqvist, D.V. Nanopoulos and F. Zwirner, CERN preprint TH.4323/85 (1985).
47. K.S. Narain, Rutherford preprint RAL-85-097 (1985).
48. J. Ellis, K. Enqvist, D.V. Nanopoulos and S. Sarkar, Phys. Lett. 167B:457 (1986);
G. Steigman, K. Olive, D. Schramm and M. Turner, Minnesota preprint UMN-TH-562 (1986).
49. L.E. Ibáñez and C. Lopez, Nucl. Phys. B233:511 (1984);
L.E. Ibáñez, C. Lopez and C. Muñoz, Nucl. Phys. B256:218 (1985).
50. S. Jones and G. Ross, Phys. Lett. 155B:69 (1984).
51. C. Kounnas, A. Lahanas, D.V. Nanopoulos and M. Quiros, Nucl. Phys. B236:438 (1984).

YUKAWA COUPLINGS BETWEEN (2,1)-FORMS

P. Candelas

Theory Group
Department of Physics
The University of Texas
Austin, Texas 78712
and
Institute for Theoretical Physics
Ellison Hall
The University of California
Santa Barbara, CA 93106

Abstract

The compactification of superstrings leads to an effective field theory for which the space-time manifold is the product of a four dimensional Minkowski space with a six dimensional Calabi-Yau space. The particles that are massless in the four dimensional world correspond to differential forms of type (1,1) and of type (2,1) on the Calabi-Yau space. The Yukawa couplings between the families correspond to certain integrals involving three differential forms. For an important class of Calabi-Yau manifolds, which includes the cases for which the manifold may be realized as a complete intersection of polynomial equations in a projective space, the families correspond to (2,1)-forms. The relation between (2,1)-forms and the geometrical deformations of the Calabi-Yau space is explained and it is shown, for those cases for which the manifold may be realized as the complete intersection of polynomial equations in a single projective space or for many cases when the manifold may be realized as the transverse intersection of polynomial equations in a product of projective spaces, that the calculation of the Yukawa coupling reduces to a purely algebraic problem involving the defining polynomials. The generalization of this process is presented for a general Calabi-Yau manifold.

Supported in part by Robert A. Welch Foundation and NSF Grants PHY 8503890, 8605978, and PHY 8217853 as supplemented by NASA.

1. Introduction

In the low energy regime superstring theory is believed to reduce to an effective ten-dimensional field theory propagating on a spacetime manifold which is the product of a four dimensional Minkowski space with a Calabi-Yau space, i.e., a compact Kähler manifold of vanishing first Chern class and complex dimension three. The massless families and antifamilies correspond to harmonic (1,1) and (2,1) forms of the internal manifold [1]. The Yukawa couplings arise as integrals of triples of harmonic forms[2]. The precise form of the integral depends on whether the three zero modes correspond to (1,1)-forms or (2,1)-forms. If they correspond to three (1,1) forms a,b,c, say, then the coupling is

$$\kappa = \int_M a \wedge b \wedge c. \tag{1.1}$$

As has been pointed out by Strominger this leads, in this case, to an interesting topological interpretation for the coupling. Through deRham cohomology each of the two-forms is associated with a four-surface. In a six (real) dimensional space three four-surfaces meet in some number of points. There exists a basis in which $\kappa(a,b,c)$ is the number of points of intersection of these surfaces. Couplings of this type have been discussed for a number of cases by Strominger [3] and it is not our purpose to dwell further on them here, rather this article is devoted to a discussion of the couplings for the case that the zero modes are related to (2,1)-forms which is the case for many manifolds of interest.

One might reasonably enquire how it is that three 3-forms can be integrated over a six-dimensional manifold. The way in which this comes about depends on the special properties of Calabi-Yau spaces. It is a crucial property of these spaces that there exists a nowhere zero holomorphic 3-form. That is a nowhere zero 3-form

$$\Omega = \frac{1}{3!} \Omega_{\mu\nu\rho}(x) dx^\mu \wedge dx^\nu \wedge dx^\rho \tag{1.2}$$

which has $\Omega_{\mu\nu\rho}$ as its only nonzero components[f1] and where the

f1. Notation: We shall use x^μ, $\mu = 1,2,3$ to denote the coordinates of M, Latin indices m,n.. will refer to a real coordinate basis and run over six values

$\Omega_{\mu\nu\rho}$ are moreover holomorphic functions of the coordinates. Given Ω it is possible to define for every (2,1)-form

$$\omega = \frac{1}{2}\,\omega_{\mu\nu\bar{\rho}}\, dx^{\mu}{}_{\wedge} dx^{\nu}{}_{\wedge} dx^{\bar{\rho}} \tag{1.3}$$

an equivalent object

$$\omega^{\mu} = \frac{1}{2}\,\bar{\Omega}^{\mu\rho\sigma}\omega_{\rho\sigma\bar{\nu}}\, dx^{\bar{\nu}}. \tag{1.4}$$

ω^{μ} is a (0,1)-form with a holomorphic vector index. It may be regarded as a (0,1)-form that takes values in the holomorphic tangent bundle T. It is straightforward to show that the original (2,1)-form $\omega_{\rho\sigma\bar{\nu}}$ is harmonic with respect to the exterior derivative d if an only if ω^{μ} is harmonic with respect to $\bar{\partial}$, the antiholomorphic part of d. The Yukawa coupling involves three such forms a^{μ}, b^{ν}, c^{ρ}, say, and takes the explicit form [2]

$$\kappa = \int \Omega_{\wedge} a^{\mu}{}_{\wedge} b^{\nu}{}_{\wedge} c^{\rho}\Omega_{\mu\nu\rho} \, . \tag{1.5}$$

The purpose of this article is to discuss the significance of the coupling in relation to deformations of the complex structure of M and to evaluate it for a number of cases of interest. We shall see that for the case that M may be realized as a complete intersection of polynomial equations in a projective space, and in a number of other cases, the evaluation of κ reduces to a purely algebraic problem.

Much of what follows is material that is already present in the literature, though some is new. The discussion of the deformation of complex structure draws heavily on the classic work of Kodaira and Spencer [4,5]. The relationship between the holomorphic three-form and the defining polynomials was known to the mathematicians but I learnt it from the articles by Witten [6] and Strominger and Witten [7]. I review these relations here in order to make the account reasonably self contained. Witten [6] also showed that the Yukawa couplings are strongly constrained by the pseudosymmetries of the manifold. Finally the fact that statements about the cohomology classes of algebraic varieties translate into algebraic statements is a fact I learnt from conversations with Bott.

The process of evaluating κ breaks naturally into two parts. The first is to understand the geometrical meaning of the integral and the second is its explicit evaluation.

2. The geometrical meaning of $H^{(0,1)}_{\bar{\partial}}(M,T)$

Let us suppose initially that M is given as a hypersurface in an (N+3) dimensional projective space P_{N+3} defined as the transverse intersection of N polynomials

$$p^{\alpha}(z) = 0 \quad , \quad \alpha = 1,\cdots,N \tag{2.1}$$

where z^A, $A = 1,\cdots,N+4$ are homogeneous coordinates for the projective space and the degrees of the polynomials p^{α} are so chosen that the hypersurface has vanishing first Chern class. If the hypersurface has vanishing first Chern class then, in virtue of Yau's theorem, we may take it to be endowed with a Ricci-flat metric g_{mn}

$$R_{mn}(g) = 0 \tag{2.2}$$

Consider an infinitesimal variation of M which takes us to another nearby Ricci-flat manifold with metric $g_{mn} + h_{mn}$

$$R_{mn}(g+h) = 0 \tag{2.3}$$

Linearization of this equation and imposition of the coordinate condition $\nabla^m h_{mn} = 0$ leads to a differential equation for h_{mn}, the Lichnerowicz equation,

$$\Delta_L h_{mn} \equiv h_{mn} + 2R_m{}^r{}_n{}^s h_{rs} = 0 . \tag{2.4}$$

The first point to note is that, owing to the special proprties of Kähler spaces the Lichnerowicz operator Δ_L does not mix modes of different type i.e., the $h_{\mu\bar{\nu}}$, $h_{\mu\nu}$, $h_{\bar{\mu}\bar{\nu}}$ parts of h_{mn} separately satisfy the equation.

First let us examine the mixed part of h_{mn}. Form from $h_{\mu\bar{\nu}}$ a (1,1)-form

$$k = ih_{\mu\bar{\nu}}dx^{\mu}{}_{\wedge}dx^{\bar{\nu}}. \tag{2.5}$$

It turns out that, in virtue of the fact that $h_{\mu\bar{\nu}}$ satisfies the Lichnerowicz equation, k satisfies the Hodge-de Rham equation

$$(dd^{+} + d^{+}d)k = 0 . \tag{2.6}$$

In other words k is harmonic. For the case we are considering the embedding space consists of a single projective space. For this case it is known that $b_{11} = 1$. The unique (1,1)-form is the Kähler-form

$$J = ig_{\mu\bar{\nu}}\, dx^{\mu}{}_{\wedge}dx^{\bar{\nu}} . \tag{2.7}$$

Thus the only metric perturbation of this type is a constant multiple of the metric itself. This perturbation corresponds simply to a change in the overall scale of M.

Of greater interest for our present purposes are the pure modes $h_{\mu\nu}$ and $h_{\bar{\mu}\bar{\nu}}$ which are, of course, related by complex conjugation so that it suffices to consider $h_{\bar{\mu}\bar{\nu}}$. Form from this a (0,1)-form that takes values in T.

$$h^{\mu} = h^{\mu}{}_{\bar{\nu}}dx^{\bar{\nu}} . \tag{2.8}$$

It may be shown that $h_{\bar{\mu}\bar{\nu}}$ satisfies the Lichnerowicz equation if and only if h^{μ} satisfies the Hodge-de Rham equation corresponding to the operator $\bar{\partial}$

$$(\bar{\partial}\bar{\partial}^{+} + \bar{\partial}^{+}\bar{\partial})h^{\mu} = 0. \tag{2.9}$$

In other words $h_{\bar{\mu}\bar{\nu}}$ satisfies the Lichnerowicz equation if and only if h^{μ} is harmonic with respect to $\bar{\partial}$. The elements of the cohomology group

$H_{\bar{\partial}}^{(0,1)}(M,T)$ are in one-one correspondence with the Ricci-flat perturbations of M. These perturbations are also in one-one correspondence with the deformations of the complex structure of M. It can happen that two manifolds are diffeomorphic, that is they are the same as real manifolds, and yet they are different as complex manifolds because there is no biholomorphic map between them. The simplest example of this phenomenon concerns two tori

$$T_1 = \{(x,y)|(x,y) \simeq (x+1,y) \simeq (x,y+1)\}$$

$$T_2 = \{(\xi,\eta)|(\xi,\eta) \simeq (\xi+1,\eta) \simeq (\xi,\eta+2)\} . \tag{2.10}$$

These manifolds are diffeomorphic as real manifolds since

$$(\xi,\eta) = (x,2y) \tag{2.11}$$

defines a C^∞ map between them. If however we set $z = x+iy$ and $\zeta = \xi+i\eta$ then ζ is not a holomorphic function of z

$$\zeta = \frac{3}{2} z - \frac{1}{2} \bar{z} \tag{2.12}$$

and it is not possible to eliminate the $\bar{z}$.

The various Ricci-flat manifolds that were discussed previously are related in just such a way. They may be smoothly deformed into each other and are in fact diffeomorphic. They are also all Kähler manifolds, being analytic submanifolds of P_{N+3}. However they have different complex structures. We know this because a Kähler manifold has a hermitean metric with only mixed components whereas the metric perturbation h_{mn} has pure parts $h_{\mu\nu}$ and $h_{\bar{\mu}\bar{\nu}}$. The complex structures of the two manifolds are different because although the pure part of h_{mn} can be removed by a transformation of coordinates it cannot be removed by a holomorphic coordinate transformation. Under a holomorphic transformation

$$x^\mu \rightarrow \xi^\mu = x^\mu + \varepsilon^\mu(x) \tag{2.13}$$

the metric perturbation transforms according to the familiar rule

$$h_{mn} \to \tilde{h}_{mn} = h_{mn} - \varepsilon^r{}_{,m} g_{rn} - \varepsilon^r{}_{,n} g_{mr}, \qquad (2.14)$$

if ε^μ is holomorphic then $h_{\mu\nu}$ and $h_{\bar{\mu}\bar{\nu}}$ are invariant.

For a space realized as a complete intersection of polynomials p^α, $\alpha = 1,\cdot\cdot,N$ in P_{N+3} the deformations of the complex structure can be simply realized as deformations of the defining polynomials p^α. Consider infinitesimal changes $p^\alpha \to p^\alpha + q^\alpha$ in the defining polynomials

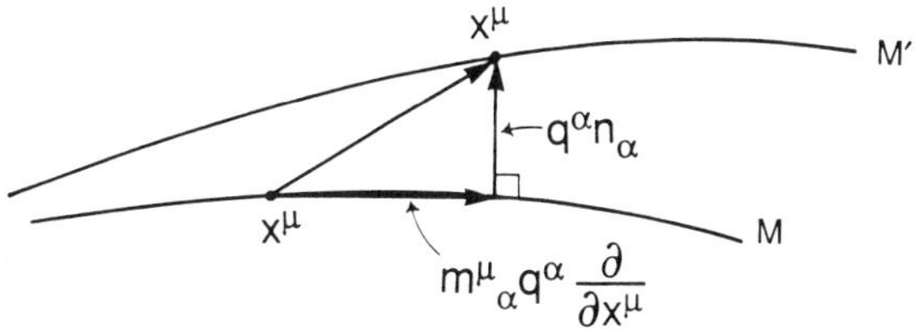

Figure 1

We may study the deformation as in General Relativity where it is common to decompose deformations into lapses and shifts [7]. The situation here is analogous except that there are now several "times" labelled by α. On M we choose a holomorphic atlas, i.e., coordinates $x^\mu{}_j$ defined in patches U_j and we choose a similar atlas for the deformed manifold M′ also labelling the points by $x^\mu{}_j$ in such a way that the point labelled by $x^\mu{}_j$ on M′ approaches that labelled by $x^\mu{}_j$ on M as the $q^\alpha \to 0$. From now on we shall largely suppress the j that labels the different coordinate patches. We join the point labelled by x^μ on the original manifold to that with same label x^μ on the perturbed manifold by a vector and decompose this vector into a normal lapse and a tangential shift which, being linear in the deformation q^α, takes the form

$$m^\mu{}_\alpha q^\alpha \frac{\partial}{\partial x^\mu} .$$

Associated with the embedding of M in the ambient space is an important geometrical tensor the extrinsic curvature or second fundamental form. Since the extrinsic curvature plays a pivotal role in what follows we shall pause briefly to recall how it arises in the study of embedded hypersurfaces.

It proves useful to choose a reciprocal set of basis vectors and one forms

$$e_\mu = \frac{\partial}{\partial x^\mu} \quad ; \quad e^\mu = dx^\mu + m^\mu{}_\alpha n^\alpha$$

$$n_\alpha = \frac{\partial}{\partial p^\alpha} - m^\mu{}_\alpha e_\mu \quad ; \quad n^\alpha = dp^\alpha. \tag{2.15}$$

With respect to this basis the lapse vector in Fig. 2 is $q^\alpha n_\alpha$.

The extrinsic curvature $\chi_{mn}{}^a$ is defined by the relation

$$\hat{\nabla}_m e_n = \nabla_m e_n + \chi_{mn}{}^a n_a \tag{2.16}$$

where a "^" is used to distinguish quantities that refer to the embedding space while quantities without the "^" are intrinsic to the surface.

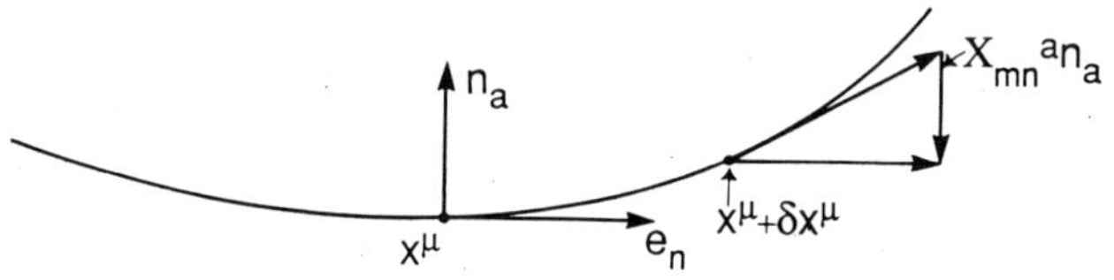

Figure 2

The difference between parallel propagation in the embedding space and parallel propagation in the hypersurface is illustrated by Fig. 3. This difference which is due to the rotation of the tangent vectors e_n measures the curvature of the surface. Alternatively one may concentrate on the rotation of the normal vectors n_a. In terms of frame rotation coefficients we have

$$\hat{\nabla}_m n_a = \hat{\Gamma}_{ma}{}^r e_r + \hat{\Gamma}_{ma}{}^b n_b \tag{2.17}$$

Now

$$\hat{\Gamma}_{mas} = \hat{g}(e_s, \hat{\nabla}_m n_a) = - \hat{g}(\hat{\nabla}_m e_s, n_a) = - \chi_{msa}. \tag{2.18}$$

Thus

$$\hat{\nabla}_m n_a = - \chi_m{}^r{}_a e_r + \hat{\Gamma}_{ma}{}^b n_b \tag{2.19}$$

which provides an alternative definition of $\chi_{mn}{}^a$.

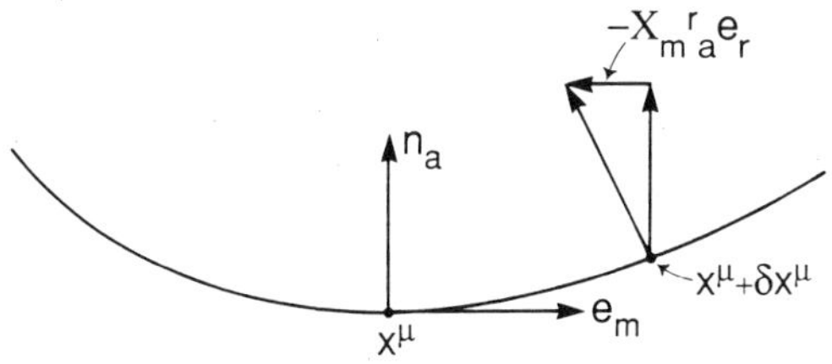

Figure 3. An alternative but equivalent way to measure the curvature of a hypersurface is to parallely propagate the normals n_a from a point x^μ to a nearby point $x^\mu+\delta x^\mu$. There it will differ from the preexisting normals by an amount proportional to the extrinsic curvature.

It is straightforward to show that for the case we are considering, of a hypersurface embedded in a Kähler manifold by holomorphic equations, that the extrinsic curvature has

$$\chi_{\mu\nu}{}^{\alpha} \quad \text{and} \quad \chi_{\bar{\mu}\bar{\nu}}{}^{\bar{\alpha}} \tag{2.20}$$

as its only nonzero components.

The fact that the extrinsic curvature relates the rotation of the tangent vectors to that of the normal vectors is important in the following. From equation (2.16) we may obtain an explicit expression for the extrinsic curvature in terms of the defining polynomials p^α.

$$\chi_{\mu\nu}{}^{\alpha} = \frac{\partial z^i}{\partial x^\mu}\frac{\partial z^j}{\partial x^\nu}\left(\frac{\partial^2 p^\alpha}{\partial z^i \partial z^j} - \hat{\Gamma}_{ij}{}^k \frac{\partial p^\alpha}{\partial z^k}\right) \tag{2.21}$$

where in this equation i, j, k run over N+3 values corresponding to choosing one of the z^A to be unity. If we endow P_{N+3} with the Fubini-Study metric

$$\hat{g}_{j\bar{k}} = \frac{1}{\sigma}\left(\delta_{j\bar{k}} - \frac{z_j z_{\bar{k}}}{\sigma}\right) \tag{2.22}$$

with

$$z_j = \delta_{j\bar{k}} z^{\bar{k}} \quad , \quad z_{\bar{k}} = \delta_{j\bar{k}} z^j \tag{2.23}$$

and

$$\sigma = \sum_{A=1}^{N+4} |z^A|^2 \ , \tag{2.24}$$

then it is easy to see that the term involving the Christoffel symbol does not contribute to $\chi_{\mu\nu}{}^{\alpha}$, moreover we may use the fact that one of the z's is a constant to write $\chi_{\mu\nu}{}^{\alpha}$ in the more symmetric form

$$\chi_{\mu\nu}{}^{\alpha} = \frac{\partial z^A}{\partial x^{\mu}} \frac{\partial z^B}{\partial x^{\nu}} \frac{\partial^2 p^{\alpha}}{\partial z^A \partial z^B} \ . \tag{2.25}$$

The extrinsic curvature may also be related, via equation (2.19), to the shift $m^{\mu}{}_{\alpha}$

$$\chi_{\bar{\lambda}}{}^{\mu}{}_{\alpha} = \frac{\partial}{\partial x^{\bar{\lambda}}} (m^{\mu}{}_{\alpha}) \ . \tag{2.26}$$

This last equation enables us to relate explicitly the deformations q^{α} of the defining polynomials to the elements of $H^{(0,1)}_{\bar{\partial}}(M,T)$.
Set

$$a^{\mu} = \chi_{\bar{\lambda}}{}^{\mu}{}_{\alpha} q^{\alpha} dx^{\bar{\lambda}} . \qquad (2.27)$$

a^{μ} is a (0,1)-form valued in T. Moreover by (2.26) we have

$$a^{\mu} = \bar{\partial}(m^{\mu}{}_{\alpha} q^{\alpha}) . \qquad (2.28)$$

Since $\bar{\partial}^2 = 0$ we see that a^{μ} is closed

$$\bar{\partial} a^{\mu} = 0. \qquad (2.29)$$

Naively one might suppose that eqn. (2.28) shows that a^{μ} is also exact, however this is in general false owing to the fact that our construction is tied to a particular coordinate patch. Since M is compact it cannot be covered by a single patch and there is no reason to suppose that where two patches overlap that the corresponding shifts transform correctly. In other words the shift $m^{\mu}{}_{\alpha}$ does not, in general, extend to a non-singular tensor field. On the other hand $\chi_{\mu\nu}{}^{\alpha}$ is a bona fide geometrical tensor defined everywhere on M since M is assumed smooth. Thus eqn. (2.27) defines a^{μ} globally.

It is of interest to enquire under what circumstances a^{μ} is exact i.e., when can we write

$$a^{\mu} = \bar{\partial} v^{\mu} \qquad (2.30)$$

with

$$v = v^{\mu} \frac{\partial}{\partial x^{\mu}} \qquad (2.31)$$

a globally defined holomorphic vector field on M. For this purpose and for later use we introduce some notation:

Set

$$p^{\alpha}{}_{A} = \frac{\partial p^{\alpha}}{\partial z^{A}} , \quad p^{\bar{\beta}} = \overline{(p^{\beta})} \tag{2.32}$$

and

$$k^{\alpha\bar{\beta}} = \sum_{A} p^{\alpha}{}_{A} p^{\bar{\beta}}{}_{\bar{A}} \tag{2.33}$$

$k^{\alpha\bar{\beta}}$ is closely related to the lapse. The perpendicular part of the metric is

$$h^{\alpha\bar{\beta}} = \sigma k^{\alpha\bar{\beta}} . \tag{2.34}$$

We denote by $k_{\alpha\bar{\beta}}$ the matrix inverse to $k^{\alpha\bar{\beta}}$ and use $k^{\alpha\bar{\beta}}$ and $k_{\alpha\bar{\beta}}$ to raise and lower polynomial indices. It is also convenient to denote complex conjugation of the embedding coordinates by raising and lowering indices. For example we set

$$dz_{A} = dz^{\bar{A}} \quad \text{and} \quad \partial^{A} = \frac{\partial}{\partial z^{\bar{A}}} . \tag{2.35}$$

This facilitates a summation convention for the embedding coordinates. With these conventions we have

$$k^{\alpha\bar{\beta}} = p^{\alpha}{}_{A} p^{\bar{\beta}A} = p^{\alpha\bar{A}} p^{\bar{\beta}}{}_{\bar{A}} . \tag{2.36}$$

Returning to the problem at hand, we can show from (2.22), (2.25) and (2.27) that

$$a^{\mu} \frac{\partial}{\partial x^{\mu}} = -\bar{\partial}\left(q^{\alpha} p_{\alpha}{}^{A} \frac{\partial}{\partial z^{A}}\right). \tag{2.37}$$

Here, again, one is tempted to conclude that we have shown a^{μ} to be exact since the quantity inside the brackets on the right hand side is globally defined on M. This is true, the catch is that

$$q^{\alpha} p_{\alpha}{}^{A} \frac{\partial}{\partial z^{A}} \tag{2.38}$$

is not in general a vector field that is tangent to M.

For a vector field

$$\nu = \nu^{A} \frac{\partial}{\partial z^{A}} \tag{2.39}$$

defined on the embedding space to be tangent to M it is necessary (and sufficient) that the quantities

$$\nu^{A} \frac{\partial p^{\alpha}}{\partial z^{A}} \tag{2.40}$$

vanish with the p^{α}. That is there must exist polynomials $f^{\alpha}{}_{\beta}$ of the appropriate degrees (a polynomial of negative degree will be understood to be zero) such that

$$\nu^{A} \frac{\partial p^{\alpha}}{\partial z^{A}} = f^{\alpha}{}_{\beta} p^{\beta} . \tag{2.41}$$

Note also that owing to the presence of the $\bar{\partial}$ operator in (2.37) there is a certain freedom in writing

$$a^{\mu} \frac{\partial}{\partial x^{\mu}} = - \bar{\partial}\left(\nu^{A} \frac{\partial}{\partial z^{A}}\right) \tag{2.42}$$

we can take

$$\nu^A = q^\alpha p_\alpha{}^A - c^A \tag{2.43}$$

where the $c^A(z)$ are any linear holomorphic functions of the z^B

$$c^A(z) = c^A{}_B z^B . \tag{2.44}$$

(The c^A have to be linear in order to have the same homogeneity degree as the first term on the right-hand side of eqn. (2.43)).

Thus $a^\mu \dfrac{\partial}{\partial x^\mu}$ is exact if an only if

$$(q^\alpha p_\alpha{}^A - c^A)\, p^\gamma{}_A = f^\gamma{}_\delta p^\delta \qquad (2.45$$

for some choice of c^A and $f^\gamma{}_\delta$. On making use of eqn. (2.36) this condition becomes

$$q^\alpha = c^A p^\alpha{}_A + f^\alpha{}_\beta p^\beta . \tag{2.46}$$

This condition admits of a simple interpretation. Let us consider the two terms on the right-hand side separately. To understand the first consider making a linear change of coordinates in the embedding space.

$$z^A \to z^A + c^A(z) = z^A + c^A{}_B z^B \tag{2.47}$$

Under such a transformation the defining polynomials transform, to linearized order, in the following way

$$p^\alpha(z) \to p^\alpha(z+c) = p^\alpha(z) + c^A p^\alpha{}_A . \tag{2.48}$$

In other words a deformation q^α which is of the form $c^A p^\alpha{}_A$ corresponds merely to a change of coordinates in the embedding space rather than to a change in the shape of M.

The second term on the right-hand side of eqn. (2.46) corresponds to the freedom to describe M either by the equations

$$p = 0 \tag{2.49}$$

or by the equivalent set, f is assumed small,

$$p + f_{\beta}p^{\beta} = 0. \tag{2.50}$$

The burden of (2.46) is therefore that a^{μ} is exact precisely when we would not wish to think of q^{α} as describing a genuine perturbation. Such a q^{α} corresponds merely to a coordinate change together with a rearrangement of the defining polynomials. In fact we would wish to identify

$$q^{\alpha} \simeq q^{\alpha} + c^{A}p^{\alpha}{}_{A} + f^{\alpha}{}_{\beta}p^{\beta} \tag{2.51}$$

for all c^{A} and $f^{\alpha}{}_{\beta}$ since the left and the right-hand sides represent the same physical deformation.

For a given choice of defining polynomials p^{α} let us define sets I^{α} of polynomials which consist precisely of the polynomials of the form (2.46).

$$I^{\alpha} = \{q^{\alpha} |\ q^{\alpha} = c^{A}p^{\alpha}{}_{A} + f^{\alpha}{}_{\beta}p^{\beta}\} \tag{2.52}$$

for all possible holomorphic choices of the c^{A} and $f^{\alpha}{}_{\beta}$. The above discussion has established the following transcription of closed forms, exact forms and cohomology classes into algebraic terms. The closed forms are in one-one correspondence with polynomials q^{α} of the appropriate degrees. Let us denote the set of polynomials of the same degree as p^{α} by C^{α} then a closed form corresponds to a vector of polynomials q^{α} with $q^{\alpha}\varepsilon C^{\alpha}$ for each α. The exact forms are in one-one correspondence with the elements of I^{α}, i.e., with vectors of polynomials $q^{\alpha}\varepsilon I^{\alpha}$. Finally the cohomology classes correspond to closed forms identified modulo exact forms i.e., to vectors of polynomials subject to the identification (2.51). These are vectors of polynomials $q^\alpha \varepsilon P^\alpha$ where we define

$$P^\alpha = C^\alpha / I^\alpha . \tag{2.53}$$

This situation is summarized in Table 1.

Table 1.

closed forms	↔	$q^\alpha \varepsilon C^\alpha$
exact forms	↔	$q^\alpha \varepsilon I^\alpha$
cohomology classes	↔	$q^\alpha \varepsilon P^\alpha$

The correspondence between cohomology classes and polynomials will allow us to transcribe operations on the cohomology classes into algebraic operations on polynomials. Before proceeding to the evaluation of the Yukawa couplings, however, we need to discuss the properties of the holomorphic three-form Ω which appears explicitly in expression (1.5).

3. A Digression on the Holomorphic 3-Form and some Jacobians

Our purpose here is to derive and relate a number of relations between different representations of the holomorphic 3-form Ω that are of importance in what follows. These relations follow from the fact that, for the class of spaces that we are presently considering, the holomorphic 3-form admits explicit representations in terms of the defining polynomials.

We shall first work in the coordinate patch

$$U_{N+4} = \{z^{N+4} \neq 0\} . \tag{3.1}$$

We set

$$z^{N+4} = \zeta , \tag{3.2}$$

with ζ an arbitrary nonzero constant, and use the other N+3 z's as coordinates in this patch. We shall use lower case latin indices to range over the values $1,2,\cdots,N+3$.

Choose i,j,k to be distinct in the first instance and set

$$\Delta^{ijk} = \varepsilon^{ijkm_1\cdots m_N} \frac{\partial p^1}{\partial z^{m_1}} \frac{\partial p^2}{\partial z^{m_2}} \cdots \frac{\partial p^N}{\partial z^{m_N}}$$

$$= \frac{\partial(z^i,z^j,z^k,p^1,\cdots,p^N)}{\partial(z^1,z^2,\cdots\cdots,z^{N+3})} . \qquad (3.3)$$

Δ^{ijk} is the Jacobian of the transformation between the coordinates $(z^1,\cdots,z^{N+3})$ and another set consisting of z^i,z^j,z^k and the N polynomials p^α. We may use (z^i,z^j,z^k) as coordinates for M. The Jacobian between this set and another such set (z^m,z^n,z^r), say, is

$$\frac{\partial(z^i,z^j,z^k)}{\partial(z^m,z^n,z^r)} = \frac{\Delta^{ijk}}{\Delta^{mnr}} \qquad (3.4)$$

in virtue of (3.3).

The holomorphic 3-form can be written[f2] in terms of Δ^{ijk}

$$\Omega = \frac{\zeta dz^i_{N+4}\, dz^j_{N+4}\, dz^k_{N+4}}{\Delta^{ijk}_{N+4}} \qquad (3.5)$$

where there is no summation over the indices i,j,k and we have appended subscripts to emphasize that what is, in reality, meant by setting $z^{N+4} = \zeta$ is that we choose coordinates

f2. In the following we shall frequently omit the "$\wedge$" symbol when exterior multiplication is clear from the context.

$$z^i_{N+4} = \frac{\zeta z^i}{z^{N+4}} , \quad i = 1,\cdots,N+3. \tag{3.6}$$

Relation (3.5) requires a certain amount of checking.

(i) Note first of all that the homogeneity degree of the right-hand side is

$$N+4 - \sum_{\alpha=1}^{N} \deg(\alpha) \tag{3.7}$$

where $\deg(\alpha)$ denotes the degree of p^α. Since z^A and λz^A represent the same point in P_{N+3} the homogeneity degree of Ω must be zero for this representation to be sensible. Thus we must have

$$\sum_\alpha \deg(\alpha) = N+4 \tag{3.8}$$

this is the condition for M to have vanishing first Chern class.

(ii) We must check that the right-hand side of (3.5) is independent of the choice of i,j,k. This is easy since we have

$$\frac{dz^i\, dz^j\, dz^k}{\Delta^{ijk}} = \frac{1}{\Delta^{ijk}} \frac{\partial(z^i,z^j,z^k)}{\partial(z^m,z^n,z^r)} dz^m\, dz^n\, dz^r \tag{3.9}$$

$$= \frac{dz^m\, dz^n\, dz^r}{\Delta^{mnr}}$$

in virtue of (3.4).

(iii) We must check that the representation is independent of our choice of z^{N+4} as the constant coordinate. To this end we transform to coordinates $z^m{}_1$, say, $m = 2,3,\ldots,N+4$ obtained by taking $z^1 = \eta$. On the set $U_1 \quad U_{N+4}$ the transformation between the two sets of coordinates takes the form

$$z^1_{N+4} = \frac{\zeta}{z^{N+4}_1} \quad , \quad z^i_{N+4} = \frac{\zeta z^i_1}{z^{N+4}_1} \quad , \quad i = 2, \quad ,N+3 \; . \tag{3.10}$$

In virtue of (ii) above it is sufficient to show that, for example

$$\frac{dz^1_{N+4}\; dz^2_{N+4}\; dz^3_{N+4}}{{}^{123}_{N+4}} = - \frac{dz^{N+4}_1\; dz^2_1\; dz^3_1}{{}^{(N+4)23}_{1}} \; . \tag{3.11}$$

(The sign change is due to the permutation that arises in going from (N+4, 1, 2, 3, N+3) to (1, N+4, 2, 3, N+3)). Given (3.10) eqn. (3.10) follows straightforwardly once account is taken of the fact that it is implicit that on the left-hand side of (3.11) the arguments of the polynomials p in ${}^{123}_{N+4}$ are the z^i_{N+4} while on the right-hand side the arguments of the p are the z^i_1.

(iv) Finally we must check that, as defined, Ω is never singular. For a singularity to occur within U_{N+4}, say, all the Δ^{ijk}_{N+4} would have to vanish simultaneously. This however would imply the vanishing of $dp^1 dp^2 \cdots dp^N$ which cannot happen if M is smooth.

There is a very beautiful construction of Ω due to Bott [8] which illuminates the properties we have described. There is a natural "form" which may be constructed from the homogeneous coordinates z^A:

$$\mu = \frac{(-1)^{N+3}}{(N+3)!} \epsilon_{A_1 A_2 \cdots A_{N+4}} z^{A_1}\, dz^{A_2} \cdots dz^{A_{N+4}} . \tag{3.12}$$

Note that in coordinates with z^{N+4} = const. this is just

$$z^{N+4}\, dz^1\, dz^2 \ldots dz^{N+3} .$$

μ is not well defined on P_{N+3} since under a scaling $z^A \to \lambda z^A$ it is not invariant, it scales as

$$\mu \to \lambda^{N+4}\mu \, . \tag{3.13}$$

This difficulty can be cured by dividing μ by the product of the N polynomials p^α. The (N+3)-form

$$\nu = \frac{\mu}{p^1 p^2 \cdots p^N} \tag{3.14}$$

is invariant under $z^A \to \lambda z^A$ in virtue of (3.8). ν is well-defined on P_{N+3} but has poles on the hypersurfaces $p^\alpha = 0$. Let us integrate ν around a contour

$$\Gamma_N = \gamma_1 \times \gamma_2 \times \cdots \times \gamma_N \tag{3.15}$$

which is the cartesian product of the N one-dimensional curves $|p^\alpha| = \delta$, with δ assumed small which wind around the hypersurfaces $p^\alpha = 0$.

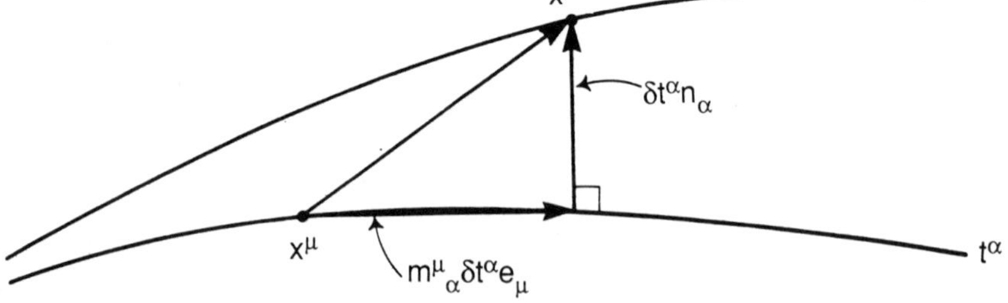

Figure 4. The contour Γ_N

Consider the integral (the limit $\delta \to 0$ is understood)

$$\int_{\Gamma_N} \frac{\mu}{p^1 p^2 \cdots p^N} = \int_{\Gamma_N} \frac{z^{N+4} dz^1 \cdots dz^{N+3}}{p^1 \cdots p^N}$$

$$= \int_{\Gamma_N} \frac{z^{N+4}}{p^1 \cdots\cdots p^N} \frac{\partial(z^1, z^2, \cdots\cdots\cdots z^{N+3})}{\partial(z^i, z^j, z^k, p^1, \cdots, p^N)}$$

$$= \int_{\Gamma_N} \frac{z^{N+4} dz^i dz^j dz^k}{\Delta^{ijk}} \frac{dp^1 dp^2 \cdots dp^N}{p^1 p^2 \cdots p^N}$$

$$= (2\pi i)^N \Omega. \tag{3.16}$$

The second line follows by changing variables of integration from the N+3 z´s to any three distinct z´s and the N polynomials p^α, and the fourth by integrating out the p^α in the limit $\delta \to 0$.

4. Computation of the Yukawa Coupling

We return now to the task of computing the coupling

$$\kappa(a,b,c) = \int_M \Omega \wedge a^\mu \wedge b^\nu \wedge c^\rho \Omega_{\mu\nu\rho} \tag{4.1}$$

between three elements of the group $H_{\bar\partial}^{(0,1)}(M,T)$.

We have seen that each of the elements of this group corresponds to a polynomial defined modulo I^α. Let us denote the polynomials corresponding to a^μ, b^ν, c^ρ by q^α, r^β, s^γ respectively.

We have the explicit relation

$$a^\mu = q^\alpha \chi_{\bar\lambda}{}^{\mu}{}_{\alpha} dx^{\bar\lambda}$$

$$= - q_{\bar\alpha} \frac{\partial z^{\bar J}}{\partial x^{\bar\rho}} g^{\mu\bar\rho} \, dp^{\bar\alpha}_{\bar J} \tag{4.2}$$

the second relation following from (2.25).

We need to form the (0,3)-form

$$\bar{\ } = a \quad b \cdot c \quad \Omega_{\mu\nu\rho}, \tag{4.3}$$

a little algebra which makes use of the relations of the previous section leads to the expression

$$\bar{\omega} = -\frac{\sigma}{|z^{N+4}|^2} q_{\bar{\alpha}} r_{\bar{\beta}} s_{\bar{\gamma}} \; z^{N+4} \, k^{1/2} \, \Delta^{\bar{I}\bar{J}\bar{K}} \, dp^{\bar{\alpha}}{}_{\bar{I}} \, dp^{\bar{\beta}}{}_{\bar{J}} \, dp^{\bar{\gamma}}{}_{\bar{K}} \tag{4.4}$$

where

$$k = \det(k_{mn}) = \det{}^2(k_{\alpha\bar{\beta}}). \tag{4.5}$$

We know on general grounds that $\bar{\omega}$ is covariant under changes of coordinates so we may multiply this equation by $\frac{|z^{N+4}|^2}{\sigma}$ and add similar equations, one for each coordinate patch, since

$$\frac{1}{\sigma}\sum_A |z^A|^2 = 1 \tag{4.6}$$

we arrive at the important expression

$$\bar{\omega} = q^{\alpha} r^{\beta} s^{\gamma} e_{ABC} \, \bar{\partial} p_{\alpha}{}^{A} \bar{\partial} p_{\beta}{}^{B} \bar{\partial} p_{\gamma}{}^{C} \tag{4.7}$$

where

$$e_{ABC} = \frac{1}{N!} \varepsilon_{ABCDM_1 \ldots M_N} \, z^D \varepsilon^{\sigma_1 \ldots \sigma_N} p_{\sigma_1}{}^{M_1} \ldots p_{\sigma_N}{}^{M_N} \tag{4.8}$$

This expression has very nice properties but is, perhaps, somewhat beset by indices. In order to appreciate what is going on let us specialize temporarily to the simplest case $P_4(5)$ of the quintic in P_4 for which there is only one polynomial so that we may omit the α indices provided we remember that

$$k^{11} = \frac{1}{k_{11}} = \bar{p}^A p_A. \tag{4.9}$$

To save a little writing we write

$$\bar{p}^A p_A = \|p\|^2 \quad \text{and} \quad \pi^A = \frac{\bar{p}^A}{\|p\|^2} \tag{4.10}$$

so that

$$\pi^A p_A = 1. \tag{4.11}$$

then expression (4.7) becomes

$$\bar{\omega} = qrs\, \varepsilon_{ABCDE} z^A \pi^B \bar{\partial}\pi^C \bar{\partial}\pi^D \bar{\partial}\pi^E \tag{4.12}$$

Now quite apart from other considerations $\bar{\omega}$ is a (0,3)-form which is closed in virtue of the fact that a^μ, b^ν and c^ρ are. For a Calabi-Yau manifold there is precisely one harmonic (0,3)-form, $\bar{\Omega}$. Thus we may write

$$\bar{\omega} = \kappa\bar{\Omega} + \bar{\partial}\bar{\nu} \tag{4.13}$$

with $\bar{\kappa}$ a constant that is determined by (a^μ, b^ν, c^ρ) or equivalently by (q, r, s) and $\bar{\nu}$ is a (0,2)-form. The exact part $\bar{\partial}\bar{\nu}$ does not contribute to the integral. Hence

$$\int \Omega \wedge \bar{\omega} = \kappa \tag{4.14}$$

which identifies the constant appearing in (4.12) as the value of the coupling.

If for some choice of q, r and s, $\bar{\omega}$ is exact then the corresponding κ is zero. We wish to enquire, therefore, when the expression (4.12) is exact. We shall now show that $\bar{\omega}$ is exact if the product qrs is of the form $E^A p_A$ with the E^A any eleventh order polynomials. This requires an identity. Consider the quantity

$$\varepsilon_{AFCDE} z^A \bar{\partial}\pi^C \bar{\partial}\pi^D \bar{\partial}\pi^E, \tag{4.15}$$

the permutation symbol has F as a free index and is contracted on the other four indices with the tensor

$$z^A \bar{\partial}\pi^C \bar{\partial}\pi^D \bar{\partial}\pi^E. \tag{4.16}$$

The latter tensor is orthogonal to p_A on all its indices, either because

$$z^A p_A = 5p = 0 \text{ on } M \tag{4.17}$$

or because

$$p_C \bar{\partial}\pi^C = \bar{\partial}(p_C \pi^C) = 0. \tag{4.18}$$

It follows that the free index F of the quantity (4.15) must be parallel to p_F. Thus we obtain the identity

$$\varepsilon_{AFCDE} z^A \bar{\partial}\pi^C \bar{\partial}\pi^D \bar{\partial}\pi^E = p_F \varepsilon_{ABCDE} z^A \pi^B \bar{\partial}\pi^C \bar{\partial}\pi^D \bar{\partial}\pi^E \tag{4.19}$$

Now our immediate aim is to show that $\bar{\omega}$ is $\bar{\partial}$-exact if qrs εI. Suppose that qrs = $E^F p_F$ then

$$\begin{aligned} \bar{\omega} &= E^F p_F \varepsilon_{ABCDE} z^A \pi^B \bar{\partial}\pi^C \bar{\partial}\pi^D \bar{\partial}\pi^E \\ &= E^B \varepsilon_{ABCDE} z^A \bar{\partial}\pi^C \bar{\partial}\pi^D \bar{\partial}\pi^E \\ &= \bar{\partial}(E^B \varepsilon_{ABCDE} z^A \pi^C \bar{\partial}\pi^D \bar{\partial}\pi^E) \end{aligned} \tag{4.20}$$

where the second line follows in virtue of the identity (4.19). We knew previously that each of the polynomials q, r, s is defined modulo I . We have learnt that the same is true of their product. The possible cohomology classes of the product qrs are the elements of

$$P_{15} = C_{15}/I . \tag{4.21}$$

It is natural at this point to enquire how many of those there are. The answer comes as a surprise on first aquaintance. There is precisely one. This is easy to see explicitly for the simplest example for which

$$p = (z^1)^5 + (z^2)^5 + (z^3)^5 + (z^4)^5 + (z^5)^5 \tag{4.22}$$

for this case I is generated by the monomials $(z^A)^4$ and it is clear that there is precisely one polynomial of fifteenth order than can be written down without writing any z^A to the fourth power. It is

$$20^5 (z^1)^3(z^2)^3(z^3)^3(z^4)^3(z^5)^3 \tag{4.23}$$

the numerical factor being inserted for later convenience. In other words

$$\dim P_{15} = 1. \tag{4.24}$$

This is clear for this special case but is true more generally. Being an integer dim P_{15} does not change under a sufficiently small variation of p and, in fact, does not change as p varies so long as the manifold defined by p is nonsingular. Let us denote the unique fifteenth order polynomial that is not in P by Q. We may take Q to be given by the determinant of the matrix of second derivatives p

$$Q = \frac{1}{5!}\varepsilon^{A_1\cdots A_5}\varepsilon^{B_1\cdots B_5} \frac{\partial^2 p}{\partial z^{A_1}\partial z^{B_1}} \cdots \frac{\partial^2 p}{\partial z^{A_5}\partial z^{B_5}} . \tag{4.25}$$

This is a more general prescription for the form of Q than (4.22).

It is shown in appendix A that with the product qrs replaced by Q the value of the integral (4.13) is unity. The burden of these remarks is that there is a unique way of writing

$$qrs \quad = \quad \kappa Q + E^A p_A \qquad (4.26)$$

with κ a constant which we identify as the Yukawa coupling. Thus the problem of computing the integral (4.15) has been reduced to a purely algebraic process. One multiplies the three polynomials q,r,s and writes the product as a multiple of Q the unique fifteenth order polynomial not in I plus a remainder that is in I . This leads to a unique expression and the coefficient of Q is the Yukawa coupling. For the simplest case with p given by (4.23) the procedure is very simple. One multiplies the polynomials q,r,s and the coefficient of the term

$$20^5 \ (z^1)^3(z^2)^3(z^3)^3(z^4)^3(z^5)^3$$

in the product is the Yukawa coupling.

Before returning from the simplest case of only one polynomial to the general case. We would like to remark on the identifications that have been made between forms and polynomials. These are summarized in Table 2.

Table 2. The correspondence between cohomology classes and sets of polynomials for the manifold $P_4(5)$. Here P_k denotes the set of homogeneous polynomials of degree k identified modulo the ideal I.

Cohomology group	Space of polynomials
$H^{(3,0)}$	P_0
$H^{(2,1)}$	P_5
$H^{(1,2)}$	P_{10}
$H^{(0,3)}$	P_{15}

The only entry that has not been explicitly discussed is the correspondence between (1,2)-forms and P_{10}. This is provided by the observation that any (1,2)-form can be expressed as a sum of the terms of the form

$$\Omega_{\mu\nu\rho}dx^{\mu}a^{\nu}b^{\rho}.$$

It follows that tenth order polynomials u(z), defined modulo I, can be identified with (1,2) forms $\nu^{(1,2)}$ via the correspondence

$$\nu^{(1,2)} = \frac{u}{\|p\|^4}\Omega_{\mu\nu\rho}\chi^{\nu}{}_{\bar{\sigma}}\chi^{\rho}{}_{\bar{\tau}}dx^{\mu}dx^{\bar{\sigma}}dx^{\bar{\tau}}$$

$$= -u\varepsilon_{ABCDE}z^{A}\pi^{B}dz^{C}\bar{\partial}\pi^{D}\bar{\partial}\pi^{E} \qquad (4.27)$$

it is straightforward to show that $\nu^{(1,2)}$ is exact if $u\varepsilon I$.

Suppose now that there are N polynomials p^{α}. We shall first enquire as to when the (0,3)-form (4.7) is exact and then we shall discuss the generalization of Table 2.

The assertion is that

$$\bar{\omega} = q^{\alpha}r^{\beta}s^{\gamma}\varepsilon_{ABC}\bar{\partial}p_{\alpha}{}^{A}\bar{\partial}p_{\beta}{}^{B}\bar{\partial}p_{\gamma}{}^{C} \qquad (4.28)$$

is exact whenever the tensor of polynomials $q^{(\alpha}r^{\beta}s^{\gamma)}$ can be written in the form

$$c^{A(\alpha\beta}p^{\gamma)}{}_{A} + f^{\alpha\beta\gamma}{}_{\delta}p^{\delta}$$

for some choice of polynomials $c^{A\alpha\beta}$ and $f^{\alpha\beta\gamma}{}_{\delta}$. We shall denote the set of all tensors of polynomials whose components $q^{\alpha\beta\gamma}$ are of degree $\deg(\alpha) + \deg(\beta) + \deg(\gamma)$ and are of this form by $I^{\alpha\beta\gamma}$. The assertion is that $\bar{\omega}$ is exact if $q^{(\alpha}r^{\beta}s^{\gamma)}\varepsilon I^{\alpha\beta\gamma}$.

The substantiation of the assertion is analogous to that for the case of one polynomial. Note first that

$$k_{F}{}^{C} = p^{\gamma}{}_{F}p_{\gamma}{}^{C} \qquad (4.29)$$

is a tensor that projects normal to M:

$$k_F{}^C\, p^\alpha{}_C = p^\alpha{}_F \quad , \quad k_F{}^C \frac{\partial z^F}{\partial x^\mu} dx^\mu = 0 . \tag{4.30}$$

Now if in expression (4.29) $q^\alpha r^\beta s^\gamma$ is replaced by $C^{F\alpha\beta}p^\gamma{}_F$ then, in virtue of the observation that

$$p^\gamma{}_F \bar{\partial} p_\gamma{}^C = \bar{\partial} k_F{}^C \tag{4.31}$$

we find that

$$\begin{aligned} \bar{\omega} &= C^{F\alpha\beta}\, e_{ABC}\, \bar{\partial} p_\alpha{}^A \bar{\partial} p_\beta{}^B \bar{\partial} k_F{}^C \\ &= \bar{\partial}\, \{ C^{F\alpha\beta} k_F{}^C e_{ABC} \bar{\partial} p_\alpha{}^A \bar{\partial} p_\beta{}^B \} \\ &- \frac{1}{(N-1)!} C^{F\alpha\beta} k_F{}^C \varepsilon_{ABCDM_1 \cdots M_N}\, \varepsilon^{\sigma_1 \cdots \sigma_N}\, p_{\sigma_1}{}^{M_1} \cdots p_{\sigma_{N-1}}{}^{M_{N-1}} z^D \bar{\partial} p_{\sigma_N}{}^{M_N}\, \bar{\partial} p_\alpha{}^A\, \bar{\partial} p_\beta{}^B \end{aligned} \tag{4.32}$$

The second term vanishes, this is seen most easily from the fact that the permutation symbol is projected orthogonal to $p_\sigma{}^A$ on the five indices (A,B,C,D,M_N). Since the $p_\sigma{}^A$, $\sigma = 1,\cdots,N$ are a set of N linearly independent vectors that span an N-dimensional space projecting the permutation symbol orthogonal to the $p_\sigma{}^A$ on more than four indices must give zero.

It is compelling that there is the following correspondence between spaces of tensors of polynomials and harmonic 3-forms:

Table 3. The correspondence between cohomology classes and spaces of polynomials

Cohomology group	Space of polynomials	Identification
$H^{(3,0)}$	1	
$H^{(2,1)}$	$p^\alpha = C^\alpha / I^\alpha$	$C^A p^\alpha{}_A + f^\alpha{}_\beta p^\beta$
$H^{(1,2)}$	$p^{\alpha\beta} = C^{\alpha\beta} / I^{\alpha\beta}$	$C^{A(\alpha} p^{\beta)}{}_A + f^{\alpha\beta}{}_\gamma p^\gamma$
$H^{(0,3)}$	$p^{\alpha\beta\gamma} = C^{\alpha\beta\gamma} / I^{\alpha\beta\gamma}$	$C^{A(\alpha\beta} p^{\gamma)}{}_A + f^{\alpha\beta\gamma}{}_\delta p^\delta$

Since there is precisely one harmonic (0,3)-form there is precisely one polynomial $Q^{\alpha\beta\gamma}$ which is not in $I^{\alpha\beta\gamma}$. Thus the rule for computing the Yukawa coupling is to take the quantity $q^{(\alpha}{}_r{}^{\beta}{}_s{}^{\gamma)}$ and to express it as a multiple of $Q^{\alpha\beta\gamma}$ plus a remainder that lies in $I^{\alpha\beta\gamma}$

$$q^{(\alpha}{}_r{}^{\beta}{}_s{}^{\gamma)} = \kappa Q^{\alpha\beta\gamma} + C^{A(\alpha\beta}p^{\gamma)}{}_A + f^{\alpha\beta\gamma}{}_\delta p^\delta \qquad (4.33)$$

It is straightforward, though tedious, to check these assertions by calculating the dimensions of P^α, $P^{\alpha\beta}$ and $P^{\alpha\beta\gamma}$ by computing directly the difference between the number of independent polynomials of the relevant degree and the number of independent polynomials in the corresponding tensor ideal. However it is simpler to defer the substantiation of Table 3 to the next section where it will be subsumed in an argument valid for all Calabi-Yau spaces.

Finally we list representative forms for each of the cohomology classes of rank 3 and give an explicit expression for $Q^{\alpha\beta\gamma}$. These expressions are useful, when M has discrete symmetries, since they permit us to relate the transformation properties of the polynomials to those of the differential forms

Table 4. Explicit expressions for representatives of third rank cohomology groups. The polynomials q^γ, $q^{\beta\gamma}$, $q^{\alpha\beta\gamma}$ belong to P^γ, $P^{\beta\gamma}$, $P^{\alpha\beta\gamma}$ respectively.

Differential form	Representation
Ω	$e_{ABC}\, dz^A\, dz^B\, dz^C$
$\omega^{(2,1)}$	$q^\gamma\, e_{ABC}\, dz^A\, dz^B\, \bar\partial p_\gamma{}^C$
$\omega^{(1,2)}$	$q^{\beta\gamma}\, e_{ABC}\, dz^A\, \bar\partial p_\beta{}^B\, \bar\partial p_\gamma{}^C$
$\bar\Omega$	$q^{\alpha\beta\gamma}\, e_{ABC}\, \bar\partial p_\alpha{}^A\, \bar\partial p_\beta{}^B \bar\partial p_\gamma{}^C$

Finally we record an explicit representative for $\bar\Omega$:

$$Q^{\alpha\beta\gamma} = p^\alpha{}_{;A_1B_1}\, p^\beta{}_{;A_2B_2}\, p^\gamma{}_{;A_3B_3}\, e^{A_1A_2A_3A_4}{}_{;B_4}\, e^{B_1B_2B_3B_4}{}_{;A_4} \qquad (4.34)$$

where

$$e^{ABCD} = \frac{1}{N!}\varepsilon^{ABCDM_1\cdots M_N}\varepsilon_{\sigma_1\cdots\sigma_N}p^{\sigma_1}{}_{M_1}\cdots p^{\sigma_N}{}_{M_N}. \qquad (4.35)$$

Note that e^{ABCD} is a polynomial in the z^A which is of degree 4 in virtue of (3.8). The expression we have given for $Q^{\alpha\beta\gamma}$ is, in fact, the only polynomial of the appropriate degree that can be constructed from the natural ingredients p^{α} and e^{ABCD}.

The considerations of this section are suggestive of more general arguments applicable to all Calabi-Yau spaces. It is to this subject that we now turn.

5. The General Case

Heretofore we have discussed the calculations of the Yukawa couplings for manifolds realized by giving N polynomial equations in a single projective space P_{N+3}. There are interesting generalizations of this class, among which are spaces realized as complete intersections of polynomials in products of projective spaces such as the interesting space given by Yau which is realized by giving the three polynomial equations

$$\sum_{A=1}^{4} x_A^3 = 0 \quad , \quad \sum_{A=1}^{4} y_A^3 = 0 \qquad (5.1)$$

$$\sum_{A=1}^{4} x_A y_A = 0$$

in $P_3 \times P_3$. Our analysis generalizes in a straightforward way to cover these examples but rather than repeat it it is simpler to pass to the general case.

The essential observation is that the Yukawa couplings are intimately related to the deformations of the complex structure of the manifold. Fortunately the theory of the deformation of complex structures has been well developed by Kodaira and Spencer and later writers [4,5]. The complex structure of a complex manifold depends on a certain number, say m, of complex parameters.[f3]

In our previous examples m is equal to the number of coefficients in the defining polynomials once account is taken of the freedom to remove some of the coefficients by choice of coordinates.

---

f3. For a general complex manifold $m \leq b_{21}$. Since there might exist solutions to the linearized perturbation equations which are not the linearization of some perturbation. For Calabi-Yau manifolds however it has been shown by Tian [9] that all the perturbations are unobstructed so that $m = b_{21}$.

Following Kodaira [5] we shall denote the parameters on which M depends by t^α, $\alpha = 1,\cdots,m$ (technically, these are the effective parameters). This is a somewhat different use of the index α from our previous use, which we hope will not confuse the reader. To emphasize the dependence of the manifold on the parameters we shall sometimes write M_t for M. The parameter space is denoted by B, in general this is itself a complex manifold but since we shall be concerned with the infinitesimal deformations of a given manifold M_0, corresponding to $t^\alpha = 0$, say, it is sufficient to take B to be a neighborhood of the origin in C^m.

The fiber bundle

$$\mathcal{M} = \{(M_t,t)\,|\,t\varepsilon B\} \tag{5.2}$$

with projection $M_t \to t$ is a complex manifold of dimension m + 3 in which the given manifold M is the embedded submanifold specified by the m holomorphic equations

$$t^\alpha(y^A) = 0 \tag{5.3}$$

with the y^A, A=1,$\cdots$,m+3, coordinates for . We may proceed in close analogy with our previous treatment. As before we may choose basis vectors and basis forms adapted to the deformations

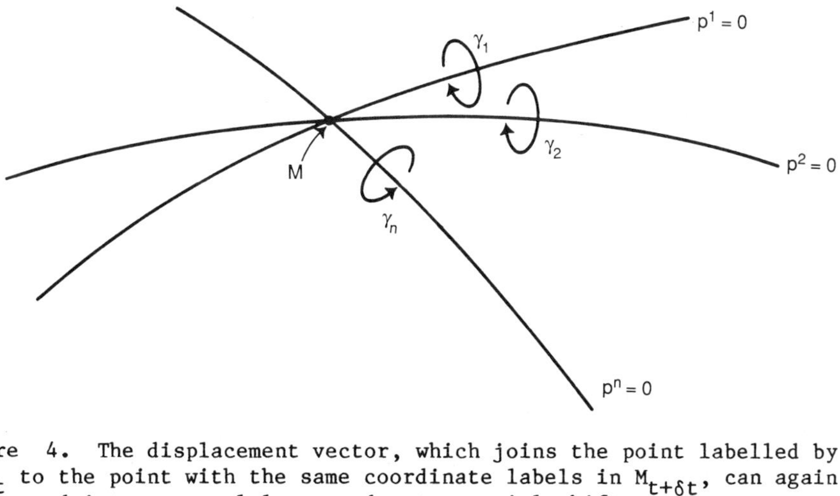

Figure 4. The displacement vector, which joins the point labelled by x^μ in M_t to the point with the same coordinate labels in $M_{t+\delta t}$, can again be decomposed into a normal lapse and a tangential shift.

$$e_\mu = \frac{\partial}{\partial x^\mu} \quad ; \quad e^\mu = dx^\mu + m^\mu{}_\alpha n^\alpha$$

$$n_\alpha = \frac{\partial}{\partial t^\alpha} - m^\mu{}_\alpha \frac{\partial}{\partial x^\mu} \quad ; \quad n^\alpha = dt^\alpha \tag{5.4}$$

There is one difference between this and the previous case which is that we may not assume that the ambient manifold $\mathcal{M}$ is Kähler. However since it is a complex manifold so we may take it to be endowed with a hermitean metric and the hermitean connection [10]. The previous analysis goes through almost unchanged except that now the extrinsic curvature

$$\chi^\alpha_{\mu\nu} = \frac{\partial y^A}{\partial x^\mu} \frac{\partial y^B}{\partial x^\nu} \left(\frac{\partial^2 t^\alpha}{\partial y^A \partial y^B} - \Gamma^C_{AB} \frac{\partial t^\alpha}{\partial y^C}\right) \tag{5.5}$$

is not now necessarily symmetric in A and B since the hermitean connection has torsion and is not symmetric in its lower indices unless $\mathcal{M}$ is in fact Kähler. This makes no difference to the analysis since the symmetry of $\chi^\alpha_{\mu\nu}$ was never used. We must however take care to represent the elements of $H^{(0,1)}_{\bar\partial}(M,T)$ as in eqn. (2.27) with the $dx^{\bar\nu}$ contracted on the first index. The elements of $H^{(0,1)}_{\bar\partial}(M,T)$ are now represented in the form

$$a^\mu = q^\alpha \chi_{\bar\nu}{}^\mu{}_\alpha dx^{\bar\nu} \tag{5.6}$$

with the q^α corresponding to variations in the parameters t^α. More precisely let us suppose that the equations

$$t^\alpha(y^A,u) = 0 \tag{5.7}$$

depend holomorphically on a parameter u. Then with a deformation of M we associate a vector

$$q^\alpha(y^A) = \frac{\partial t^\alpha}{\partial u}(y^A,0). \tag{5.8}$$

We are again interested in the question as to when the form (5.6) is exact. A short calculation reveals that we may write

$$a^{\mu}\frac{\partial}{\partial x^{\mu}} = \bar{\partial}\left(q^{\alpha}t_{\alpha}{}^{D}\frac{\partial}{\partial y^{D}}\right) \tag{5.9}$$

where we employ a notation similar to that of Sect. 2. The $\bar{\partial}$-operator refers to

$$t^{\alpha}{}_{A} = \frac{\partial t^{\alpha}}{\partial y^{A}} \tag{5.10}$$

and the α-indices are raised and lowered by means of the normal metric

$$h^{\alpha\bar{\beta}} = t^{\alpha}{}_{A}t^{\bar{\beta}A}. \tag{5.11}$$

a^{μ} is exact if and only if

$$(q^{\alpha}t_{\alpha}{}^{D} - c^{D})\frac{\partial t^{\gamma}}{\partial y^{D}} = 0 \tag{5.12}$$

on $t^{\beta} = 0$ for some holomorphic $c^{D}(y)$ i.e., if and only if

$$q^{\alpha} = c^{D}\, t^{\alpha}{}_{D} + f^{\alpha}{}_{\beta}\, t^{\beta} \tag{5.13}$$

for some holomorphic $c^{D}(y)$ and $f^{\alpha}{}_{\beta}(y)$.

Let us return now to a consideration of the product of three elements of $H^{(0,1)}_{\bar{\partial}}(M,T)$

$$\bar{\omega} = a^{\mu}{}_{\wedge}\, b^{\nu}{}_{\wedge}\, c^{\rho}\Omega_{\mu\nu\rho} \tag{5.14}$$

and set

$$\bar{\omega}^{\mu\nu\rho} = a^{[\mu}{}_{\wedge}\, b^{\nu}{}_{\wedge}\, c^{\rho]} \tag{5.15}$$

and

$$\|\Omega\|^2 = \frac{1}{3!}\,\Omega_{\mu\nu\rho}\,\bar{\Omega}^{\mu\nu\rho} \tag{5.16}$$

so that

$$\bar{\omega} = \bar{\omega}^{\mu\nu\rho}\Omega_{\mu\nu\rho} \tag{5.17}$$

and

$$\bar{\omega}^{\mu\nu\rho} = \frac{1}{3!\,\|\Omega\|^2}\,\bar{\Omega}^{\mu\nu\rho}\bar{\omega}. \tag{5.18}$$

Although it might seem at first sight improbable the quantity

$$\frac{\bar{\Omega}^{\mu\nu\rho}}{\|\Omega\|^2}$$

is a holomorphic tensor field

$$\bar{\partial}\left(\frac{\bar{\Omega}^{\mu\nu\rho}}{\|\Omega\|^2}\right) = 0. \tag{5.19}$$

The most explicit way to see this is to recall that Ω is holomorphic and nowhere zero. Thus in local coordinates

$$\Omega_{\mu\nu\rho} = f(x)\varepsilon_{\mu\nu\rho} \tag{5.20}$$

with $f(x)$ holomorphic and nowhere zero. It follows that

$$\frac{\bar{\Omega}^{\mu\nu\rho}}{\|\Omega\|^2} = \frac{1}{f(x)}\varepsilon^{\mu\nu\rho}. \tag{5.21}$$

Since f is nonzero we verify (5.19). With this observation in hand it is strightforward to check that, given two forms $\bar{\omega}$ and $\bar{\omega}^{\mu\nu\rho}$ related by (5.17) that $\bar{\omega}$ is closed if and only if $\bar{\omega}^{\mu\nu\rho}$ is and $\bar{\omega}$ is exact if and only if $\bar{\omega}^{\mu\nu\rho}$ is.

We enquire first when $\bar{\omega}^{\mu\nu\rho}$ is exact. If a^μ, b^ν, c^ρ correspond to perturbations q^α, r^β, s^γ then

$$\bar{\omega}^{\mu\nu\rho} = q^\alpha r^\beta s^\gamma \, \chi_{\bar{\kappa}}{}^{\mu}{}_{\alpha} \, \chi_{\bar{\lambda}}{}^{\nu}{}_{\beta} \, \chi_{\bar{\sigma}}{}^{\rho}{}_{\gamma} \, dx^{\bar{\kappa}} \, dx^{\bar{\lambda}} \, dx^{\bar{\sigma}} \, . \tag{5.22}$$

If we replace the product $q^\alpha r^\beta s^\gamma$ by any holomorphic tensor $q^{\alpha\beta\gamma}$ of the appropriate degree then the corresponding (0,3)-form $\bar{\omega}^{\mu\nu\rho}$, and hence $\bar{\omega}$, is still closed. By an analysis that is precisely analogous to the discussion of the deformations q^α we find the corresponding (0,3)-form is exact if and only if

$$q^{\alpha\beta\gamma} = C^{A(\alpha\beta}t^{\gamma)}{}_{A} + f^{\alpha\beta\gamma}(y,t) \tag{5.23}$$

with the $f^{\alpha\beta\gamma}$ holomorphic functions that vanish with t^α. In a similar way it is possible to relate (1,2)-forms to symmetric tensors $q^{\alpha\beta}$ subject to an appropriate tensor ideal. These results are summarized in Table (5).

Table 5. The correspondence between cohomology groups and spaces of functions for a general Calabi-Yau space.

Type of form	Representative	Tensor ideal
(3,0)	1	
(2,1)	q^α	$c^A t^\alpha{}_A + f^\alpha$
(1,2)	$q^{\alpha\beta}$	$C^{A(\alpha}t^{\beta)}{}_A + f^{\alpha\beta}$
(0,3)	$q^{\alpha\beta\gamma}$	$C^{A(\alpha\beta}t^{\gamma)}{}_A + f^{\alpha\beta\gamma}$

We know on general grounds that there is precisely one cohomology class of (0,3)-forms. Thus there exists a tensor $Q^{\alpha\beta\gamma}$, say, which represents $\bar{\Omega}$ which is such that an arbitrary tensor $q^{\alpha\beta\gamma}(y)$ is equal to some multiple of $Q^{\alpha\beta\gamma}(y)$ modulo the tensor ideal. We may take $Q^{\alpha\beta\gamma}$ to be given by the expression

$$Q^{\alpha\beta\gamma} = \chi_{\mu_1\nu_1}{}^{\alpha} \, \chi_{\mu_2\nu_2}{}^{\beta} \, \chi_{\mu_3\nu_3}{}^{\gamma} \, \frac{\bar{\Omega}^{\mu_1\mu_2\mu_3}\bar{\Omega}^{\nu_1\nu_2\nu_3}}{\|\Omega\|^4} \, . \tag{5.24}$$

$Q^{\alpha\beta\gamma}$ is holomorphic in virtue of our earlier remarks concerning the quantity $\bar{\Omega}^{\mu\nu\rho}/\|\Omega\|^2$. We may check that, as defined, $Q^{\alpha\beta\gamma}$ does not correspond to an exact form i.e., it is not in the tensor ideal, by showing that the integral

$$\int_M \Omega \wedge \bar{\omega}$$

with $\bar{\omega}$ the corresponding (0,3)-form, is nonzero. This is straightforward

$$\bar{\omega} \;=\; Q^{\alpha\beta\gamma}\, \chi_{\bar{\mu}_1}{}^{\nu_1}{}_{\alpha}\, \chi_{\bar{\mu}_2}{}^{\nu_2}{}_{\beta}\, \chi_{\bar{\mu}_3}{}^{\nu_3}{}_{\gamma}\, \Omega_{\nu_1\nu_2\nu_3}\, dx^{\bar{\mu}_1}\, dx^{\bar{\mu}_2}\, dx^{\bar{\mu}_3} \tag{5.25}$$

and we have the elementary identity

$$dx^{\bar{\mu}_1}\, dx^{\bar{\mu}_2}\, dx^{\bar{\mu}_3} \;=\; \frac{\Omega^{\bar{\mu}_1\bar{\mu}_2\bar{\mu}_3}}{\|\Omega\|^2}\, \bar{\Omega} \, . \tag{5.26}$$

Thus

$$\bar{\omega} \;=\; Q^{\alpha\beta\gamma}\, \bar{Q}_{\alpha\beta\gamma}\, \|\Omega\|^2\, \bar{\Omega} \tag{5.27}$$

moreover since

$$\Omega \wedge \bar{\Omega} \;=\; \|\Omega\|^2\, g^{1/2} d^6x \tag{5.28}$$

we find that

$$\int_M \Omega \wedge \bar{\omega} \;=\; \int_M Q^{\alpha\beta\gamma}\bar{Q}_{\alpha\beta\gamma}\, \|\Omega\|^4 g^{1/2} d^6x \, . \tag{5.29}$$

The right-hand side of this equality is the integral of a positive real quantity and so cannot vanish.

Finally, the relation that determines the Yukawa coupling is

$$q^{(\alpha}{}_r{}^{\beta}{}_s{}^{\gamma)} \;=\; \kappa Q^{\alpha\beta\gamma} + C^{A(\alpha\beta}{}_t{}^{\gamma)}{}_A + f^{\alpha\beta\gamma} \, . \tag{5.30}$$

6. Symmetries and Pseudosymmetries

Of particular importance for application to the compactification of superstring theories are manifolds M endowed with discrete isometry groups. We shall not attempt a detailed discussion of the implications of such symmetries but will instead make some observations for the case that M may be realized as the transverse intersection of polynomials p^{α} in P_{N+3}. Even in this case we have very little to add to the detailed discussion of Witten [6] who first pointed out that the symmetries and pseudosymmetries of M greatly restrict the possible structure of the couplings $\kappa(a,b,c)$. Indeed for the case of $P_4(5)$ with the simplest quintic (4.24). Witten showed purely on the basis of the pseudosymmetries of the manifold that the coupling between the quintics q, r and s vanishes unless the product qrs contains the term

$$(z^1)^3(z^2)^3(z^3)^3(z^4)^3(z^5)^3 \ .$$

The ratios of nonzero couplings $\kappa(q,r,s)$ i.e., those for which the product qrs does contain the above term are not determined purely on symmetry grounds. Nevertheless it is surprising that it is possible to go so far purely on the basis of symmetries especially since the procedures we have developed in this article has a priori nothing to do with symmetries. The simple but essential observation is that the polynomial $Q^{\alpha\beta\gamma}$ which represents $\bar{\Omega}$ is covariant under the symmetry transformation, a

consequence of this is that noncovariant polynomials are in the tensor ideal $I^{\alpha\beta\gamma}$ and hence correspond to exact forms. More precisely suppose that M has a group Δ of discrete isometries that are induced by a group of $\hat{\Delta}$ of isometries of the embedding space. Then for each $g\varepsilon\hat{\Delta}$ there are unitary matrices $L^A{}_B$ and $\ell^{\alpha}{}_{\beta}$ such that the action of g is represented by

$$z^A \rightarrow L^A{}_B z^B \tag{6.1}$$

and

$$p^{\alpha}(Lz) \ = \ \ell^{\alpha}{}_{\beta} p^{\beta}(z) \tag{6.2}$$

($\ell^{\alpha}{}_{\beta}$ must be unitary to preserve the normal metric $k^{\alpha\bar{\beta}}$). For the simplest cases $\ell^{\alpha}{}_{\beta}$ is just the identity $\delta^{\alpha}{}_{\beta}$ and for these cases we wish to show that for the purposes of computing the coupling it suffices to consider the symmetric part $(q^{(\alpha_r\beta_s\gamma)})_0$ of the product $q^{(\alpha_r\beta_s\gamma)}$. For an arbitrary polynomial ϕ we may make a decomposition into a symmetric and a non symmetric part by writing

$$\phi(z) = \phi_0(z) + \phi_1(z) \tag{6.3}$$

with

$$\phi_0(z) = \frac{1}{|\hat{\Delta}|} \sum_{g\varepsilon\hat{\Delta}} \phi(Lz) \tag{6.4}$$

where $|\hat{\Delta}|$ denotes the order of $\hat{\Delta}$. Now it is easy to see from the explicit representation (4.35) that $Q^{\alpha\beta\gamma}$ is invariant i.e., that

$$Q^{\alpha\beta\gamma} = Q_0^{\alpha\beta\gamma} \tag{6.5}$$

it is also clear that if a polynomial $R^{\alpha\beta\gamma}(z)$ is in the ideal then so is $R^{\alpha\beta\gamma}(Lz)$ for each $g\varepsilon\hat{\Delta}$. Hence if for some polynomial $q^{\alpha\beta\gamma}$ we may write

$$q^{\alpha\beta\gamma} = \kappa Q^{\alpha\beta\gamma} + R^{\alpha\beta\gamma} \tag{6.6}$$

with $R^{\alpha\beta\gamma}$ a remainder in the tensor ideal then

$$q_0^{\alpha\beta\gamma} = \kappa Q^{\alpha\beta\gamma} + R_0^{\alpha\beta\gamma} \tag{6.7}$$

so the symmetric part of $q^{\alpha\beta\gamma}$ leads to the same value of the coupling. In particular the non-symmetric part $q_1^{\alpha\beta\gamma}$ is in the tensor ideal.

Finally we note that representation (4.35) also affords a simple transformation rule for the holomorphic 3-form, relaxing the assumption that $\ell^{\alpha}{}_{\beta}$ is the identity, we have

$$\Omega \to \overline{\det(\ell^{\alpha}{}_{\beta})}\ \Omega \tag{6.8}$$

if det $(\ell^{\alpha}{}_{\beta})$ is not unity then the low energy theory has an R-symmetry.

7. The Normalization Matrix

The elements of the cohomology group $H_{\bar{\partial}}^{(0,1)}(M,T)$ correspond to the families of light particles in the low energy effective theory. An important question of normalization arises. The kinetic terms of the low energy effective Lagrangian will not be diagonal and canonically normalized if a basis for $H_{\bar{\partial}}^{(0,1)}(M,T)$ is chosen arbitrarily. In order to diagonalize the kinetic terms one must diagonalize the matrix

$$N(a^{\mu},b^{\nu}) = \int a \wedge \bar{b} \tag{7.1}$$

where a and b are the (2,1)-forms corresponding to a^{μ} and b^{ν}. The computation of this matrix is an important problem about which little is known. In terms of deformations q^{α},r^{β} we have

$$N(q,r) \propto \int q^{\alpha} r^{\bar{\beta}} \chi^{\mu\nu}{}_{\alpha} \chi_{\mu\nu\bar{\beta}} \; . \tag{7.2}$$

For the special cases of manifolds endowed with large symmetry or pseudosymmetry groups much information about the normalization maxtrix is available on the basis of symmetry alone. However even for the simplest manifolds it does not seem to be known how to determine $N(q^{\alpha},r^{\bar{\beta}})$ completely short of numerical computation.

The reason that the evaluation of $N(q^{\alpha},r^{\beta})$ is difficult seems to be that whereas, for the case of a complete intersection of polynomials in P_{N+3} say, the coupling $\kappa(q^{\alpha},r^{\beta},s^{\gamma})$ are polynomial functions of the coefficients of the defining polynomial p^{α} this is not true of the normalization matrix. It seems that $N(q^{\alpha},r^{\beta})$ depends transcendentally[f4] on the coefficients of p^{α}.

There is an alternative and intriguing way to pose the problem of calculating the normalization matrix it is the following. Consider for definiteness the manifold $P_4(5)$ corresponding to a quintic in P_4. We know from our previous discussion that (2,1)-forms are in correspondence with quintic polynomials in P_5 and (1,2)-forms are in correspondence with tenth order polynomials in P_{10}. Thus every quintic q defines a

f4. I am grateful to P. Griffiths for this observation.

(2,1)-form and hence, by complex conjugation, also a (1,2)-form. The question is to which tenth order polynomial does q correspond?

Acknowledgements

It is a pleasure to acknowledge fruitful conversations with R. Bott, P. Griffiths and R. O. Wells, Jr. The author wishes to thank the Santa Barbara Institute for Theoretical Physics where part of this work was performed.

Appendix: Direct Evaluation of the Coupling for $P_4(5)$

The purpose of this appendix is to record the fact that it is possible, at least for the case of $P_4(5)$, to evaluate the coupling $\kappa(q,r,s)$ directly. Ultimately this is what one must do if one wishes to fix its overall normalization.

We begin by setting

$$\bar{\omega} = a^\mu b^\nu c^\rho \Omega_{\mu\nu\rho} \tag{A1}$$

as in Sect.5 and extending the domain of integration to the whole of P_4 by means of a δ-function

$$\kappa = -\frac{1}{2i}\int_{P_4} \Omega\bar{\omega}\,dp\,d\bar{p}\,\delta^{(2)}(p) \;, \tag{A2}$$

where $\delta^{(2)}(p)$ means $\delta(\mathrm{Re}p)\delta(\mathrm{Im}p)$. To proceed we observe that

$$\delta^{(2)}(p)d\bar{p} = \frac{1}{\pi}\bar{\partial}\left(\frac{1}{p}\right) \tag{A3}$$

and also that

$$\Omega dp = \frac{1}{4!}\epsilon_{ABCDE} z^A dz^B dz^C dz^D dz^E$$

$$= z^5 dz^1 dz^2 dz^3 dz^4 \tag{A4}$$

$$= z^5 d^4 z$$

which follow from (3.5). The second equality holding in the coordinate patch U_5 hence

$$\kappa = \frac{1}{2\pi i.4!}\int_{P_4} \bar{\omega}\bar{\partial}\left(\frac{1}{p}\right) z^5 d^4 z \ . \tag{A5}$$

The next maneuver is to integrate by parts. From eqn (4.4) we have the expression, valid in U_5,

$$\frac{|z^5|^2 \bar{\omega}}{\sigma} = z^5 \frac{qrs}{\| p \|^8} \varepsilon^{\bar{i}\bar{j}\bar{k}\bar{\ell}} \bar{p}_{\bar{I}} d\bar{p}_{\bar{J}} d\bar{p}_{\bar{K}} d\bar{p}_{\bar{\ell}} \tag{A6}$$

where the indices have the range 1,..., 4. It follows that

$$\bar{\partial}\left(\frac{|z^5|^2 \bar{\omega}}{\sigma}\right) = \frac{qrs}{\| p \|^{10}} z^5 p_5 \bar{\tau} \tag{A7}$$

with

$$\tau = \varepsilon^{ABCDE} p_A dp_B dp_C dp_D dp_E \tag{A8}$$

by adding four similar expressions we find

$$\bar{\partial}\bar{\omega} = \frac{5pqrs}{\| p \|^{10}} \bar{\tau} \ . \tag{A9}$$

Notice that there is an explicit factor of p in this expression which cancels the $\frac{1}{p}$ in the integrand. So far we have

$$\kappa = -\frac{5}{2\pi i}\int_{P_4} \frac{qrs}{\|p\|^{10}} z^5 d^4 z\ \bar{\tau}\ . \tag{A10}$$

It is possible to transform the integral to an integral over the nine sphere S^9. This is, of course, a consequence of the fact that S^9 is the Hopf fibration of P_4 but it may be understood in less sophisticated language as a consequence of using the scaling freedom associated with the homogenous coordinates to enforce the coordinate condition $\sigma = 1$ rather than the more usual condition of taking one of the z^A to be a constant. To effect the transformation insert unity in the guise of

$$1 = \frac{i}{2\pi}\int d\eta\, d\bar{\eta}\ \frac{|z^5|^2}{|\eta|^4}\ \delta\left(\sigma - \frac{|z^5|^2}{|\eta|^2}\right) \tag{A11}$$

into the integral and make the change of variable

$$z^k \to \frac{z^k z^5}{\eta}\ , \quad k = 1,\ldots,4 \tag{A12}$$

noting that if a function $F(z^k, z^5)$ is homogeneous of degree m then

$$F\left(\frac{z^k z^5}{\eta}\ ,\ z^5\right) = \left(\frac{z^5}{\eta}\right)^m F(z^k, \eta)\ . \tag{A13}$$

Finally rename the variable $\eta \to z^5$. We find

$$\kappa = \frac{-5}{(2\pi)^2}\int_{C_5} d^5 z\ \bar{\tau}\ d\bar{z}^5\ \frac{qrs}{\|p\|^{10} z^5}\ \delta(\sigma - 1) \tag{A14}$$

where $d^5 z = dz^1 dz^2 dz^3 dz^4 dz^5$. We wish now to eliminate the specific reference to $\bar{z}^5$ in the integrand. To this end note that

$$\tau\frac{dz^5}{z^5} = \varepsilon^{ABCDE}\, p_A dp_B dp_C dp_D dp_E \frac{1}{z^5}\frac{\partial z^5}{\partial p_F} dp_F$$

$$= \varepsilon^{ABCDE}\, \varepsilon_{BCDEF}\, p_A \frac{1}{z^5}\frac{\partial z^5}{\partial p_F} d^5p \qquad \text{(A15)}$$

$$= 4!\frac{1}{z^5}\, p_A \frac{\partial z^5}{\partial p_A} d^5p$$

$$= 3!\, d^5p$$

where $d^5p = dp_1dp_2dp_3dp_4dp_5$ and the last equality follows from the fact that each z^A is a function homogeneous of degree 1/4 of the p_B. There is a further identity which is useful and which may be established analogously

$$\tau\frac{\partial\sigma}{\sigma} = 3\, d^5p. \qquad \text{(A16)}$$

Thus

$$\tau\frac{dz^5}{z^5} = \tau\frac{\partial\sigma}{\sigma}\, . \qquad \text{(A17)}$$

This identity permits us to perform the σ-integration in (A14)

$$\kappa = \frac{-5}{(2\pi)^2}\int_{S_9} \frac{qrs}{\|p\|^{10}}\, d^5z\, \bar{\tau}\, . \qquad \text{(A18)}$$

We have brought the integral to a standard form to which we may apply the Bochner-Martinelli theorem [11]. This leads to a drastic simplication

$$\kappa = \frac{-5}{(2\pi)^2}\int_{\Gamma_5} \frac{qrs\, d^5z}{p_1p_2p_3p_4p_5} \qquad \text{(A19)}$$

where the contour of integration $\Gamma_5 = \gamma_1 \times \gamma_2 \times .. \times \gamma_5$ is the direct product of five 1-dimensional contours that wind around the five curves $p_A = 0$.

We may now recover our previous results in virtue of Cauchy's theorem. First note that we may take the product qrs to be defined modulo I since a shift qrs $\rightarrow$ qrs + $E^A p_A$ has no effect on the integral. For each term of the point $E^A p_A$ that is in the ideal the factor p_A cancels one of the poles in the denominator the corresponding γ_A is contractible to a point. Hence the contribution vanishes. Thus the only integral that needs to be computed is

$$\int_{\Gamma_5} \frac{Q d^5 z}{p_1 p_2 p_3 p_4 p_5} = \int_{\Gamma_5} \frac{d^5 p}{p_1 p_2 p_3 p_4 p_5} = (2\pi i)^5 \tag{A20}$$

where the first equality is a consequence of the fact that

$$Q = \det\left(\frac{\partial^2 p}{\partial z^A \partial z^B}\right) = \frac{\partial(p_1, p_2, \ldots, p_5)}{\partial(z^1, z^2, \ldots, z^5)} \tag{A21}$$

is the Jacobian of the transformation from the p_A to the z^A.

References

1. P. Candelas, G. Horowitz, A. Strominger and E. Witten, Nucl. Phys. B258(1985) 46.

2. A. Strominger and E. Witten, Commun. Math. Phys. 101, 341 (1985).

3. A. Strominger, Phys. Rev. Lett. 55 (1985) 2547, and in "Unified String Theories", M. Green and D. Gross Eds, World Scientific, 1986.

4. K. Kodaira and D. C. Spencer, Ann. Math. 67 (1958) 328, Ann. Math 67 (1958) 403, Ann. Math. 71 (1960) 43.

5. K. Kodaira, Complex Manifolds and Deformations of Complex Structure, Springer-Verlag 1985.

6. E. Witten, Nucl. Phys. B258 (1985) 75.

7. R. Arnowitt, S. Deser and C. W. Misner, "The Dynamics of General Relativity" in Gravitation: An Introduction to Current Research, L. Witten Ed. Wiley, New York 1962.

8. M. F. Atiyah, R. Bott and L. Garding, Acta Mathematica 131, 145 (1973).

9. G. Tian, "Smoothness of the Universal Deformation Space of Compact Calabi-Yau Manifolds and its Peterson-Weil Metric" preprint, Department of Mathematics, University of California, San Diego.

10. S.Kobayashi and K. Nomizu, Foundations of Differential Geometry, Interscience, New York, 1963.

11. P. Griffiths and J. Harris, Principles of Algebraic Geometry, Interscience, New York, 1978.

THE POLYAKOV APPROACH AND DIVERGENCES IN OPEN SUPERSTRINGS

C.P. Burgess†

Institute for Advanced Study
Princeton, NJ 08540, USA

INTRODUCTION

String theories have recently re-emerged, pheonix-like, into the mainstream of particle physics. The principal reasons for this revival are threefold:

i) Superstrings are believed to be renormalizable (possibly even finite), unitary, theories that include gravity as well as the usual gauge and matter interactions [1,2,30,33,34].

ii) Strings 'miraculously' satisfy very stringent requirements of mathematical consistency [2,3,4]. These requirements arise in the form of anomaly-cancellation conditions and are sufficiently restrictive that only a handful (Type I, Types IIa, IIb, and heterotic with gauge groups $E_8 \times E_8$ or $Spin(32)/Z_2$) of candidate string theories survive.

iii) Finally, the Type I and heterotic strings appear to have the potential of producing a realistic phenomenology at energies low enough (*i.e.* $E \ll M_P$) to be accessible to experiment [5]. (For a review see L. Ibáñez, this volume.)

In total, superstrings offer the appealing prospect of providing a (more or less) unique theory of physics at the Planck scale from which (hopefully) definite low-energy predictions may be made. In order to explore the potential of these theories, two questions must be answered.

i) Firstly, *are* all of the superstrings really renormalizable (or finite) and unitary quantum theories of gravity as claimed? Related to this is the question of whether the known strings remain consistent beyond one-loop order.

ii) What are the allowed vacuum states, and what are their low-energy spectra? Apart from permitting an exploration of their phenomenology, a resolution of this question would indicate whether the known strings are really distinct rather than being versions of the same string theory expanded about different vacua.

Several techniques are available to address these issues. One approach is to construct a covariant, second-quantized string field theory [6]. This would allow the application of techniques that are familiar from point-particle field theories. This is not the route followed in this paper. The alternative is to extract the information available in the first-quantized version of the superstring. Although possibly a less

† Research supported in part by an N.S.E.R.C. fellowship and DOE contract DE-AC02-76ER02220

fundamental approach, this is the framework in which virtually all of the presently known properties of string theories have been obtained.

Even within the formalism of the first-quantized string, there are two competing approaches. These differ in the features of string theory that are made manifest. The Green-Schwarz (GS) covariant superstring [7] keeps all of the spacetime (*i.e.* ten-dimensional) symmetries intact, at the expense of world-sheet supersymmetry. The Neveu-Schwarz-Ramond (NSR) formalism [8], on the other hand, keeps the two-dimensional supersymmetry and ten-dimensional Lorentz-invariance manifest but does not have manifest spacetime supersymmetry. Both versions are covariant with respect to world-sheet reparameterizations. The GS approach is particularly convenient in light-cone gauge calculations because of its manifest spacetime supersymmetry, but is intractable in a covariant gauge [7]. For covariant calculations the NSR string is at present easiest to use.

The Polyakov approach [9,10,11] consists of the application of path integral techniques to the covariant quantization of the NSR string. This is the framework within which the remainder of this paper is presented. It is particularly well suited to addressing the problems listed earlier, especially those concerning the quantum properties of strings, such as divergences and anomalies. It is a technique that has been fruitful on the issue of the string vacuum as well, through the connection between the string equations of motion and the quantum properties of two-dimensional sigma models [12].

In this paper the Polyakov technique is applied to the identification of the one-loop divergences that appear in superstring theories. Particular attention is paid to the Type I ($SO(32)$) string. There are two reasons for focussing on this string. Firstly, the Type I string is an open, unoriented string, while the other four on the short list given above are closed and oriented. This makes the perturbative expansion more complicated for the Type I string than for its alternatives. In fact, the graphs that appear in the perturbative expansions for the other strings form a subset of those needed for the Type I string. The result is that there are more potential divergences to deal with in the Type I string and their cancellation is more delicate. The divergence analysis for the other strings generally follows as a corollary to the Type I results.

The second, more predatory, reason for paying particular attention to the Type I string is related to the first. It, together with the heterotic string, is among the most attractive for phenomenological purposes, so it would be particularly interesting if it could be shown to be anomalous. The absence of world-sheet anomalies for the heterotic string at higher loops is understood [4], while the understanding of divergence and anomaly cancellation in the Type I string is more tenuous. Following the philosophy of pushing the weakest link the hardest, it is therefore of interest to better understand divergence cancellations and their connection with anomalies for open superstrings.

The rest of the paper is organized as follows: The following two sections are devoted to a description of the Polyakov formalism with the Type I string in mind. It is fairly self-contained but requires some exposure to the quantization of the free superstring. Section 2 sets up the Polyakov apparatus in the simpler context of the bosonic string and section 3 extends the formalism to the superstring. This is followed, in sections 4 and 5, by an application of these techniques to a discussion of the divergences encountered in one-loop string graphs, and to a demonstration of their cancellation for the Type I string with gauge group $SO(32)$.

The conventions used in what follows are those of refs. [13,14]. In euclidean-signature two-dimensional space the gamma-matrices are: $\gamma_1 = \tau_1$, $\gamma_2 = -\tau_2$, and $\gamma_3 = i\gamma_1\gamma_2 = \tau_3$ where the τ_k denote the Pauli matrices. Left-handed spinors satisfy $\frac{1}{2}(1+\gamma_3)\psi_+ = \psi_+$ and Majorana spinors obey $\psi^* = \gamma_1\psi$ in this basis. Greek indices run from 1 to 2 and upper-case latin indices run from 1 to D, the dimension of spacetime. The curvature conventions are $R^\alpha_{\beta\gamma\delta} = \partial_\delta\Gamma^\alpha_{\beta\gamma} + \ldots$, and $R_{\alpha\beta} = R^\gamma_{\alpha\gamma\beta}$. In two dimensions $R_{\alpha\beta\gamma\delta} \equiv \frac{1}{2}R(g_{\alpha\gamma}g_{\beta\delta} - g_{\alpha\delta}g_{\beta\gamma})$. In complex coordinates $w = \frac{1}{2}(\sigma^1 + i\sigma^2)$, with metric $\delta_{\alpha\beta}$, the euclidean Dirac action $\overline{\psi}\not{\partial}\psi$ is $\psi_+\partial^*_w\psi_+ + \psi_-\partial_w\psi_-$.

BOSONIC STRING S-MATRIX ELEMENTS à la POLYAKOV

The Polyakov approach to the calculation of string scattering amplitudes is most easily described for the bosonic string and then extended to the superstring theories. The presentation here follows that used by refs. [11,13].

The starting point for the Polyakov path-integral calculations in string theories is the expression for S-matrix elements. Suppose the particle states are labelled $|k, a>$ in which k^M denotes the particle's D-momentum and a represents the remainder of the quantum numbers necessary to specify the state. The amplitude for the scattering of E such particles is given by [9,10,11]:

$$\mathcal{S} = \prod_{n=0}^{E} \left(\frac{1}{(2\pi)^{(D-1)/2}\sqrt{2|k_n^0|}} \right) \sum_{\mathcal{M}} \mathcal{S}_{\mathcal{M}}(k_1, \ldots, k_E) \tag{2.1a}$$

$$\mathcal{S}_{\mathcal{M}} = < \prod_{n=1}^{E} \mathcal{V}_n(k_n, a_n) >_{\mathcal{M}} \tag{2.1b}$$

$$< \mathcal{O} >_{\mathcal{M}} = Z_{\mathcal{M}}^{-1} C_{\mathcal{M}} \int \mathcal{D}x^M \mathcal{D}g_{\alpha\beta} e^{-S(x,g)} \mathcal{O} \tag{2.1c}$$

$$Z_{\mathcal{M}} = \int \mathcal{D}x^M \mathcal{D}g_{\alpha\beta} e^{-S(x,g)} \tag{2.1d}$$

The integration variables are the string coordinate, $x^M(\sigma)$, and the world-sheet metric, $g_{\alpha\beta}(\sigma)$. The action appearing in (2.1c) is the bosonic free string action:

$$S(x, g) = \frac{T}{2} \int_{\mathcal{M}} d^2\sigma \sqrt{g} g^{\alpha\beta} \eta_{MN} \partial_\alpha x^M \partial_\beta x^N \tag{2.2}$$

The dimensionful constant T is the string tension, often denoted $1/2\pi\alpha'$. This action is invariant with respect to two-dimensional reparameterizations, D-dimensional Poincaré transformations, and Weyl transformations, defined by: $x^M \to x^M$, $g_{\alpha\beta} \to \Omega^2(\sigma) g_{\alpha\beta}$. There is one local operator, $\mathcal{V}(k, a) = \int d^2\sigma V(\sigma; k, a)$, corresponding to each external particle type. It represents the amplitude for producing a string initial or final state for which only one oscillation mode (corresponding to the particle type of interest) is excited. They are constructed such that the complete path integral shares the symmetries of the classical action.

The sum in (2.1a) is over all choices of world sheet appropriate to the string theory considered. This corresponds to a sum over all compact euclidean two-surfaces, in which two surfaces related by coordinate or Weyl transformations are not regarded as distinct. The contribution of each surface to the total amplitude is weighted by the constant, $C_{\mathcal{M}}$, appearing in (2.12). To calculate the connected part of the S-matrix it suffices to sum over connected world-sheets. There are four cases to consider, distinguished by the choice of closed vs. open, and oriented vs. unoriented strings. For closed strings the sum is, by definition, only over surfaces without boundary. If the string is open, then the sum must also include surfaces with boundary. Similarly, an oriented string includes only orientable surfaces in the sum, while an unoriented string includes both orientable and unorientable world sheets.

In general, any compact, connected euclidean two-surface is equivalent to a sphere with h handles, b discs removed, and c 'cross-caps' [13,15]. (A cross-cap consists of removing a disc and identifying antipodal points on the edge of the resulting hole.) The sum over surfaces therefore boils down to a sum over $h, b, c = 0, 1, 2, \ldots$. A closed string includes only surfaces with $b = 0$ and an oriented string keeps only those with $c = 0$. The Euler number for any such surface is $\chi(\mathcal{M}) = 2 - b - c - 2h$.

In order to complete the specification of the rules for computing graphs it is necessary to construct the vertex operators appropriate to any given particle type, and to give the constants, $C_{\mathcal{M}}$, that weight each graph. There are two methods for doing so. The first would appeal to an underlying formulation, such as a string field theory, in which a derivation of the Feynman rules would supply the operators and factors consistent with principles like unitarity and Lorentz invariance. The alternative [11] is to take eqs. (2.1) and (2.2) as the starting point and to derive the vertex operators by imposing the requirement that their inclusion in eq. (2.1) not spoil the symmetries of the action (2.2). In particular, the conformal anomalies are required to cancel. This determines the critical dimension ($D = 26$ or 10 for the bosonic or superstring respectively) and the anomalous dimension of the various vertex operators. The vertex function for each particle type is completely determined up to normalization. These normalizations and the constants, $C_{\mathcal{M}}$, are then fixed up to a common overall scaling by forcing the resulting S-matrix to be unitary.

The indeterminacy of the overall normalization of these factors corresponds to the freedom of rescaling the string coupling constant, λ. Its appearance is constrained by requiring consistency with unitarity. If the coupling constant is defined so that a specific open-string vertex operator (such as the gauge boson) is proportional to λ, then so are all of the other open-string vertices. The closed-string vertices are then $o(\lambda^2)$ and $C_{\mathcal{M}} = o(\lambda^{-2\chi(\mathcal{M})})$ [13]. For closed strings there are no open-string vertices and so it is customary to define the closed-string coupling by $\lambda_c = \lambda^2$ so all vertices are $o(\lambda_c)$ and $C_{\mathcal{M}} = o(\lambda_c^{-\chi(\mathcal{M})})$ [11]. The ordering of manifolds, $\mathcal{M}$, by powers of string coupling is then: 1) ($h = c = b = 0$): Sphere (S_2)—$o(\lambda^{-4})$; 2) ($h = c = 0, b = 1$): Disc (D_2) and ($h = b = 0, c = 1$): Real Projective Plane (RP_2)—$o(\lambda^{-2})$; 3) ($c = b = 0, h = 1$): Torus (T_2), ($h = b = 0, c = 2$): Klein bottle (K_2), ($h = c = 0, b = 2$): Cylinder (C_2), and ($h = 0, c = b = 1$): Möbius Strip (M_2)—$o(1)$; etc.

The vertex operators for the lowest-lying open, oriented bosonic string states that are obtained in this fashion are (up to normalization) [11,16]:

I) Open string sector:

$$
\begin{aligned}
&M^2 = -2\pi T\text{: [Tachyon] } \mathcal{V}_T(k) = \lambda M_A \epsilon^A \int_{\partial\mathcal{M}} \sqrt{g} : \exp(ik\cdot x) : \\
&M^2 = 0\text{: [Gauge boson] } \mathcal{V}_A(k,a) = \lambda \int_{\partial\mathcal{M}} \sqrt{g} T_A \epsilon^A_M(k,a) t^\alpha : \partial_\alpha x^M \exp(ik\cdot x) : \\
&M^2 = 2\pi T\text{: ....}
\end{aligned}
\tag{2.3}
$$

Here colons denote normal-ordering, t^α is the unit tangent to the boundary of $\mathcal{M}$, T_A is a matrix in the fundamental representation of the gauge group $U(N)$, $Sp(N)$, or $O(N)$. No other choice of gauge group is consistent with unitarity [17,18]. For unoriented open strings world-sheet parity invariance of the path integral rules out the $U(N)$ option. The gauge representation of the tachyon is fixed by the $N\times N$ matrix, M_A, upon which the gauge group is represented by $[T,M] \sim M$. Unitarity implies that they must satisfy various algebraic identities [18]. *e.g.* for $U(N)$ M_A is an arbitrary hermitian matrix and for $O(N)$ it is an arbitrary symmetric matrix. The polarization tensor, $\epsilon_M(k,a)$, is transverse: $k\cdot\epsilon(k,a) = 0$.

II) Closed string sector:

$$
\begin{aligned}
&M^2 = -8\pi T\text{: [Tachyon] } \mathcal{V}_T(k) = \lambda^2 \int_{\mathcal{M}} \sqrt{g} : \exp(ik\cdot x) : \\
&M^2 = 0 \text{ [Dilaton] } \mathcal{V}_D(k) = \lambda^2 \int_{\mathcal{M}} \sqrt{g} \epsilon^{(d)}_{MN} : \big[g^{\alpha\beta} \partial_\alpha x^M \partial_\beta x^N \\
&\qquad\qquad + A\eta^{MN} R(\sigma) \big] \exp(ik\cdot x) : \\
&\qquad\text{[Graviton] } \mathcal{V}_G(k,a) = \lambda^2 \int_{\mathcal{M}} \sqrt{g} \epsilon^{(g)}_{MN}(k,a) g^{\alpha\beta} : \partial_\alpha x^M \partial_\beta x^N \exp(ik\cdot x) : \\
&\qquad\text{[Skew Tensor] } \mathcal{V}_B(k,a) = \lambda^2 \int_{\mathcal{M}} \epsilon^{(b)}_{MN}(k,a) \varepsilon^{\alpha\beta} : \partial_\alpha x^M \partial_\beta x^N \exp(ik\cdot x) : \\
&M^2 = 8\pi T \text{ ....}
\end{aligned}
\tag{2.4}
$$

The polarization tensors, ϵ_{MN}, are transverse on all indices. Those for the graviton and dilaton are symmetric and that for the skew tensor is antisymmetric. The graviton's is also trace-free. $R(\sigma)$ denotes the world-sheet scalar curvature and A is a specific constant determined by the requirement that the path integral be Weyl invariant [19]. $\varepsilon^{\alpha\beta}$ is the completely antisymmetric tensor density on the world sheet.

For closed strings the vertex functions that couple only to the boundary of the world sheet obviously drop from the spectrum. Similarly for unoriented strings all vertex functions that are odd under world-sheet parity must be omitted. This is because the integral of a pseudoscalar on an unorientable surface is not well defined.

The integral over the metric in eq. (2.1c) is easily simplified. There are three field variables in a symmetric tensor, $g_{\alpha\beta}$, in two dimensions. There are also three parameters available in the reparametrization and Weyl symmetries. It is therefore always possible, locally, to choose the gauge, $g_{\alpha\beta} = \delta_{\alpha\beta}$, in which the metric is trivial. On the sphere, disc and projective plane this gauge may also be chosen globally [13,15]. On all of the other world sheets such a gauge is not globally possible.

Consider the torus as an example. Taking the torus as the unit square on which the string coordinate and metric are periodic, the question is whether it is possible to reach the gauge $g_{\alpha\beta} = \delta_{\alpha\beta}$ from an arbitrary metric *through coordinate and Weyl transformations that are all themselves periodic*. Although Weyl transformations may always be used to arrange that $g_{11} = 1$, for example, it may not be possible to choose a coordinate transformation to go the rest of the way. This is because the transformation law of the metric involves derivatives of the coordinate transformation parameter which must be inverted to determine the transformation required to take a given metric to the trivial one. In general the relevant differential operator has zero modes and so cannot be inverted [20] subject to the required boundary conditions.

As a result there is, in general, a t-parameter family of metrics, $\hat{g}_{\alpha\beta}(\tau^i), i = 1, ..., t$ to which any metric is gauge-equivalent. These parameters are the Teichmüller parameters of the world sheet. This family includes, but is not exhausted by, the trivial metric $\delta_{\alpha\beta}$. To see what this family is on the torus, consider starting at the trivial metric and examining which small fluctuations are not pure gauge. Expanding all fluctuations and transformation parameters in a Fourier series shows that it is the constant (σ-independent) mode of the metric that cannot be gauged away. Using Weyl invariance to set $g_{11} = 1$ allows the most general gauge-fixed metric to be written:

$$\hat{g}_{\alpha\beta} = \begin{pmatrix} 1 & x \\ x & x^2 + y^2 \end{pmatrix} \tag{2.5}$$

in which the Teichmüller parameters, x and y, are constants. The metric is parameterized in this peculiar way because the quantities appearing in the functional integral then naturally occur in the complex combination $\tau = x + iy$.

A similar analysis [13] on the other spaces that appear at the same order in perturbation theory as does the torus (*i.e.* K_2, C_2, and M_2) shows that the most general gauge-fixed metric is given in this case by eq. (2.5) with $x = 0$.

In general, then, the functional integral over the metric degenerates into an ordinary integration over the Teichmüller parameters. There is generally a Jacobian, $J(\tau^i)$, [9,20] that must be included in changing from the $g_{\alpha\beta}$ variables to the τ^i's. The functional integral in eq. (2.1c) becomes:

$$< \mathcal{O} >_{\mathcal{M}} = Z_{\mathcal{M}}^{-1} C_{\mathcal{M}} \int d^t\tau J(\tau) \int \mathcal{D}x^M e^{-S(x,\hat{g}(\tau))} \mathcal{O} \tag{2.6}$$

Since the vertex functions are polynomials in x^M times an exponential whose argument is linear in x^M, the remaining functional integral over x^M is gaussian and may be done explicitly. The integral over the zero mode $x^M = constant$ ensures momentum conservation, although some care must be taken to ensure that the normalization used for the zero modes is the same as that used for the rest of the modes. The result is (up to irrelevant constant factors that are eventually cancelled by $Z_{\mathcal{M}}$ in eq. (2.1c)):

$$< \mathcal{O} >_{\mathcal{M}} = C_{\mathcal{M}}(2\pi)^D \delta^D(k_1 + ... + k_E) \int d^t\tau \left(\frac{V_{\mathcal{M}}(\tau)}{2\pi}\right)^{D/2} J(\tau)[\det{}'(-\Box)]^{-D/2}$$

$$\left(\prod_{n=1}^{E} \int d\sigma_n\right) \exp\left[\frac{1}{2T} \sum_{m,n=1}^{E}{}' k_m \cdot k_n N_{\mathcal{M}}(\sigma_m, \sigma_n)\right]$$

$$\times \{ \text{ polynomial in } k_n^M, \partial_\alpha N_{\mathcal{M}}\}$$

$$\equiv C_{\mathcal{M}} \int d^t\tau \mathcal{F}(\tau) \left(\prod_{n=1}^{E} \int d\sigma_n\right) \mathcal{J}_{\mathcal{M}}(\sigma_n, \tau) \tag{2.7}$$

This last line defines the functions $\mathcal{F}$ and $\mathcal{J}$ in terms of the corresponding quantities in the line above. The functional determinants are defined by zeta-function regularization. $N_{\mathcal{M}}(\sigma, \sigma')$ denotes the x^M-propagator on the world sheet of interest. This satisfies Neumann conditions on surfaces with boundary. The polynomial in momentum and the derivatives of $N_{\mathcal{M}}$ arise from evaluating the contractions between the x^M's in the different vertex functions making up $\mathcal{O}$. The functional determinants and Neumann functions for the various spaces appearing up to one loop in string perturbation theory are given explicitly in ref. [13]. $V_{\mathcal{M}}(\tau) = \int_{\mathcal{M}} \sqrt{g}$ denotes the volume of the surface $\mathcal{M}$.

All that remains is to evaluate the remaining finite-dimensional integrals for the spaces of interest and to sum their contributions to get the S-matrix. This is particularly simple on the tree-level spaces, S_2, D_2 and RP_2, for which there are no Teichmüller integrations.

SUPERSTRING AMPLITUDES à la POLYAKOV

All of the above formalism can be extended to superstring theories. The basic new feature is the imposition of world-sheet supersymmetry in addition to the symmetries noted earlier. The Type I (II) theory is obtained by making the action and vertex operators of the open, unoriented (closed, oriented) bosonic string invariant under one left-handed and one right-handed supersymmetry [22]. Both strings therefore contain a supermultiplet of fields, $\{x^M, \psi^M_\pm, F^M\}$, all of which carry a spacetime vector index, coupled to the world-sheet supergravity multiplet consisting of zweibein and gravitino fields, $\{e_\alpha{}^a, \chi^\alpha_\pm\}$. ψ^M is a world-sheet spinor and the auxiliary field, F^M, is a scalar.

One might expect there to be two more string theories of this type: a supersymmetric, closed, unoriented string or an open, oriented superstring. These alternatives are thought not to be unitary because of their low-energy limits [22]. The closed, unoriented superstring becomes $N = 1$, ten-dimensional supergravity in this limit, which is known to have gravitational anomalies [3]. The oriented, open superstring becomes $N = 2$ supergravity coupled to $N = 1$ super-Yang-Mills theory in the low-energy limit. Apart from having anomalies, this theory couples one of its gravitino fields to an unconserved supercurrent and so is expected to be sick.

The heterotic string is constructed in a more subtle way [2]. Starting with the closed, oriented bosonic string, only symmetrize with respect to *right – handed* supersymmetry transformations. This introduces a left-handed field ψ_+^M, a right-handed gravitino, χ_-^α and auxiliary fields. As it stands this theory has two-dimensional reparameterization and conformal anomalies but these may be cancelled by adding a set of 32 right-handed world-sheet fermions that transform as spacetime scalars, together with their superpartners (auxiliary fields) under the right-handed supersymmetry. This string is not pursued in what follows since divergence cancellation in the Type I string is of principal interest.

Some of the features of the Polyakov method applied to superstrings that are not shared by bosonic strings are summarized in the following paragraphs:

(i) Spin Structures and Boundary Conditions:

In the same way that the sum over world sheets can be thought of as a sum over the allowed boundary conditions for the two-dimensional field, x^M, it is necessary, in superstring theory, to sum over all possible boundary conditions allowed for the world-sheet fermions [23]. Since they must respect two-dimensional supersymmetry, a choice of boundary conditions for the field ψ^M determines those for the remaining fermionic fields. The contribution of each choice of boundary condition to the S-matrix is a priori weighted by some constants, $C_{\mathcal{M},(a)}$, whose values are to determined, up to coupling-constant dependence, by unitarity. Here (a) schematically denotes the fermion boundary condition under consideration.

For the open superstring $\psi_+^M = \pm\psi_-^M$ is chosen on the boundaries of the world sheet. Moving past a Ramond (R) sector, open-string vertex on the boundary flips the sign in the boundary condition. The corresponding vertex operator cannot, therefore, be a single-valued function of the two-dimensional field variables. It can either be thought of as being proportional to a double-valued "spin field" [24], or as being defined on a two-valued covering surface with the vertex position acting as a 'branch point'. The boundary condition is unchanged when a NS-sector vertex is passed.

In general additional 'periodicity conditions' must also be chosen to completely specify a spinor field theory on a two-surface. For every independent non-contractable, orientation-preserving [1] loop of $\mathcal{M}$ [13], a condition of the general form $\psi_\pm^M(2\pi) = P_{(\pm)}{}^M{}_N \psi_\pm^N(0)$ must be chosen. For orientation-reversing loops the same kind of condition is necessary but relates $\psi_\pm(2\pi)$ to $\psi_\mp(0)$. The complete specification of these condition is called the choice of a spin structure. If the boundary and periodicity conditions are to respect the entire D-dimensional Lorentz group, the matrix $P^M{}_N$ must be $\pm\delta^M{}_N$. In this case the choice of spin structure boils down to a choice of a sign for all independent non-contractable loops. Notice that left- and right-handed fermions need not have the same spin structure.

Just as for open-string vertices, closed-string vertex functions need not be single-valued functions of the two-dimensional fields [24]. It is convenient to think of the position of the vertex function as a puncture in the world sheet about which a spin structure must be chosen. If the operator corresponds to a closed-string NS-type particle, then the vertex is single-valued and the periodicity condition for fermions

[1] A loop is said to be orientation-preserving (-reversing) if a set of basis vectors keeps (doesn't keep) its orientation when parallel-propagated about the loop.

about a loop encircling the puncture allows the loop to be continuously shrunk to a point. It is the R-type operators that are multivalued (or sit at branch points).

The necessity for these multivalued operators is due to the existence of interactions that couple the NS and R sectors together. Since the two sectors satisfy different conditions on the world sheet, the operator that describes their interaction *must* change these conditions and so cannot be a single-valued function of the fields. In order to avoid such complications only scattering amplitudes involving NS-type particles are considered here.

(ii) World-Sheet Symmetries:

As is the case for any theory that is both locally supersymmetric and Weyl invariant, the resulting superstring theories automatically enjoy a further symmetry called super-Weyl invariance. Under this symmetry the gravitino transforms as $\delta\chi_\alpha = \gamma_\alpha\eta$, for some spinor $\eta(\sigma)$, with all other fields invariant. This symmetry can, and will, always be used to ensure that $\gamma\cdot\chi = 0$. The supersymmetry transformations then locally allow the gravitino to be gauged to zero. Just as was the case for the metric, it may not be possible to reach this gauge from an arbitrary field configuration through supersymmetry transformations that respect the boundary conditions at hand [13,25]. The modes of the gravitino that cannot be so removed are the super-Teichmüller modes. There are none on the sphere, disc or projective plane [13,25]. On the torus and other spaces with Euler number zero, these modes are constants and only appear for some of the spin structures. The gravitino path integral degenerates to the integral over a finite number of Grassmann variables. There is a Jacobian factor, $\tilde{J}^{-1}(\tau)$, coming from the transformation from χ_α to the super-Teichmüller variables.

(iii) Anomalies:

Being intrinsically chiral, the heterotic string is rife with potential world-sheet anomalies. Anomalies in infinitessimal reparameterizations and Weyl transformations cancel in the critical dimension regardless of the boundary conditions. The vertex operators must have the correct anomalous dimension, to ensure that the entire path integral is conformally invariant. The same is true for the Type I and II strings in the critical dimension.

Having ensured invariance with respect to infinitessimal reparameterizations, the next step is to ensure that there are no 'global' [4,26] reparameterization anomalies that would spoil invariance with respect to those transformations not homotopic to the identity. Since the transformation properties of the path integral under these 'big diffeomorphisms' are sensitive to the boundary conditions satisfied by the fields, the condition for the absence of anomalies gives information about what fermion boundary conditions are allowed. It is this condition (when the world sheet is a torus) that selects the gauge groups $E_8\times E_8$ or $Spin(32)/Z_2$ for the heterotic string [2,4]. For the Type I and II theories with the $SO(D-1,1)$-invariant spin structures the same arguments give no information beyond the requirement that $D = 10$.

For more complicated boundary conditions, however, the absence of these anomalies can give nontrivial information.

(iv) Spin Structure Sums:

For some world sheets, such as the torus, there are big diffeomorphisms that change the boundary conditions of the spinor fields. Requiring that the path integral be invariant with respect to these transformations is a condition that fixes the relative size of the constants $C_{\mathcal{M},(a)}$ for the spin structures so related [23]. This condition is simpler to use, when available, than directly using unitarity to fix these constants. If the left- and right-handed sectors of the string are required to be separately invariant (up to an overall phase) then the resulting relation among the $C_{\mathcal{M},(a)}$'s implies the GSO projection [27] that ensures the spacetime supersymmetry in the operator approach [23].

(v) Scattering Amplitudes:

For the superstring the analogues of eqs. (2.1) are:

$$\mathcal{S} = \prod_{n=0}^{E} \left(\frac{1}{(2\pi)^{(D-1)/2}\sqrt{2|k_n^0|}} \right) \sum_{\mathcal{M}} \mathcal{S}_{\mathcal{M}} \tag{3.1a}$$

$$\mathcal{S}_{\mathcal{M}} = < \prod_{n=1}^{E} \mathcal{V}_n(k_n, a_n) >_{\mathcal{M}} \tag{3.1b}$$

$$< \mathcal{O} >_{\mathcal{M}} = \int \mathcal{D}x^M \mathcal{D}F^M \mathcal{D}g_{\alpha\beta} \sum_a Z_{\mathcal{M},(a)}^{-1} C_{\mathcal{M},(a)} \int \mathcal{D}\psi_{(a)}^M \mathcal{D}\chi_\alpha^{(a')} e^{-S(x,\psi,F,g,\chi)} \mathcal{O} \tag{3.1c}$$

$$Z_{\mathcal{M},(a)} = \int \mathcal{D}x^M \mathcal{D}F^M \mathcal{D}g_{\alpha\beta} \mathcal{D}\psi_{(a)}^M \mathcal{D}\chi_\alpha^{(a')} e^{-S(x,\psi,F,g,\chi)} \tag{3.1d}$$

The fermionic integration in the normalization factor is defined to omit any zero mode integrations. The supersymmetric version of the action (2.2) appropriate to Type I and II strings is:

$$S(x,\psi,F,g,\chi) = \frac{T}{2} \int d^2\sigma \sqrt{g}\, \eta_{MN} \Big[g^{\alpha\beta} \partial_\alpha x^M \partial_\beta x^N + \overline{\psi}^M \partial\!\!\!/ \psi^N - F^M F^N - (\overline{\chi}_\alpha \gamma^\beta \gamma^\alpha \psi^M) \partial_\beta x^N - \frac{1}{8} (\overline{\psi}^M \psi^N)(\overline{\chi}_\alpha \gamma^\beta \gamma^\alpha \chi_\beta) \Big] \tag{3.2}$$

Anticipating the imposition of the super-Weyl gauge condition discussed earlier, terms in eq. (3.2) that vanish when $\gamma \cdot \chi = 0$ have been dropped.

The vertex functions for the NS sector are single-valued and are most easily found by supersymmetrizing the bosonic string vertices. For some purposes, such as tree graphs, and one-loop graphs involving a sufficiently small number of external lines, it is sufficient to know those terms in the vertex functions that do not involve the world-sheet gravitino. These are given here for the graviton and Type I gauge boson:

$$\mathcal{V}_A(k,a) = \lambda T_A \epsilon_M^A(k) \int_{\partial\mathcal{M}} \sqrt{g} t^\alpha : \Big[\partial_\alpha x^M + \frac{i}{2} k_N (\overline{\psi}^M \gamma_\alpha \psi^N) \Big] e^{ik\cdot x} : \quad + \chi_\alpha\text{-terms} \tag{3.3}$$

$$\mathcal{V}_G(k,a) = \lambda^2 \epsilon_{MN}(k) \int_{\mathcal{M}} \sqrt{g} : \Big[g^{\alpha\beta} \partial_\alpha x^M \partial_\beta x^N + \overline{\psi}^M \partial\!\!\!/ \psi^N - F^M F^N$$

$$
- ik_P(\overline{\psi}^P\gamma^\alpha\psi^M)\partial_\alpha x^N + ik_P(\overline{\psi}^P\psi^M)F^N - \frac{i}{2}k\cdot F(\overline{\psi}^M\psi^N)
$$
$$
-\frac{1}{4}k_Pk_Q(\overline{\psi}^P\psi^Q)(\overline{\psi}^M\psi^N)\Big] e^{ik\cdot x} : +\chi_\alpha \text{ -terms} \tag{3.4}
$$

Since the Type I string is open and unoriented, the matrices T_A, generate either $Sp(N)$ or $O(N)$.

The metric and gravitino integrations greatly simplify. If the (Grassmann) super-Teichmüller variables are denoted p^i, $i = 1, ..., \tilde{t}$, then the functional integral (3.1c), after gauge-fixing becomes:

$$
< \mathcal{O} >_{\mathcal{M}} = \int d^t\tau J(\tau) \int \mathcal{D}x^M \mathcal{D}F^M \sum_a Z^{-1}_{\mathcal{M},(a)} C_{\mathcal{M},(a)} \tilde{J}^{-1}_{(a')}(\tau)
$$
$$
\int \mathcal{D}\psi^M_{(a)} d^{\tilde{t}}p_{(a')} e^{-S(x,\psi,F,\hat{g}(\tau),p^i)}\mathcal{O} \tag{3.5}
$$

Once the metric and gravitino are in their gauge-fixed forms, and the trivial integral over the super-Teichmüller parameters are performed, the remaining functional integrals over x^M, ψ^M and F^M are gaussian and can be done explicitly.

Define the function (including zero-mode normalization factors):

$$
\mathcal{H}_{\mathcal{M},(a)}(\tau) = V_{\mathcal{M}}^{-n_{(a)}/2}\tilde{J}^{-1}(\tau)[\det{}'(\nabla\!\!\!/)]^{D/2} \tag{3.6}
$$

where $V_{\mathcal{M}}$ is defined following eq. (2.7) and $n_{(a)}$ is the number of zero modes for the field ψ^M with spin structure (a). ∇ is the world-sheet covariant derivative, constructed using the gauge-fixed metric. Because all world-sheet anomalies vanish, the determinant of $\nabla\!\!\!/$ is well defined. The path integral becomes:

$$
< \mathcal{O} >_{\mathcal{M}} = \int d^t\tau \sum_a Z^{-1}_{\mathcal{M}} C_{\mathcal{M},(a)} \mathcal{F}_{\mathcal{M}}(\tau)\mathcal{H}_{(a)}(\tau) \int \left(\prod_{n=1}^{E} d\sigma_n\right)
$$
$$
\exp\left[\frac{1}{2T}\sum_{m,n=1}^{E}{}' k_m\cdot k_n N_{\mathcal{M}}(\sigma_m,\sigma_n)\right]\Big\{ \text{ polynomial in } \partial N_{\mathcal{M}}, G_{\mathcal{M},(a)}, k^M\Big\}
$$
$$
\equiv \int d^t\tau \sum_{(a)} Z^{-1}_{\mathcal{M}} C_{\mathcal{M},(a)} \Lambda^{(a)}_{\mathcal{M}}(\tau) \int_{\mathcal{N}} \left(\prod_{n=1}^{E} d\sigma_n\right) \mathcal{I}_{\mathcal{M}}(\sigma_n,\tau) \tag{3.7}
$$

Here $G_{\mathcal{M},(a)}(\sigma,\sigma')$ denotes the Dirac propagator for the two-surface and spin structure of interest. The polynomial terms in the propagators, their derivatives and the external momenta come, as before, from evaluating contractions involving the fields appearing in the vertex functions, as well as from terms that multiplied the super-Teichmüller modes in the action. The functions $\mathcal{F}$ and $\mathcal{H}$ are as defined in eqs. (2.7) and (3.6). They are irrelevant constants for the tree-level spaces and are calculated explicitly for the one-loop surfaces in ref. [13,14] as are the relevant propagators. Λ is defined as the combination $\Lambda = \mathcal{F}\mathcal{H}$. The last line of eq. (3.7) defines $\mathcal{I}$.

DIVERGENCES ARISING FROM VERTEX FUNCTION INTEGRATIONS

Any divergences in the final expressions for the amplitude, eqs. (2.7) and (3.7), must come when the finite-dimensional integrals over the positions of the vertex functions or the Teichmüller variables are performed. These divergences can appear in two ways: in the integrations over the Teichmüller parameters, or over the positions of the vertex functions. Since the integrand does not diverge for finite, nonzero values of the Teichmüller parameters this integral can only diverge at the boundaries of the integration region. These are the infinities that cancel in the Type I superstring if the gauge group is $O(32)$.They are discussed in more detail in section 5.

The second source of potential infinities is the integration over the positions of the vertex functions. All of the surfaces considered are compact and so these integrals can only diverge when the integrand blows up. This occurs when some of the vertex functions approach one another. The purpose of the present section is to sketch the singular behaviour of the amplitude (3.7) coming from this part of the vertex-function integration region. The analysis follows that of refs. [28,29]. Since this is not the divergence whose cancellation in the Type I string is the principal purpose of this paper, many of the details of the arguments are omitted.

The coincident behaviour of the Neumann propagator for any two-surface is given in complex coordinates, $w = \sigma^1 + i\sigma^2$, with the Weyl gauge $ds^2 = dw\, dw^*$, by:

$$N_{\mathcal{M}}(w, w') \sim \frac{1}{4\pi} \log |w - w'|^2 + ... \qquad (4.1a)$$

for $w \to w'$ with w' in the interior of $\mathcal{M}$. If w' should lie on the boundary, $\partial\mathcal{M}$, with coordinates chosen to put the boundary near w' on the real axis, then the coincident limit is:

$$N_{\mathcal{M}}(w, w') \sim \frac{1}{4\pi} \left(\log |w - w'|^2 + \log |w^* - w'|^2\right) + ... \qquad (4.1b)$$

It is therefore straightforward to identify the divergent contributions to eq. (3.7) coming from that part of the integration region in which any n vertex functions approach one another with the Teichmüller parameter held fixed. Because of the exponential term in eq. (3.7) the resulting singularity is momentum dependent, diverging when the n external momenta are chosen to put an intermediate particle on shell. This gives part of the mass-shell discontinuity of the S-matrix corresponding to the exchange of an intermediate state between the n coincident vertex functions and the $E - n$ remaining vertices. The rest of the discontinuity comes from the singularity that arises when the other $E - n$ vertices approach one another with the original n kept apart. As pointed out in ref. [28] the divergent behaviour ocurring when the original n vertices come together *and* the remaining $E - n$ also approach one another gives the shift in the position of the lower order discontinuity due to radiative corrections to the mass of the exchanged state.

If the n vertex functions correpond to closed-string states, then the singular behaviour is given by the factorized diagram obtained by 'factoring out' a sphere from the original surface. Alternatively, if the vertex functions that approach one another represent open-string states, the singularity corresponds to factoring out a disc. In either case the divergent part of the amplitude is proportional to a reduced amplitude on the original space involving the emission of a closed (or open) string intermediate state multiplied by the production amplitude for this state with the remaining vertices on the sphere (or disc).

Choosing the momenta of the n external states that are brought together such that no exchanged particle is put on shell, ensures that the integral in eq. (3.7) does *not* diverge as the vertex functions approach one another. The only bona fide infinities that can arise in the vertex-function integrations are therefore those that are momentum-independent. They occur when it is not possible to choose momenta to avoid the mass-shell singularity of the factorized diagram. As advertised earlier this happens when *all* or *all but one* of the vertex functions approach one another. In these two cases momentum conservation requires that the exchanged particle be on shell, giving a momentum-independent divergence in the amplitude. From the form of the factorized amplitude it is easy to determine conditions under which these infinities vanish.

If all but one of the vertex functions approach one another, the divergence is proportional to the two-point function, evaluated on the original surface, for the particle associated with the lone vertex. This two-point function has the interpretation of a radiative correction to the mass of the particle in question and vanishes for particles whose masses are protected by a symmetry (such as massless particles like the graviton or gauge boson) [28]. It therefore need not be considered when investigating divergences in amplitudes only involving these particles. Such infinities can a priori appear in amplitudes containing massive external states, starting at one loop (torus) for the closed, oriented superstrings, and at tree level (disc and projective plane), for the Type I string.

This same factorization argument shows that the divergence that corresponds to bringing all of the vertex functions together is proportional to the zero momentum dilaton tadpole, evaluated on the surface in question. This vanishes for the sphere by virtue of its conformal symmetries. Potential divergences of this type on the disc and projective plane are not considered here since their cancellation is ensured once the one-loop graphs are shown to be finite. For superstrings (in flat ten-dimensional spacetime) the one-loop surfaces give vanishing contributions. The simplest way to see this is to recognize that the dilaton vertex operator evaluated at zero momentum is given by the graviton vertex operator with polarization tensor $\epsilon_{MN} \sim \eta_{MN}$, apart from world-sheet curvature terms that are irrelevant for the one-loop spaces. When $k^M = 0$, this is equal to the string action appearing in eq. (3.2). Since the string action is proportional to the string tension, T, the zero momentum dilaton tadpole is proportional to the derivative of the vacuum energy with respect to T. The equal spacetime fermi-bose count of the superstring ensures vanishing vacuum energy, [2] so this tadpole must vanish graph-by-graph for each of the superstring theories at one-loop.

Clearly the only one-loop divergences of worry for the superstring are therefore those that arise in the Teichmüller integrations. These are addressed in the next section.

---

[2] For explicit calculations of the flat-space vacuum energy via the Polyakov path integral for the various strings see refs. [13,14,31].

DIVERGENCES IN TEICHMÜLLER INTEGRATIONS

Recall that the four spaces appearing in one-loop graphs are the torus, T_2, the Klein bottle, K_2, the cylinder, C_2, and the Möbius strip, M_2. Big diffeomorphisms of the surface need not preserve the gauge-fixed metric and so induce an action in the space of Teichmüller parameters. Only those values of the Teichmüller variables that are not related by these 'modular transformations' are to be considered physically distinct and so it suffices to integrate over the quotient of Teichmüller space divided by the modular group. This space for the four one-loop surfaces is:

$$\begin{aligned} \tau = x + iy; \quad & y > 0, \quad |\tau| > 1, \quad |x| < \frac{1}{2} \quad (T_2) \\ \tau = iy; \quad & y > 0 \qquad\qquad (K_2, C_2, M_2) \end{aligned} \tag{5.1}$$

The integrand is well behaved for all finite τ so the only potential divergences in the τ-integration come from the boundaries $|\tau| \to \infty$ (for all four spaces) and $|\tau| \to 0$ (for K_2, C_2 and M_2). For the torus the limits $|\tau| \to 0$ or ∞ are related by the modular transformation $\tau \to -1/\tau$ [32] and so are not independent. This is reflected by the omission of the former limit in the space of moduli. The key difference between the remaining spaces and the torus is that their group of big diffeomorphisms does not induce a change in the Teichmüller parameter and so no regions of Teichmüller space are excluded by considerations of modular invariance.

A second striking fact is that the Teichmüller and moduli spaces for K_2, C_2 and M_2 are identical. This is a prerequisite for the cancellation of Teichmüller divergences among these spaces.

a) Consider first the $|\tau| \to \infty$ limit common to all four spaces. In order to identify the divergent behaviour of the Teichmüller integral of eqs. (2.1) or (3.1) it is necessary to know the asymptotic form of the integrand as $|\tau| \to \infty$. The asymptotic forms of the functional determinants and propagators are found in ref. [29]. The main observation is that the scalar propagator for the torus, given (up to an irrelevant, additive constant) by:

$$N_T(w, w') = \frac{1}{4\pi} \log \left| \frac{\vartheta_1(w - w'|\tau)}{\vartheta_1'(0|\tau)} \right|^2 - \frac{1}{2} \left(\frac{[Im\,(w - w')]^2}{Im\,\tau} \right) \tag{5.2}$$

approaches in the $y \to \infty$, $w - w'$ fixed, limit a nonsingular limit given by:

$$N_T(w, w') \to \frac{1}{4\pi} \log \left| \sin[\pi(w - w')] \right|^2 - \frac{1}{2} \left(\frac{[Im\,(w - w')]^2}{Im\,\tau} \right) + o(e^{2i\pi\tau}) \tag{5.3}$$

The same is true for the propagators on the other three spaces, since they are obtained from eq. (5.2) by the method of images.

The asymptotic behaviour of the functional determinants therefore determine the convergence of the Teichmüller integration for large y and fixed u_{mn}.

In the open superstring inspection of the results of ref [29] shows that for Ramond boundary conditions these determinants fall off at least as fast as y^{-6} for large y, thereby ensuring convergence.

The Neveu-Schwarz sector at first appears to diverge exponentially, just as does the bosonic string. For example the functional determinants for the spin structures of the Klein bottle behave asymptotically like:

$$\begin{aligned}
\Lambda_K^{(+,1)}(y) &\to \frac{\sqrt{2}}{4}\left(\frac{\pi}{2}\right)^{-D/2}(1+y^{-2})^{-1/2}y^{-2-D/2}[1+o(e^{-6\pi y})] \\
\Lambda_K^{(+,2)}(y) &\to \frac{1}{16}\left(\frac{\pi}{4}\right)^{-D/2}y^{-1-D/2}[1+o(e^{-6\pi y})] \\
\Lambda_K^{(-,1)}(y) &\to \frac{1}{4}\pi^{-D/2}y^{-1-D/2}e^{\pi y(D-2)/8}[1-(D-2)e^{-\pi y}+o(e^{-2\pi y})] \\
\Lambda_K^{(-,2)}(y) &\to \frac{1}{4}\pi^{-D/2}y^{-1-D/2}e^{\pi y(D-2)/8}[1+(D-2)e^{-\pi y}+o(e^{-2\pi y})]
\end{aligned} \tag{5.4}$$

This exponential divergence cancels when the spin sum is performed. This cancellation comes about because the leading term in the asymptotic form of the Dirac propagator as $y \to \infty$ is the same for both choices of Neveu-Schwarz boundary conditions. All terms in the divergent part of the integrand therefore factor out of the spin sum except for those coming from the functional determinants themselves. The resulting cancellation is then a consequence of the Jacobi identity $\sum \pm\vartheta^4 = 0$ that ensures supersymmetry in the NSR formalism. Since this potential exponential divergence can be traced to the tachyon of the Neveu-Schwarz sector, the cancellation corresponds in the operator formalism to omitting the tachyon via the GSO projection onto odd fermion number [23,27]. After the cancellation of this exponential divergence the remaining terms fall off as y^{-6} again ensuring convergence.

b) Turn now to the limit of small $|\tau|$. Only K_2, C_2 and M_2 need be considered. The scalar propagator on the torus in this limit is given by:

$$N_T(w,w') \to \frac{1}{4\pi}\log\left|\sin[\pi\frac{(w-w')}{\tau}]\right|^2 + \frac{1}{2}\left(Im\ \frac{(w-w')^2}{\tau} - \frac{[Im\ (w-w')]^2}{Im\ \tau}\right) + o(e^{-2i\pi/\tau}) \tag{5.5}$$

If $|\tau| \to 0$ with u_{mn} fixed, the propagators become singular and the exponential factor of the integrand (2.7) or (3.7) behaves as:

$$\prod_{m,n=1}^{E}{}' \left|e^{-\pi[Im\ (u_{mn}^2/\tau)-(Im\ u_{mn})^2/Im\ \tau]}\sin[\pi\frac{u_{mn}}{\tau}]\right|^{k_m\cdot k_n/4\pi T} \tag{5.6}$$

giving at best a momentum-dependent singularity rather than a divergence.

The limit that leads to a momentum-independent divergence is $\tau \to 0$, u_{mn}/τ fixed. This is the limit that is related, for the torus, by modular invariance to the limit that was considered earlier: $y \to \infty$, u_{mn} fixed. Comparison of eqs. (5.3) and (5.5) shows that the propagators approach the same limit as they did when $y \to \infty$ so the convergence is again dictated by the asymptotic behaviour of the functional determinants.

For the Type I superstring, for two of the four spin structures on each surface the divergence appears to be exponential, but this exponential behaviour disappears when the sum over spin structures is performed. After cancellation of the exponentially diverging terms, the subleading behaviour of these amplitudes is $\sim y^2$, so their integral converges.

The only boundary conditions for which the $y \to 0$ limit is singular are $(+,1)$ on K_2, and $(++)$ on C_2 and M_2. See ref. [14] for the spin structure notation. These three choices of boundary conditions behave asymptotically as y^{-2} and so their integral diverges linearly as $y \to 0$. This is the infinity that must cancel amongst these three graphs. Their asymptotic form for small y is given by:

i) Klein Bottle:

$$\Lambda_K^{(+,1)}(y) \to \frac{\sqrt{2}}{4}\left(\frac{\pi}{2}\right)^{-D/2}(1+y^2)^{-1/2}y^{-2}[1+o(e^{-6\pi/y})] \tag{5.7}$$

ii) Cylinder: $\Lambda_C = 2^{-D}\Lambda_K$.

iii) Möbius Strip:

$$\Lambda_M^{(++)}(y) \to \frac{\sqrt{2}}{2}\pi^{-D/2}(1+y^2)^{-1/2}y^{-2}[1+o(e^{-2\pi/y})] \tag{5.8}$$

In order to exhibit the divergence it is convenient to regulate all three graphs by cutting them off at $y = \epsilon$. Because the singularity is only linear in ϵ^{-1} there are no subleading infinities so only the leading, small-y behaviour of the integrand is necessary to determine the divergent part of the amplitude.

In the absence of any open-string external states (graviton-dilaton scattering) all three graphs contribute so the divergent part of eq. (3.7) is given by:

$$< \mathcal{O}_c >_\infty = \int_\epsilon dy \sum_{\mathcal{M}=K,C,M} C_{\mathcal{M},(++)}\Lambda^{(++)}_{\mathcal{M},\infty}(y) \int_{\mathcal{M}} \left(\prod_{n=1}^{E} d^2 w_n\right) \mathcal{I}_{\mathcal{M},\infty}(w_n, y) \tag{5.9a}$$

In this expression the Klein Bottle boundary condition $(+,1)$ is denoted $(++)$ for convenience. Λ_∞ denotes the leading term in the small-y expansion of the function $\Lambda = \mathcal{F}\mathcal{H}$ (defined after eq. (3.7)) given explicitly in (5.7) and (5.8). Similarly $\mathcal{I}_\infty$ denotes the function $\mathcal{I}$ of eq. (3.7) evaluated using the propagators evaluated in the limit $y \to 0$ with $v_{mn} \equiv u_{mn}/iy$ fixed. Finally, the subscript 'c' of $\mathcal{O}_c$ is intended to emphasize that only closed-string vertex operators appear.

For amplitudes that also involve external gauge bosons only the cylinder and Möbius strip contribute and the divergent part is:

$$< \mathcal{O}_{o,c} >_\infty = \int_\epsilon dy \sum_{\mathcal{M}=C,M} C_{\mathcal{M},(++)}\Lambda^{(++)}_{\mathcal{M},\infty}(y) \int_{\mathcal{N}} \left(\prod_{n=1}^{E} dw_n\right) \mathcal{I}_{\mathcal{M},\infty}(w_n, y) \tag{5.9b}$$

c) Divergence Cancellation:

i) Closed-String External Lines Only:

Consider first the case in which none of the external states come from the open string. In this case the conditions under which eq. (5.9a) vanishes are required.

The key point is that, as is shown below, the integral, $A_{red} \equiv \int (\prod dw_n) \mathcal{I}_{\mathcal{M},\infty}$, is independent of the space $\mathcal{M}$ under consideration. This allows the complicated part of the divergent amplitude to be factored out of the sum over $\mathcal{M}$. The condition for finiteness is then that the remaining sum:

$$\Delta_c \equiv \sum_{\mathcal{M}=K,C,M} C_{\mathcal{M},(++)} \Lambda^{(++)}_{\mathcal{M},\infty} \tag{5.10a}$$

be zero.

To establish the equivalence, for differing $\mathcal{M}$, of the integral over $\mathcal{I}$ notice that the propagators on the three spaces are all identical in the limit that $y \to 0$ with $v_{nm} = u_{nm}/iy$ fixed. (The fact that additive, w-independent constants may be freely dropped from the scalar propagator because momentum conservation keeps any such term from contributing to the $\mathcal{S}$-matrix—see eq. (3.7)—is required in establishing this result.) The limiting form of the scalar propagator is given explicitly by:

$$N_{K,C,M}(w,w') \to \frac{1}{4\pi} \log \left| \sin \left[\pi \frac{(w-w')}{iy} \right] \right|^2 + \frac{1}{2y} Re\,(w+w') \tag{5.11}$$

The common factor, $A_{red}(k_1, ..., k_E)$, is a reduced amplitude, evaluated on the disc with a puncture in it, involving the original external states. The reduced amplitude is calculated on the punctured disc because this is the region that the Klein bottle, cylinder and Möbius strip reduce to in the limit that $y \to 0$. Around the puncture the world-sheet fermions satisfy periodic boundary conditions. See ref. [29] for details.

The explicit form for Λ_∞ given above and the Chan-Paton [17] factors, $C_{C,(++)} = N^2 C_{K,(++)}$ and $C_{M,(++)} = \mp N C_{K,(++)}$, for gauge groups $SO(N)$ and $Sp(N)$ respectively, allow eq. (5.9a) to be written:

$$\begin{aligned} < \mathcal{O}_c >_\infty &= A_{red}(k_1, ..., k_E) \Delta_c \int_\epsilon \frac{dy}{y^2} \\ \Delta_c &= C_{K,(++)} \frac{\sqrt{2}}{4} (2\pi)^{-5} (2^D + N^2 \mp 2^{D/2+1} N) \\ A_{red}(k_1, ..., k_E) &= \int_{D_2} \left(\prod_{n=1}^{E} d^2 w_n \right) \mathcal{I}_\infty (v_n) \end{aligned} \tag{5.12a}$$

In the last of eqs. (5.12a) D_2 represents the disc and $v_n = w_n/iy$.

Clearly the divergence is proportional to $(N \mp 32)^2$ and so vanishes only for the gauge group $SO(32)$ as claimed.

ii) Open- (and Closed-) String External Lines:

The argument here is much the same as just used. As before the intregration over $\mathcal{I}$ factors out of the expression (5.9b) giving a reduced amplitude, A_{red}, that now involves open-string vertices on the boundary of the disc.

The only new feature is that because some of the vertices must lie on the boundary of the graph in question, the requirement that $v_{mn} = u_{mn}/iy$ be fixed (or, equivalently, $|w_n - w_m| \to 0$) requires all vertex functions to approach some point on $\partial\mathcal{M}$.

On the cylinder this is impossible if there are open-string vertex functions on *both* boundaries. These (non-planar) open-string diagrams therefore do not contribute to this divergence. Furthermore, the cylinder graph with all open-string vertices on the same boundary contributes *twice* to the divergent part of the amplitude (5.9*b*); once for each of the boundaries that all of the vertex functions can be on.

The gauge-boson scattering analogue of eq. (5.10*a*) is:

$$\Delta_o = 2C_{C,(++)}\Lambda^{(++)}_{C,\infty} + C_{M,(++)}\Lambda^{(++)}_{M,\infty} \tag{5.10b}$$

and so eq. (5.9*b*) becomes:

$$\begin{aligned} < \mathcal{O}_o >_\infty &= A_{red}(k_1, ..., k_E)\Delta_o \int_\epsilon \frac{dy}{y^2} \\ \Delta_o &= C_{K,(++)}\frac{\sqrt{2}}{4}(2\pi)^{-5}(2N \mp 2^{D/2+1}) \end{aligned} \tag{5.12b}$$

Again finiteness requires gauge group $SO(32)$.

CONCLUSIONS

In this paper the Polyakov approach to the calculation of string S-matrix elements is reviewed and applied to the exhibition of the potential divergences that can occur in Type I superstring amplitudes. For one-loop amplitudes the various divergences are shown not to contribute to S-matrix elements involving arbitrary numbers of external gravitons, dilatons and gauge bosons when the gauge group is $SO(32)$.

The advantage of the Polyakov path integral approach is its manifest covariance, which allows the infinities to be easily identified and displayed. The divergences can be traced to two sources in the Polyakov formula for the S-matrix. Some potential divergences arise when the integrations over the positions of the vertex functions are performed. These can also appear in the Type II and heterotic strings. It is argued that because they are proportional to either the zero-momentum dilaton tadpole or the radiative mass shift of one of the scattered particles they vanish at one loop for amplitudes only involving massless external states. Factorization implies the absence of the corresponding divergences for the lower-order spaces, the disc and RP_2 [30], for gauge group $SO(32)$, once the one-loop graphs have been proven finite.

A divergence that only arises for the Type I string comes from the integration over the Teichmüller parameters of the one-loop two-surfaces. In particular the Klein bottle, cylinder and Möbius strip all have one real Teichmüller variable, y, that must be integrated from zero to infinity. This integral diverges at the $y \to 0$ end for one choice of spin structures on each of these three spaces. The divergence factorizes into the product of a common reduced amplitude calculated on the disc times numerical factors coming from the asymptotic form of the functional determinants for each space. The divergent parts, when summed over these three graphs, cancel when the gauge group is $SO(32)$.

This result includes that of earlier authors [30,33] who demonstrated in light-cone gauge that four- and five-point gauge-boson scattering amplitudes are finite for gauge group $SO(32)$. A similar result holds [34] for the n-point gauge amplitudes for restricted external states.

A great simplification of the Polyakov computation relative to those done previously is the absence of subleading divergences $\sim \log \epsilon$. Presumably their absence here is a reflection of the manifest symmetry of the Polyakov approach.

Some care must be taken with the regulation of the divergences that cancel amongst the various one-loop graphs. The prescription used here is the most natural from the Polyakov point of view. The Klein bottle, cylinder and Möbius strip are all related in their definitions to a torus with points identified under the action of some involution, as is discussed in ref. [13]. In this approach the Teichmüller variable on the quotient space is inherited from that on the torus. This common torus modular parameter, $\tau = iy$, is here cut of at a lower limit ϵ.

When this regularization is transcribed into the language of light-cone gauge operators, such as in the appendix of [13,14], it corresponds to differing choices of cutoff for the different operator contributions to the amplitude. Explicitly, the torus, Klein bottle, cylinder and Möbius strip cutoffs in the light-cone approach, when translated from the path integral formalism, are given by $\epsilon_T = \epsilon$, $\epsilon_K = \epsilon/2$, $\epsilon_C = 2\epsilon$ and $\epsilon_M = \epsilon$ respectively. This corresponds to the results of refs. [30,33,34] where it was found that in light-cone computations different cutoffs were required for the cylinder and Möbius strip in order to ensure finiteness.

As noted elsewhere [30,33,34], there appears to be an ambiguity in the choice of regularization corresponding to the freedom to redefine Teichmüller variables differently for the three spaces before imposing the cutoff. In string theory there should be no such freedom to choose regularizations independently for different graphs. Indeed if such a choice were available the divergences discussed here could be arranged to cancel for *any* gauge group by suitably choosing the cutoffs. As in all questions of the relation between different two-surfaces in the Polyakov approach, unitarity must dictate the proper cutoff for the Klein bottle, cylinder and Möbius strip given that for the torus. Because of the connection between these divergences and spacetime anomalies in the Type I string, and the fact that the low-energy, field theory limit of the Type I string is anomaly free only for gauge group $SO(32)$ [3] it is expected that the regularization used here is the choice that is consistent with unitarity.

Although all divergences cancel for one-loop amplitudes involving massless bose particles in the Type I string, additional arguments beyond those used here are required to ensure that all amplitudes are finite. In particular the analysis should be extended to include spacetime fermions and higher loops. If all massless amplitudes are finite then the finiteness of the massive ones should follow by factorization of the massless ones.

Acknowledgements

The author would like to thank Tim Morris for collaboration on some of the work reported here, as well as C.S. Lam, D.X. Li and Mark Mueller for useful discussions.

References

1) M.B. Green and J.H. Schwarz, Nucl. Phys. **B181** (1976) 502; **B198** (1982) 252; **B198** (1982) 441; and Phys. Lett. **109B** (1982) 444; M.B. Green, J.H. Schwarz, and L. Brink, Nucl. Phys. **B198** (1982) 474; E. Martinec, Phys. Lett. **171B** (1986) 189.

2) D.J. Gross, J. Harvey, E. Martinec, and R. Rohm, Phys. Rev. Lett. **54** (1984) 502; Nucl. Phys. **B256** (1985) 253; and Nucl. Phys. **B267** (1986) 75.

3) L. Alvarez-Gaumé and E. Witten, Nucl. Phys. **B234** (1983) 269; M.B. Green and J.H. Schwarz, Phys. Lett. **149B** (1984) 117; and Nucl. Phys. **B255** (1985) 93; C.M. Hull and E. Witten, Phys. Lett. **160B** (1985) 398.

4) E. Witten, in "Geometry, Anomalies and Topology", edited by W.A. Bardeen and A.R. White, World Scientific (1985).

5) P. Candelas, G.J. Horowitz, A. Strominger and E. Witten, Nucl. Phys. **B258** (1985) 46.

6) E. Witten, "Interacting Field Theory of Open Superstrings" Princeton preprint (1986); Nucl. Phys. **B268** (1986) 253; T. Yoneya, to appear in the proceedings of the 7th Workshop on Grand Unification, Toyama, Japan (1986); H. Hata, K. Itoh, T. Kugo, H. Kunitomo and K. Ogawa, Kyoto preprints; Phys. Lett. **175B** (1986) 138; Phys. Lett. **172B** (1986) 186, 195; A. Neveu and P.C. West, Phys. Lett. **168B** (1986) 192; S.P. de Alwis and N. Ohta, Phys. Lett. **174B** (1986) 383; N.P. Chang, H.Y. Guo, Z. Qiu and K. Wu, CCNY-HEP 86/5 (1986); A. Jevicki, Phys. Lett. **169B** (1986) 359; J. Lykken and S. Raby, LA-UR-1334 (1986).

7) M.B. Green and J.H. Schwarz, Nucl. Phys. **B218** (1983) 43; Nucl. Phys. **B243** (1984) 285; M.B. Green, J.H. Schwarz and L. Brink, Nucl. Phys. **B219** (1983) 437.

8) P. Ramond, Phys. Rev. **D3** (1971) 2415; A. Neveu and J.H. Schwarz, Nucl. Phys. **B31** (1971) 86; Phys. Rev. D4 (1971) 1109; J.-L. Gervais and B. Sakita, Nucl. Phys. **B34** (1971) 477.

9) A.M. Polyakov, Phys. Lett. **103B** (1981) 207, 211.

10) D.B. Fairlie and H.B. Nielsen, Nucl. Phys. **B20** (1970) 637; C.S. Howe, B. Sakita, and M.A. Virasoro, Phys. Rev. **D2** (1970) 2857; O. Alvarez, Nucl. Phys. **B216** (1983) 125; D. Freidan, E. Martinec and S. Shenker, Nucl. Phys. **B271** (1986) 93.

11) S. Weinberg, Phys. Lett. **156B**(1985), 309;

12) E.S. Fradkin and A.A. Tseytlin, Phys. Lett. **158B** (1985) 316; Phys. Lett. **160B** (1985) 69; and Nucl. Phys. **B261** (1985) 1; C.G. Callan, E.J. Martinec and M.J. Perry, Nucl. Phys. **B262** (1985) 593; C.G. Callan, I. R. Kebanov, and M.J. Perry, "String Theory Effective Actions" Princeton preprint (1986); B.E. Fridling and A. Jevicki, Phys. Lett. **174B** (1986) 75; C.M. Hull and E. Witten, Phys. Lett. **160B** (1985) 398; A. Sen, SLAC-PUB-3794 (1985); S.P. de Alwis, Texas preprint UTTG-15-86 (1986).

13) C.P. Burgess and T.R. Morris, "Open and Unoriented Strings à la Polyakov", I.A.S. preprint (1986).

14) C.P. Burgess and T.R. Morris, "Open Superstrings à la Polyakov", I.A.S. preprint (1986).

15) M. Schiffer and D.C. Spencer, "Functionals of Finite Riemann Surfaces" Princeton University Press (1954).

16) E. del Guidice, P. DiVecchia and S. Fubini, Ann. Phys. **70** (1972) 378.

17) J.Paton and H.M.Chan, Nucl. Phys. **B10**(1969)519

18) J.H. Schwarz, in the proceedings of the Johns Hopkins Workshop, (1982); N. Marcus and A. Sagnotti, Phys. Lett. **119B** (1982) 97.

19) S. de Alwis, Phys. Lett. **168B** (1986) 59; C.G. Callan and Z. Gan, Nucl. Phys. **B272** (1986) 647.

20) O. Alvarez, Nucl. Phys. **B216** (1983) 125; and in the proceedings of the Workshop on Unified String Theories, Santa Barbara (1985).

21) L. Brink, P. DiVecchia and P. Howe, Phys. Lett. **65B** (1976) 471; S. Deser and B. Zumino, Phys. Lett. **65B** (1976) 369.

22) J.H. Schwarz, Phys. Rep. **89C** (1982) 223.

23) N. Seiberg and E. Witten, "Spin Structures in String Theory" Princeton-IAS preprint, (1986).

24) D. Friedan, S. Shenker and E. Martinec, Phys. Lett. **160B** (1985) 55; J. Cohn, D. Friedan, Z. Qiu and S. Shenker, Chicago preprint EFI 85-90-Rev. (1986).

25) G. Moore, P. Nelson and J. Polchinski, Phys. Lett. **169B** (1986) 47; E.D' Hoker and D.H. Phong, Columbia preprint CU-TP-340 (1986);

26) E. Witten, Comm. Math. Phys. **100** (1985) 197.

27) F. Gliozzi, J. Scherk and D.I. Olive, Phys. Lett. **65B** (1976) 282; Nucl. Phys. **B122** (1977) 253.

28) S. Weinberg, Texas preprint UTTG-22-85 (1985).

29) C.P. Burgess, "Finiteness and the Type I Superstring", IAS preprint (1986).

30) M.B. Green and J.H. Schwarz, Phys. Lett. **151B** (1985) 21.

31) J.Polchinski, Comm. Math. Phys. **104** (1986) 37; E. D'Hoker and D.H.Phong, Nucl. Phys. **B269** (1986) 205; E.D' Hoker and D.H.Phong, Columbia preprint CU-TP-340 (1986).

32) J. Shapiro, Phys. Rev. **D5** (1972) 1945.

33) H. Yamamoto, Y. Nagahami and N. Nakazawa, Hiroshima preprint RRK 86-20; P.H. Frampton, P. Moxhay, Y.J. Ng, North Carolina preprint IFP-256-UNC (1986); R. Potting and J. Shapiro, (to appear) Phys. Rev. **D**;L. Clavelli, (to appear) Phys. Rev. **D**.

34) P.H. Frampton, P. Moxhay, Y.J. Ng, Phys. Rev. Lett. **55** (1985) 2107; L. Clavelli, Phys. Rev. **D33** (1986) 1098.

ON THE EVALUATION OF SUPERSTRING ANOMALIES

R.B. Mann

Dept. of Physics
University of Toronto
Toronto, Ontario
Canada M5S 1A7

Abstract

The current divergences of the parity violating part of the one-loop graph in $2n$ dimensions with $n+1$ external gauge particles in type I superstring theories are evaluated using a variety of regulating techniques. In contrast to the point field theory case, the current divergences are not equivalent to one another but depend upon the scheme used, vanishing in some schemes and diverging in others. However in all schemes the anomaly is proportional to $N + 32\ell$.

Introduction

The recent interest in superstring theories as candidate "Theories of Everything" is primarily founded upon the results reported by Green and Schwarz [1,2] concerning the parity violating part of the one-loop amplitude

$$\int \frac{d^{2n}p}{(2\pi)^{2n}} Tr[\Delta V(1)\Delta \cdots \Delta V(n+1)(\frac{1}{2}(1+\Gamma_{2n+1}))] \qquad (1)$$

in $D = 2n$ dimensions. For $n = 5$ $(D = 10)$, which is needed to avoid ghosts in the covariant formalism (or to preserve Lorentz invariance in the light-cone formalism), Green and Schwarz found that such amplitudes in general rendered the theory anomalous, except in the special case of SO(32) in type I superstring theory [1] (they subsequently showed [2] in the low energy effective theory resulting from the superstring that $E_8 \otimes E_8$ was another possible anomaly-free theory; this led to the construction of the heterotic string [3]). In their computation [1], Green and Schwarz used a Pauli-Villars regulator and a Gaussian regulator, obtaining differing results for the form of the anomaly (although cancellation of anomalies occurred for SO(32) in each case).

It is well-known in point-particle field theory that the evaluation of the anomaly crucially depends upon the well-known fact that naïve shifts of integration variable in more than logarithmically divergent integrals are not permitted [4,5,6]. Indeed, it is possible to make use of this fact to evaluate the anomaly entirely in terms of such loop momentum routing ambiguities without imposing any *a-priori* symmetry requirements such as Bose-symmetry or vector current conservation [7,8]. In $2n$ dimensions the vector-current and axial-vector-current divergences are (graph-by-graph) uniquely computable in terms of surface terms which may be parametrized as a linear combination of the external momenta; the anomaly arises because there are at most n independent parameters available to satisfy $n+1$ current conservation relations. Such arbitrariness in the loop momentum routing has been shown to be equivalent to ambiguities inherent in the Pauli-Villars technique [9], as well as in conventional dimensional regularization (CDR), regularization by dimensional reduction (RDR) [8], and the aforementioned Gaussian technique. These shift-of-integration variable techniques have also been shown to be useful in evaluating loop diagrams in supersymmetric theories [10].

This paper describes the results of the evaluation of the current divergences of the graphs associated with the parity violating part of the amplitude (1) in the type I superstring theory using the above techniques. In analogy with the point-particle case, the current divergences of a given graph (both the planar and the non-orientable graphs are separately considered) depend upon differing ambiguities inherent in each scheme. Unlike the familiar point particle case, however, the answers are not equivalent but depend upon the regulating technique employed, although for any scheme freedom from anomalies is guaranteed provided the group is SO(32). The relationship between this result and the results of reference [1] is discussed. In particular, it is pointed out that the Pauli-Villars technique replaces the evaluation of the current divergences of the parity violating graph with the evaluation of another graph. In the pre-regulating shift-of-integration variable method [10], the current divergences are multiplied by ground-state expectation values which *vanish* as a consequence of the equality of two adjacent Koba-Nielsen variables [11]. Since this is true for any graph in the amplitude, all current divergences associated with the parity-violating part of the amplitude vanish using this technique. In CDR and RDR the current divergences become infinite as the regulating parameters approach their limiting values. In the Gaussian technique the anomaly arises as a consequence of a delicate cancellation between a zero due to a cancelled propagator and an infinity due to the vanishing radius of the inner loop of the annulus. Using this last technique, it is possible to understand the anomaly as arising due to a discontinuity in the parameters that regulate the momentum integration and the string oscillator trace.

In order to understand the situation in string theory, it is useful to review the situation in the point-particle case [12]. The VVA triangle graph, $S_{\lambda\mu\nu}$ may be written as

$$S_{\lambda\mu\nu} = \int \frac{d^4k}{(2\pi)^4} Tr[\gamma_5\gamma_\lambda \, \not{k}^{-1}\gamma_\mu(\not{k}+\not{q}_1)^{-1}\gamma_\nu(\not{k}+\not{q}_1+\not{q}_2)^{-1}] \tag{2}$$

where q_1 and q_2 are the two external momenta flowing into the graph. The full amplitude consists of the sum of this graph and the associated crossed graph.

The above integral is linearly divergent, and various regulating schemes may be used to evaluate it. The usual approach is to impose symmetry requirements in the full amplitude, *ie.* vector current conservation and Bose-symmetry and then compute the remaining axial-vector-current divergence. The non-zero result is the usual chiral anomaly; it is scheme-independent. This scheme-independence in fact extends to each graph $S_{\lambda\mu\nu}$ in the amplitude without imposing any symmetry requirements.

The general form of the current divergences of $S_{\lambda\mu\nu}$ is

$$q_0^\tau S_{\tau\mu\nu} = \frac{-X_1}{(2\pi)^2}\epsilon_{\mu\rho_1\nu\rho_2}q_1^{\rho_1}q_2^{\rho_2} \tag{3a}$$

$$q_1^\tau S_{\mu\tau\nu} = \frac{X_2}{(2\pi)^2}\epsilon_{\mu\rho_1\nu\rho_2}q_1^{\rho_1}q_2^{\rho_2} \tag{3b}$$

$$q_2^\tau S_{\mu\nu\tau} = \frac{-X_3}{(2\pi)^2}\epsilon_{\mu\rho_1\nu\rho_2}q_1^{\rho_1}q_2^{\rho_2} \tag{3c}$$

Each method has *a-priori* distinct ambiguities associated with the evaluation of $S_{\lambda\mu\nu}$. In the pre-regulating technique, there is an ambiguity in the routing of the loop-momentum which arises due to the linear divergence of $S_{\lambda\mu\nu}$ [7]. In the Gaussian scheme, one inserts (after Wick rotation) a factor of $e^{-\eta k^2}$ to regulate the graph; there is an ambiguity in the choice of Gaussians one may introduce. In the Pauli-Villars scheme, there is an ambiguity in choice of the introduction of the massive regulators [9]. In CDR there is an ambiguity in the location of the (non-anticommuting) γ_5 [7], and in RDR the ambiguity is in the ordering of matrices in the trace since trace-cyclicity is abandoned [7]. Despite these widely differing ambiguities, the current divergences of each scheme are equivalent, *ie.*, there is a $1-1$ mapping between the X-parameters in each scheme. Superficially, each method has a 3-fold ambiguity inherent in the redefinition of $S_{\lambda\mu\nu}$ that each method necessarily imposes. However the X-parameters obey $X_1 + X_2 + X_3 = 1$ in any scheme; this leads to a unique scheme independent result for the chiral anomaly [12]. In fact, the alternating sum of current divergences in (3 a-c) yields

$$q_0^\tau S_{\tau\mu\nu} - q_1^\tau S_{\mu\tau\nu} + q_2^\tau S_{\mu\nu\tau} = \frac{-1}{(2\pi)^2}\epsilon_{\mu\rho_1\nu\rho_2}q_1^{\rho_1}q_2^{\rho_2} \tag{3d}$$

The right hand side is half the usual Chern-Pontryagin invariant [12].

To summarize, the current divergences of any graph in the VVA-amplitude in point field theory are equivalent in any of the above regulating schemes (independent of the imposition of any physical constraints) despite the widely differing ambiguities inherent in each scheme. This result generalizes to any (even) number of dimensions [8]. In fact, it is possible to show that each graph in the V^nA amplitude is finite in $2n$ dimensions [9]. Hence any regulating scheme must yield the same result for the V^nA graph apart from the appearance of the arbitrary (but finite) X-parameters which depend upon the scheme employed and which arise due to the linear divergence of the graph.

The anomaly in (open) superstring theory arises from the current divergences of the parity violating 1-loop graphs [1]. There are three types of graph that occur in the amplitude: planar (in which all incoming/outgoing particles attach to one boundary of the world-sheet annulus), non-planar (in which incoming/outgoing particles attach to both boundaries) and the Möbius strip, or non-orientiable diagram. The anomaly arises due to contributions from the first and last types of graph [1]. As in point-field theory there are $n+1$ emissions in each graph, and each graph is linearly divergent. In the open string case in 10 dimensions ($n = 5$) the result of Green and Schwarz, using the Pauli-Villars scheme (with vector-current conservation imposed), is that [1]

$$k_6 \cdot \Gamma = -(N+32l)\frac{(4\pi)^5}{(i\sqrt{2})^4}\epsilon(\xi,k)\int \prod_{i=1}^{5} d\nu_i \theta(\nu_{i+1}-\nu_i) < 0|\prod_{i=1}^{6} V_0(k_j, e^{2\pi i\nu_j})|0> \tag{4}$$

where the conventions regarding commutators, vertices, and propagators are the same as in [1]:

$$V(i) \equiv V(k_i, \xi_i, 1) \tag{5a}$$

$$V(k, \xi, z) = \xi \cdot \Gamma(z) V_0(k, z) \tag{5b}$$

$$V_0(k, z) = \; : exp[ik \cdot X(z)] : \tag{5c}$$

$$\Gamma^\mu(z) = \gamma^\mu + i\sqrt{2}\gamma_{2n+1} \sum_{n=1}^{\infty} (d_n^\mu z^{-n} + d_{-n}^\mu z^n) \tag{5d}$$

$$\Gamma_{2n+1} = \gamma_{2n+1}\, exp(i\pi \sum_{n=1}^{\infty} d_{-n} \cdot d_n) = \gamma_{2n+1}\Gamma_d \tag{5e}$$

$$X(z) = x^\mu - ip^\mu ln(z) + i\sum_{n=1}^{\infty} \frac{1}{n}(\alpha_n^\mu z^{-n} - \alpha_{-n}^\mu z^n) \tag{5f}$$

$$F_0 = \frac{1}{i\sqrt{2}}\, \not{p} + \gamma_{2n+1} \sum_{n=1}^{\infty} (\alpha_{-n} \cdot d_n + \alpha_n \cdot d_{-n}) \tag{5g}$$

$$L_0 = F_0^2 = \frac{1}{2}p^2 + N \tag{5h}$$

$$N = \sum_{n=1}^{\infty} (\alpha_{-n} \cdot \alpha_n + n d_{-n} \cdot d_n) \tag{5i}$$

$$\epsilon(\xi, k) = \epsilon_{\alpha_1 \rho_1 \cdots \alpha_5 \rho_5} \xi_1^{\alpha_1} \cdots \xi_5^{\alpha_5} k_1^{\rho_1} \cdots k_5^{\rho_5} \tag{5j}$$

The k's are external momenta and the ξ's are the polarization vectors of the external states. Freedom from anomalies is guaranteed provided the right-hand side of (4) vanishes; this is possible if $N + 32l = 0$, where N is the dimension of the group, and $l = (-1, 0, 1)$ depending on whether the group is orthogonal, unitary or symplectic. Hence the only anomaly free group in the open string case is SO(32) [1].

Green and Schwarz also evaluated the anomaly using Gaussian regulators. In this case they obtained a different form for the anomaly for the right-hand side of eq. (4), although the overall coefficient of $N + 32l$ was the same (hence the SO(32) anomaly-free result is not altered). It is of interest, therefore, to see how the anomaly arises using other regulating schemes, and to see to what extent the previously mentioned point-field theory results are carried over to the string case. In analogy with the point-field theory case, the current divergences of a given graph in the amplitude will be considered without imposing any physical constraints such as current conservation. The results are summarized here; details will appear in a forthcoming paper [13].

Consider first the parity violating part of the planar graph in $2n$ dimensions. It is given by the following expression:

$$T = \int d^{2n}p Tr[\prod_{i=1}^{n+1} \frac{1}{F_0} V(i) \Gamma_{2n+1}] \tag{6}$$

The r-th current divergence of (6) shall be denoted by T_r; it is obtained by replacing ξ_r by k_r and then using the relation [1]

$$V(k, k, 1) = (i\sqrt{2})[F_0, V_0(k, 1)]. \tag{7}$$

Results for the evaluation of T_r using a variety of techniques will now be summarized.

Pauli-Villars

In the Pauli-Villars scheme, each graph T in the amplitude is replaced by $T \to T - T^M$ where T^M is formed from T by replacing the fermion propagators F_0 by 'massive' propagators $F_0 - iM$. Since

$$\Gamma_{2n+1}\frac{1}{F_0 - iM} = -\frac{1}{F_0 + iM}\Gamma_{2n+1}$$

there is an ambiguity associated with the introduction of the massive propagators that is identical to that in point-field theory [9]. This ambiguity may be parametrized as

$$T^R = T - T^M \tag{8}$$

where T^R is the regulated graph, and

$$T^M = \sum_{j=1}^{n+1} P_j \int d^{2n}pTr[\prod_{i=1}^{j} \frac{1}{F_0 - iM}V(i) \prod_{l=j+1}^{n+1} \frac{1}{F_0 + iM}V(l)\Gamma_{2n+1}] \tag{9}$$

Each of T and T^M are linearly divergent; however T^R is required to be free of divergences. This forces a constraint on the P's that is the same as in point field theory [9]:

$$\sum_{j=1}^{n+1} P_j = -1 \tag{10}$$

Evaluation of T_r proceeds in a manner similar to that in [1]; the result is

$$T_r = (-1)^r P_r \frac{(-4\pi)^n}{(i\sqrt{2})^{n-1}} \epsilon_r(\xi, k) \times \int \prod_{i=1}^{n} d\nu_i \theta(\nu_{i+1} - \nu_i) < 0| \prod_{j=1}^{n+1} V_0(k_j, e^{2\pi i \nu_j})|0 > \tag{11}$$

where

$$\epsilon_r(\xi, k) = \epsilon_{\alpha_1\rho_1\cdots\alpha_{r-1}\rho_{r-1}\alpha_r\rho_r\cdots\alpha_n\rho_n}\xi_1^{\alpha_1}\cdots\xi_{r-1}^{\alpha_{r-1}}\xi_{r+1}^{\alpha_r}\cdots\xi_{n+1}^{\alpha_{n+1}}k_1^{\rho_1}\cdots k_n^{\rho_n}$$

and where

$$\nu_{n+1} \equiv 1 \ .$$

This result is completely analogous to the point-field theory result in [9]. Each Möbius strip diagram gives a similar contribution. Requiring vector-current conservation in the full amplitude fixes the n independent V's, yielding the result of reference [1]. Alternatively, the alternating sum of current divergences in the full amplitude yields

$$\sum_{r=0}^{n} \Gamma_r = (N + 2^n l)\frac{(-4\pi)^n n!}{(i\sqrt{2})^{n-1}} \epsilon_r(\xi, k) \times \int \prod_{i=1}^{n} d\nu_i \theta(\nu_{i+1} - \nu_i) < 0| \prod_{j=1}^{n+1} V_0(k_j, e^{2\pi i \nu_j})|0 > \tag{12}$$

which only vanishes in the case SO(32) when $n = 5$.

Pre-regularization

The integral in (6) is linearly divergent; hence there is an ambiguity associated with the arbitrariness in the routing of loop momentum in such an integral [1,5,7]. Rewriting (6) as

$$T = \int d^{2n}p' Tr[\prod_{i-1}^{n+1} \frac{1}{F_0} V(i)\Gamma_{2n+1}] \tag{13}$$

parametrizes this ambiguity: the quantity p' in the integration measure is related to p by $p = p' + s$; s reflects the arbitrariness inherent in the graph of the routing of the external momenta through the loop. The most general form of s is

$$s^\mu = \sum_{i=1}^{n} A_i q_i^\mu$$

The r-th current divergence T_r is given by the difference of two linearly divergent integrals, each of which has a cancelled propagator beside the r-th vertex operator. Such an expression is ill-defined because the product of two V_0's is ill defined [1,14]. In order to deal with this, a regulator of the form $f(\eta, N)$ such that $f(0, N) = 1$ needs to be inserted between these two vertices, with the $\eta \to 0$ limit taken at the end. This gives

$$T_r = \frac{(-1)^{r+1}(4\pi)^n}{n(i\sqrt{2})^{n-1}} \epsilon_r(\xi, k)$$

$$\times \int \prod_{i=1}^{n-1} d\nu_i \theta(\nu_{i+1} - \nu_i)\{(A_{r+1} - A_r) < 0| \prod_{j=1}^{r} V_0(k_j, e^{\frac{2\pi i \nu_j}{1+\eta}}) \prod_{l=r+1}^{n+1} V_0(k_l, e^{2\pi i \nu_{l-1}})|0>$$

$$+(A_r - A_{r-1}) < 0| \prod_{j=1}^{r-1} V_0(k_j, e^{\frac{2\pi i \nu_j}{1+\eta}}) \prod_{l=r}^{n+1} V_0(k_l, e^{2\pi i \nu_{l-1}})|0>\} \tag{14}$$

with $v_n \equiv 1$ in this case. This expression vanishes in the $\eta \to 0$ limit because this forces the equality of two adjacent Koba-Nielsen variables in the vacuum expectation values [14,15]. Note that if the zero-slope limit is taken before the $\eta \to 0$ limit is taken this reduces to the usual result for the r-th current divergence of the chiral anomaly graph [8], apart from a constant which arises due to the differing normalizations of the F_0 and L_0 propagators. An analysis of the first and $n+1$-th current divergences is slightly different because F_0 must be anticommuted through Γ_{2n+1}; however the end result is the same: in the $\eta \to 0$ limit these current divergences also vanish. The evaluation of current divergences in the Möbius strip diagram also yields vanishing results using this technique. The current divergences of the full amplitude therefore vanish for any group. If the full amplitude is computed before taking the $\eta \to 0$ limit, the result is proportional to $N + 2^n l$, and so vanishes only for SO(32) when $n = 5$ for $\eta \neq 0$.

Continuous Dimension Techniques

One can consider regulating the integral in (6) using CDR: by generalizing the vector algebra and loop integration to 2ω dimensions, and taking the limit $\omega \to n$ at the end of the calculation, where ω approaches n from above. This allows one to naively shift the variable of integration [7]. This leads automatically to a vanishing anomaly, as in point-field theory. In point-field theory the solution proposed by 't Hooft and Veltman to this problem is to employ a non-anticommuting γ_{2n+1} [16]; as previously mentioned, this yields results equivalent to other schemes [7,8]. Employing a non-anticommuting Γ_{2n+1} yields an ambiguity similar to that in point-field theory, since

$$F_0\Gamma_{2n+1} = -\Gamma_{2n+1}F_0 - \frac{2}{i\sqrt{2}}\Gamma_{2n+1}\not{A}$$

where the q_μ's are the components of the momentum p for $n \leq \mu \leq 2\omega$. Hence there is an ambiguity in the choice of location of Γ_{2n+1}, which may be parametrized as:

$$T(2\omega) = \sum_{j=1}^{n+1} C_j \int d^{2\omega}p Tr\left[\prod_{i=1}^{j} \frac{1}{F_0} V(i) (\frac{1}{F_0} V(j) \Gamma_{2n+1}) \prod_{l=j+1}^{n+1} \frac{1}{F_0} V(l)\right] \tag{15}$$

where $\sum_{i=1}^{n+1} C_i = 1$ must be taken so that the regulated graph is equal to its unregulated value if $\omega \to n$ is taken before the calculation is performed.

As in the point-field theory case [7,8], one can show that only terms proportional to q^2 contribute to the current divergences in the $\omega \to n$ limit. There is no cancelled propagator in these terms, so no further regularization is needed. The result is:

$$T_r = (-1)^r C_r \frac{2^n(-2\pi)^\omega}{(i\sqrt{2})^{n-1}} \epsilon_r(\xi, k) \times \int \prod_{i=1}^{n} d\nu_i \theta(\nu_{i+1} - \nu_i) \int_0^\infty d\nu (n-\omega) \nu^{n-\omega-1} < 0| Tr[e^{-\frac{4\pi^2 N}{\nu}} \prod_{j=1}^{n+1} V_0(k_j, e^{2\pi i \nu_j}) \Gamma_d] |0> \tag{16}$$

If the zero-slope limit is taken before the $\omega \to n$ limit is taken, then the CDR result of ref [8] is recovered, *ie.* the method employed here correctly reproduces the point-field theory result. However in the full string theory this expression diverges in the limit $\omega \to n$.

The analysis of the Möbius strip diagram yields similar results. As with the previous case, the axial-vector-current divergence of the full amplitude $\Gamma(2\omega)$ is proportional to $(N + 2^n \ell)$, and so vanishes for $SO(32)$ in 10 dimensions if the $\omega \to n$ limit is not taken.

Employing the RDR scheme [7,8,17] leads to similar problems in the evaluation of T_r. Furthermore, RDR is plagued by a further inconsistency associated with the abandonment of trace cyclicity; see ref. [8] for details.

Gaussian Regulators

In this method each graph in the amplitude is regulated by inserting a factor of $e^{-B\eta L_o}$, where B is a positive constant and the $\eta \to 0$ limit is taken at the end of the calculation. There is an ambiguity in the choice of location of such regulators in each graph; to parametrize this choice a factor $e^{-B_i \eta L_o}$ may be inserted in the ith propagator yielding

$$T = \int d^{2n}p Tr\left[\prod_{j=1}^{n+1} \frac{e^{-B_i \eta L_o}}{F_o} V_o(j) \Gamma_{2n+1}\right] \tag{17}$$

For the regularization to be effective, the sum of the $B's$ must be a positive constant; this may be taken to be unity without loss of generality. The calculation of T_r proceeds as in [1]; the result for T_r is

$$T_r = \frac{i(-4\pi)^n}{(i\sqrt{2})^{n-1}} \varepsilon_r(\xi, k) \left[\int_{R_r} - \int_{R_r+1}\right] \prod_{i=1}^{n} d\nu_i < 0| \prod_{j=1}^{n+1} V_o(k_j, e^{2\pi i \nu_j}) |0> \tag{18}$$

where the R_r region of integration is given by

$$R_r = \{(\nu_1 ... \nu_n) | B_i(\nu_r - \nu_{r-1}) \leq B_r(\nu_i - \nu_{i-1})\} \ . \tag{19}$$

The axial-vector-current divergence T_{n+1} is given by an expression identical to (18) except the integrals are $[\int_{R_{r+1}} + \int_{R_1}]$. These expressions are non-vanishing, and yield the results of [1] when all the B's are equal and where vector-current conservation is imposed in the full amplitude. The result differs from the Pauli-Villars case (11) since the regions of integration in parameter space differ. In the zero-slope limit the r-th current divergence of (17) yields an ambiguity structure for the point-field theory case equivalent to those found in [8]. The Möbius strip diagram is evaluated in a similar manner. As in the other cases, the current-divergences of the the full amplitude are proportional to $N + 2^n \ell$.

Conclusions

Unlike the point-field theory case, the current divergences of the open superstring anomly graph are inequivalent: different methods yield different results in the absence of constraints imposed by vector current conservation. Even if these constraints are imposed, the resultant axial-vector current divergences (i.e. the anomalies) are inequivalent in different methods, although each method yields a result proportional to $(N+32\ell)$ in 10 dimensions. Furthermore certain methods (CDR and RDR) yield indeterminate values for the current divergences, whereas others (pre-regularization) give zero. All methods reproduce the ambiguity structure obtained in [8] for the chiral anomaly graph when the zero-slope limit is taken.

One can understand how the finite non-vanishing results of the Pauli-Villars and Gaussian techniques are obtained in the following way. In the former case, as in point-field theory [1], the effect of T^M is to replace T_r with an evaluation of a graph with n vector emissions and one pseudoscalar emission. This graph is free of cancelled propagators and is always finite for any non-zero M now matter how large. In the Gaussian case the effect of the regulator is to cancel a zero which arises from the cancelled propagator with an infinity that arises as the inner radius of the loop shrinks to zero in the integration over all values of this radius.

It is possible to generalize the Gaussian method to parametrize this effect. Specifically, if a regulator $e^{-\eta p^2 - \eta' N}$ is chosen so that

$$\frac{1}{F_0} \rightarrow \frac{e^{-\eta p^2 - \eta' N}}{F_0}$$

in the graph (16), then an evaluation of Tr gives

$$\mathrm{T_r} \sim \epsilon_r(\xi, \mathrm{k}) \frac{\eta}{\eta'} \int_{\mathrm{R_r}} \prod_{1=1}^{\mathrm{n}} \mathrm{d}\nu_\mathrm{i} < 0|\mathrm{T_r}\left[\mathrm{e}^{\frac{4\pi^2 \mathrm{N}}{\eta'}} \prod_{\mathrm{j}=1}^{\mathrm{n}+1} \mathrm{V_o}(\mathrm{k_j}, \mathrm{e}^{2\pi \mathrm{i}\nu_\mathrm{i}})\right] \Gamma_\mathrm{d}|0> \tag{20}$$

From this it is clear that the final value for T_r depends upon the ordering of the limits $\eta \rightarrow 0$, $\eta' \rightarrow 0$. If $\eta \rightarrow 0$ first T_r vanishes; if $\eta' \rightarrow 0$ first T_r is infinite. Only for η/η' finite is a non-zero result obtained, leading to the superstring anomaly found in [1]. From this point of view, the superstring anomaly arises due to a discontinuity in the Gaussian regulators.

Two further comments should be made. Suzuki [18] has evaluated superstring anomalies using a proper time method; this is equivalent to the Pauli-Villars scheme as he shows [18]. Clavelli, Cox and Harms [19] have directly evaluated the graph T in 10 dimensions and have shown it to be finite using a formalism different from that employed here. In point-field theory, the finiteness of the chiral anomaly graph has been shown [9] to imply the current-divergence equivalence between schemes found in ref. [8]. The analogous lack of equivalence found here in the string case therefore needs to be reconciled with the results of ref. [8]. Presumably the discrepancy is a consequence of the different formalisms used. Work on this area is in progress.

Acknowledgements

I would like to thank Professor M. Green for helpful discussions. This work was supported by the Natural Sciences and Engineering Research Council of Canada.

References

1. M. Green and J. Schwarz, Nucl. Phys. **B255** (1985) 93.
2. M. Green and J. Schwarz, Phys. Lett. **B149** (1984) 117.
3. D. Gross, J. Harvey, E. Martinec and R. Rohm, Nucl. Phys. **B256** (1985) 253.
4. S. Adler in *Lectures on Elementary Particles and Quantum Field Theory*, eds. S. Deser, M. Grisaru, and H. Pendleton (MIT Press, Cambridge, Mass. 1970) p.1; J.S. Bell and R. Jackiw, Nuovo Cim. **60A** (1969) 47.
5. J.M. Jauch and F. Rohrlich, *The Theory of Photons and Electrons* (Springer, Berlin and N.Y., 1976) pp. 457-460; R.E. Pugh, Can.J.Phys. **47** (1969) 1263; V. Elias and R.B. Mann, Nuc.Phys. **B219** (1983) 524.
6. V. Elias, G. McKeon and R.B. Mann, Phys.Rev.**D28** (1983) 1978; Can. J. Phys. **63** (1985) 1498.
7. V. Elias, G. McKeon and R.B. Mann, Nucl. Phys. **B229** (1983) 487.
8. A.M. Chowdhury, G. McKeon and R.B. Mann, Phys. Rev. **D33** (1986) 3090.
9. R.B. Mann, Nucl. Phys. **B266** (1986) 125.
10. V. Elias, G. McKeon, R.B. Mann and S.B. Phillips, Phys. Lett. **B133** (1983) 83.
11. Z. Koba and H.B. Nielsen, Nucl.Phys. **B10** (1969) 633.
12. R.B. Mann, *Proc. of XIVth Int. Colloquium on Group Theoretical Methods in Physics.*, p. 543, ed. Y.M. Cho (World Scientific, 1986).
13. R.B. Mann, Toronto preprint (to appear).
14. J. Scherk, Rev. Mod. Phys. **47** (1975) 123.
15. A. Neveu, J. Schwarz and C.B. Thorn, Phys. Lett **B35** (1971) 529.
16. G't Hooft and M. Veltman, Nuc. Phys. **B44** (1972) 189.
17. H. Nicolai and P. Townsend, Phys. Lett. **B93** (1980) 111.
18. H. Suzuki, KEK preprint 86-2, April 1986.
19. L. Clavelli, P. Cox and B. Harms, UA-HEP 861 preprint (1986).

ON THE COVARIANT QUANTIZATION OF ANOMALOUS GAUGE THEORIES

C.-M. Viallet

L.P.T.H.E. Universite Pierre et Marie Curie, Tour 16, 1°étage
4, Place Jussier
75252 Paris Cedex 05, France

The standard model of gauge theory for weak and electromagnetic interactions describes fermions as having a definite chirality : left and right chiralities are so different that for example the left handed part of the electron and its right-handed part (each of which is a Weyl fermion) do not have the same quantum numbers (weak hypercharge for example) [1,2], and thus it is natural to try to quantize gauge theories with Weyl fermions.

Unfortunately the presence of Weyl fermions in a gauge theory is known to produce, at the quantum level, non-gauge-invariant interactions. This is the phenomenon of anomalies [3].

The presence of anomalies in the theory is a disease which one usually avoids by carefully choosing the fermion content, to force mutual cancellations between the various non-invariances possibly produced by each fermion [4,5].

Another cancellation mechanism is to add extra fields in the theory, e.g. a group valued field with definite transformation as introduced by Wess and Zumino [6], or other fields coming from a larger theory and a judicious choice of the group of symmetry [7].

Our point, described in a work of O. Babelon, F. Schaposnik and the author [8] (see also the work of K. Harada and T. Tsutsui [9]) is to show that the use of lagrangian formalism and functional integral formalism leads, in the case of anomalous gauge theories -with no forced cancellations- to reconsider the quantization rules [10].

Our motivation is - 1) The idea promoted by Faddeev and Shatashvili [11,12] in the context of hamiltonian formalism, together with the need of explicit Lorentz invariance.

- 2) The results obtained by Polyakov in a similar problem for the conformal anomaly in the context of string theory [13].

We thus adopt the lagrangian formalism and use the functional integral

$$Z = \int \mathcal{D}A \, \mathcal{D}\bar{\psi} \, \mathcal{D}\psi \; e^{-S}.$$

with $S = \int \mathcal{L}\, d^4x$, $\mathcal{L} = \mathcal{L}_{fermion} + \mathcal{L}_{YM} = \bar{\psi}_L (i\partial\!\!\!/ + A\!\!\!/)\psi_L + \frac{1}{4} \mathrm{tr}\, F_{\mu\nu}^2$.

where ψ_L is a left-handed Weyl fermion : $\psi_L = \frac{1-\gamma_5}{2} \psi_L$.

The outcome is that if one treats properly the functional integration, then the group valued field of Wess-Zumino is not an external field and it is present in the theory. Actually the Wess-Zumino action emerges and one recovers independence on the choice of gauge condition.

In other words, the gauge part of the gauge potential, which is irrelevant at the classical level [14] acquires a different status in the quantum theory (Exactly the same phenomenon happens for two dimensional metrics in [13]). The core of the discussion is the correct account of the number of degrees of freedom of the quantum theory, which may differ from the one for the classical theory.

1. We first recall some results [15] on pure gauge theory, coming from the geometrical analysis of the Faddeev-Popov determinant [10].

2. We then recall results [16] on the fermionic measure in the case where Weyl fermions are present.

3. With this in hand we reconsider the full functional integral, including integration over gauge potentials and integration over the fermionic degrees of freedom, without changing the starting point -we pretend to study Yang-Mills theory- but merely keeping in mind all features of the objects we write down. In the absence of anomalies we recover the standard results.

4. We give some explicit formulae for the 2-dimensional case.

I. THE GEOMETRICAL CONTENT OF THE USUAL QUANTIZATION RULE OF GAUGE THEORY

In usual Yang-Mills theory, in the lagrangian approach, one writes a functional integral over the space of gauge potentials (pure Yang-Mills)

$$Z_{YM} = \int \mathcal{D}A\, e^{-S_{YM}}$$

This integral is a priori an integral over the whole space $\mathcal{E}$ of gauge potentials

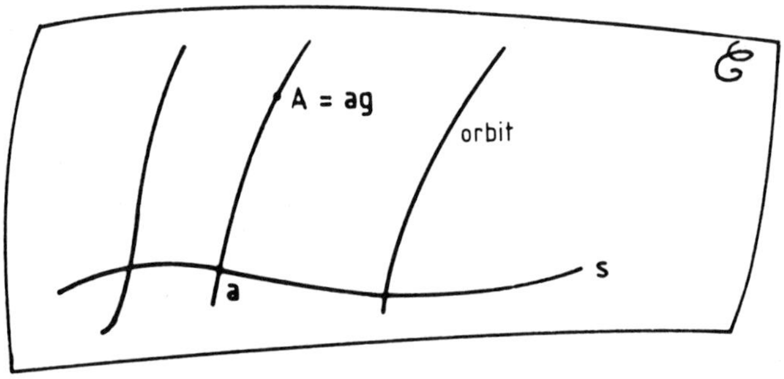

The invariance of $S_{YM} = \frac{1}{4}\int \mathrm{tr}\, F_{\mu\nu}^2$ under gauge transformations means that $S_{YM}(A)$ is constant on any orbit of the group $\mathcal{G}$ of gauge transformations. This invariance is a trivial source of divergence for the integral Z_{YM} and leads to choosing a gauge section $\mathcal{S}$ and restricting the integral to $\mathcal{S}$. Such a choice of $\mathcal{S}$ is the choice of a unique representative in each orbit.

The choice of $\mathcal{S}$ may be done by imposing a differential equation on the gauge potential, eg $\partial_\mu A^\mu = 0$. The independence on the choice of $\mathcal{S}$ persists only through the presence of an additional term in the measure :

the Faddeev-Popov determinant, and it manifests itself in the Becchi-Rouet-Stora invariance of the total action S_{YM} + S gauge fixing + S Fad-Pop [17].

The origin of this invariance is best explained by the geometrical meaning of the Faddeev-Popov determinant [15] : The volume element $\mathcal{D}A$ has to be written in the system of coordinates determined by the choice of the section $\mathcal{S}$. Any point A in $\mathcal{E}$ is given coordinates a point $a \in \mathcal{S}$ and a group element g so that A is the gauge transformed a^g of a by g.

The volume element $\mathcal{D}A$ may be written $\mathcal{D}A = \mathcal{D}\mu(a) \cdot \rho(a) \cdot \mathcal{D}g$ (1)

where : $\mathcal{D}\mu(a)$ is a natural volume element on the quotient space $\mathcal{E}/\mathcal{G}$ [14].

$\mathcal{D}g$ is an invariant volume element on the group of gauge transformations.

$\rho(a)$ is a factor depending only on the orbit ; it is actually the scale of this orbit. Its value is the determinant of the covariant laplacian constructed with a [18].

It was shown in [15] that the Faddeev-Popov determinant is just the product $\rho(a) \cdot \mathcal{D}\mu(a)$.

The situation may be pictured as follows, in a plane x,y.

Suppose one integrates over the trapezium shown below a function f(x,y) which is constant along the y axis.

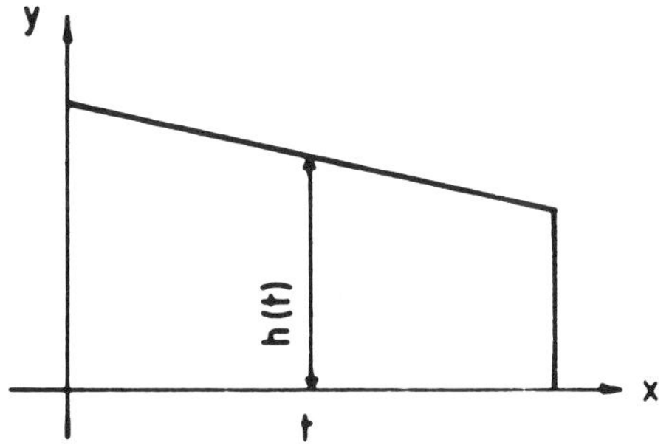

One has $I = \int dx\,dy\; f(x,y) = \int h(x)\, dx\; f(x,y)$, where the height h(t) is the scale of the orbit, and dx is the integration element over the quotient space.

The factor $\rho(a)$ in (1) is nothing but the factor h(x).

The presence of this factor is a remnant of the integration over the gauge degrees of freedom. In this sense we say that in the usual procedure, the functional integral -for the lagrangian approach- is an integral over the whole space of gauge potentials, rather than over the orbit space [19].

Note that the situation is essentially different in the hamiltonian formalism where the group of gauge transformations has to be taken away [20,14].

II. THE FERMIONIC MEASURE

In the presence of Weyl fermions, the fermionic measure $\mathcal{D}\bar{\psi}\mathcal{D}\psi\, e^{-S_{fer}(A,\psi)}$ which has to be properly defined, is not gauge invariant [16, and Fujikawa's lecture in this Workshop]:

In a gauge transformation $\gamma : \quad \psi \to \psi' = \gamma^{-1}\psi$. (each fermion in its representation), one has

$$\mathcal{D}\bar{\psi}'\,\mathcal{D}\psi'\, e^{-S_{fer}(A,\psi')} = \mathcal{D}\bar{\psi}\,\mathcal{D}\psi\, e^{-S_{fer}(A\gamma^{-1},\psi)} \cdot J(A\gamma^{-1},\gamma) \quad (2)$$

where J is the jacobian of the transformation, and depends on the regularization scheme. Suppose we use ζ function regularization, we may define:

$$\det_L (i\partial\!\!\!/ + A\!\!\!/) = \exp\left(\frac{-d\zeta(s,\Delta(A))}{ds}\Big|_{s=0} \right)$$

ζ being the ζ-function for the operator

$\Delta(A) = i\partial\!\!\!/ + A\!\!\!/\left(\frac{1-\gamma_5}{2}\right)$ acting on Dirac fermions [21].

We then have :

$J(B,\gamma) = \dfrac{\det_L (i\partial\!\!\!/ + C\!\!\!/)}{\det_L (i\partial\!\!\!/ + B\!\!\!/)}$ with $C = B\gamma$ (gauge transformed of B by γ)

The jacobian J has an immediate cocycle property (see R. Jackiw's exposé) :

$$J(a,g) = J(a,\gamma)\,.\,J(a\gamma,\gamma^{-1}g) \quad (3)$$

The non invariance of the fermionic measure leads us to reconsider the Faddeev-Popov procedure.

III. THE FULL FUNCTIONAL INTEGRAL

We start from $$Z = \int \mathcal{D}A\, \mathcal{D}\bar{\psi}\mathcal{D}\psi\, e^{-S_{fer}(A,\psi) - S_{YM}(A)} \quad (4)$$

We coordinatize the space of A's by a gauge section s (local coordinatization)

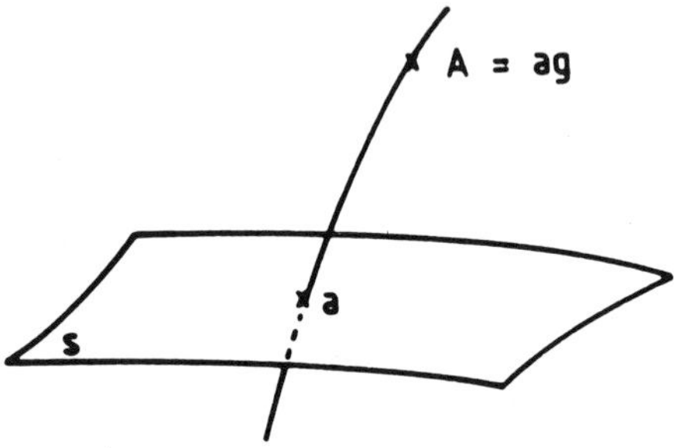

A has coordinates a,g.

Thus (4) may be written, by the change $\psi \to g^{-1}\psi$.

$$Z = \int \mathcal{D}\mu(a)\, \rho(a)\, \mathcal{D}g\, \mathcal{D}\bar{\psi}\,\mathcal{D}\psi\,.J(a,g).\, e^{-S(a,\psi)} \quad (5)$$

or

$$Z = \int \underbrace{\delta(\text{gauge fixing})\, \Delta^{FP}(a)}\, \mathcal{D}a\, \mathcal{D}g\, \mathcal{D}\bar{\psi}\mathcal{D}\psi\, J(a,g)\, e^{-S}$$

usual part containing gauge fixing + Faddeev-Popov determinant Δ^{FP}.

Clearly if the Jacobian J was 1, then the integration over g would factor out as was first used in [10]. In our case, J is not 1, and the integration over g has to be performed. Actually J = exp $(WZ(a,g))$ where $WZ(a,g)$ is the Wess-Zumino action as introduced in [6]. See also [12], but notice that we do not have to introduce an extra g field by hand : it arises spontaneously.

<u>Claim</u> : Z is independent on the choice of section $\mathcal{S}$.
Suppose we choose another section $\mathcal{S}'$ in $\mathcal{E}$.

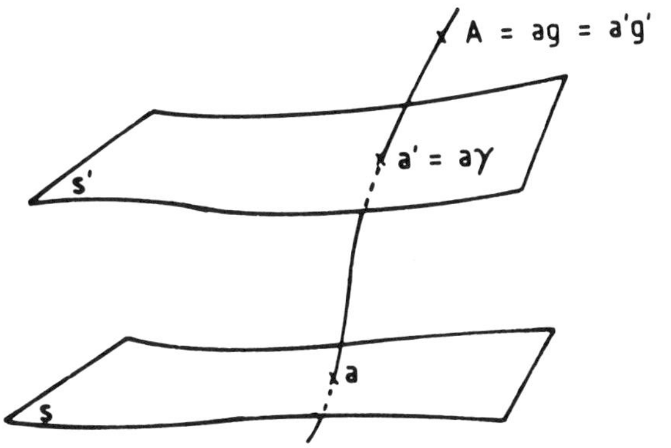

Any point A in $\mathcal{E}$ is gauge related to some point a' in $\mathcal{S}'$ and some point a in $\mathcal{S}$. The two points a and a' are related by some gauge transformation γ : $a' = a\gamma$. Of course γ is <u>not</u> constant over the gauge section.

Our claim is a mere consequence of the independence of the value of the integral Z on the coordinatization we choose.

It may however be detailed as follows :
We have to compare

$$Z = \int \mathcal{D}\mu(a)\ \rho(a)\ \mathcal{D}g\ \mathcal{D}\bar{\psi}\,\mathcal{D}\psi\ \ J(a,g)\ e^{-S(a,\psi)}$$

and

$$Z' = \int \mathcal{D}\mu(a')\ \rho(a')\ \mathcal{D}g'\ \mathcal{D}\bar{\psi}'\,\mathcal{D}\psi'\ \ J(a',g')\ e^{-S(a',\psi')}$$

In Z' we perform the changes of variables :

$$\psi' = \gamma^{-1}\psi \quad \text{and} \quad g' = \gamma^{-1}g\,.$$

Noting that

$$\rho(a) = \rho(a')\,, \qquad \mathcal{D}\mu(a) = \mathcal{D}\mu(a') \qquad \text{and} \qquad \mathcal{D}g = \mathcal{D}g' \qquad \text{, we get}$$

$$Z' = \int \mathcal{D}\mu(a)\ \rho(a)\ \mathcal{D}g\ \ \mathcal{D}\bar{\psi}\,\mathcal{D}\psi\ \ J(a\gamma,\gamma^{-1}g)\cdot J(a,\gamma)\ e^{-S(a,\psi)}$$

and thus : $\quad Z' = Z\,.$
by using the cocycle property (3).

We have recovered the independence on the choice of gauge.

IV. AN EXAMPLE : THE 2-DIMENSIONAL MODEL

We use Minkowski space and parametrize the gauge potentials by two group valued fields U and V such that :

$$A_+ = i\, U^{-1} \partial_+ U .$$
$$A_- = i\, V \partial_- V^{-1} .$$

The gauge "condition" $A_- = 0$ yields $a = UV$ and $g = V^{-1}$.

Then

$$Z = \int \mathcal{D}\mu(a)\ \rho(a)\ \mathcal{D}g\ \mathcal{D}\psi\ \exp\left(i\, S(a,\psi) + i\,\phi(g) - \frac{1}{16\pi} \mathrm{tr} \int (a^{-1}\partial_- a)(g\partial_- g^{-1}) \right)$$

where

$$\phi(g) = \frac{1}{16\pi}\, \mathrm{tr} \int d^2x\ \partial_\mu g^{-1} \cdot \partial^\mu g - \Gamma(g)$$

and

$$\Gamma(g) = \frac{-i}{8\pi}\, \mathrm{tr} \int d^2x \int_0^1 dt\ \varepsilon^{\mu\nu}\, g_t^{-1} \partial_t g_t\ g_t^{-1} \partial_\mu g_t\ g_t^{-1} \partial_\nu g_t ,$$

with g_t some interpolating family between 1 and g.

V. CONCLUSION

The analysis we have performed is valid in all dimensions, but is clearly regularization dependent and the choice of a preferred regularization is necessary.

For example in 4 dimensions, fourth order derivatives are present in the kinetic energy term of the g field and the resulting action for g has to be studied in detail. Moreover, the canonical (hamiltonian) interpretation may be absent for this field. This is a result of forcing Lorentz invariance to be present all the way. This corroborates the results of [22], see also G. Semenoff exposé in this Workshop, and also [23] . The question which remains open here is the problem of unitarity of the theory and the exact status of the g field as a physical field.

Finally if various spinors of both chiralities in different representations are present, not all directions in the group of gauge transformations are anomalous and only an element of $\mathcal{G}/\mathcal{H}$ arises in (4) (if $\mathcal{H}$ is the subgroup of $\mathcal{G}$ which is anomaly free), an a finer analysis is necessary.

REFERENCES

1. See for example, E.S. Abers & B.W. Lee, Phys. Reports 9C(1973).
2. T.D. Lee & C.N. Yang, Phys. Rev. 105(1957)1671.
3. S. Adler, Lectures at Brandeis Summer School 1970, Deser, Grisaru & Pendleton eds, MIT Press.
4. C. Bouchiat, J. Iliopoulos & P. Meyer, Phys. Lett. 38B(1972)519.
5. D. Gross & R. Jackiw, Phys. Rev. D6(1972)477.
6. J. Wess & B. Zumino, Phys. Lett. 37B(1971)95.
7. M.B. Green & J.H. Schwarz, Phys. Lett. 149(1984)117.
8. O. Babelon, F.A. Schaposnik & C.M. Viallet, "Quantization of Gauge Theories with Weyl Fermions", preprint PAR LPTHE 86/31, to appear in Physics Letters.
9. K. Harada & T. Tsutsui, "On the Path-Integral Quantization of Anomalous Gauge Theories", preprint TIT/HEP 94 (1986).
10. L.D. Faddeev & V.N. Popov, Phys. Lett. 25B(1967)29.
11. L.D. Faddeev, Phys. Lett. 145B(1984)81.
12. L.D. Faddeev, S.L. Shatashvili, Phys. Lett. 167B(1986)225.

13. A.M. Polyakov, Phys. Lett. 103B(1981)207.
14. O. Babelon, C.M. Viallet, Comm. Math. Phys. 81(1981)515.
15. O. Babelon, C.M. Viallet, Phys. Lett. 85B(1979)246.
16. K. Fujikawa, Phys. Rev. Lett. 42(1979)1195.
K. Fujikawa, Phys. Rev. D21(1980)2848.
17. C. Becchi, A. Rouet & R. Stora, Ann. Phys. (N.Y.) 98(1976)287.
18. A.S. Schwarz, Comm. Math. Phys. 64(1979)233.
19. C.M. Viallet, Lectures at the XXII Karpacz Winter School of Theoretical Physics (1986), to appear, World Scientific, A. Jaczyk ed.
20. L.D. Faddeev, Theor. Math. Phys. 1(1969)3.
21. L. Alvarez-Gaumé & P. Ginsparg, Nucl. Phys. B243(1984)449.
22. A. Niemi & G. Semenoff, Phys. Rev. Lett. 56(1986)1019.
23. R. Jackiw & R. Rajaraman, Phys. Rev. Lett. 54(1985)1219; R. Rajaraman, Phys. Lett. 154B(1985)305; 162B(1985)148; L. Alvarez-Gaumé, S. Della Pietra, V. Della Pietra, Phys. Lett. 166B(1986)177; J.G. Halliday, E. Rabinovici, A. Schwimmer, M. Chanowitz, Nucl. Phys. B268(1986)413; M. Chanowitz, Phys. Lett. 171B(1986)280.

QUANTUM ADIABATIC PHASES AND CHIRAL GAUGE ANOMALIES

Gordon W. Semenoff †

Department of Physics
University of British Columbia
Vancouver, British Columbia, Canada V6T 2A6.

I. INTRODUCTION

The quantum adiabatic phase [1,2] has recently emerged as a universal element in the topological analysis of various quantum mechanical problems. It can be understood as an Ahoronov-Bohm effect on the parameter space of a quantum mechanical system and has seen numerous applications, notably to the analysis of corrections to semiclassical quantization [2,3,4], the statistics of quasiparticles such as vortices in two dimensional systems [5,6,7], adiabatic effective actions, the quantum Hall effect [8] and chiral anomalies in the Hamiltonian, Schrödinger picture of gauge theories [9]. Furthermore, some of its predicted interference phenomena have recently found good agreement with experiment [10,11]. In this review we shall give a simple account of the origin of the quantum adiabatic phase, describe how the concept is used to construct adiabatic effective actions and discuss chiral gauge anomalies within this framework.

II. QUANTUM ADIABATIC PHASE

We shall review an idea which has been known for some time and has been recently popularized in the theoretical physics literature by Berry and Simon [1]. It concerns the problem of adiabatic transport of wavefunctions on the parameter space of a quantum mechanical system. This transport is natural in the adiabatic limit and the associated $\mathcal{U}(1)$ holonomy reflects the nontrivial topology of a wavefunction viewed as a section of a Hilbert line bundle with base manifold the parameter space. We shall show in the following Sections that this holonomy determines the leading terms in adiabatic effective actions and in a chiral gauge theory it can obstruct the implementation of gauge invariance, giving rise to projective representations of the gauge group and chiral anomalies.

Consider a Hamiltonian $\mathcal{H}(p, q, \lambda)$ which is specified by a set of external parameters λ taking values on a parameter space Λ. We solve the quantum mechanics described by $\mathcal{H}$ for each value of the parameters

† Supported in part by NSERC of Canada.

$$\mathcal{H}(\lambda)\ \psi_\omega(\lambda)\ =\ \omega(\lambda)\ \psi_\omega(\lambda) \tag{2.1}$$

and consider a curve $\mathcal{C} = \{\lambda^\tau : 0 \le \tau \le \mathrm{T}\}$ such that a particular eigenvalue of the Hamiltonian $\omega(\lambda^\tau)$ is nondegenerate and evolves continuously for all values of τ. We further assume that $\psi_\omega(\lambda^\tau)$ is single valued on Λ.

The dimension of the subspace of Λ in which $\mathcal{C}$ can lie follows from the Wigner-Von Neuman theorem [12] which states that for a complex Hermitian differential operator it is necessary to adjust three parameters to make a generic eigenvalue degenerate. This implies that a given eigenvalue is degenerate on a subspace of Λ of codimension 3. If $\mathcal{H}$ is complex $\mathcal{C}$ must be taken in the 3 dimensional subspace where $\omega(\lambda)$ is nondegenerate. If $\mathcal{H}$ is real or exhibits certain discrete symmetries, the dimension of this subspace is reduced to 2.

We consider the adiabatic evolution of the quantum state which solves the time dependent Schroedinger equation

$$[\ i\frac{\partial}{\partial\tau}\ -\ H(\lambda^\tau)\ ]\Psi(\tau)\ =\ 0 \tag{2.2}$$

with the boundary condition $\Psi(0) = \psi_\omega(\lambda^0)$. Equation (2.2) has the adiabatic solution

$$\Psi(T)\ =\ e^{-i\int_0^T d\tau\omega(\tau)+i\gamma(T)}\ \psi_\omega(\lambda^T) \tag{2.3}$$

where the first term in the exponential is the standard dynamical adiabatic phase and the second term is a nondynamical contribution given by

$$\gamma(T) = \int_0^T d\tau < \psi_\omega(\lambda^\tau)|i\frac{\partial}{\partial\tau}|\psi_\omega(\lambda^\tau) >= \int_0^T d\tau\ \dot\lambda < \psi_\omega(\lambda^\tau)|\frac{i\partial}{\partial\lambda^\tau}|\psi_\omega(\lambda^\tau) > \tag{2.4}$$

$$=\ \int_0^T d\tau\ \dot\lambda\ \mathcal{A}(\tau) \tag{2.5}$$

where the second equality follows from the fact that the states depend on τ only through their implicit dependence on λ.

This construction has an intrinsic $\mathcal{U}(1)$ gauge structure. The requirement that the wavefunctions $\psi_\omega(\lambda)$ solve equation (2.1) leaves their phase arbitrary and it can be redefined in a λ-dependent way,

$$\psi_\omega(\lambda) \to e^{i\chi(\lambda)}\psi_\omega(\lambda)\ . \tag{2.6}$$

Under this redefinition $\mathcal{A}$ transforms like a $\mathcal{U}(1)$ gauge potential

$$\mathcal{A}_\omega(\tau)\ \to\ \mathcal{A}_\omega(\tau)\ -\ \frac{\partial\chi[\lambda^\tau]}{\partial\lambda} \tag{2.7}$$

and the phase integral like

$$\gamma(T)\ \to \gamma(T)\ -\ \chi(\lambda^T)\ +\ \chi(\lambda^0) \tag{2.8}$$

This implies that for open paths γ is always trivial and can be transformed away. However, for closed paths, *i.e.* where $\lambda^T = \lambda^0$ we require the transformation (2.6) to be single valued on Λ,

$$\chi(\lambda^T) = \chi(\lambda^0) + 2\pi n \tag{2.9}$$

and $\gamma \; mod \; 2\pi n$ is $\mathcal{U}(1)$ gauge invariant. It has nontrivial physical consequences and could for example give rise to observable interference effects in real adiabatic processes.

Using equation (2.4) for a direct calculation of γ would require a set of basis wavefunctions which are globally single valued on Λ. (This requirement is a partial $\mathcal{U}(1)$ gauge fixing to single valued wavefunctions.) It is furthermore generally difficult to choose a good $\mathcal{U}(1)$ gauge in which to perform computations. For this reason it is usually more convenient to compute the $\mathcal{U}(1)$ curvature tensor

$$\mathcal{F}_{ij} = (\delta\mathcal{A})_{ij} = -i\sum_{\omega'}\{ < \psi_\omega|\frac{\partial}{\partial\lambda_i}|\psi_{\omega'} >< \psi_{\omega'}|\frac{\partial}{\partial\lambda_j}|\psi_\omega > - \\ - < \psi_\omega|\frac{\partial}{\partial\lambda_j}|\psi_{\omega'} >< \psi_{\omega'}|\frac{\partial}{\partial\lambda_i}|\psi_\omega > \} \tag{2.10}$$

and to deduce the $\mathcal{U}(1)$ connection

$$\mathcal{A} = (\delta^{-1}\mathcal{F}) + \delta\chi[\lambda] \tag{2.11}$$

Then γ is given by

$$\gamma = \oint \mathcal{A} = \int\int_D \mathcal{F}\, d\lambda \wedge d\lambda \tag{2.12}$$

where the integration in the final term is over a disc on Λ which subtends $\mathcal{C}$. The curvature tensor in equation (2.10) depends only on the off-diagonal elements of the derivative operator and can be evaluated using the formula

$$< \psi_\omega|\frac{\partial}{\partial\lambda}|\psi_{\omega'} > = \frac{1}{\omega' - \omega} < \psi_\omega|\frac{\partial\mathcal{H}}{\partial\lambda}|\psi_{\omega'} >$$

so that

$$\mathcal{F}_{ij} = \sum_{\omega'}\frac{i}{(\omega-\omega')^2}\{< \psi_\omega|\frac{\partial\mathcal{H}}{\partial\lambda_i}|\psi_{\omega'} >< \psi_{\omega'}|\frac{\partial\mathcal{H}}{\partial\lambda_j}|\psi_\omega > - \\ - < \psi_\omega|\frac{\partial\mathcal{H}}{\partial\lambda_j}|\psi_{\omega'} >< \psi_{\omega'}|\frac{\partial\mathcal{H}}{\partial\lambda_i}|\psi_\omega >\} \tag{2.13}$$

Notice that $\mathcal{F}$ vanishes if the Hamiltonian is real, *i.e.* since

$$< \psi_\omega|\frac{\partial\mathcal{H}}{\partial\lambda}|\psi_{\omega'} >=< \psi_{\omega'}|\frac{\partial\mathcal{H}^*}{\partial\lambda}|\psi_\omega > \quad ,$$

if $\mathcal{H} = \mathcal{H}^*$ the terms in (2.13) cancel. Furthermore, $\mathcal{F}$ is well defined only where $\omega(\lambda)$ is nondegenerate. Thus if the Hamiltonian is real, and if $\mathcal{C}$ can be subtended by a disc which encounters no degeneracies of $\omega(\lambda)$, γ vanishes. If $\mathcal{C}$ encloses degeneracies $\mathcal{A}$ is locally flat, *i.e.* $\mathcal{F} = 0$, but is not globally a pure gauge.

This can be seen by adding a complex perturbation to $\mathcal{H}$ to produce a three dimensional region of Λ where ω is nondegenerate surrounding the point where it is degenerate. Then consitency requires that

$$\oint \mathcal{A} = \int\int_{D^+} \mathcal{F} \; mod \; 2\pi = \int\int_{D^-} \mathcal{F} \; mod \; 2\pi \tag{2.14}$$

where D^+ and D^- are discs which subtend the loop and avoid the degeneracy point on opposite sides. This implies that

$$\int\int_{D^+ + D^-} \mathcal{F} = 2\pi n \tag{2.15}$$

i.e. the surface integral of $\mathcal{F}$ on a sphere which surrounds a degeneracy point yields $2\pi n$. It is straightforward to show by example that for a single degeneracy, $n = 1$ and that n is the number of degeneracy points enclosed by $D_+ + D^-$.

When $n \neq 0$, analogous to the case of the Dirac magnetic monopole [13] of a $U(1)$ gauge theory, $\mathcal{A}$ cannot be defined globally on the sphere but must exist only in local patches with the appropriate $\mathcal{U}(1)$ transition functions. If this sphere and the degeneracy point are intersected by a plane on Λ where the Hamiltonian is real then $\mathcal{F} = 0$ on the circle of intersection. We can choose the transition region to contain this circle and the sphere is divided into two hemispheres D^+ and D^- with connections $\mathcal{A}^+$ and $\mathcal{A}^-$ respectively and $\mathcal{A}^+ - \mathcal{A}^- = \delta\chi$ along the equator. Equation (2.15) implies that for a line integral along the equator,

$$\oint \mathcal{A}^+ - \oint \mathcal{A}^- = \oint \delta\chi = 2\pi n \tag{2.16}$$

Furthermore, uniqueness of the adiabatic phase requires that

$$\oint \mathcal{A}^+ = -\oint \mathcal{A}^- \quad mod\ 2\pi \tag{2.17}$$

and we conclude that $\oint \mathcal{A}^{\pm} = \pi n \quad mod\ 2\pi$. Thus if $\mathcal{H}$ is real

$$e^{i\gamma_C} = (-1)^{deg_C} \tag{2.18}$$

where deg_C is the number of degeneracies enclosed by C.

Thus, when $\mathcal{H}$ is real, degeneracies are analogous to vortex line singularities in a $U(1)$ gauge theory with $\frac{1}{2}$ of a magnetic flux quantum and (2.18) is the ensuing Aharonov-Bohm phase in parameter space. Furthermore, if $\mathcal{H}$ is complex, by (2.15) the degeneracies are analogous to Dirac magnetic monopoles with $\mathcal{U}(1)$ *magnetic* flux 2π.

III. ADIABATIC EFFECTIVE ACTION

We shall now examine how the adiabatic phase appears in the construction of effective actions [3,4]. Consider a quantum system whose dynamical variables can be divided into two sets, one which we call fast variables (p,q) and the other which we call slow variables (π,χ) and with Hamiltonian of the form

$$\mathcal{H} = \mathcal{H}_1(\pi,\chi) + \mathcal{H}_2(p,q,\chi) \ . \tag{3.1}$$

The effective action is obtained by eliminating the fast variables from the partition function,

$$Z = \int_{p.b.c.} [d\pi][d\chi][dp][dq] \ \exp\left(\int_0^\beta d\tau \{ip\dot{q} + i\pi\dot{\chi} - \mathcal{H}(p,q,\pi,\chi)\} \right) \ . \tag{3.2}$$

We first treat χ as a fixed external variable, quantize the subsystem governed by the Hamiltonian $\mathcal{H}_2(p,q,\chi)$,

$$\mathcal{H}_2(p,q,\chi)\,|\omega,\chi> = \omega(\chi)\,|\omega,\chi> \tag{3.3}$$

and construct the propagator

$$Z[\chi] = \mathrm{tr}_{p,q}\{\mathcal{T}e^{-\int_0^\beta d\tau\, \mathcal{H}_2(p,q,\chi^\tau)}\} \tag{3.4}$$

where $\mathcal{T}$ denotes time ordering and $\chi^\beta = \chi^0$. We shall evaluate (3.4) to leading order in time derivatives of χ^τ. We begin by dividing the time into infinitesimal slices of length ϵ,

$$Z[\chi] = \lim_{\epsilon\to 0} \mathrm{tr}_{p,q}\{\prod_n e^{-\epsilon\mathcal{H}_2(p,q,\chi^{\tau_n})}\} \tag{3.5}$$

where $\tau_n = n\epsilon$. We evaluate the trace in the basis of wavefunctions $|\omega,\chi^\tau>$.

$$Z[\chi] = \lim_{\epsilon\to 0} \sum_\omega < \omega,\chi^\beta| \prod_n e^{-\epsilon\mathcal{H}_2(p,q,\chi^{\tau_n})}\, |\omega,\chi^0> \tag{3.6}$$

and insert the complete sets of states

$$I = \sum_{\omega'} |\omega',\chi^{\tau_n}> < \omega',\chi^{\tau_n}| \tag{3.7}$$

The adiabatic approximation consists of neglecting the contributions to the intermediate state sum from all states except $|\omega,\chi^{\tau_n}>$, the state which occurs at the endpoints of the trace. It is easy to see that all other contributions are of higher orders in time derivatives of χ^τ, consider

$$\begin{aligned}
&\sum_{\omega'} |\omega',\chi^{\tau_{n+1}}> < \omega',\chi^{\tau_{n+1}}|\, e^{-\epsilon\mathcal{H}_2(p,q,\chi^{\tau_n})}\, |\omega,\chi^{\tau_n}> \\
&= e^{-\epsilon\omega(\chi^{\tau_n})} \sum_{\omega'} |\omega',\chi^{\tau_{n+1}}> < \omega',\chi^{\tau_{n+1}}|\omega,\chi^{\tau_n}> \\
&= e^{-\epsilon\omega(\chi^{\tau_n})} \sum_{\omega'} |\omega',\chi^{\tau_{n+1}}> \;(\delta_{\omega,\omega'} - \epsilon < \omega',\chi^{\tau_n}|\frac{\partial}{\partial\tau}|\omega,\chi^{\tau_n}>) \\
&= e^{-\epsilon\omega(\chi^{\tau_n})+i\epsilon\mathcal{A}_\omega(\chi^{\tau_n})}(|\omega,\chi^{\tau_n}> - \epsilon\sum_{\omega'\neq\omega} \frac{|\omega',\chi^{\tau_n}>< \omega',\chi^{\tau_n}|}{\omega - \omega'}\frac{\partial\mathcal{H}}{\partial\tau_n}|\omega,\chi^{\tau_n}>)\;.
\end{aligned} \tag{3.8}$$

Iterating this proceedure, we have

$$Z[\chi] = \sum_\omega e^{-\int_0^\beta d\tau\omega(\chi^\tau)+i\oint\mathcal{A}_\omega}\{1 + O(<\frac{\partial\mathcal{H}}{\partial\tau}>^2)\} \tag{3.9}$$

Corrections are at least of second order in time derivatives since in (3.8) if we had included a step to a state with $\omega' \neq \omega$ periodicity would force us to consider a transition back to ω at another time slice, thus involving at least two matrix elements to the time derivative of the $\mathcal{H}$. Therefore, to linear order in time derivatives of χ^τ the effective action is

$$S_{eff}[\chi] = -\ln Z[\chi] = -\ln(\sum_\omega e^{-\int_0^\beta d\tau\omega(\tau)}) + \frac{\sum i\oint\mathcal{A}_\omega\, e^{-\int_0^\beta d\tau\omega(\tau)}}{\sum e^{-\int_0^\beta d\tau\omega(\tau)}} \tag{3.10}$$

At low temperature, $\beta \to \infty$, the leading term in each series dominates and the effective action reduces to

$$S_{eff}[\chi] = \int d\tau \; \omega_<[\chi(\tau)] + i\int d\tau \; \mathcal{A}_{\omega_<}[\chi(\tau)] \tag{3.11}$$

where $\omega_<$ is the smallest eigenvalue of $\mathcal{H}_2$. (Without loss of generality we can assume that it is positive.) Equations (3.10) and (3.11) yield the effective action to linear order in time derivatives of the slow variables. This formula has many potential applications, one of which we shall explore in the following Section.

IV. ANOMALIES

In this section we shall examine the relationship between quantum holonomy and gauge theory anomalies [9]. The anomalies which we consider can be viewed as the impossibility of defining the dynamics of fermions in a background gauge field such that the locality and Lorentz and gauge invariance of the full quantum theory is preserved.

We shall first consider the chiral Schwinger model with classical Hamiltonian

$$\mathcal{H} = \int \{\frac{1}{2}E^2 + \Psi^\dagger h \Psi\} = \mathcal{H}_1[E] + \mathcal{H}_2[\Psi, \Psi^\dagger, A] \tag{4.1}$$

where $h = i\frac{\partial}{\partial x} + A(x)$ and gauge constraint

$$\mathcal{G}(x) = \frac{\partial E}{\partial x} + \Psi^\dagger(x)\Psi(x) \sim 0 \tag{4.2}$$

and examine its adiabatic effective action by treating the fermions as fast variables and the gauge fields as slow variables. If the fermionic dynamics were gauge invariant, we would expect that the propagator

$$Z[A, u] = \mathrm{tr}\{\mathcal{T} e^{-\int_0^\beta d\tau \mathcal{H}_2(\tau)}\} \tag{4.3}$$

with the twisted boundary condition $A(\beta, x) = A(0, x) + du_1(x)$ is invariant under the static gauge transform

$$A(x) \to A(x) + du_2(x) \tag{4.4}$$

To find the adiabatic limit of (4.3) we construct the eigenmodes of $\mathcal{H}_2$ by first solving the single particle problem

$$[\, i\frac{\partial}{\partial x} + A(x)\,]\psi_\epsilon(x) = \epsilon\, \psi_\epsilon(x) \tag{4.5}$$

We work on a circle of radis 2π and choose periodic boundary conditions. (We assume that $u_1(x)$ and $u_2(x)$ are periodic.) The wavefunctions are

$$\psi_n(x) = \frac{1}{\sqrt{2\pi}} e^{-inx - iax + i\int_0^x A(x)dx} \tag{4.6}$$

where $a = \frac{1}{2\pi}\int_0^{2\pi} A(x)dx$ and the eigenvalues are

$$\epsilon_n = n + a \tag{4.7}$$

The second quantized fermion field operator is defined by

$$\Psi(x) = \sum_{\epsilon} \alpha_\epsilon \, \psi_\epsilon(x) \tag{4.8}$$

where $\{\alpha_\epsilon, \alpha'_\epsilon\} = \delta_{\epsilon,\epsilon'}$ and the Fock vacuum by

$$\alpha_\epsilon|0> = 0\,, \qquad \epsilon > 0 \tag{4.9a}$$

$$\alpha_\epsilon^\dagger|0> = 0\,, \qquad \epsilon < 0 \tag{4.9b}$$

Excited states are constructed by operating creation operators on the Fock vacuum. Then $\mathcal{H}_2 = \sum_\epsilon \epsilon : \alpha_\epsilon^\dagger \alpha_\epsilon :$ is diagonal and $|0, A>$ is its ground state.

The $\mathcal{U}(1)$ connection and curvature are

$$\mathcal{A}(x) = i < 0|\frac{\partial}{\partial A(x)}|0> \tag{4.10a}$$

$$\mathcal{F}(x,y) = \sum_\sigma \{\frac{i}{\omega_\sigma} < 0|\frac{\delta \mathcal{H}_2}{\delta A(x)}|\sigma >< \sigma|\frac{\delta \mathcal{H}_2}{\delta A(y)}|0> \; - \; (x \leftrightarrow y)\,\} \tag{4.10b}$$

respectively, where σ denote excited states, ω_σ their energies and $\mathcal{H}_2$ is defined in (4.1). Using (4.1), (4.8) and (4.10) we can write the formula for the $\mathcal{U}(1)$ curvature as

$$\mathcal{F}(x.y) = \sum_{\epsilon\epsilon'} \frac{-i}{(\epsilon - \epsilon')^2} \psi_\epsilon^\dagger(x)\psi_{\epsilon'}(x)\psi_{\epsilon'}^\dagger(y)\psi_\epsilon(y)\frac{1}{2}(\text{sign}(\epsilon') \; - \; \text{sign}(\epsilon)) \tag{4.11}$$

The summation in (4.11) is linearly divergent. We regulate by cutting off all frequency summations with functions of the frequencies $f(\epsilon)$ such that $f(0) = 1$ and $f(\infty) = 0$ and $f \to 1$ in some well defined limit. Then (4.10) can be evaluated as

$$\mathcal{F}(x,y) = -\frac{i}{(2\pi)^2}\sum_{n \neq 0} \frac{e^{-in(x-y)}}{n} = -\frac{1}{4\pi} sign(x-y) \tag{4.12}$$

We can deduce the $\mathcal{U}(1)$ connection

$$\mathcal{A}(x) = \frac{1}{8\pi}\int dy \; sign(x-y)A(y) \tag{4.13}$$

and the adiabatic effective action from (3.11),

$$S_{eff} = (\text{static vacuum contribution}) + \frac{i}{8\pi}\int dxdy \dot{A}(x) sign(x-y)A(y) \tag{4.14}$$

This efffective action is not invariant under the gauge transformation (4.4),

$$\Delta S_{eff} = \frac{i}{4\pi}\int u_1 du_2 \tag{4.15}$$

and the propagator (4.3) transforms by a phase,

$$Z[A + du_2, u_1] = e^{-\frac{i}{4\pi}\int u_1 du_2} Z[A, u_1] \tag{4.16}$$

This gauge variance cannot be cancelled by adding local counterterms to the action. It can however be understood if the gauge symmetry is realized projectively on the Fock states and if the background field time evolution operator $\mathcal{T}e^{-\int \mathcal{H}_2}$ is gauge invariant. The appropriate transformation of the Fock states would be

$$e^{i\int u(x)\Psi^{\dagger}(x)\Psi(x)}|F, A+du> \to e^{\frac{i}{4\pi}\int u(x)A(x)}|F, A> \tag{4.17}$$

To see that this is the case, it is necessary to show that the Fock states are eigenstates of the gauge generator. Indeed, it is straightforward to show that the Fock vacuum, regarded as an implicit functional of $A(x)$ is an eigenstate of $\mathcal{G}(x)$, *i.e.* that all off-diagonal matrix elements of $\mathcal{G}(x)$ vanish.

$$<\sigma, A|\mathcal{G}(x)|0, A> = 0 \quad ; \sigma \neq 0 \; . \tag{4.18}$$

This follows from the formula (2.13) for off-diagonal matrix elements of the derivative operator by noting that they cancel those of the charge operator. Similar considerations show that all Fock states are eigenstates of $\mathcal{G}(x)$ and that their phases can be redefined in such a way that they all have the same eigenvalue which we shall denote $g(x, A)$. Then

$$g(x, A) = <0, A|\mathcal{G}(x)|0, A> = -\partial_x \mathcal{A}(x) + <0, A|\Psi^{\dagger}(x)\Psi(x)|0, A> \tag{4.19}$$

The vacuum matrix element of the charge density operator is readily evaluated once we choose a normal ordering and regularization prescription. We shall use the Dirac commutator for normal ordering and, analogously to $\mathcal{F}$, we cut off diverging frequency sums with a function of the spectrum, f. Then the vacuum matrix element of the charge density operator is

$$<0, A|\frac{1}{2}[\Psi^{\dagger}(x), \Psi(x)]|0, A> = -\frac{1}{2}\lim_{f\to 1}\sum_{\omega}\psi_{\omega}^{\dagger}(x)\psi_{\omega}(x)\mathrm{sign}(\omega)\, f(\omega)$$

$$= \frac{1}{2\pi}(a - [a] + \frac{1}{2}) \tag{4.20}$$

and

$$g(x, A) = i\frac{A(x)}{4\pi} + \frac{1}{2\pi}(a - [a] + \frac{1}{2})$$

where $[a]$ is the largest integer less than a. Notice that the charge density (4.20) is invariant under translations and both large ($a \to a +$ *integers*) and small ($A \to A + dv$, $a \to a$) gauge transformations.

The gauge symmetry is realized projectively. The extended algebra of the gauge generators is

$$[\mathcal{G}(x), \mathcal{G}(y)]\; |F, A> = (\mathcal{G}(x)g(y, A) - \mathcal{G}(y)g(x, A))\; |F, A>$$

$$= [\delta_{\mathcal{G}}\, g](x, y, A)\; |F, A> = \frac{i}{2\pi}\delta'(x-y)\; |F, A> \tag{4.21}$$

where $\delta_{\mathcal{G}}$ is the coboundary operator [14] with the known Schwinger term. Since $\delta_{\mathcal{G}}^2 = 0$ the Schwinger term in (4.21) is automatically an infinitesimal 2-cocycle and the gauge algebra is associative. Furthermore, the Schwinger term is a coboundary.

However, as we shall discuss in the following, local cohomology is too restrictive to allow its removal.

As a consequence of (4.17), expectation values of gauge invariant operators such as the electric field $\frac{\delta}{i\delta A(x)}$ are not gauge invariant.

$$< 0, A + du|\frac{i\delta}{\delta A(x)}|0, A + du > = < 0, A|\frac{i\delta}{\delta A(x)}|0, A > + \frac{1}{4\pi}u(x) \tag{4.22}$$

One might try to compensate by introducing an external $\mathcal{U}(1)$ connection and define the electric field operator as

$$\mathcal{E}(x) = -i\frac{\delta}{\delta A(x)} - \mathcal{A}(x) \tag{4.23}$$

This operator has gauge invariant expectation values and if we use it to construct the full Hamiltonian

$$\mathcal{H} = \int dx \frac{\mathcal{E}^2}{2} + \mathcal{H}_2 \tag{4.24}$$

$\mathcal{H}$ has gauge invariant matrix elements in the Fock space. If we construct full states of the quantum theory as superpositions of the direct products of Fock states and invariant gauge field wavefunctionals, the eigenvalues of $\mathcal{H}$ and consequently the time evolution would be gauge invariant. (Recall that we have already shown that $\mathcal{H}_2$ generates gauge invariant time evolution of the fermionic variables.)

We would regard (4.24) as arising from the canonical quantization of the Lagrangian

$$\mathcal{L} = \int \{-\frac{1}{4}F^2 + \Psi^\dagger(i\frac{\partial}{\partial t} + A_0 + \mathrm{h})\Psi + \dot{A}\mathcal{A}\} \tag{4.25}$$

with the external $\mathcal{U}(1)$ connection $\mathcal{A}$ which appears as a counterterm to cancel the gauge variant terms in the adiabatic effective action. This yields a gauge invariant theory where the gauge symmetry is realized projectively. If we further modify the Gauss' law constraint by the redefinition $\mathcal{G}(x) \to \mathcal{G}(x) - g(x, A)$ which would follow canonically from the Lagrangian

$$\mathcal{L} = \int \{-\frac{1}{4}F^2 + \Psi^\dagger(i\frac{\partial}{\partial t} + A_0 + h)\Psi + \dot{A}\mathcal{A} - A_0 g(x, A)\} \tag{4.26}$$

we obtain invariant dynamics with a faithful representation of the gauge symmetry.

However, since the $\mathcal{U}(1)$ connection is neither local nor Lorentz invariant, we would expect (4.26) to violate at least one of these symmetries. Indeed, the quantum mechanics of the Chiral Schwinger model with this modification has been solved and shown not to be Lorentz invariant [15].

A $3 + 1$ dimensional gauge theory exhibits essentially the same structure. The gauge symmetry can be shown to be realized projectively. The ensuing anomalous commutator has been computed [9] and shown to be cohomologous to that suggested by Faddeev [14]. Taking the induced connection $\mathcal{A}$ into account yields a gauge invariant Hamiltonian and unitary, gauge invariant time evolution. However, we would conjecture that Lorentz invariance is also broken by that quantization.

In conclusion, we have seen how anomalies arise from an induced connection on the gauge field configuration space. Anomalies could be cancelled if this connection could be compensated in a Lorentz invariant way. It would be interesting to examine this possibility further [16].

References

1. M.V. Berry, Proc. Roy. Soc. Lond. Ser392, 45 (1984); B. Simon, Phys. Rev. Lett. 51, 2167 (1983); L. Schiff, "Quantum Mechanics" (McGraw-Hill, New York 1955) pg. 290.
2. F. Wilczek and A. Zee, Phys. Rev. Lett. 52, 2111 (1984). J. Moody, A. Shapere and F. Wilczek, Phys. Rev. Lett. 56, 893 (1986); R. Jackiw, Phys. Rev. Lett. 56 2779 (1986).
3. M. Kuratsuji and S. Iida, Phys. Lett. 111A, 220 (1985).
4. A. Niemi and G. Semenoff, Phys. Rev. Lett. 55, 227 (1985).
5. D. Arovas, R. Schrieffer and F. Wilczek, Phys. Rev. Lett. 53, 772 (1984).
6. D. Arovas, R. Schrieffer, F. Wilczek and A. Zee, Nucl. Phys. B251[FS13], 117 (1985).
7. D. Haldane and Yong-Shi Wu, Phys. Rev. Lett. 55, 2287 (1985).
8. G. Semenoff and P. Sodano, Phys. Rev. Lett. 57, 1195 (1986).
9. P. Nelson and L. Alvarez-Gaume, Comm. Math. Phys. 99, 103 (1985); A. Niemi and G. Semenoff, ref. 2 and Phys. Rev. Lett. 56, 1019 (1986); H. Sonoda, Phys. Lett. 156B, 220 (1985); Nucl. Phys. B266, 440 (1986); A. Niemi, G. Semenoff and Yong-Shi Wu, Nucl. Phys. B276, 173 (1986).
10. M. V. Berry, Bristol Preprint, 1986 R. Chiao and Yong-Shi Wu, Phys. Rev. Lett. 57, 933 (1986).
11. A. Tomita and R. Chiao, Phys. Rev. Lett. 57, 937 (1986); R. Chiao, private communication.
12. J. Von Neumann and E. Wigner, Z. Phys. 30, 467 (1929).
13. T.T. Wu and C.N. Yang, Nucl. Pyhs. B107, 365 (1976).
14. L. Faddeev, Phys. Lett. B145, 81 (1984); L. Faddeev and S. Shatashvili, Teor. Math. Fiz. 60, 206 (1984); J. Mickelsson, Comm. Math. Phys. 97, 361 (1984); I. Frenkel and I. Singer, unpublished.
15. A. Niemi and G. Semenoff, Phys. Lett. in press (1986).
16. See B. Grossman, Rockefeller University Report, 1986.

PREREGULARIZATION AND THE AMBIGUITY STRUCTURE OF THE JACOBIAN FOR CHIRAL SYMMETRY TRANSFORMATIONS

Victor Elias

Department of Applied Mathematics
University of Western Ontario
London, Ontario N6A 5B9
Canada

INTRODUCTION

Preregularization is a procedure for evaluating non-dimensionally-continued field theoretical amplitudes by constraining ambiguous terms within such amplitudes to absorb any Ward-identity violating terms that may arise. Usually these ambiguous terms correspond to some parametrizable arbitrariness in the loop momenta percolating into shift-of-integration-variable surface terms peculiar to integer-dimensional spacetime.[1,2,3] Indeed, since the proposal of preregularization, originally motivated to retain a spacetime structure consistent with supersymmetry,[4] the procedure has been shown to be a "regularization-free" method for analyzing anomalies perturbatively. Preregularization enables a distinction to be made between the manifest finiteness of anomaly-generating amplitudes[5] and the Ward-identity ambiguities associated with such amplitudes. In particular, preregularization has been shown to be applicable to the VVA triangle anomaly,[2,4,6,7] two dimensional anomalies,[8,9] six-and-higher-dimensional anomalies,[9,10] a perturbative demonstration of the Adler Bardeen Theorem,[11] the supercurrent anomaly,[12] and (as discussed by Mann elsewhere in this volume) superstring anomalies.[13]

In this talk, I will first review briefly the preregularization of the VVA triangle graph and the decoupling of surface terms from infinite bounds of integration. I will then discuss the path-integral approach to anomalies and establish that ambiguities in the fermionic-measure Jacobian for chiral-symmetry transformations may be expressed in terms of shift-of-integration variable surface terms. Finally, I will show that such Jacobian ambiguities are resolved through the imposition of gauge invariance, in much the same way that similar ambiguities are resolved in the perturbative chiral anomaly.

THE CHIRAL ANOMALY AND SURFACE TERMS

Consider the triangle graph and cross-graph amplitudes listed below in figure 1. After momentum conservation is imposed at each vertex, the internal loop-momentum in each graph is still arbitrary up to an overall constant, denoted by s_1 for the first graph and s_2 for the second graph. The overall amplitude from both graphs is then given by

$$T_{\mu\rho\sigma} = \frac{-ie^2}{(2\pi)^4} \int d^4r \left\{ \mathrm{Tr}\left[\gamma_\rho \frac{1}{\gamma\cdot(r+q+s_1)-m} \gamma_\sigma \frac{1}{\gamma\cdot(r+q+s_1-k)-m} \right.\right.$$

$$\left. \times \gamma_\mu\gamma_5 \frac{1}{\gamma\cdot(r+s_1-k)-m}\right]$$

$$+ \mathrm{Tr}\left[\gamma_\rho \frac{1}{\gamma\cdot(r+s_2+q)-m} \gamma_\mu\gamma_5 \frac{1}{\gamma\cdot(r+s_2)-m}\right.$$

$$\left.\left. \times \gamma_\sigma \frac{1}{\gamma\cdot(r+s_2-k)-m}\right]\right\} \quad (1)$$

To find the axial vector current divergence, contract $-q^\mu$ into (1) and use the identity

$$\frac{1}{\gamma\cdot(t+q)-m} (\gamma\cdot q\, \gamma_5) \frac{1}{\gamma\cdot t-m}$$

$$= \frac{1}{\gamma\cdot(t+q)-m} (2m\,\gamma_5) \frac{1}{\gamma\cdot t-m}$$

$$+ \gamma_5 \frac{1}{\gamma\cdot t-m} + \frac{1}{\gamma\cdot(t+q)-m} \gamma_5 \quad (2)$$

in order to obtain

$$-q^\mu T_{\mu\rho\sigma} = -2m\, T_{5\rho\sigma} +$$

$$\left[-ie^2(2\pi)^{-4}\right] \int d^4r \left\{ \left[\mathrm{Tr}\ \gamma_\rho [\gamma\cdot(r+q+s_1)-m]^{-1} \gamma_\sigma\gamma_5 [\gamma\cdot(r+s_1-k)-m]^{-1} \right.\right.$$

$$\left. + \mathrm{Tr}\ \gamma_\rho [\gamma\cdot(r+q+s_1)-m]^{-1} \gamma_\sigma [\gamma\cdot(r+q+s_1-k)-m]^{-1}\ \gamma_5 \right]$$

$$+ \left[\mathrm{Tr}\ \gamma_\rho\gamma_5 [\gamma\cdot(r+s_2)-m]^{-1} \gamma_5 [\gamma\cdot(r+s_2-k)-m]^{-1} \right.$$

$$\left.\left. + \mathrm{Tr}\ \gamma_\rho [\gamma\cdot(r+s_2+q)-m]^{-1} \gamma_5\gamma_\sigma [\gamma\cdot(r+s_2-k)-m]^{-1} \right]\right\} \quad (3)$$

The first and fourth traces in (3) correspond (after a trivial anticommutation of γ_5 through γ_σ) to the difference of integrands that are identical under the integration variable shift $r+s_1 \rightarrow r+s_2$ (or $r \rightarrow r' = r + s_2 - s_1$).

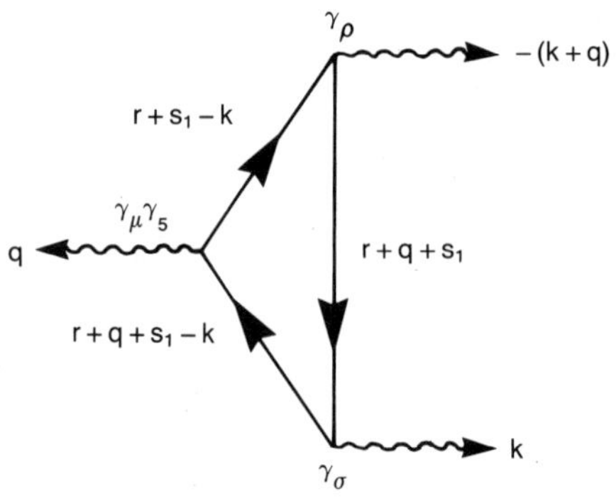

Fig. 1a

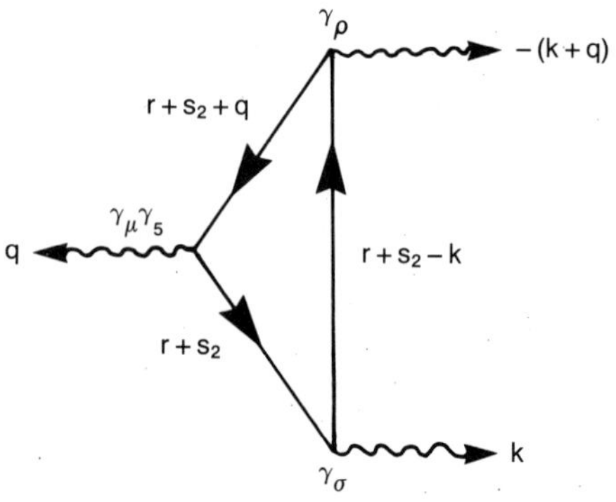

Fig. 1b

Similarly, the second and third traces of (3) are also the difference of identical integrands upon application of the integration variable shift $r+q+s_1 \to r+s_2$ (or $r \to r' = r+(s_2-s_1)-q$). However, one cannot make such naive variable shifts in linearly divergent 4-dimensional integrals without inclusion of surface terms proportional to $r'-r$:[1,2,3]

$$\int \frac{d^4r\, r_\mu}{[r'^2-\mu^2]^2} - \int \frac{d^4r'\, r_\mu}{[r'^2-\mu^2]^2}$$

$$= \frac{i\pi^2}{2}(r'-r)_\mu \tag{4}$$

Both such terms in (3) involve the difference of s_1-s_2, which can be parametrized as an arbitrary linear combination of the external momenta q and k

$$(s_1-s_2)_\rho = Aq_\rho + Bk_\rho \tag{5}$$

The arbitrary coefficient B enters into the divergence of the axial-vector current, obtained entirely from surface terms in evaluating (3):

$$-q^\mu T_{\mu\rho\sigma} + 2m\, T_{5\rho\sigma} = \frac{ie^2}{8\pi^2}\varepsilon_{\rho\tau\sigma\eta}(1-B)q^\tau k^\eta \tag{6}$$

Hence, the right-hand side of (6) represents a possible departure from the naive axial-vector-current Ward identity, rectifiable by choosing B=1. However, corresponding vector-current divergences are also sensitive to loop-momentum ambiguities:[6]

$$k^\sigma T_{\mu\rho\sigma} = \frac{-ie^2}{8\pi^2}\varepsilon_{\mu\tau\rho\eta}(1+A)q^\tau k^\eta \tag{7}$$

$$-(q+k)^\rho T_{\mu\rho\sigma} = \frac{ie^2}{8\pi^2}\varepsilon_{\mu\tau\sigma\eta}(2-A+B)q^\tau k^\eta \tag{8}$$

Preregularization entails resolving the ambiguities in (6), (7) and (8) by imposing gauge invariance (hence vector current conservation) on ambiguous amplitudes. Thus, for (7) and (8) to vanish, A = -1, B = -3, in which case (6) corresponds to the usual chiral anomaly in momentum space.

Note that the surface term (4) is not an artifact of regularization; indeed the surface term is completely independent of how one chooses to parametrize the infinite bounds of the improper momentum-space integrals. To demonstrate this point, consider the following difference between linearly divergent Euclidean space integrals

$$Cp_\mu \equiv \int_{-\infty}^{\infty} \frac{d^4k(k-p)_\mu}{[(k-p)^2]^2} - \int_{-\infty}^{\infty} \frac{d^4k(k+p)_\mu}{[(k+p)^2]^2} \tag{9}$$

The denominators can be expanded in four-dimensional Legendre polynomials[14]

$$\frac{1}{[(k\mp p)^2]^2} = \frac{1}{|k^2||p^2|}\sum_{\ell=0}^{\infty}\sum_{s=0}^{\infty}(\ell+1)(\pm 1)^\ell P_\ell(\hat{p}\cdot\hat{k})R^{\ell+2s+2} \tag{10}$$

where

$$R = \begin{pmatrix} |p|/|k| \ , & |k|\geq|p| \\ |k|/|p| \ , & |p|\geq|k| \end{pmatrix} \tag{11}$$

and where

$$\int d\Omega_{\hat{k}}\ P_\ell(\hat{p}\cdot\hat{k})P_m(\hat{p}\cdot\hat{k}) = 2\pi^2\ \delta_{\ell m} \tag{12}$$

If p^μ is contracted into (9), the relations $p^\mu k_\mu = |p||k|P_1(\hat{p}\cdot\hat{k})/2$ and $p^\mu p_\mu = (p^2)P_0(\hat{p}\cdot\hat{k})$ may be used to perform all angular integrals via (12), so as to obtain

$$Cp^2 = 4\pi^2 \int_0^\infty \frac{|k|d|k|}{|p^2|} \sum_{s=0}^\infty \left(|k|\,|p|R^{2s+3} - |p^2|R^{2s+2}\right) \tag{13}$$

One can see from the definition of R (11) that the integrand of (13) vanishes except for values of $|k|$ smaller than $|p|$. Consequently, one finds that the surface term is insensitive to the infinite upper bound in (13):[3]

$$Cp^2 = 4\pi^2 \int_0^{|p|} \frac{|k|d|k|}{|p^2|} \sum_{s=0}^\infty \left(\frac{|k|^{2s+4}}{|p|^{2s+2}} - \frac{|k|^{2s+2}}{|p|^{2s}}\right)$$

$$= 2\pi^2 \sum_{s=0}^\infty \left(\frac{1}{s+3} - \frac{1}{s+2}\right) = -\pi^2, \tag{14}$$

a result consistent with the Euclidean space version of (4).

PARAMETRIZATION OF INDETERMINACY IN THE PATH-INTEGRAL JACOBIAN

Consider the Lagrangian for massless fermion ψ coupled to an external field $V_\mu \equiv -igV_\mu^a T^a([T_a,T_b] = if_{abc}\,T_c)$:

$$L = i\bar{\psi}\not{D}\psi + \frac{1}{2g^2}\,\mathrm{Tr}\,F_{\mu\nu}F^{\mu\nu} \tag{15}$$

$$D_\mu \equiv \partial_\mu + V_\mu \tag{16}$$

$$F_{\mu\nu} \equiv \partial_\mu V_\nu - \partial_\nu V_\mu + [V_\mu, V_\nu]\ . \tag{17}$$

The generating functional is given by

$$Z[V] \equiv \int d\psi\, d\bar{\psi}\, \exp[-i \int d^4x\ L(x)] \tag{18}$$

Under the chiral transformation

$$\delta\psi = i\ c(x)\gamma_5\psi \tag{19}$$

the variation of the Lagrangian is given by

$$\delta L = (-\partial_\mu c(x))\bar{\psi}\gamma_\mu\gamma_5\,\psi \tag{20}$$

Nevertheless $\delta Z = 0$ as Z is a functional of the external field only. Thus a chiral transformation (19) on (18) yields

$$\begin{aligned} Z[V] &\to J(V,c) \int d\psi\, d\bar{\psi}\, \exp[-i \int d^4x(L(x) + \delta L(x))] \\ &= Z[V] \end{aligned} \tag{21}$$

where J[V,c] is the Jacobian associated with the altered fermionic measure generated by (19).[15] The chiral Ward identity may be realized from (21) by requiring $\delta Z|_{c=0} = 0$:

$$0 = \frac{1}{Z[V]} \left.\frac{\delta Z[V]}{\delta c(x)}\right|_{c=0} = \left.\frac{\delta J[V,c]}{\delta c(x)}\right|_{c=0}$$

$$+ \frac{J[V,0]}{Z[V]} \frac{\delta}{\delta c(x)} \left[\int d\psi\, d\bar{\psi} \exp\left[-i \int d^4x (\mathcal{L}(x) + \delta\mathcal{L}(x)) \right] \right]_{c=0} \tag{22}$$

Using (20) and the fact that $J[V,0] = 1$, the second term on the right hand side of (22) is precisely the definition of $\langle \partial_\mu j^{\mu 5} \rangle_0$. In Fujikawa's seminal work,[15,16] a path integral approach to the anomaly involves direct evaluation of $J[V,c]$ by expanding the fermion fields $\psi(x)$ in an eigenbasis $\phi_k(x)$ of a hermitian operator:[15]

$$J[V,c] = \exp\left[-2i \sum_k \mathrm{Tr}_{\mathrm{Dirac}} \int d^4x\, \phi_k^+(x) c(x) \gamma_5 \phi_k(x) \right] \tag{23}$$

Naively, (23) is indeterminate:

$$\mathrm{Tr}_{\mathrm{Dirac}} \sum_k \phi_k^+(x) \gamma_5 \phi_k(x) = \delta^4(x-x) \mathrm{Tr}_{\mathrm{Dirac}} \gamma_5$$

$$= \infty \cdot 0 \tag{24}$$

This indeterminacy is successfully resolved by utilizing eigenstates of $(i)\not{D}(x)$ and by regulating the large eigenvalues of this operator[15,17]

$$\mathrm{Tr}_{\mathrm{Dirac}} \sum_k \phi_k^+(x) \gamma_5 \phi_k(x) = \lim_{\substack{M\to\infty \\ y\to x}} \mathrm{Tr}_{\mathrm{Dirac}} \sum_k \phi_k^+(y) \exp\left[-\left(\frac{(i)\not{D}(x)}{M} \right)^2 \right] \phi_k(x)$$

$$= \lim_{\substack{M\to\infty \\ y\to x}} \mathrm{Tr}_{\mathrm{Dirac}} \gamma_5 \exp\left[(-) - \left(\frac{4D_\mu(x) D^\mu(x) + [\gamma^\mu, \gamma^\nu][D_\mu(x), D_\nu(x)]}{4M^2} \right) \right] \delta^4(x-y) \tag{25}$$

Eigenstates of a γ-contracted non-commuting operator ($[D_\mu, D_\nu] = F_{\mu\nu}$) are essential in order to bring additional γ-matrix structure into the Dirac trace; had eigenstates been taken of the (commuting)operator $\gamma\cdot\partial$, the Jacobian would remain trivial (J=1) as a consequence of the tracelessness of γ_5. It is worth noting that overall gauge invariance suggests use of the correct choice of hermitian operator $[(i)\not{D}]$, just as gauge invariance resolved the indeterminacy of the VVA triangle discussed in the previous section.

It is of some interest to establish whether this parallel between path integral and perturbative treatments of the anomaly can be placed in closer correspondence. In particular, is regularization of $J[V,c]$ really necessary, and can the indeterminacy of $J[V,c]$ be parametrized in the absence of gauge invariance? In other words, is it possible to preregularize the Jacobian $J[V,c]$ and utilize integration variable ambiguities to absorb violations of gauge invariance?

To answer these questions, consider evaluation of $\delta J/\delta c|_0$ from (22) directly, through the introduction of external sources η and η':[17]

$$-Z[V]\left.\frac{\delta J[V,c]}{\delta c(x)}\right|_{c=0}$$

$$= \frac{\delta}{\delta c(x)}\left(\int d\psi\, d\bar{\psi}\, \exp[\int d^4x(L+\Delta L)]\right)$$

$$= \int d\psi\, d\bar{\psi}\left(\partial_\mu[\bar{\psi}(x)\gamma^\mu\gamma_5\,\psi(x)]\right)\sum_{n=0}^{\infty}\frac{(-ig)^n}{n!}$$

$$\times\left\{\prod_i^n \int \bar{\psi}(z_i)\not{V}^a(z_i)T^a\psi(z_i)dz_i\right\}\exp[i\int d^4z\bar{\psi}(z)\not{\partial}\psi(z)]$$

$$= \int d\psi\, d\bar{\psi}\left(\partial_\mu\left[-\frac{\delta}{\delta\eta(x)}\gamma^\mu\gamma_5\frac{\delta}{\delta\bar{\eta}(x)}\right]\right)\sum_{n=0}^{\infty}\frac{(-ig)^n}{n!}$$

$$\times\left\{\prod_i^n\int dz_i\left(-\frac{\delta}{\delta\eta(z_i)}\right)\not{V}(z_i)\frac{\delta}{\delta\bar{\eta}(z_i)}\right\}\exp[\int d^4z(\bar{\psi}(z)i\not{\partial}\psi(z)$$

$$\left.+\bar{\eta}(z)\psi(z)+\bar{\psi}(z)\eta(z))]\right|_{\eta=\bar{\eta}=0} \qquad (26)$$

One can complete the square in the exponential, via

$$\int d^4z[\bar{\psi}(z)i\not{\partial}\,\psi(z)+\bar{\eta}(z)\psi(z)+\bar{\psi}(z)\eta(z)]$$

$$= \int d^4z\, d^4z'[\tilde{\bar{\psi}}(z)\not{K}(z,z')\tilde{\psi}(z')-\bar{\eta}(z)\not{K}^{-1}(z,z')\eta(z')], \qquad (27)$$

$$K^\mu(z,z')\equiv i\,\delta(z-z')\frac{\partial}{\partial z'_\mu}, \qquad \int d^4z\,\not{K}(z',z)\not{K}^{-1}(z,z'')\equiv\delta^4(z'-z''), \quad (28)$$

so as to obtain

$$\frac{Z[V]}{Z[0]}\left.\frac{\delta J}{\delta c(x)}\right|_{c=0} = -\left[\partial_\mu\left(-\frac{\delta}{\delta\eta(x)}\gamma^\mu\gamma_5\frac{\delta}{\delta\bar{\eta}(x)}\right)\sum_{n=0}^{\infty}\frac{(-ig)^n}{n!}\right.$$

$$\times\left\{\prod_i^n\int d^4z_i\left(-\frac{\delta}{\delta\eta(z_i)}\right)\not{V}(z_i)\frac{\delta}{\delta\bar{\eta}(z_i)}\right\}$$

$$\left.\left.\exp[-\int d^4z\int d^4z'\,\bar{\eta}(z)\not{K}^{-1}(z,z')\eta(z')]\right]\right|_{\eta=\bar{\eta}=0}, \qquad (29)$$

where [for η and η' sufficiently well-behaved at infinity]

$$Z(0) = \int d\tilde{\psi}\, d\tilde{\bar{\psi}}\ \exp[\int d^4z\ \tilde{\bar{\psi}}(z)\ i\not{\partial}\,\tilde{\psi}(z)]$$

$$= \int d\psi\, d\bar{\psi}\ \exp[\int d^4z\ \bar{\psi}(z)\ i\not{\partial}\,\psi(z)]\ . \qquad (30)$$

An expansion of (18) analogous to (26) yields

$$\frac{Z[V]}{Z[0]} = \left[\sum_{n=0}^{\infty} \frac{(-ig)^n}{n!} \left\{\prod_i^n \int d^4z_i \left(-\frac{\delta}{\delta\eta(z_i)}\right) \not{V}(z_i) \frac{\delta}{\delta\bar{\eta}(z_i)}\right\}\right.$$

$$\left.\left. \times \exp\left[-\int d^4z \int d^4z' \, \bar{\eta}(z)\not{K}^{-1}(z,z')\eta(z')\right]\right]\right|_{\eta=\bar{\eta}=0} . \qquad (31)$$

From substitution of (31) into (29), one can see that $\delta J/\delta c|_0$ is equal to those contributions to the right hand side of (29) not entailing factorization of the integrals over z_i for each value of n in the summation. For example, the n=0 contribution to $\frac{\delta J}{\delta c}$ is just proportional to

$$\mathrm{Tr}\left[\frac{\partial}{\partial x^\mu} \not{K}^{-1}(x,x)\gamma^\mu\gamma_5\right] = 0 .$$

The n=1 contribution to the right hand side of (29) that factorizes an integral over z_1,

$$ig \, \mathrm{Tr}\left[\frac{\partial}{\partial x^\mu} \not{K}^{-1}(x,x)\gamma^\mu\gamma^5\right] \mathrm{Tr}\left[\int d^4z_1 \not{V}(z_1)\not{K}^{-1}(z_1,z_1)\right] ,$$

corresponds to multiplication of the (vanishing) n=0 contribution to $\frac{\delta J}{\delta c}$ by the next-to-lead contribution to Z[V]/Z[0]. The n=1 contribution that does not factorize an integral over z_1 vanishes by Dirac algebra:

$$ig \int d^4z_1 \frac{\partial}{\partial x^\mu} \int \frac{d^4k}{(2\pi)^4} \int \frac{d^4q}{(2\pi)^4} \frac{1}{k^2q^2} e^{i(k-q)(x-z_1)}$$

$$\times \mathrm{Tr}[\not{k}\not{V}(z_1)\not{q}\gamma^\mu\gamma_5] = 0 \qquad (32)$$

Consequently the lowest order contribution to $\delta J/\delta c|_0$ arising from the right hand side of (29) is given by the n=2 term whose z_1 and z_2 integrals do not factorize from the remainder of the expression:

$$\left.\frac{\delta J}{\delta c(x)}\right|_{c=0} = + g^2 \frac{\partial}{\partial x^\mu} \int d^4z_1 \int d^4z_2 \, \mathrm{Tr} \, \not{K}^{-1}(x,z_1)\not{V}(z_1)$$

$$\times \not{K}^{-1}(z_1,z_2)\not{V}(z_2)\not{K}^{-1}(z_2,x)\gamma_\mu\gamma_5$$

$$+ \mathcal{O}(g^3)$$

$$= -g^2 \frac{\partial}{\partial x^\mu} \int d^4z_1 \int d^4z_2 \int \frac{d^4k_1}{(2\pi)^4} \int \frac{d^4k_2}{(2\pi)^4} \int \frac{d^4k_3}{(2\pi)^4} e^{ik_1\cdot(x-z_1)}$$

$$\times e^{ik_2\cdot(z_1-z_2)} e^{ik_3\cdot(z_2-x)} \mathrm{Tr} \frac{\not{k}_1}{k_1^2} \not{V}(z_1) \frac{\not{k}_2}{k_2^2} \not{V}(z_2) \frac{\not{k}_3}{k_3^2} \gamma_\mu\gamma_5 + \mathcal{O}(g^3)$$

(33)

Let the Fourier transforms of the external fields in (33) be defined by

$$V_\tau(u) \equiv \int \frac{d^4k'}{(2\pi)^2} e^{ik'u} \mathcal{V}_\tau(k') , \quad \mathcal{V}_\tau(k') \equiv \int \frac{d^4y}{(2\pi)^2} e^{-ik'y} V(y) \tag{34}$$

To parametrize the indeterminacy of J[V,c] in (33), it is useful to re-express (33) in terms of the transform variables k' and k" associated with the Fourier transforms of the external fields $V(z_1)$ and $V(z_2)$ respectively. Integration over z_1 and z_2 constrains $k' = k_1 - k_2$ and $k'' = k_2 - k_3$. Consequently, the most general linear relabeling of the integration variables k_1, k_2, and k_3 into k', k" and a third variable (k) will necessarily contain two arbitrary parameters, denoted here by α and β:

$$k_1 \equiv k + (1+\alpha)k' + \beta k''; \; k_2 \equiv k + \alpha k' + \beta k''; \; k_3 \equiv k + \alpha k' + (\beta-1)k'' \tag{35}$$

The Jacobian of this transformation is unity, regardless of α and β. Substitution of (35) into (33) yields an ambiguous expression,

$$\left.\frac{\delta J}{\delta c(x)}\right|_{c=0} = -\frac{g^2}{(2\pi)^8}\frac{\partial}{\partial x^\mu}\int d^4k' \int d^4k'' \, e^{i(k'+k'')x}$$

$$\times \left[\mathrm{Tr} \int d^4k \left\{ \frac{[\not k+(1+\alpha)\not k'+\beta\not k'']}{[k+(1+\alpha)k'+\beta k'']^2} \gamma\cdot\mathcal{V}(k') \frac{[\not k+\alpha\not k'+\beta\not k'']}{[k+\alpha k'+\beta k'']^2} \gamma\cdot\mathcal{V}(k'') \right.\right.$$

$$\left.\left. \times \frac{[\not k+\alpha\not k'+(\beta-1)\not k'']}{[k+\alpha k'+(\beta-1)k'']^2} \gamma^\mu\gamma_5 \right\}\right] , \tag{36}$$

corresponding identically to the VVA triangle graph (fig. 1 with k' + k" = q) for massless fermions with a momentum-contracted axial-vector-current vertex. Indeed (36) is precisely the relationship between J[V,c] and quantum corrections to $\langle j_{\mu 5}\rangle_0$ needed to ensure that a non-trivial chiral-symmetry Jacobian might not conspire with well-known quantum corrections to $\langle j_{\mu 5}\rangle_0$ so as to cancel the anomaly.[18] Moreover, we see that α and β, chosen to parametrize the indeterminacy of J[V,c], correspond to ambiguity in the internal loop momentum of Fig. 1a equal to $s_1 = \alpha k' + (\beta-1)k''$.

PREREGULARIZATION OF THE JACOBIAN

Equation (36) may be evaluated using the methods of two sections previous to this. The answer one obtains upon integrating over k,

$$\left.\frac{\delta J}{\delta c(x)}\right|_{c=0} = \frac{4ig^2}{(2\pi)^8}\int d^4k' \int d^4k'' \, e^{i(k'+k'')x} \, \mathcal{V}_\nu(k')\mathcal{V}_\mu(k'')$$

$$\times \left[-\frac{\pi^2}{2} \varepsilon_{\rho\nu\sigma\mu} (-\beta+\alpha) k'_\rho k''_\sigma \right]$$

$$= -\frac{ig^2(\beta-\alpha)}{8\pi^2} \varepsilon_{\rho\nu\sigma\mu} [\partial_\rho V_\nu(x)][\partial_\sigma V_\mu(x)], \tag{37}$$

corresponds to those terms on the right hand side of the equation

$$-\langle \partial_\mu j^{\mu 5}(x)\rangle_0 = \frac{-ig^2}{16\pi^2}(\beta-\alpha)F^{\sigma\nu}(x)\tilde{F}_{\sigma\nu}(x) \tag{38}$$

that are bilinear in vector fields. Of course, (38) is an ambiguous expression; a choice of $\beta=\alpha$ removes entirely the anomalous term from the chiral current. To resolve this ambiguity (without appealing to the perturbative approach of two sections previous), one must regard the Jacobian to be both a functional of the chiral symmetry-transformation parameter c(x) (19) as well as a functional (in principle) of the quantity $\Lambda(x)$ parametrizing local gauge-transformations on the external field V(x):

$$V_\mu(x) \to V_\mu(x) + \partial_\mu\Lambda(x) + \ldots. \tag{39}$$

One can then establish that[19]

$$\left.\frac{\delta J[V,c,\Lambda]}{\delta\Lambda(x)}\right|_{\Lambda=0}$$

$$\sim \partial_\mu c(x)\int d^4k'\int d^4k''\, e^{i(k'+k'')x}\, \mathrm{Tr}\int d^4k$$

$$\times\left[\left\{\frac{(\not k+(1+\alpha)\not k'+\beta\not k'')}{[k+(1+\alpha)k'+\beta k'']^2}\,\not k'\,\frac{(\not k+\alpha\not k'+\beta\not k'')}{[k+\alpha k'+\beta k'']^2}\,\gamma\cdot V(k'')\,\frac{(\not k+\alpha\not k'+(\beta-1)\not k'')}{[k+\alpha k'+(\beta-1)k'']^2}\,\gamma^\mu\gamma_5\right\}\right.$$

$$\left.+\left\{\frac{(\not k+(1+\alpha)\not k'+\beta\not k'')}{[k+(1+\alpha)k'+\beta k'']^2}\,\gamma\cdot V(k')\,\frac{(\not k+\alpha\not k'+\beta\not k'')}{[k+\alpha k'+\beta k'']^2}\,\not k''\,\frac{(\not k+\alpha\not k'+(\beta-1)\not k'')}{[k+\alpha k'+(\beta-1)k'']^2}\,\gamma^\mu\gamma_5\right\}\right]$$

$$= -2\pi^2\,\varepsilon_{\rho\nu\sigma\mu}\,\partial_\mu c(x)\int d^4k'\int d^4k''\, e^{i(k'+k'')x}\, k'_\rho k''_\sigma[(\beta-1)V_\nu(k'')$$

$$+ (\alpha+1)V_\nu(k')] \tag{40}$$

The last line of (40) is obtained from the previous line first by using the identity

$$\frac{\not k+\not q}{[k+q]^2}\,\not q\,\frac{\not k}{k^2} = \frac{\not k}{k^2} - \frac{\not k+\not q}{(k+q)^2} \tag{41}$$

to transform traces of six γ-matrices into traces of four γ-matrices, and then by seeing that the (Euclidean space) integrals over k only generate shift-of-integration-variable surface terms:

$$\varepsilon_{\rho\nu\sigma\mu}\int d^4k\,\frac{(k+q)_\rho(-K)_\sigma}{(k+q)^2(k+q-K)^2}$$

$$= \varepsilon_{\rho\nu\sigma\mu}(-K_\sigma)\int d^4k\int_0^1 dz\,\frac{(k+q-Kz)_\rho}{[(k+q-Kz)^2+K^2z(1-z)]^2}$$

$$= \varepsilon_{\rho\nu\sigma\mu}(-K_\sigma)\int_0^1 dz\left[\int\frac{d^4t\; t_\rho}{[t^2+K^2z(1-z)]^2} + \frac{\pi^2}{2}(q-Kz)_\rho\right] = -\frac{\pi^2}{2}\,\varepsilon_{\rho\nu\sigma\mu}q_\rho K_\sigma \tag{42}$$

The final line of (40) vanishes provided $(\beta-\alpha) = 2$, a result which when substituted into (38) yields the usual chiral anomaly.[20]

ACKNOWLEDGEMENTS

The results presented in the final two sections of this paper are the product of collaborative research with G. McKeon, R.B. Mann, T.N. Sherry, T. Steele and T. Treml.[19] I am also grateful to the Natural Sciences and Engineering Research Council of Canada for financial support of this work.

REFERENCES

1. J.M. Jauch and F. Rohrlich, The Theory of Photons and Electrons (Springer, Berlin and N.Y., 1976) pp. 457-460; R.E. Pugh, Can. J. Phys. 47, 1263 (1969).
2. V. Elias, G. McKeon and R.B. Mann, Phys. Rev. D28, 1978 (1983).
3. V. Elias, G. McKeon and R.B. Mann, Can. J. Phys. 63, 1498 (1985).
4. V. Elias, G. McKeon, R.B. Mann and S.B. Phillips, Phys. Lett. B133, 83 (1983).
5. R.B. Mann, Nucl. Phys. B266, 125 (1986).
6. V. Elias, G. McKeon and R.B. Mann, Nucl. Phys. B229, 487 (1983).
7. F.B. Little, V. Elias, G. McKeon and R.B. Mann, Phys. Rev. D32, 2707 (1985).
8. V. Elias, G. McKeon and R.B. Mann, Phys. Rev. D27, 3027 (1983).
9. V. Elias, R.B. Mann, A.M. Chowdhury, G. McKeon, S. Samant and S.B. Phillips, Can. J. Phys. 63, 1453 (1985).
10. A.M. Chowdhury, R.B. Mann and G. McKeon, Phys. Rev. D33, 3090 (1986).
11. A.M. Chowdhury and G. McKeon, Phys. Rev. D (in press).
12. A.M. Chowdhury, G. McKeon, V. Elias and R.B. Mann, Phys. Rev. D34, 619 (1986).
13. R.B. Mann, Univ. of Toronto Preprint UTPT/86/13.
14. A.E. Terrano, Phys. Lett. B93, 424 (1980).
15. K. Fujikawa, Phys. Rev. Lett. 42, 1195 (1979).
16. K. Fujikawa, Phys. Rev. D21, 2848 (1980) and Phys. Rev. D25, 285 (1984).
17. Our Euclidean space conventions[$\gamma_\mu = \gamma_\mu^+$, $\bar{\psi} = \psi^+$, $\gamma_5 = \gamma_1\gamma_2\gamma_3\gamma_4$, Tr $\gamma_\mu\gamma_\nu\gamma_\rho\gamma_\sigma\gamma_5 = 4\varepsilon_{\mu\nu\rho\sigma}$] differ from those of refs. 15 and 16, which employ antihermitian γ-matrices. For our conventions, Fujikawa's eigenbasis are eigenstates of $i\not{D}$-hence, the parenthetical factor of i.
18. The relationship between perturbative and path integral approaches to the anomaly has also been considered by R. Delbourgo and G. Thompson, Phys. Rev. D32, 3300 (1985).
19. V. Elias, G. McKeon, T. Steele, T.N. Sherry, R.B. Mann and T.F. Treml, Univ. of Western Ontario Preprint UWO/86/137.
20. A difference in relative sign with ref. 15 is accounted for by differing γ_5 conventions.

OPERATOR REGULARIZATION

D.G.C. McKeon

Department of Applied Mathematics
University of Western Ontario
London, Ontario N6A 5B9
Canada

T.N. Sherry

Department of Mathematical Physics
University College Galway
Galway, Ireland

ABSTRACT

Operator regularization is introduced in conjunction with a perturbative expansion of Schwinger to compute Green's functions in the background field formalism. Feynman diagrams do not arise, no explicit divergences occur, and no symmetries of the theory are broken by the regulating procedure.

We would like to outline a procedure for calculating Green's functions in quantum field theory that may best be called "operator regularization". This work was done over the past year, in part with Subhash Rajpoot and Sanjiv Samant.

Let us begin by first sketching the usual procedure for computing Green's functions through Feynman diagrams in the background field method. If L is the classical effective action, and the field ϕ_i are split into a background part f_i and a quantum part h_i,

[1] $\phi_i = f_i + h_i$

then the Euclidean generating functional is given by

[2] $Z[f_i,J_i] = \int dh_i \exp \int dx[\, L(f_i+h_i) + J_i h_i].$

If L is of the form

$$[3] \quad L = \frac{1}{2!} h_i M_{ij}(f) h_j + \frac{1}{3!} a_{ijk}(f) h_i h_j h_k + \frac{1}{4!} b_{ijk\ell} h_i h_j h_k h_\ell$$

and if

$$M_{ij}(f) = M_{ij}^{(0)} + M_{ij}^{(1)}(f)$$

then the usual Feynman diagram expansion is generated by the following way of re-expressing eq. [2]

$$[4] \quad Z[f_i, J_i] = \sum_{n=0}^{\infty} \frac{1}{n!} \left[\int dx \left(\frac{1}{2!} M_{ij}^{(1)}(f) \frac{\delta^2}{\delta J_i \delta J_j} + \frac{1}{3!} a_{ijk}(f) \frac{\delta^3}{\delta J_i \delta J_j \delta J_k} + \frac{1}{4!} b_{ijk\ell} \frac{\delta^4}{\delta J_i \delta J_j \delta J_k \delta J_\ell} \right) \right]^n \exp\left(-\tfrac{1}{2} \int dx \; J_i M_{ij}^{(0)^{-1}} J_j \right)$$

Divergences that arise in this expansion must first be regulated through insertion of a regulating parameter into the initial Lagrangian. In theories such as supersymmetric gauge theories it proves difficult to do this in a way that is consistent with the initial symmetries present in the theory. When the regulating parameter approaches its limiting value, the resulting divergences must be removed from the theory through a renormalization of the physical parameters in the initial Lagrangian.

Operator regularization is a technique for computing Green's functions that is distinct in several respects from the usual procedure sketched above. In this approach:

(1) The functional integral is preformed prior to expanding in powers of the background field.

(2) Operators ($\det A, A^{-1}$) are regulated, not the action.

(3) Green's Functions are evaluated by using an expansion in powers of the background field devised by Schwinger.

Three features of operator regularization are:

(1) Symmetries of the theory are not broken.

(2) Feynman diagrams do not arise.

(3) As an added bonus, no explicit ultraviolet divergences are ever encountered.

In our approach, the first step consists of combing eqs. [2] and [3] to write the generating functional as

$$[5] \quad Z[f_i, h_i] = \operatorname{sdet}^{-\frac{1}{2}}(M_{ij}(f)) \sum_{n=0}^{\infty} \frac{1}{n!} \left[\int dx \left(\frac{1}{3!} a_{ijk}(f) \frac{\delta^3}{\delta J_i \delta J_j \delta J_k} + \frac{1}{4!} b_{ijk\ell} \frac{\delta^4}{\delta J_i \delta J_j \delta J_k \delta J_\ell} \right) \right]^n \exp -\tfrac{1}{2} \int dx \left(J_i M_{ij}^{-1}(f) J_j \right)$$

In eq. [5], the supermatrix M is separated into Bose-Bose etc. blocks

$$[6] \quad M = \begin{pmatrix} M_{bb} & M_{bf} \\ M_{fb} & M_{ff} \end{pmatrix}$$

and sdet M and M^{-1} have two representations

$$[7a] \quad \operatorname{sdet} M = \det M_{bb} \det^{-1}\left(M_{ff} - M_{fb} M_{bb}^{-1} M_{bf}\right)$$

$$[7b] \quad \operatorname{sdet} M = \det^{-1} M_{ff} \det\left(M_{bb} - M_{bf} M_{ff}^{-1} M_{fb}\right)$$

$$[8a] \quad M^{-1} = \begin{pmatrix} \left(M_{bb} - M_{bf} M_{ff}^{-1} M_{fb}\right)^{-1} & 0 \\ -M_{ff}^{-1} M_{fb}\left(M_{bb} - M_{bf} M_{ff}^{-1} M_{fb}\right)^{-1} & 1 \end{pmatrix} \begin{pmatrix} 1 & -M_{bf} M_{ff}^{-1} \\ 0 & M_{ff}^{-1} \end{pmatrix}$$

$$[8b] \quad M^{-1} = \begin{pmatrix} 1 & -M_{bb}^{-1} M_{bf}\left(M_{ff} - M_{fb} M_{bb}^{-1} M_{bf}\right)^{-1} \\ 0 & \left(M_{ff} - M_{fb} M_{bb}^{-1} M_{bf}\right)^{-1} \end{pmatrix} \begin{pmatrix} M_{bb}^{-1} & 0 \\ -M_{fb} M_{bb}^{-1} & 1 \end{pmatrix}$$

For an operator A, the key to regulating det A and A^{-1} is the equation

$$[9] \quad \ln A = -\lim_{s\to 0} \frac{d^m}{ds^m}\left\{\frac{s^{m-1}}{m!} A^{-s}\right\} \qquad m = 1,2,\ldots$$

In eq. [9], we choose m equal to the number of loops. To one loop order,

$$[10a] \quad \det A = \exp[\operatorname{tr} \ln A]$$

$$= \exp\left[-\operatorname{tr} \lim_{s\to 0} \frac{d}{ds} A^{-s}\right]$$

and at m(>1) loop order

$$[10b] \quad A^{-1} = \frac{d}{dA} \ln A$$

$$= \lim_{s\to 0} \frac{d^m}{ds^m}\left\{\frac{s^m}{m!} A^{-s-1}\right\} .$$

The next step is to write

$$[11] \quad A^{-\lambda} = \frac{1}{\Gamma(\lambda)} \int_0^\infty dt\ t^{\lambda-1} e^{-At}$$

in eq. [10] ($\lambda = s,\ s + 1$).
In eq. [11] we now preform the Schwinger expansion

$$[12] \quad \operatorname{tr} e^{-(A_0+A_1)t} = \operatorname{tr}\left\{e^{-A_0 t} - t\, e^{-A_0 t} A_1 + \frac{t^2}{2}\int_0^1 du\ e^{-(1-u)A_0 t} A_1\, e^{-uA_0 t} A_1 + \ldots\right\}$$

and similarly for $e^{-(A_0+A_1)t}$.

Evaluation of functional traces etc. is now most easily done in momentum space where

$$[13]\quad \langle p|x\rangle = \exp(ip\cdot x)/(2\pi)^{n/2} \qquad (p \equiv -i\partial)$$

For example, consider ϕ_6^3 theory. The generating functional is to one loop order

$$[14]\quad Z^{(1)}[f,0] = \det^{-\frac{1}{2}}(p^2+\lambda f) = \exp \tfrac{1}{2}\,\zeta'(0)$$

where $\zeta(s)$ is the usual ζ-function. Eqs. [10a], [11] and [12] imply

$$[15]\quad \zeta(s) = \frac{1}{\Gamma(s)}\int_0^\infty dt\; t^{s-1}\; \mathrm{tr}\Big[e^{-p^2t} - t\,e^{-p^2t}(\lambda f) + \frac{t^2}{2}\int_0^1 du\; e^{-(1-u)p^2t}(\lambda f)e^{-up^2t}(\lambda f) + \ldots\Big]$$

The two point function can be read off eq. [15], and after computing the functional traces, we obtain the finite result

$$[16]\quad \zeta'_{ff}(0) = \int \frac{d^6p}{(4\pi)^3}\, f(p)f(-p)\left[\frac{p^2}{12}\left(\ell n\; p^2 - \frac{8}{3}\right)\right]$$

To two loop order, the regulated generating functional in ϕ_6^3 is

$$[17]\quad Z^{(2)}[f,0] = \lim_{s\to 0} \frac{d^2}{ds^2}\left[\frac{s^2}{2!}\,\frac{\lambda^2}{2!3!}\int d^6x\, d^6y\left(\langle x|(p^2+\lambda f)^{-s-1}|y\rangle\right)^3\right]$$

The Schwinger expansion has been applied to eq. [17] in order to determine the two point function. Again, no ultraviolet divergences are encountered.

We have also applied our approach to ϕ_4^4 two loop effective potential, the two and three point functions in four and n dimensional QED, the two loop vacuum polarization in QED, the anomalous divergence of the spinor current in N=1 super Yang-Mills theory, and the two point function in the massless Wess-Zumino superfield model. In all cases, finite symmetry preserving results are recovered.

QED in a covariant gauge (non-covariant gauges still present an outstanding problem), using eq. [7b], has the one loop generating functional

$$[18]\quad Z^{(1)} = \frac{\det(\not{p}-e\not{V})}{\det^{\frac{1}{2}}\left[p^2\delta_{\mu\nu} - \left(1-\frac{1}{\alpha}\right)p_\mu p_\nu - e^2\bar{\chi}\gamma_\mu(\not{p}-e\not{V})^{-1}\gamma_\nu\chi - e^2\bar{\chi}\gamma_\nu(\not{p}-e\not{V})^{-1}\gamma_\mu\chi\right]}$$

(Using eq. [7a] gives results related to results obtained from eq. [18] by a finite renormalization.)

The spinor self energy that follows from eq. [18] upon applying operator regularization in 4-2ε dimensions can be derived from $\zeta'_{\bar{\chi}\chi}(0)$ where

$$[19]\quad \zeta_{\bar{\chi}\chi}(s) = \frac{e^2}{8\pi^2}\int d^np\, \frac{s}{s+\varepsilon}\Big[\alpha + \alpha\varepsilon(1-\gamma+\ell n\,4\pi - \ell n\, p^2) + s\left(\tfrac{3}{2} + \alpha\,\ell n\,\alpha - \alpha\,\ell n\, p^2\right)\Big]\bar{\chi}(p)\not{p}\chi(-p)\;.$$

We see from eq. [19] that

$$[20]\quad \lim_{\varepsilon\to 0}\left(\lim_{s\to 0}\frac{d}{ds}\zeta_{\bar{\chi}\chi}(s)\right)$$

reproduces the results of evaluating the appropriate Feynman diagram. After renormalizing so as to excise the pole at $\varepsilon=0$ in eq. [20], we obtain a result related to

$$[21]\quad \lim_{s\to 0}\frac{d}{ds}\left(\lim_{\varepsilon\to 0}\zeta_{\bar{\chi}\chi}(s)\right)$$

by a finite renormalization.

The anomalous Green's functions <VVA> and <AAA> follow from the generating functional

$$[22]\quad Z = \det^{\frac{1}{2}}(\not{p} - e\not{V} - ig\not{A}\gamma_5)$$

In eq. [22] the factor of i is needed to maintain Hermiticity in Euclidean space, but destroys compact $U_A(1)$ gauge invariance.

The computation of <VVA> and <AAA> from eq. [22] using operator-regularization has the following features:

(1) In contrast to the Feynman diagram approach, the two calculations are completely distinct.

(2) The V vertex is automaticly transverse and the A vertex is automaticly anomalous.

(3) In n dimensions, no divergences occur, and γ_5 need only anticommute with all γ_μ (the 't Hooft Veltman prescription gives no extra contribution).

We are currently applying our approach to other anomalies, and to non-renormalizable interactions, such as quantum gravity. Preliminary calculations indicate that the one loop graviton correction to the spinor propagator is finite in our approach.

Papier 1524

AN ANALYTIC REGULARIZATION FOR SUPERSYMMETRY ANOMALIES

H.C. Lee

Theoretical Physics Branch
Chalk River Nuclear Laboratories
Atomic Energy of Canada Research Company
Chalk River, Ontario K0J 1J0, Canada
and
Department of Applied Mathematics
University of Western Ontario
London, Ontario N6A 5B9, Canada

Q. HoKim

Physics Department
Laval University
Quebec, Quebec G1K 7P4, Canada

1. INTRODUCTION TO ANALYTIC REGULARIZATION

In the originally devised form of dimensional regularization [1-4], a theory in D (= integer) dimensions is defined as the limit $\varepsilon \to 0$ of a theory in $d \equiv D + 2\varepsilon$ dimensions. The method is incompatible with supersymmetry [5-8], since the algebras for vectors and spinors in a supersymmetric theory depend differently on D, so that supersymmetry is manifestly broken in the method.

In a modification, known as dimensional reduction [9], the algebra of Dirac matrices (which act only on spinors) is kept in the N-dimensional space, but the Lorentz algebra lives in d-dimensional space. This approach is basically inconsistent [9,10], but except for anomalies it can be made with special provisions to work for most purposes.

Dimensional regularization also fails when used to compute axial anomalies [11-15] since a totally antisymmetric tensor cannot be realized in a nonintegral-dimensional space. There is another more technical reason. Because anomalies (in perturbation theory) are closely related to surface integrals [17-19] - not surprising since in the topological approach [15] anomalies are associated with boundaires of manifolds - they are always expressible as finite differences of pairs of divergent Feynman integrals. But the nature of analytic continuation employed by dimensional regularization is such that these differences are exact cancellations so that the method always (incorrectly) gives zero values for anomalies.

Several modifications have been devised for dimensional regularization for the purpose of making it useful for the computing anaomalies [1,20-23].

The modification usually centres on the properties of the totally antisymmetric tensor and Dirac matrix (γ_5 when D=4) and requires special care for its application, thus tarnishing the elegance of the method in some way.

Another misconception that gives doubts to the applicability of dimensional regularization in computing anomalies is the belief that shift of integration variables in Feynman integrals is not allowed (a notion not inconsistent with the association between anomalies and surface integrals). In fact shift of integration variables is permitted, provided it is done consistently (in a way explained later). This is important because the shift operation is essential for an analytic approach.

Our goal is to find a regularization method based on analytic continuation that (i) preserves supersymmetry and (ii) is useful for computing anomalies. It goes without saying that the method must also preserve gauge invariance, Lorentz invariance, and so on. The method, which shall be called analytic regularization for reasons that will become obvious, is based on the observations that: (i) In (the perturbative approach to) field theory divergences that need to be regulated occur only in Feynman integrals; (ii) All amplitudes involving Feynman integrals and transforming as tensor of the Lorentz group can be expressed as sums of terms that are products of known Lorentz tensors and Lorentz invariant Feynman integrals. Thus in the proposed method only (Lorentz) invariant (Feynman) integral are regulated.

A method of regulating only invariant integrals entails a retreat from the more ambitious "global" approach of dimensional regularization, in which a theory is regulated. As mentioned earlier, the retreat is necessary if the objective is to preserve supersymmetry.

There is an earlier (and different) version of analytic regularization [24-26] often referred to as Speer's method. This approach is also global: propagators having the form $(k^2 + m_i^2)^{-1}$ is replaced by the expression $(k^2 + m_i^2)^{-\lambda_i}$, and the original theory is viewed as the limit $\lambda_i \to 1$ of the theory with the new propagators. In principle the method calls for as many parameters λ_i as there are species of particles, a procedure which manifestly breaks supersymmetry. The method has had some success in verifying supersymmetric Ward identities at the one-loop level, provided the different λ_i's are identified as a single parameter [27].

As a vehicle for testing our method, we consider the N=1 supersymmetric Yang-Mills theory [5-8] in the Wess-Zumino gauge, and use the method to verify the anomalous supersymmetric identity [28-35]

$$\delta(\mathcal{S}\cdot\gamma) = 2i\bar{\varepsilon}\ \mathrm{Tr}(\mathcal{T}) + 3\bar{\varepsilon}\ \gamma_5(\partial\mathcal{J}_5) \qquad (1)$$

at the one-loop order. In the equations δ denotes the supersymmetric transformation, ε is a constant spinor field, $\mathcal{S}$ is the supersymmetry current, $\mathcal{T}$ is the energy-momentum tensor and $\mathcal{J}_5$ is the axial-vector current. The non-vanishing quantities $\mathcal{S}\cdot\gamma$, $\mathrm{Tr}(\mathcal{T})$ and $\partial\mathcal{J}_5$ are anomalous since they respectively signal the violations of conformal invariance, energy-momentum conservation and chiral invariance due to quantum fluxuation. Eq. (1) is therefore an unbroken supersymmetric relation among anomalies. It will be shown that our method neatly separates the task of evaluating the three anomalies and the task of making manifest the supersymmetric relations connecting them.

2. N=1 SUPERSYMMETRIC YANG-MILLS THEORY [5-8]

We restrict ourselves to working in the Wess-Zumino gauge [5]. Then the Lagrangian is

$$\mathcal{L} = -\frac{1}{4} F_{\mu\nu}F^{\mu\nu} + \frac{i}{2}\bar{\chi}\not{\mathcal{D}}\chi + \frac{1}{2}D^2 \tag{2}$$

which is invariant under the supersymmetric transformation

$$\delta F_{\mu\nu} = i\bar{\varepsilon}(\mathcal{D}_\mu\gamma_\nu - \mathcal{D}_\nu\gamma_\mu)\chi \tag{3}$$

$$\delta\chi = -\frac{i}{2} F^{\mu\nu}\sigma_{\mu\nu}\varepsilon + iD\gamma_5\varepsilon \tag{4}$$

$$\delta D = \bar{\varepsilon}\gamma_5\not{\mathcal{D}}\chi \tag{5}$$

Here $F_{\mu\nu}$, χ, $\mathcal{D}_\mu$, D and ε are respectively the Yang-Mills field tensor, spinor field, covariant derivative, sclar field and a constant spinor field. At the tree-level,

$$\mathcal{S}_\mu = \bar{\chi}\gamma_\mu\sigma_{\nu\lambda}F^{\nu\lambda} \tag{6}$$

$$\mathcal{T}_{\mu\nu} = -F_{\mu\sigma}F^\sigma_\nu - g_{\mu\nu}\mathcal{L} + \frac{i}{4}\bar{\chi}(\gamma_\mu\mathcal{D}_\nu + \gamma_\nu\mathcal{D}_\mu)\chi \tag{7}$$

$$\mathcal{J}_{5\mu} = \frac{1}{2}\bar{\chi}\gamma_\mu\gamma_5\chi \tag{8}$$

and (1) is trivially satisfied, since the symmetries corresponding to those currents are conserved classically, so that

$$\mathcal{S}\cdot\gamma = \mathrm{Tr}(\mathcal{T}) = \partial\mathcal{J}_5 = 0 \tag{9}$$

Define the amplitudes

$$S_{\mu\alpha}(p,q,r) = \langle\mathcal{S}_\mu(p)A_\alpha(q)\chi(r)\rangle \tag{10}$$

$$T_{\mu\nu\alpha\beta}(p,q,r) = \langle\mathcal{T}_{\mu\nu}(p)A_\alpha(q)A_\beta(r)\rangle \tag{11}$$

$$J_{5\mu\alpha\beta}(p,q,r) = \langle\mathcal{J}_{5_\mu}(p)A_\alpha(q)A_\beta(r)\rangle \tag{12}$$

We reserve as indices the middle Greek letter for the current vertices and early letters for external vector fields. Using power counting, it is easy to show that all on-shell (i.e. $\not{\mathcal{D}}\chi = 0$) amplitudes bilinear in spinors for either $\mathcal{T}$ or $\mathcal{J}_5$ vanish identically. The conservation of supersymmetry current, energy-momentum and gauge invariance dictates that all divergences of the three amplitudes in (10)-(12) must vanish at the tree-level. However, at the one-loop level, the divergences

$$p^\mu S^{(1)}_{\mu\alpha},\ q^\alpha S^{(1)}_{\mu\alpha},\ p^\mu T^{(1)}_{\mu\nu\alpha\beta},\ \cdots,\ r^\beta J^{(1)}_{5_{\mu\alpha\beta}} \tag{13}$$

may not vanish. Anomalies would result only if some of these divergences cannot be absorbed into local counter-terms $\Delta\mathcal{S}_\mu$, $\Delta\mathcal{T}_{\mu\nu}$ and $\Delta\mathcal{J}_{5\mu}$. The renormalized currents at the one-loop level are

$$\mathcal{S}^{\,\mathrm{ren}}_\mu = \mathcal{S}^{(1)}_\mu + \Delta\mathcal{S}_\mu \tag{14}$$

and similar expressions for $\mathcal{T}_{\mu\nu}$ and $\mathcal{J}_{5\mu}$. It will be shown that the counter-terms are uniquely determined (for $\mathcal{S}_\mu$, modulo terms not contributing to $\Delta\mathcal{J}\cdot\gamma$ and for $\mathcal{T}_{\mu\nu}$, modulo terms not contributing to $\mathrm{Tr}(\Delta\mathcal{T})$) by the nonvanishing divergences in (13) and the requirement that gauge invariance be preserved. The true anomalies are then revealed in the form of

$$\mathcal{J}^{\mathrm{ren}}\cdot\gamma \neq 0, \quad \mathrm{Tr}(\mathcal{T}^{\mathrm{ren}}) \neq 0, \quad \partial\mathcal{J}_5^{\mathrm{ren}} \neq 0$$

Our task is to find an analytic regularization that preserves the supersymmetric relation among these anomalies.

3. ANALYTIC REGULARIZATION, TENSOR REDUCTION AND INVARIANT INTEGRALS

Given a rank-m tensor $T_{\mu_1\cdots\mu_m}(p_1\cdots,p_n)$ being an n-point function with external momenta p_i it is always possible to expand [36-37] the tensor in terms of known tensors $O^{(\ell)}_{\mu_1\cdots\mu_m}(p_1,\cdots,p_\mu)$ transforming under the Lorentz group with coefficients being Lorentz invariant functions A_ℓ of the p_i's. Thus, using an obvious short hand

$$T_{\{\mu\}}(p_i) = \sum_\ell A_\ell(p_i)\, O^{(\ell)}_{\{\mu\}}(p_i)\ , \qquad \{\mu\} = \mu_1\cdots\mu_m \tag{15}$$

In perturbation theory, the computation of $T_{\{\mu\}}(p_i)$ may involve multi-loop integrals, which determine the invariant functions A_ℓ in (15). In any case, (15) implies that for any n-point function it is sufficient to evaluate only Feynman integrals with Lorentz invariant integrands. (This statement needs to be modified in a straightforward manner for non-covariant gauges such as the light-cone gauge [37].) A simple example of an expansion such as (15) is

$$\int d^Dk \frac{\not{k}k_\mu}{[k^{2m}(k-p)^{2n}]} = A_1 p^2\gamma_\mu + A_2\not{p}p_\mu \tag{16}$$

Using Lorentz invariance it is easily derived that

$$A_1 = \int \frac{d^Dk}{[\cdots]}\left\{\frac{1}{D-1}(x-y^2)\right\} \tag{17}$$

$$A_2 = \int \frac{d^Dk}{[\cdots]}\left\{\frac{1}{D-1}(Dy^2-x)\right\} \tag{18}$$

where $x = k^2/p^2$ and $y = k\cdot p/p^2$. It is a general property of such expansions that the expressions in the curly brackets in (17) and (18) are independent of the denominator $[\cdots]$.

By using the relation

$$k\cdot p = \frac{1}{2}\left[k^2 + p^2 - (k-p)^2\right] \tag{19}$$

the invariant integrals A_ℓ can further be brought to "canonical" form

$$A_\ell = \sum_{i,j} C_{\ell,ij}\int d^Dk(k^2)^{m_i}\left[(k-p)^2\right]^{n_i} \tag{20}$$

Specifically, from (16), (17) and (19)

$$A_1 = \frac{1}{4(D-1)} \{2[-m+1,-n] + 2[-m+1,-n+1] + 2[-m,-n+1]$$

$$-[-m+2,-n] - [-m,-n] + [-m,-n+2]\} \tag{21}$$

where

$$[m_i,n_i] \equiv (p^2)^{1-m-n-m_i-n_i} \int d^Dk(k^2)^{m_i}[(k-p)^2]^{n_i} \tag{22}$$

A similar expression can be derived for A_2.

Although the example of (16) typifies a one-loop calculation of a two-point function, the expansion (16) and the reduction of invariant integrals to canonical form are general. The reduction is particularly simple at the one-loop order; (canonical) invariant integrals for n-point functions in massless theories have at most n factors of quadratures in the integrand, so that they can be classified by a set of n exponents. The classifications of integrals appearing in theories with massive particles is more complicated, but we will not discuss it here.

From now on we shall specify to what is sufficient to tackle (1), namely one-loop, three-point integrals in D=4 dimensions. Furthermore, without loss of generality, we use the on-shell condition

$$q^2 = r^2 = 0 \tag{23}$$

where q and r are momenta appearing in (10)-(12). Then, from $p+q+r = 0$,

$$2q\cdot r = -2p\cdot q = -2p\cdot r = p^2 \tag{24}$$

which means that aside from the exponents m_i, the invariant integrals can only be functions of a single variable, p^2.

There are three classes of integrals

$$I_1(D,\ell_1,\ell_2,\ell_3) \equiv \int d^Dk[(k-q)^2]^{\ell_1}[(k+p)^2]^{\ell_2}(k^2)^{\ell_3} \tag{25}$$

$$I_2(D,\ell_1,\ell_2,\ell_3) = \int d^Dk[(k-r)^2]^{\ell_1}[(k+p)^2]^{\ell_2}(k^2)^{\ell_3} \tag{26}$$

$$I_3(D,\ell_1,\ell_2,\ell_3) = \int d^Dk(k^2)^{\ell_1}[(k-q)^2]^{\ell_2}[(k+r)^2]^{\ell_3} \tag{27}$$

where on the left-hand sides the fact that the I_i's are functions of p^2 is not made explicit. By simple power counting, the integrals are ultraviolet divergent when

$$D/2 + \ell_1 + \ell_2 + \ell_3 \geq 0$$

so that regularization is needed.

We regulate the integrals by analytic continuation in the most general possible way [38]. Let ω, λ_1, λ_2, λ_3 be continuous (for our purposes real is sufficient) variable and define

$$M_1(\omega,\lambda_1,\lambda_2,\lambda_3) \equiv \int d^{2\omega}k[(k-q)^2]^{\lambda_1}[(k+p)^2]^{\lambda_2}(k^2)^{\lambda_3} \tag{28}$$

then there will be regions in the $\omega\lambda_1\lambda_2\lambda_3$-hyperplane ($\omega\lambda$-plane for short) in which the integral on the right-hand side of (28) is well-defined. Then in principle a close-form representation for the integral (in that region) exists, which we call M_1. We now define

$$I_1(D,\ell_1,\ell_2,\ell_3) \overset{\text{def}}{=} \lim_{\substack{2\omega\to D \\ \lambda_i\to\ell_i}} M_1(\omega,\lambda_1,\lambda_2,\lambda_3) \tag{29}$$

The other two classes of integrals I_2 and I_3 are similarly defined. Because (we assume, and it turns out to be so) M_1 is an analytic function of ω and the λ_i's, divergences in the original integral I_1 will appear in M_1 as poles in the $\omega\lambda$-plane. The method differs from dimensional regularization through the presence of generalized exponents λ_i. Since there are more than one parameter, the regularization is not completely defined until the limiting process in (29) is specified. We defer the choosing of the limit to the end of the computation.

Elsewhere [36,37] it has been shown that the multi-parameter continuation has allowed our method to render a rigorous treatment of certain integrals (e.g. tadpoles) undefinable in dimensional regularization and to analytically separate ultraviolet from infrared and mass singularities.

With the on-shell conditions (23) and (24), and using Euler's representation for exponentiating the factors in the integrand in (28) and standard dimensional regularizations technique [1,4,38] for the k-integration one derives [39] the close-form representation for M_1

$$M_1(\omega,\lambda_1,\lambda_2,\lambda_3) = \frac{(-\pi)^{\omega}(p^2)^{\alpha_1}\Gamma(-\alpha_1)\Gamma(\alpha_1-\lambda_2)\Gamma(\alpha_1-\lambda_3)}{\Gamma(-\lambda_2)\Gamma(-\lambda_3)\Gamma(2\alpha_1-\lambda_1-\lambda_2-\lambda_3)} \tag{30}$$

where $\alpha_1 \equiv \omega+\lambda_1+\lambda_2+\lambda_3$. Similar calculation shows that both M_2 and M_3 are identical to the right-hand side of (30). We therefore suppress the subscript on M henceforth. Now write

$$\omega = D/2 + \varepsilon, \qquad \lambda_i = \ell_i + \sigma_i \tag{31}$$

with the small continuous parameter ε and the σ_i's being of the same order. Then

$$I_{1,2,3}(D,\ell_1,\ell_2,\ell_3) \overset{\text{def}}{=} \lim_{\varepsilon,\sigma_i\to 0} M(\tfrac{D}{2}+\varepsilon,\ \ell_1+\sigma_1,\ \ell_2+\sigma_2,\ \ell_3+\sigma_3) \tag{32}$$

where, as usual, the limit $\varepsilon\to 0$ means ignoring terms of $\mathcal{O}(\varepsilon)$ and higher. To complete the regularization we still need to specify the approach to the origin in the $\varepsilon\sigma_i$-hyperplane.

4. PRESERVATION OF SUPERSYMMETRY

The method described in the last section clearly preserves Lorentz invariance and leaves unchanged the Dirac algebra. It does not manifestly break gauge invariance or supersymmetry, but neither is it guaranteed that such symmetries are preserved. Since properties of such symmetries are not transparently transmitted to Feynman integrals, it is difficult to directly assess whether our method indeed preserves them. Normally Ward identities and identities such as (1) are used for the purpose of making such assessments. This we shall do in the next section. Here we show how we can, by constraining the small parameters ε, σ_i expect our regularization to preserve the symmetries.

We begin with the supposition that if the representation (32) retains all the symmetries of the original integrals in (25)-(27), then the regularization should preserve all symmetries of the theory, as far as three-point and lower functions are concerned. The symmetries of the original integrals are exposed by considering the following changes of integration variable, under which the integrals should be invariant: (i) $k \to -k-p$ in (25) and (26); (ii) $k \to -k+q$ in (25) and (27); $k \to k+r$ in (26); and $k \to k-r$ in (27). Thus we have

$$I(D,\ell_1,\ell_2,\ell_3) = I(D,\ell_1,\ell_3,\ell_2) = I_i(D,\ell_1,\ell_2,\ell_3), \quad i = 1,2,3 \tag{33}$$

For the right-hand side of (32) to reflect this symmetry we must therefore have

$$\sigma_2 = \sigma_3 \tag{34}$$

Conversely, unless there exists symmetries among the integrals I_i not covered by (33), we expect the constraint (34) on the representation (32) to be sufficient to render our regularization to preserve all symmetries of the theory, including supersymmetry and gauge invariance (that are not broken by anomalies).

Summarizing (30)-(34), we arrive at a symmetry preserving representation for (on-shell) three-point invariant integrals [39]

$$I(D,\ell_1,\ell_2,\ell_3) \overset{\text{def}}{=} \lim_{\varepsilon_i \to 0} \frac{(-\pi)^{D/2}(p^2)^{a_1+\varepsilon_1}\Gamma(-a_1-\varepsilon_1)\Gamma(a_1-\ell_2+\varepsilon_0)\Gamma(a_1-\ell_3+\varepsilon_0)}{\Gamma(-\ell_2+\varepsilon_0-\varepsilon_1)\Gamma(-\ell_3+\varepsilon_0-\varepsilon_1)\Gamma(D+\ell_1+\ell_2+\ell_3+\varepsilon_2)} \tag{35}$$

where

$$a_1 = D/2 + \ell_1 + \ell_2 + \ell_3 \tag{36}$$

is the "ultraviolet index", and for reasons to be made clear we have expressed the small parameters in terms of

$$\varepsilon_0 = \varepsilon + \sigma_1 + \sigma_2$$

$$\varepsilon_1 = \varepsilon + \sigma_1 + 2\sigma_2$$

$$\varepsilon_2 = 2\varepsilon + \sigma_1 + 2\sigma_2 \tag{37}$$

Note that in (35) the crucial invariance under $\ell_2 \leftrightarrow \ell_3$ is manifest.

In using our method, it is vitally important to realize that integrals having only two non-vanishing exponents (i.e. a two-point integral) should be viewed as a special case of three-point integrals. For example

$$I(D,\ell_1,\ell_2,0) = \lim M(D/2+\varepsilon,\ \ell_1+\sigma_1,\ \ell_2+\sigma_2,\ \sigma_2)$$

$$\neq \lim M(D/2+\varepsilon,\ \ell_1+\sigma_1,\ \ell_2+\sigma_2,\ 0)$$

This is especially important in the case of computing anomalies which are surface terms, for which a higher degree of rigor is required than in tasks such as the computation of β-functions.

5. PROPERTIES OF THE ANALYTIC REPRESENTATION

The ultraviolet index a_1 in (36) is so called because, by power counting, its value dictates the ultraviolet behavior of the integral. Similarly the values $a_1-\ell_2$ and $a_1-\ell_3$ determine the infrared behavior. Thus, the representation yields simple poles of $\mathcal{O}(1/\varepsilon_1)$ for ultraviolet divergences and poles of $\mathcal{O}(1/\varepsilon_0)$ for infared divergences. The delineation of these two types of poles makes clear why it is unnecessary in this method to use artificially massive propagators in integrals for the purposes of isolating infrared divergences.

When either ℓ_2 or ℓ_3 is nonnegative, the representation has "zeroes" of $\mathcal{O}(\varepsilon_0-\varepsilon_1)$. The limit when all the σ_i's vanish exactly corresponds to dimensional regularization. In that case $\varepsilon_0 = \varepsilon_1$ and the aforementioned zeroes become exact, thus providing a rigorous derivation of the vanishing of tadpole integrals in that method.

In addition to simple poles, the representation also admits double poles of $\mathcal{O}((1/\varepsilon_0)^2)$ which originate from terms such as $(1/\varepsilon_0)\ \ell n q^2$ in the limit $q^2 \to 0$. As expected, in the calculation of anomalies all simple and double ε_0-poles cancel among themselves.

The choice of linear paths (recall ε, σ_1 and σ_2 are by definition of the same order) on the $\varepsilon\sigma$-plane for the limit in (35) is constrained by the possible occurence of $\mathcal{O}(1/\varepsilon_0)$ and $\mathcal{O}(1/\varepsilon_1)$ poles such that paths defined by $\varepsilon_0 = 0$ and $\varepsilon_1 = 0$ are not allowed; any other path is permissible.

Note that, because of the canonicalization of invariant integrals(see paragraph leading to (21)), the exponents ℓ_i in (35) are no more directly related to propagators, and even less to that of any species of particle. This makes clear the difference between our method and the analytic method of Speer, and gives reason to why a necessary (but not sufficient) condition for Speer's method [8] to work is to dissociate the exponents from particle species.

6. ANOMALOUS SUPERSYMMETRY IDENTITY

We now compute one-loop contributions to the three amplitudes (10), (11) and (12). The four types of graphs that occur in the calculation are given in Figure 1; (a) in the calculation of $J^{(1)}_{5\mu\alpha}$, (a) and (b) in $S^{(1)}_{\mu\alpha\beta}$ and

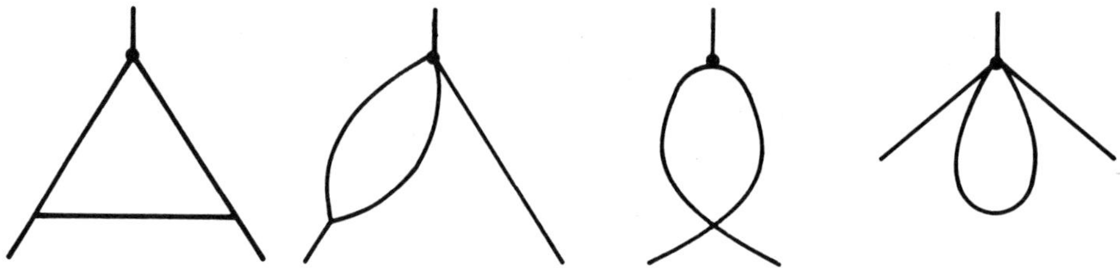

(a) (b) (c) (d)

all four in $T^{(1)}_{\mu\nu\alpha\beta}$. These graphs are evaluated following the procedure described in section 3 so that each of the (amputated) amplitudes is reduced to a sum of known operators with coefficients being sums of invariant integrals. The tensor reduction is straightforward but can be tedious; here it is carried out with the aid of the algebraic computer program SCHOONSCHIP [40]. The results, with the on-shell conditions (23)

$$\not{\mathcal{D}}\chi = \mathcal{D}_\mu F^{\mu\nu} = 0 \tag{38}$$

incorporated, are [39]

$$p^\mu J^{(1)}_{5\mu\alpha\beta} = -16iI_u\ \varepsilon_{\alpha\beta\rho\sigma}p^\rho q^\sigma \tag{39}$$

$$q^{\alpha}J^{(1)}_{5\mu\alpha\beta} = -i(4I_u + 8I_x)\varepsilon_{\mu\beta\rho\sigma}p^{\rho}q^{\sigma} \tag{40}$$

$$p^{\mu}S^{(1)}_{\mu\alpha} = (3I_u - 2I_x)p^2\gamma_{\alpha} - (6I_u - 8I_x)\not{q}r_{\alpha} \tag{41}$$

$$q^{\alpha}S^{(1)}_{\mu\alpha} = 2I_x(p^2\gamma_{\mu} - 4\not{q}r_{\mu}) \tag{42}$$

$$S^{(1)}_{\mu\alpha}\gamma^{\mu} = 0 \tag{43}$$

$$p^{\mu}T^{(1)}_{\mu\nu\alpha\beta} = -(\frac{3}{2}I_u + I_y)p^2p_{\nu}g_{\alpha\beta} + 2(I_u + I_y)p_{\nu}r_{\alpha}q_{\beta} + (\frac{5}{4}I_u - I_y)\,p^2(g_{\nu\alpha}q_{\beta} + g_{\nu\beta}r_{\alpha}) \tag{44}$$

$$q^{\alpha}T^{(1)}_{\mu\nu\alpha\beta} = \frac{7}{4}I_up^2g_{\mu\nu}q_{\beta} + \frac{1}{2}I_u(q_{\mu}r_{\nu} + q_{\nu}r_{\mu})q_{\beta} - \frac{15}{4}I_up^2(g_{\mu\beta}q_{\nu} + g_{\nu\beta}q_{\mu}) - (10I_u + 4I_y)r_{\mu}r_{\nu}q_{\beta} + (\frac{5}{2}I_u + I_y)p^2(g_{\mu\beta}r_{\nu} + g_{\nu\beta}r_{\mu}) \tag{45}$$

$$T^{(1)\mu}_{\mu\alpha\beta} = 0 \tag{46}$$

where (in units of g^2C_2)

$$I_u = (2\pi)^{-4}\left[(0,0,-1) - (-1,0,0)\right] \tag{47}$$

$$I_x = (2\pi)^{-4}\left[(0,-1,-1) + 2(1,-1,-1) - \frac{7}{2}(0,0,-1) + \frac{3}{2}(-1,0,0)\right] \tag{48}$$

$$I_y = (2\pi)^{-4}\left[(-1,0,-1) - 2(-1,1,-1) + 2(-1,0,0)\right] \tag{49}$$

with (ℓ_1,ℓ_2,ℓ_3) being the shorthand of $I(4,\ell_1,\ell_2,\ell_3)$ as defined in (35). The full expressions for $J^{(1)}_{5\mu\alpha\beta}$, $S^{(1)}_{\mu\alpha}$ and $T^{(1)}_{\mu\nu\alpha\beta}$ are lengthy and will not be given here.

We make some remarks concerning the calculation. (i) For a given amplitude, say $S^{(1)}_{\mu\alpha}$, the internal consistency for the tensor reduction (to invariant integrals) can be checked as follows. One can first reduce $S^{(1)}_{\mu\alpha}$ then contract it separately with p^{μ} and q^{α}, or one can directly and separately reduce $p^{\mu}S^{(1)}_{\mu\alpha}$ and $q^{\alpha}S^{(1)}_{\mu\alpha}$. The two sets of results should be identical. This internal consistency is further verified by the equalities derived from (39), (42), (44) and (45),

$$q^{\alpha}(p^{\mu}S^{(1)}_{\mu\alpha}) = p^{\mu}(q^{\alpha}S^{(1)}_{\mu\alpha}) \quad ; \quad q^{\alpha}(p^{\mu}T^{(1)}_{\mu\nu\alpha\beta}) = p^{\mu}(q^{\alpha}T^{(1)}_{\mu\nu\alpha\beta})$$

(ii) The calculation made extensive use of the symmetry relations (33). A significant consequence is that the value of each graph in Fig. 1 is independent of which internal line is assigned to carry the integration momentum k_{μ}. This is not usually the case, especially when a non-analytic regularization method is used, with the consequential "route-momentum ambiguity" further adding to the myth of anomalies in perturbation theory.

(iii) The tadpole graph (d) in Fig. 1 does not have a vanishing value, but contributes terms proportional to (0,0,-1) to both (44) and (45). Among all integrals this integral is actually the most important one, being directly proportional to the anomalies.

The non-vanishing divergents (39)-(42), (44) and (45) represent anomalies only if they cannot be removed by local counterterms. It can be shown that [39] gauge invariance (all contraction with q^α must vanish) demands the constraint

$$I_x = 0 \tag{50}$$

with which the counterterms become uniquely determined. They are

$$(\Delta J_5)_{\mu\alpha\beta} = 4i\ I_u \varepsilon_{\mu\alpha\beta\sigma}(q-r)^\sigma \tag{51}$$

$$(\Delta S)_{\mu\alpha} = -6i\ I_u \gamma_5 \gamma^\sigma \varepsilon_{\mu\alpha\sigma\beta} q^\beta \tag{52}$$

$$(\Delta T)^\mu_{\mu\alpha\beta} = -12\ I_u (q\cdot r g_{\alpha\beta} - r_\alpha q_\beta) \tag{53}$$

In terms of fields, the renormalized currents ($\mathcal{S}^{ren} = \mathcal{S}^{(1)} + \Delta\mathcal{S}$, etc.) satisfy

$$\partial\mathcal{J}_5^{ren} = -2\ I_u {}^*F_{\mu\nu} F^{\mu\nu} \tag{54}$$

$$\mathcal{S}^{ren}\cdot\gamma = 6\ I_u\ \bar{\chi}\ \sigma^{\mu\nu}\gamma_5 {}^*F_{\mu\nu} \tag{55}$$

$$\mathrm{Tr}(\mathcal{T}^{ren}) = 3\ I_u\ F_{\mu\nu}F^{\mu\nu} \tag{56}$$

where ${}^*F_{\mu\nu} = \frac{1}{2}\varepsilon_{\mu\nu\rho\sigma}F^{\rho\sigma}$ is the dual of $F_{\mu\nu}$. A comparison of (52) and (53) with (43) and (46) reveals that the two anomalies $\mathcal{S}^{ren}\cdot\gamma$ and $\mathrm{Tr}(\mathcal{T}^{ren})$ arise entirely from counterterms. The supersymmetry transformation of the right-hand side of (55) gives

$$(\delta\bar{\chi})\sigma^{\mu\nu}F_{\mu\nu} = i\bar{\varepsilon}F_{\mu\nu}F^{\mu\nu} - \bar{\varepsilon}\gamma_5\ {}^*F_{\mu\nu}F^{\mu\nu} \tag{57}$$

$$\bar{\chi}\ \sigma^{\mu\nu}(\delta F_{\mu\nu}) = 2i\bar{\chi}\sigma^{\mu\nu}(\bar{\varepsilon}\mathcal{D}_\mu\gamma_\nu\chi) = 0 \tag{58}$$

to within a total derivative (because $\bar{\chi}\not{\mathcal{D}} = \not{\mathcal{D}}\chi = 0$). Thus, from (54)-(58)

$$\delta(\mathcal{S}^{ren}\gamma) - 2i\bar{\varepsilon}\ \mathrm{Tr}(\mathcal{T}^{ren}) - 3\bar{\varepsilon}\gamma_5(\partial\mathcal{J}_5^{ren}) = I_u[0] \equiv 0 \tag{59}$$

That is, the identity (1) is satisfied algebraically as a result of supersymmetry, independently of the value of I_u.

7. LIMIT FOR ANALYTIC REPRESENTATION

The magnitude of the anomalies (54)-(56) is determined by I_u, subject to the constraint (50). From (35), (47) and (48), we have

$$I_u = \frac{1}{32\pi^2}\left(\frac{\varepsilon_0-\varepsilon_1}{\varepsilon_1}\right) \tag{60}$$

$$I_x = \frac{1}{32\pi^2}\frac{1}{\varepsilon_1}(\varepsilon_0 + 3\varepsilon_1 - 2\varepsilon_2) \tag{61}$$

which are undetermined until the limiting path in (35) is chosen. The ellusive nature of these integrals, a reflection of the fact that anomalies depend on finite differences of divergent integrals, is now transparently displayed.

If we take the limit corresponding to dimensional regularization, namely letting $\sigma_1 = \sigma_2 = 0$ so that $\varepsilon_0 = \varepsilon_1 = \varepsilon_2/2 = \varepsilon$. Then $I_x = I_u = 0$. That is, the constraint (50) is satisfied but the anomalies are also incorrectly given vanishing values. (Because $(-1,0,0) \sim (\varepsilon_0-\varepsilon_1)^2/\varepsilon_1 \sim \mathcal{O}(\varepsilon_i)$, I_u depends only on (0,0,-1), the integral intimately related to the tadpole graph Fig. 1(d). The indistinguishability of ultraviolet and infrared singularities ($\varepsilon_0 = \varepsilon_1$), the vanishing of tadpoles and the vanishing of anomalies are closely related shortcomings of dimensional regularization.)

If the integrals are indeterminate, we can ask whether a path in the $\varepsilon\sigma$-plane exists that satisfies $I_x = 0$ and yields the known value [17]

$$I_u = -\frac{1}{16\pi^2} \tag{62}$$

From (60)-(62), it must satisfy

$$\varepsilon_1 = -\varepsilon_0 = \varepsilon_2 \tag{63}$$

The limiting path, from (37), is therefore

$$\varepsilon \equiv 0 \tag{64}$$

$$2\sigma_1 + 3\sigma_2 = 0 \tag{65}$$

The first condition is remarkable. It states that a supersymmetry preserving analytic regularization can be extended for the computation of anomalies only if the number of dimensions is not changed. Note that, had we restricted ourselves to $\varepsilon=0$ from the outset, then the constraint $I_u=0$ alone would have yielded (65), from which (62) and (63) would have followed. With (64) and (65), the anomalies are (in units of g^2C_2).

$$\partial\mathcal{J}_5^{ren} = \frac{1}{8\pi^2}\,{}^*F_{\mu\nu}F^{\mu\nu}$$

$$\mathcal{J}^{ren}\cdot\gamma = -\frac{3}{8\pi^2}\bar{\chi}\,\sigma^{\mu\nu}F_{\mu\nu}$$

$$\mathrm{Tr}(\mathcal{T}^{ren}) = -\frac{3}{16\pi^2}F_{\mu\nu}F^{\mu\nu}$$

We close with some final remarks. We suspect $\varepsilon \equiv 0$ is a general constraint for an analytic regularization if it is to be used to compute anomalies unambiguously. But this can only be confirmed by further studies. Our method differs from that of Speer's because there is not a fixed relation between the exponents ℓ_i in (35) and any set of propagators in the theory. We do not believe an analytic regularization (even with $\varepsilon \equiv 0$) without adhering to the tensor reduction procedure described in section 3 will in general work; at least not simply, and certainly not for anomalies. We believe our method can in principle be generalized to higher (more than three-point) functions and to multi-loop order. The procedure for tensor reduction is loop-independent, and it may not be too far-fetched to assume that a representation M preserving all the symmetries of I exists. These should be sufficient for the existence of an analytic regularization that preserves all symmetries of the theory. In practise, the derivation of M is not always easy.

REFERENCES

1. G. t'Hooft and M. Veltman, Nucl. Phys. B44(1972)189.
2. C.G. Bellini and J.J. Giambiagi, Nuovo Cimento B12(1972)20.
3. J.F. Ashmore, Nuovo Cimento 4(1972)289.
4. See also the review; G. Leibbrandt, Rev. Mod. Phys. 47(1975)849.
5. J. Wess and B. Zumino, Nucl. Phys. B70(1974)39.
6. A. Salam and J. Strathdee, Nucl. Phys. B76(1974)477.
7. B. deWit and D.Z. Freedman, Phys. Rev. D12(1975)2286.
8. See also the review, M.F. Sohnius, Phys. Rep. 128(1985)39; and the book, J. Wess and J. Bagger, "Supersymmetry and Supergravity" (Princeton University Pres, 1983).
9. W. Siegel, Phys. Lett. 84B(1979)193; Phys. Lett. 94B(1980)37.
10. L.V. Ardeev and A.A. Vladimirov, Nucl. Phys. B214(1983)62.
11. S. Adler, Phys. Rev. 177(1969)2426.
12. J.S. Bell and R. Jackiw, Nuovo Cimento 60A(1969)47.
13. S. Adler and W.A. Bardeen, Phys. Rev. 182(1969)1517.
14. S.B. Treiman, R. Jackiw, B. Zumino and E. Witten, "Current Algebra and Anomalies" (World Scientific, Singapore, 1985).
15. W.A. Bardeen and A.R. White, "Symposium on Anomalies, Geometry and Topology" (World Scientific, Singapore, 1985).
16. K. Fujikawa, contribution to this volume.
17. J.M. Jauch and F. Rohrlich, "The Theory of Photons and Electrons" (Springer-Verlag, Berlin, 1976) pp: 457-460.
18. V. Elias, G. McKeon and R.B. Mann, Phys. Rev. D28(1983)1978; Nucl. Phys. B229(1983)487.
19. V. Elias, contribution to this volume.
20. D. Ahyeampong and R. Delbourgo, Nuovo Cimento 17A(1973)578; 19A(1974)219.
21. P.M. Chanowitz, M. Furman and I. Hinchliffe, Nucl. Phys. B159(1979)225.
22. Y. Fujii, N. Ohta and H. Taniyuchi, Nucl. Phys. B177(1981)297.
23. P.H. Frampton and T.W. Kephart, Phys. Rev. D28(1983)1010.
24. C.G. Bellini, J.J. Giambiagi and a.G. Dominquez, Nuovo Cimento 31(1964)550.
25. E. Speer, J. Math. Phys. 9(1968)1404; 15(1974)1.
26. P. Breitenlohner and H. Mitter, Nucl. Phys. B7(1968)443; Nuovo Cimento 10A(1972)655.
27. S. Kummar and Y. Fujii, Prog. Theo. Phys. 69(1983)653.
28. S. Ferrara and B. Zumino, Nucl. Phys. B87(1975)207.
29. L. Abbott, M. Grisara and H. Schnitzer, Phys. Rev. D16(1977)2995.
30. T.E. Clark, O. Piguet and K. Sihold, Nucl. Phys. B143(1978)445.
31. H. Nicolai and P.K. Townsend, Phys. Lett. 93B(1980)111.
32. P. Majamdar, E. Poggio and H. Schnitzer, Phys. Rev. D21(1980)2203.
33. O. Piguet and K. Sibold, Nucl. Phys. B196(1982)428.
34. N.K. Nielsen, Nucl. Phys. B247(1984)157; B252(1985)401.
35. I.N. McArthur and H. Osborn, Nucl. Phys. B268(1986)573.
36. H.C. Lee and M.S. Milgram, J. Math. Phys. 26(1985)1793.
37. H.C. Lee and M.S. Milgram, Phys. Rev. 55(1986)2172; Nucl. Phys. B268(1986)543.

38. H.C. Lee and M.S. Milgram, Phys. Lett. 133B(1983)320; Ann. Phys. (NY) 157(1984)408; H.C. Lee, Chi. J. Phys. 23(1985)90.
39. H.C. Lee, Q. Ho-Kim and F.Q. Liu, CRNL preprint, TP-86-VI-11.
40. M. Veltman, "SCHOONSCHIP" (Univ. Michigan, 1984).

NON-SYMMETRIC COSET SPACES WITH TORSION-
ALTERNATIVE COMPACTIFICATIONS OF 10-d SUPERSTRINGS*

Brian P. Dolan, David C. Dunbar,
Alfredo B. Henriques† and R. Gordon Moorhouse

Department of Physics & Astronomy, University of Glasgow
Glasgow, G12 8QQ, Scotland

ABSTRACT

The equations of motion of a modified 10-dimensional supergravity Lagrangian coupled to supersymmetric Yang-Mills are solved using non-symmetric 6-dimensional coset spaces with torsion. Four-dimensional Minkowski space is obtained, together with some realistic gauge groups and particle spectra.

The motivation for the following work is to try to find compactification schemes for modified 10-d supergravity which naturally split the 10 dimensions into 4+6. This splitting is obtained by assuming that the three index tensor, H, is non-zero.

The starting point is the 10-d supergravity Lagrangian coupled to supersymmetric Yang-Mills of Chapline and Manton[1], which is a 10-form in differential form language

$$\mathcal{L} = \frac{1}{2} R(*1) - \frac{1}{4} \exp(2\sigma)\, H \wedge * H$$
$$+ \frac{1}{2} \exp(\sigma)\, \mathrm{tr}\,(F \wedge * F) - d\sigma \wedge * d\sigma$$

+ Fermionic terms.

In the following it is assumed that the fermionic background fields vanish and only the bosonic fields are considered. These are

The scalar field (dilation), σ

The Yang-Mills two form $F = dA + A \wedge A$

The three form $H = dB + \mathrm{tr}\,(A \wedge F - \frac{1}{3} A \wedge A \wedge A) = dB - \omega_3^{Y-M}$

Orthonormal one-forms $e^A \quad A = 0, 1, \ldots, 9$

* Presented by B.P. Dolan. Work partly supported by NATO grant #RG85/0128.
† Permanent address, CFMC, Lisbon.

The Einstein scalar, $\mathcal{R}$, is obtained from the curvature two-forms

$$R_{AB} = d\omega_{AB} + \omega_{AC} \wedge \omega^{C}{}_{B}$$

where ω_{AB} are the connection one-forms, related to the orthonormal one -forms via the torsion, T^A

$$de^A + \omega^A{}_B \wedge e^B = T^A$$

The Hodge * operator maps p-forms into 10-p forms

$$*(e^{A_1} \wedge \dots \wedge e^{A_p}) = \frac{1}{(10-p)!} \varepsilon^{A_1 \dots A_p}{}_{B_1 \dots B_{10-p}} e^{B_1} \wedge \dots \wedge e^{B_{10-p}}$$

The gauge group could be $E_8 x E_8$, $SO(32)/Z_2$ or any other group compatible with superstring formalism[2].

It has been shown[3] (and see B. de Wit's lecture in this volume) that there exists no solution of the equations of motion of $\mathcal{L}$ that consists of a compact, six-dimensional, internal space and a maximally symmetric 4-d space-time. However, superstring arguments imply[4] that terms higher order in derivatives should be added to $\mathcal{L}$ in order to reproduce the full superstring Lagrangian. A term, quartic in derivatives, which is a ghost free modification of the Einstein scalar was suggested by Zweibach[5].

$$\frac{1}{4} \exp(\sigma)\, R_{AB} \wedge R_{CD} \wedge *(e^A \wedge e^B \wedge e^C \wedge e^D)$$

The modification

$$H = dB - \omega_3^{Y-M} \rightarrow H = dB - \omega_3^{Y-M} + \omega_3^{LORENTZ}$$

$$\Rightarrow \quad dH = tr(R \wedge R) + tr(F \wedge F)$$

was also suggested by Green and Schwarz[2], to avoid gravitational anomalies. We shall add these terms to the Lagrangian, $\mathcal{L}$, and endeavour to find solutions of the resulting equations of motion with six internal, compact dimensions and 4-d Minkowski space-time. This is similar to the procedure of Candelas et al.,[4] where Calabi-Yau spaces were suggested as internal manifolds.

In order to split off six dimensions as being special, we shall look for an ansatz in which $H \neq 0$. A natural choice might be to make H proportional to the volume element of three space, but one would expect this to lead to very high curvatures in four dimensions, as the only natural length scale is the Planck length, so we shall avoid this. Instead, H will be taken to be non-zero only on the internal, six-dimensional manifold. Apart from group manifolds $[U(1)]^6$, $SU(2)x[U(1)]^3$ and $SU(2)xSU(2)$, three candidates for a 6-dimensional compact space with non-zero H are

Sp(4)/SU(2)xU(1), SU(3)/U(1)xU(1) and G_2 /SU(3) $\approx S^6$, all non-symmetric coset spaces.

Let S and R be two Lie groups with $R \subset S$. The set of left cosets of S with respect to R can be considered as a differentiable manifold, S/R, of dimension (dim S - dim R). Following the notation of J. Strathdee's lecture in this volume[6], let $\bar{a}$ = 1,..., dim R; $\hat{a}$ = 1,..., dim S; a = 1,..., dim S - dim R).

Let $Q_{\bar{a}}$ generate R and $Q_{\hat{a}}$ generate S. Then the imbedding R → S is symmetric if

$$[Q_a, Q_b] = C_{ab}{}^{\bar{c}} Q_{\bar{c}} \qquad \text{(no terms with } Q_c\text{)}.$$

The imbedding is non-symmetric if the structure constants $C_{ab}{}^{c} \neq 0$. Non-symmetric coset spaces allow a natural definition of a non-zero H as

$$H = h C_{abc}\, e^a \wedge e^b \wedge e^c$$

where h is a real constant and e^a are orthonormal one-forms on S/R. Non-symmetric coset spaces also admit non-zero torsion which we shall use as an extra degree of freedom to avoid a cosmological constant on 4-d space-time.

Let y be co-ordinates which label points on S/R and $L(y) \in S$ be a choice of a single element of each distinct left coset in S/R. Since S/R is a differentiable manifold, we can choose L to depend smoothly and differentiably on y. We can construct a metric and connection on S/R as follows:

Define the S-Lie algebra valued one forms

$$\begin{aligned} e(y) &= \left(-\tfrac{1}{\lambda}\right) L^{-1}(y)\, d L(y) \\ &= e^a Q_a + e^{\bar{a}} Q_{\bar{a}} \end{aligned}$$

where d is the exterior derivative on S/R.

e^a can be used as orthonormal one-forms to construct a metric on S/R

$$g = \sum_{a=1}^{\dim S/R} e^a \otimes e^a$$

$e^{\bar{a}}$ can be used to give a connection for a non-zero Yang-Mills background field on S/R, in the Lie algebra of R.

$$\begin{aligned} A &= -\lambda e^{\bar{a}} Q_{\bar{a}} \\ F &= dA + A \wedge A \end{aligned}$$

The above metric can be shown to be Einstein and F satisfies the Yang-Mills equations in this metric and gives a diagonal energy momentum tensor, compatible with Einstein's equations.

From the definition of e(y) above, it follows that

$$de^{\hat{a}} = \frac{\lambda}{2} C_{\hat{b}\hat{c}}{}^{\hat{a}} e^{\hat{b}} \wedge e^{\hat{c}}$$

where λ is a free parameter, and

$$F = -\frac{\lambda^2}{2} C_{ab}{}^{\bar{c}} (e^a \wedge e^b) Q_{\bar{c}}$$

A torsion field can be defined on S/R as follows

$$T^a = \frac{\lambda}{2}(1-\beta) C_{bc}{}^a e^b \wedge e^c$$

$$\Rightarrow \quad \omega^{ab} = \frac{\lambda}{2} \beta C^{ab}{}_c e^c + \lambda C^{ab}{}_{\bar{c}} e^{\bar{c}}$$

Note $T^a = 0$ for any symmetric coset space, since $C_{ab}{}^c = 0$ by definition. β is a free, real, parameter.

β = 1 is the case of zero torsion (canonical connection of the 1st kind in ref. [7]).

β = 0 is the canonical connection of the 2nd kind, which leads to holonomy group R.

Other values of β are of interest, e.g. $\beta = 1 \pm \sqrt{5}$ leads to a Ricci flattening connection[8] in all three cases while other values of β remove the conformal anomaly from the σ-model approach[9].

In general, for all three coset spaces under study

$$\mathcal{R} = \lambda^2 (2 + \beta - \tfrac{1}{2}\beta^2)$$

and, for large enough β, $\mathcal{R}$ becomes negative, allowing the 4-d space-time cosmological constant to vanish.

Having discussed the geometry of non-symmetric coset spaces, we now make the following ansatz for the fields.

σ = const. (in fact, σ scales out of everything leaving us with an overall degree of freedom - the length scale, which is not fixed dynamically).

$$H = h\, C_{abc}\, e^a \wedge e^b \wedge e^c$$

This choice of H automatically satisfies the H equation

$d * H = 0$

using the Jacobi identity.

F and e^a are chosen as above, thus the Yang-Mills equations are automatically satisfied. This leaves the Einstein equation, the dilaton equation and the dH Bianchi identity. It can be shown that, with 4-d space-time being Minkowski and for our ansatz, the dilaton equation follows automatically by tracing the Einstein equations. Thus it is sufficient merely to solve the Einstein equations. These provide two constraints, one from the internal, 6 dimensional, Einstein equations and one from the 4-d space-time equations.

The generalised Bianchi identity

$$dH = \mathrm{tr}(R \wedge R) + \mathrm{tr}(F \wedge F)$$

is compatible with the above ansatz for the fields on S/R (using the Jacobi identity) leading to one constraint which fixes h in terms of λ and β. We now have two polynomial equations (the Einstein equations) in two unknowns, λ and β, which can be solved. For example, in the case of $G_2/SU(3) \approx S^6$, there is one solution

$$\beta = -2.13, \quad \lambda^2 = 3.3$$

Note that this value of β gives negative $\mathcal{R}$.

We now look at some possible residual gauge groups and fermion spectra in 4-dimensions, taking the original Yang-Mills group to be $E_8 \times E_8$. If the starting group is G, then the residual gauge group in 4 dimensions is the centraliser of R in G[10] (with some subtleties when R contains U(1) factors, but Witten[11] has shown that, with the presence of the Chern-Simons terms in H, these U(1)'s are broken also). The question of fermion spectra on these manifolds has been discussed in refs. [8] and [12] and more specifically for SU(3)/U(1)xU(1) by Pilch and Schellekens[13] and for $G_2/SU(3)$ by Fogleman et al[14], in the last supersymmetry on $G_2/SU(3)$ is also discussed.

(i) $G_2/SU(3)$: Giving F a non-zero background value in SU(3) as in the above ansatz breaks the $E_8 \to SU(3) \times E_6$. The adjoint rep. of E_8 decomposes as

$$248 \to (1,8) + (78,1) + (27,3) + (\overline{27},\overline{3})$$

as in Candelas et al[4].

Using the index theorem, the imbalance of massless chiral fermions is

$$n_+ - n_- = -\frac{i}{(2\pi)^3}\frac{1}{6}\int_{S^6} \mathrm{tr}(F \wedge F \wedge F)$$

Explicit calculation with the above ansatz for F yields the value unity for this integral giving only one family, unless the G_2 group of isometries can be used to give a horizontal symmetry.

(ii) SU(3)/U(1) xU(1): There appear to be many possible residual fermion spectra, depending on the way that the U(1)xU(1) is imbedded. Pilch and Schellekens[13] have used group theory arguments to construct a model with three generations in 16's of SO(10).

iii) Sp(4)/SU(2) xU(1): Chapline and Grossman[12] have discussed a scheme with the following symmetry breaking pattern

$$E_8 \rightarrow SU(5) \times SU(5) \rightarrow SU(5) \times SU(3)_F \times SU(2) \times U(1)_F$$

with three flavours of $(\bar{5} + 10)$'s of SU(5) in four dimensions.

In conclusion, by adding R^2 terms to the Chapline-Manton Lagrangian, and including torsion on the internal manifold, compactifications of modified 10-dimensional supergravity coupled to supersymmetric Yang-Mills have been achieved, compatible with the equations of motion. Four dimensional Minkowski space with one of the three non-symmetric coset spaces in six dimensions are the possible geometries. E_6, SO(16) or SU(5) are possible residual gauge groups, leading to some realistic chiral fermion spectra in four dimensions.

REFERENCES

1. G.F. Chapline & N.S. Manton, Phys. Lett. 120B(1983)105.
2. M.B. Green & J.H. Schwarz, Phys. Lett. 149B(1984)117; D.J. Gross, J.A. Harvey, E. Martinec & R. Rohn, Nucl. Phys. B256(1985)253. Narain, Rutherford Laboratory Preprint.
3. D.Z. Freedman, G. Gibbons & P.C. West, Phys. Lett. 124B(1983)491.
4. P. Candelas, G.T. Horowitz, A. Strominger & E. Witten, Nucl. Phys. B258(1985)46.
5. B. Zweibach, Phys. Lett. 156B(1985)315.
6. A. Salam & J. Strathdee, Ann. Phys. (NY) 141(1982)316.
7. S. Kobayashi & K. Nomizu, "Foundations of Differential Geometry" (Vol. II) (1969) Wiley.
8. D. Lüst, "Compactification of Ten-Dimensional Superstring Theories over Ricci Flat Coset Spaces". CalTech preprint CALT-68-1329, to appear in Nucl. Phys. B.
9. L. Castellani & D. Lüst, "Superstring Compactification on Homogeneous Coset Spaces with Torsion", CalTech preprint CALT-68-1353.
10. P. Forgacs & N.S. Manton, Comm. Math. Phys. 72(1980)15.
11. E. Witten, Phys. Lett. 149B(1984)351.
12. G. Chapline & B. Grossman, Phys. Lett. 143B(1984)161.
13. K. Pilch & A.N. Schellekens, Phys. Lett. 164B(1985)31.
14. G. Fogleman, K.S. Viswanathan & B. Wong, "Superstring Compactification on S^6 with Torsion", Simon Fraser preprint.

DIRECT COMPACTIFICATION OF HETEROTIC STRINGS, MODULAR INVARIANCE AND THREE FAMILIES OF CHIRAL FERMIONS

Da-Xi Li

Department of Physics, McGill University
3600 University Street
Montreal, Quebec, Canada H3A 2T8

Superstring theories[1,2] are the most promising candidates for the unified theory including gravity. To make contact with low energy physics one must compactify the ten dimensional space-time into the four dimensional one. This is usually carried out within the context of a d=10 field theory,[3] but this approach may have problems.[4] Recent calculations[5] of the Beta function of the nonlinear Sigma-model in four-loop level showed that Calabi-Yau manifold may not be the solution for the compactification, because of the conformal anomalies arising from the four-loop level. It is much more desirable to carry out the compactification directly with strings. Compactifying string theories to four dimensions via tori has been studied.[6] But it is impossible to get chiral fermions in this way. Dixon, Harrey, Vafa and Witten[7] considered compactifying the d=10 heterotic strings on a six dimensional orbifold. They found 36 families of chiral fermions. But this orbifold has 27 singularities.

In ref. 8, we proposed to compactify directly the $E_8 x E_8'$ heterotic strings from 26 dimensions to 4 dimensions via a 22-dimensional manifold $U=T^{22}/G$, where $T^{22}=T^{16} x T^6$, and T^{16} and T^6 are the maximal tori of $E_8 x E_8'$ and SU(3)xSU(3)xSU(3) respectively. G acts on T without fixed points, so that U is free of singularities. Several examples were analyzed in ref. 8. The modular invariance and one-loop finiteness of this theory have been proved in an example through an explicit calculation of the one-loop amplitudes.[9]

This talk is based on the work carried out in cooperation with C.S. Lam. In this talk, we discuss the problem of the direct compactification of the heterotic strings further, derive conditions of modular invariance for the more general cases, and construct a new U manifold to compactify the heterotic strings to four dimensions with three families of chiral fermions and $SU(3)_3 \times SU(2)_L \times U(1)^3$ gauge symmetry.[10] The E_8' symmetry is broken down to $SU(3)' \times SU(4)' \times U(1)'^3$. The $SU(4)'$ group can play the role of the technicolor group to break the $SU(2)_L \times U(1)^3$ symmetry.

The $E_8 \times E_8'$ heterotic string consists of a ten dimensional (d=10) right moving superstring and a d=26 left moving boson string, with the latter compactified and to be considered as a string living in ten dimensional space-time with an accompanying internal space given by the 16 dimensional torus T^{16}. To compactify the ten dimensional space-time into a four dimensional physical space-time with an associated six dimensional internal manifold M, the usual procedure calls for two separate compactifications: from 26 to 10, and then from 10 to 4 dimensions. In ref. 8, we have investigated the possibility of combining the compactifications, so that the 22=16+6 dimensional internal space U is no longer the direct product of T^{16} and M. For this purpose it is convenient to regard the superstring as a 26 dimensional string which is dormant in the extra 16 dimensions.

If the manifold M is curved then we must look for string theories in curved space. This is very difficult. Alternatively one can retain the original string theories but compactify them in a flat manifold M, though this usually leads to non-chiral theories. Unlike M, the 22 dimensional manifold U mixes in internal quantum numbers in a non-trivial way. As we shall see in the following examples, this enables us to keep chirality in a flat manifold U.

We shall show below examples how such U can be constructed, how it can be used to break the gauge symmetry and obtain chiral fermions, and how the massless states can be determined in such a compactified theory.

Imagine first compactifying the 10 dimensional heterotic string by a six dimensional torus T^6. Then one can show that the 22 dimensional space $T=T^{22}=T^{16} \times T^6$ has a discrete symmetry group G which acts on T without fixed points. The manifold U we are after is then obtained as U=T/G. Since G acts without fixed points, U does not have singularities.

Let us use the index I(1-16) to label coordinates in T^{16}, b(1-3) and $\bar{b}$(1-3) to label the complex coordinates in T^6 ($Z^b=X^{b+2}+iX^{b+5}$, $\bar{Z}^b=Z^{\bar{b}}=X^{b+2}-iX^{b+5}$, $Z^b \approx Z^b+R_b \approx Z^b+R_b\exp(i2\pi/3)$), and β(1-2) to label the transverse coordinates in the four dimensional space-time. The index labelling all these 24 coordinates together will be denoted by M, and the index B will represent $(b,\bar{b},\beta)$. $S^b(\sigma,\tau)$ and $S^{\bar{b}}(\sigma\ \tau)$ (b=0,1,2,3) will denote the fermionic string coordinates.

The group G is a discrete group generated by a twist operator Q which contains rotations on T^6 and a shift on T^{16}.

Since $U=T^{22}/G$, it is simpler to consider the strings on T^{22} first and then project them onto G invariant states. Because every state is a combination of a left and a right moving state, we must first find out the G eigenvalues λ and $\tilde{\lambda}$ of the right and left moving states, and then combine them to make $\lambda\tilde{\lambda}=1$.

Since X^M, $Q^j X^M Q^{-j}$ in T^{22} are identified in U, closed strings in U may satisfy any one of the following boundary conditions in T^{22}:

$$X^M(\sigma+\pi,\tau)=Q^j X^M(\sigma,\tau)Q^{-j} \ , \tag{1}$$

$$S^b(\sigma+\pi,\tau)=Q^j S^b(\sigma,\tau)Q^{-j} \ .$$

Therefore besides the untwisted sector (j=0), we should also consider the solutions in the twisted sectors (j≠0). Besides the G invariant requirement, the twisted sectors should also satisfy the usual physical constraint on closed strings,

$$N+C_R=\tilde{N}+P_I'^2/2+C_L \tag{2}$$

where N, $\tilde{N}$ are the occupation numbers of the right and left oscillators. The quantum numbers of the oscillators, P_I', and the constants C_R and C_L can be determined as in refs. 7 and 8.

To construct a U manifold with three chiral fermion families we need to "twist" the T^{22} more than once. In our model $G=G_1xG_2xG_3xG_4$, and $U=T^{22}/G$.[10]

G_i is generated by a twisted operator Q_i which will be defined by the parameters r_i^b, and q_i^I:

$$Q_i Z^b Q_i^{-1}=\exp(i2\pi r_i^b)Z^b$$

$$Q_i X^I Q_i^{-1}=X^I+\pi q_i^I$$

$$Q_i X^\beta Q_i^{-1}=X^\beta \tag{3}$$

$$Q_i S^b Q_i^{-1}=\exp[i\pi(r_i^0+r_i^1+r_i^2+r_i^3-2r_i^b)]S^b$$

To be clear, we describe the construction of the theory in four steps. First consider the theory in manifold $U=T^{22}/G_1$.

$$\vec{r}_1=(0,\tfrac{1}{3},\tfrac{1}{3},-\tfrac{2}{3}),$$

$$3\vec{q}_1=\varepsilon_6+\varepsilon_7-2\varepsilon_8+\varepsilon_6'+\varepsilon_7'-2\varepsilon_8' \tag{4}$$

ε's are the unit vectors in T^{16}. The root vectors of E_8 are[11] $e_1=\frac{1}{2}(\varepsilon_1-\varepsilon_2-\varepsilon_3-\varepsilon_4-\varepsilon_5-\varepsilon_6-\varepsilon_7+\varepsilon_8)$, $e_2=\varepsilon_1+\varepsilon_2$, $e_3=\varepsilon_2-\varepsilon_1$, $e_4=\varepsilon_3-\varepsilon_2$, $e_5=\varepsilon_4-\varepsilon_3$, $e_6=\varepsilon_5-\varepsilon_4$, $e_7=$

$\varepsilon_6-\varepsilon_5$, $e_8=\varepsilon_7-\varepsilon_6$, $\varepsilon'_I=\varepsilon_{I+8}$. Since $3\vec{q}_1$ is a lattice vector, $G^3=1$, so that the order of G is 3.

The G_1 invariance requirement breaks the $E_8 xE'_8$ to $SU(3)xE_6 xSU'(3)xE'_6$. This can be shown from the spectrum. First let us inspect the untwisted sector.

For the untwisted sector (j =0), the solution of the equation of motion is that of the heterotic string given in ref. 2. The momentum of the right moving bosonic string will be called p', with the right moving oscillators collectively denoted by a'_K. Similarly the left moving momenta and oscillators will be denoted by p and a_K respectively. A quarter of the square of the mass for the right (left) moving strings are given by $N'+(p')^2/2$ ($N+p^2/2+C$). These two masses must be equal. For massless states both of them must vanish. Here N', N are the occupation numbers of the right and left oscillators, in the twisted sectors ($j \neq 0$), the solution has the same form though p, k, and C take on different values as $p^I=\Lambda^I + jq^I$, $k=n \pm \frac{1}{3}$, and $C = -2/3$. Moreover, $p^b=0$ and the c.m. positions (z_0^b, z_0^b) may take on 27 discrete values corresponding to the 27 "fixed points" of G when restricted to T^6.

Although the string now comes in three sectors, it does not mean that the states are tripled. Only states in T invariant under G are allowed states in U.

To survive the G_1 invariant projection, a right moving mode with eigenvalue λ_1 of G_1 must be combined with a left-moving mode of eigenvalue λ_1^{-1}. For the right mover, the massless modes are:

$$\begin{aligned}
\lambda_1=1:&\quad (1, \tfrac{1}{2})+(-1, -\tfrac{1}{2})\\
\lambda_1=\alpha:&\quad (\tfrac{1}{2},0)+(\tfrac{1}{2},0)+(\tfrac{1}{2},0)\\
\lambda_1=\alpha^2:&\quad (-\tfrac{1}{2},0)+(-\tfrac{1}{2},0)+(-\tfrac{1}{2},0)
\end{aligned}\tag{5}$$

where the numbers in (,) denote the four dimensional helicities, and $\alpha=e^{i2\pi/3}$. The massless modes in the $SU(3)xE_6 xSU(3)'xE_6'$ representations are:

$$\begin{aligned}
\tilde{\lambda}_1=1:&\quad (8,1;1,1)+(1,78;1,1)+(1,1;8,1)+(1,1;1,78)\\
\tilde{\lambda}_1=\alpha^2:&\quad (3,27;1,1)+(1,1;3,27)\\
\tilde{\lambda}_1=\alpha:&\quad (\overline{3},\overline{27};1,1)+(1,1;\overline{3},\overline{27})
\end{aligned}\tag{6}$$

For simplicity, we will not write down the $E_8 xE'_8$ singlet zero modes which will form the gravity multiplet. Combining together the left and right moving states to make $\lambda_1\tilde{\lambda}_1=1$, the allowed massless states are as follows. Besides the gauge supermultiplets of $SU(3)xE_6 xSU(3)'xE'_6$, there are three families of (3,27;1,1)+(1,1;3,27) with helicities (1/2,0) and their antiparticles.

We should also consider the twisted sectors. The twisted sectors are crucial to the modular invariance of the theory. Modular invariance is required for the consistency and finiteness of the closed string theories. The requirement of modular invariance restricts the ways of compactification of closed string theories. The necessary and sufficient conditions in orbifold compactification have been obtained.[12] Here we rewrite them in our notation for the bosonic formulation:

$$\sum_b r_i^b=0 \qquad \text{mod } 1$$

$$\sum_{I=1 \text{ to } 8} q_i^I=0 \qquad \text{mod } 2 \tag{7}$$

$$\sum_{I=9 \text{ to } 16} q_i^I=0 \qquad \text{mod } 2$$

$$n(C_{iL}-C_{iR}+\frac{1}{2}q_i^2)=0 \quad \begin{cases} \text{mod } 2 \text{ if } n \text{ is even} \\ \text{mod } 1 \text{ if } n \text{ is odd} \end{cases} \tag{8}$$

(n is the order of the group G_i)

Condition (8) is just the mass match condition of the twisted sectors. Our choices of the r_i^b, q_i^I satisfy all those conditions, so that the theory is modular invariant. The vacua of the twisted sectors may transform nontrivially under the group G. To determine the spectrum in the twisted sectors, we should know how the vacua transform under the G transformation. Let $|j\vec{q}_1>$ denote the "tachyon-like" state with internal momentum $j\vec{q}_1$. Then in general (j=0,±1)

$$G_1|j\vec{q}_1>_j=\exp(-i2\pi(jq_1^2+\psi_j))|j\vec{q}_1>_j \tag{9}$$

Since $G_1^3=1$, we must require

$$3(jq_1^2+\psi_j)=0 \quad \text{mod } 1 \tag{10}$$

The phase factor can also be determined by the requirement of modular invariance. Through an explicit evaluation of the determinant appearing in the measure for the string in orbifold, it can be shown that to ensure modular invariance we must require

$$\frac{1}{2}j^2(r_1^2-\sum_b|r_1^b|+q_1^2)+j\psi_j=0 \qquad \text{mod } 1 \tag{11a}$$

$$jk(r_1^2-\sum_b|r_1^b|+q_1^2)+j\psi_k+k\psi_j=0 \qquad \text{mod } 1 \tag{11b}$$

We will show (11a) and (11b) in the Appendix. The LHS of (11a) and (11b) are the phase factors under the modular transformation $\tau\to\tau+1$ and $\tau\to -1/\tau$, respectively. In (11b), let j=0, k=1, we have $\psi_0=0$. This is as expected that the vacuum in the untwisted sector is invariant under the G transformation. Since $C_{1L}=-1-(r_1^2-\sum_b|r_1^b|)/2$, combining (10) and (11b) we obtain the mass match condition (8). From (11a) and (11b), we find

$$\psi_1=1/3,\ \psi_{-1}=-1/3 \tag{12}$$

The zero modes in the j=1 sector are $27(\bar{3},1,\bar{3},1)$ with helicities $(0,\frac{1}{2})$. The factor 27 is present because the c.m. position of the Z^b in the twisted sectors may take 27 discrete values corresponding to the 27 "fixed points" of G, when restricted to T^6. From (12), we find that they are invariant under the G_1 transformation. The particles from the j=-1 sector are the antiparticles of that in j=1. Those $27(\bar{3},1,\bar{3},1)$ also cancelled the SU(3) anomalies from the untwisted sector alone. In terms of $\underline{27}$ of E_6, there are 9 families of chiral fermions. The G_2,G_3,G_4 invariance will reduce the number of the chiral fermions and the gauge symmetry.

Group G_2 is defined by

$$\vec{r}_2=(0,1/3,-1/3,0),$$

$$\vec{q}_2=\frac{1}{3}(\varepsilon_1+\varepsilon_2-2\varepsilon_3+\varepsilon_1'-\varepsilon_2') \tag{13}$$

G_2 breaks the gauge group to SU(3)xSU(3)xSU(3)xSU(3)xSU(3)'xSU(6)'xU(1)'. We can write the zero modes of the right movers with their G_1,G_2 eigenvalues as follows.

$$\begin{aligned}
&\lambda_1=1,\lambda_2=1: && (1,1/2)+(-1/2,-1)\\
&\lambda_1=\alpha,\lambda_2=1: && (1/2,0)\\
&\lambda_1=\alpha,\lambda_2=\alpha: && (1/2,0)\\
&\lambda_1=\alpha,\lambda_2=\alpha^2: && (1/2,0)\\
&\lambda_1=\alpha^2,\lambda_2=1: && (0,-1/2)\\
&\lambda_1=\alpha^2,\lambda_2=\alpha^2: && (0,-1/2)\\
&\lambda_1=\alpha^2,\lambda_2=\alpha: && (0,-1/2)
\end{aligned} \tag{14}$$

The zero modes of left movers with suitable G_1,G_2 eigenvalues such that they can combine the right movers to form G_1 and G_2 invariant states are as follows.

$$\begin{aligned}
&\tilde{\lambda}_1=1,\tilde{\lambda}_2=1: && (8,1,1,1;1,1)+(1,8,1,1;1,1)+(1,1,8,1;1,1)+(1,1,1,8;1,1)\\
& && +(1,1,1,1;8,1)+(1,1,1,1;1,35)+(1,1,1,1;1,1)\\
&\tilde{\lambda}_1=\alpha^2,\tilde{\lambda}_2=1: && (3,1,\bar{3},\bar{3};1,1)+(1,1,1,1;3,15)\\
&\tilde{\lambda}_1=\alpha^2,\tilde{\lambda}_2=\alpha^2: && (3,3,3,1;1,1)+(1,1,1,1;3,6)\\
&\tilde{\lambda}_1=\alpha^2,\tilde{\lambda}_2=\alpha: && (3,\bar{3},1,3;1,1)+(1,1,1,1;3,6)\\
&\tilde{\lambda}_1=\alpha,\tilde{\lambda}_2=1: && (\bar{3},1,3,3)+(1,1,1,1;\bar{3},\overline{15})\\
&\tilde{\lambda}_1=\alpha,\tilde{\lambda}_2=\alpha: && (\bar{3},\bar{3},\bar{3},1;1,1)+(1,1,1,1;\bar{3},\bar{6})
\end{aligned} \tag{15}$$

$\tilde{\lambda}_1=\alpha, \tilde{\lambda}_2=\alpha^2$: $(\bar{3},3,1,\bar{3};1,1)+(1,1,1,1;\bar{3},\bar{6})$

Combining these states with the right movers to form $\lambda_1\tilde{\lambda}_1=1$, $\lambda_2\tilde{\lambda}_2=1$ states, we find that besides super Yang-Mills multiplet, there are three families of chiral fermions of $(3,3,3,1;1,1)+(3,\bar{3},1,3;1,1)+(3,1,\bar{3},\bar{3};1,1)+(1,1,1,1;3,6)$ $+(1,1,1,1;3,6)+(1,1,1,1;3,15)$ and their antiparticles.

Now we have 8 twisted sectors. We need to find the phase factors of the vacua of these twisted sectors. To ensure modular invariance, the conditions for the phase factors in the case of $G=G_1xG_2xG_3xG_4$ are similar to eq. (11). In the (j_1,j_2,j_3,j_4) sector, the boundary condition for closed strings are:

$$X(\sigma+\pi)=Q_1^{j_1}Q_2^{j_2}Q_3^{j_3}Q_4^{j_4}X(\sigma)Q_4^{-j_4}Q_3^{-j_3}Q_2^{-j_2}Q_1^{-j_1} \tag{16}$$

where $j_1,j_2=0,\ \pm1$; $j_3,j_4=0,1$.

The phase factor under the G_1 transformation is

$$G_i|\vec{q}_{j_1j_2j_3j_4}>_{j_1j_2j_3j_4}=\exp[2\pi i(\vec{q}_i\vec{q}_{j_1j_2j_3j_4}+\psi^{(i)}_{j_1j_2j_3j_4}]$$
$$|\vec{q}_{j_1j_2j_3j_4}>_{j_1j_2j_3j_4} \tag{17}$$

where

$$\vec{q}_{j_1j_2j_3j_4}=j_1\vec{q}_1+j_2\vec{q}_2+j_3\vec{q}_3+j_4\vec{q}_4 \tag{18}$$

Then the conditions similar to (11a) and (11b) are:

$$\frac{1}{2}\gamma'^{2}_{j_1j_2j_3j_4}-\frac{1}{2}\sum_b|\gamma'^{b}_{j_1j_2j_3j_4}|+\frac{1}{2}\vec{q}^{\,2}_{j_1j_2j_3j_4}+\sum_i j_i\psi^{(i)}_{j_1j_2j_3j_4}=0 \bmod 1 \tag{19a}$$

$$\vec{\gamma}'_{j_1j_2j_3j_4}\,\vec{\gamma}'_{\mu_1\mu_2\mu_3\mu_4}-\frac{1}{2}\sum_b[\gamma^b_{j_1j_2j_3j_4}\mathrm{Sign}(\gamma'^{b}_{\mu_1\mu_2\mu_3\mu_4})$$
$$+\gamma^b_{\mu_1\mu_2\mu_3\mu_4}\mathrm{Sign}(\gamma'^{b}_{j_1j_2j_3j_4})]$$
$$+\vec{q}_{j_1j_2j_3j_4}\cdot\vec{q}_{\mu_1\mu_2\mu_3\mu_4}+\sum_i(\mu_i\psi^{(i)}_{j_1j_2j_3j_4}+j_i\psi^{(i)}_{\mu_1\mu_2\mu_3\mu_4})=0 \quad \bmod 1 \tag{19b}$$

where

$$r^b_{j_1j_2j_3j_4}=4j_1r_1^b+4j_2r_2^b+j_3r_3^b+j_4r_4^b \tag{19c}$$

$$r'^{b}_{j_1j_2j_3j_4}=r^b_{j_1j_2j_3j_4}-\text{integer part of } r^b_{j_1j_2j_3j_4} \tag{19d}$$

$$\mathrm{Sign}(x)=\begin{cases}1 & \text{if } x\geq 0\\ -1 & \text{if } x<0\end{cases}$$

We find the solution

$$\psi^{(1)}_{0100}=\psi^{(1)}_{1000}=\psi^{(1)}_{1100}=-\frac{1}{3},\ \psi^{(1)}_{1-100}=\psi_{1000}=\frac{1}{3}$$

$$\psi^{(2)}_{1,-100}=\frac{2}{9},\ \psi^{(2)}_{0100}=-\frac{2}{9},\ \psi^{(2)}_{1100}=\frac{4}{9} \tag{20}$$

satisfies conditions (19a) and (19b).

Now we can use the G_1 and G_2 invariant projection to determine the massless particles from the twisted sectors. The particles from sector $(-j_1,-j_2)$ are always the antiparticles of those from sector (j_1,j_2). We find the massless particles with helicities (0,1/2) in the representations of SU(3)xSU(3)xSU(3)xSU(3)xSU(3)'xSU(6)' are

$$\begin{aligned}
&\text{sector } (1,0): && \text{none}\\
&\text{sector } (0,1): && 9\{(\bar{3},1,\bar{3},1;1,1)+(1,1,\bar{3},\bar{3};1,1)\}\\
&\text{sector } (1,1): && 9\{(1,1,3,1;1,6)+(1,1,1,\bar{3};\bar{3},1)\}\\
&\text{sector } (1,-1): && 9(1,1,1,3;1,\bar{6})
\end{aligned} \tag{21}$$

The G_3 invariant requirement breaks the supersymmetry and breaks the gauge group further to $SU(3)xSU(3)xSU(2)xSU(2)xU(1)^2xSU(3)'xSU(4)'xSU(2)'xU(1)'^2$. The number of chiral fermions in each family is reduced from 27 to 16. G_3 is defined by a 360° rotation in the Z^b plane and a shift $\pi\vec{q}_3$ in T^{16} as in ref. 13:

$$\begin{aligned}
&Q_3\ Z^b\ Q_3^{-1}=Z^b\\
&Q_3\ X^I\ Q_3^{-1}=X^I+\pi q_3^I\\
&Q_3\ S^b\ Q_3^{-1}=\exp(i2\pi r_3'^b)S^b\\
&\vec{q}_3=\varepsilon_4-\varepsilon_4'\\
&r_3'^b=1/2, \qquad (b=0,1,2,3)
\end{aligned} \tag{22}$$

The chiral fermions in the untwisted sector in the representations of SU(3) xSU(3)xSU(2)xSU(2)xSU(3)'xSU(2)'xSU(6)' are $(3,3,2,1;1,1,1)+(3,\bar{3},1,2;1,1,1)+(3,1,2,1;1,1,1)+(3,1,1,2;1,1,1)$, which are three families of 16 chiral fermions, and $(1,1,1,1;3,4,2)+(1,1,1,1;3,4,1)+(1,1,1,1;3,4,1)$, which are the chiral fermions in the "hidden sector" (E_8' sector).

There are also massless bosons. Besides the gravity multiplet and the gauge particles, there are spin zero bosons of $(3,3,1,1;1,1,1)+(3,\bar{3},1,1;1,1,1)+(3,1,2,1;1,1,1)+(3,1,1,2;1,1,1)$.

Using eq. (19) to determine $\psi^{(i)}_{j_1 j_2 j_3 0}$, we find, after the G_1, G_2 and G_3 invariant projections, the surviving chiral fermions in the twisted sectors are:

$$\begin{aligned}
&\text{sector } (0,1,0,0): && 9\{(\bar{3},1,2,1;1,1,1)+(1,1,2,1;1,1,1)+(1,1,1,2;1,1,1)\}\\
&\text{sector } (1,1,0,0): && 9\{(1,1,1,1;1,1,2)+(1,1,2,1;1,4,1)+(1,1,1,2;\bar{3},1,1)\}\\
&\text{sector } (1,-1,0,0): && 9\{(1,1,1,1;1,1,2)+(1,1,1,2;1,4,1)\}\\
&\text{sector } (0,0,1,0): && (\bar{3},1,1,1;3,1,1)+(3,1,1,1;\bar{3},1,1)+(3,1,1,1;3,1,1)
\end{aligned}$$

$+(\bar{3},1,1,1;1,6,1)+(\bar{3},1,1,1;1,1,2)+(\bar{3},1,1,1;1,1,2)$

$+(1,3,1,1;\bar{3},1,1)+(1,\bar{3},1,1;\bar{3},1,1)+(1,1,1,2;\bar{3},1,1)$

$+(1,1,2,1;\bar{3},1,1)$

sector (1,1,1,0): $9\{(3,1,1,1;1,1,1)+(1,1,1,1;3,1,1)\}$

There are no tachyons in any of the twisted sectors.

To break the gauge symmetry further we need G_4. G_4 is defined by

$$\vec{r}_4=(0, \frac{1}{2}, \frac{1}{2}, 0),$$

$$\vec{q}_4= \frac{1}{4}(\varepsilon_1+\varepsilon_2+\varepsilon_3+\varepsilon_4-\varepsilon_5-\varepsilon_6-\varepsilon_7-\varepsilon_8+3\varepsilon_1'-\varepsilon_2'+\varepsilon_3'+\varepsilon_4'+\varepsilon_5'+\varepsilon_6'+\varepsilon_7'+\varepsilon_8') \tag{23}$$

The G_4 invariant requirement breaks the gauge symmetry further to $SU(3)_F$x $SU(3)_C$x$SU(2)_L$x$U(1)^3$x$SU(3)'$x$SU(4)'$x$U(1)'^3$. The chiral fermions in the untwisted sector with the representations of $SU(3)_F$x$SU(3)_C$x$SU(2)_L$x$SU(3)'$x $SU(4)'$ are $(3,3,2;1,1)+(3,\bar{3},1;1,1)+(3,\bar{3},1;1,1)+(3,1,2;1,1)+(3,1,1;1,1)$, which have exactly the same quantum numbers as that in the standard model, and $(1,1,1;3,4)+(1,1,1;3,4)+(1,1,1;3,4)$, which are from the "hidden sector".

So we end up with a string theory in four dimensions, whose effective field theory in the low energy limit is very close to the standard model. The $SU(4)'$ group can play the role of the technicolor group[14] to break the $SU(2)_L$x$U(1)^3$ to $U(1)_{em}$ at the mass scale M_W. This is possible because from renormalization group analysis, the broken scale $\Lambda_{SU(4)'}$ of $SU(4)'$ is much larger than $\Lambda_{SU(3)_C}$ of $SU(3)_C$, and the chiral fermions in the twisted sectors with both $SU(4)'$ and $SU(2)_L$x$U(1)^3$ quantum numbers can be the bridge introducing the $SU(4)'$ force to break $SU(2)_L$x$U(1)^3$, as in the usual technicolor model. The $SU(3)'$ can be broken by the same mechanism at the same scale. There are no chiral fermions with both $SU(3)_F$ and $SU(4)'$ index, hence the $SU(3)_F$ will not be broken by the $SU(4)'$ force. That protects the chiral fermions with $SU(3)_F$ index from getting masses of the order M_W, while the $SU(3)_F$ singlet chiral fermions from the twisted sectors will get masses of the order M_W when the $SU(2)_L$x$U(1)^3$ and $SU(3)'$ are broken. This explains why at energy much lower than M_W we can only see three families of the standard quarks and leptons. Because there are so many fermions coupled to $SU(3)_F$, the β-function of the coupling constant α_F of $SU(3)_F$ is a large positive number, so at low energy, α_F is very small. The number of the chiral fermions coupled to $SU(3)_C$ is the same as that in the standard model, the coupling constant α_C of $SU(3)_C$ will be large at low energy, and $SU(3)_C$ can act as a technicolor group to break the $SU(3)_F$.

In conclusion, we have shown how to compactify the heterotic strings directly from 26 to 4 dimensions via a smooth manifold T^{22}/G. The conditions of modular invariance have been derived. A realistic string model in four dimensions has been constructed. It is modular invariant and tachyon free. In the low energy limit, the gauge group is $SU(3)_F xSU(3)_C xSU(2)_L xU(1)^3 xSU(3)'xSU(4)'xU(1)^3$. The local $SU(3)_F$ group can be broken as in the paper of Candelas et al.[3] The SU(4)' group can play the role of the technicolor group to break the $SU(2)_L xU(1)^3$. The massless fermions are three families of standard quarks and leptons. They can get mass when the global chiral $SU(3)_F$ group is broken by the $SU(3)_C$ force and/or the string radiative effect.

A more detailed study will be published elsewhere.

I thank C.S. Lam, H. Guo, B. Sharp, K.C. Wali, and M. Walton for discussions. This research is supported in part by the Natural Sciences and Engineering Research Council of Canada and the Quebec Department of Education.

APPENDIX: MODULAR INVARIANCE

A crucial test of the theory is modular invariance. It can be checked at one-loop level. World sheet tori contribute to the string one-loop amplitudes. The conformally inequivalent world sheet tori can be described by a complex parameter τ. Besides the conformal invariance, the string theory has also the "large diffeomorphism" symmetry. The "large diffeomorphism" symmetry requires the one-loop amplitudes, or in the path integral formalism, the measure of the string coordinates, to be invariant under the modular transformation

$$\tau \to (A\tau+B)/(C\tau+D) \qquad \text{(A1)}$$

where A, B, C, D are integers, and AD-BC=1.
This invariance is not trivially true. It constrains the way of compactification. The constraints to the compactification parameters for the orbifold compactification have been obtained.[12] They are eqs. (7) and (8) in the text. The phase factors of the vacua can also be obtained from the requirement of modular invariance. Here we derive the conditions of modular invariance by calculating the measure of the string coordinates.

The determinant appearing in the measure of the string coordinates in the path integral formulation is the same as the partition function in the operator formulation.[9] In the usual heterotic strings, it is

$$D_o(\tau)=Tr(-1)^F q^H \qquad \text{(A2)}$$

where $q=e^{i2\pi\tau}$ and Tr means sum (or integrate) over all the momenta, trace over all the oscillators. H is the Hamiltonian and F is the fermion number operator. For strings on T^{22}/G, we can calculate the partition function in a similar way in T^{22} first, then do the G-invariant projection. The contributions from the twisted sectors shall also be added.

The G-invariant projection can be done by introducing a projection operator

$$P_r = \frac{1}{h'} \sum_{g \varepsilon G} g \tag{A3}$$

where h' is the order of group G. When G is a cyclic group and h'=2h+1, then

$$P_r = \frac{1}{h'} \sum_{\mu=-h}^{h} g^{\mu} \tag{A4}$$

If h' is even, h'=2h, then

$$P_r = \frac{1}{h'} \sum_{\mu=-h+1}^{h} g^{\mu} \tag{A5}$$

We will consider the case of h'=2h+1 first. In this case there are 2h twisted sectors which will be denoted by an integer j, $h \geqslant j \geqslant -h$. Then the measure is

$$D(\tau) = \sum_{j=-h}^{h} \mathrm{Tr}\{P_r (-1)^F q^{H_j}\} \tag{A6}$$

where H_j is the Hamiltonian of the j-th sector in T^{22}. From (A4),

$$D(\tau) = \sum_{j=-h}^{h} \sum_{\mu=-h}^{h} \mathrm{Tr}\{g^{\mu}(-1)^F\ q^{H_j}\}$$

$$= \sum_{j,\mu} D_{j\mu}(\tau) \tag{A7}$$

where

$$D_{j\mu}(\tau) = \mathrm{Tr}\{g^{\mu}(-1)^F\ q^{H_j}\} \tag{A8}$$

We can write $D_{j\mu}(\tau)$ as

$$D_{j,\mu}(\tau) = e^{-i2\pi\mu\psi_j} \prod_{b=0 \text{ to } 3} \{DF^b_{j,\mu}(\tau)\ DB^b_{j,\mu}(\tau)\ \tilde{\tilde{DB}}{}^b_{j,\mu}(\tau)\} \cdot$$

$$\cdot \prod_{I=1 \text{ to } 16} \{DI^I_{j,\mu}(\tau)\} \tag{A9}$$

where ψ_j is the phase factor of the j-th sector's vacuum as defined in eq. (9); $DF^b_{j,\mu}(\tau)$ are the contributions from the fermionic coordinates; $DB^b_{j,\mu}(\tau)$ and $\tilde{\tilde{DB}}{}^b_{j,\mu}(\tau)$ are the contributions from the ten-dimensional right- and left-moving bosonic coordinates, respectively; $DI^I_{j,\mu}(\tau)$ are the contributions from the bosonic coordinates in T^{16}. Under modular transformation, the arising phase factors of $DB^b_{j,\mu}(\tau)$ and $\tilde{\tilde{DB}}{}^b_{j,\mu}(\tau)$ always cancel each other. Hence to study the properties under modular transformation, we need only consider $DF^b_{j,\mu}(\tau)$ and $DI^I_{j.\mu}(\tau)$.

In the operator formalism, we find

$$\prod_b DF^b_{j,\mu}(\tau) = \prod_b \{e^{i\pi\tau(j^2\gamma^{b^2} - |j\gamma^b| + \frac{1}{6})} \cdot (1 - e^{2\pi i[|j\gamma^b| + \mu\gamma^b \mathrm{Sign}(j\gamma^b)]})$$

$$\cdot \prod_{n=1}^{\infty} [(1-e^{2\pi i[(n+j\gamma^b)\tau+\mu\gamma^b]})(1-e^{2\pi i[(n-j\gamma^b)\tau-\mu\gamma^b]})]\}$$

$$=\prod_b\{f^{-1}(\tau)e^{-i\pi\tau/12}\, e^{i\pi[\tau j^2\gamma^{b^2}+\mu\gamma^b \mathrm{Sign}(j\gamma_b)]}\theta_1(j\gamma^b\tau+\mu\gamma^b|\tau)\} \qquad \text{(A10)}$$

where

$$f(\tau)=\prod_{n=1}^{\infty}(1-e^{i2\pi n\tau}) \qquad \text{(A11)}$$

and $\theta_1(v|\tau)$ is the Jacobi theta function.[15] Using the properties of $\theta_1(v|\tau)$ and $f(\tau)$[15], we find

$$\prod_b DF^b_{j,\mu}(\tau)=\{\prod_b DF_{j,\mu-j}(\tau+1)\}e^{-i\pi\sum_b[j^2\gamma^{b^2}-|j\gamma^b|]}\varepsilon^4 \qquad \text{(A12)}$$

$$\prod_b DF^b_{j,\mu}(\tau)=\{\prod_b DF_{\mu,-j}(\tfrac{1}{-\tau})\}e^{-i\pi\sum_b[2j\mu\gamma^{b^2}-\mu\gamma^b \mathrm{Sign}(j\gamma^b)-j\gamma^b \mathrm{Sign}(\mu\gamma^b)]}\varepsilon^4 \qquad \text{(A13)}$$

where ε is a phase factor satisfying $\varepsilon^{12}=1$.

The contributions from T^{16} coordinates can also be calculated in the operator formalism.

$$\prod_{I=1\text{ to }16} DI^I_{j,\mu}(\tau)=f^{-16}(\bar\tau)e^{-i4\pi\bar\tau/3}\sum_{P_I\varepsilon\Lambda_{16}}\prod_I e^{i\pi[(P_I+jq_I)^2\bar\tau-2(P_I+jq_I)q_I\mu]}$$

$$=f^{-16}(\bar\tau)e^{-i\frac{4}{3}\pi\bar\tau}\prod_{I=1\text{ to }8}\{\sum_{\substack{\lambda_1=0,\frac{1}{2}\\ \lambda_2=0,\frac{1}{2}}}\sum_{n_I} e^{i\pi[(n_I+jq_I+\lambda_1)^2\bar\tau-2(n_I+jq_I+\lambda_1)q_I\mu+\lambda_2]}$$

$$\cdot \prod_{I=9\text{ to }16}\{\sum_{\substack{\lambda'_1=0,\frac{1}{2}\\ \lambda'_2=0,\frac{1}{2}}}\sum_{n_I'} e^{i\pi[(n_{I'}+jq_{I'}+\lambda'_1)^2\bar\tau-2(n_{I'}+jq_{I'}+\lambda'_1)q_{I'}\mu+\lambda'_2]}\}$$

$$=f^{-16}(\bar\tau)e^{-i\frac{4}{3}\pi\bar\tau}[\sum_{i=1,2,3,4}\prod_{I=1\text{ to }8}\theta_i(jq_I\bar\tau+\mu q_I|\bar\tau)]$$

$$\cdot[\sum_{i=1,2,3,4}\prod_{I'=9\text{ to }16}\theta_i(jq_{I'}\bar\tau+\mu q_{I'}|\bar\tau)] \qquad \text{(A14)}$$

where $\theta_i(v|\tau)$ (i=1,2,3,4) are the Jacobi theta functions.[15] Then we find

$$\prod_I DI^I_{j,\mu}(\tau)=[\prod_I DI^I_{j,\mu-j}(\tau+1)]e^{-i\pi\sum_I j^2q_I^2}\bar\varepsilon^{16} \qquad \text{(A15)}$$

$$\prod_I DI^I_{j,\mu}(\tau)=[\prod_I DI^I_{\mu,j}(-\tfrac{1}{\tau})]e^{-i\pi\sum_I 2j\mu q_I^2}\bar\varepsilon^{16} \qquad \text{(A16)}$$

Combining (A9), (A12), (A13), (A15), and (A16), we find

$$D_{j,\mu}(\tau)=D_{j,\mu-j}(\tau+1)e^{-i\pi\{\sum_b(j^2\gamma^{b^2}-|j\gamma^b|)+\sum_I q_I^2+2j\psi_j\}} \tag{A17}$$

$$D_{j,\mu}(\tau)=D_{\mu,-j}(-\frac{1}{\tau})e^{-i2\pi\{\sum_b[j\mu\gamma^{b^2}-\frac{1}{2}j\gamma^b\mathrm{Sign}(\mu\gamma^b)-\frac{1}{2}\mu\gamma^b\mathrm{Sign}(j\gamma^b)]+\sum_I j\mu q_I^2+j\psi_\mu+\mu\psi_j\}} \tag{A18}$$

To ensure modular invariance, the phase factors in the right handed sides of eq. (A17) and (A18) must vanish. Then we obtain the conditions of eqs. (11a) and (11b) in the text:

$$\sum_b \frac{1}{2}(j^2\gamma^{b^2}-|j\gamma^b|)+\frac{1}{2}\sum_I q_I^2+j\psi_j=0 \quad \mathrm{mod}\ 1 \tag{11a}$$

$$\sum_b[j\mu\gamma^{b^2}-\frac{1}{2}j\gamma^b\mathrm{Sign}(\mu\gamma^b)-\frac{1}{2}\mu\gamma^b\,\mathrm{Sign}(j\gamma^b)]+\sum_I \mu j q_I^2+j\psi_\mu+\mu\psi_j=0 \tag{11b}$$

For the case of $\bar{h}=2h$, we can find the same result in the same way. The only difference is that in Eq. (A9), the sum now is from $(-h+1)$ to h.

When group $G=G_1xG_2xG_3xG_4$ and G_i are cyclic groups, as in the model in the text, we can derive the conditions eq. (19) as above. The only thing we should be careful of now is that in twisted sector (j_1,j_2,j_3,j_4), the γ^b in the above procedure is now replaced by:

$$\gamma^b_{j_1j_2j_3j_4}=4j_1\gamma_1^b+4j_2\gamma_2^b+j_3\gamma_3^b+j_4\gamma_4^b \tag{19c}$$

To reach the result, we need $|\gamma^b|<1$, this condition is not always satisfied by $\gamma^b_{j_1j_2j_3j_4}$, that is why we need to use

$$\gamma'^b_{j_1j_2j_3j_4}=\gamma^b_{j_1j_2j_3j_4}-\text{integer part of }\gamma^b_{j_1j_2j_3j_4} \tag{19d}$$

After this replacement, we can derive the conditions in (19) as above.

REFERENCES

1. J. Schwarz, Physics Reports 9:223 (1982); M. B. Green, Surveys in High Energy Physics 3:127 (1983).
2. D. J. Gross, J. A. Harvey, E. Martinec, and R. Rohm, Nucl. Phys. B256:253 (1985) and B267:75 (1986)
3. P. Candelas, G.T. Horowitz, A. Strominger, and E. Witten, Nucl. Phys. B258:46 (1985); A. Strominger and E. Witten, Comm. Math. Phys. 101:341 (1985); A. Strominger, preprints NSF-ITP-85-105 and 109; M. Dine, V. S. Kaplunovsky, M. Mangano, C. Nappi, and N. Seiberg, Nucl. Phys. B259:549 (1985); J. D. Breit, B. A. Ovrut, and G. C. Segre, Phys. Lett. 158B:33 (1985); R. I. Nepomechie, Y. S. Wu, and A. Zee, Phys. Lett. 158B:311 (1985).
4. M. Dine and N. Seiberg, Phys. Rev. Lett. 55:366 (1985); V. S. Kaplunovsky, Phys. Rev. Lett. 55:1036 (1985).

5. M. T. Grisaru, A. van de Ven and D. Zanon, preprints HUTP-86/A020 (BRX-TH-196); HUTP-86/A026 (BRX-TH-198); HUTP-86/A027 (BRX-TH-199); M. D. Freeman and C. N. Pope, Imperial College preprint, Imperial/TP/ 85-86/16 (1986).
6. M. B. Green, J. H. Schwarz and L. Brink, Nucl. Phys. B198:474 (1982); F. Englert and A. Neveu, Phys. Lett. 163B:349 (1985).
7. L. Dixon, J. A. Harvey, C. Vafa, and E. Witten, Nucl. Phys. B261:651 (1985); Princeton preprint, 1986.
8. C. S. Lam and Da-Xi Li, Direct Compactification of Heterotic Strings from 26 to 4 Dimensions via T^{22}/G, McGill preprint (Dec. 1985) (to be published in Comm. in Theor. Phys.).
9. Da-Xi Li, One-Loop Finiteness of the String Theory on the T^{22}/G Manifold, McGill preprint (1986); C. S. Lam and Da-Xi Li, Phys. Rev. Lett. 56:2575 (1986).
10. Da-Xi Li, String Theory in Four Dimensions with Three Families of Chiral Fermions and Standard Gauge Symmetry, McGill preprint (1986).
11. N. Bourbaki, 'Groupes et algebras de lie' (Hermann); R. Slansky, Phys. Rep. 79:1 (1981).
12. C. Vafa, Modular Invariance and Discrete Torsion on Orbifolds, Harvard preprint HUTP-86/A011.
13. L. Dixon and J. A. Harvey, String theories in ten dimensions without space-time supersymmetry, Princeton preprint, 1986; L. Alvarez-Gaume, P. Ginsparg, G. Moore, and C. Vafa, Harvard preprint HUTP-86/A013.
14. E. Farhi and L. Susskind, Phys. Rep. 74:277 (1981).
15. A. Erdelyi et al., Higher Transcendental Functions, Vol. 2 (McGraw-Hill, New York, 1953) Ch. 13.

SUPERSTRING COMPACTIFICATION ON S^6 WITH TORSION

K.S. Viswanathan, G. Fogleman* and Brenden Wong**

Department of Physics
Simon Fraser University
Burnaby, British Columbia, Canada V5A 1S6

ABSTRACT

We show that if space-time supersymmetry is required and if internal manifolds with torsion not due to the three index field strength H_{pqr} are considered, the almost complex manifold S^6 emerges as a natural candidate for superstring compactification.

INTRODUCTION

Superstring theories[1,2] appear to be good candidates for a mathematically consistent theory of quantum gravity. Anomaly cancellations[3] and the likelihood of finiteness[4] have been demonstrated for the specific gauge groups SO(32) and $E_8 \times E_8$. Hopes are that these theories are phenomenologically realistic as well. In the low energy limit, the Type I superstring theories (and the heterotic string theories[5]) reduce to N = 1 supergravity coupled to super Yang-Mills theory in ten dimensions.[6]

Candelas, Horowitz, Strominger, and Witten (CHSW) have studied vacuum configurations at the tree level for the N = 1, d = 10 supergravity super Yang-Mills theory.[7] Requiring these configurations to have N = 1 supersymmetry in ten dimensions and a geometry consistent with $M^4 \times K$, where M^4 is a four dimensional maximally symmetric space and K is a compact six dimensional internal space, they find that the supersymmetry transformation laws require that M^4 be Minkowski spacetime and K a Calabi-Yau (CY) manifold. In their treatment they set $H_{pqr} = 0$ where H_{pqr} is the totally antisymmetric field strength of the two form potential B_{MN} plus Chern-Simons terms. Furthermore, the number of fermion generations was shown to be one half the Euler characteristic of K. The latter follows from the embedding of the spin connection in an SU(3) subgroup of the gauge group. For the phenomenologically preferred $E_8 \times E_8$ theory, for instance, the residual gauge

---

* Permanent address: Department of Physics and Astronomy, San Francisco State Univ., San Francisco, California.

**Current address: Stanford Linear Accelerator Center, Stanford, California.

group is the $E_6 \times E_8$. A large number of six dimensional CY manifolds (Kähler manifolds with SU(3) holonomy) have been studied[7,8] which give varying numbers of generations. In general, simply connected CY manifolds give much too many generations. Furthermore, CY manifolds do not possess isometries so harmonic expansions cannot be applied in order to study the effective four dimensional theory.

Our motivation in this paper is to seek alternatives to compactification on CY spaces and in particular to seek compactification of the heterotic string (or, more correctly, its field theory limit) on coset spaces. Chapline and Slansky[9] have pointed out that three nonsymmetric coset spaces, $G_2/SU(3)$, $SU(3)/U(1)\times U(1)$, and $SP(4)/SU(2)\times U(1)$, can be compatible with N = 1 supersymmetry. Chapline and Grossman[10] considered compactification on $SP(4)/SU(2)\times U(1)$. Two of these manifolds, $G_2/SU(3)$ and $SP(4)/SU(2)\times U(1)$, are known to admit almost complex structures. The feasibility of coset space compactification of N = 1, d = 10 supergravity coupled to super Yang-Mills theory has recently been discussed by Brian Dolan[11] and by Dieter Lüst.[12] Castellani and Lüst[13] find that both $G_2/SU(3)$ and $SU(3)/U(1)\times U(1)$ are promising candidates for superstring compactification. In these two cases the β-function of the corresponding σ-model, as well as the central charge of the Virasoro algebra, are found to vanish at each order of perturbation theory for a particular choice of torsion and radius of the manifold. These two models therefore provide a perturbative solution of the classical string equations. In the discussions[11-13] above the torsion is taken to be proportional to the structure constants of the corresponding Lie algebra. In the $G_2/SU(3)$ case the holonomy group is SO(6) so it would appear that N = 1 supersymmetry will not be preserved in four dimensions. The manifold is, however, Ricci flat with this torsion. Lüst,[12] in fact, looked for nonsymmetric coset manifolds which are Ricci flat while Dolan[11] looked for solutions to the equations of motion. C. Hull[14] has also remarked on the possibility of compactification on the almost complex manifold $G_2/SU(3)$.

In this paper we analyze the prospects for compactification on $G_2/SU(3)$ realized as an almost complex structure by considering the restrictions imposed by requiring N = 1 supersymmetry. In the analysis of CHSW, H_{pqr} is set to zero and therefore the case of an internal manifold with no torsion is considered. In our discussion we relax this requirement and suggest that the torsion on K may be induced by nonvanishing fermion bilinears and hence may be a nonperturbative effect. A similar suggestion has been made (for a different reason) by Foda and Halayel-Neto.[15] Some of these condensates, such as the gluino condensates, will break supersymmetry.[16] If a bilinear in the gravitino field ψ_m acquires a vacuum expectation value, then it could induce torsion on K by adding a contorsion term to the spin connection. Thus, in this article, we consider the possibility of K with torsion which is, however, not due to H_{pqr}. We will set H_{pqr} to zero for convenience. It will be shown that K must admit an almost complex structure J_n^m whose torsion is, in general, nonvanishing. Furtherfore, the two form $J_n^m R_m^n$, where R_m^n is the curvature two form, must vanish. We then explain how S^6, equipped with a connection which renders it isomorphic to the coset space $G_2/SU(3)$, satisfies the above requirements. The question of how torsion is determined will be seen to be addressed automatically.

ANALYSIS

We now repeat the CHSW analysis for an internal manifold K with torsion. The appropriate equations to analyze are the supersymmetric transformation laws for fermions:

$$\delta\psi_\mu = \nabla_\mu \varepsilon + \frac{\sqrt{2}}{32}\, e^{2\phi}(\gamma_\mu \gamma_5 \otimes H)\varepsilon$$

$$\delta\psi_m = \nabla_m \varepsilon + \frac{\sqrt{2}}{32}\, e^{2\phi}(\gamma_m H - 12\, H_m)\varepsilon$$

$$\delta\lambda = \sqrt{2}\,(\gamma^m \nabla_m \phi)\varepsilon + \frac{1}{8}\, e^{2\phi}\, H\varepsilon \qquad (1)$$

$$\delta\chi^a = -\frac{1}{4}\, e^{\phi}\, F^a_{mn}\, \gamma^{mn}\, \varepsilon\,.$$

Our notation is the same as that in CHSW. These equations are not quite complete as we have ignored terms on the right hand side in (1) that involve bilinears in Fermi fields. As remarked earlier we consider only the possibility of nonvanishing bilinears in the gravitino field and these do not occur in the transformation laws. Thus it is relevant to analyze the vanishing of the set of transformations in (1) even in the case of non-zero torsion. We assume that the covariant derivative is with respect to a connection that has a nonvanishing torsion. The integrability condition

$$[\nabla_m, \nabla_n]\varepsilon = -\frac{1}{4}\, R_{mnpq}\, \gamma^{pq}\, \varepsilon + T_{mn}{}^{p} \nabla_p \varepsilon \qquad (2)$$

is used. R_{mnpq} is the curvature tensor and in the presence of torsion does not have the full symmetry properties that it has in the torsion-free case. We also assume, although it is not necessary, that $H_{pqr} = 0$. The analysis is similar to that of CHSW up to the point where we find that M^4 is Minkowski space and the requirement $H_{pqr} = 0$ leads to $\nabla_m \eta = 0$ where η is a six dimensional spinor part of ε. The last condition, that is the existence of a covariantly constant spinor, implies that

$$R^p{}_{qmn}\, \gamma^q_p\, \eta = 0. \qquad (3)$$

If K is Kähler and therefore torsion free, then this condition is equivalent to Ricci flatness. For our case, where K has torsion, K may not be Ricci flat. Now the tensor $J^m_n = -i\eta^{+}\gamma^m \gamma_n \gamma\eta$, where $\gamma = \frac{1}{6!}\, g^{1/2}\, \varepsilon_{mnpqrs}\gamma^{mnpqrs}$, satisfies

$$J^m_p\, J^p_n = -\delta^m_n\,, \qquad (4)$$

so that K must admit an almost complex structure J^m_n. Since η is covariantly constant, it is obvious that J^m_n has vanishing covariant derivative. The Niejenhuis tensor, or torsion of the almost complex structure, is defined by[17]

$$N^m_{np} = J^q_p(\partial_q J^m_n - \partial_n J^m_q) - J^q_n(\partial_q J^m_p - \partial_p J^m_q)\,. \qquad (5)$$

If $N^m_{np} = 0$ then (5) may be integrated and a complex structure constructed from the solutions to give a complex manifold. If N^m_{np} does not vanish then J^m_n is not integrable and there is no corresponding complex structure. Let an affine connection be defined on K, then (5) may be rewritten as

$$\begin{aligned} N^m_{np} &= J^q_p(\nabla_q J^m_n - \nabla_n J^m_q) - J^q_n(\nabla_q J^m_p - \nabla_p J^m_q) \\ &\quad - T^m_{np} + T^m_{qr}\, J^q_n\, J^r_p + J^m_q(J^r_p\, T^q_{rn} - J^r_n\, T^q_{rp}), \end{aligned} \qquad (6)$$

where the torsion tensor of the affine connection is defined as usual by $T^m_{np} = \Lambda^m_{np} - \Lambda^m_{pn}$. Now we see from (6) the difference that emerges between the CHSW case $T^m_{np} = 0$ and our case $T^m_{np} \neq 0$. For the former case the demand that torsion be zero means that the Kähler condition $\nabla_m J_p^n = 0$ also requires a complex manifold. For our case, we must still demand that $\nabla_m J_p^n = 0$ since this follows from the definition of J^m_n in the physical theory. However, the presence of torsion indicates that J^m_n will not satisfy the integrability condition $N^m_{np} = 0$ in general so that J^m_n does not correspond to any underlying complex structure of K. It is not surprising that the torsion tensors N^m_{np} of the almost complex structure and T^m_{np} of the affine connection are in fact related by the theorem[18] which states that every almost complex manifold admits an almost complex affine connection (that is, an affine connection with respect to which the almost complex structure is covariantly constant) such that $N^m_{np} = -4T^m_{np}$. Hence we find the nice result that the torsion is fixed by geometric considerations. Furthermore, by premultiplying (3) by $-i\eta^+\gamma$, we find that the curvature tensor defined as usual in terms of the affine connection must satisfy

$$J^p_q \, R^q_{\ pmn} = 0 \ . \tag{7}$$

For a Riemannian manifold, (7) is a necessary and sufficient condition for the holomony group of K to be contained in $O(6) \cap SL(3,C) = SU(3)$.[18] (In the Kähler manifold case, the identity $R_{np} = R^m_{\ nmp} = \frac{1}{2} J^r_s R^s_{\ rqp} J^q_n$ shows that (7) is equivalent to Ricci flatness, $R_{np} = 0$.) Summarizing, we find that a candidate manifold for compactification should admit an almost complex structure which is nonintegrable as well as satisfy (7), leading to SU(3) holonomy. We now turn to the specific example of S^6 as one such possible candidate.

$G_2/SU(3)$ AS AN ALMOST COMPLEX STRUCTURE

It is well known that S^6 may be considered as a hypersurface in the space of imaginary Cayley numbers E^7 spanned by elements of unit modulus. The subspace of E^7 orthogonal to a given point x on the hypersurface may be identified with the tangent space T_x of S at x. An almost complex structure (which is an endomorphism of T_x) is now induced locally by the (Cayley number) vector product $x \wedge y$ where y is in the orthogonal subspace to x.[19] In a neighbourhood of the north pole, the almost complex structure is given explicitly by $(m,n=1,\ldots,6)$

$$J^m_n = A^m_n - A^m_o \, x^n/x^o \ , \tag{8}$$

where $x^o = (1-x^m x^m)^{1/2}$ and the matrix A is

$$\begin{pmatrix} A^0_0 \cdots A^0_6 \\ \vdots \quad\quad \vdots \\ A^6_0 \cdots A^6_6 \end{pmatrix} = \begin{pmatrix} 0 & -x^2 & x^1 & -x^4 & x^3 & -x^6 & x^5 \\ x^2 & 0 & -x^0 & x^5 & -x^6 & -x^3 & x^4 \\ -x^1 & x^0 & 0 & -x^6 & -x^5 & x^4 & x^3 \\ x^4 & -x^5 & x^6 & 0 & -x^0 & x^1 & -x^2 \\ -x^3 & x^6 & x^5 & x^0 & 0 & -x^2 & -x^1 \\ x^6 & x^3 & -x^4 & -x^1 & x^2 & 0 & -x^0 \\ x^5 & -x^4 & -x^3 & x^2 & x^1 & x^0 & 0 \end{pmatrix} \tag{9}$$

The almost complex affine connection is given in terms of J^m_n as

$$\Lambda^m_{np} = \Gamma^m_{np} + \frac{1}{2} J^m_{p;q} J^q_n \tag{10}$$

where ; refers to covariant differentiation with respect to the Riemannian connection Γ^m_{np}. In fact, J^m_n and Λ^m_{np} are invariant under the automorphism group of Cayley numbers G_2. G_2 acts transitively on S^6 and, since under the SU(3) subgroup the $\underline{7}$ of G_2 decomposes as $1+3+\bar{3}$, it follows that S^6 endowed with Λ^m_{np} in (10) may be identified with the coset space $G_2/SU(3)$. The following results are easily verified to be true at the north pole (and hence everywhere since G_2 acts transitively on S^6):

$$\begin{aligned} J^m_{n;p} + J^m_{p;n} &= 0 \\ J^m_{q;n} J^q_p + J^m_q J^q_{p;m} &= 0 \quad . \end{aligned} \tag{11}$$

By use of (11), explicit computation then shows that indeed $J^p_q R^q_{pmn} = 0$. Hence we have shown that $G_2/SU(3)$ is an almost complex (but not complex with the above J^m_n) manifold with vanishing two-form $J^p_q R^q_p$ (known as the Ricci form for Kähler manifolds). The torsion is given from (10) by $T^m_{np} = \frac{1}{2}(J^m_{p;q} J^q_n - J^m_{n;q} J^q_p)$. This torsion reduces the usual O(7) isometries of S^6 since Killing's equation is modified in the presence of torsion.

CONCLUSIONS

To conclude, requiring unbroken supersymmetry in four dimensions but allowing for intrinsic torsion on K selects $G_2/SU(3)$ as a natural candidate for superstring compactification on a coset manifold endowed with an almost complex structure. On the other hand, as shown in references 11-13, requiring Ricci flatness also allows $G_2/SU(3)$ as a possible manifold, although it is realized differently from above. In Lüst's case,[12] the holonomy group is SO(6) and supersymmetry is broken. In our case we cannot say whether the β-function vanishes or not because we have not investigated the corresponding σ-model with nonvanishing intrinsic torsion. σ-models studied in the literature have torsion through the Wess-Zumino term and it is not clear to us how to introduce torsion otherwise. One nice feature of coset spaces is that the G_2 isometries of the internal space may be expected to manifest themselves in the Kaluza-Klein sense as four dimensional gauge degrees of freedom. This effect could give rise to a rich structure and extend the possibilities for phenomenology. For example, these extra symmetries could possibly give a mechanism for the replication of generations of particles (the Euler characteristic for S^6 is 2, naively giving one generation of fermions). Certainly S^6 is not unique in the sense that it admits a nonintegrable, almost complex, structure although it is undoubtedly the simplest such example. In fact, it is known that the algebraic properties of Cayley numbers induce such a structure on any closed oriented hypersurface of R.[7,20] In any case, $G_2/SU(3)$ and perhaps the other two nonsymmetric coset spaces deserve more study.

ACKNOWLEDGEMENTS

The work reported here has been supported in part by an NSERC Operating Grant to one of us (K.S.V.).

REFERENCES

1. M.B. Green and J.H. Schwarz, Nucl. Phys. B181:502 (1981); Nucl. Phys. B198:252 (1982); Nucl. Phys. B198:444 (1982); M.B. Green, J.H. Schwarz and L. Brink, Nucl. Phys. B198:474 (1982).
2. For reviews see: J.H. Schwarz, Phys. Rep. 89:223 (1982); "Lectures on Superstring Theory," Caltech Preprint CALT-668-1247; M.B. Green, Surveys in High Energy Phys. 3:127 (1983); L. Brink, "Superstrings," Ahrenshoop Sympos., pg. 234 (1984).
3. M.B. Green and J.H. Schwarz, Phys. Lett. 149B:117 (1984); Nucl. Phys. B255:93 (1985).
4. M.B. Green and J.H. Schwarz, Phys. Lett. 149B:117 (1984).
5. D.J. Gross, J. Harvey, E. Martinec and R. Rohm, Phys. Rev. Lett. 54:46 (1985).
6. A.H. Chamseddine, Nucl. Phys. B185:403 (1981); E. Bergshoeff, M.M. de Roo, B. de Wit and P. van Niewenhuizen, Nucl. Phys. B195:97 (1982); G.F. Chapline and N.S. Manton, Phys. Lett. B120:105 (1983).
7. P. Candelas, G. Horowitz, A. Strominger and E. Witten, Nucl. Phys. B258:46 (1985).
8. A. Strominger and E. Witten, Commun. Math. Phys. 101:341 (1985).
9. G.F. Chapline and R. Slansky, Nucl. Phys. B209:461 (1982).
10. G.F. Chapline and B. Grossman, Phys. Lett. 143B:161 (1984).
11. B. Dolan, These proceedings.
12. D. Lüst, CALT-68-1329 (1986), to appear in Nucl. Phys. B.
13. L. Castellani and D. Lüst, Caltech Preprint CALT-68-1353 (1986).
14. C. Hull, These proceedings.
15. O. Foda and J.A. Helayel-Neto, Class. Quant. Grav. 3:607 (1986).
16. M. Dine, R. Rohm, N. Seiberg and E. Witten, Phys. Lett. 156B:55 (1985).
17. S.S. Chern, "Complex Manifolds Without Potential Theory," D. van Nostrand (1967).
18. S. Kobayashi and K. Nomizu, "Foundations of Differential Geometry," Vol. II, Wiley-Interscience (1969).
19. A. Frohlicher, Math. Ann. 129:50 (1955); LT. Fukami and S. Ishihara, Tohoku. Math. J. 7:151 (1955).
20. E. Calabi, Trans. Amer. Math. Soc. 87:407 (1958).

SUPERSTRING COSMOLOGY AT LATE TIMES AND TIME VARIATION OF FUNDAMENTAL COUPLING CONSTANTS

Yong-Shi Wu

Department of Physics, University of Utah
Salt Lake City, UT 84112, USA

ABSTRACT

The variation of Newton's gravitational constant G at the present time is studied in the framework of superstring cosmology in an attempt to confront superstring prediction with observations. It turns out that the present value of $\dot{G}/G$ critically depends on the shape of the potential for the size of internal space. If the potential is almost flat, as in perturbation theory to all orders, $|\dot{G}/G|$ is estimated to be in the range $1 \times 10^{-10} - 1 \times 10^{-12}$ yr.$^{-1}$. If the potential has a minimum with finite curvature due to unknown nonperturbative effects, $\dot{G}/G$ will become unobservably small. Thus the improvement of the measurement of $\dot{G}/G$ would discriminate between the two situations. Complications in superstring unified theories for time variation of other coupling constants are also pointed out, which invalidate the assumptions made in previous analyses of data.

1. INTRODUCTION AND CONCLUSIONS

Recently superstring theories[1], especially the $E_8 \times E_8$ heterotic string[2], have attracted a lot of attention. They appear to be promising candidates of a consistent quantum theory unifying all known interactions including gravity. A crucial problem is to find observational tests for superstring theories. Of course, this is a very hard task, considering the tremendously huge gap between the Planck scale ($\sim 10^{19}$ GeV) that string theories are dealing with and the energy scale that now we can reach ($\sim 10^2$GeV). Thus any attempt to confront predictions of superstring theories with observational data is urgently welcome. In such an attempt, very recently we have suggested[3] to consider the time variation of fundamental coupling constants, such as Newton's gravitational constant G and the fine-structure constant α. Some materials are published here the first time.[4]

The idea that time variation of fundamental constants may provide a connection between cosmology and particle physics can be traced back to Dirac.[5] Unfortunately, his proposal for $\dot{G}/G$ seems not supported by observations[6]; and his original observation of the coincidence of two Large Numbers can be understood in a way not related to the time variation of G at all.[7]

Although the time variation of coupling constants looks conceptually attractive when one observes that our Universe is expanding, there seems no way to deal with it in standard Friedmann cosmology. Generally speaking, Kaluza-Klein (KK) theories, in which space-time is enlarged to 4+n dimensions, can provide an approach to the problem.[8] This is because in such theories the coupling constants in our four dimensional world are related to the size of the compact internal space formed by the n extra spatial dimen sions. To see this, let us consider the following action in 4+n dimensions:

$$\int d^4x \sqrt{-\det g_{MN}} \left\{ \frac{1}{2\kappa^2_{4+n}} R - \frac{1}{4 g^2_{4+n}} F^2 \right\} \tag{1.1}$$

For massless fluctuations around the vacuum we have

$$g_{MN} = \begin{pmatrix} g_{\mu\nu}(x) & 0 \\ 0 & g_{mn}(y) \end{pmatrix} \tag{1.2}$$

$$F_{MN} = \begin{pmatrix} F_{\mu\nu}(x) & 0 \\ 0 & F_{mn}(y) \end{pmatrix} \tag{1.2'}$$

where $g_{mn}(y)$ and $F_{mn}(y)$ are the internal backgrounds in vacuum. (Here we assume that the internal space has no isometry so that in 4 dimensions no gauge fields arise from gravity in higher dimensions.) Substituting (1.2) and (1.2') into (1.1) and integrating over the internal coordinates y, we obtain the action in 4 dimensions:

$$\int d^4x \sqrt{-\det g_{\mu\nu}} \left\{ \frac{1}{2\kappa^2_4} R^{(4)} - \frac{1}{4 g^2_4} F^2 + \Lambda^{(4)} \right\} \tag{1.3}$$

with the coupling constants[9]

$$\kappa^2_4 = \kappa^2_{4+n}/V_n \quad , \quad g^2_4 = g^2_{4+n}/V_n \tag{1.4}$$

where V_n is the volume of internal space. (By the way, the cosmological constant $\Lambda^{(4)}$ receives contributions from g_{mn} and F_{mn}.) Eq. (1.4) is crucial, showing that the time variation of coupling constants κ_4 and g_4 is dynamically determined by the higher dimensional cosmology, since the latter would cause time variation of the internal space and, therefore, of its volume. However, a generic KK approach suffers from the arbitrariness in choosing the internal backgrounds g_{mn} and F_{mn}, including the dimension n.

Superstring theories provide a more appropriate framework for studying the time variation of coupling constants. As is well-known,[1,2] superstring theories are consistent only in ten dimensions, fixing $n = 6$. As in usual KK theories, after we do dimensional reduction from ten to four, the coupling constants are inversely proportional to the volume of internal space. To dynamically determine the time evolution of both ordinary 3-space and internal 6-space one needs input for the background fields in internal space in the cosmological equations of motion. Another virtue of superstring theories is that certain classes of internal metrics and internal massless bosonic backgrounds are preferred by string compactification and phenomenological considerations (see below). All these make superstring theories more predictive than KK theories.

How does the quantum effects affect the above essentially classical picture? The size of the internal space, R_6, is expected to be close to or one or two orders larger than the Planck length ($\sim 10^{-33}$cm). Although in the classical equations of motion, there is no potential which would hold R_6 fixed, one expects that the quantum effects in the internal space may induce one. Indeed, in a generic KK approach all the following mechanisms can contribute to such a potential $V(R_6)$:

1) gravitational (one-loop) Casimir effect;[10]
2) one-loop quantum (Casimir) effect for spin-0 massless bosons;[11]
3) nonvanishing internal pressure due to bosonic backgrounds;[12]
4) nonvanishing scalar curvature of the internal space;[13]
5) a fine-tuned cosmological constant in higher dimensions;[13]

Some models have been discussed in KK theories with several of the above mechanisms which lead to constant R_6 in the radiation-dominated era.[12-14] However, it is amazing to notice that because of their supersymmetric nature, superstring theories provide neither of the above mechanisms, even up to all finite orders in perturbation theory. This is the content of a non-renormalization theorem[15] proved by Witten recently. At the one-loop level, the situation is obvious in virtue of the boson-fermion cancellation within a supermultiplet. (The vanishing of internal scalar curvature is because the $N = 1$ supersymmetry at low energies after compactification requires the internal space to be Ricci-flat.[16]) Witten's nonrenormalization theorem asserts that the same is true, i.e. $V(R_6)$ is flat, up to all orders in perturbation theory. So far, the study of nonperturbative supersymmetry-breaking effects[17,18], including such as world-sheet instantons, also failed to produce a potential V with a minimum at finite R_6, whose existence is expected by the conventional wisdom. As we will see, the time variation of coupling constants critically depends on the shape of this potential. Thus, we may view the theoretical predictions on time variation of coupling constants as a consequence of internal vacuum backgrounds and of the potential for the size of the internal space and, therefore, use the observations on the former as a test for the latter.

We emphasize that we are only interested in cosmology at late times since most of the relevant observations are referred to the present (rather than early) universe. Many problems appearing in the early universe, such as why the cosmological constant vanishes, do not appear in the late universe which we know about much better. As the first step, we will consider stable time-dependent vacuum solutions, which physically correspond to a very late universe which has become sufficiently dilute due to previous expansion. A remarkable result is the following. Even if the potential for R_6 is flat, i.e. there is no force to hold it, it turns out that with each previously suggested superstring vacuum background in the internal space, there is, as an exact solution, an open universe ($k = -1$) with ordinary three-space expanding and internal six-space fixed in size. Furthermore this configuration is stable against time-dependent perturbations, while the usual static Minkowski space-time is not. This result, allowing no asymptotic time variation of coupling constants, seem to be encouraging for obtaining in superstring cosmology non-vanishing but tolerably small time variations at present times.

Indeed, after including cosmic matter and doing perturbation theory around the asymptotic solution, we are able to estimate the present time variation of coupling constants. If the potential for R_6 is flat, the present value of $\dot{G}/G$ for an open universe is given by

$$(\dot{G}/G)_o = (q_o - \frac{13}{8}\,\Omega_o H_o^2 t_o^2)/t_o \tag{1.5}$$

where H_o is the Hubble constant, t_o the age of the universe, q_o the deceleration parameter and $\Omega_o \equiv 8\pi G_o \rho_o / 3H_o^2$ the density parameter. Here ρ_o is the density in ordinary 3-space and the subscript o denotes the present value of the quantity. We estimate $(\dot{G}/G)_o$ to be in the range

$$|(\dot{G}/G)_o| \approx 1 \times 10^{-11\pm 1}\ \mathrm{yr}^{-1} \tag{1.6}$$

which overlaps the present observational upper bound[19]

$$|\dot{G}/G| \leq 1 \times 10^{-11}\ \mathrm{yr}^{-1} \tag{1.7}$$

However, if the potential really has a minimum at finite R_6, $(\dot{G}/G_o)$ will be much suppressed and become unobservably small. So an improvement on the measurements of $\dot{G}/G$ will discriminate between the two situations.

Here we concentrate on $\dot{G}/G$, since theoretically it is independent of the dilaton field and experimentally extracting it from data is simple and direct. Some remarks about time variation of other coupling constants in superstring theories are given at the end of the paper. An important point is that the time dependence of R_6 will lead to the variation in the grand unified coupling constant at momentum R_6^{-1}, which, in turn, through the renormalization group equation, gives rise to the variation in almost every coupling constant and mass measured at low energies, though with different rates. This feature invalidates the usual assumption made in previous analyses of data that only one coupling constant, e.g. α, is time-varying. So the re-analysis of data becomes both necessary and complicated.

In summary, superstring theories provide an appropriate framework for studying time variation of fundamental coupling constants. The present value of $\dot{G}/G$ critically depends on the shape of the potential arising from quantum effects for the size of internal space. Further improvement in measuring $\dot{G}/G$ would give us important information about the shape of the potential. If the potential is almost flat or has no minimum for finite R_6, probably we are on the edge of observing $\dot{G}/G$. We encourage that old data be re-analyzed and new experiments be done. Especially new clever ideas for precise short-time laboratory experiments would be most welcome.

2. COSMOLOGICAL EQUATIONS OF MOTION AND ANSATZ

In this section we will show that compared with usual KK cosmology, superstring cosmology are more definite in many aspects.

The extra dimensions are fixed to be six by the consistency of the theory. The content of the massless sector is also fixed to be N = 1 supergravity and N = 1 super Yang Mills multiplets (in d = 10). So the massless bosonic states include the metric g_{MN}, the Kolb-Ramond antisymmetric field B_{MN}, the dilaton ϕ and the Yang-Mills potential A_M^a (with gauge group SO(32) or $E_8 \times E_8$). At low energies their interactions are described by the effective action in the field-theory ($\alpha' \to 0$) limit, which can be derived, in principle, from string theory and turns out to be the usual Chapline-Manton action[20] modified by string considerations. First, there are terms quadratic in the curvature tensor. Although they have not been explicitly calculated, there are good reasons to argue that these terms just form the so-called Gauss-Bonnet term, i.e.

$$\frac{1}{4}\,\phi^{-3/4}\,(R_{MNPQ}^2 - 4\,R_{MN}^2 + R) \tag{2.1}$$

This is inferred either from the ghost-free nature of string theories[21] or from the supersymmetrization of the H^2 term in the action[22] [see (2.4)]. Secondly, the gauge invariant strength H_{MNP} of the field B_{MN} is given by (using 3-forms)

$$H = dB + \frac{1}{30}\,\mathrm{Tr}(AF - \frac{1}{3}\,A^3) - \mathrm{tr}(\omega R - \frac{1}{3}\,\omega^3) \tag{2.2}$$

Especially, the addition of the second term by Green and Schwarz[1] is due to the anomaly-free condition and can be derived in the non-linear sigma model approach to string theories.[23] This leads to the Bianchi identity

$$dH = \frac{1}{30} \mathrm{Tr} F_\Lambda F - \mathrm{tr} R_\Lambda R \tag{2.3}$$

which is a constraint on these bosonic backgrounds.

Therefore, the effective action for massless bosonic fields is

$$S = \int d^{10}x \sqrt{-g} \left\{ R - \frac{3}{4} \kappa_{10}^2 \phi^{-3/2} H_{MNP}^2 - \frac{9}{16} \frac{1}{\kappa_{10}^2} (\phi^{-1} \partial_M \phi)^2 \right.$$
$$\left. - \frac{1}{4} \phi^{-3/4} \left[\frac{1}{30} \mathrm{Tr}\, F_{MN}^2 - (R_{MNPQ}^2 - 4 R_{MN}^2 + R^2) \right] + \mathcal{L}_f \right\} \tag{2.4}$$

So the classical equations of motion for these massless bosonic backgrounds in ten dimensions are given by

$$R_{AB} - \frac{1}{2} g_{AB} R = \left[\frac{9}{2} \phi^{-3/2} (H_{AMN} H_B{}^{MN} - \frac{1}{6} g_{AB} H_{MNP}^2) + 9 \nabla^M (\phi^{-3/2} H_{APQ} R_{MB}{}^{PQ}) \right] \kappa_{10}^4$$
$$+ \frac{9}{8} \phi^{-2} \left[\partial_A \phi \partial_B \phi - \frac{1}{2} g_{AB} (\partial_M \phi)^2 \right] + \frac{1}{30} \kappa_{10}^2 \phi^{-3/4} (\mathrm{Tr} F_{AM} F_B{}^M - \frac{1}{4} g_{AB} \mathrm{Tr} F_{MN}^2)$$
$$+ \frac{1}{2} \kappa_{10}^2 \phi^{-3/4} \left[\frac{1}{2} g_{AB} (R_{MNPQ} - 4 R_{MN}^2 + R^2) - 2 R R_{AB} \right.$$
$$\left. + 4 R_{AM} R_B + 4 R_{AMBN} R^{MN} - 2 R_A{}^{MNP} R_{BMNP} \right] + \kappa_{10}^2 T_{AB} \ . \tag{2.5}$$

$$\nabla_M (\phi^{-3/2} H^{MNP}) = 0 \ . \tag{2.6}$$

$$D_M (\phi^{-3/4} F^{MPa}) + 9 \kappa_{10}^2 (\phi^{-3/2} F_{MN}{}^a H^{MNP}) = 0 \ . \tag{2.7}$$

$$6 \nabla_M (\phi^{-2} \partial^M \phi) + 6 \phi^{-3} (\partial_M \phi)^2 + 6 \kappa_{10}^4 \phi^{-5/2} H_{MNP}^2$$
$$+ \kappa_{10}^2 \phi^{-7/4} \left[\frac{1}{30} \mathrm{Tr} F_{MN}^2 - (R_{MNPQ}^2 - 4 R_{MN}^2 + R^2) = 0 \right. \ . \tag{2.8}$$

where $A,B,M,N, = 0,1,\cdots,9$; $F_{MN}{}^a$ are the Yang-Mill strengths. ∇_M is the ordinary covariant derivative in curved space-time and D_M is the doubly covariant derivative. T_{AB} is the thermal energy-momentum tensor, and we have neglected the effects of matter on other bosonic backgrounds.

To obtain the cosmological equations of motion, we assume as in usual KK cosmology that the metric in ten dimensions is of the form

$$g_{MN} = \begin{pmatrix} -1 & & \\ & R_3^2(t)\, \tilde{g}_{ij}(x) & \\ & & R_6{}^2(t)\, \tilde{g}_{mn}(y) \end{pmatrix} \tag{2.9}$$

where $i,j = 1,2,3$; $m,n = 4,5,\cdots,9$; $R_3(t)$, $R_6(t)$ are the scale factors. $\tilde{g}_{ij}(x)$ is maximally symmetric in ordinary 3-space.

The time-independent part of the internal metric, $\tilde{g}_{mn}(y)$, and the vacuum backgrounds for other massless bosonic fields in internal space are to be chosen according to string theory considerations. The consistency and/or finiteness up to three loops of the quantized nonlinear sigma model describing the string propagation in curved backgrounds prefer to select $\tilde{g}_{mn}(y)$ to be respectively Ricci-flat, Kähler Ricci-flat and hyper-Kähler Ricci-flat, if the sigma model has $N = 1,2,4$ supersymmetry.[24] To have low-energy $N = 1$ supersymmetry in four dimensions one must choose[16]

$\tilde{g}_{mn}(y)$ is Calabi-Yau, i.e. Ricci-flat 6-d Kähler (2.10) (2.10) if, for simplicity, one has assumed all components of H vanish:

$$H_{MNP} = 0 \ (M.N.P = 0,\cdots,9) \tag{2.11}$$

To preserve gauge invariance in four dimensions, Yang-Mills fields survive only in the internal space. There they are set equal to the spin-connection of $\tilde{g}_{mn}(y)$ with a suitable embedding into SO(32) or $E_8 \times E_8$:

$$\begin{aligned} F_{MN}{}^{\alpha\beta} &= \tilde{R}_{mn}{}^{\alpha\beta} \text{ if } M = m,\ N = n \\ &= 0 \quad , \quad \text{otherwise} \end{aligned} \tag{2.12}$$

where α,β are internal vielbein indices. This is the simplest and natural way to satisfy the Bianchi identity (2.3) which becomes, under ansatz (11),

$$\text{tr } R_\Lambda R = \frac{1}{30} \text{Tr} F_\Lambda F \tag{2.3'}$$

Finally, from the action (2.4) it is easy to see that the background value of ϕ plays the role of coupling constants in ten dimensions. (The gauge coupling constant g_{10} has been absorbed by rescaling ϕ[25].) One usually takes

$$\phi = \text{const.} \tag{2.13}$$

It is easy to verify that together with the Minkowski metric in four dimensions the above backgrounds (2.10)-(2.13) form a static vacuum solution.[16]

If we have accepted (2.10)-(2.13) as the static vacuum for local phenomena, then in the cosmological case of eq. (2.9) we should keep them as much as possible, since in the universe at late times local physics at each instant should be nearly the same. Therefore, in accordance to our philosophy of making only necessary and minimal modifications, we will assume that the ansatz (2.10)-(2.13) are still valid for a cosmological vacuum configuration. In particular, eq. (2.13) now implies that the coupling constants in ten dimensions do not vary in time.

The key observation here is that in the cosmological case with (2.9), the Kalb-Ramond eq. (2.6), Yang-Mill eq. (2.7) and Bianchi identity (2.3') are still satisfied by the ansatz (2.10)-(2.13). This is because of the conformal invariance of these equations[26], and the fact that $g_{mn}(y,t) = R_6(t)^2 \tilde{g}_{mn}(y)$ in eq. (2.9) can be viewed as a conformal transformation in the internal space. Thus, we are left with the Einstein equation (2.5) and the dilaton equation (2.8); they are a set of four highly nonlinear differential equations for two unknown functions $R_3(t)$ and $R_6(t)$. (The detailed form can be found in ref. 3.)

We notice that actually we can relax the assumption (2.10) so that $\tilde{g}_{mn}(y)$ is Ricci-flat. Alternatively, the assumption (2.11) can be altered such that both the internal components of H[27,28] and an appropriate gluino condensate become nonvanishing, but their contributions to the cosmological constant cancel[28]. All the equations for $R_3(t)$ and $R_6(t)$ are the same as before, but we would lose N = 1 supersymmetry in four dimensions.

Also we remark that with all these ansatz, there is no internal energy-momentum tensor arising from nonvanishing Kalb-Ramond and Yang-Mills backgrounds. This is a feature of superstring theories.

3. COSMOLOGICAL VACUUM SOLUTIONS AND THEIR STABILITY

Let us first consider the vacuum solutions with $T_{AB} = 0$. In a universe that expands forever, a stable cosmological vacuum solution corresponds to a very late universe in which matter has become extremely dilute.

Here we emphasize the importance of studying the stability problem. This is because in superstring theories the terms quadratic in the curvature tensor lead to higher-derivative terms in the equations of motion which, though normally very small, may destabilize the solution with them neglected. An example for such instability is the following: Consider

$$-\epsilon\ddot{y} + \dot{y} + y = 0 \qquad (0 < \epsilon << 1) \tag{3.1}$$

If we neglect ϵ, then $y(t) \sim e^{-t}$. However, if we keep ϵ, $y(t) \sim e^{t/\epsilon}$ for very large t. Similar things may happen here. It has been shown that in a k = 0, radiation-dominated universe, the R^2_{ABCD} term would destabilize the ordinary Friedmann solution; however, the Gauss-Bonnet term would not.[29] Here we will discuss the vacuum case with k arbitrary.

As we said, the Einstein equations and dilaton equation are a set of four highly nonlinear differential equations for $R_3(t)$ and $R_6(t)$. Fortunately, we have been able to find two solutions by trying the power-law form:

$$\text{(i)} \quad R_3(t) = t + t_o, \quad R_6(t) = \text{const.} \quad \text{for } k = -1 \tag{3.2}$$

$$\text{(ii)} \quad R_3(t) = \text{const.}, \quad R_6(t) - \text{const.} \quad \text{for } k = 0 \tag{3.3}$$

The solution (ii) is just the static vacuum proposed in ref. 16. The solution (i) represents a cosmological vacuum in a late universe.

To study stability of the solutions we consider time-dependent perturbations in $R_3(t)$ and $R_6(t)$. For the solution (3.2) we write

$$R_3(t) = t + r_3(t) \quad , \quad R_6(t) = R_{60} + r_6(t) \tag{3.4}$$

Substituting (3.4) in Eqs. (2.5) and (2.8) we obtain, to first-order,

$$\ddot{r}_3 + 2t^{-1}\dot{r}_3 = 0 \quad , \quad \ddot{r}_6 + 3t^{-1}\dot{r}_6 = 0 \quad , \quad R_{60}\ddot{r}_3 + 2t\ddot{r}_6 = 0 \tag{3.5}$$

with the solution

$$r_3(t) = \alpha_3 - \beta_3\, t^{-1} \quad , \quad r_6(t) = \alpha_6 + \frac{\beta_3 R_{60}}{6}\, t^{-2} \tag{3.6}$$

So, the solution (3.2) is critically stable. However, the static solution (3.3) is critically unstable, since perturbations around it satisfy

$$\ddot{r}_3 = 0 \quad , \quad \ddot{r}_6 = 0 \tag{3.7}$$

Some remarks about our cosmological vacuum configuration (3.2) plus (2.10)-(2.13) are in order. For this configuration the action (2.4) vanishes so it has a zero cosmological constant. And it is expected to lead to the same physics as the static vacuum for phenomena occuring in regions which are small compared to both cosmological length and time scales. Whether our three-space expanding vacuum leads to a supersymmetry breaking on the cosmological time scale remains to be seen. It is amusing to note that in eq. (3.2) the time-dependence of the scale factor in three-space, namely $R_3(t) \approx t$, coincides with that of $R_3(t)$ at late times in the standard Friedmann cosmology with $k = -1$ (see ref. 6). This result is nontrivial. Finally, the ansatz (2.10)-(2.13) plus stability considerations lead to the prediction f an open universe ($k = -1$) with vanishing deceleration parameter ($q \equiv -R_3 \ddot{R}_3/\dot{R}_3^2 = 0$). Conversely, cosmological implications may be used to restrict the choice for internal vacuum backgrounds.

In summary, as a consequence of superstring internal vacuum backgrounds, such as the ansatz (2.10)-(2.13) or their variations, our time-dependent vacuum (3.2) both incorporates the expansion of the universe and predicts the time non-variation of fundamental constants in a very late universe. This seems encouraging for obtaining non-vanishing but tolerably small time variations in the present universe even in the absence of $V(R_6)$, which we will discuss in next section.

4. $\dot{G}/G$ IN THE ABSENCE OF $V(R_6)$

In this section we include the effect of cosmic matter. We assume

$$T_{AB} = \text{diag.}\ (\rho, p\tilde{g}_{ij}, p'\tilde{g}_{mn}) \tag{4.1}$$

In the matter-dominant era, $p = p' = 0$[30]. The conservation of T_{AB} gives

$$\rho(t)R_3{}^3(t)R_6{}^6(t) = \text{const.} \tag{4.2}$$

(We have normalized $\tilde{g}_{mn}(y)$ such that $\int d^6y \sqrt{\det \tilde{g}} = 1.$) Using above ansatz and assumptions and neglecting the Gauss-Bonnet term which is of order $\kappa_{10}^2/t^2 \propto (t_p/t)^2$ (where $t_p \sim 10^{-43}$ sec is the Planck time), the dilaton eq. (2.8) is satisfied and the Einstein equations (2.5) become

$$\frac{\ddot{R}_3}{R_3} + 2\frac{\ddot{R}_6}{R_6} = -\frac{7}{24}\frac{\kappa_{10}^2\rho_o}{R_6^6}\left[\frac{R_3(t_o)}{R_3(t)}\right] \tag{4.3}$$

$$\frac{2k}{R_3^2} + \frac{\ddot{R}_3}{R_3} + \frac{2\dot{R}_3^2}{R_3^2} + 6\frac{\dot{R}_3\dot{R}_6}{R_3R_6} = \frac{1}{8}\frac{\kappa_{10}^2\rho_o}{R_6^6}\left[\frac{R_3(t_o)}{R_3(t)}\right]^3 \tag{4.4}$$

$$\frac{\ddot{R}_6}{R_6} + 5\frac{\dot{R}_6^2}{R_6^2} + 3\frac{\dot{R}_3\dot{R}_6}{R_3R_6} = \frac{1}{8}\frac{\kappa_{10}^2\rho_o}{R_6^6}\left[\frac{R_3(t_o)}{R_3(t)}\right]^3 \tag{4.5}$$

where we have used $R = (1/4)\kappa_{10}^2\rho$; ρ_o is the present matter density.

For the $k = -1$ case, the stable cosmological vacuum solution (3.2) can be viewed as the large-t asymptotic solution. Now we assume ρ_o can be treated as small quantities and use perturbation theory to solve eqs. (4.3)-(4.5). Define $r_3(t)$ and $r_6(t)$ by

$$R_3(t) = (t + c)\ [1 + r_3(t)] \quad , \quad R_6(t) = R_{60}\ [1 + r_6(t)] \tag{4.6}$$

and recall that the gravitational constant in four dimensions κ_4^2 is given by

$$\kappa_4^2(t) \equiv 8\pi G(t) = \kappa_{10}^2/R_6^6(t) \tag{4.7}$$

Treat $r_3(t)$ and $r_6(t)$ as small quantities. Up to first order eqs. (4.3)-(4.5) are reduced to linear differential equations with κ_4^2 replaced by $\kappa_4^2(t_o) \equiv 8\pi G_o$. Introducing $\Omega_o = \kappa_4^2\rho_o/3H_o^2$, the solution is given by

$$r_3(t) = -\frac{5}{8}\,\Omega_o H_o^2 t_o^3\,\frac{\ell n(ht)}{t} - \frac{b}{t^2} \tag{4.8}$$

$$r_6(t) = -\frac{3}{8}\,\Omega_o H_o^2 t_o^3\,\frac{1}{t} + \frac{b}{6t^2} \tag{4.9}$$

where h and b are integration constants and we have made the change $t + c \rightarrow t$. $(\dot{G}/G)_o = -6\dot{r}_6(t_o)$ is given by eq. (1.5). The parameter t_o may differ from the age of the universe by a factor of order one, which we will neglect in the following estimation.

Astronomical observations have produced quite diverse values for the cosmological parameters Ω_o, H_o, q_o and t_o. (For details, see ref. 31.) The most "satisfactory" set of parameters recommended by ref. 31, i.e., $(\Omega_o, q_o, H_o) = (0.05, 0.025, 67\ \text{km}\cdot\text{sec}^{-1}\ \text{Mpc}^{-1})$ and $t_o = 1.6 \times 10^{10}$yr, gives

$$(\dot{G}/G)_o = -3.6 \times 10^{-12}\ \text{yr}^{-1} \tag{4.10}$$

The extreme sets are given by $(\Omega_o, q_o, H_o) = (0.05, -0.925, 100\ \text{km}\cdot\text{sec}^{-1}\cdot\text{Mpc}^{-1})$ and $(1, 0.5, 40\ \text{km}\cdot\text{sec}^{-1}\cdot\text{Mpc}^{-1})$. Correspondingly,

$$(\dot{G}/G_o) = -7.1 \times 10^{-11}\ \text{yr}^{-1} \text{ or } -1.2 \times 10^{-11}\ \text{yr}^{-1} \tag{4.11}$$

Thus we estimate the range for $(\dot{G}/G)_o$ as given by eq. (1.6).

Rigorously speaking, if Ω_o is close to 1, the above perturbation calculation breaks down. One needs to use computer for solving eqs. (4.3)-(4.5), but this would not change the estimation (1.6). The same is expected to be true for $k = 0$ or $k = +1$ cases. The key point here is that eq. (4.5) with $\rho_o \neq 0$ does not allow $\dot{R}_6 = 0$. Thus $\dot{G}/G = -6\dot{R}_6/R_6 \neq 0$. Since t_o is the only relevant cosmological time scale, $(\dot{G}/G)_o$ must be proportional to $1/t_o$ with coefficient of order unit or, probably, one to two orders lower.

In fact, the $k = 0$ case can be exactly solved without using perturbation theory. Let us try the power ansatz for both $R_3(t)$ and $R_6(t)$:

$$R_3(t) = \alpha_3 t^{\beta_3} \quad , \quad R_6(t) = \alpha_6 t^{\beta_6} \tag{4.12}$$

where $\alpha_3, \alpha_6, \beta_3$ and β_6 are constants to be determined. It is easy to find

$$\beta_3 = \beta_6 = 2/9 \quad , \quad \alpha_3^3 \alpha_6^6 = \frac{9}{16} \kappa_{10} \rho_o R_3(t_o)^3 \tag{4.13}$$

$$(\dot{G}/G)_o = - 6\dot{R}_6(t_o)/R_6(t_o) = - (12/9) t_o^{-1} \tag{4.14}$$

$$= - 8.4 \times 10^{-11} \text{ yr}^{-1} \tag{4.14'}$$

It is just of the order of t_o^{-1}, as we expected! This value is too big to be compatible with the experimental limit (1.7). We also notice that $R_6 \sim t^{2/9}$ does not tend to a finite R_{60} as $t \to \infty$, and this is consistent with the instability of the vacuum solution (3.3).

We emphasize that the results in this section are obtained in the absence of a nontrivial potential $V(R_6)$ for the size of internal space. We have seen that in this case, the time variation of coupling constants at the present time is, generally speaking, rather "large": either of the order of or one or two orders lower than, the inverse age of the Universe $1/t_o$.

5. $\dot{G}/G$ IN THE PRESENCE OF $V(R_6)$

Although the efforts so far to break supersymmetry nonperturbatively in superstring theories have failed[17,18] to produce a potential $V(R_6)$ which has a minimum at finite R_6, it is still desirable to see how sensitive $\dot{G}/G$ is to the shape of the potential $V(R_6)$.

So let us assume that by some yet unknown nonperturbative quantum effects, there is a potential $V(R_6)$ in the effective action which has a minimum at R_{60} ($\neq 0$ and ∞). The time evolution in the early universe would have brought the present value of R_6 to the neighborhood of R_{60}; in particular, for the k = -1 case, the asymptotic constant value of R_6 in eq. (4.6) must coincide with this minimum. Furthermore, since $V(R_{60})$ contributes a cosmological constant in 4 dimensions in a late universe, we assume it is zero. Thus around R_{60} we have

$$V(R_6) = \frac{1}{2} \mu^2 \left(\frac{R_6 - R_{60}}{R_{60}}\right)^2 + \cdots \tag{5.1}$$

For the k = -1 case, the asymptotic solution (3.2) is still valid for very large t. For finite t, when we do perturbation around (3.2), a new term $\mu^2 r_6(t)$ must be added to the Einstein equation (4.5):

$$\ddot{r}_6 + \frac{3\dot{r}_6}{t} + \mu^2 r_6 = \frac{1}{8} \kappa_{40}^2 \rho_o \frac{t_o^3}{t^3} \tag{5.2}$$

where we have used eqs. (4.6). Assuming $(\mu t)^2 \gg 1$, the solution is

$$r_6(t) = \frac{3}{8}\Omega_o H_o^2 t_o^3 \frac{1}{\mu^2 t^3} + t^{-3/2} A \cos(\mu t + \delta) \tag{5.3}$$

The second term is oscillatory and vanishes after being averaged over the period $2\pi/\mu$. (This averaging is necessary for a macroscopic determination of the coupling constants, and is consistent with $(\mu t)^2 >> 1$.) Thus,

$$\bar{r}_6(t) = \frac{3}{8}\Omega_o H_o^2 t_o^3 \frac{1}{\mu^2 t^3} \tag{5.4}$$

So,

$$\left(\frac{\dot{G}}{G}\right)_o = -6\dot{\bar{r}}_6(t_o) = \frac{27}{4}\Omega_o H_o^2 t_o^2 \cdot \frac{1}{t_o} \cdot \frac{1}{(\mu t_o)^2} \tag{5.5}$$

We notice that on the r.h.s. of eq. (5.5), the factor in front of $1/(\mu t_o)^2$ is of the same order as the r.h.s. of eq. (1.5). So the factor $1/(\mu t_o)^2$ represents a suppression factor in the presence of a minimum in $V(R_6)$ with finite curvature μ. Since

$$1/(\mu t_o)^2 = (10^{-32}\text{eV}/\mu)^2 \tag{5.6}$$

a very tiny mass $\mu >> 10^{-32}$eV would make $(\dot{G}/G)_o$ unobservably small. In this way, we see that the present rate of the time variation of G is very sensitive to the curvature at the minimum of the induced quantum potential $V(R_6)$. (The results of the last section can be viewed as the results for the case $(\mu t)^2 << 1$.) Therefore, <u>the improvement of the present upper limit on $(\dot{G}/G)_o$ by two orders of magnitude would tell us whether the quantum potential $V(R_6)$ is flat or not around R_{60}</u>.

There is a physical interpretation of R_6. In local phenomena, the size of internal space may also depend on the points in ordinary 3-space, so actually R_6 is a local field in our 4-dimensional world. It can be shown by dimensional reduction that in fact it is the well-known Brans-Dicke scalar field[32]. The parameter μ can be interpreted as the mass of this field. The conventional wisdom favors a not very small mass μ for it, since the field would compete with gravitons in producing long-ranged interactions and would have been observed if it is massless. However, the coupling of this field to matter might be anomalously weak; if so, a flat potential is not in conflict with the non-observation of it so far.

7. REMARKS ON TIME VARIATION OF OTHER CONSTANTS

We have been concentrating on $\dot{G}/G$. As for the time variation of particle-physics constants, such as α, strong and weak coupling constants or particle masses, the following remarks are in order.

First, the time dependence of R_6 will lead also to the variation in the grand unified coupling constant at momentum R_6^{-1}

$$\alpha_{GUT}(R_6^{-1}) \equiv g_{GUT}^2(R_6^{-1})/4\pi = \phi^{3/4} R_6^{-6}/4\pi \tag{7.1}$$

which, in turn, gives rise to the variation in almost every coupling constant and mass measured at low energies. This is an important feature of unified string theories, in contrast to the usual assumption in previous analyses of data that only the quantity considered is varying alone.

Second, there is a renormalization-group (RG) running of coupling constants[33], which relates those measured at low energy $\mu(<<R_6^{-1})$ to $\alpha_{GUT}(R_6^{-1})$ calculated from eq. (7.1) as follows:

$$\alpha_i^{-1}(\mu) = \alpha_i^{-1}(R_6^{-1}) - \frac{1}{\pi}\sum_j C_{ij}\left[\ell n\,\frac{1}{m_j R_6} + \theta(\mu - m_j)\ell_n\,\frac{m_j}{\mu}\right] \tag{7.2}$$

where i = 1,2,3 correspond to $U(1)_{em}$, $SU(2)_W$ and $SU(3)_c$; the sum is over j = leptons, quarks, gluons, $W^{\pm}$, etc. The C_{ij} are well-known numbers depending on the spin and group-representation of the jth particle. If one neglects the variation of the second term, then

$$\frac{\dot{\alpha}_i(\mu)}{\alpha_i(\mu)} = \frac{\alpha_i(\mu)}{\alpha_{GUT}(R_6^{-1})}\cdot\frac{\dot{\alpha}_{GUT}(R_6^{-1})}{\alpha_{GUT}(R_6^{-1})} = \frac{\alpha_i(\mu)}{\alpha_{GUT}(R_6^{-1})}\cdot\frac{\dot{G}}{G} \tag{7.3}$$

For α this is two orders of magnitude lower than $\dot{G}/G$. Thus, though both G and atomic clocks vary in the time defined by the metric (2.10), the latter probably varies more slowly. Similarly there is also RG running of particle masses which depends very much on whether there are heavy families[34]. Also according to eq. (7.1), the time dependence of the background ϕ, which might arise upon appropriate modification of our ansatz, would lead to an extra contribution to $\dot{\alpha}_{GUT}/\alpha_{GUT}$. It might be important to include this effect in considering the variation of, e.g. α over a very long period.[35]

Because of the above complications, which were neglected in the previous analyses of observations on, e.g., $\dot{\alpha}/\alpha$, we do not think we can trust the experimental upper limits obtained in these analyses. Particularly in view of the uncertainties in the renormalization group running from the compactification scale down to the low-energy scale, a complete analysis of experimental data on time variation of particle-physics constants, which is consistent with superstring unified theories, would be extremely complicated. Nevertheless, we still think it is worthwhile.

Acknowledgement

The author would thank the organizer, Dr. H.C. Lee, for the kind invitation and warm hospitality. The work was in part supported by U.S. NSF Grant No. PHY-8405648.

References

1. J.H. Schwarz, Phys. Rep. 89 (1982) 223; M.B. Green, Surv. High Energy Phys. 3 (1983) 127. M.B. Green and J.H. Schwarz, Phys. Lett. 149B (1984) 117.
2. D.J. Gross, J.A. Harvey, E. Martinec, and R. Rohm, Phys. Rev. Lett. 54 (1984), 502; Nucl. Phys. B256 (1985) 253.
3. Y.S. Wu and Z. Wang, U.U. - U.S.U. preprint (1986); Y.S. Wu and Z. Wang, Phys. Rev. Lett. (to be published).
4. Y.S. Wu (in preparation).
5. P.A.M. Dirac, Nature (London) 139 (1937) 323; Proc. Roy. Soc. London, Ser. A165 (1938) 199.
6. S. Weinberg, Gravitation and Cosmology, part five, John Wiley & Sons, 1972.
7. R.H. Dicke, Nature 192 (1961) 442.
8. A. Chodos and S. Detweiler, Phys. Rev. D21 (1980) 2167; P. Freund, Nucl. Phys. B209 (1982) 146; S. Randjbar-Daemi, A Salam and J. Strathdee, Phys. Lett. 135B (1984) 388; M. Gleiser, S. Rajpoot and J.G. Taylor, Ann. of Phys. (N.Y.) 160 (1985) 299.
9. Y.S. Wu, Acta Physica Sinica, 29 (1980) 395; P. Freund, Phys. Lett. 120B (1983) 335; S. Weinberg, Phys. Lett. 125B (1983) 265.
10. T. Applequist and A. Chodos, Phys. Rev. Lett. 50 (1983) 141.
11. D. Candelas and S. Weinberg, Nucl. Phys. B237 (1984) 397.
12. F.S. Accetta, M. Gleiser, R. Holman and E.W. Kolb, Fermilab-pub. 86/38-A (1986).
13. S. Randjbar-Daemi, A. Salam and J. Strathdee, in [8].
14. G.F. Chapline and G.W. Gibbons, Phys. Lett. 135B (1984) 43; D. Bailin, A. Love and C.G. Voyonakis, Phys. Lett. 142B (1984) 344.
15. E. Witten Nucl. Phys. B268 (1986) 79.
16. P. Candelas, G.T. Horowitz, A. Strominger and E. Witten,, Nucl. Phys. B256 (1985) 46.
17. R. Rohm, Nucl. Phys. B237 (1984) 553.
18. M. Dine, N. Seilberg, X.G. Wen nd E. Witten, Princeton preprint (1986).
19. R.W. Hellings et al. Phys. Rev. Lett. 51 (1983) 1609.
20. A.H. Chamseddine, Cucl. Phys. B185 (1981) 403; E. Bergshoeff, M. deRoo, B. deWit, and P. Van Nieuwenhuizen, Nucl. Phys. B195 (1982) 97; G.F. Chapline and N.S. Manton, Phys. Lett. 120B (1983) 105.
21. D.J. Gross and E. Witten,, unpublished; B. Zwiebach, Phys. Lett. 156B (1985) 315; B. Zumino, UCB-PTH-85/12, LBL-19302 preprint (1985).
22. L. Romoms and N. Warner, CALT preprint 68-1291 (1985); S. Cecotti, S. Ferrara, L. Girardello and M. Porrati, CERN preprint TH4253185 (1985); D. Deser, in Proceedings of the Workshop on Supersymmetry and its

Applications, Cambridge 1985 (Cambridge University Press, Cambridge, MASS).

23. C. HULL and E. Witten, Phys. Lett. 160B (1985) 398; A. Sen. Phys. Lett. 166B (1986) 300.

24. L. Alvarez-Gaume and D.Z. Freedman Phys. Rev. D22 (1980) 846; Comm. Math. Phys. 80 (1981) 443; L. Alvarez-Gaume, D.Z. Freedman and S. Mukhi, Ann. Phys. 134 (1981) 85; C. Hull, Nucl. Phys. B260 (1985) 182; L. Alvarez-Gaume and P. Ginsparg, Comm. Math. Phys. 102 (1985) 311; L. Alvarez-Gaume, S. Coleman and P. Ginsparg, preprint HUTP-85/A037.

25. M. Dine and N. Seiberg, Phys. Rev. Lett. 55 (1985) 366.

26. C.N. Yang, Phys. Rev. D16 (1977) 330; S.S. Chern and J. Simons, Proc. Natl. Acad. Sci. (U.S.A.) 68 (1971) 791; A. Avez, Proc. Natl. Acad. Sci. (U.S.A.) 66 (1970) 265.

27. R. Nepomechie, Y.S. Wu and A. Zee, Phys. Lett. 158B (1985) 311.

28. M. Dine, R. Rohm, N. Seiberg and E. Witten, Phys. Lett. 156B (1985) 55.

29. D. Bailin, A. Love and D. Wong, Phys. Lett. 165B (1985) 270; D. Bailin and A. Love, Phys. Lett. 163B (1985) 135.

30. E.W. Kolb, D. Lindley and D. Seckel, Phys. Rev. D30 (1984) 1205.

31. M. Rowan-Robinson, The Cosmological Distance Ladder, (W.H. Freeman & Company, New York, 1985) p. 281.

32. C. Brans and R.H. Dicke, Phys. Rev. 124 (1961) 925.

33. H. Georgi, H. Quinn and S. Weinberg, Phys. Rev. Lett. 33, (1974) 451.

34. J. Bagger, S. Dimopoulos and E. Masso, Nucl. Phys. B253 (1985) 397.

35. F.H. Dyson, Phys. Rev. Lett. 22 (1967) 1291, and in The Fundamental Constants and Their Time Variation; eds. by A. Salam and E. Wigner (Cambridge University Press, 1972); A. Shlyakhter, Nature (London) 264 (1976) 340.

THE GROUND STATE OF STRINGY GRAVITY

J.D. Gegenberg

Department of Mathematics and Statistics
University of New Brunswick, P.O. Box 4400
Fredericton, N.B. E3B 5A3

If the heterotic string theory is the correct "theory of everything," then the physics of our world at energies much lower than 10^{19} Gev should be governed, to a good approximation, by an appropriately modified version of N = 1 supergravity in ten dimensions coupled to a supersymmetric $E_8 \times E_8$ gauge field.[1,2] In fact, there is reason to believe that to lowest order in an appropriate string parameter, the low energy effective theory is the Chapline-Manton theory,[3] modified so as to arrange for the cancellation of gauge, gravitational and mixed anomalies.[1] This modified Chapline-Manton theory will be abbreviated here as MCM theory.

In order that the MCM theory be compatible with the physics of our world, it must admit a ground state field configuration satisfying the following requirements:

1. The ten dimensional space-time is a product manifold $M^4 \times B^6$, where M^4 is a four dimensional Lorentzian space-time whose metric $g_{\mu\nu}(x)$ is maximally symmetric, i.e. the curvature tensor on M^4 is given by

$$R_{\mu\nu\pi\sigma} = \Lambda\,(g_{\mu\pi}g_{\nu\sigma} - g_{\mu\sigma}g_{\nu\pi}), \tag{1}$$

with $|\Lambda| << 1$. The internal space, B^6, is compact (or at least of finite volume) with mean radius approximately equal to the Planck length ($\sim 10^{-32}$ cm).

2. The gauge group of the full theory must be broken to SU(3) x SU(2) x U(1) at the appropriate scale.

3. The configuration of gauge and gravitational fields (including metric, possible torsion and topology of B^6) must admit the existence of at least three generations of four dimensional chiral fermions.

4. The ground state configuration will be a stable solution of the MCM field equations.

5. A more controversial requirement is that the dimensionally reduced theory have N = 1 supersymmetry. This may provide for a solution of the "gauge hierarchy" problem.[4]

By using requirement 5 above, Candelas, et. al.[4] (CHSW) found candidate ground states of MCM potentially satisfying the other requirements 1 - 4. In particular, the CHSW ground state has a flat M^4, and an internal space B^6 which is Ricci-flat, Kähler and which has SU(3) holonomy. Many examples of such manifolds, called Calabi-Yau spaces, are known to exist,[5] but no explicit metrics have been constructed to date. By imbedding the spin connection of B^6 into one of the E_8 gauge fields, the gauge group $E_8 \times E_8$ is broken to $E_6 \times E_8$. The physics of elementary particles at low energies emerges by breaking E_6 into, say, SU(3) x SU(2) x U(1), but this is controversial - see for example the contribution of L. Ibañez in this volume.

Not withstanding its elegance and potential physical relevance, the CHSW ground state ansatz has some troublesome aspects. First, it now seems to be the case that the CHSW field configuration is compatible with string theory only to order $(\alpha')^3$, where α' is essentially the inverse string tension. See the contribution of D. Zanon in this volume, wherein the $O[(\alpha')^4]$ corrections are shown to prohibit ten dimensional Ricci-flat space-times. Second, the CHSW ansatz rules out any Kaluza-Klein-like geometrical origin for gauge fields. This is because Calabi-Yau spaces have no isometries.[4] Finally, the ansatz admits thousands of distinct Calabi-Yau spaces, only some of which are physically relevant, with no mechanisms for selecting one over the others. In addition, if B^6 is allowed to have torsion, then non-Calabi-Yau ground states are possible,[6] again with no mechanisms selecting among them, or indeed for selecting internal spaces with or without torsion.

One problem which the above discussion suggests is the question of the uniqueness of the CHSW ansatz if one drops the requirement that four dimensional physics has one unbroken supersymmetry at or near the Planck scale, but still requires that the internal space B^6 be torsion-free. My investigation of this question has led me to propose the following conjecture: There are no physically realistic ground state solutions of the MCM field equations (including terms proportional to the square of the curvature) for which B^6 is torsion-free and not Ricci-flat.

I have not succeeded in proving the conjecture for the general case, but I will show that obvious counter-examples do not exist. To begin with, the MCM action, including a phenomonological term in order to partially compensate for the absence of all the terms to $O[(\alpha')^2]$, and the corresponding field equations will be displayed. It will then be shown that if one requires the internal space to be an Einstein space, i.e., that $R_{ij} = b\gamma_{ij}$, then $b<<1$ follows from $|\Lambda|<<1$. In the above γ_{ij} is the metric on B^6, R_{ij} the corresponding Ricci tensor and b a non-negative constant whose inverse is the square of the size of B^6. Hence, only if $b = 0$ exactly is it possible for B^6 to be very small and M^4 to be very large.

The purely bosonic part of the MCM action, including $(\text{curvature})^2$ terms in the Gauss-Bonnet (ghost-free) form and a phenomonological term depending on the dilaton field can be written as:[7]

$$\hat{S} = \frac{1}{2K^2} \int_{M^{10}} d^{10}x \ \sqrt{-\hat{g}} \ \{\hat{R} - 2V(\phi) - 2\hat{g}^{AB} \ \phi_{,A} \ \phi_{,B} + \\ - \frac{1}{2} \alpha' e^{\phi} \ [\frac{1}{30} \mathrm{Tr}(\hat{F}_{AB} \ \hat{F}^{AB}) - \hat{L}] + \\ - \frac{1}{12} (\alpha' e^{\phi})^2 \ \hat{H}_{ABC} \ \hat{H}^{ABC}\}. \tag{2}$$

The ten dimensional gravitational constant K is related to the Regge slope parameter α' (which is essentially the inverse of the string tension) and the gauge coupling constant g by $K^2 = \alpha' g^2$. The ten dimensional metric $\hat{g}_{AB}$ has signature +8. Capital latin indices have the range 0 - 9. The dilaton field is denoted by ϕ and $V(\phi)$ is an unspecified phenomonological term. $\hat{R}$ is the ten dimensional curvature scalar. The conventions on the curvature tensors are $\hat{R}^A_{BCD} = \hat{\Gamma}^A_{BD,C} - \ldots$ and $\hat{R}_{BD} \equiv \hat{R}^A_{BAD}$. The field stregths of the $E_8 \times E_8$ gauge fields are $\hat{F}_{AB}$. The trace Tr is over the (hermitian) adjoint representation of $E_8 \times E_8$. $\hat{L}$ is the Gauss-Bonnet term:

$$\hat{L} \equiv \hat{R}_{ABCD} \ \hat{R}^{ABCD} - 4\hat{R}_{AB} \ \hat{R}^{AB} + \hat{R}^2. \tag{3}$$

$\hat{H}_{ABC}$ is a totally anti-symmetric tensor field constructed from an anti-symmetric potential $\hat{B}_{BC}$, the metric and the gauge fields by:

$$\hat{H}_{ABC} \equiv \hat{B}_{[AB,C]} - \frac{1}{30} \hat{\omega}^{(YM)}_{ABC} + \hat{\omega}^{(L)}_{ABC} \tag{4}$$

The Chern-Simons forms $\hat{\omega}^{(YM)}_{ABC}$ and $\hat{\omega}^{(L)}_{ABC}$ are inserted in order that the anomalies cancel. The conventions are such that

$$\hat{\omega}^{(YM)}_{[ABC,D]} = \mathrm{Tr}(\hat{F}_{[AD} \ \hat{F}_{BC]}),$$

and

$$\hat{\omega}^{(L)}_{[ABC,D]} = \hat{R}^{EF}_{[AD} \ \hat{R}_{BC]EF}.$$

The field equations are obtained by varying S with respect to the metric $\hat{g}^{AB}$ to get Einstein-like equations, with respect to the gauge potential of $\hat{F}_{AB}$ to get Yang-Mills-like equations, with respect to $\hat{B}_{AB}$ and with respect to ϕ. The result is:

$$\hat{E}_{AB} \equiv \hat{G}_{AB} - 2\hat{D}_{AB} + \hat{g}_{AB} V(\phi) - \frac{\alpha' e^{\phi}}{2} [\frac{1}{2} \hat{g}_{AB} \hat{L} +$$

$$+ 2(\frac{1}{30} \hat{Y}_{AB} - \hat{L}_{AB})] - \frac{1}{4} (\alpha' e^{\phi})^2 \hat{S}_{AB} +$$

$$- 2\alpha' e^{\phi} [\hat{R}_{ACBD}(\phi^{|CD} + \phi^{|C} \phi^{|D}) + 2\hat{R}_{C(A} \phi_{|B)}{}^{|C} +$$

$$+ \hat{g}_{AB} \hat{R}^{CD} \phi_{|CD} + \hat{G}_{AB} \phi_{|C} \phi^{|C}] +$$

$$+ \frac{1}{6} \hat{\nabla}_E [(\alpha' e^{\phi})^2 \hat{H}_{(A}{}^{CD} R^{E}{}_{B)CD}] = 0, \tag{5}$$

$$\hat{J}^{A} \equiv \hat{D}_B(\alpha' e^{\phi} \hat{F}^{BA}) - \frac{1}{12} (\alpha' e^{\phi})^2 \hat{F}_{BC} \hat{H}^{BCA} = 0, \tag{6}$$

$$\hat{K}^{AB} \equiv \hat{\nabla}_C [(\alpha' e^{\phi})^2 \hat{H}^{ABC}] = 0, \tag{7}$$

$$\hat{D} \equiv \hat{g}^{AB} \hat{\nabla}_A \hat{\nabla}_B \phi + \frac{1}{8} \alpha' e^{\phi} [\hat{L} - \frac{1}{30} \mathrm{Tr}(\hat{F}_{AB} \hat{F}^{AB})] +$$

$$- \frac{1}{24} (\alpha' e^{\phi})^2 \hat{H}_{ABC} \hat{H}^{ABC} + \frac{d}{d\phi} V(\phi) = 0. \tag{8}$$

The symbols $\hat{\nabla}_A$ and $\hat{D}_A$ denote, respectively, covariant derivatives with respect to $\hat{g}_{AB}$ and gauge covariant derivatives, i.e. $\hat{D}_B = \hat{\nabla}_B + [\hat{A}_B,]$, where $\hat{A}_B$ is the gauge potential, defined so that $\hat{F}_{AB} = 2\hat{A}_{[B,A]} + [\hat{A}_A, \hat{A}_B]$. The energy-momentum tensors of the gauge field $\hat{F}_{AB}$ and anti-symmetric tensor field $\hat{H}_{ABC}$ and the dilaton field ϕ are, respectively,

$$\hat{Y}_{AB} \equiv \mathrm{Tr}(\hat{F}_{AC} \hat{F}_B{}^C - \frac{1}{4} \hat{g}_{AB} \hat{F}_{CD} \hat{F}^{CD}), \tag{9}$$

$$\hat{S}_{AB} \equiv \hat{H}_{ACD} \hat{H}_B^{CD} - \frac{1}{6} \hat{g}_{AB} \hat{H}_{CDE} \hat{H}^{CDE}, \tag{10}$$

$$\hat{D}_{AB} \equiv \phi_{,A} \phi_{,B} - \frac{1}{2} \hat{g}_{AB} \phi_{,C} \phi_{,D} \hat{g}^{CD}. \tag{11}$$

Finally, the symmetric tensor $\hat{L}_{AB}$ (roughly, the energy-momentum tensor of the Gauss-Bonnet term) is defined as:

$$\hat{L}_{AB} = \hat{R}_{ACDE} \hat{R}_B^{CDE} - 2\hat{R}^{CD} \hat{R}_{ACBD} -$$

$$- 2\hat{R}_{AC} \hat{R}_B^C + \hat{R} \hat{R}_{AB}. \tag{12}$$

Besides the usual Bianchi identities satisfied by $\hat{R}_{ABCD}$ and $\hat{F}_{AB}$, there is the following restrictive condition which is the "curl" of equation (4):

$$\hat{\nabla}_{[D} \hat{H}_{ABC]} = \frac{1}{30} \mathrm{Tr}(\hat{F}_{[AD} \hat{F}_{BC]}) - \hat{R}^{EF}_{[AD} \hat{R}_{BC]EF}. \tag{13}$$

It should be noted here that the ten dimensional space-time is assumed to be torsion-free. However, if $\hat{H}_{ABC} \neq 0$, one can construct a non-Riemannian connection from the Christoffel symbols of $\hat{g}_{AB}$ and from $\hat{H}_{ABC}$.[8]

The first ground state ansatz I will examine here has a direct product metric.

$$\underline{\hat{g}} \equiv \hat{g}_{AB} d\hat{x}^A \otimes d\hat{x}^B = g_{\mu\nu}(x) dx^\mu \otimes dx^\nu + \gamma_{ij}(y) dy^i \otimes dy^j, \tag{14}$$

where $\mu, \nu, \ldots = 0, 1, 2, 3$ and $i, j, \ldots = 4, 5, \ldots, 9$.
The space-time metric $g_{\mu\nu}(x)$ has signature +2 and is maximally symmetric, i.e. satisfies equation (1). The Riemannian metric $\gamma_{ij}(y)$ on the internal space will be assumed to be a compact Einstein space

$$R_{ij} = b\, \gamma_{ij}, \tag{15}$$

where the constant $b \geq 0$. Furthermore, it will be required that $\hat{H}_{ABC} = 0$, ϕ = constant, $\hat{F}_{\mu A} = 0$ and that the spin connection on B^6 (constructed from the γ_{ij} metric) is imbedded in the gauge field $\hat{F}_{ij}$, so that

$$R_{iklm} R_j^{\;klm} = \frac{1}{30} \mathrm{Tr}(\hat{F}_i^{\;k} \hat{F}_{jk}), \tag{16}$$

where indices are raised and lowered by γ_{ij}.

With the above in place, the field equations $\hat{E}_{\mu\nu} = 0$, $\hat{E}_{ij} = 0$, $\hat{D} = 0$, $\hat{J}^i = 0$ reduce to:

$$3(b + \Lambda) + 3\alpha b(b + 6\Lambda) - V = 0, \tag{17}$$

$$2(b + 3\Lambda) + \alpha(b^2 + 12b\Lambda + 6\Lambda^2) - V = 0, \tag{18}$$

$$3\alpha(b^2 + 12b\Lambda + 2\Lambda^2) + 2dV/d\phi = 0, \tag{19}$$

$$D_j F^{ij} = 0, \tag{20}$$

where $\alpha \equiv \alpha' e^\phi$. The other field equations hold identically. From (17) and (18) it follows that

$$2\alpha = \frac{3\Lambda - b}{b^2 + 3b\Lambda - 3\Lambda^2}. \tag{21}$$

On physical grounds, the size of B^6 must be much smaller than M^4, so that $b \gg |\Lambda|$. Using this in (21) forces $\alpha < 0$, i.e. negative string tension. Only if $b = \Lambda = 0$ is it possible to find potentially physically realistic solutions of (17) - (20), and this is essentially the CHSW ground state configuration. It is worth noting that this result is not affected by the phenomonological term. The same problem occurs in the pure Einstein-Hilbert theory (i.e. in the limit as $\alpha' \to 0$), as has been known for a long time.[9] In a slightly different context, Müller-Hoissen has investigated spontaneous compactification of higher dimensional pure gravity with "dimensionally continued Euler forms" in the action.[10] But he achieved potentially physically realistic compactifications at the price of losing any connection to string theory, since his solutions require negative string tension.

Under current investigation is another ground state ansatz, motivated by the work of Freund and Rubin on eleven dimensional supergravity.[11] The idea is to split B^6 into the product of Riemannian three-manifolds, $B_1^3 \times B_2^3$, by choosing the non-zero components of $\hat{H}_{ABC}$, namely $\hat{H}_{i_1j_1k_1}$ and $\hat{H}_{i_2j_2k_2}$, to be proportional, respectively, to the volume three-forms $\eta_{i_1j_1k_1}$ on B_1^3 and $\eta_{i_2j_2k_2}$ on B_2^3. The outcome appears to be that B_1^3 and B_2^3 are necessarily spheres with radii approximately equal to $|\Lambda|^{-\frac{1}{2}}$.

References

1. M.B. Green and J. H. Schwarz, Phys. Lett. B149, 117 (1984); B151, 21 (1985).
2. D. Gross, J. Harvey, E. Martinec and R. Rohm, Nucl. Phys. B256, 253 (1985).
3. F. Chapline and N. S. Manton, Phys. Lett. B120, 301 (1983).
4. P. Candelas, G. Horowitz, A. Strominger and E. Witten, Nuc. Phys. B256, 46 (1985).
5. S.-T. Yau, Proc. Natl. Acad. Sci. 74, 1798 (1977).
6. G. Fogelman, K. S. Viswanathan and B. Wong, "Superstring Compactification on S^6 with Torsion", Simon Fraser University Preprint, 1985.
   B. P. Dolan, A. B. Henriques and R. G. Moorhouse, Phys. Lett. B166, 392 (1986);
   R. I. Nepomechie, Y.-S. Wu and A. Zee, Phys. Lett. B158, 311 (1985).
7. R. E. Kallosh, Phys. Lett. B159, 111 (1985).
8. I. Bars, "Compactification of Superstrings and Torsion", USC Preprint 85/15, 1985.
9. M. J. Duff in An Introduction to Kaluza-Klein Theories, ed. H. C. Lee (World Scientific, Singapore, 1984).
   J. D. Gegenberg and G. Kunstatter, Class. Quant. Grav. 3, 379 (1986).
10. F. Müller-Hoissen, Phys. Lett. B163, 106 (1985).
11. P. G. O. Freund and M. A. Rubin, Phys. Lett. B97, 233 (1980).

THE FUNCTIONAL MEASURE IN QUANTUM FIELD THEORY

David J. Toms

Department of Theoretical Physics
University of Newcastle upon Tyne
Newcastle upon Tyne, NE1 7RU
England

1. INTRODUCTION

One of the great legacies left to the physics community by Albert Einstein is the idea that physical predictions should be independent of the choice of coordinates. A natural way to ensure this is to formulate a theory in a way which is manifestly covariant. This is what is done for instance in the general theory of relativity. Today most physicists would not even contemplate founding a theory based on a preferred coordinate system or frame. The lack of a preferential frame is often exploited in calculations since, for example, derivations of certain results are often simpler if one adopts a coordinate system suitable for the symmetries of the problem. In general relativity it is often convenient to use a local orthonormal frame in which the metric tensor becomes just that for flat spacetime. One of the main purposes of this talk is to discuss how these ideas carry over into quantum field theory, particularly in relation to the functional measure.

An obvious question to ask is "Why should we be concerned with the functional measure?" Most discussions of functional integral methods are very cavalier in their treatment of the measure. However, the measure contains interesting physics. Fujikawa[1-6] has shown us how the existence of all known anomalies are linked to the non-invariance of the measure under the appropriate transformation. (For example chiral anomalies are due to the fact that the measure is not chirally invariant.) Also, Polyakov[7,8] has shown how the critical dimension for strings can be understood as a conseqence of a careful treatment of the measure.

There has been a certain amount of disagreement in the literature over exactly what should be taken as the functional measure. (See Refs. 1-6, 9, 10 for a guide to some of the literature.) For example, if we consider a scalar field $\varphi_0(x)$ on a curved background with a Riemannian metric $g_{\mu\nu}(x)$, Fujikawa[2,4-6] claims that the basic variable in the theory is not $\varphi_0(x)$ but rather $\tilde{\varphi}(x)=[-g(x)]^{\frac{1}{4}}\varphi_0(x)$. This is based upon the observation that the measure $\prod_x d\varphi_0(x)$ is not invariant under general coordinate transformations, but that $\prod_x d\tilde{\varphi}(x)$ is. On the other hand, Unz[9,10] has analysed the problem using a phase space analysis, and claims that the measure is $\prod_x \{(-g^{00})^{1/2} d\tilde{\varphi}(x)\}$. In this talk I wish to show that in contrast to Fujikawa, we can choose whatever variables we like, including the usual scalar fields provided that sufficient care is taken in defining the measure. In addition, I will discuss why the analysis of Unz[9,10] is incorrect and describe the proper phase space approach. This talk is based on Ref. 11 where more details can be found.

2. CONFIGURATION SPACE ANALYSIS

Before discussing the functional measure in quantum field theory, it is helpful to initially consider the analogous procedure for a finite dimensional system. Let V be a real vector space of dimension n with u, $v \in V$. Assume that V is equipped with an inner produced defined by

$$(u,v) \quad = \quad \sum_{i,j=1} u^i E_{ij}\, v^j \tag{1}$$

where u^i, v^i are the components of u, v with respect to some basis in V, and E_{ij} are the components of the metric for V with respect to this same basis. The measure which is invariant under a change of basis in V is

$$d\mu(v) \quad = \quad \left|\det E_{ij}\right|^{1/2} d^n v \tag{2}$$

where $d^n v = \prod_{i=1}^{n} dv^i$. If we had chosen to work in a local orthonormal basis, then $E_{ij} = \delta_{ij}$ and the measure in (2) would be simply $d\mu(v) = \prod_{i=1}^{n} dv^i$.

Now consider a real bosonic field $\varphi^i(x)$. Define an inner product which is invariant under general coordinate transformations by

$$(\varphi,\psi) \quad = \quad \int d^N x\, \varphi^i(x)\, E_{ij}(x)\psi^j(x) \tag{3}$$

where $E_{ij}(x)$ plays the role of the metric in function space. By analogy with the finite dimensional case above, the proper functional measure should be taken to be

$$d\mu[\varphi] = \prod_x [\det E_{ij}(x)]^{1/2} \prod_{i=1}^{n} d\varphi^i(x) \tag{4}$$

where $\det E_{ij}(x)$ means a determinant over the indices i and j (as opposed to a functional determinant). The idea of using the invariant metric in function space to define the functional measure was suggested by DeWitt[12] and used within the context of quantum gravity. He also emphasized the need for the functional measure to be invariant under general coordinate transformations[13].

As an explicit example, consider a real scalar field $\varphi_0(x)$. We may form a scalar field density of weight w by taking

$$\varphi(x) = [-g(x)]^{-\frac{w}{2}} \varphi_0(x). \tag{5}$$

Fujikawa's[2,4,6] variables correspond to the special case $w = -\frac{1}{2}$. The inner product (4) which is generally coordinate invariant is easily seen to be

$$(\varphi,\psi) = \int d^N x [-g(x)]^{w+\frac{1}{2}} \varphi(x)\, \psi(x), \tag{6}$$

from which we see that the role of the function space metric is played by $[-g(x)]^{w+\frac{1}{2}}$. The invariant functional measure for densities of weight w should therefore be taken as

$$d\mu[\varphi] = \prod_x [-g(x)]^{\frac{w}{2}+\frac{1}{4}} d\varphi(x). \tag{7}$$

Because the spacetime metric is held constant here, it is seen from (5) that the measure in (7) is equivalent to

$$d\mu[\varphi] = \prod_x [-g(x)]^{\frac{1}{4}} d\varphi_0(x) \tag{8}$$

This proves that the measure is independent of the weight chosen for the variables.

If the usual scalar field variables $\varphi_0(x)$ are chosen, it is seen from (8) that the functional measure is not simply $\prod_x d\varphi_0(x)$ as may be naively supposed. The viewpoint taken here is that $\prod_x d\varphi_0(x)$ is no more expected to be invariant than is $d^n x$ expected to be the invariant volume element in curved spacetime. Both of these expressions neglect the fact that there may be a nontrivial metric.

If Fujikawa's choice of variable ($w = -\frac{1}{2}$) is made, it is clear from (6) that the metric in function space is simply the identity. Thus, choosing scalar field densities of weight $-\frac{1}{2}$ is analogous to choosing a local orthonormal basis in the finite dimensional case. Although this is often a convenient choice for calculational purposes, it is in no way compulsory. We are free to choose whatever variables we like so long as we include the appropriate metric factor in the functional measure.

3. PHASE SPACE ANALYSIS

Once again it is helpful to consider first of all a finite-dimensional system before proceeding to the field theoretical case. Consider a system with a Lagrangian

$$L = \frac{1}{2} E_{ij} \dot{q}^i \dot{q}^j - V(q) \tag{9}$$

where E_{ij} is a non-singular matrix. The canonical momentum is

$$p_i = \frac{\partial L}{\partial \dot{q}^i} = E_{ij} \dot{q}^j \tag{10}$$

and the Hamiltonian is

$$H(q,p,t) = \frac{1}{2}(E^{-1})^{ij} p_i p_j + V(q) \tag{11}$$

where

$$(E^{-1})^{ik} E_{kj} = \delta^i{}_j \tag{12}$$

From (11) we see that the volume element in momentum space is

$$d\mu(p) = (\det E^{-1})^{\frac{1}{2}} d^n p = (\det E)^{-\frac{1}{2}} d^n p \tag{13}$$

where $d^n p = \prod_{i=1}^{n} dp_i$. From (9), and the discussion of the preceding section the invariant volume element in configuration space is

$$d\mu(q) = (\det E)^{\frac{1}{2}} d^n q \tag{14}$$

Thus the invariant volume element in phase space is

$$d\mu(p,q) = d^n p \, d^n q, \tag{15}$$

where the factors of detE cancel between (13) and (14). For a quantum mechanical path integral based on this system, the measure should be taken as

$$d\mu[p,q] = \prod_t \left(\prod_{i=1}^{n} dp_i(t)\right)\left(\prod_{i=1}^{n} dq^i(t)\right). \tag{16}$$

This was first written down by Feynman[14]. The result of (16) is usually directly transcribed to quantum field theory; however, it turns out to be incorrect for quantum field theory in curved spacetime.

A careful discussion of canonical quantization in curved spacetime, and its general covariance is described by Fulling[15]. The canonical formalism requires that one of the coordinates, say x^o, be chosen as the time coordinate, and that for each t, the hypersurface x^o=t be a Cauchy hypersurface for the region of spacetime covered by the coordinate system. The behaviour of quantitites under a change of coordinates which involve only those on the Cauchy hypersurface or else a transformation of the local time scale with the coordinates on the Cauchy hypersurface held fixed may be considered.

The Lagrangian for a free, massless scalar field density of weight w is

$$L[\varphi,\dot{\varphi},t] = -\frac{1}{2}\int d^{N-1}x[-g(x)]^{\frac{1}{2}} g^{\mu\nu}\partial_\mu[(-g)^{\frac{w}{2}}\varphi] \times \partial_\nu[(-g)^{\frac{w}{2}}\varphi] \tag{17}$$

where the integration extends over the (N-1)-dimensional Cauchy hypersurface. The momentum canonically conjugate to $\varphi(x)$ is defined by

$$\pi(x) = -[-g(x)]^{w+\frac{1}{2}} g^{0\nu}(x)\,\nabla_\nu\varphi(x). \tag{18}$$

The Hamiltonian density is

$$\begin{aligned}\mathcal{H} = & -\tfrac{1}{2}(g^{oo})^{-1}(-g)^{-w-\frac{1}{2}}\pi^2 - (g^{00})^{-1}g^{oi}\pi\nabla_i\varphi \\ & -\tfrac{w}{2}[\partial_o \ln(-g)]\pi\varphi - \tfrac{1}{2}(g^{oo})^{-1}(-g)^{w+\frac{1}{2}} g^{oi}g^{oj}\nabla_i\varphi\nabla_j\varphi \\ & +\tfrac{1}{2}(-g)^{w+\frac{1}{2}} g^{ij}\nabla_i\varphi\nabla_j\varphi .\end{aligned} \tag{19}$$

It is clear from (19) that the momentum space measure for the field is

$$d\mu[\pi] = \prod_x \left\{[-g^{00}(x)]^{-\frac{1}{2}}[-g(x)]^{-\frac{w}{2}-\frac{1}{4}}d\pi(x)\right\}. \quad (20)$$

The factor of $[-g^{00}(x)]^{-\frac{1}{2}}[-g(x)]^{-\frac{w}{2}-\frac{1}{4}}$ which followed from the coefficient of π^2 in (19) is seen to be the analogue of $(\det E^{-1})^{1/2}$ in (13). When (20) is combined with the configuration space measure (7) it gives

$$d\mu[\pi,\varphi] = \prod_x \left\{[-g^{00}(x)]^{-\frac{1}{2}} d\pi(x)\, d\varphi(x)\right\} \quad (21)$$

as the invariant measure in phase space. It is straightforward to verify that (21) is invariant under arbitrary changes in the local time scale, as well as under changes in the coordinates on the spacelike Cauchy hypersurface. The presence of $[-g^{00}(x)]^{-1/2}$ which is not present in the naive transcription of (16) to field theory, is crucial for this invariance.

Integration over $\pi(x)$ in the functional integral leads directly to the configuration space functional integral based on the measure (7). Without the explicit factor of $[-g^{00}(x)]^{-1/2}$ present in (21) this will not be true as explicit factors of g^{00} will appear in the configuration space measure. This is the situation encountered by Unz[9,10] who starts off with the non-covariant phase space measure $\prod_x d\pi(x)\; d\varphi(x)$ and not surprisingly ends up with a non-invariant measure in configuration space.

4. DISCUSSION

The analysis above has focussed on the nature of the functional measure in quantum field theory, particularly in relation to Fujikawa's[1-6] work. It was demonstrated that the choice of variables for the functional integral are not uniquely specified by the demand that the functional measure be generally coordinate invariant, in contrast to claims of Fujikawa. The problem of defining the functional measure was related to the simpler one of defining a metric in function space to give an invariant inner product. This approach applies equally well to phase space, the correct form of the functional measure for a scalar field density of arbitrary weight being given in (21). The reason for Unz's[9,10] results being incorrect was traced to the naive transcription of results valid for finite dimensional systems to quantum field theory.

Because Fujikawa's[2] treatment of conformal anomalies was based on his particular choice of variable, it might be wondered what happens if densities of arbitrary weight are chosen. This problem is analysed in Ref. 11 where it is shown that the conformal anomaly may be understood as arising from a non-unitary change of bases in Hilbert space. The Jacobian which arises in this non-unitary transformation is independent of the weight chosen for the field variables and leads to the usual form for the conformal anomaly. The extension of these results to other types of fields is also presented in Ref. 11.

Finally, I wish to make a few remarks on the relation of this approach to Vilkovisky's[16] formulation of the effective action. Fields should be viewed as coordinates in function space. In the preceding sections it was assumed that they are also sections of some vector bundle so that expression such as (3) make sense. This will not be true in general; however, it still makes sense to define an inner product in function space between vectors. It was in fact done in the case of quantum gravity by DeWitt[17]. Specifying an inner product gives a metric in function space. As Vilkovisky[16] has discussed, this metric can be used to construct a connection and give a formulation of the effective action which is invariant under arbitrary field redefinitions. All of this supposes that the functional measure is also invariant under field redefinitions. This is guaranteed to be the case if it is defined as described in sec. 2 above. This gives the whole approach to quantization by the background-field method and the effective action a very geometrical feel.

REFERENCES

1. K. Fujikawa, Path-integral measure for gauge-invariant fermion theories, Phys. Rev. Lett. 42 : 1195 (1979).
2. K. Fujikawa, Comment on chiral and conformal anomalies, Phys. Rev. Lett 44: 1733 (1980).
3. K. Fujikawa, Path integral for gauge theories with fermions, Phys. Rev. D 21 : 2848 (1980).
4. K. Fujikawa, Energy-momentum tensor in quantum field theory, Phys. Rev. D 23 : 2262 (1981).
5. K. Fujikawa, Path integral measure for gravitational interactions, Nucl. Phys. B226 : 437 (1983).
6. K. Fujikawa, Path integral quantization of gravitational interactions - local symmetry properties, in: "Quantum Gravity and Cosmology", H. Sato and T. Inami, ed., World Scientific, Singapore (1986).
7. A. Polyakov, Quantum geometry of bosonic strings, Phys. Lett. 103B : 207 (1981).
8. A. Polyakov, Quantum geometry of fermionic strings, Phys. Lett. 130B : 211 (1981).
9. R.K. Unz, Path integration and the functional measure, SLAC-PUB-3 656, unpublished (1985).

10. R.K. Unz, The functional measure in Kaluza-Klein theories, Phys. Rev. D 32 : 2539 (1985).
11. D.J. Toms, The functional measure for quantum field theory in curved spacetime, University of Newcastle upon Tyne report NCL 86 - TP (1986).
12. B.S. DeWitt, Quantum gravity: the new synthesis, in: "General Relativity", S.W. Hawking and W. Israel, ed., Cambridge University Press, Cambridge (1979).
13. B.S. DeWitt, Covariant quantum geometrodynamics, in "Magic Without Magic", J. Klauder, ed., W.H. Freeman, San Francisco (1972).
14. R.P. Feynman, An operator calculus having applications in quantum electrodynamics, Phys. Rev. 84 : 108 (1951).
15. S.A. Fulling, Ph.D. Thesis, Princeton University, unpublished (1972).
16. G. Vilkovisky, The gospel according to DeWitt, in "Quantum Theory of Gravity", S.M. Christensen, ed., Adam Hilger, Bristol (1984).
17. B.S. DeWitt, Quantum theory of gravity. I. The canonical theory, Phys. Rev. 160 : 1113 (1967).

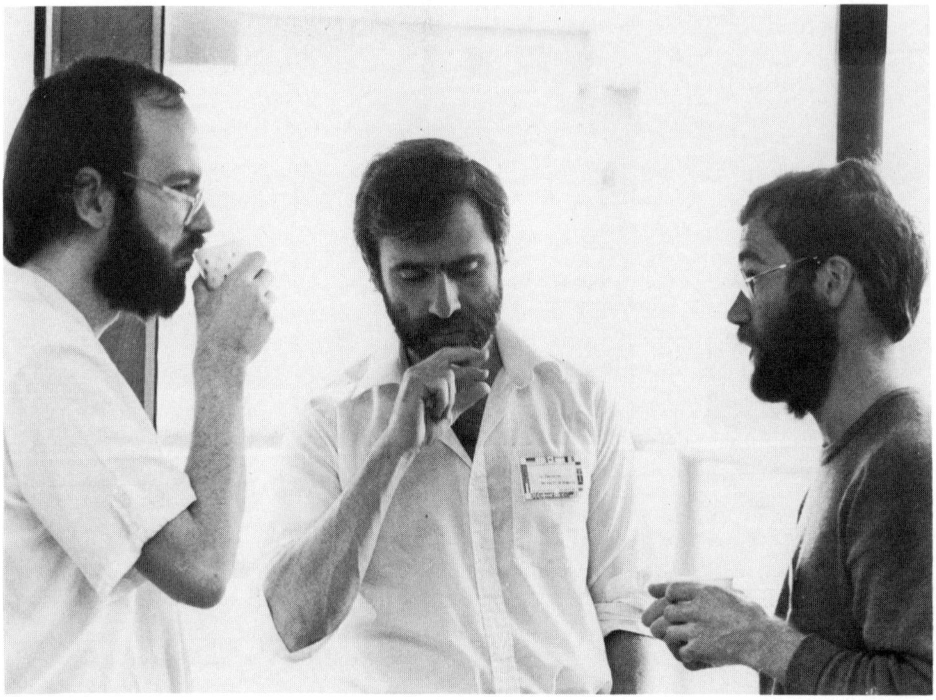

VILKOVISKY'S UNIQUE EFFECTIVE ACTION:

AN INTRODUCTION AND EXPLICIT CALCULATION*

Gabor Kunstatter

Physics Department
University of Winnipeg
Winnipeg, Manitoba
CANADA R3B 2E9

While I was walking on the stair
I met a man who wasn't there.
He wasn't there again today
I wish that he would stay away.
author unknown

I. INTRODUCTION

The effective action is an important tool in modern quantum field theory. In particular it can be used to probe the non-perturbative vacuum structure of a theory. Examples of this are provided by the Coleman-Weinberg[1] mechanism in scalar QED, self-consistent dimensional reduction in Kaluza-Klein theory[2-12], and β-function techniques for evaluating the field equations for the target space metric in non-linear σ-models[13-14]. In non-perturbative applications, it is generally necessary to evaluate the effective action off-shell. It is well known, however, that the off-shell effective action as usually defined is both gauge dependent and reparametrization dependent. Reparametrization dependence means that under a field redefinition $\phi \to \phi'$, the new effective action $\Gamma'[\phi']$ does not equal the original one $\Gamma[\phi]$ evaluated at the same physical field ϕ. In terms more appropriate to the following discussion, one might say that the effective action is not a scalar function of the configuration space coordinates.

Although gauge dependence of the effective action has been extensively analyzed in the past[9,10,15-29], reparametrization dependence has not received as much attention in the literature. In a recent paper[11], it was shown that the one-loop effective action depends on the parametrization in complex ϕ^4 theory, scalar QED and quantum gravity. Although physical quantities are not affected by reparametrization in the first two cases, the prediction for the radius of the internal dimension in the self-consistent dimensional reduction of Kaluza-Klein theory does in fact depend on the choice of parametrization[11].

The solution to the problem of gauge and parametrization dependence has been known for some time: Vilkovisky has introduced[28] a natural

---

*Based on a talk given at the NATO Advanced Research Workshop on Super Field Theories, Simon Fraser University, July 24- August 5, 1986.

definition for a unique effective action which is manifestly a scalar function of the configuration space coordinates. It is also precisely equal to the standard expression on-shell, so that S-matrix elements are not affected. Moreover, Vilkovisky's extension of the formalism to gauge theories yields an expression which is both background gauge and gauge parameter independent*. The purpose of this talk is to provide an introduction to Vilkovisky's unique effective action and to illustrate its usefulness with an explicit calculation.

The standard definition of the effective action will be briefly reviewed in section II, followed by three examples of reparametrization dependence[11]. In section III, Vilkovisky's formalism will be outlined, and the calculation of the unique effective action for scalar QED using a non-standard parametrization will be presented in some detail. Although the gauge parameter independence of Vilkovisky's definition has been checked for scalar QED by Fradkin and Tseytlin[29], this has not been done with regard to parametrization dependence. As expected, the formalism works. Section IV concludes with discussion and prospects for future work.

II. The Ordinary Effective Action: The Problem

The standard discussion of the effective action starts from the generating functional W[J] for connected Green's functions**:

$$\langle 0|0\rangle_J = \exp \frac{i}{\hbar} W[J]$$

$$= \int [D\phi] \exp\frac{i}{\hbar} (I[\phi] + \phi^i J_i) \tag{2.1}$$

The effective action is then defined via a functional Legendre transformation:

$$\Gamma[\Phi] = W[J] - \Phi^i J_i \tag{2.2}$$

where Φ^i is the normalized vacuum expectation value of the field ϕ^i in the presence of the source J_i

$$\Phi^i = \frac{\delta W[J]}{\delta J_i}$$

$$= \int [D\phi]\, \phi^i \exp\frac{i}{\hbar} (I[\phi] + \phi^i J_i)/\langle 0|0\rangle_J \tag{2.3}$$

From Eqs. (2.2) and (2.3) it follows identically that

$$\frac{\delta\Gamma}{\delta\Phi^i} = - J_i \tag{2.4}$$

---

*It is important to stress the distinction between background gauge invariance of the effective action, which requires $\Gamma[U\Phi]=\Gamma[\Phi]$ for an arbitrary gauge transformation U, and gauge parameter independence, which demands that Γ be independent of the gauge choice used to regulate the functional integration over quantum fluctuations. (I am grateful to Roman Jackiw for conversations on this point)

**ϕ^i denotes the complete set of quantum fields in the theory. The label i represents both discrete indices and space-time coordinates. For example in scalar QED $\phi^i = \{\phi(x), A_\mu(x)\}$, while in general relativity $\phi^i = \{g_{\mu\nu}(x)\}$. Summation over repeated indices is implied and includes integration over coordinates with the appropriate measure.

so that the quantum corrected vacuum field equations for the background field are $\delta\Gamma/\delta\Phi^i = 0$.

An alternative, but formally equivalent definition of the effective action will be useful in the following discussion. It can be obtained by exponentiating Eq.(2.2) and substituting $-\delta\Gamma/\delta\Phi^i$ for J_i in the resulting expression. This yields the following functional differential equation for $\Gamma[\Phi]$:

$$\exp\frac{i}{\hbar}\,\Gamma[\Phi] = \exp\frac{i}{\hbar}\,(W[J] + \Phi^i\,\frac{\delta\Gamma}{\delta\Phi^i})$$
$$= \int[D\phi]\exp\frac{i}{\hbar}\,(\, I[\phi] + \phi^i\, J_i + \Phi^i\frac{\delta\Gamma}{\delta\Phi^i})$$
$$= \int[D\phi]\exp\frac{i}{\hbar}\,(\, I[\phi] + (\Phi^i - \phi^i)\cdot\frac{\delta\Gamma}{\delta\Phi^i}) \tag{2.5}$$

It is now straightforward to derive the loop expansion of the effective action. First expand the classical action in a functional Taylor series about the background Φ:

$$I[\phi] = I[\Phi] + \frac{\delta I}{\delta\Phi^i}\,(\sqrt{\hbar}\eta^i) + \frac{1}{2}\,\hbar\,\frac{\delta^2 I}{\delta\Phi^i\delta\Phi^j}\eta^i\eta^j + \ldots \tag{2.6}$$

where we have defined the quantum fluctuations: $\sqrt{\hbar}\eta^i \equiv \phi^i - \Phi^i$. Substituting Eq.(2.6) into (2.5) and noting that $I - \Gamma$ is necessarily of order $\hbar$, we get the standard expression[27] for the one-loop effective action, namely

$$\Gamma[\Phi] = I[\Phi] + \frac{i\hbar}{2}\,\ln\,\mathrm{Det}\left|\frac{\delta^2 I}{\delta\Phi^i\delta\Phi^j}\right| + O(h^2) \tag{2.7}$$

where Det denotes a functional determinant.

The above discussion has avoided two important technical details concerning the effective action: the correct choice of functional measure and the issue of parametrization dependence. (The latter is in fact a more general case of the off-shell gauge-parameter dependence[28]). It turns out that both these questions are related and that their resolution depends on the realization that one must consider the configuration space as a manifold[28]. Like the classical action $I[\phi]$, the effective action must therefore be defined as a scalar function on that manifold. For a recent discussion of the measure problem see reference[30]. Before discussing the solution to the issue of the parametrization dependence of the effective action, we will briefly illustrate the nature of the problem using three simple examples, first treated in ref. 11.

The simplest example of a field theory in which parametrization dependence poses a potential problem is massless complex ϕ^4 theory. The standard calculation[27] starts with the classical lagrangian:

$$\pounds = \frac{1}{2}\,\partial_\mu\phi^*\partial^\mu\phi - \frac{\lambda}{4!}(\phi^*\phi)^2 \tag{2.8}$$

where $\phi = \phi_1 + i\phi_2$. The quantum fluctuations about a background $\Phi = \Phi_1 + i\Phi_2$ are then defined by:

$$\eta_1 = \phi_1 - \Phi_1;\ \eta_2 = \phi_2 - \Phi_2 \tag{2.9}$$

yielding the following one-loop effective potential[27]:

$$V(\rho_0) = \frac{\lambda}{4!}\rho_o^4 + \frac{\hbar\lambda}{64\pi^2}\left(\frac{5}{18}\right)\rho_o^4\left(\ln\left(\frac{\lambda\rho_o^2/2}{M^2}\right) - \frac{25}{6}\right) \tag{2.10}$$

where $\rho_o = (\Phi^*\Phi)^{1/2}$ and the factor of $\hbar$ is retained only as a book-keeping device for keeping track of the loop expansion. In Eq.(2.10), the mass has been renormalized to zero at $\rho_o=0$, while λ is the coupling constant renormalized at $\rho_o=M$.

Alternatively, one could have performed the following invertible (except at $\rho=0$) field redefinition prior to quantization:

$$\phi_1 = \rho\cos\chi; \qquad \phi_2 = \rho\sin\chi \tag{2.11}$$

The quantum fluctuations would then be defined by

$$v = \rho - \rho_o; \qquad \gamma = \chi - \chi_o \tag{2.12}$$

where

$$\rho_o \exp(i\chi_o) = \Phi \tag{2.13}$$

In this case, the same renormalization conditions as before yield the following one-loop effective potential:

$$V'(\rho_o) = \frac{\lambda}{4!}\rho_o^4 + \frac{\hbar\lambda^2}{64\pi^2}\left(\frac{1}{4}\right)\rho_o^4\left(\ln\left(\frac{\lambda\rho_o^2/2}{M^2}\right) - \frac{25}{6}\right) \tag{2.14}$$

Thus the one-loop effective potential depends explicitly on the parametrization. Note, however, that consistent truncation of the loop expansion requires $\lambda^2 \ll \lambda$, so that both V and V' have only one minimum, namely at $\rho_o=0$. The on-shell effective potential is therefore invariant under the reparametrization.

The second example of parametrization dependence is somewhat less trivial. Consider the classical lagrangian for scalar Q.E.D. with gauge fixing term added:

$$\begin{aligned}\pounds = &-\frac{1}{4}F_{\mu\nu}F^{\mu\nu} + \frac{1}{2}\partial_\mu\phi^*\partial^\mu\phi - \frac{\lambda(\phi^*\phi)^2}{4!}\\ &-e\varepsilon_{ab}\partial_\mu\phi_a\phi_b A^\mu + \frac{1}{2}e^2\phi^*\phi A_\mu A^\mu - \frac{1}{2\alpha}(\partial_\mu A^\mu)^2\end{aligned} \tag{2.15}$$

The one-loop effective potential has been calculated by Jackiw[27] in terms of the fluctuations

$$\eta_1 = \phi_1 - \Phi_1 \tag{2.16a}$$
$$\eta_2 = \phi_2 - \Phi_2 \tag{2.16b}$$
$$a_\mu = A_\mu - 0 \tag{2.16c}$$

The answer depends explicitly on the gauge parameter α:

$$V(\Phi_1,\Phi_2) = \frac{\lambda\rho_o^4}{4!} + \frac{\hbar}{64\pi^2}\left(3e^2 + \frac{5}{18}\lambda^2 - \frac{2}{3}\alpha e^2\lambda\right)\rho_o^4\left(\ln\left(\frac{\lambda\rho_o^2/2}{M^2}\right) - \frac{25}{6}\right) \tag{2.17}$$

where $\rho_o^2 = \Phi_1^2 + \Phi_2^2$ as before.

We now perform the field redefinitions (2.11) on the complex scalar field while leaving the vector potential unchanged. This is different from the unitary gauge calculation of Dolan and Jackiw[26] in which the vector potential was redefined in order to trivialize the dependence of the classical lagrangian on the gauge degrees of freedom. The lagrangian after the field redefinition is:

$$\begin{aligned}\pounds = &-\frac{1}{4}F_{\mu\nu}F^{\mu\nu} + \frac{1}{2}\partial_\mu\rho\partial^\mu\rho + \frac{1}{2}\rho^2\partial_\mu\chi\partial^\mu\chi\\ &+ e\rho^2 A^\mu\partial_\mu\chi + \frac{1}{2}e^2\rho^2 A_\mu A^\mu - \frac{\lambda}{4!}\rho^4\\ &-\frac{1}{2\alpha}(\partial_\mu A^\mu)^2\end{aligned} \tag{2.18}$$

The Jacobian of the transformation from (ϕ_1,ϕ_2) to (ρ,χ) is

$$\left|\frac{\partial(\phi_1\phi_2)}{\partial(\rho,\chi)}\right| = \mathrm{Det}\left(\rho(x)\delta^4(x-y)\right) \tag{2.19}$$

After a straightforward calculation, one finds the following one-loop effective potential[11]:

$$V = \frac{\delta m^2}{2}\rho_o^2 + \frac{(\lambda+\delta\lambda)}{4!}\rho_o^4 + i\hbar\int\frac{d^4k}{(2\pi)^4}\ln(\rho_o)$$
$$-\frac{i\hbar}{2}\int\frac{d^4k}{(2\pi)^4}\ln\left((k^2-e^2\rho_o^2)^3(k^2-\lambda\rho_o^2/2)\right)$$
$$-\frac{i\hbar}{2}\int\frac{d^4k}{(2\pi)^4}\ln(\rho_o^2k^4/\alpha) + i\hbar\int\frac{d^4k}{(2\pi)^4}\ln(k^2) \tag{2.20}$$

Surprisingly, the gauge parameter dependence in the above expression is trivial. It is also important to note that the Jacobian contribution represented by the third term in Eq.(2.20) is required to cancel a quartic divergence coming from the inverse propagator of the longitudinal photon mode (fifth term in Eq.(2.20)[see also ref.26].

The final expression for the effective potential after Wick rotation and renormalization is[11]:

$$V = \frac{\lambda\rho_o^4}{4!} + \frac{\hbar}{64\pi^2}\left(3e^2 + \frac{\lambda^2}{4}\right)\rho_o^4\left(\ln\frac{\lambda\rho_o^2/2}{M^2} - \frac{25}{6}\right) \tag{2.21}$$

The co-efficient of the λ^2 term differs from the standard result[27] in the same family of gauges. Physical quantities, however, are still not affected. As discussed in Coleman and Weinberg's original paper[1], a minimum to the effective potential away from $\rho_o=0$ exists providing that λ is of the same order as e^4. Consistency of the loop expansion therefore demands that the λ^2 term be neglected when compared to the e^4 term. Similarly the gauge dependent contribution found by Jackiw[27] (cf. Eq.(2.17)) is also irrelevant to the order of the calculation. Thus in scalar QED, the reparametrization dependence, like the gauge parameter dependence, does not seem to pose any real problems. In principle, one must nonetheless always be aware of the fact that the off-shell effective potential is parametrization dependent.

The final example of parametrization dependence of the effective potential is perhaps the most important. Since it is discussed in some detail in H.P. Leivo's contribution to these proceedings[31] and in ref.[11], I will merely mention it briefly. In general, self-consistent dimensional reduction involves a calculation of the off-shell effective action in 4+n dimensional quantum gravity with non-zero cosmological constant and compactified internal dimensions. A non-perturbative minimum is then found to the one-loop corrected equations of motion by fine-tuning the bare cosmological constant so that solutions with flat four dimensional space times are allowed. In this case, the issue of parametrization dependence is of utmost importance. Although a natural way to proceed with the calculation is to quantize the fluctuations about the full 4+n-dimensional metric, it is also possible to first express the theory in terms of the lower dimensional fields by imposing the Kaluza-Klein ansatz, which for n=1 can be written[23]:

$$g_{AB} = \phi^s\begin{vmatrix} g_{\mu\nu} + \phi A_\mu A_\nu & \phi A_\mu \\ \phi A_\nu & \phi \end{vmatrix} \tag{2.22}$$

where $\{A,B = 0,1,2,3,5\}$ and $\{\mu,\nu = 0,1,2,3,\}$. In five-dimensional Kaluza-Klein theory, Eq.(2.22) represents a parametrization and not a restrictive ansatz providing that the fields are allowed to depend on all five coordinates. The Weyl exponent, s, has been introduced in order to provide a one-parameter family of parametrizations. Classically, it is completely undetermined (except for s = -1 which defines a degenerate parametrization and is therefore forbidden.)

In ref.[11] it was shown that the predictions of self-consistent dimensional reduction depend on whether the effective potential is calculated in terms of fluctuations of the reduced fields $\{g_{\mu\nu}, A_\mu, \phi\}$ or in terms of fluctutations of the five-dimensional metric*. In fact, some values of the Weyl exponent yield an effective potential which is not pure real. Thus, the parametrization dependence appears to render the predictions of self-consistent dimensional reduction meaningless. In the following section we will present an introduction to Vilkovisky's unique effective action, which in principle provides a method for determining a unique prediction for the radius of the internal dimensions in self-consistent dimensional reduction.

III. Vilkovisky's Unique Effective Action: The Solution

The solution to the problem of parametrization dependence of the effective action lies in the realization that the configuration space, $\mathcal{C}$, of a field theory must be treated as a manifold. Any viable definition of the effective action must yield a scalar function of the coordinates of that manifold. It is clear that $\Gamma[\Phi]$ defined in Eq.(2.5) is not a scalar, because it depends explicitly on the quantity $\eta^i = \phi^i - \Phi^i$, which is the difference between the coordinates of two different points in configuration space. The key, therefore, is to replace η^i by a geometrical quantity which is equal to η^i in the special case that one is considering Cartesian coordinates in a flat configuration space. Such a geometrical quantity can be defined as follows: Consider the geodesic joining the two points Φ^i and ϕ^i in $\mathcal{C}$. This will be unique providing that ϕ is in some suitable neighbourhood** of Φ^i. Let $x^i(\tau)$ denote the coordinates along the geodesic, given as functions of an affine parameter τ, such that $x^i(0)=\phi^i$ and $x^i(\tau_f) = \Phi^i$. These coordinates are solutions to the infinite dimensional analogue of the geodesic equation:

$$\frac{d^2x^i}{d\tau^2} + \Gamma^i_{jk}\frac{dx^j}{d\tau}\frac{dx^k}{d\tau} = 0 \tag{3.1}$$

We can now define the following vector at each point along the geodesic:

$$\sigma^i(\tau) = \tau dx^i/d\tau \tag{3.2}$$

It is possible to expand x^i about $\tau=\tau_f$ in a Taylor series:

$$x^i(0) = x^i(\tau_f) + \left.\frac{dx^i}{d\tau}\right|_{\tau_f}(-\tau_f) + \frac{1}{2}\left.\frac{d^2x^i}{d\tau^2}\right|_{\tau_f}(-\tau_f)^2 + \frac{1}{6}\left.\frac{d^3x^i}{d\tau^3}\right|_{\tau_f}(-\tau_f)^3 + 0(\tau_f^4).$$

*This is not too surprising in light of earlier results[8,9,10] which showed that the predictions were gauge dependent.

**An interesting question concerns the extension of this definition to topologically non-trivial configuration spaces. This question is not relevant to the following discussion, however, since validity of the truncated loop expansion requires that only fluctutations in the neighbourhood of the background contribute to the effective action.

$$= x^i(\tau_f) - \sigma^i - \frac{1}{2}\Gamma^i_{jk}\sigma^j\sigma^k$$

$$+ \frac{1}{6}(\Gamma^i_{jk,\ell} - 2\Gamma^i_{nj}\Gamma^n_{k\ell})\sigma^j\sigma^k\sigma^\ell + O(\sigma^4) \tag{3.3}$$

where we have used Eq.(3.1) and defined the two-point function[32]

$$\sigma^i \equiv \sigma^i(\Phi, \phi) = \sigma^i(\tau_f) \tag{3.4}$$

Eq.(3.3) can be inverted to yield the following perturbative expansion for $\sigma^i(\Phi,\phi)$:

$$\sigma^i(\Phi,\phi) = -\eta^i - \frac{1}{2}\Gamma^i_{jk}\eta^j\eta^k$$

$$-\frac{1}{6}(\Gamma^i_{jk,\ell} + \Gamma^i_{jn,}\Gamma^n_{k\ell})\eta^j\eta^k\eta^\ell + O(\eta^4) \tag{3.5}$$

$\sigma^i(\Phi,\phi)$ is precisely the required geometrical quantity. It is covariant under a change of function space coordinates at the point Φ (it lies in the tangent space of $\mathcal{C}$ at Φ) and it is a scalar under coordinate changes at ϕ. Vilkovisky's unique, scalar effective action can now be defined as follows[28]:

$$\tilde{\Gamma}[\Phi] = \int[D\phi]\exp\frac{i}{h}\left(I[\phi] + \sigma^i(\Phi,\phi)\frac{\delta I}{\delta\Phi}\right) \tag{3.6}$$

Note that this procedure is different from an expansion of Eq.(2.5) in normal coordinates [see ref. 14].

The next important question concerns the choice for the connection Γ^i_{jk} in the geodesic equation (3.1). Vilkovisky imposes the following reasonable criteria[28]:

(1) The connection should be determined from the classical action by a universal rule. (This eliminates the possibility of the type of quantization ambiguity discussed by Hull[14].)

(2) $\tilde{\Gamma} = I$ for free field theories.

(3) The connection must be ultralocal (i.e. contain only δ-functions; no derivatives of δ-functions). This condition is necessary for the S-matrix to be well-defined[28].

As shown in detail by Vilkovisky, these assumptions lead uniquely to the most natural choice for the configuration space connection, namely the Christoffel symbol constructed from the configuration space metric. The correct choice for the configuration space metric is discussed in general by Vilkovisky[28]. Here we will merely illustrate the procedure with two explicit examples.

First, however, we note that the above approach also automatically solves the problem of the correct function space measure to be used in the definition of the effective action. It must be a scalar density of weight 1 constructed from the configuration space metric G,

$$[D\phi] = \Pi_x d\phi(x)(\det G)^{1/2} \tag{3.7}$$

The loop expansion of $\tilde{\Gamma}$ can now be derived exactly as in the previous section to yield:

$$\tilde{\Gamma}[\Phi] = I[\Phi] - \frac{i\hbar}{2}\ln\operatorname{Det}G + \frac{i\hbar}{2}\ell n\operatorname{Det}\tilde{D}^{-1}_{ij} \tag{3.8}$$

where $\tilde{D}^{-1}_{ij}$ is the modified inverse propagator:

$$\tilde{D}^{-1}_{ij} = \frac{\delta^2 I}{\delta\Phi^i\delta\Phi^j} - \Gamma_{ij}^{k}[\Phi]\,\frac{\delta I}{\delta\Phi^k} \tag{3.9}$$

For example, consider the Lagrangian in Eq.(2.8). In terms of the real fields $\{\phi^a\} = \{\phi^1,\phi^2\}$, the configuration space metric in an obvious notation is:

$$G_{\phi_a(x)\phi_b(y)} = \delta_{ab}\delta^4(x-y) \tag{3.10}$$

In order to understand the origin of this metric, compare the kinetic term in Eq.(2.8), namely

$$1/2\int d^3x d^3y(\partial\phi_a(x)/\partial t)(\partial\phi_b(y)/\partial t)\delta_{ab}\delta^4(x-y)$$

to the analogous term in nonrelativistic mechanics:

$$m/2\ \ (dx^i/dt)(dx^j/dt)g_{ij}$$

Since the metric in Eq.(3.10) is constant with respect to the fields, its associated Christoffel symbol is zero. In this case, therefore, $\tilde{\Gamma}[\Phi]=\Gamma[\Phi]$.

After the field redefinition in Eqs(2.11), the Lagrangian is

$$\pounds = \frac{1}{2}\,\partial_\mu\rho\partial^\mu\rho + \frac{1}{2}\,\rho^2\partial_\mu\chi\partial^\mu\chi - \frac{\lambda\rho^4}{4!} \tag{3.11}$$

The configuration space metric is

$$G'_{\rho(x)\rho(y)} = \delta^4(x-y) \tag{3.12a}$$

$$G'_{\chi(x)\chi(y)} = \rho^2\delta^4(x-y) \tag{3.12b}$$

$$G'_{\chi(x)\rho(y)} = G'_{\rho(x)\chi(y)} = 0 \tag{3.12c}$$

whose determinant is

$$(\mathrm{Det}G')^{1/2} = \mathrm{Det}\left|\rho(x)\delta^4(x-y)\right|. \tag{3.13}$$

The functional measure in this parametrization is therefore precisely the one used in section 2.

The Christoffel symbol of G' is non-vanishing, although the associated curvature tensor is of course zero. The only non-zero components of the Christoffel symbol are

$$\left\{ {\rho(x) \atop \chi(y)\chi(z)} \right\} = \frac{1}{2}\int d^4x' G'^{\rho(x)\rho(x')}\left\{\frac{\delta G'_{\chi(y)\rho(x')}}{\delta\chi(z)} + \frac{\delta G'_{\chi(z)\rho(x')}}{\delta\chi(y)} - \frac{\delta G'_{\chi(y)\chi(z)}}{\delta\rho(x')}\right\}$$

$$= -\ \rho(x)\ \delta^4(x-y)\delta^4(y-z) \tag{3.14}$$

$$\left\{ {\chi(x) \atop \chi(y)\rho(z)} \right\} = \frac{1}{2\rho(x)}\,\delta^4(x-y)\delta^4(y-z), \tag{3.15}$$

where no summation is implied in the final expressions.

The modified inverse propagator is therefore

$$\tilde{D}^{-1}_{ij} = D^{-1}_{ij} - \int dx^4 \left\{ {\rho(x) \atop \phi^i\phi^j} \right\} \frac{\delta I}{\delta\rho(x)}\Bigg|_{\rho_0}$$

$$= D^{-1}_{ij} + \frac{\lambda}{3!}\int d^4x \begin{Bmatrix}\rho(x)\\ \phi^i\phi^j\end{Bmatrix} \rho^3(x)\Big|_{\rho_o} \tag{3.16}$$

In particular

$$\tilde{D}^{-1}_{\chi(x)\chi(y)} = D^{-1}_{\chi(x)\chi(y)} - \frac{\lambda\rho o^4}{6}\,\delta^4(x-y)$$

$$= -\,\rho_o{}^2\Big[\Box + \frac{\lambda\rho o^2}{6}\Big]\delta^4(x-y) \tag{3.17}$$

$$\tilde{D}^{-1}_{\rho(x)\rho(y)} = D^{-1}_{\rho(x)\rho(y)} = -(\Box + \frac{\lambda\rho o^2}{2})\;\delta^4(x-y) \tag{3.18}$$

Finally, we find the effective potential for Vilkovisky's unique effective action to be

$$\tilde{V}(\rho o) = \frac{\lambda+\delta\lambda^2}{4!}\,\rho_o{}^4 + \frac{\delta\mu^2}{2}\rho_o{}^2 - \frac{i\hbar}{2}\int\frac{d^4k}{(2\pi)^4}\ln\Big[(k^2-\frac{\lambda\rho_o{}^2}{2})(k^2-\frac{\lambda\rho_o{}^2}{6})\Big] \tag{3.19}$$

which yields precisely Eq.(2.10) after renormalization. The modified effective potential is indeed reparametrization independent.

We now turn to the application of Vilkovisky's method to scalar QED. The metric on the space of fields $\{\phi_a, A_\mu\}^*$ that one would naively extract from Eq.(2.14) is

$$G_{\phi_a(x)\phi_b(y)} = \delta_{ab}\delta^4(x-y)$$

$$G_{A_\mu(x)A_\nu(y)} = -g^{\mu\nu}\delta^4(x-y)$$

$$G_{\phi_a(x)A_\mu(y)} = G_{A_\mu(x)\phi_a(y)} = 0 \tag{3.20}$$

which again has a vanishing Christoffel symbol. The target space metric implied by the Lagrangian (2.18) after the field redefinition (2.11) is:

$$G_{\rho(x)\rho(y)} = \delta^4(x-y)$$

$$G_{\chi(x)\chi(y)} = \rho^2\delta^4(x-y)$$

$$G_{A_\mu(x)A_\nu(y)} = -g^{\mu\nu}\delta^4(x-y) \tag{3.21}$$

The only non-vanishing components of the Christoffel symbol of this metric are in the ρ and χ directions and are equal to those given in Eqs(3.14) and (3.15) above.

In the case of gauge theories, however, the choice of connection in Eq.(3.9) must be modified substantially because the target space is <u>not</u> the physical configuration space. The group of gauge transformations must be factored out[33]. In particular, the action in Eq.(2.17) is invariant under the following transformations:

$$\delta\rho(x) = 0 \tag{3.22a}$$
$$\delta\chi(x) = e\lambda(x) \tag{3.22b}$$
$$\delta A_\mu(x) = -\,\partial_\mu\lambda(x) \tag{3.22c}$$

---

*For lack of a better term we shall refer to this space of fields as the target space.

More generally, the classical action for a gauge theory is invariant under transformations of the fields:

$$\delta\phi^i = \lambda^\alpha K^i{}_\alpha \tag{3.23}$$

so that

$$\delta I = \lambda^\alpha \frac{\delta I}{\delta\phi^i} K^i{}_\alpha = 0 \tag{3.24}$$

for arbitrary λ^α. The $K^i{}_\alpha$ are the generators of the gauge transformations. In the above expressions and much of the following we again revert to the condensed notation in which the indices $i, j, k, \ldots$ and $\alpha, \beta, \gamma, \ldots$ represent both discrete labels and the spacetime coordinates. In the case of QED, we find by inspection of Eqs.(3.22a-c) that

$$K^{\rho(x)}_{\bar{x}} = 0 \tag{3.25a}$$

$$K^{\chi(x)}_{\bar{x}} = e\delta^4(x-\bar{x}) \tag{3.25b}$$

$$K^{A\mu(x)}_{\bar{x}} = -\partial_\mu\delta^4(x-\bar{x}) \tag{3.25c}$$

where $\partial_\mu\delta^4(x-y)$ will always denote partial differentiation with respect to the first argument of the δ-function.

Since the physical configuration space is the target space modulo the group of gauge transformations[33], the configuration space metric must "ignore" translations in unphysical (i.e. pure gauge) directions. Such a metric can be defined as follows[34]:

$$\gamma_{ij} = G_{ij} - K^k{}_\alpha K^l{}_\beta N^{\alpha\beta} G_{ki} G_{lj} \tag{3.26}$$

where $N^{\alpha\beta}$ is the inverse of

$$N_{\alpha\beta} = K^k{}_\alpha K^l{}_\beta G_{kl} \tag{3.27}$$

It is straightforward to show that

$$\gamma_{ij} K^i{}_\alpha = 0 \tag{3.28}$$

for all α. Thus the metric γ only measures displacements in physical directions, as required. An additional requirement that must be satisfied is that the generators $K^j{}_\alpha$ be Killing vectors of the target space metric G_{ij}. We leave it as an exercise for the reader to verify that the following Killing equation is satisfied by the metric and generators defined in Eqs(3.21) and (3.25a-c) respectively

$$D_i K_{j\alpha} + D_j K_{i\alpha} = 0 \tag{3.29}$$

In the above, D_i denotes covariance differentiation with respect to the Christoffel symbol of the target space metric.

The configuration space connection $\Gamma^i{}_{jk}$ now also follows in a straightforward manner*. The only necessary condition is that the connection be compatible with the physical metric; i.e. that it preserve the metric under parallel transport:

---

*I am grateful to Steve Carlip for pointing out this simple derivation to me.

$$\nabla_i \gamma_{jk} = 0 \tag{3.30}$$

where ∇ denotes covariant differentiation with respect to $\Gamma^i{}_{jk}$. By adding and subtracting appropriate cyclic permutations of Eq.(3.30), one finds that:

$$\gamma_{ij}\Gamma^j_{kl} = [i;kl] \tag{3.31}$$

where $[i;kl]$ denotes the Christoffel symbol of the first kind:

$$[i;kl] = \frac{1}{2}(\gamma_{ki,l} + \gamma_{li,k} - \gamma_{lk,i}) \tag{3.32}$$

Since γ_{ij} is degenerate, Eq.(3.31) does not uniquely determine $\Gamma^i{}_{jk}$. In fact if Γ^i_{jk} is a solution to Eq.(3.31) then so is $\Gamma^i_{jk} + K^i_\alpha X^\alpha_{jk}$, for arbitrary $X^\alpha{}_{jk}$. As shown by Vilkovisky [28], any term in $\Gamma^i{}_{jk}$ which is proportional to the Killing vectors does not contribute to the unique effective action. This is rather obvious to one-loop (c.f. Eq.(3.9)) if one recalls Eq.(3.24).

Multiplying both sides of Eq.(3.31) by G^{ni} and using the definition (3.26) of γ_{ij} one finds:

$$\begin{aligned}(\delta^n_j - G^{ni}K_{i\alpha}K_{j\beta}N^{\alpha\beta})\Gamma^j_{k\ell} &= G^{ni}[i;k\ell] \\ &= \frac{1}{2}G^{ni}[G_{ki,\ell} + G_{\ell i,k} - G_{k\ell,i} \\ &- (K_{k\alpha}K_{i\beta}N^{\alpha\beta})_{,\ell} - (K_{\ell\alpha}K_{i\beta}N^{\alpha\beta})_{,k} + (K_{k\alpha}K_{\ell\beta}N^{\alpha\beta})_{,i}].\end{aligned} \tag{3.33}$$

By noting the following identities**:

$$K_{i\beta,j} - K_{j\beta,i} = D_jK_{i\beta} - D_iK_{j\beta} \tag{3.34}$$

$$\begin{aligned}N^{\alpha\beta}{}_{,i} &= -N^{\alpha\gamma}N^{\beta\delta}N_{\gamma\delta,i} \\ &= -2N^{\alpha(\gamma}N^{\delta)\beta}D_{(i}K_{j)\delta}K^j{}_\gamma\end{aligned} \tag{3.35}$$

Eq.(3.33) can be reduced to:

$$\Gamma^i{}_{jk} = \left\{ {i \atop jk} \right\} + T^i{}_{jk} + K^i{}_\alpha H^\alpha{}_{jk} \tag{3.36}$$

where

$$T^i{}_{jk} = -2B^\alpha{}_{(j}D_{k)}K^i_\alpha + K^\rho_\alpha D_\rho K^i_\beta B^\alpha_{(j}B^\beta_{k)} \tag{3.37}$$

$$B^\alpha_k = N^{\alpha\beta}K_{k\beta}, \tag{3.38}$$

and

$$H^\alpha_{ij} = B^\alpha{}_k\Gamma^k_{ij} - N^{\alpha\beta}(D_{(i}K_{j)\beta} - N_{\beta\gamma,(i}B^\gamma{}_{j)}) \tag{3.39}$$

Since terms proportional to $K^i{}_\alpha$ do not contribute to the unique effective action, the last term in Eq.(3.36) may be neglected. Vilkovisky proved[28] that Eqs.(3.5,3.6,3.36-3.39) lead to a reparametrization and gauge parameter independent effective action providing that the effective action is constructed to be gauge

---

**Round brackets around indices denote symmetrization: e.g. $D_{(i}K_{j)\beta} \equiv \frac{1}{2}(D_iK_{j\beta} + D_jK_{i\beta})$.

independent by using DeWitt's background field method[35]. Here we will merely show that the technique does in fact work for the case of scalar QED. Fradkin and Tseytlin[29] have verified the gauge parameter independence of the scalar QED unique effective action using the standard parametrization. In their calculation, the target space Christoffel symbol was identically zero, but $T^i{}_{jk}$ in Eq.(3.36) was non-zero. In the following, we will give explicit details of the calculation of Vilkovisky's effective action for scalar QED using the non-standard parametrization of Eq.(2.18). In this case, both the Christoffel symbol and the gauge part of the connection contribute to yield the same answer as that of Fradkin and Tseytlin[29]. To the best of our knowledge, this is the first calculation which actually checks both the reparametrization and gauge parameter invariance of Vilkovisky's formalism.

The ordinary inverse propagator derivable from the Lagrangian in Eq.(2.18) is

$$D^{-1}_{A_\mu(x)A_\nu(y)} = (\Box g_{\mu\nu} - (1-1/\alpha)\partial_\mu\partial_\nu + e^2\rho_o^2 g_{\mu\nu})\delta^4(x-y) \tag{3.40a}$$

$$D^{-1}_{\rho(x)\rho(y)} = -(\Box + \lambda\rho_o^2/2)\delta^4(x-y) \tag{3.40b}$$

$$D^{-1}_{\chi(x)\chi(y)} = -\rho_o^2\,\Box\,\delta^4(x-y) \tag{3.40c}$$

$$D^{-1}_{\chi(x)A_\mu(y)} = -e\rho_o^2\,\partial_\mu\delta^4(x-y) = -D^{-1}_{A_\mu(x)\chi(y)} \tag{3.40d}$$

The only non-trivial contribution to $\tilde{D}^{-1}_{ij}$ from the Christoffel symbol is

$$-\left\{{\rho(z) \atop \chi(x)\chi(y)}\right\}\frac{\delta I}{\delta\rho(z)}d^4z\Big|_{\rho_o} = -\frac{\lambda\rho_o{}^4}{6}\delta^4(x-y) \tag{3.41}$$

It now remains to calculate the contribution from the gauge part of the connection T^i_{jk}. We will need the following quantities:

$$\begin{aligned}N_{\bar{x}\bar{y}} &= \int d^4x d^4y\{G_{\chi(x)\chi(y)}K^{\chi(x)}_{\bar{x}}K^{\chi(y)}_{\bar{y}} + G_{A_\mu(x)A_\nu(y)}K^{A_\nu(x)}_{\bar{x}}K^{A_\nu(y)}_{\bar{y}}\}\\ &= \int d^4x d^4y[\rho_o^2\delta^4(x-y)e^2\delta^4(x-\bar{x})\delta^4(y-\bar{y})\\ &\quad - g^{\mu\nu}\delta^4(x-y)(-\partial_\mu\delta^4(x-\bar{x}))(-\partial_\nu\delta^4(y-\bar{y}))]\\ &= (\Box + e^2\rho_o^2)\delta^4(\bar{x}-\bar{y})\end{aligned} \tag{3.42}$$

$$\begin{aligned}B^{\bar{x}}_{A_\mu(x)} &= \int d^4y d^4\bar{y}\, N^{\bar{x}\bar{y}}G_{A_\mu(x)A_\nu(y)}K^{A_\nu(y)}_{\bar{y}}\\ &= +\int d^4y d^4\bar{y} N^{\bar{x}\bar{y}}(-g^{\mu\nu})\,\delta^4(x-y)\,(-\partial_\nu\delta^4(y-\bar{y}))\\ &= g^{\mu\nu}\frac{\partial N^{x\bar{x}}}{\partial x^\nu}\end{aligned} \tag{3.43}$$

$$\begin{aligned}B^{\bar{x}}_{\chi(x)} &= \int d^4\bar{y}\, N^{\bar{x}\bar{y}}\,(\rho^2\delta^4(x-y))e\delta^4(y-\bar{y})\\ &= e\rho^2 N^{\bar{x}x}\end{aligned} \tag{3.44}$$

$$B^{\bar{x}}_{\rho(x)} = 0 \tag{3.45}$$

For the sake of brevity we will illustrate explicitly the calculation of one component of $T^i{}_{jk}$, and merely quote the results for the remainder

$$T^{\rho(z)}_{\chi(x)\chi(y)} = - B^{\alpha}_{\chi(x)}D_{\chi(y)}K^{\rho(z)}_{\alpha} - B^{\alpha}_{\chi(y)}D_{\chi(x)}K^{\rho(z)}_{\alpha} + \frac{1}{2} K^{p}_{\alpha}D_{p}K^{\rho(x)}_{\beta}(B^{\alpha}_{\chi(x)}B^{\beta}_{\chi(y)}+B^{\beta}_{\chi(x)}B^{\alpha}_{\chi(y)}) \tag{3.46}$$

But

$$K^{p}_{\bar{x}} D_{p}K^{\rho(z)}_{\bar{y}} = \int d^4y(K^{\chi(y)}_{\bar{x}} D_{\chi(y)}K^{\rho(z)}_{\bar{y}} + K^{A_\mu(y)}_{\bar{x}}D_{A_\mu(y)}K^{\rho(z)}_{\bar{y}})$$

$$= \int d^4y\left[e\delta^4(\bar{y}-\bar{x}) \int d^4\bar{z}\left[-\rho(z)\delta^4(z-\bar{z})\delta^4(\bar{z}-z)e\delta^4(\bar{z}-\bar{y})\right]\right.$$

$$= -e^2\rho_o\delta^4(z-\bar{y})\delta^4(\bar{x}-\bar{y}) \text{ (no summation over } \bar{y}) \tag{3.47}$$

and

$$D_{\chi(y)}K^{\rho(z)}_{\bar{x}} = -e\rho(z)\delta^4(z-\bar{x})\delta^4(\bar{x}-y) \quad \text{(no summation over } \bar{x}) \tag{3.48}$$

so that

$$T^{\rho(z)}_{\chi(x)\chi(y)} = e^2\rho_o^3 \{\delta^4(z-y)N^{zx} + \delta^4(z-x)N^{zy} -e^2\rho_o^2N^{xy}N^{xz}\} \tag{3.49}$$

The calculation of the remaining components of T^i_{jk} is completely analogous. The only non-zero components which contribute to $\tilde{\Gamma}$ are*

$$T^{\rho(z)}_{A_\mu(x)A_\nu(y)} = -e^4\rho_o^5\partial^\mu N^{yx}\partial^\nu N^{zx} \tag{3.50}$$

and

$$T^{\rho(z)}_{A_\mu(x)\chi(y)} = e\rho_o\left[\partial^\mu N^{xz}\delta^4(y-z) -e^2\rho_o^2 \ \partial^\mu N^{xz}N^{zy}\right] \tag{3.51}$$

Using Eqs.(3.40a-3.40d), Eq.(3.41), Eqs.(3.49-3.51) and the fact that

$$N^{\bar{x}\bar{y}} = (N_{\bar{x}\bar{y}})^{-1} = \int\frac{d^4k}{(2\pi)^4} \frac{e^{ik(\bar{x}-\bar{y})}}{(-k^2+e^2\rho_o^2)} \tag{3.52}$$

we are finally led to the following modified momentum space inverse propagators:

$$\tilde{D}^{-1}_{A_\mu A_\nu} = (-k^2+e^2\rho_o^2)g^{\mu\nu} + \frac{\left[(1-1/\alpha)-\lambda e^2\rho_o^4\right]k^\mu k^\nu}{6N^4} \tag{3.53a}$$

$$\tilde{D}^{-1}_{A_\mu\chi} = ik^\mu e\rho_o^2 \left[1+ \frac{\lambda\rho_o^2}{6N^2} - \frac{\lambda e^2\rho_o^4}{6N^4}\right] \tag{3.53b}$$

$$\tilde{D}^{-1}_{\chi\chi} = \rho_o^2 \left[k^2 - \frac{\lambda\rho^2{}_o}{6} + \frac{\lambda e^2\rho_o^4}{3N^2} - \frac{\lambda e^4\rho_o^6}{6 N^4}\right] \tag{3.53c}$$

$$\tilde{D}^{-1}_{\rho\rho} = k^2 - \frac{\lambda\rho^2{}_o}{2}, \tag{3.53d}$$

---

*In the following, differentiation is always with respect to the first argument: e.g. $\partial^\nu N^{zx} \equiv g^{\nu\rho} \frac{\partial}{\partial z^\rho}N^{zx}$

where

$$N^2 \equiv (-k^2 + e^2\rho_o^2) \tag{3.54}$$

After Wick rotation, the one-loop effective potential is:

$$\tilde{V}_{eff}(\rho o;\lambda,e^2) = \frac{\lambda\rho_o^4}{4!} - \frac{\hbar}{2}\int\frac{d^4k}{(2\pi)^4}\ell n\rho_o - \hbar\int\frac{d^4k}{(2\pi)^4}\ell n(k^2)$$
$$+ \frac{\hbar}{2}\int\frac{d^4k}{(2\pi)^4}\ell n\left\{\frac{\rho_o k^4}{\alpha}(k^2 + \frac{\lambda\rho_o^2}{2})(k^2 + e^2\rho_o{}^2)(k^2 + m_+^2)(k^2 + m_-^2)\right\} \tag{3.55}$$

where $m_\pm^2$ are the roots to the quadratic (in k^2) equation:

$$(k^2)^2 + (2e^2\rho_o{}^2 + \frac{\lambda\rho_o{}^2}{6})\ k^2 + e^4\rho_o{}^4 = 0 \tag{3.56}$$

In Eq.(3.55) the second term arises from the functional measure, and is again necessary to cancel a quartic divergence in $\ell n\det\mathcal{D}_{ij}^{-1}$. As expected, the gauge parameter dependence is indeed trivial. The final expression for the renormalized unique effective potential is:

$$\tilde{V}_{eff} = \frac{\lambda\rho_o{}^4}{4!} + \frac{\hbar}{64\pi^2}(3e^4 + \frac{5\lambda^2}{18} + \frac{2}{3}\lambda e^2)\left\{\ell n(\frac{\rho_o{}^2}{M^2}) - \frac{25}{6}\right\}\rho_o^4 \tag{3.57}$$

This is precisely the same answer as obtained by Fradkin and Tseytlin[29]. Note that it is not equal to the ordinary effective potential calculated in terms of either parametrization in section II.

IV. CONCLUSIONS

We have seen that Vilkovisky's formalism does indeed solve the problem of reparametrization dependence of the off-shell effective action, at least to one loop in the case of complex ϕ^4 theory and scalar QED. In these examples, however, it was already known that parametrization dependence does not result in any ambiguity of physical observables providing that the loop expansion is handled correctly.

The most interesting example of parametrization dependence occurs in the case of self-consistent dimensional reduction. In this case it is not clear that the truncated loop expansion provides a reliable approximation to the full effective action[12]. Vilkovisky's formalism should in principle provide a unique prediction for the radius of the internal dimensions and for the effective fine structure constant. This calculation, which is currently in progress[36], is interesting for two reasons. First it will provide an explicit check of Vilkovisky's formalism in the case of quantum gravity. Secondly it should reveal whether the gauge parameter and reparametrization dependence of earlier results were due to the fact that the extrema were gauge artifacts, or simply a consequence of the fact that the incorrect (i.e. ordinary) definition of the effective action was used. In either case, it is clear that Vilkovisky's formalism is an essential tool for probing the vacuum structure of quantum gravity.

ACKNOWLEDGEMENTS

I am grateful to Steve Carlip, Chris Hull, Roman Jackiw, Peter Leivo and David Toms for many helpful discussions. This work was supported by the Natural Sciences and Engineering Research Council of Canada.

REFERENCES

1. S. Coleman & E. Weinberg, Phys. Rev. D7(1973)1888.
2. S. Weinberg, Phys. Letts. B126(1983)265; P. Candelas and S. Weinberg, Nucl. Phys. B237(1984)397.
3. A. Chodos & E. Myers, Ann. Phys. (N.Y.) 156(1984)412.
4. M.H. Sarmadi, Nucl. Phys. B263(1986)187.
5. C.R. Ordonez & M.A. Rubin, Nucl. Phys. B260(1985)465.
6. S.R. Huggins & D.J. Toms, Phys. Lett. 153B(1985)247; Nucl. Phys. B263(1986)433.
7. D.J. Toms, Proc. 1985 CAP Summer Workshop on Quantum Field Theory, Can. J. Phys. 64(1986)644.
8. E.J. Copeland & D.J. Toms, Nucl. Phys. B255(1985)201.
9. G. Kunstatter & H.P. Leivo, Phys. Lett. 166B(1986)321; Nucl. Phys. B (in press).
10. S. Randjbar-Daemi & M.H. Sarmadi, Phys. Lett. 151B(1985)343.
11. G. Kunstatter & H.P. Leivo, "On the Reparametrization Dependence of the Effective Potential", Univ. Winnipeg preprint, June (1986).
12. G. Kunstatter & D.J. Toms, in Proc. 1985 CAP Summer Workshop on Quantum Field Theory", Can. J. Phys. 64(1986)641.
13. D. Friedan, Phys. Rev. Lett. 45(1980)1057.
14. C. Hull, Proc. NATO Advanced Workshop on Superfield Theories, (this volume) and references therein.
15. N.K. Nielsen, Nucl. Phys. B101(1975)173.
16. I.J.R. Aitchison & C.M. Fraser, Ann. Phys. 156(1984)1.
17. R. Fukuda & T. Kugo, Phys. Rev. D13(1976)3469.
18. P. Mansfield, Nucl. Phys. B267(1986)575.
19. E. Cohler & A. Chodos, Phys. Rev. D30(1984)492.
20. O. Yasuda, Phys. Lett. 137B(1984)52.
21. S. Ichinose, Phys. Lett. 152B(1986)56.
22. T. Inami & O. Yasuda, Phys. Lett. 133B(1983)180.
23. G. Kunstatter, H.C. Lee & H.P. Leivo, Phys. Rev. D33(1986)1018.
24. I. Antoniadis, J. Iliopoulos & T.N. Tomaras, Nucl. Phys. B267(1986)496
25. I. Antoniadis & E.T. Tomboulis, Phys. Rev. D33(1986)2756.
26. L. Dolan & R. Jackiw, Phys. Rev. D9(1974)2904.
27. R. Jackiw, Phys. Rev. D9(1974)1686.
28. G. Vilkovisky, "The Gospel According to DeWitt" in Quantum Theory of Gravity, ed. S.M. Christensen (Adam Hilger, Bristol, 1983); Nucl. Phys. B234(1984)125.
29. E.S. Fradkin & A.A. Tseytlin, Nucl. Phys. B234(1984)274; (1984)509.
30. D.J. Toms, "The Functional Measure for Quantum Field Theory in Curved Spacetime", Univ. of Newcastle preprint (1986); Proc. NATO Advanced Research Workshop on Superfield Theories (this volume).
31. H.P. Leivo, Proc. NATO Advanced Research Workshop on Superfield Theories (this volume).
32. B.S. DeWitt, Relativity Groups & Topology, (Gordon Breach, N.Y. 1963); Dynamical Theory of Group & Fields, (Gordon Breach, N.Y. 1965).
33. M. Daniel, C.M. Viallet, Rev. Mod. Phys. 52(1980)175.
34. B.S. DeWitt, Phys. Rev. 160(1967)1113.
35. B.S. DeWitt, in Quantum Gravity II, eds. C.J. Isham, R. Penrose & D. Sciama (Oxford University Press, 1981).
36. S.R. Huggins, G. Kunstatter, H.P. Leivo & D.J. Toms (in preparation).

GAUGE AND PARAMETRIZATION DEPENDENCE
IN KALUZA-KLEIN THEORY

H.P. Leivo

Department of Physics
University of Toronto
Toronto, Ontario, Canada

INTRODUCTION

A desideratum in physics is to exercise control over the approximations one is forced to make; in particular, when these are not founded upon physical arguments, one must take care that they nevertheless reflect the properties of the underlying theory as faithfully as possible. For this to be the case, one must surely demand some measure of unanimity in the predictions following from application of a given approximation in various guises. For example, the ubiquitous loop expansion is an unphysical one, based not on the smallness of coupling constants, but rather on topological features of diagrams or equivalently on the introduction of a unit counting parameter into the generating functional of Green functions. The diagrammatic viewpoint reveals that the loop expansion deals with patterns of propagation of fluctuations on some background; indeed the one-loop approximation to the effective action, in particular, is determined by the operator appearing in the term quadratic in fluctuations in the expansion of the tree-level action. The question which naturally arises is what effect the identification of the fluctuations, or more generally the choice of parametrization of the fields involved, has on the loop expansion.

In fact, in evaluating effective actions one finds[1] that the one-loop contributions calculated with different parametrizations are in general different, at least off shell, i.e., when the background does not satisfy the classical equations of motion.* The off-shell effective potential for gauge-theories is also known in general to depend on the choice of gauge fixing[3], a fact which is related to the parametrization dependence[4,5].

An important application of the effective potential is to the production of dynamical symmetry breaking by radiative corrections, as in the familiar Coleman-Weinberg[6] approach. There one is interested not in perturbative corrections to properties calculated at a classical solution, but in seeking nonperturbative solutions far from the classical one. It is in this context that the problems of gauge and parametrization dependence arise.

* See also the article by G. Kunstatter in this volume[2] and references therein.

The off-shell gauge-choice and parametrization dependence have recently been addressed by Vilkovisky, who has proposed a new definition of the effective action[4,5]. It differs from the standard one in that the coupling of the fluctuations to the source is non-linear, and hence explicitly mixes orders in the loop expansion, which it thereby redefines in just such a way as to yield a unique expression to a given order, independent of the choice of parametrization or of gauge. There is more to it, of course; in fact the procedure rests on an elegant geometrization of the configuration space, rather than the details of loop expansions.

Vilkovisky's definition is discussed in detail in the contribution of G. Kunstatter to this volume[2]. As illustrated there, at times, as in Φ^4 theory, the parametrization dependence of the effective potential presents no problems, because there are no nonperturbative solutions to the effective equations of motion. Commonly, as in the case of massless scalar QED[3,6,7], a reinterpretation of the approximation allows one to push the parametrization and gauge dependence to higher orders. In either case, application of the new formalism cleans things up, but does not alter the physical predictions.

Since Vilkovisky's approach imports tools and concepts familiar to the relativist into the quantum field theory arsenal, it is fitting that a case in which the formalism may truly be essential is gravity itself. The effective potential for higher-dimensional gravity on a compactified Kaluza-Klein background[8–11] has been used to predict the radius of internal spherical dimensions in the presence of a flat four-dimensional space-time. The off-shell gauge-parameter and parametrization dependence of the effective potential in this approach, known as self-consistent dimensional reduction[9–16], has been shown[1,15] to intrude into the physical predictions, such as they are: the radius and associated effective gauge-coupling and fine-structure constants.

This talk will be devoted to pointing out the nature of the problems in self-consistent dimensional reduction; whether the resolution is truly to be found in the formalism proposed by Vilkovisky is presently under investigation[17].

GAUGE-PARAMETER DEPENDENCE OF SELF-CONSISTENT DIMENSIONAL REDUCTION

Let us consider the simplest example of self-consistent dimensional reduction, which starts with the action for pure gravity with cosmological constant in 5 dimensions*,

$$I = \frac{1}{16\pi\bar{G}} \int d^5x \sqrt{-\bar{g}}(\bar{R} + 2\lambda). \tag{1}$$

With the assumption that the topology is fixed to be $R^4 \times S^1$, it suffices to evaluate the effective action for a background metric parametrized as

$$\bar{g}^{\circ}_{AB} = \begin{pmatrix} g^{\circ}_{\mu\nu} + \kappa^2\phi_o A^{\circ}_{\mu}A^{\circ}_{\nu} & \kappa\phi_o A^{\circ}_{\mu} \\ \kappa\phi_o A^{\circ}_{\nu} & \phi_o \end{pmatrix} \tag{2}$$

* Our conventions are the following: the metric has signature -++++; Latin indices $A,B = 0,1,2,3,5$, Greek indices $\mu,\nu = 0,1,2,3$; a bar denotes higher-dimensional quantities; the final coordinate $x^5 = y$ is periodic, $0 \le y \le 2\pi R$ and the radius of the compact dimension thus $r = R\sqrt{\phi_o}$; $\hbar = c = 1$.

with ϕ_o constant and the other fields restricted to be y-independent. The one-loop contribution to the effective action is given by the functional formula

$$\Gamma^{(1)}[\bar{g}^o_{AB}] = \frac{1}{2}\ln \mathrm{Det}\, D^{-1}[\bar{g}^o_{AB}], \tag{3}$$

where $D^{-1}[\bar{g}^o_{AB}]$ is the wave operator or inverse propagator for the part of the action (1) quadratic in the fluctuations

$$\bar{h}_{AB} = \begin{pmatrix} h_{\mu\nu} + s\psi\eta_{\mu\nu} & \sqrt{\phi_o}a_\mu \\ \sqrt{\phi_o}a_\nu & (s+1)\phi_o\psi \end{pmatrix} \tag{4}$$

of $\bar{g}_{AB}$ about the background $\bar{g}^o_{AB}$.* To first order in the curvature, one obtains[12–15] a result of the form

$$\Gamma^{(1)} = \int d^4x\sqrt{-g^o}\left[-\frac{A(z)}{r^4} + \frac{B(z)}{r^2}R(g^o) + \frac{C(z)}{r^2}\kappa^2\phi_o F^o_{\mu\nu}F^{o\,\mu\nu}\right], \tag{5}$$

where A(z), B(z) and C(z) are functions of the dimensionless parameter $z = \sqrt{2\lambda r^2}$.

The above approximation to the effective action can be written in the form

$$\Gamma = I + \Gamma^{(1)} = -\int d^4x\sqrt{-g^o}V_{\text{eff}}(r;\lambda) + \frac{1}{16\pi G}\int d^4x\sqrt{-g^o}R(g^o)$$
$$-\frac{1}{4}\int d^4x\sqrt{-g^o}F^o_{\mu\nu}F^{o\,\mu\nu}, \tag{6}$$

providing the following identifications are made:

$$V_{\text{eff}}(r;\lambda) = -\frac{2\pi r}{16\pi\bar{G}}2\lambda + \frac{A(z)}{r^4}, \tag{7}$$

$$\frac{1}{16\pi G} = \frac{2\pi r}{16\pi\bar{G}} + \frac{B(z)}{r^2}, \tag{8}$$

and

$$\kappa^{-2} = \left(\frac{2\pi r}{16\pi\bar{G}} - \frac{4C(z)}{r^2}\right)\phi_o. \tag{9}$$

The first terms in these expressions are the purely geometrical ones arising from the standard Kaluza-Klein reduction of the action (1). The desired flat vacuum solutions $M^4 \times S^1$ to the dynamics determined by

$$\frac{\delta\Gamma}{\delta\bar{g}^o_{AB}} = 0 \tag{10}$$

evaluated at $g^o_{\mu\nu} = \eta_{\mu\nu}$ and $A^o_\mu = 0$ are given by solving for r_o and λ_o the equations

* The peculiar parametrization with a Weyl parameter s is explained in the next section.

$$\left.\frac{\partial V_{\text{eff}}}{\partial r}\right|_{r_o;\lambda_o} = 0 \tag{11}$$

and

$$V_{\text{eff}}(r_o;\lambda_o) = 0. \tag{12}$$

The latter requires the effective four-dimensional cosmological constant to vanish in order to admit the flat space-time portion M^4; this is the meaning of self-consistency. These equations combine to yield a single equation to locate a solution z_o:

$$z_o\frac{dA(z_o)}{dz} - 5A(z_o) = 0. \tag{13}$$

Equation (11) then yields r_o and hence the value of λ required; it reads

$$r_o^3 = 8\overline{G}\left[\frac{A(z_o)}{z_o^2}\right] \tag{14}$$

and gives

$$\lambda_o = \frac{z_o^2}{2r_o^2} = \frac{1}{8}\left[\frac{\overline{G}A(z_o)}{z_o^5}\right]^{-\frac{2}{3}}. \tag{15}$$

One notes that only the effective potential is required in order to locate self-consistent solutions. The induced Einstein-Hilbert and Maxwell terms B(z) and C(z) do contribute to the effective one-loop quantities predicted: the gravitational constant (8) and the fine structure constant and electric charge given by

$$4\pi\alpha_f = e^2 = \phi_o\frac{\kappa^2}{r_o^2}. \tag{16}$$

To calculate A(z), one adds to the full action the background gauge fixing term

$$I_{GF} = \frac{-1}{16\pi\overline{G}}\int d^5x\sqrt{-\overline{g}^o}\frac{1}{2\alpha}G_A G_B \overline{g}^{o\,AB}, \tag{17}$$

with

$$G_A = \overline{g}^{o\,CD}(\nabla^o_C\overline{g}_{AD} - \frac{1}{2}\nabla^o_A\overline{g}_{CD}), \tag{18}$$

and takes into account the associated Faddeev-Popov ghost contribution, obtaining eventually

$$\frac{A(z;\alpha)}{r^4} = V^{(1)}(r;\lambda,\alpha) + V_{FP}(r), \tag{19}$$

where

$$V^{(1)} = \frac{1}{2}\sum_{n=-\infty}^{\infty}\int\frac{d^4k}{(2\pi)^4}\ln\det\tilde{D}_{ij}^{(n)\,-1}(k). \tag{20}$$

Evaluation of the determinant yields the simple form

$$\det \tilde{D}_{ij}^{(n)\,-1}(k) = \text{const.} \times (s+1)^2\left(\frac{k^2}{\alpha} + \frac{m_n^2}{\alpha} + 2\lambda\right)^5 (k^2 + m_n^2 + 2\lambda)^{10} \tag{21}$$

where $m_n = n/r$. Except for the special case $s = -1$, which corresponds to a degenerate parametrization (cf. (23)), the dependence on the Weyl parameter in (21) may be absorbed into the irrelevant non-vanishing constant.

Putting everything together, one finds[15], finally, that not only does the location z_o, λ_o of a self-consistent solution depend on the choice of gauge-parameter, but so do the ratio of the radius to the 5-dimensional Planck length $\bar{L}_p = \bar{G}^{1/3}$ and also the physical predictions based solely on the effective potential, i.e., neglecting B(z) and C(z). The prospect is under investigation that the α-dependence of the induced one-loop effects might be precisely such as to compensate for that of the effective potential, so as to render gauge-parameter independent the full one-loop values of e^2 and the ratio of the radius to the effective four-dimensional Planck length $L_p = G^{1/2}$. As this would be rather curious, with the ratio of four- and five-dimensional Planck lengths itself α-dependent, one may provisionally conclude that self-consistent dimensional reduction, at least when based on the standard effective action, does present a problem of gauge-dependence in physical quantities.

PARAMETRIZATION DEPENDENCE OF SELF-CONSISTENT DIMENSIONAL REDUCTION

Self-consistent dimensional reduction also exhibits parametrization dependence, the other, more general symptom which Vilkovisky's effective action seeks to cure. In the previous section the higher dimensional metric was parametrized as such in determining the fluctuations. In the original calculation[8] of the Casimir energy of the fifth dimension in Kaluza-Klein theory without a cosmological constant, the fluctuations of the metric were identified somewhat differently. Rather than writing

$$\bar{g}_{AB} = \bar{g}^{\circ}_{AB} + \bar{h}_{AB} \tag{22}$$

and considering the $\bar{h}_{AB}$ (or linear combinations, as in (4)), one can first apply the Kaluza-Klein Ansatz with a so-called Weyl factor to give

$$\bar{g}_{AB} = \left(\frac{\phi}{\phi_o}\right)^s \begin{pmatrix} g_{\mu\nu} + \phi A_\mu A_\nu & \phi A_\mu \\ \phi A_\nu & \phi \end{pmatrix} \tag{23}$$

and then consider the fluctuations about $M^4 \times S^1$ defined by

$$g_{\mu\nu} = \eta_{\mu\nu} + h'_{\mu\nu}$$

$$A_\mu = 0 + \frac{1}{\sqrt{\phi_o}} a'_\mu \tag{24}$$

$$\phi = \phi_o(1 + \psi')$$

These primed fluctuations are related non-linearly to the unprimed ones appearing in (4), which were chosen so as to give

$$h'_{\mu\nu} = h_{\mu\nu} + \text{higher order terms in } (h,\ a,\ \psi) \quad ,\ \text{etc.} \tag{25}$$

The Weyl factor, with parameter s, provides a continuous family of parametrizations. It is only in the presence of a cosmological constant, however, that the choice of parametrization affects the one-loop effective potential. Then the flat background is off shell and the terms linear in fluctuations consequently appearing in the expansion of the classical action produce new quadratic terms when the substitutions (25) are made.

Restricting attention now to a fixed choice of gauge-parameter, $\alpha = 1$, one finds[1], instead of the determinant (21),

$$\det \tilde{D}_{ij}^{(n)\,-1}(k) = \text{const.} \times (s+1)^2 (k^2 + m_n^2 + 2\lambda)^9 (k^2 + m_n^2)^4$$
$$\times (k^2 + m_n^2 + 2\lambda_1)(k^2 + m_n^2 + 2\lambda_2) \tag{26}$$

where

$$\lambda_1 = \frac{10s + 1}{s + 1}\lambda, \tag{27}$$

$$\lambda_2 = \frac{-20s + 6}{6(s + 1)}\lambda. \tag{28}$$

The s-dependence is no longer trivial. In fact, for certain ranges of s, one or both of λ_1 and λ_2 becomes negative, yielding an imaginary contribution to the effective potential. Even when the effective potential is real, the s-dependence is also present in the physical quantities predicted, just as is the case with gauge-parameter dependence.

CONCLUSIONS

It is important to understand the nature of the dependence upon unphysical parameters of supposedly physical quantities predicted in one-loop self-consistent dimensional reduction, as exhibited for the simplest model with a single compact dimension. At present there are more questions than answers. Are such self-consistent solutions merely gauge- or parametrization artefacts, or do they point to a need for revision of the apparatus used to derive them, such as self-consistent dimensional reduction itself or the definition of the effective action upon which it rests? Even if Vilkovisky's formalism yields unique one-loop predictions, are they the physically correct ones? After all, for other, simpler theories the unique effective action does not alter the physical predictions, but merely in a sense calibrates the machinery of effective actions to the loop expansion. If the loop expansion itself is inappropriate for self-consistent dimensional reduction[16], the new effective action may be irrelevant in this context. Calculations are underway which aim to shed some light on these issues[17].

ACKNOWLEDGMENTS

I would like to thank Gabor Kunstatter and David Toms for helpful discussions. This work was supported by the Natural Sciences and Engineering Research Council of Canada.

REFERENCES

1. G. Kunstatter and H.P. Leivo, preprint (June 1986).
2. G. Kunstatter, Vilkovisky's Unique Effective Action, in: Proceedings of the NATO Advanced Research Workshop on Super Field Theories, H. C. Lee, ed., Plenum, New York (1986) (this volume).
3. R. Jackiw, Phys. Rev. D9:1686 (1974).
4. G. A. Vilkovisky, The Gospel According to DeWitt, in: "Quantum Theory of Gravity", S. M. Christensen, ed., Adam Hilger, Bristol (1983); Nucl. Phys. B234:125 (1984).
5. E. S. Fradkin and A.A. Tseytlin, Nucl. Phys. B234:509 (1984).
6. S. Coleman and E. Weinberg, Phys. Rev. D7:1888 (1973).
7. N. K. Nielsen, Nucl. Phys. B101:173 (1975); I.J.R. Aitchison and C.M. Fraser, Ann. Phys. 156:1 (1984).
8. T. Appelquist and A. Chodos, Phys. Rev. Lett. 50:141 (1983); Phys. Rev. D28:772 (1983).
9. P. Candelas and S. Weinberg, Nucl. Phys. B237:397 (1984).
10. A. Chodos and E. Myers, Ann. Phys. 156:412 (1984); Can. J. Phys. 64:633 (1986).
11. M. H. Sarmadi, Nucl. Phys. B263:187 (1986); S. Randjbar-Daemi and M. H. Sarmadi, Phys. Lett. 151B:343 (1985).
12. S. R. Huggins and D.J. Toms, Phys. Lett. 153B:247 (1985); Nucl. Phys. B263:433 (1986).
13. D. J. Toms, Can. J. Phys. 64:644 (1986).
14. E. J. Copeland and D. J. Toms, Nucl. Phys. B255:201 (1985).
15. G. Kunstatter and H.P. Leivo, Phys. Lett. 166B:321 (1986);Nucl. Phys.B (in press).
16. G. Kunstatter and D. J. Toms, Can. J. Phys. 64:641 (1986).
17. S. R. Huggins, G. Kunstatter, H. P. Leivo and D. J. Toms (in preparation).

NEW ALGEBRAIC CANONICAL STRUCTURES OF INTEGRABILITY IN 2-D FIELD THEORIES

Jean Michel Maillet

Laboratoire de Physique Théorique et Hautes Energies
Paris, France
Laboratoire associé au CNRS UA 280, Université
Pierre et Marie Curie, Tour 16, 1er étage, 4 place Jussieu
75252 Paris Cedex 05, France

INTRODUCTION

Considerable progress has been made in the last decade in the elaboration of non perturbative methods allowing the exact resolution of a large class of two dimensional models in field theory and in statistical mechanics both at classical and quantum levels. Underlying these exiting developments of the notion of complete integrability in systems with an infinite number of degrees of freedom, the concept of new algebraic structures (The Yang-Baxter algebras and equations) has emerged[1]. In fact, these algebras usually provide us with the complete canonical structure of such models of field theory, leading in particular to an infinite set of conserved charges in involution, as a signature of the complete integrability.

In this lecture we will be concerned with the notion of integrability in classical two-dimensional field theories. Hence, we will be interested mainly in the Hamiltonian formalism for such models and in the canonical algebraic structures responsible for their complete integrability that can be determined and used in this approach.

At the classical level, the starting point for the integrable structure of a two dimensionnal field theory is the existence of a matrix linear differential system (Lax pair)[1]

$$\partial_x \Psi(\underline{x},\lambda) = L(\underline{x},\lambda)\,\Psi(\underline{x},\lambda) \;;\; \partial_t \Psi(\underline{x},\lambda) = M(\underline{x},\lambda)\,\Psi(\underline{x},\lambda) \qquad (1)$$

where L and M are two traceless NxN matrices depending locally on the fields $\varphi(\underline{x})$ and their derivatives and on a complex spectral parameter λ such that the equations of motion of the fields are represented by the compatibility conditions of (1) (identically verified in λ) :

$$\partial_t L - \partial_x M + [L, M] = 0 \qquad (2)$$

It must be noted that the existence of a Lax pair (1) is an highly non trivial fact and has crucial consequences for the theory considered. In particular, it is possible to construct an infinite set of conserved quantities for it from particular matrix solutions $T(x,y,\lambda)$ of the differential linear system (1) verifying (the t dependence being understood) :

$$\partial_x T(x,y,\lambda) = L(x,\lambda)\,T(x,y,\lambda) \qquad (3)$$

$$T(y,x,\lambda) = T(x,y,\lambda)^{-1}, \quad T(x,x,\lambda) = \mathbb{1}$$

and $$\partial_t T(x,y,\lambda) = M(x,\lambda)\,T(x,y,\lambda) - T(x,y,\lambda)\,M(y,\lambda)$$

In fact, let us, for example, put the theory in a box $[-L, +L]$ with periodic boundary conditions, then we have :

$$\partial_t\{tr(T(L,-L,\lambda)^n)\} = 0 \tag{4}$$

leading, through a power series expansion in λ, to an infinite set of conserved charges. Moreover, the monodromy matrix $T(\lambda)$, obtained as an infinite volume limit of $T(x,y,\lambda)$ ($x \to +\infty$, $y \to -\infty$)with proper regularisation, provide us, when a mass scale is present in the theory, not only with an infinite series of conserved charges, but also, through its off-diagonal elements, with variables varying linearly with time. In fact, the classical inverse scattering method 1 applied to such a theory allows us to define in this case a canonical transformation from the original fields and momentum of the theory to action-angles variables constructed from the monodromy matrix $T(\lambda)$, hence leading to the complete integrability of the theory [1] .

In addition, exact classical solutions of the equations of motion (2) (soliton solutions) can be obtained using this method [1] .

Hence, the monodromy matrix $T(\lambda)$ appears to be the fundamental object to study in such theories, and in order to extract the canonical integrable structure of them, we will be interested at first, in the algebra under Poisson brackets of two monodromy matrices defined by (3), from which the algebra of matrix elements of $T(\lambda)$, $T(\mu)$ can be obtained [1] .

THE ALGEBRA OF MONODROMY MATRICES

The algebra of monodromy matrices can be derived from the definition(3) and the knowledge of the Lax matrices algebra. For (x,x',y,y') all differents we have [1,2]

$$\{T(x,y,\lambda) \overset{\otimes}{,} T(x',y',\mu)\} = \int_y^x dz \int_{y'}^{x'} dz' \; T(x,z,\lambda) \otimes T(x',z',\mu) \,\cdot \tag{5}$$
$$\cdot\{L(z,\lambda) \overset{\otimes}{,} L(z',\mu)\} \,.\, T(z,y,\lambda) \otimes T(z',y',\mu)$$

where we have used tensor product notations :

$$(A \otimes B)_{ij,k\ell} = A_{ik} B_{j\ell} \quad ; \quad \{A \overset{\otimes}{,} B\}_{ij,k\ell} = \{A_{ik}, B_{j\ell}\}$$

Therefore, the Poisson bracket algebra (5) is determined from the algebra of L matrices. For the class of ultralocal type models, for which these Poisson brackets contain only the distribution $\delta(z-z')$, it writes [1] :

$$\{L(z,\lambda) \overset{\otimes}{,} L(z',\mu)\} = \delta(z-z')\left[r(\lambda,\mu) \,,\, L(z,\lambda)\otimes 1\!\!1 + 1\!\!1 \otimes L(z,\mu)\right] \tag{6}$$

where $r(\lambda,\mu)$ is a numerical matrix depending on the spectral parameters and and acting in the tensor product of spaces 1 and 2. Using now the remarquable linear algebraic structure (6) of ultralocal type models, it is easy to evaluate explicitly and in a close way the algebra (5), valid in this case also for the case $x = x'$ and $y = y'$:

$$\{T(x,y,\lambda) \overset{\otimes}{,} T(x,y,\mu)\} = [r(\lambda,\mu), T(x,y,\lambda)\otimes T(x,y,\mu)] \tag{7}$$

and hence :

$$\{tr(T(x,y,\lambda)^m) \,,\, tr(T(x,y,\mu)^n)\} = 0 \tag{8}$$

In order to be consistent, the algebras (6) and (7) have to verify the antisymmetry of Poisson brackets and Jacobi identity, leading to the following constraints on the r-matrix :

$$P\, r(\lambda,\mu)\, P = -r(\mu,\lambda) \tag{9}$$

where $P_{ij,k\ell} = \delta_{i\ell}\delta_{jk}$, and :

$$[r_{23}(\mu,\eta), r_{12}(\lambda,\mu)] + [r_{23}(\mu,\eta), r_{13}(\lambda,\eta)] + \tag{10}$$
$$+ [r_{13}(\lambda,\eta), r_{12}(\lambda,\mu)] = 0$$

with the notations $r_{12}(\lambda,\mu) = r(\lambda,\mu)\otimes 1\!\!1_3$

This quadratic equation is the classical Yang-Baxter equation for r-matrices. It has to be noted that the structure given by eqs(6,9,10) is completely similar to those of usual Lie algebras, $r(\lambda,\rho)$ playing the role of matrix structure constants of it.

Using now the remarkable quadratic Yang-Baxter algebra (9) and its infinite volume limit, one can extract the complete integrable canonical structure of such a theory [1] .

Moreover, a quantum R-matrix formalism has been developed for these ultra local type models and applied with great success to the exact resolution (exact mass spectrum and S matrix) of many two dimensional models of this class like Non linear Schrödinger, Sine-Gordon or Toda theories (see [1] for a review).

However this scheme can be applied only to models having the ultralocality property, ie, when in the r.h.s of (6), only the $\delta(z-z')$ distribution is present in the algebra of L matrices. Unfortunately, some very interesting models like the O(N) non linear σ-models [3] , the principal Chiral fields [4] or the Complex Sine-Gordon theory [4,5] do not share this property and are called non ultralocal, due to the presence of first derivatives of $\delta(z-z')$ in the Lax matrices Poisson brackets.

A more general method has been constructed recently [2] which take into account these more general class of integrable models, at the classical canonical level as a first step. It leads to new Yang-Baxter type algebraic structures which contain as a particular case the usual one.

THE EXTENDED YANG-BAXTER TYPE ALGEBRAIC STRUCTURES [2]

Let us now consider a large class of two dimension field theories having an associated Lax pair for which the Poisson bracket algebra writes:

$$\begin{aligned} \{ L(z,\lambda) \overset{\otimes}{,} L(z',\mu) \} = {} & A(z,\lambda,\mu)\,\delta(z-z') + \\ & + B(z,z',\lambda,\mu)\,\partial_{z'}\delta(z-z') + C(z,z',\lambda,\mu)\,\partial_z\delta(z-z') \end{aligned} \tag{11}$$

This class includes, as particular examples, the O(N) non linear σ models, the principal Chiral fields and the Complex Sine Gordon models. Then, it is possible to show that the canonical algebraic structure (11) can be recasted in the following linear form [2] .

$$\begin{aligned} \{ L(z,\lambda) \overset{\otimes}{,} L(z',\mu) \} = {} & \delta(z-z')\Big(\big[r(z,\lambda,\mu), L(z,\lambda)\otimes \mathbb{1} + \mathbb{1}\otimes L(z,\mu) \big] \\ & + \big[s(z,\lambda,\mu), \mathbb{1}\otimes L(z,\mu) - L(z,\lambda)\otimes \mathbb{1} \big] \Big) \\ & - \delta'(z-z')\Big(r(z,\lambda,\mu) + s(z,\lambda,\mu) - r(z',\lambda,\mu) + s(z',\lambda,\mu) \Big) \end{aligned} \tag{12}$$

where we have defined two matrices r and s depending on the spectral parameters λ and μ ,and possibly on the fields of the theory,which can be computed in terms of A, B and C. The remarkably linear form (12) allows us to evaluate explicitly and in a compact way the monodromy matrix algebra given by (5) [2] . The antisymmetry of these canonical bracket algebras holds through the relations :

$$\begin{aligned} \mathcal{P}\, r(x,\lambda,\mu)\,\mathcal{P} &= -\, r(x,\mu,\lambda) \\ \mathcal{P}\, s(x,\lambda,\mu)\,\mathcal{P} &= s(x,\mu,\lambda) \end{aligned} \tag{13}$$

Therefore, it follows from these two different behavior that the algebra(12) involves two type of matrix structure constants, the r and s matrices, in contrast with the ultralocal case (B=C=0) where s vanishes identically. In addition these two matrices (r and s) could now depend on the fields of the theory and hence be dynamical quantities. Further, as a consistency condition, we have to require the Jacobi identity to hold. It can be shown that it results in a new extended Yang-Baxter type equation for the r and s matrices [2] :

$$\begin{aligned} &\left[(r+s)_{13}(x,\lambda,\eta) \,,\, (r-s)_{12}(x,\lambda,\mu) \right] + \left[(r+s)_{23}(x,\mu,\eta) \,,\right. \\ &\left.(r+s)_{12}(x,\lambda,\mu) \right] + \left[(r+s)_{23}(x,\mu,\eta) \,,\, (r+s)_{13}(x,\lambda,\eta) \right] + \\ &+ H_{1,23}(x,\lambda,\mu,\eta) - H_{2,13}(x,\mu,\lambda,\eta) = 0 \end{aligned} \tag{14}$$

where we have taken into account the possible dynamical nature of r and s by the H term defined by :

$$\left\{ L_1(x,\lambda) \overset{\otimes}{,} (r+s)_{23}(y,\mu,\eta) \right\} = \delta(x-y)\, H_{1,23}(x,\lambda,\mu,\eta) \tag{15}$$

These new extended Yang-Baxter type algebraic structures (12-15) have been shown to hold for the various non ultralocal type models cited before [2,6]. It must be noted that these new algebras reduce to the usual ultralocal one's (6-10) when s =0 and $\partial_x r = 0$, as it should be.

As a result we obtain the monodromy matrix algebras in the equal points limits (weak algebras, see [2] for a detailed discussion) ($x > y > z$). For equal intervals we have :

$$\begin{aligned} \left\{ T(x,y,\lambda) \overset{\otimes}{,} T(x,y,\mu) \right\} &= r(x,\lambda,\mu) . T(x,y,\lambda) \otimes T(x,y,\mu) \\ &- T(x,y,\lambda) \otimes T(x,y,\mu) . r(y,\lambda,\mu) \end{aligned} \tag{16}$$

And, for adjacent intervals, we obtain, due to the non ultralocality :

$$\left\{ T(x,y,\lambda) \overset{\otimes}{,} T(y,z,\mu) \right\} = T(x,y,\lambda) \otimes 1\!\!1 . s(y,\lambda,\mu) . 1\!\!1 \otimes T(y,z,\mu) \tag{17}$$

For a theory defined in a box [-L, L] with periodic boundary conditions we obtain with $T_L(\lambda) = T(L, -L, \lambda)$:

$$\begin{aligned} \left\{ T_L(\lambda) \overset{\otimes}{,} T_L(\mu) \right\} &= \left[r(L,\lambda,\mu) \,,\, T_L(\lambda) \otimes T_L(\mu) \right] \\ &+ (T_L(\lambda) \otimes 1\!\!1) . s(L,\lambda,\mu) . (1\!\!1 \otimes T_L(\mu)) \\ &- (1\!\!1 \otimes T_L(\mu)) . s(L,\lambda,\mu) . (T_L(\lambda) \otimes 1\!\!1) \end{aligned} \tag{18}$$

As a consequence we obtain an infinite abelian subalgebra of commuting charges (as in the ultralocal case) :

$$\left\{ tr(T_L^n(\lambda)) \,,\, tr(T_L^m(\mu)) \right\} = 0 \tag{19}$$

Using these new structure it has been possible, in the particular example of the Complex Sine Gordon theory, to construct the action-angle variable for such a theory, leading to its complete integrability [2] .

NEW EXTENDED YANG-BAXTER STRUCTURE AND KAC MOODY ALGEBRAS [6]

It has been shown that the canonical algebraic structure of integrability for ultralocal type models (6-10) can be constructed from graded Lie algebras [7]. More recently a similar construction has been done in the more general situation of non ultralocal type models by using graded Kac-Moody algebras [6], the central extension being related with the new s matrix in (12). Here, we want to sketch the procedure described in [6].

Let us consider a semi-simple Lie algebra G with generators X_o and structure constants C_{ab}^c:

$$G: \qquad [X_a, X_b] = C_{ab}^c X_c$$

And let us define the graded Lie algebra of function F(λ) with values in G and with generators X_a^m, $m \in \mathbb{Z}$:

$$\tilde{G}: \qquad [X_a^m, X_b^n] = C_{ab}^c X_c^{m+n}$$

We then distinguish two subalgebras $\tilde{G}_+$ and $\tilde{G}_-$ for $m \geqslant 0$ and $m < 0$, redefining $m \longrightarrow -1 - m$ in $\tilde{G}_-$.
It is then possible to define associated graded Poisson brackets algebras g_+ and g_- :

$$\begin{aligned} g_+: \qquad & \{\xi_a^m, \xi_b^n\} = C_{ab}^c \xi_c^{m+n} \\ g_-: \qquad & \{\xi_a^m, \xi_b^n\} = C_{ab}^c \xi_c^{m+n+1} \end{aligned} \tag{20}$$

where m and n are positive integers.
However, we wish now to introduce the space (x) as an additional index in order to generate the algebra (12). For ultralocal type models the space index enters dragonally in the algebra, ie, only with $\delta(x - y)$ distributions [7]. In our more general models we will consider central extension of the current algebras (20) :

$$\begin{aligned} \tilde{g}_+: \quad & \{\xi_a^m(x), \xi_b^n(y)\} = C_{ab}^c \xi_c^{m+n}(x)\, \delta(x-y) - K_{ab} B_+(m+n)\, \delta'(x-y) \\ \tilde{g}_-: \quad & \{\xi_a^m(x), \xi_b^n(y)\} = C_{ab}^c \xi_c^{m+n+1}(x)\, \delta(x-y) + K_{ab} B_-(m+n)\, \delta'(x-y) \end{aligned} \tag{21}$$

where $B_\pm$ are functions of positive integers and Kab is an invertible Cartan-Killing matrix associated to the basis X_a of G.
We then define two functions (for $\tilde{g}_+$ and $\tilde{g}_-$ respectively) by their series expansion in λ :

$$\tilde{g}_\pm: \quad f(\lambda)^{-1} = f_0 + \sum_{m \geqslant 0} B_\pm(m)\, \lambda^{\mp(m+1)} \tag{22}$$

and we obtain the corresponding Lax matrices satisfying (12) :

$$L(x,\lambda) = K^{a,b} L_a(x,\lambda) \cdot X_b$$

where
$$L_a(x,\lambda) = -f(\lambda) \sum_{m \geqslant 0} \xi_a^m(x)\, \lambda^{-m-1} \qquad \tilde{g}_+$$

and
$$L_a(x,\lambda) = f(\lambda) \sum_{m \geqslant 0} \xi_a^m(x)\, \lambda^m \qquad \tilde{g}_-$$

The r and s matrices are in this case :

$$r(\lambda,\mu) = \frac{1}{2}\,\Pi\,\frac{f(\lambda)+f(\mu)}{\lambda-\mu}$$
$$s(\lambda,\mu) = \frac{1}{2}\,\Pi\,\frac{f(\lambda)-f(\mu)}{\lambda-\mu} \qquad (23)$$

where $\Pi = K^{ab}\, X_a \otimes X_b$ and K^{ab} is the inverse matrix of K_{ab}.
It is easy to verify that such r and s matrices satisfy the relation(13) and the new Yang-Baxter equation (14) with H = 0 since here r and s are numerical matrices.
It has been shown in [6] that the principal chiral fields theories are particular example of this construction for $f(\lambda) = \gamma^2 \lambda^2/\lambda^2-1$

CONCLUSIONS AND PERSPECTIVES

In this lecture we have sketch how extended Yang-Baxter type algebras, which are able to describe the algebraic structures of integrability of two-dimensional integrable field theories of non ultralocal type, can be constructed at the classical canonical level. Moreover, a purely algebraic construction in term of graded Kac Moody algebras has been given for a particular class of them containing, as an example, the principal chiral field models.

The next step, of great importance in this program, will be to generalize these extended structures at the quantum level. This problem is under investigation now, the canonical approach developed here at the classical level being of great help to get first insight in it.

References

1. Surveys where many references to original works can be found are:
L.D. Faddeev, in Les Houches Lectures (1982), Recent Advances in Field Theory & Statistical Mechanics, eds. J.B. Zuber & R. Stora (North Holland, 1984).
P.P. Kulish & E.K. Sklyanin, in Tvärmine Lectures, eds. J. Hietarinta & C. Montonen, Springer Lectures in Physics (1982) Vol. 151.
R.J. Baxter, Exactly Solved Models in Statistical Mechanics (Academic Press, 1982).
2. J.M. Maillet, Nucl. Phys. B269(1986)54.
J.M. Maillet, Phys. Lett. 162B(1985)137.
3. M. Lüscher, Nucl. Phys. B135(1978)1.
M. Lüscher & K. Pohlmeyer, Nucl. Phys. B137(1978)46.
4. K. Pohlmeyer, Comm. Math. Phys. 46(1976)207.
H. Eichenherr & M. Forger, Nucl. Phys. B155(1979)381.
V.E. Zakharov & A.M. Mikhailov, Sov. Phys. JETP 47(1978)1017.
E. Brezin, C. Itzykson, J. Zinn-Justin, J.B. Zuber, Phys. Lett. 82B(1979)442.
H.J. de Vega, Phys. Lett. 87B(1979)233.
5. F. Lund & T. Regge, Phys. Rev. D14(1976)1524.
F. Lund, Phys. Rev. Lett. 38(1977)1175.
B.S. Getmanov, JETP Lett. 25(1977)119.
H.J. de Vega & J.M. Maillet, Phys. Rev. D28(1983)1441.
6. J.M. Maillet, Phys. Lett. 167B(1986)401.
7. A.A. Belavin & V.G. Drinfeld, Funkts. Anal. 16(1982)1 and 17(1983)69.
N.Y. Reshetikhin & L.D. Faddeev, Teor. Mat. Fiz. 56(1983)33.

B.R.S. ALGEBRA AND ANOMALIES

Michel Talon

L.P.T.H.E. Université Paris VI (U.A. 280), Tour 16, 1°étage
4, place Jussieu 75005 Paris - France

I. INTRODUCTION

It is convenient for discussion of anomalies to use geometric notations. Let us recall (see for example [1]) that the gauge potential A is naturally a 1-form on a principal bundle P(M,G), with base space M (the space-time) and <u>structure group</u> G (a Lie group), with values in the Lie algebra $\mathcal{G}$ of G. Usually, if P is trivial (no instanton), choosing a section of P, A appears as a 1-form on space-time M, with values in $\mathcal{G}$, behaving under change of section in the "non-covariant" way :

$$A \rightarrow A^{g} = g^{-1} A g + g^{-1} dg$$

where $x \in M \mapsto g(x) \in G$ is a "gauge transformation", i.e. a change of section of P. The curvature F of A is the covariant derivative of A on P, and, in some section :

$$F = dA + A^2 = dA + 1/2\, [A, A]$$

is also a 2-form with values in $\mathcal{G}$. It behaves in a "covariant" way under a gauge transformation :

$$F \rightarrow F^{g} = g^{-1} F g$$

Similarly, the set of all gauge transformations (set of maps $x \mapsto g(x)$) is a group under pointwise multiplication, and can be given the structure of an infinite dimensional Lie group. It is called the <u>gauge group</u>. There is a natural 1-form χ on this Lie group : the Maurer-Cartan form. An <u>infinitesimal</u> gauge transformation ξ is of the form $g(x) = 1 + \epsilon \xi(x)$ with ϵ very small, and $\xi(x) \in \mathcal{G}$. Then χ is the unique left-invariant 1-form on the gauge group such that, for infinitesimal transformations ξ :

$$\chi(x) \cdot \xi = \xi(x)$$

So χ is a field on space-time, with values in $\mathcal{G}$, also called <u>the Faddeev-Popov field</u> (note that χ anticommutes with odd degree forms).

If δ is the exterior differential on the gauge group, the Maurer-Cartan equation says :

$$\delta \chi = - \chi^2 = - 1/2\, [\chi, \chi]$$

and this is the B.R.S. transformation of the anticommuting Faddeev-Popov field. Moreover, an infinitesimal gauge transformation on A and F is given by :

$$\delta_\xi A = (\delta A)\cdot\xi \qquad \delta_\xi F = (\delta F)\cdot\xi$$

with

$$\delta A = -(d\chi + \chi A + A\chi) = -d\chi - [A,\chi]$$
$$\delta F = -\chi F + F\chi = [F,\chi]$$

where the second form always displays the Lie algebraic structure of all these operations. Note that δ augments the χ degree by 1, and that the B.R.S. transformations [2] of χ,A,F are naturally recovered in this way. The 2-form $d\chi$ of "ghost degree" 1, and space degree 1 appears in δA. It is defined by :

$$(d\chi)(x)\cdot\xi = d\xi(x) = dx^\mu\,\partial_\mu\,\xi(x)$$

Note that all these objects are natural geometric objects on the cartesian product P(M,G) x gauge group. At the point (p,g) of this product, we can consider the forms χ, $A(p)^g$, $F(p)^g$ and the differentials d (along P), δ (along gauge group), and d+δ (total). This obviously reproduces the formulas for $\delta\chi$, δA, δF and makes clear that:

$$d^2 = \delta^2 = (d+\delta)^2 = d\delta + \delta d = 0$$

This is also easily checked by direct computation.

Let us now consider basic abstract symbols A^α, χ^α (anticommuting) and F^α, $d\chi^\alpha$ (commuting). Here α = 1,2, ..., dim $\mathfrak{g}$. These symbols may be replaced by the components of A, χ,F,dχ in some base of $\mathfrak{g}$. By doing sums and products these symbols generate a graded algebra with two degrees (space-time degree and ghost degree) and two antiderivations d and δ. Since :

$$dA = F - 1/2\,[A,A] \qquad \Rightarrow \qquad dF = [F,A]$$

the whole structure only depends on the Lie algebra structure of $\mathfrak{g}$ (a similar idea was introduced by A. Weil in 1949, leading to the Weil algebra and the Weil homomorphism [3]). The generated algebra is clearly isomorphic to :

$$\Lambda\mathfrak{g}^* \otimes S\mathfrak{g}^* \otimes \Lambda\mathfrak{g}^* \otimes S\mathfrak{g}^*$$

By analogy with the Weil algebra $\Lambda\mathfrak{g}^*\otimes S\mathfrak{g}^*$ generated by A^α, F^α with one antiderivation d (see [3]), we call it the B.R.S. algebra. Its properties are studied in more detail in [4], [5] . Notably, it is shown here that the B.R.S. algebra is universal in the following sense : let us assume some identity to be valid between our differential forms A,F, χ, dχ. More precisely we assume this identity to be of algebraic, local nature, and non trivially true (that is, a form of space degree greater than the dimension of space-time is trivially zero), but valid for generic values of the fields.
(EXAMPLE : Tr F^2 = d Tr(AF - 1/3 A^3) is true in dimension 4, generically, non trivial, and involves only algebraic expressions in the components A^α, F^α of the fields). Then this identity has to be true in the abstract B.R.S. algebra, and so must be consequence of the Lie algebra structure of $\mathfrak{g}$.

Finally, we mention briefly the discussion of gravitational anomalies [6] , which is completely similar. The principal bundle is here the bundle of orthonormal frames on some riemannian variety. Then G is the orthogonal group, and $\mathfrak{g}$ the algebra of antisymmetric matrices. Of course A has to be replaced by the "spin connection" ω, and F by the curvature R. Then everything comes out like the gauge case. As a matter of fact, local Lorentz anomalies are so obtained, which can look like true gravitational anomalies, but are not [6]. No real diffeomorphism anomaly has ever been found.

II. ANOMALIES

We shall adopt the presentation of Wess-Zumino [7]. Using conservation laws derived from gauge invariance, other definitions of anomalies can be given, leading to covariant, non consistent anomalies [8]. The classical action L(A) is gauge invariant : $\delta L(A) = 0$.
Going to the quantum theory, L is replaced by $\Gamma(A)$ and perhaps :

$$\delta \Gamma(A) \neq 0$$

This is the anomaly.

More precisely, in the full classical theory, one has :

$$L = L(A, \chi, \psi, \cdots)$$

$$L = \int_M \left(-\frac{1}{4} F_{\mu\nu}^2 + \frac{1}{2}(\partial_\mu A^\mu)^2 + \bar{\chi} M \chi + \bar{\psi}\, i \not{D}_A \psi + \cdots \right)$$

where χ is the Faddeev-Popov field, and :

$$\delta(\partial_\mu A^\mu) = M\chi$$

while ψ is, for example, a Weyl fermion and D_A the covariant derivative. Such an L is B.R.S. invariant under appropriate definition of $\delta\bar{\chi}$, [2]. In the quantum theory L is replaced by $\Gamma(A)$: the generating function for 1-particle-irreducible graphs [9] :

$$\Gamma(A) = L(A) + \hbar\, \Gamma^{(1)}(A) + \hbar^2\, \Gamma^{(2)}(A) + \cdots$$

Of course $\Gamma(A)$ is divergent and has to be renormalized. This means adding some local polynomial of the fields $A, \chi, \psi, \ldots$ and their derivatives of low enough degree [9], integrated over space-time M, to $\Gamma(A)$. Now there is a quantum perturbative anomaly if $\delta\Gamma \neq 0$, for any allowed renormalization of Γ. It is shown in [9] that true anomalies can only depend on A and χ (with ghost degree 1) :

$$\delta\Gamma = a(\chi, A)$$

where a is a local polynomial of the fields, integrated over M. Since $\delta^2=0$, one has the constraint on the anomaly a:

$$\delta a(\chi, A) = 0$$

and this is the Wess-Zumino consistency condition [7].

Using the above discussion, one usually considers only a quantized fermion ψ in an external classical gauge field A. Then

$$L = L(\psi, A) = \int_M \bar{\psi}\, i \not{D}_A \psi$$

Going to the quantum theory, fermions are integrated out, leaving a quantum $\Gamma(A)$. Then, the anomaly is given by [7] :

$$\delta\Gamma(A) = a(\chi, A)$$

and, of course $\delta a = 0$. This chiral anomaly is related to the index of the family of Dirac operators $\not{D}_A$, see [6], and this gives an efficient way to compute $\mathfrak{a}$, for a given fermionic content of the theory. Note that, taking into account the definition of exterior differentiation, if ξ and η are two infinitesimal gauge transformations, one has :

$$(\delta a)\cdot(\xi,\eta) = \delta_\xi\, a(\eta,A) - \delta_\eta\, a(\xi,A) - a([\xi,\eta],A)$$

and so, $\delta \mathfrak{a} = 0$ is equivalent to

$$\{ [\delta_\xi , \delta_\eta] - \delta_{[\xi,\eta]} \}\ \Gamma(A) = 0$$

as originally discussed in [7].

Of course, $\Gamma(A)$ is a general, non-local, non-polynomial functional of A, but it follows from renormalization theory that its variation $\delta\Gamma(A) = \mathfrak{a}(\chi, A)$ is a local, polynomial functional of A and its derivatives (integrated over M). Moreover, if under renormalization

$$\Gamma(A) \to \Gamma(A) + \mathcal{B}(A)$$

where $\mathcal{B}$ is local, polynomial then $\mathfrak{a} \to \mathfrak{a} + \delta\mathcal{B}$ and so $\mathfrak{a}$ is determined only up to the δ of some local polynomial function of A integrated over M. Let us write :

$$a(\chi, A) = \int_M Q^1_m(\chi, A)$$

where Q^1_m is a local polynomial of A and χ of space-time degree m = dim M and ghost degree 1. Then :

$$\delta\mathfrak{a} = 0 \quad \text{if and only if} \quad \delta Q^1_m + dQ^2_{m-1} = 0$$

$$\mathfrak{a} = \delta\mathcal{B} \quad \text{if and only if} \quad Q^1_m = \delta L^o_m + dL^1_{m-1}$$

for similar local polynomials Q^2_{m-1}, L^o_m, L^1_{m-1} (or, at least, we shall admit these equivalences). Finally, we have to solve $\delta Q + dQ' = 0$ (we say that Q is a δ-cocycle modulo d), modulo trivial solutions of the form $Q = \delta L + dL'$ (we say that Q is a δ-coboundary modulo d), i.e. we have to find $H^1_m(\delta/d)$, (the δ-cohomology modulo d in degrees 1,m), and all this in the class of local polynomials of A, χ and their derivatives.

Recently, L. Faddeev has shown that Schwinger terms obey a similar cohomological formulation [10], [11]. Looking at gauge theories in the Hamiltonian approach, he shows that the symmetry is realized projectively and so the Gauss law constraints do not obey naïve commutators : a Schwinger term is present. This type of anomaly is defined on space (t=0) of dimension m-1 and contains two gauge variations, so is given by a Q^2_{m-1}. The natural constraint on projective representations implies that :

$$\delta \int_{t=0} Q^2_{m-1} = 0 ,$$

while if $\int_{t=0} Q^2_{m-1} = \delta \int_{t=0} L^1_{m-1}$ all projective phases can be cancelled. Finally Schwinger terms are given by $H^2_{m-1}(\delta/d)$, at least possible Schwinger terms. Faddeev has shown in [11] by a concrete calculation that such a term is indeed present. Then R. Jackiw has generalized this analysis to 3-cocycles (see Jackiw's lecture at this conference), and it is of interest to compute $H^K_{m+1-K}(\delta/d)$ for all K.

In principle, all these polynomials could be algebraic polynomials of A, χ and their derivatives :

EXAMPLE : $(F_{\mu\nu})^2$, $(D_\mu F_{\nu\rho})^2$ with symmetrization on $\mu \leftrightarrow \nu$ are δ-cocycles. But no true solution of $\delta Q + dQ' = 0$ (with $Q' \neq 0$) is known outside the restricted class of polynomials in the differential forms A,

dA, χ, $d\chi$ (involving only 1st derivatives of fields and automatic anti-symmetrizations). All known anomalies are of this type :

(EXAMPLE : Adler-Bardeen's anomaly :

$$Q_4^1 = \mathrm{Tr}\ \chi d\,(A\,dA + 1/2\ A^3) \qquad .)$$

and so can be universally projected on all types of space-time M (natural covariance of differential forms).

We shall simply assume Q, Q', L, L' to be of this type and proceed to compute $H(\delta/d)$. By the universal property of B.R.S. algebra, our problem is then equivalent to the calculation of $H(\delta/d)$ for the B.R.S. algebra, and this only depends on the Lie algebra $\mathcal{G}$. The topological problems involved in the structure of the gauge groups are thus irrelevant for our study. We shall see that $H(\delta/d)$ is expressed in terms of invariant polynomials and invariant forms of the Lie algebra $\mathcal{G}$.

III. COHOMOLOGY OF δ

The B.R.S. algebra is generated by the formal elements A^α, χ^α, F^α, $d\chi^\alpha$ ($\alpha = 1,\ldots,\dim\mathcal{G}$). Since $F^\alpha = dA^\alpha + 1/2\ C^\alpha_{\beta\gamma} A^\beta A^\gamma$ it is also generated by (A^α, dA^α, χ^α, $d\chi^\alpha$). Each subalgebra (A^α, dA^α), ..., has no d-cohomology because the d-cocycles $(dA^\alpha)^n$ are also d-coboundaries $d(A^\alpha (dA^\alpha)^{n-1})$. By Kunneth theorem, their tensor product (i.e. the B.R.S. algebra) is also d-trivial :

$$H(d) = H_0^0\,(d) = \mathbb{R}$$

(of course, constants are d-cocycles, but not d-coboundaries).

Similarly, the B.R.S. algebra is generated by A^α, δA^α, χ^α, F^α because δA^α contains $d\chi^\alpha$. The part (A^α, δA^α) is δ-trivial and δ is internal to (χ^α, F^α) which is the only relevant part for the calculation of $H(\delta)$. So, we are left with the algebra $\Lambda\mathcal{G}^* \otimes S\,\mathcal{G}^*$, generated by χ^α and F^α, on which δ acts by :

$$\delta\chi^\alpha = -1/2\ C^\alpha_{\beta\gamma}\,\chi^\beta\chi^\gamma \qquad \delta F^\alpha = C^\alpha_{\beta\gamma}\,F^\beta\,\chi^\gamma$$

In other words, this algebra is the algebra of exterior forms on $\mathcal{G}$, with values in the $\mathcal{G}$-module $S\mathcal{G}^*$ of polynomials on $\mathcal{G}$. Then δ reduces on χ^α to the formula for exterior differential on forms, and on F^α to the $\mathcal{G}$-action on the $\mathcal{G}$-module. Finally $H(\delta)$ is the Lie algebra cohomology of $\mathcal{G}$ with values in the $\mathcal{G}$-module $S\mathcal{G}^*$

$$H(\delta) = H(\mathcal{G}\,,\,S\mathcal{G}^*)$$

In particular $H^\ell_K(\delta) = 0$ if K is odd, because F^α is of space degree 2.

We shall assume the Lie algebra $\mathcal{G}$ to be reductive, i.e. the direct product of a semi-simple Lie algebra and an abelian algebra, or equivalently the Lie algebra of a compact Lie group. Then, it is known (Hochschild-Serre) that for each $\mathcal{G}$-module $\mathcal{M}$ with invariant subspace $\mathcal{M}^I$:

$$H(\mathcal{G}\,,\,\mathcal{M}) = H(\mathcal{G}\,,\mathbb{R}) \otimes \mathcal{M}^I$$

Moreover $H(\mathcal{G},\mathbb{R}) \simeq \mathcal{I}_\Lambda(\mathcal{G})$: the algebra of invariant exterior forms on $\mathcal{G}$. (It may be shown that a left invariant form on a compact Lie group G is also right invariant if and only if it is closed, and if a biinvariant form is exact, it vanishes). Here $\mathcal{M}^I = \mathcal{I}_S(\mathcal{G})$: the algebra of invariant polynomials on $\mathcal{G}$. Finally :

$$H(\delta) = \mathcal{I}_\wedge(\mathcal{G}) \otimes \mathcal{I}_s(\mathcal{G})$$

For reductive Lie algebras, much more is known on $\mathcal{I}_\wedge(\mathcal{G})$ and $\mathcal{I}_s(\mathcal{G})$ (work of Koszul, Cartan, Chevalley, Weil). Among invariant forms, we can choose r(= rank of $\mathcal{G}$) forms of odd degree, called primitive forms, which generate $\mathcal{I}_\wedge(\mathcal{G})$. So, if P_o is the vector space of primitive forms,

$$\dim P_o = \text{rank}(\mathcal{G}) \qquad\qquad \mathcal{I}_\wedge(\mathcal{G}) = \wedge P_o$$

We shall give here a table of degrees of primitive forms for various simple Lie groups (for a product, simply juxtapose corresponding primitive forms):

SU(N)	3, 5, 7, ..., 2N-1
SO(2N+1), Sp(2n)	3, 7, 11, ..., 4N-5, 4N-1
SO(2N)	3, 7, 11, ..., 4N-5, 2N-1
G_2	3, 11
F_4	3, 11, 15, 23
E_6	3, 9, 11, 15, 17, 23
E_7	3, 11, 15, 19, 23, 27, 35
E_8	3, 15, 23, 27, 35, 39, 47, 59

For each degree (2K-1) the primitive form is of type $\mathrm{Tr}\,\chi^{2K-1}$ (in some representation).

For each primitive form (and only for primitive forms), there is a chain of equations of the type introduced by Stora : let $\omega(\chi)$ be a primitive form ; one can choose an invariant polynomial $(\tau\omega)(F)$, called a transgression of ω such that :

$$Q_o^{2K-1} = \omega(\chi) \rightarrow \left\{ \begin{array}{rcl} \tau\omega(F) &=& d\, Q^{o}_{2K-1}(A,F) \\ 0 &=& \delta Q^{o}_{2K-1} + d\, Q^{1}_{2K-2} \\ \hline 0 &=& \delta Q_1^{2K-2} + d\, Q_o^{2K-1} \\ 0 &=& \delta Q_o^{2K-1} \end{array} \right.$$

and the Q_K^{ℓ} are polynomials of χ, $d\chi$, A, F. Usually, for $\omega(\chi) = \mathrm{Tr}\,\chi^{2K-1}$ one chooses :

$$(\tau\omega)(F) = C^{te}\ \mathrm{Tr}\, F^K$$

Finally, the polynomials $\tau\omega$ for $\omega \in P_o$ also generate $\mathcal{I}_s(\mathcal{G})$ and are algebraically independant, so :

$$\mathcal{I}_s(\mathcal{G}) = S\,\tau P_o$$

Our result is :

$$\boxed{H(\delta) = \wedge P_o \otimes S\,\tau P_o}$$

EXAMPLE : G = SU(3). Then $H(\delta)$ is the algebra of polynomials in the odd

Tr χ^3 and Tr χ^5 and the even TrF^2 and TrF^3.

IV. COHOMOLOGY OF δ MODULO d

We have to solve $\delta Q + dQ' = 0$ in the B.R.S. algebra, modulo Q such that $Q = \delta L + dL'$. In fact $H(\delta/d)$ is related to $H(\delta)$ by an exact couple (Massey) :

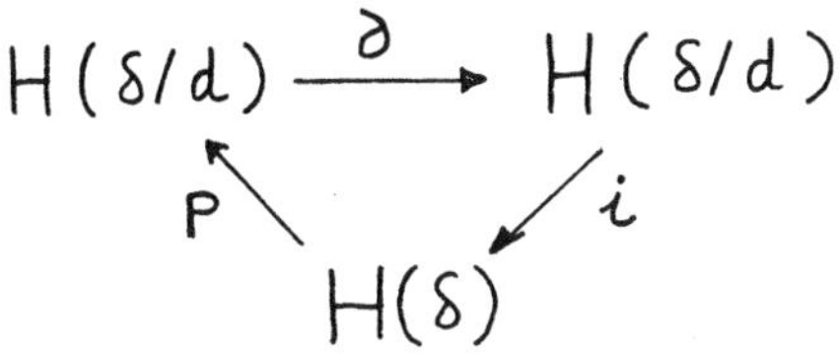

where, if $\delta Q + dQ' = 0$ one sets :

$$\partial([Q]_{H(\delta/d)}) = [Q']_{H(\delta/d)}$$
$$i([Q]_{H(\delta/d)}) = [dQ]_{H(\delta)} \quad \text{and if } \delta Q = 0$$
$$p([Q]_{H(\delta)}) = [Q]_{H(\delta/d)}$$

Using triviality of d, it is easy to show that these maps are well defined, and all 3 corners are exact :

$$\mathrm{Ker}\ \partial = \mathrm{Im}\ p \qquad \mathrm{Ker}\ p = \mathrm{Im}\ i \qquad \mathrm{Ker}\ i = \mathrm{Im}\ \partial$$

EXAMPLE : Assume $p([Q]_{H(\delta)}) = 0$, so $\delta Q = 0$ and

$Q = \delta L + dL' \Rightarrow d\delta L' = 0 \Rightarrow \delta L' = dM$

(triviality of d) so L' is a δ-cocycle modulo d. But

$$[Q]_{H(\delta)} = [dL']_{H(\delta)} = i([L']_{H(\delta/d)}), \quad \text{so} \quad \mathrm{Ker}\, p = \mathrm{Im}\, i$$

REMARK : Since $H^0_0(d) \neq 0$ constants are excluded from $H(\delta)$ and $H(\delta/d)$ so that everything works well.

Now, we can use the standard trick [12] : iterate the exact couple. Since $H(\delta)$ has only even space dimension, we have

$$\partial : H_{odd}(\delta/d) \xrightarrow{\sim} H_{even}(\delta/d)$$

and, taking $F_0 = H_{even}(\delta/d)$, $E_0 = H(\delta)$, $p_0 = p$, $i_0 = i \circ \partial^{-1}$ well defined on $H_{even}(\delta/d)$ one gets the exact triangle :

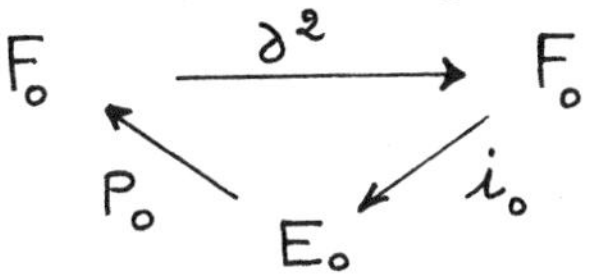

Put $F_1 = \partial^2 F_0$ and define $d_0 : E_0 \to E_0$ by $d_0 = i_0 \circ p_0$. Then $d_0^2 = 0$ and so, define:

$$E_1 = H(E_0, d_0)$$

It is easy to construct $i_1 : F_1 \to E_1$ and $p_1 : E_1 \to F_1$ which together with $\delta^2 : F_1 \to F_1$ form another exact couple. By induction, one then defines an rth derived exact couple $(F_r, E_r, i_r, p_r, d_r)$ and $E_{r+1} = H(E_r, d_r)$, $d_r = i_r \circ p_r$.

Here, $F_o = H_{even}(\delta/d)$, while $E_o = \Lambda P_o \otimes S\tau P_o$ and one is able to show that $E_r = \Lambda P_r \otimes S\tau P_r$ where P_r is the vector space of primitive forms of degree $\geq 2r+1$ (In E_o, E_r, the constants $\Lambda^o P_r \otimes S^o \tau P_r$ are excluded). For the proof, we refer to [4] or [13]. In particular, for r large enough $E_r = 0$, so the iteration stops. Moreover, d_r has also a very simple expression : On E_r, d_r is defined as an antiderivation such that $d_r\omega = \tau\omega$ if ω is a primitive form of degree (2r+1), d_r vanishing on all other generators. Finally i_r and p_r have equally simple expressions.

Note that $F_o \simeq \mathrm{Ker}\,\delta^2 \oplus \mathrm{Im}\,\delta^2 \simeq \mathrm{Im}\, p_o \oplus F_1$ and by induction :

$$F_o = H_{even}(\delta/d) \simeq \bigoplus_r p_r(E_r)$$

Knowing E_r and p_r our problem is solved. Moreover :

$$p_r(E_r) \simeq \frac{E_r}{\mathrm{Ker}\, p_r} \simeq \bigoplus_{K\geq 0} \frac{E^{r+K}_{r+K}}{\mathrm{Im}\, d_{r+K}}$$

(see [4] or [13] for the proof ; E^r_r is the part of E_r containing at least one ω of degree 2r+1 or its transgression $\tau\omega$). This reduces the determination of $H(\delta/d)$ to a simple enumeration, explained for important examples in [14] . Note that all independent elements are so obtained.

To write their explicit form, it is necessary to apply δ^{-2r} to an element of $p_r(E_r)$ because of the isomorphisms implicit in the above discussion.

EXAMPLE : $F_o \simeq p_o(E_o) \oplus F_1 \simeq p_o(E_o) \oplus p_1(E_1) \oplus F_2$

An element of $p_1(E_1) \subset F_1$ is associated with an element in F_o via $\delta^{-2} : F_1 \to F_o$, or more precisely via some choice of δ^{-2}.

A convenient way to do that is to use the "generalized transgression lemma" (in [4], [13] this lemma is the key to the proof of the above results): let $X \in E_r$. One can choose $X = \prod_i (\tau\xi_i)(F)\,\omega_o(X) \cdots \omega_n(X)$ where ξ_i, ω_j are primitive forms of degree $\geq 2r+1$. By the usual transgression trick and the "Russian formula" of Stora (see [15], [16]), there are invariant $L_p(A,F)$ such that :

$$(\tau\omega_p)(F) = dL_p(A,F) = (d+\delta)\,L_p(A+X,F)$$

Then, computing :

$$(d+\delta)\left\{\prod_i(\tau\xi_i)(F)\,L_o(A+X,F)\cdots L_n(A+X,F)\right\} =$$

$$\sum_{p=0}^{n}(-1)^p \prod_i(\tau\xi_i)(F)(\tau\omega_p)(F)L_o(A+X,F)\cdots\hat{L}_p\cdots L_n(A+X,F)$$

and expanding in decreasing δ degree yields :

$$\delta X = 0$$
$$\delta Q_1 + d X = 0$$
$$- - - - - - - - - -$$
$$\delta Q_{2r+1} + d Q_{2r} = 0$$
$$\delta Q_{2r+2} + d Q_{2r+1} = d_r X$$

where d_r is the antiderivation described above. Then $Q_1, \ldots, Q_{2r}$ are δ-cocycles modulo d of a new type, and, in particular, one can choose:

$$\partial^{-2r}\left([X]_{H(\delta)} \right) = \left[Q_{2r} \right]_{H(\delta/d)}$$

This obviously gives a practical way to compute these new δ-cocycles modulo d (alluded to in [15]). An $X \in E_r^r$ is in the Kernel of p_r precisely when $d_r X = 0$, i.e. when the above construction degenerates, so everything comes out quite nicely. This is applied in particular to the case $G = U(1)^N$ in [14], leading to an interesting result.

V. CONCLUSION

As we have seen, the B.R.S. algebra is a natural algebraic object in which one can state and solve the problem of consistency conditions for anomalies. The general solution in this class can be obtained by developing products of secondary Chern-Weil polynomials. All independent solutions are classified by looking at a new cohomology (similar to Spencer cohomology) defined on invariant polynomials and forms of the structure Lie algebra.

Moreover, the elements of the B.R.S. algebra also have natural geometric interpretation. Replacing A^α, F^α, χ^α, $d\chi^\alpha$ by the corresponding components of the fields, to each element of the B.R.S. algebra is associated a differential form (real-valued) on P(M,G)x gauge group. This is similar to the Weil homomorphism.

If one is only interested in $H^1_n(\delta/d)$ and $H^2_n(\delta/d)$, i.e. Adler anomaly and Schwinger term, it is sufficient to note that, for semi-simple groups (even with one U(1) factor) $H^1(\delta) = 0$ and $H^2(\delta) = 0$. Then Adler anomalies and Schwinger terms are in one to one correspondence with invariant polynomials (such as TrF^n) by the usual procedure :

$$I_{2n+2}(F) = d\, Q^0_{2n+1}(A, F)$$
$$0 = \delta Q^0_{2n+1} + d Q^1_{2n} \quad \leftarrow \text{Adler}$$
$$0 = \delta Q^1_{2n} + d Q^2_{2n-1} \quad \leftarrow \text{Schwinger}$$
$$- - - - - - - - - - - - - -$$

If two U(1) factors are present, $H^2(\delta) \neq 0$ and new Schwinger terms appear For general G, $H^3(\delta) \neq 0$ and 3-cocycles have to be obtained by the above procedure.

Of course, this is appropriate for the analysis of perturbative anomalies. As already stated by Wess-Zumino [7], and confirmed in the beautiful analysis of Singer [17], the anomalies produced by regularization of the fermionic determinant have a form constrained only by the Lie algebra of the structure group. The topology of the gauge group plays a role in the value of the coefficient of this anomaly (Novikov, Witten's quantification of this coefficient). On the contrary, this topology completely affects the existence of non perturbative anomalies [18].

REFERENCES

[1] C.M. Viallet, Lectures at 1986 Karpacz Winter School.

[2] C. Becchi, A. Rouet, R. Stora, Ann. Phys. N.Y. 98 (1976) 287.

[3] H. Cartan in Colloque de Topologie, Bruxelles (1950), Masson.

[4] M. Dubois-Violette, M. Talon, C.M. Viallet, Comm. Math. Phys. 102 (1985) 105.

[5] M. Dubois-Violette, Lectures at 1986 Karpacz Winter School.

[6] L. Alvarez-Gaumé, P. Ginsparg, Ann. Phys. N.Y. 161 (1985) 423.

[7] J. Wess, B. Zumino, Phys. Lett. 37B (1971) 95.

[8] W. Bardeen, B. Zumino, Nucl. Phys. B244 (1984) 421.

[9] B. Lee, Lectures at 1975 Les Houches Summer School, North Holland.

[10] L. Faddeev, Phys. Lett. 145B (1984) 81.

[11] L. Faddeev, S. Shatashvili, Phys. Lett. 167B (1986) 225.

[12] S. Mac Lane, Homology, Springer

[13] M. Talon, Lecture at 1985 Cargese Summer School, Plenum

[14] M. Dubois-Violette, M. Talon, C.M. Viallet, Ann. Inst. H. Poincaré 44 (1986) 103.

[15] R. Stora, Lectures at 1976 Cargese Summer School, Plenum.

[16] B. Zumino, Lectures at 1983 Les Houches Summer School, North Holland.

[17] I. Singer in Asterisque 1985 (S.M.F. Lyon, Juin 1984).

[18] E. Witten, Phys. Lett. 117B (1982) 324.

BRST CURRENT ALGEBRA DERIVATION OF THE HIGHER COCYCLES

Bernard Grossman

Rockefeller University

1230 York Avenue, New York, N.Y. 10021

INTRODUCTION

Two topics of much current research in mathematical physics are the BRST quantization of gauge theories and the understanding of anomalies in gauge theories. While a large part of the stimulus for this research is the revival of interest in string theory, much remains to be learned about gauge theories. Here, we shall present a new extension of BRST symmetry in the form of a local symmetry and demonstrate that this leads to an understanding, in the context of a field theory, of the higher cocycles that have been mathematically associated with chiral anomalies.[1] We obtain, furthermore, a unification of many approaches to the chiral anomaly. Namely, once the BRST symmetry is gauged, we can examine BRST current algebra. Anomalies and the cocyles associated with them are consequences of this current algebra. Moreover, the BRST charge derived from the BRST current is not nilpotent.

In the usual case of Yang-Mills theory, with Lie-algebra valued 1-form potential A, ghost c, antighost $\bar{c}$, auxiliary field b, and matter Ψ with ghost numbers 0,1, -1,0,0 respectively, we have the following transformation under the BRST operator S

$$\begin{aligned} SA &= -Dc \\ Sc &= -\frac{1}{2}[c,c] \\ S\bar{c} &= 0 \\ Sb &= 0 \\ S\Psi &= -c\Psi \end{aligned}$$

As a consequence of closure of the algebra and the Jacobi identity, we have $S^2=0$. In addition for differential $d=dx^{\mu}\partial_{\mu}$, we have $d^2=0$ and $Sd+dS=0$. S acts as a differential operator in a direction perpendicular to d and graded by the sum of the ghost and Lorentz degree of forms.

Talk delivered at Super Field Theories at Nato Advanced Research Workshop, July 25-30, 1986, to be published by Plenum Press, ed. by H.C. Lee.

If we define curvatures

$$F = (d+A)^2$$

and

$$\tilde{F} = (d+S+A=c)^2$$

the horizontality condition on the curvature $F=\tilde{F}$ is easily seen to be equivalent to the BRST transformations.

The above BRST operator S defines a global symmetry with anti-commuting constant ξ

$$\delta_\xi \Phi = \xi S \Phi$$

We would like to make ξ local. This requires the introduction of a ghost number -1 gauge field α_μ, to compensate for $\partial_\mu \xi$, along with its ghost number 0 ghost field λ.

Defining the local BRST S (which reduces to the previous S for $\alpha=0$, $\lambda=1$), we have the following transformation laws:

$$SA' = dc' + [A',c']$$
$$Sc' = -\frac{1}{2}[c',c']$$
$$Sc = \lambda b$$
$$Sb = 0$$
$$S\alpha = d\lambda$$
$$S\lambda = 0$$

where $A'=A-\alpha c$, $c'=\lambda c$. Once again, the above BRST transformations are easily seen to be nilpotent and to be equivalent to the horizontality condition for the curvature

$$F(A',c') = \tilde{F}(A',c')$$

Moreover the invariant polynomials in the curvature that determine the Chern classes are also horizontal. This fact has important consequences for the current algebra.

Given the above local BRST transformation, one can construct very easily an invariant Lagrangian, L, with or without matter fields. Differentiating this Lagrangian with respect to α, the source of the BRST current, one obtains the form of the BRST current

$$J_\mu = \left.\frac{\delta L}{\delta \alpha_\mu}\right|_{\alpha=0}$$

Integration of J_0 over space yields the BRST charge Q, commutation with which determines the BRST transformations.

In gauge theories with anomalies, the BRST current algebra has interesting consequences. The anomaly can be calculated by n-point functions in the external gauge current or with external BRST current. Both are related to the Chern-Simons polynomial of, degree d+1, T_{d+1}, that is obtained from the invariant Chern polynomial, P^{inv}_{d+2}, of degree d+2 (d=dimension of space-time). Because of the horizontality condition

$$P^{inv}_{d+2}(F(A',c') = P^{inv}_{d+2}(\tilde{F}(A',c'))$$

we have

$$(d+S)\ T_{d+1}(A'+c',\tilde{F})=dT_{d+1}(A',F)$$

By the standard descent procedure

$$S\ T_{d+1} = -\ d\Delta_d^{\ 1}$$

where $\Delta_d^{\ 1}$ is the standard anomaly. However, $\Delta_d^{\ 1}$ is now dependent on α. By expanding in a Taylor series, one obtains not only the above anomaly, but also all the higher cocycles. Moreover, since these are obtained by differentiating with respect to α, one easily obtains that the presence of such gauge anomalies is equivalent to the violation of BRST Noether current conservation. By the standard technique of removing a derivative inside a time order product, one obtains the anomalous commutation relations involving the BRST current as well as the failure of the BRST charge to have square zero. For example, in two dimensions, the above procedure relates

$$<T(\partial_\mu A^\mu,\ A_\nu)<$$

to

$$<T(\partial_\mu J^\mu_{BRST},\ J^{BRST}_\nu,\ \bar{c},\ \bar{c})>$$

Removing the derivative from inside the time-order product gives the anomalous equal-time commutation relation

$$\{J_0,\ J_\mu\}_{ETC} \neq 0$$

Setting μ=0 and integrating over space yields $Q^2\neq 0$.

References

1. The local BRST for gauge theories, gravity and string theories was first derived by L. Baulieu, C. Becchi, and R. Stora (in preparation). The derivation of the higher cocycles appears in L. Baulieu, B. Grossman and R. Stora Rockefeller preprint RU86/B/160 (to be published in Physics Letters B). See these papers for further references.

SUPERCONFORMAL ALGEBRAS

A. Van Proeyen†

Instituut voor theoretische fysica, K.U. Leuven
Celestijnenlaan 200 D
B-3030 Leuven, Belgium

INTRODUCTION

The construction of locally supersymmetric actions involves usually long calculations. These can be facilitated by making effective use of a maximal amount of symmetry. The largest nontrivial extension of the Poincare algebra is the conformal algebra. Even for constructing actions which only exhibit super-Poincare invariance, one can start by constructing a superconformal invariant one, and then impose gauge choices to break the extra symmetries.[1] It turns out that this way of proceeding is efficient. Moreover it clarifies the structure of the theory. In 2 dimensions the importance of the superconformal invariance is well known. String theories owe their consistency to conformal invariance.

It is then clear that it is important for the construction of general models to know which are the super-extensions of the conformal algebras. In the next section we repeat the superconformal algebras in dimensions higher than 2, and the finite dimensional d=2 superconformal algebras. In section 3 we present new results[2] on N-extended infinite dimensional superconformal algebras (d=2). Section 4 shows some features of the superconformal gauge theory in 2 dimensions, when one starts from the infinite dimensional algebras.[3]

FINITE DIMENSIONAL SUPERCONFORMAL ALGEBRAS

The conformal algebra in d dimensions is SO(d,2). In 2 dimensions, however, the conformal algebra is infinite dimensional. SO(2,2) is then a finite dimensional subalgebra. In this section we consider d>2 or d=2 with only the finite dimensional conformal algebra. The superconformal algebras which we will look for are these with the following three characteristics.

1) They contain the conformal algebra SO(d,2) which is built from translations P, Lorentz generators M, dilatations D and special conformal generators K.

2) They are extended with supersymmeteries. These are generators which are in the spinor representation of the SO(d-1,1) generators M. It turns out that for each supersymmetry Q one needs a second supersymmetry S in the commutator [Q,K]. So we have introduced 2N spinors of SO(d-1,1).

†Bevoegdverklaard navorser, N.F.W.O., Belgium

3) To satisfy the Jacobi identities with 3 spinor generators one needs in general to introduce more genertors in the anticommutator $\{Q,S\}$. These extra generators form a bosonic algebra, which we demand to commute with SO(d,2).

The last requirement is motivated by the Coleman-Mandula theorem[4] which states that in d=4 all scattering amplitudes are trivial if there are (bosonic) symmetries of the S matrix noncommuting with the conformal algebra. We have two remarks on this theorem. First, it considers symmetries between on-shell states. In the applications of the superconformal algebra which we have in mind, some of these symmetries are broken or could act only between off-shell states. Secondly, as far as we know, this theorem has only been proved in 4 dimensions. On the other hand, we have up to now only succeeded in using superalgebras satisfying 3) for constructing sensible theories.

For further reference we repeat the Lie superalgebras of 'classical type'[5]. These are the simple superalgebras where the 'defining representation' of the bosonic algebra in the fermionic generators is completely reducible.

Table 1: Lie superalgebras of classical type.

Name	Range	Bosonic algebra	Defining representation	Number of generators
SU(n|m)	m≠n	SU(m)+SU(n)+U(1)	(m,n) + (m,n)	$m^2+n^2-1, 2mn$
SU(m|m)	m=2,3,...	SU(m)+SU(m)	$2(m,\bar{m})$	$2(m^2-1), 2m^2$
Osp(m|n)	m=1,2,... n=2,4,...	SO(m)+Sp(n)	(m,n)	$\frac{1}{2}(m^2-m+n^2+n), mn$
D(2,1,α)	$0<\alpha\leq 1$	SU(2)+SU(2)+SU(2)	(2,2,2)	9,8
F(4)		$\overline{SO(7)}$+SU(2)	(8,2)	21,16
G(3)		G_2+SU(2)	(7,2)	14,14
P(m-1)	m=3,4,...	SU(m)	$(m\times m)_S+(m\times m)_A$	m^2-1, m^2
Q(m-1)	m=3,4,...	SU(m)	Adjoint	m^2-1, m^2-1

Note that (algebra) ismorphisms SU(2|1) ≃ OSp(2|2) and D(2,1,1) ≃ OSp(4|2).

Now we can select the algebras which satisfy the three requirements mentioned above. Observing the isomorphisms SO(3,2) ≃ Sp(4) and SO(4,2) ≃ SU(2,2), we obtain[6] table 2 listing superconformal algebras with $3\leq d\leq 6$. In these dimensions one finds solutions with a general extension N, except

Table 2: Superconformal algebras for $3\leq d\leq 6$.

dimension	superalgebra	bosonic algebra
d=3	OSp(N|4)	Sp(4) + SO(N)
d=4	SU(2,2|N)	SU(2,2) + SU(N) (+ U(1) if N≠4)
d=5	F(4)	SO(5,2) + SU(2)
d=6	OSp(6,2|N)	SO(6,2) + Sp(N) (N even)

when d=5. In the latter case one could try explicitly to find generators which one can introduce in $\{Q,S\}$, commuting with the conformal algebra, such that the Jacobi identities can be satisfied.[7]

$$\{Q_{ai}, S^{bj}\} = -\frac{1}{2}\,\delta_i{}^j(\delta_a{}^b D + \Gamma_{\mu\nu a}{}^b M^{\mu\nu}) + \delta_a{}^b X_i{}^j \qquad (1)$$

where X are unknown bosonic generators, and the spinors satisfy a symplectic Majorana condition.[8] We suppose that all supersymmetries square to translations with the symplectic metric Ω:

$$\{Q_a{}^i, Q_b{}^j\} = \frac{1}{2}\Omega^{ij}\Gamma^{\mu}{}_{ab}P_{\mu} \tag{2}$$

We then parametrize the [X,Q] commutator as

$$[X_i{}^j, Q_{ak}] = A_i{}^j{}_k{}^{\ell} Q_{a\ell} \tag{3}$$

Two different sectors in the [Q,Q,S] Jacobi identity determine

$$4\, A_{[i}{}^k{}_{j]}{}^{\ell} = -9\, \delta_{[i}{}^k\, \delta_{j]}{}^{\ell} = -\delta_{[i}{}^k\, \delta_{j]}{}^{\ell} + 4\Omega_{ij}\Omega^{k\ell} \tag{4}$$

which is contradictory except when N=2, in which case one can then obtain the full algebra.[7] This confirms the group theory argument that in d=5 we have only a N=2 superconformal algebra with the properties mentioned above. We will encounter a similar situation in the infinite dimensional superconformal algebra in d=2, where under some assumptions[2] (see section 3) only N=1, 2 and 4 are possible.[9]

In dimensions higher than 6 one cannot get this type of superconformal algebra. One could relax condition 3 and also allow new generators in {Q,Q}. In this way one gets superconformal algebras in higher dimensions.[10] E.g. one gets for d=10 the superalgebra OSp(1|32), for d=11 OSp(1|64) and for d=12 there is SU(64|1). However the superconformal structure of d=10 supergravity[11] has, as far as we know, no relation to this algebra. It makes use of structure functions depending on fields. So far, this is not unusual in supersymmetric gauge theories[12] or in supergravity, and some general features have been studied by Sohnius.[13] Usually there exists a 'linearized' limit. This means that if we set the fields equal to zero or to some vacuum expectation value, one obtains a consistent algebra. In the algebra of Bergshoeff et al.[11] one has

$$\{Q_a, Q_b\} = -\frac{1}{2}\Gamma^{\mu}{}_{ab}P_{\mu} - \frac{21}{16}\Gamma^{\mu}{}_{ab}\,\bar{\lambda}\Gamma_{\mu}Q + \frac{1}{1280}\Gamma^{\mu\nu\rho\sigma\tau}_{ab}\,\bar{\lambda}\Gamma_{\mu\nu\rho\sigma\tau}Q$$

$$+ \text{ other field-dependent terms}$$

$$\{Q_a, S_b\} = -\frac{1}{2}C_{ab}D - \frac{1}{2}\Gamma_{ab}{}^{\mu\nu}M_{\mu\nu} + \text{field-dependent terms} \tag{5}$$

λ is a spinor field of the theory which under S transformations behaves as

$$[S_a, \lambda_b] = C_{ab} \qquad \text{or} \qquad \delta_S(\eta)\,\lambda = \eta \tag{6}$$

This makes it already clear that we cannot go to a 'linearized' limit. In the [Q,Q,S] Jacobi identity the λ terms are necessary. In the case of the OSp(1|32) algebra one does not have the λ terms but {Q,Q} contains a 5-index symmetry $Z_{[\mu\nu\rho\sigma\tau]}$ and {Q,S} contains a 4-index symmetry $U_{[\mu\nu\rho\sigma]}$.

There are also 4-dimensional superconformal algebras for which the bosonic subalgebra is not of the type SO(4,2) + G. These algebras[14] also contain extra bosonic generators in {Q,Q}. It is the algebras OSp(4,4|N) which therefore exist for even values of N. They are proposed for including central charges in the superconformal framework. However, so far the only successful way to do so starts from a field dependent algebra based on the generators of SU(2,2|2) and a bosonic algebra.[15]

The d=2 conformal algebra splits in 2 mutually commuting parts, as $SO(2,2) = SU(1,1) + SU(1,1)$. The superalgebras are also direact sums of two superextensions of $SU(1,1) \simeq SO(2,1) \simeq Sp(2)$. For each sector there are now several possibilities[6] listed in table 3 with the bosonic subgroup, the dimension of the algebra commuting with SU(1,1), and the number of Q (or S) supersymmetries. Observe that there is one N=1, 2 and 3 algebra as $SU(1|1,1) \simeq OSp(2|2)$, but then e.g. for N=4 there are different possibilities $D^1(2,1,\alpha)$ with $D^1(2,1,1) \simeq OSp(4|2)$ and there is $SU(2|1,1)$. We will come back to this in the next section.

Table 3: Finite dimensional d=2 superconformal algebras.

superalgebra	bosonic algebra	dim G	N
$OSp(N\|2)$	$O(N) + Sp(2)$	$\frac{1}{2}N\ (N-1)$	N
$SU(m\|1,1)\quad (m\neq 2)$	$SU(1,1) + SU(m) + U(1)$	m^2	2m
$SU(2\|1,1)$	$SU(1,1) + SU(2)$	3	4
$OSp(4\|2m)$	$SU(1,1) + SU(2) + Sp(2m)$	m(2m+1)+3	4m
G(3)	$G_2 + SU(1,1)$	14	7
F(4)	$SO(7) + SU(1,1)$	21	8
$D^1(2,1,\alpha)$	$SU(1,1) + SU(2) + SU(2)$	6	4

INFINITE DIMENSIONAL SUPERCONFORMAL ALGEBRAS

The infinite dimensional superconformal algebra in d=2 splits also in two identical parts, and we can again restrict ourselves to one of them. This is the Virasoro algebra. The SU(1,1) subalgebra is generated by L_{-1}, L_0 and L_1 corresponding to $\frac{1}{2}(P_2+iP_1)$, $\frac{1}{2}(M-D)$ and $\frac{1}{2}(K_2-iK_1)$ of the previous section. The Q^i and S^i supersymmetries are now the $G^i_{-1/2}$ and $G^i_{1/2}$ components of the family G^i_r, $r = \pm 1/2, \pm 3/2, \ldots$ (Neveu-Schwarz fermionic operators). They are dimension 3/2 conformal operators. A conformal operator X_m of dimension D satisfies

$$[L_n, X_m] = (n(D-1) - m)\ X_{n+m} + \text{terms with other operators} \tag{7}$$

Note that this allows 2D-1 operators $X_{D-1},\ldots,X_{-(D-1)}$ in the finite dimensional subalgebra. The anticommutator of fermionic operators is at this point unknown.

$$\{G_r^{\ i}, G_s^{\ j}\} = X_{rs}^{\ \ ij} \tag{8}$$

Both subindices of X run over an infinite range. Jacobi identities determine:[2]

$$\begin{aligned}
[L_n, X_{rs}^{\ \ ij}] &= (\tfrac{1}{2}n-s)\ X_{r,s+n}^{\ \ \ \ ij} + (\tfrac{1}{2}n-r)\ X_{r+n,s}^{\ \ \ \ ij} \\
[X_{rs}^{\ \ ij}, G_t^{\ k}] &= G_{r+s+t}^{\ \ \ \ \ell}[(s-r)O^{[ij]k}{}_{\ell} + (2t-r-s)\ O^{(ij)k}{}_{\ell}]
\end{aligned} \tag{9}$$

To arrive at the latter we assumed that those $G_r^{\ i}$ are the only fermionic generators, that they are all independent and non-zero. $O^{ijk}{}_{\ell}$ are constants which have to satisfy a set of equations. Further also the X operators are constrained:

$$O^{ijk}{}_h X^{h\ell}{}_{rs} = Y_{r+s}{}^{ijk\ell} + (r-s)\, V_{r+s}{}^{ijk\ell} + [\tfrac{1}{2}(r^2-s^2)-(r^2+s^2-\tfrac{1}{2})] Z_{r+s}{}^{ijk\ell} \quad (10)$$

where Y, V and Z are conformal operators of dimension resp. 2, 1 and 0 which satisfy symmetry relations on their upper indices. If the $N^3 \times N$ matrix $O^{ijk}{}_\ell$ has rank N then all operators X are determined in terms of Y, V and Z. If the rank of O is R<N then there are $(N-R)^2$ operators left with 2 subindices running over the infinite range. For N=2 all solutions of the mentioned equations have been obtained.[2] There is one simple algebra[16] to which we come back shortly, and further several other ones, the most remarkable is a solvable algebra obtained by having as only nonzero constant $O^{111}{}_2$. Apart from commutators with L_n the nonzero (anti)commutators are

$$\{G_r{}^1, G_s{}^1\} = X_{rs}$$

$$\{G_r{}^1, G_s{}^2\} = Y_{r+s} + [\tfrac{1}{2}(r^2-s^2)-(r^2+s^2-\tfrac{1}{2})]\ Z_{r+3}$$

$$[X_{rs}, G_t{}^1] = (2t-r-s) G_{r+s+t}{}^2$$

$$[X_{r,m-r}, X_{s,n-s}] = 2(n-m)\ Y_{m+n} + (m^3-m-n^3+n)\ Z_{m+n} \quad (11)$$

All these algebras still contain a finite dimensional subalgebra spanned by L_{-1}, L_0, L_1, $G_{-1/2}$, $G_{1/2}$, $X_{-1/2,-1/2}$, $X_{1/2,-1/2}$ and $X_{1/2,1/2}$.

If the Virasoro operators L occur in the anticommutator of the fermionic operators (as we usually see supersymmetries as 'square roots' of translations), then O must have rank N, and so the X operators split into dimension 2, 1 and 0. Moreover, the latter ones are then restricted to 'central extensions':

$$Z_n{}^{ij} = z^{ij}\ \delta_n \quad (12)$$

The finite dimensional subalgebra contains now instead of the X operators mentioned above: Y_{-1}, Y_0, Y_1 and V_0. The Y generators in this list do not commute with the L operators. If we want, as in the previous section, that all extra bosonic symmetries commute with the space-time symmetries, then we are led to demand that the Virasoro operators L are the only dimension 2 operators. If the remaining algebra is simple it must be in table 3.

If we look for infinite dimensional superconformal algebras with L_n as only dimensional 2 operator we can derive that

$$O^{[ij]k}{}_\ell = -2\ b^{k[i}\ \delta^{j]}{}_\ell + \phi^{ijk}{}_\ell \quad (13)$$

where b^{ij} is a symmetric matrix which is from now on used to reduce indices, and $\phi^{ijk}{}_\ell$ is antisymmetric in 3 indices, and even in 4 indices if the last one is raised. The algebra contains now

$$\{G_r{}^i, G_s{}^j\} = -2\ b^{ij}(L_{r+s} + (c/6)(r^2-1/4)\delta_{r+s}) + (r-s)\ V_{r+s}{}^{ij}$$

$$[V_n{}^{ij}, G_r{}^k] = -O^{[ij]k}{}_\ell\ G_{n+r}{}^\ell$$

$$[V_m{}^{ij}, V_n{}^{k\ell}] = 2\ O^{[ij][k}{}_h\ V_{m+n}{}^{\ell]h} + \tfrac{1}{3}\,cm\ \delta_{m+n}\ O^{[ij](kl]} \quad (14)$$

V^{ij} is antisymmetric. The Jacobi identity $[V^{ij}, V^{kl}, G^u]$ implies

$$O^{[ij]k}{}_h \, O^{[\ell h]u}{}_v - (k \leftrightarrow \ell) = O^{[ij]u}{}_h \, O^{[kl]h}{}_v - (ij \leftrightarrow kl) \tag{15}$$

From the antisymmetry of the last commutator in (14) we get a condition on V^{ij}:

$$O^{[ij][k}{}_h V^{\ell]h} = - O^{[k\ell][i}{}_h V^{j]h} \tag{16}$$

A second condition follows from the Jacobi identity $[V_m{}^{ij}, G^k, G^\ell]$. This is trivial for m=0 (the only one occurring in the finite dimensional subalgebra), but for m≠0 it implies

$$O^{[ij](k}{}_h V_n{}^{l)h} = -b^{kl} V_n{}^{ij} \tag{17}$$

A consequence of this is (first commuting with G and then a contraction of indices)

$$2(2-N)\, b^{k[i} b^{j]\ell} + (N-4)\phi^{ijk\ell} - \phi^{iju}{}_v \, \phi^{klv}{}_u = 0 \tag{18}$$

Another important consequence from (17) is that the number of independent dimension 1 operators (dim G) is N-1.

At this point we can make one of the following two assumptions: 1) Each fermionic operator G is a 'square root' of L, i.e. the matrix b in (14) is nonsingular, and can therefore be redefined to δ^{ij}. 2) The finite dimensional subalgebra is simple (which is necessary for the full algebra to be simple). In both cases one arrives only at an N = 1, 2 or 4 algebra. The proof using assumption 1) is in Gastmans et al.[3] and follows the line of Schwarz and Ramond.[9] If we use assumption 2) we examine table 3 using the above mentioned result dim G = N - 1. We only find then OSp(1|2), OSp(2|2) ≃ SU(1|1,1) and SU(2|1,1).

To illustrate the difference between the finite and infinite dimensional algebras we first solve (15) and (16) for O. Those are already solved by (13) with ϕ=0 and no more conditions on V. These are the algebras OSp(N|2). For the infinite dimensional algebras, we should also have (18) which implies that this solution is only good for N=2. The first extra solution, because of the antisymmetry of ϕ, can only occur at N=4.

$$b^{ij} = \delta^{ij} \qquad \phi^{ijkl} = - x\, \varepsilon^{ijkl} \tag{19}$$

This solves (15) and (16) for arbitrary values of x. SO(4) ≃ SU(2) × SU(2) corresponds to the fact that the selfdual and antiselfdual parts of V^{ij} mutually commute. In the algebra they now occur with weights (1±x): e.g.

$$[V^+{}_0{}^{ij}, V^+{}_0{}^{k\ell}] = (1+x)\, 4\, \delta^{[i}{}_{[k} V^+{}_0{}^{j]}{}_{\ell]}$$
$$[V^+{}_0{}^{ij}, G_r{}^k] = (1+x)\, \delta^{k[i} G_r{}^{j]} \tag{20}$$

This algebra is $D^1(2,1,(1-x)/(1+x))$. Inequivalent algebras are obtained in the range $0 \le x \le 1$. The case x=0 is OSp(4|2). For x=1 the algebra is not simple. We can then impose self-duality on V to arrive at the simple algebra SU(2|2). For the infinite dimensional algebras the extra condition (17) has the consequence (see (18)) x = ±1. So now we have only SU(2|1,1) and V must be (anti)selfdual.

GAUGE THEORY OF INFINITE DIMENSIONAL ALGEBRAS

It is also possible to start a conformal calculus from the infinite dimensional algebras.[3] In fact this has some nice features. For simplicity we will consider here just the Virasoro algebra. Recall first what was done for the gauge theories of the conformal algebras in d>2. The transformation on space-time is then

$$\delta x_\mu = \xi_\mu(x) = a_\mu + \omega_{[\mu\nu]}x^\nu + \Lambda_D x_\mu + \Lambda_K{}^\nu(2x_\nu x_\mu - g_{\mu\nu}x^2) \tag{21}$$

The constants a_μ, $\omega_{\mu\nu}$, Λ_D and $\Lambda_K{}^\nu$ are parameters for translations P^a, Lorentz rotations M^{ab}, dilatations D and special conformal transformations K^a. For the gauge theory we allow these parameters to be independent functions of x. We introduce gauge fields for each of them $e_\mu{}^a$, $\omega_\mu{}^{ab}$, b_μ and $f_\mu{}^a$. By imposiing constraints on the curvatures, one can replace the translations in the algebra by general coordinate transformations. These can be solved for ω and f. On the other hand b can be chosen to vanish as as gauge choice for K. So one is left with the vierbein e as gauge field and general coordinate transformations, local Lorentz transformations and dilatations as gauge symmetries.

For d=2 the Killing vector (21) is

$$\xi = \xi^0 + \xi^1 = \sum_{n=-\infty}^{\infty} \xi_n z^{n+1} \qquad z = x^0 + x^1$$
$$\bar\xi = \xi^0 - \xi^1 = \sum_{n=-\infty}^{\infty} \bar\xi_n \bar z^{n+1} \qquad \bar z = x^0 - x^1 \tag{22}$$

where ξ and $\bar\xi$ are the constant parameters corresponding to L and $\bar L$, which in the gauge theory are independent functions of x. The transformtions of a conformal field of dimension D is (we restrict ourselves further to the ξ transformations)

$$\delta\phi(z) = \sum \xi_n L_n \phi(z) = \int \frac{dw}{2\pi i}\, \xi(w)\ T(w)\ \phi(z)$$
$$= \xi(z)\ \phi'(z) + \xi'(z)\ D\phi(z) + \text{other terms with } \xi \text{ and its derivatives} \tag{23}$$

(this corresponds to eq. (7)). In these transformations occur $\xi(z)$ and its derivatives $\xi'(z)$, $\xi''(z)$, ···, which are thus some combinations of the ξ_n parameters. In the gauge theory they become independent functions of z and $\bar z$, which we call $\xi^{-1}(z,\bar z)$, $\xi^0(z,\bar z)$, $\xi^1(z,\bar z)$, ···

$$\xi(z) = \sum_{-\infty}^{\infty} \xi_n z^{n+1} \quad \to \quad \xi^{-1}(z,\bar z) = \sum_{-\infty}^{\infty} \xi_n(z,\bar z)\ z^{n+1}$$
$$\xi'(z) = \sum (n+1)\xi_n z^n \quad \to \quad \xi^0(z,\bar z) = \sum (n+1)\xi_n(z,\bar z) z^n \tag{24}$$

We introduce then gauge fields $h^n{}_z$ and $h^n{}_{\bar z}$ and for the $\bar\xi$ transformations $\bar h^n{}_z$ and $\bar h^n{}_{\bar z}$, where n = -1, 0, 1, 2, ···. The four n=-1 fields correspond to the zweibein $e_\mu{}^a$ (which should be invertible), the n=0 fields to b and ω, and the n=1 fields to f. As constraints we put now all curvatures zero.

$$R^{n}(z,\bar{z}) = \bar{R}^{n}(z,\bar{z}) = 0 \qquad n = -1, 0, 1, \cdots \tag{25}$$

These constraints can be solved for $h_{\bar{z}}^{n+1}$ and $\bar{h}_{z}^{n+1}$. Also as in the gaugings of the finite dimensional algebra we can impose gauge choices for L_{n+1} and $\bar{L}_{n+1}$ transformations for $n \geq 0$

$$h_{z}^{n} = \bar{h}_{\bar{z}}^{n} = 0 \qquad \text{for} \qquad n = 0, 1, 2, \cdots \tag{26}$$

The final result is the same as in the gaugings of the finite dimensional conformal algebra: only the zweibein remains as an independent field and we have general coordinate transformations $(\xi^{-1},\bar{\xi}^{-1})$, local Lorentz transformations and dilatation $(\xi^{0},\bar{\xi}^{0})$. In two dimensions these can be used to gauge the zweibein to a constant.

In the gauging of the finite dimensional conformal algebra in d=2,[17] the curvature constraints do not determine $f_{\mu}{}^{a}$ completely. There are two remaining components which however disappear in all theories. This is now clear by the gauging of the infinite dimensional algebra as presented above. Also, in this way the gauge theory is elegant by the simplicity of the constraints (25).

CONCLUSIONS

We first review finite dimensional superconformal algebras. They are summarized in tables 2 and 3. In d=10 a superconformal calculus has been started, which is not based on a usual type of superalgebra.[11] For d=2 there are several classes of finite dimensional superconformal algebras. A superconformal tensor calculus has been developed based on the superalgebras OSp(N|2).[17] All the mentioned superconformal algebras have the following characteristics:

1) They are simple superalgebras with as bosonic part a direct sum containing as one of the terms the conformal algebra SO(d,2): SO(2,2) = SU(1,1) + SU(1,1), and we can consider one factor only.

2) The fermionic generators are in the spinor representation of SO(d-1,1).

In d=2 the full conformal algebra is infinite dimensional (the Virasoro algebra). We[2] have investigated the conditions for infinite dimensional superalgebras, with only 'dimension 3/2' fermionic operators (condition A). We have given all solutions for N=2. Then we have adopted a second condition (B), that L is the only dimension 2 operator. This requirement can be based on the finite dimensional subalgebra, which is always present in these infinite dimensional algebras. Demanding the structure 1) for this algebra leads to the condition B. However, as mentioned, there was no firm ground for this requirement before. Moreover now it is only based on the subalgebra if no suitable generalisation of the Coleman-Mandula theorem exists. Even with conditions A and B there exist superconformal algebras for all N. There was in the literature a theorem[9] that only N=1, 2 and 4 are possible. To prove this we need one of the two following conditions: C1: All supersymmetries square to the Virasoro algebra. C2: The superalgebra is simple. It is not clear whether condition B is necessary when we use C2. On the other hand, it is also not obvious that the superalgebras which can be useful are simple or semisimple.

One can also consider the gauge theories of infinite dimensional algebras. We used the Virasoro algebra to illustrate some features. However this can be extended to superconformal algebra, and becomes the basis of a tensor calculus.[3]

REFERENCES

1. M. Kaku, P. Townsend and P. van Nieuwenhuizen, Phys. Rev. D17(1978)3179.
   B. deWit, in "Supersymmetry and Supergravity 1982", eds. S. Ferrara, J.G. Taylor, and P. van Nieuwenhuizen (World Scientific, 1983).
   A. Van Proeyen, in "Supersymmetry and Supergravity 1983", ed. B. Milewski (World Scientific, 1983).
2. R. Gastmans, A. Servrin, W. Troost and A. Van Proeyen, preprint Leuven, KUL-TF-86/6 and in the proceedings of the Berkeley conference, 1986.
3. J.W. van Holten, Acta Physica Polonica, proceedings of the Zakopane school, to be published. A. Sevrin and J.W. van Holten, forthcoming publication.
4. S. Coleman and J. Mandula, Phys. Rev. 159(1967)1251.
5. V.G. Kac, Commun. Math. Phys. 53(1977)31.
6. W. Nahm, Nucl. Phys. B135(1978)149.
7. J. Joos, licenciaatsthesis, K.U. Leuven, 1986.
8. E. Cremmer, in "Superspace and Supergravity", eds. S.W. Hawking and M. Rocek (Cambridge University Press, 1981).
9. P. Ramond and J.H. Schwarz, Phys. Lett. B64(1976)75.
10. J.W. van Holten and A. Van Proeyen, J. Phys. A15(1982)3763.
11. E. Bergshoeff, M. de Roo and B. de Wit, Nucl. Phys. B217(1983)489.
12. B. de Wit and D.Z. Freedman, Phys. Rev. D12(1975)2286.
13. M.F. Sohnius, Z. Physik C18(1983)229.
14. S.P. Bedding, Nucl. Phys. B236(1984)368; Phys. Lett. 157B(1985)183.
15. B. de Wit, J.W. van Holten and A. Van Proeyen, Phys. Lett. 95B(1980)51.
16. M. Ademollo et al., Nucl. Phys. B111(1976)77; B114(1976)297.
17. E. Vergshoeff, E. Sezgin and H. Nishino, Phys. Lett. 166B(1986)141.
    P. van Nieuwenhuizen, IJMPA 1(1986)155.
    M. Hayashi, S. Nojiri and S. Uehara, Kyoto preprint RRK 86-2.
    J.W. van Holten, preprint NIKHEF Amsterdam, 1986.

SUPERMANIFOLDS AND SUPER RIEMANN SURFACES*

Jeffrey M. Rabin

Enrico Fermi Institute and Department of Mathematics
The University of Chicago
Chicago, IL 60637

ABSTRACT

The theory of super Riemann surfaces is rigorously developed using Rogers' theory of supermanifolds. The global structures of super Teichmüller space and super moduli space are determined. The super modular group is shown to be precisely the ordinary modular group. Super moduli space is shown to be the gauge-fixing slice for the fermionic string path integral.

1. Introduction

The theory of Riemann surfaces [1-3] has recently become an important mathematical tool for string theorists. The world sheet of a bosonic string is in fact a Riemann surface, and the g-loop contribution to an amplitude in string theory can be expressed as an integral over the moduli space of Riemann surfaces of genus g [4-6]. Such a representation of the amplitudes allows one to use powerful techniques from algebraic geometry to study their analytic properties, investigate their finiteness, and even to compute them in terms of theta functions [7-9].

Since it is the superstring rather than the bosonic string which is physically relevant and possibly finite, there is great interest in generalizing the algebraic geometry to include the fermionic coordinates of superspace [10,11]. Several authors have described a notion of "super Riemann surface" which is appropriate for this purpose [12-15]. However, the intuitive concept of a supermanifold in the physics literature is not a sufficient foundation for a theory of super Riemann surfaces because it describes only the local geometry of superspace and not its global topology. The theory of super Riemann surfaces, like that of Riemann surfaces, makes use of several topological constructions including universal covering spaces, quotient spaces, and homotopy groups. Therefore one requires a theory of

* Research supported by the NSF (PHY 83-01221) and DOE (DE-AC02-82-ER-40073).

supermanifolds which is sufficiently rigorous to give meaning to these constructions and to ensure that they have their usual mathematical properties. In this paper I will develop the theory of super Riemann surfaces using as foundation the theory of supermanifolds due to A. Rogers [16]. Other supermanifold theories exist [17], but Rogers' is both the most general and the closest to the physicist's intuitive view of superspace as a manifold with some anticommuting coordinates. It also has the great advantage that a supermanifold is in fact an ordinary manifold, so that the topological constructions have their standard meanings. Because of its extreme generality, Rogers' theory includes many topologically exotic supermanifolds which are not physically useful. This can be viewed as a disadvantage of the theory which could be avoided by using a less general one, or as an advantage in that it makes explicit the properties which are actually assumed in physical applications.

In Section 2 I review the connection between bosonic strings and Riemann surfaces, specifically the identification of the gauge-fixing slice for the functional integral with the moduli space of Riemann surfaces. This motivates the definition of a super Riemann surface, which must be such that the super moduli space coincides with the gauge-fixing slice for the fermionic string path integral. Section 3 develops Rogers' theory of supermanifolds, which allows one to discuss topological properties of super Riemann surfaces rigorously. It also clarifies some obscure points in the physics literature, such as whether the commuting coordinates of superspace are ordinary numbers or have nilpotent parts (the latter is true), and explains why the anticommuting dimensions must be topologically trivial in physical applications. Section 4 proves the basic properties of super Riemann surfaces, beginning with the relation between a super Riemann surface and an ordinary Riemann surface with spin structure, and continuing through the determination of the dimension and global structure of super moduli space. As a concrete example of this abstract discussion, the case of super tori (genus 1) is worked out explicitly. Section 5 is a brief survey of results which can be obtained with more advanced techniques, and open problems. Most of the work described here was done with Louis Crane [18,19].

2. Strings and Riemann Surfaces

Polyakov's closed bosonic string theory [20] is a theory of maps $X^\mu(\sigma^a)$ from the two-dimensional world sheet of the string with metric h^{ab} into Euclidean spacetime R^{26}, with action

$$S = \int d^2\sigma \sqrt{h}\; h^{ab}\; \partial_a X^\mu\, \partial_b X_\mu . \tag{2.1}$$

Quantization involves functional integration over the fields h^{ab} and X^μ. At g loop order we integrate over world sheets of genus g. The integral over X^μ is Gaussian, so only the integration over metrics is nontrivial. However, the action (2.1) has a large gauge symmetry: it is invariant under reparametrizations or diffeomorphisms of the world sheet as well as conformal rescalings of the metric $h^{ab} \to \Omega^2 h^{ab}$. To prevent overcounting of equivalent field configurations, we seek a gauge-fixing slice in the space of metrics which contains exactly one representative of each gauge equivalence class of metrics, and integrate over this slice. The slice will be a realization of the space of metrics modulo diffeomorphisms and conformal rescalings.

This slice is in fact the moduli space of Riemann surfaces of genus g. A Riemann surface is a two-dimensional surface with a given complex structure: a covering by local coordinate charts so that each point in a given chart is assigned a complex coordinate z and the transition functions relating the coordinates in overlapping charts are analytic. Two Riemann surfaces are considered equivalent if they are related by an analytic diffeomorphism. It turns out that not all complex structures on a given surface are equivalent, and in fact the set of all complex structures on a surface of genus g is itself a complex manifold (except at some singular points) of complex dimension 0 for $g=0$, 1 for $g=1$, and $3g-3$ for $g>1$. This manifold is called the moduli space for genus g. It is a topologically complicated space, and it is useful to define a related but topologically trivial space called Teichmüller space. The fundamental group or first homotopy group $\pi_1(M)$ of a surface M is generated by $2g$ noncontractible closed curves, but a specific set of $2g$ generators can be chosen in many different ways. A Riemann surface together with a choice of generators for its fundamental group is called a marked Riemann surface and defines a point in Teichmüller space. Each Riemann surface is represented by a single point of moduli space but by a discrete infinity of points in Teichmüller space. Moduli space can be viewed as a quotient of Teichmüller space by a discrete group which identifies all points representing the same Riemann surface. This group is the modular, or mapping class, group for genus g.

What is the connection between gauge equivalence classes of metrics and Riemann surfaces? A metric on an orientable surface determines a Riemann surface structure in the following way. Any metric in two dimensions is locally conformally flat. This means that the surface can be covered by charts such that the metric in each chart is conformal to the flat metric, $h^{ab} = \Omega^2 \delta^{ab}$. When two such charts overlap, their coordinates are conformally related. If σ^a are coordinates in such a chart, define $z = \sigma^1 + i\sigma^2$. Since a conformal transformation is an analytic function of z, this set of charts defines a Riemann surface structure. It can be shown that two metrics determine the same Riemann surface iff they are related by diffeomorphisms and conformal rescalings, and conversely that any Riemann surface is defined by some metric [3]. Therefore the gauge-fixing slice is precisely the moduli space. Questions about the behavior of the string integrand on the gauge-fixing slice can be translated into questions about the geometry of moduli space, or into questions about the action of the modular group on Teichmüller space.

The fermionic string can be formulated as a superfield $X^\mu(\sigma^a, \theta^\alpha)$ coupled to two-dimensional supergravity [21]. The action is

$$S = \int d^2\sigma \, d^2\theta \; sdet \; E \; \nabla_\alpha X^\mu \, \nabla^\alpha X_\mu, \tag{2.2}$$

where E_M^A is the super vierbein and ∇_α the corresponding covariant derivative. Again, quantization requires functional integration over X^μ and E_M^A, and we seek a gauge-fixing surface for the super reparametrization and superconformal symmetries of the action. It is known that any supergravity geometry in two dimensions is locally superconformal to flat superspace [21], meaning that coordinates can be found in which the super vierbein takes the form,

$$E_M{}^a = \Omega e_M{}^a, \tag{2.3a}$$

$$E_M{}^\alpha = \sqrt{\Omega}\, e_M{}^\alpha - i e_M{}^a\, \gamma_a^{\alpha\beta}\, D_\beta \sqrt{\Omega}. \tag{2.3b}$$

Here $\Omega(\sigma,\theta)$ is the conformal factor and $e_M{}^A$ and D_A are the flat super vierbein and covariant derivative. These local coordinates define complex coordinates by $z = \sigma^1 + i\sigma^2$, $\theta = \theta^1 + i\theta^2$. In this way we obtain a complex supermanifold of dimension (1,1) whose transition functions are superconformal maps. Such a supermanifold is a super Riemann surface (SRS). The moduli space of super Riemann surfaces should be the gauge-fixing surface for the functional integral, because supergravity geometries related by reparametrizations and superconformal transformations define the same super Riemann surface. This is the motivation for the rigorous theory of SRS's to be developed in Section 4.

3. Supermanifolds

A supermanifold should be a space having both even and odd coordinates. I will discuss the case of immediate interest in which there is one even complex coordinate z and one odd complex coordinate θ, but generalization is easy.

If z and θ are to be coordinates rather than just symbols, they must be able to assume values in some number system so that different values of the coordinates can label different points. We take as this number system a Grassmann algebra B_L with L generators $v_1, v_2, \cdots, v_L$ obeying $v_i v_j = -v_j v_i$. The most general even and odd elements of this algebra have the form,

$$\begin{aligned} z &= z_0 + z_{ij} v_i v_j + \cdots, \\ \theta &= \theta_i v_i + \theta_{ijk} v_i v_j v_k + \cdots, \end{aligned} \tag{3.1}$$

where the coefficients $z_{ij\cdots l}$ and $\theta_{ij\cdots l}$ are ordinary complex numbers, and all subscripts are in increasing order, $i<j<\cdots<l$. The coefficient z_0 is called the body of z, and the soul of z is $z - z_0$ [22]. θ has no body and is pure soul. The fact that there are 2^L independent complex coefficients in the expansions (3.1) suggests the following idea. Take an ordinary complex manifold M of dimension 2^L. In each chart it has 2^L complex coordinates, which we call the body coordinate z_0 and the soul coordinates θ_i, z_{ij}, $\cdots$. Then the series (3.1) define one even and one odd complex Grassmann coordinate in each chart, and different values of these coordinates do indeed label different points of M. Obviously this can be done on any complex manifold M. But we will now demand that the transition functions of M have special analytic properties when expressed in terms of the Grassmann coordinates. Not every M will admit coordinates with these properties; the ones that do will be called complex supermanifolds.

The transition functions relating coordinates in overlapping charts have the form $\tilde{z} = \tilde{z}(z,\theta)$, $\tilde{\theta} = \tilde{\theta}(z,\theta)$. We require these functions to have the properties normally assumed of superfields in physics. They must take the form,

$$\tilde{z} = f(z) + \theta\zeta(z), \quad \tilde{\theta} = \psi(z) + \theta g(z), \tag{3.2}$$

where the B_L-valued component functions f,ζ,ψ,g are required to have Taylor expansions in powers of the soul of z, for example,

$$f(z) = f(z_0) + (z - z_0)f'(z_0) + \cdots, \tag{3.3}$$

with $f(z_0)$ analytic. Such functions will be called superanalytic. Notice that the series (3.3) terminates because $z - z_0$ is nilpotent, so convergence is not a problem. If we were discussing real rather than complex supermanifolds, $f(z_0)$ would be required to be smooth rather than analytic, and the transition functions would then be called supersmooth or G^∞.

Eq. (3.3) clears up a confusing point in the literature. The even coordinates of superspace are always treated as ordinary numbers despite the fact that they clearly cannot be: z cannot be an ordinary number both before and after a supersymmetry transformation $\tilde{z} = z + \theta\eta$, $\tilde{\theta} = \theta + \eta$. The present formalism resolves this problem: only the body of z is an ordinary number, not z itself. But then why can superfields be manipulated as if their even arguments were ordinary numbers? The answer is that according to (3.3) a superanalytic function is completely determined when its components are known for soulless values of z. Functions of z are in 1-1 correspondence with functions of z_0 and this correspondence is preserved by algebraic manipulations. In other words, a Grassmann-valued function of z_0 can always be analytically continued to a function of z with the same algebraic properties.

One expects on physical grounds that an ordinary manifold M_0 can be obtained from a supermanifold M by "throwing away" all the soul coordinates. If M is a supergravity superspace, M_0 should be physical spacetime; if M is the super world sheet of a fermionic string, M_0 should be the physical world sheet. We will see that M_0, called the body of M, does not generally exist unless the topology of M is restricted.

In each chart of M, consider the surfaces of constant z_0. When charts overlap we have

$$\begin{aligned}\tilde{z} &= f(z) + \theta\zeta(z)\\ &= f(z_0) + (z - z_0)f'(z_0) + \cdots + \theta\zeta(z).\end{aligned} \tag{3.4}$$

Since every term on the right is pure soul except the first, $\tilde{z}_0$ depends only on z_0. This means that a surface of constant z_0 is also a surface of constant $\tilde{z}_0$, so these surfaces fit together smoothly to give the leaves of a foliation, called the soul foliation. Every point of M lies on exactly one constant body surface. A topological space M_0 is obtained by identifying all points of M which lie on the same surface. Unfortunately M_0 need not be a smooth manifold or even Hausdorff.

As a simple example, consider the Euclidean plane with coordinates denoted x_0 and θ_1. This is a real, rather than complex, supermanifold over the trivial Grassmann algebra B_1, with real Grassmann coordinates $x = x_0$ and $\theta = \theta_1 v_1$. The leaves of the soul foliation are the lines of constant x_0 and the body can be thought of as the x_0 axis. The quotient space of the plane by the group of integer translations along the coordinate axes is a torus, which is also a supermanifold. The leaves are now circles going around the torus in the θ_1 direction, and the body is a circle in the x_0 direction. The quotient of the plane by the group of integer translations along two axes having irrational slope, however, is a torus whose soul foliation consists of spirals each of which is dense in the entire torus! In this case M_0 is a non-Hausdorff topological space, and no body manifold exists. Intuition suggests that only supermanifolds with bodies can be physically relevant.

Physics imposes further restrictions on the topology of the leaves of the soul foliation. These restrictions follow from the theorem that any G^∞ function on a supermanifold must be constant along any compact leaf of the soul foliation [18]. In string theory this would imply that the map $X^\mu(\sigma,\theta)$ must send each compact leaf to a point. But then there is no embedding of the super world sheet in any flat superspace, so the super world sheet cannot represent a string moving in spacetime if it contains compact leaves. To prove the theorem, let F be a real G^∞ function on M. Expand F in terms of the basis of B_L as $F = F_0 + F_i v_i + \cdots$, and consider any of the coefficient functions $F_{ij \cdots l}$ on the compact leaf. If it is nonconstant then it necessarily achieves a maximum on the leaf. However, one can see from Eqs. (3.2) and (3.3) that a G^∞ function is always a polynomial function of the soul coordinates. In fact it is linear in each soul coordinate separately: because each soul coordinate appears together with at least one v_i which squares to zero, no higher power of a soul coordinate can appear. Elementary calculus (the second derivative test) shows that such a function cannot have maxima, but only saddle points. Therefore each $F_{ij \cdots l}$ is constant on the leaf. Evidently the supermanifolds of interest in physics should contain no compact leaves and should have bodies. Because there is no complete classification of supermanifolds, it is not known whether these requirements eliminate all possibilities for nontrivial topology in the soul directions, but examples of additional possibilities have not been found.

The simplest way to guarantee that a supermanifold M will have a body and be free of compact leaves is to demand that it have the DeWitt topology [22]. This means that each coordinate chart must be the Cartesian product of an open set in the z_0 plane with the entire complex planes of the soul coordinates. The leaves of a DeWitt supermanifold are complex vector spaces C^k, $k = 2^L - 1$, with trivial topology, and a body always exists. A set of charts for the body is obtained by projecting the charts of M onto the z_0 plane, and the transition functions for the body are the bodies $\tilde{z}_0 = f_0(z_0)$ of the transition functions of M, Eq. (3.2). M is then a fiber bundle over M_0 with fiber C^k. All supermanifolds used in physics are implicitly assumed to have the DeWitt topology. Historically this was due to the absence of any definition permitting any other topology, but as we have seen there are good physical reasons for the choice.

Finally, there are some technical points connected with the size of the Grassmann algebra B_L. Nothing in the mathematical theory depends on the value of L, which should be thought of as a large but finite integer. However, for physical applications it is necessary to take the $L \to \infty$ limit. For example, Green's functions containing more than L fermionic fields vanish identically if L is finite, because the product of more than L odd elements in B_L is zero. The limit is well understood and is fully discussed in [23].

Another technical problem which arises for finite L is an ambiguity in the definition of the derivative ∂_θ. It should be defined so that $\partial_\theta[f(z) + \theta\zeta(z)] = \zeta(z)$, but ζ is ambiguous because adding a multiple of $v_1 v_2 \cdots v_L$ to ζ does not change $\theta\zeta$. The ambiguity can be removed by insisting that the components $f(z_0), \zeta(z_0)$ of all G^∞ functions take values in the subalgebra B_{L-1} obtained by deleting the generator v_L. This method of correcting the problem is necessary if ∂_θ is to obey the Leibniz rule [23].

We now have a theory of supermanifolds which is fully rigorous and gives us control over their topological properties, yet which reduces in practical calculations to the physicist's standard calculus of superfields. We can now develop the theory of SRS's on a firm mathematical basis.

4. From Super Riemann Surfaces to Super Moduli Space

Let M be a complex supermanifold of dimension (1,1). In each chart we have the flat covariant derivative

$$D = D_\theta = \partial_\theta + \theta\partial_z, \quad D^2 = \partial_z, \tag{4.1}$$

as well as the flat frame field

$$e^z = dz + \theta d\theta, \quad e^\theta = d\theta. \tag{4.2}$$

(My convention for superforms is that $d\theta$ commutes with itself but anticommutes with dz and with θ.) The transition functions have the superanalytic form,

$$\tilde{z} = f(z) + \theta\zeta(z), \quad \tilde{\theta} = \psi(z) + \theta g(z). \tag{4.3}$$

Following Eq. (2.3a) we define a superconformal map as a superanalytic transformation (4.3) under which e^z is multiplied by a function Ω. A simple calculation shows that (4.3) is superconformal iff

$$\zeta = g\psi, \quad g^2 = f' + \psi\psi', \tag{4.4}$$

and one finds that $\Omega = (D\tilde{\theta})^2$. We say that M is a super Riemann surface if its transition functions are superconformal,

$$\begin{aligned} \tilde{z} &= f + \theta\psi\sqrt{f'}, \\ \tilde{\theta} &= \psi + \theta\sqrt{f' + \psi\psi'}. \end{aligned} \tag{4.5}$$

Unless otherwise stated, all SRS's are assumed to have the DeWitt topology.

The body M_0 of a SRS M is a Riemann surface whose transition functions are $\tilde{z}_0 = f_0(z_0)$. Then $f'_0(z_0)$ are the transition functions for the tangent bundle of M_0 (holomorphic tangent vectors transform by this factor under a change of coordinates), and a choice of signs for the square roots $\sqrt{f'_0(z_0)}$ defines a spin structure on M_0: a consistent transformation law for spinors under changes of coordinates. Recall that there can be many inequivalent spin structures on a nonsimply connected manifold [24,25]. Roughly speaking, spinors can be chosen to be either periodic or antiperiodic around each of the $2g$ noncontractible closed curves on a surface of genus g, so that there are 2^{2g} spin structures. Because the body of $\sqrt{f'(z)}$ is $\sqrt{f'_0(z_0)}$, such a choice of signs is implicit in Eqs. (4.5). Therefore a SRS determines a particular spin structure on its body. Conversely, given a Riemann surface M_0 with a particular spin structure, a SRS M can be constructed in a canonical way. The charts for M are the Cartesian products of the charts of M_0 with the entire complex planes of the soul coordinates. If $f(z_0)$ are transition functions for M_0 then the transition functions of M have $\psi(z) = 0$ and $f(z)$ the Grassmann analytic continuation of $f(z_0)$, with the signs of the square roots in Eqs. (4.5) chosen according to the given spin structure. A SRS constructed in this way will be called canonical, and we will see that not all SRS's are of this type.

We now begin the task of classifying all SRS's, with the goal of obtaining a picture of the super moduli space which is the gauge-fixing surface for the fermionic string path integral. If M is a SRS, construct its universal covering space $\hat{M}$. This is a simply connected manifold which locally looks just like M. The standard theory of covering spaces [1] shows that $\hat{M}$ is also a SRS, and $M = \hat{M}/G$ where G is a group of superconformal transformations isomorphic to $\pi_1(M)$. Furthermore, since M is a bundle with topologically trivial fibers, any noncontractible loops in M are due to the topology of M_0, meaning that $\pi_1(M) = \pi_1(M_0)$. The classification problem now splits into two parts: classifying all simply connected SRS's $\hat{M}$, and determining the groups G which can act on them.

A simply connected SRS $\hat{M}$ has a simply connected Riemann surface $\hat{M}_0$ as its body. According to the classical uniformization theorem [1] there are only three simply connected Riemann surfaces: the complex plane C, the upper half plane U, and the Riemann sphere C^* (thought of as the plane plus the point at infinity). Each of these surfaces has a unique spin structure, so exactly one canonical SRS can be constructed over each of them, to be denoted SC, SU, and SC^*. I claim that these are the only simply connected SRS's. To verify that a given SRS $\hat{M}$ is one of these three, one would examine its transition functions. If all the functions $\psi(z)$ are zero and all the $f(z_0)$'s are soulless, the SRS is canonical and must be one of these three. If these conditions do not hold, the SRS may still be canonical, because the transition functions can always be changed by redefining the coordinates in the charts. The question becomes whether some superconformal coordinate redefinitions can set to zero all the ψ's and the souls of the f's. If so, the SRS was canonical but this fact was obscured by a poor original choice of coordinates. Such questions about coordinate redefinitions can be answered by the methods of sheaf cohomology

[26]. In the present case they show that the desired coordinate redefinitions can always be found, which proves that SC, SU, and $SC*$ are indeed the only simply connected SRS's [19].

To complete the classification we must determine the groups G of superconformal automorphisms of the simply connected SRS's. Begin with $SC*$. The body of a superconformal automorphism of $SC*$ must be a conformal automorphism of $C*$, namely a Möbius transformation

$$\tilde{z}_0 = \frac{a_0 z_0 + b_0}{c_0 z_0 + d_0}. \tag{4.6}$$

The soul of the superconformal automorphism is restricted only by the condition that it be well defined at $z_0 = -d_0/c_0$, which is mapped to the point at infinity by (4.6). The behavior of a map at infinity is studied by using the transition functions of $SC*$ to switch from $(\tilde{z}, \tilde{\theta})$ to new coordinates $(-1/\tilde{z}, \tilde{\theta}/\tilde{z})$ which must be finite. This is true only for maps of the form,

$$\tilde{z} = \frac{az+b}{cz+d} + \theta\frac{\gamma z+\delta}{(cz+d)^2},$$
$$\tilde{\theta} = \frac{\gamma z+\delta}{cz+d} + \frac{\theta}{cz+d}(1+\frac{1}{2}\delta\gamma). \tag{4.7}$$

The transformations (4.7) form the supergroup of superconformal automorphisms of $SC*$. It has three independent even parameters (since a choice of normalization can impose the constraint $ad - bc = 1$) and two odd ones, and will be denoted $SPL(2,C)$. Its super Lie algebra is the subalgebra of the Neveu-Schwarz algebra generated by L_0, $L_{\pm 1}$, and $G_{\pm 1/2}$ [12,27].

The groups of superconformal automorphisms of SC and SU are much larger than $SPL(2,C)$. These SRS's do not contain the point at infinity, so finiteness at this point is no longer required. Any superconformal map whose body is a Möbius transformation is an automorphism of these SRS's. For example,

$$\tilde{z} = z + 1 + \theta\eta z^n, \quad \tilde{\theta} = \theta + \eta z^n, \tag{4.8}$$

is an automorphism of SC which does not belong to $SPL(2,C)$ for $n>1$.

The classification theorem for SRS's states that any SRS is a quotient of SC, SU, or $SC*$ by a group G of superconformal automorphisms. Such a quotient space does not automatically have the DeWitt topology, however. If G_0 is the group of Möbius transformations which are the bodies (4.6) of elements of G, then G_0 must act properly discontinuously on the body of $\hat{M}$ to get the DeWitt topology. This means that each point of $\hat{M}_0$ must have a neighborhood which does not intersect any of its images by elements of G_0. (Roughly speaking, G_0 should not have fixed points or points which are "nearly fixed".) Supersymmetry provides an example of what goes wrong if this condition is not met. The supersymmetry transformation

$$\tilde{z} = z + \theta\delta, \quad \tilde{\theta} = \theta + \delta, \tag{4.9}$$

fixes the body of every point of SC, transforming only the soul. Taking the quotient of SC by this transformation will identify points with the same body but different souls. This curls up the soul fibers, violating the DeWitt topology.

To complete the argument that the moduli space of SRS's is the gauge-fixing slice for the fermionic string, it must be shown that every SRS admits a supergravity geometry. This is in fact only true if the group G is a subgroup of $SPL(2,C)$, so the extra automorphisms of SC and SU have no physical relevance. A supergravity geometry on M can be defined by a frame field on $\hat{M}$ which is invariant under G up to a phase, since $U(1)$ is the tangent space group of two-dimensional supergravity. On SC such a frame field is given by (4.2), which is invariant up to a phase under a subgroup of $SPL(2,C)$ to be determined below but not under any other superconformal transformations. On SU the frame field

$$E^z = (\mathrm{Im}\, z + \frac{1}{2}\theta\bar{\theta})^{-1}\,(dz + \theta d\theta),$$

$$E^\theta = (\mathrm{Im}\, z + \frac{1}{2}\theta\bar{\theta})^{-1/2}\, d\theta + \frac{1}{2}(i\theta - \bar{\theta})(\mathrm{Im}\, z + \frac{1}{2}\theta\bar{\theta})^{-3/2}\,(dz + \theta d\theta), \tag{4.10}$$

is invariant up to a phase only under a subgroup $SPL(2,R)$ obtained by restricting the even parameters a, b, c, d of $SPL(2,C)$ to be real and the odd ones to obey $\bar{\gamma} = i\gamma$, $\bar{\delta} = i\delta$. SRS's which admit these frame fields will be called metrizable, so that the gauge-fixing slice for the fermionic string is the moduli space of metrizable SRS's.

To illustrate these ideas I will now work out the classification of super tori, SRS's whose bodies have genus 1. A super torus is obtained as the quotient of SC by a subgroup G of $SPL(2,C)$ which is isomorphic to the fundamental group of a torus. Therefore G has two commuting generators of the form (4.7). The generators leave the flat super vierbein invariant up to a phase only if $c = 0$, $a^2 = 1$, and $\gamma = 0$; and if G_0 acts properly discontinuously as well then $b_0 \neq 0$. The generators then take the form,

$$\tilde{z} = z + ab + \theta\delta, \quad \tilde{\theta} = a(\theta + \delta), \tag{4.11}$$

where $a = \pm 1$. We represent the two generators by the ordered triples $A = (a, b, \delta)$ and $A' = (a', b', \delta')$. The choice of signs for a and a' determines one of the four spin structures on the torus. The commutator of the generators is

$$A'^{-1}A^{-1}A'A = [1, (aa' + a + a' - 1)\delta\delta', (1 - a')\delta - (1 - a)\delta']. \tag{4.12}$$

Note that this commutator is a pure soul transformation. This means that if we dropped the requirement that the generators commute, the quotient space would not have the DeWitt topology.

As is true for Riemann surfaces, the groups G and $G^q = q^{-1}Gq$ represent equivalent SRS's for any $SPL(2,C)$ element q. This is shown by finding a superconformal diffeomorphism relating SC/G and SC/G^q. Starting from a point x in SC/G, choose any point y lying above it in SC, map y to $q^{-1}y$, and project down to SC/G^q. This map is readily seen to be superconformal, invertible, and independent of the choice of y. By an appropriate choice of q it is possible to conjugate the generators into a standard form with $b' = 1$ and $\delta' = 0$. If the spin structure is trivial, so that $a = a' = 1$, then the commutator (4.12) automatically vanishes, and the group is completely described by the two parameters b and δ. These parameters describe super tori with a distinguished choice of generators of $G = \pi_1(M)$, so they give global coordinates on super Teichmüller, rather than super moduli, space. It would appear that this space is a supermanifold of complex dimension (1,1), but this is not quite true. Conjugation by the $SPL(2,C)$ element $q: \tilde{z} = z,\ \tilde{\theta} = -\theta$ flips the sign of δ without changing b, showing that both signs of δ describe the same super torus. Except for the singular points with $\delta = 0$, the coordinates b and δ cover super Teichmüller space twice, showing it to be a super orbifold.

For the nontrivial spin structures, a and a' are not both 1; assume $a' = -1$. Then the commutator (4.12) vanishes only if $\delta = 0$, leaving just one free parameter b. The global structure of super Teichmüller space is therefore as follows. It has four disconnected pieces representing the four spin structures. One piece is a complex super orbifold of dimension (1,1), and the other three pieces are complex supermanifolds of dimension (1,0). Its body is four copies of the ordinary Teichmüller space of the torus, a one-dimensional complex manifold with coordinate b_0. Points of the body represent supertori of the canonical type, since their group parameters are soulless.

Having chosen the group G representing a super torus, we can change from A and A' to a new pair of generators. By conjugating these generators into the standard form, we can identify new values of b and δ which represent the same super torus. Such points of super Teichmüller space which represent the same SRS are related by an element of the super modular group, and the quotient of super Teichmüller space by this group is the super moduli space. For example, we can choose A'^{-1} and A as generators of G rather than A and A'. For the trivial spin structure this leads to the transformation $b \to -1/b,\ \delta \to i\,\delta b^{-3/2}$ on super Teichmüller space. In this way the action of any super modular transformation on super Teichmüller space can be worked out. One finds in particular that the nontrivial spin structures mix under super modular transformations [25]. The structure of the super modular group is also easy to determine: because G is isomorphic to $\pi_1(M_0)$, changing the choice of generators for G is the same as changing the choice of generators for $\pi_1(M_0)$. But this is precisely what the ordinary modular group of a torus does. Therefore the super modular group is isomorphic to the ordinary modular group. The super moduli space will be a complex super orbifold whose body is a four-sheeted cover of ordinary moduli space.

The results for genus $g>1$ are similar (SC^* is the only metrizable SRS of genus 0). A metrizable SRS M of genus $g>1$ is the quotient of SU by a subgroup G of $SPL(2,R)$ having $2g$ generators. The generators are described by $6g$ even and $4g$ odd parameters. By a conjugation, two odd parameters can be set to zero and three even parameters can be fixed. The group relation

$$q_1 q_2 q_1^{-1} q_2^{-1} \cdots q_{2g-1} q_{2g} q_{2g-1}^{-1} q_{2g}^{-1} = 1 \tag{4.13}$$

obeyed by the generators q_i of the fundamental group of any surface then eliminates three more even and two more odd parameters. The only other restriction on the parameters comes from the requirement that G_0 act properly discontinuously on U, and this affects only the bodies of the parameters. Super Teichmüller space therefore has the DeWitt topology, since its soul coordinates are unrestricted. It has $6g-6$ real even and $4g-4$ real odd dimensions, and comes in 2^{2g} disconnected pieces. Each piece is a super orbifold, because flipping the signs of all the odd coordinates together describes the same SRS. Note that because the parameters of $SPL(2,R)$ are real, this construction only gives real coordinates on super Teichmüller space. The space itself is actually a complex super orbifold, but more sophisticated methods are needed to prove this. The super modular group is again isomorphic to the ordinary modular group and the body of super moduli space is a 2^{2g}-sheeted cover of ordinary moduli space.

5. Conclusions and Open Problems

The results of this paper provide a rigorous foundation for the mathematical theory of super Riemann surfaces and for their applications to superstrings. The connection between SRS's and ordinary Riemann surfaces with spin structure was explained, and the gauge-fixing slice for the fermionic string path integral was shown to be the super moduli space of metrizable DeWitt SRS's. The global structure of super Teichmüller space was determined and the super modular group was shown to be the ordinary modular group. The parameters of the group representing a SRS provide local coordinates on super moduli space. These results constitute the "elementary" theory of SRS's.

More powerful results can be obtained by generalizing more advanced techniques from Riemann surface theory, specifically the theory of quasiconformal maps [2,3,19]. With these methods one can show that super Teichmüller space for genus $g>1$ actually has a complex structure. The complex structure comes from an embedding of super Teichmüller space in a certain space of superanalytic functions. A set of complex coordinates is provided by the parameters of a different kind of group representing the SRS, a Schottky group.

Many questions about SRS's remain open. As explicitly as possible, how does the super modular group act on the odd coordinates of super Teichmüller space, and on the spin structures? A proof of finiteness for superstring amplitudes via algebraic geometry probably depends on answering this question. First, finiteness is known to depend on cancellations between the contributions of different spin structures when these contributions are summed to give a modular invariant result. Second, finiteness probably depends on the existence of a compactification of super moduli space by adding points at infinity representing SRS's with certain singularities [9]. Such a compactification of ordinary moduli space is known to exist [28], but generalization of the proof depends on controlling the behavior of spin structures on singular Riemann surfaces. A possible approach to this problem is to study the boundary of the space of metrizable DeWitt SRS's within the larger space of all SRS's. Little is known about this larger space. A related problem is to describe SRS's in terms of period matrices. For Riemann surfaces the entries of these matrices are the integrals of analytic 1-forms around noncontractible loops, and they can serve as coordinates on moduli space. Modular transformations are conveniently described in terms of their effect on a period matrix, and the same should be true for super modular transformations. Finally, an abstract formulation of string theory in terms of the geometry of moduli space has been proposed [29]. The corresponding formulation for superstrings would be of interest.

References

[1] H.M. Farkas and I. Kra, *Riemann Surfaces,* Springer-Verlag, New York, 1980.

[2] L. Bers, Bull. London Math. Soc. **4** , 257 (1972).

[3] C.J. Earle, in *Discrete Groups and Automorphic Functions,* ed. W.J. Harvey, Academic Press, London, 1977.

[4] O. Alvarez, Nucl. Phys. **B216** , 125 (1983).

[5] J. Polchinski, Commun. Math. Phys. **104** , 37 (1986).

[6] E. D'Hoker and D.H. Phong, Nucl. Phys. **B269** , 205 (1986).

[7] G. Moore, "Modular forms and two-loop string physics", Harvard preprint HUTP-86/A038 (1986).

[8] A.A. Belavin and V.G. Knizhnik, Phys. Lett. **168B** , 201 (1986).

[9] R. Catenacci, M. Cornalba, M. Martellini, and C. Reina, Phys. Lett. **172B** , 328 (1986).

[10] E. D'Hoker and D.H. Phong, Nucl. Phys. **B278** , 225 (1986).

[11] G. Moore, P. Nelson, and J. Polchinski, Phys. Lett. **169B** , 47 (1986).

[12] D. Friedan, in the Proceedings of the Workshop on Unified String Theories, ed. D. Gross and M. Green, World Press, Singapore, 1986.

[13] E. Martinec, "Conformal field theory on a (super)Riemann surface", Princeton preprint (1986).

[14] M.A. Baranov and A.S. Shvarts, JETP Lett. **42** , 419 (1986).

[15] P. Nelson and G. Moore, Nucl. Phys. **B274** , 509 (1986).

[16] A. Rogers, J. Math. Phys. **21** , 1352 (1980).

[17] D. Leites, Russ. Math. Surv. **35** , 1 (1980).

[18] J.M. Rabin and L. Crane, Commun. Math. Phys. **100** , 141 (1985); **102** , 123 (1985).

[19] L. Crane and J.M. Rabin, "Super Riemann surfaces: uniformization and Teichmüller theory", University of Chicago preprint EFI 86-25, submitted to Commun. Math. Phys.

[20] A.M. Polyakov, Phys. Lett. **103B** , 207 (1981).

[21] P.S. Howe, J. Phys. A **12** , 393 (1979).

[22] B.S. DeWitt, *Supermanifolds,* Cambridge University Press, Cambridge, 1984.

[23] A. Rogers, Commun. Math. Phys. **105** , 375 (1986).

[24] R. Geroch, J. Math. Phys. **9** , 1739 (1968).

[25] L. Alvarez-Gaumé, G. Moore, and C. Vafa, "Theta functions, modular invariance, and strings", Harvard preprint HUTP-86/A017 (1986).

[26] K. Kodaira, *Complex Manifolds and Deformation of Complex Structures,* Springer-Verlag, New York, (1986).

[27] A. Neveu and J.H. Schwarz, Nucl. Phys. **B31** , 86 (1971).

[28] S. Wolpert, Ann. Math. **118** , 491 (1983).

[29] D. Friedan and S. Shenker, Phys. Lett. **175B** , 287 (1986).

NON-LINEAR REALIZATION OF HEAVY FERMIONS*

York-Peng Yao

Department of Physics
The University of Michigan
Ann Arbor, Michigan 48109, U.S.A.

I want to discuss a piece of work which is still going on. Eduardo Flores and Herbert Steger are collaborating with me on this.[1] The motivation for us to look into the present problem is as follows: The pattern of masses of the various fermion families seems to indicate that in the fourth family, if in fact there is one more family, the mass of the up-member quark will be very large. In fact, perhaps the top quark in the third family is already heavier than W and Z. If so, our analysis should apply. Let me then generically call this heavy fermion top.

Now, in the standard model, with only one Higgs' doublet, the way to give a large mass to a fermion is to raise its Yukawa coupling to the Higgs. There are many questions we can ask in this situation:

We know that if we just raise the Yukawa couplings of all members of a multiplet so as to give large masses to all of them, we do not get decoupling. Instead, we have the residual Wess-Zumino terms in the effective low energy theory.[2] We are going one step further: what will be the situation when we remove only a member of a multiplet? In a gauged theory, we have a U(1) anomaly and the theory is no more renormalizable. We want to find out how serious is this non-renormalizability.

* Work partially supported by U.S. Department of Energy.

My assumption here is that the Yukawa coupling H is much larger than the SU(2) X U(1) couplings, $g_{1,2}$ and the other Yukawas, but not so large that we need to discard perturbation. I want to investigate the low energy physics. In other words, when M_t is the largest scale in a problem, compared with all other masses and external momenta. We want to find out $\mathcal{L} \to \mathcal{L}_{eff}$.

Here, I want to deal with a simplified model, in which we have no gauge fields. We have a Higgs' doublet and one quark family.
The gauged theory will be presented elsewhere.

$$\mathcal{L}_L = -(\bar{t}, \bar{b})_L \gamma^\mu \frac{1}{i} \partial_\mu \binom{t}{b}_L - \bar{t}_R \gamma^\mu \frac{1}{i} \partial_\mu t_R - \bar{b}_R \gamma_\mu \frac{1}{i} \partial^\mu b_R - \partial_\mu \bar{\phi}\, \partial^\mu \phi$$

$$- (H (\bar{t}, \bar{b})_L \binom{\phi^0}{\phi^-} t_R + h.c.) - (h (\bar{t}, \bar{b})_L \binom{-\phi^+}{\phi^{0+}} b_R + h.c.)$$

$$- \frac{\lambda}{2} (\bar{\phi}\, \phi)^2 - \mu^2 (\bar{\phi}\, \phi)$$

Here, $(\bar{t}, \bar{b})_L$ transforms as a left-handed doublet, so does $\binom{\phi^0}{\phi^-}$. $\binom{-\phi^+}{\phi^{0+}}$ is the charge conjugated doublet. t_R and b_R are iso-singlets.

$$\binom{\phi^{0\prime}}{\phi^-} = e^{i\vec{\tau}\cdot\vec{\delta\alpha}} \binom{\phi^0}{\phi^-} \quad , \quad \binom{t}{b}_L' = e^{i\vec{\tau}\cdot\vec{\delta\alpha}} \binom{t}{b}_L$$

Let us casually look at the formal solutions at the tree level. We have for the mass of t-quark $M_t = Hv$, $v = \langle\phi^0\rangle = \langle\phi^{0+}\rangle$,
Then

$$\binom{t_R}{t_L} = \frac{1}{M_t^2+p^2} \begin{pmatrix} -\gamma\cdot p & M_t \\ M_t & -\gamma\cdot p \end{pmatrix} \begin{pmatrix} -H (\phi^+ b_L + (\phi^{0+} - v)\, t_L) \\ -H (\phi^0 - v)\, t_R + h\, \phi^+ b_R \end{pmatrix} + \binom{t_{R_0}}{t_{L_0}}$$

Where t_{R_0} and t_{L_0} are solutions of the homogeneous equation

$\begin{pmatrix} \gamma\cdot p & M_t \\ M_t & \gamma\cdot p \end{pmatrix} \binom{t_{R_0}}{t_{L_0}} = 0$. There are similar equations for b and ϕ. By iterating these equations to obtain $t = t(t_0, b_0, \phi_0)$ etc. we obtain all all the tree diagrams.

Here, we are interested in the situation when t_L and t_R cannot be produced. We should set $t_{L_0} = 0$, $t_{R_0} = 0$ for $M \gg p$, m or $H \gg h$, λ in which case, we have

$$\binom{t_R}{t_L} = \begin{pmatrix} 0 & 1 \\ 1 & 0 \end{pmatrix} \begin{pmatrix} -(\phi^+ b_L + (\phi^{0+} - v)\, t_L)\, / v \\ -(\phi^0 - v) t_R / v \end{pmatrix} + O(\tfrac{1}{H})$$

or $\quad t_R = 0 \quad , \quad t_L = -(\phi^+/\phi^{0+})b_L \equiv (t_L)_{\text{non-linear}}$

The last is a constraint $(\phi^{0+}, \phi^+)\binom{t}{b}_L = 0$, which is a SU(2) invariant statement. To put it differently, under a SU(2) rotation $e^{i\vec{\tau}.\delta\vec{\alpha}}$, we can check explicitly that

$$(\bar{t}_L)'_{\text{non-linear}} = (\bar{t}_L)_{\text{non-linear}} + i\,\delta\alpha_3\,(\bar{t}_L)_{\text{non-linear}} + (-i\,\delta\alpha_1 + \delta\alpha_2)\bar{b}_L$$

which shows that we have a non-linear realization of SU(2).

Note that we can expand $\frac{1}{\phi^0} = \frac{1}{v}(1 - \frac{\phi^0 - v}{v} + \frac{(\phi^0 - v)^2}{v^2} \pm \cdots)$ and make it meaningful.

We can show that the Lagrangian $(\underset{\sim}{t}_L = -(\underset{\sim}{\phi}^+/\underset{\sim}{\phi}^{0+})\underset{\sim}{b}_L)$

$$\mathcal{L}_{N.L.} = -\bar{\underset{\sim}{b}}_L\gamma^\mu \frac{1}{i}.\partial_\mu \underset{\sim}{b}_L - \bar{\underset{\sim}{b}}_R\,\gamma^\mu \frac{1}{i}\,\partial_\mu \underset{\sim}{b}_R - \bar{\underset{\sim}{t}}_L\gamma^\mu \frac{1}{i}\,\partial_\mu\,\underset{\sim}{t}_L$$

$$- (h(\bar{\underset{\sim}{b}}_L\phi^{0+} - \bar{\underset{\sim}{t}}_L\underset{\sim}{\phi}^+)\underset{\sim}{b}_R + \text{h.c.}) - \frac{\lambda}{2}(\bar{\underset{\sim}{\phi}}\,\underset{\sim}{\phi})^2 - \mu^2\,\bar{\underset{\sim}{\phi}}\,\underset{\sim}{\phi}$$

$$- \partial_\mu\bar{\underset{\sim}{\phi}}\,\partial^\mu\underset{\sim}{\phi} + (\bar{\eta}_R\,\underset{\sim}{b}_R + \bar{\eta}_L\underset{\sim}{b}_L + \underset{\sim}{\phi}\,J_\phi^+ + \text{h.c.})$$

reproduces all the tree results with only ϕ and b as external lines to the accuracy of $O(\frac{1}{H})$.

Let us turn to the problems of quantum corrections.[3] We have the generating functional

$$\lim_{M_t\to\infty} e^{iW(J,\eta)} = \lim_{M_t\to\infty} \int d\phi\, d\bar{\phi}\, db\, d\bar{b}\, db\, dt\, d\bar{t}$$

$$\cdot e^{i\int d^4x\{\mathcal{L}_L + (\bar{\eta}_R\,b_R + \bar{\eta}_L\,b_L + \phi J_\phi^+ + \text{h.c.})\}}$$

naively

$$= \int d\phi\, d\bar{\phi}\, db\, d\bar{b}\, e^{i\int d^4x\{\mathcal{L}_{\text{eff}} + (\bar{\eta}_R\,b_R + \bar{\eta}_L\,b_L + \phi\,J_\phi^+ + \text{h.c.})\}}$$

Note that we have made a subtle interchange of integration and large mass limit. This is a rather dangerous interchange, because the external momenta of a loop with a top quark inside may be some internal momenta of some bigger loops. The machinery which justifies this interchange after some rearrangement is the Zimmermann's algebraic identity.[4] In words, there are over-subtractions supplied by an algebraic identity which gives meaning to the interchange.

We shall do loop expansion to evaluate $\mathcal{L}_{eff}$. We have seen that the solutions with $\mathcal{L}_{N.L.}$ gives us all the tree graphs, correct to order $O(\frac{1}{H})$. This is a good point around which to do quantum fluctuation. We write $b_L = B_L + \underset{\sim}{b}_L$, $b_R = B_R + \underset{\sim}{b}_R$, $\phi = \Phi + \underset{\sim}{\phi}$, $t_L = T_L + \underset{\sim}{t}_L$, $t_R = T_R$. Quantities with small case letters and ~ underlined are solutions of the non-linear Lagrangian. Quantities which are capitalized such as B, Φ and T's are quantum fluctuations. Note that because $\underset{\sim}{b}$, $\underset{\sim}{\phi}$ and $\underset{\sim}{t}$ are <u>not exact</u> solutions of the Lagrangian $\mathcal{L}_L$, we have terms linear in fluctuations. Specifically, we have

$$\mathcal{L}_L = \mathcal{L}_{N.L.} + \{ \bar{B}_R \left(\frac{\delta \mathcal{L}_L}{\delta \bar{b}_R} + \eta_R\right)_{L \to N.L.} + \bar{B}_L \left(\frac{\delta \mathcal{L}_L}{\delta \bar{b}_L} + \eta_L\right)_{L \to N.L.}$$

$$+ \bar{T}_L \frac{\delta \mathcal{L}_L}{\delta b_L} \Big|_{L \to N.L.} + \bar{T}_R \frac{\delta \mathcal{L}_L}{\delta \bar{b}_R} \Big|_{L \to N.L.}$$

$$+ \Phi\left(\frac{\delta \mathcal{L}_L}{\delta \phi} - \partial_\mu \frac{\delta \mathcal{L}_L}{\delta \partial_\mu \phi} + J_\phi^{\ +}\right)_{L \to N.L.} + h.c.\}$$

$$+ \text{ higher orders in fluctuations.}$$

Upon using the equations of the non-linear model,

$$\frac{\delta \mathcal{L}_{N.L.}}{\delta \underset{\sim}{\bar{b}}_R} + \eta_R = 0, \quad \frac{\delta \mathcal{L}_{N.L.}}{\delta \underset{\sim}{\bar{b}}_L} + \frac{\delta \underset{\sim}{\bar{t}}_L}{\delta \underset{\sim}{\bar{b}}_L} \frac{\delta \mathcal{L}_{N.L.}}{\delta \underset{\sim}{\bar{t}}_L} + \eta_L = 0$$

$$\frac{\delta \mathcal{L}_{N.L.}}{\delta \underset{\sim}{\phi}} - \partial_\mu \frac{\delta \mathcal{L}_{N.L.}}{\delta \partial_\mu \underset{\sim}{\phi}} + \frac{\delta \underset{\sim}{\bar{t}}_L}{\delta \underset{\sim}{\phi}} \frac{\delta \mathcal{L}_{N.L.}}{\delta \underset{\sim}{\bar{t}}_L} + J_\phi^{\ +} = 0$$

we have

$$e^{iW} = e^{i\int d^4x \, \mathcal{L}_{N.L.}} \int d\Phi \, d\bar{\Phi} \, dB \, d\bar{B} \, dT \, d\bar{T} \; e^{i\int d^4x \, \mathcal{L}'}$$

where

$$\mathcal{L}' = \{(\bar{B}_L \underset{\sim}{\phi}^- + \bar{T}_L \underset{\sim}{\phi}^0) \frac{1}{\underset{\sim}{\phi}^0}(-\gamma^\mu \tfrac{1}{i} \partial_\mu \underset{\sim}{t}_L + h \, \underset{\sim}{\phi}^+ \underset{\sim}{b}_R)$$

$$+ (\Phi^0 \underset{\sim}{\bar{t}}_L + \phi^+ \underset{\sim}{b}_L) \frac{1}{\underset{\sim}{\phi}^0} (-\gamma^\mu \tfrac{1}{i} \partial_\mu \underset{\sim}{t}_L + h \, \underset{\sim}{\phi}^+ \underset{\sim}{b}_R) + h.c. \}$$

$$+ \text{ higher orders in fluctuations}$$

As we noted before, $\binom{\phi^o}{\phi^-}$ and $\binom{t}{b}_L$ transform like iso - doublets. If we make a simultaneous change of variables

$$\binom{T}{B}_L' = e^{i\delta\vec{\alpha}.\vec{\tau}} \binom{T}{B}_L , \qquad \binom{\Phi^o}{\Phi^-}' = e^{i\delta\vec{\alpha}.\vec{\tau}} \binom{\Phi^o}{\Phi^-}$$

we can easily see that terms which are higher order in fluctuations are all iso - invariant. The combinations $(\bar{B}_L \phi^- + \bar{T}_L \phi^o)$ and $(\Phi^o \bar{t}_L + \bar{\Phi}^- b_L)$ are also iso-singlets.

Then, the only term which is apparently non-invariant is

$$x = \frac{1}{\phi^o} (-\gamma^\mu \frac{1}{\dagger} \partial_\mu t_L + h \phi^+ b_R)$$

However, if we use the equations of motion, we have under SU(2), $\delta x = - (i\delta\alpha_1 + \delta\alpha_2) \frac{1}{\phi^o} \eta_L$ which has not one particle pole. Hence, W is in fact <u>SU(2) invariant on-shell</u>.

The conclusion of this discussion is that we should end up with a SU(2) invariant S-matrix. For the Greens function generating functional W, we have monomials built up from SU(2) invariant quantities, which are formed by $\binom{t}{b}_L$, b_R and $\binom{\phi^o}{\phi_-}$ and powers of x. I shall call these effective vertices.

The coefficients which multiply these monomials depend on powers of M_t and $\ell n\, M_t$. We need a power counting procedure to determine what this dependence can be and when we can drop terms.

By definition a diagram has to have some heavy fermion lines to generate effective vertices. Also, the structure of the interaction is such that we do not have power infrared singularity due to light quarks and bosons. This means M_t is the scale of the integral in inverse powers.

Let us examine a set of diagrams in which all the vertices carry the coupling H. Let n_F be the number of light external fermions and n_B external bosons. Then, the dimension of the integral is $4 - \frac{3}{2} n_F - n_B$.

Let V be the number of vertices. We have for the diagram an integrated expression

$$H^V{}_I = \frac{M_t{}^V}{(v)^V} M_t{}^{4-3/2n_F-n_B}\{f_o(\ln M_t) + f_1(\ln M_t)\frac{\gamma\cdot p}{M_t} + \ldots + f_N(\ln M_t)\left(\frac{\gamma\cdot p}{M_t}\right)^N\}$$

in which we terminate the series when $4 - \frac{3}{2}\ n_F - n_B + V = N$.

Now we use the identities (L=loop) $3\ i_F + 2i_B + \frac{3}{2}n_F + n_B = 4V$, $i_F + i_B = L - 1 + V$, and $2\ (2i_B + n_B) = 2i_F + n_F$ to obtain $N = 2(L+1) - \frac{1}{2}n_F$, which is a relation between loop order, number of external lines and the maximum number of derivatives we need to expand, such that the neglected terms are $\frac{1}{M_t{}^2}$ or smaller.

There is something quite noticeable in this relation. The number of external boson lines do not appear. This means that we can have any arbitrary number of them. One way to restrict the work further is to put a further restriction on the system, namely: $\lambda \sim H >> g$, h. This will correspond to heavy top and heavy Higgs. Another relation ensues: $\bar{\phi}\ \phi = v^2$. In this case, the non-trivial SU(2) invariants built with scalars have to have derivatives. The number of effective vertices is reduced. However, the more general case can be done, with the aid of Schoonschip.

In conclusion, I have described the program we are following. We have derived the one loop effective Lagrangian which will be published. We are working on the gauged SU(2) x U(1) model. Once we have eff here, we shall have all the top quark effects at a glance. This will yield some interesting physics.

The next step is to improve the penturbative results by summing up series. How this can be related to new divergences of the effective theory is an intriguing issue.[5]

References

1. To be published.
2. E. D'Hoker & E. Farhi, Nucl. Phys. B 241(1984)109; Y.-C. Chao, to be published.
3. The development here parallels that in T. Appelquist & C. Bernard, Phys. Rev. D 23(1981)425; R. Akhoury & Y.-P. Yao, Phys. Rev. D 25 (1982)3361.
4. Y. Kazawa & Y.-P. Yao, Phys. Rev. D 25(1982)1605.
5. M. Veltman, private communication.

SUPERSYMMETRY BREAKING IN R x S^3

Diptiman Sen

Physics Department
Carnegie - Mellon University
Pittsburgh, Pennsylvania 15213
U.S.A.

1. Introduction

The idea of supersymmetry has attracted physicists ever since its inception. If supersymmetry plays a role in the unified theory of all forces, it must be dynamically broken in nature.

To derive the physical properties of a quantum field theory, it is often useful to replace it by an approximate model which has a discrete number of degrees of freedom. One way of doing this is to define the theory in a finite volume. The usual choice is a rectangular box, but this breaks continuous rotational symmetry. A space with considerably more symmetry is the surface of a four-dimensional sphere S^3 [1].

In sect. 2, we briefly discuss the Lagrangians of spin 0, $\frac{1}{2}$ and 1 fields. In sect. 3, the two fundamental supersymmetric multiplets are considered, and the supersymmetry algebra is exhibited. A new term appears in the superalgebra, which can be identified with the well-known R-operator [2]. The role played by the R-operator is one of the most interesting features of theories in curved space. It is found that only R-invariant theories have supersymmetric actions in R x S^3.

In sect. 4, we study supersymmetry breaking breaking. This question can be answered in many theories once an index is defined [3]. The definition of the Witten index here is different from that in flat space, due to the unusual superalgebra. We then find that supersymmetry breaking can be ruled out in many theories simply by index considerations.

Throughout this paper, we quote and use results from ref. 4, which contains details of the calculations.

2. Bosonic and Fermionic Lagrangians in R x S^3

The space-time R x S^3 is the product of a flat time direction, a derivative along which is denoted ∂_0, and curved spatial directions, derivatives along which will be denoted by ∂_i [4], i=1,2,3. There exists a mass scale in S^3, the inverse radius ρ^{-1}, which we set equal to unity wherever possible. Due to the curvature, the spatial derivatives do not commute

$$[\partial_i , \partial_j] = 2 \epsilon_{ijk} \partial_k \qquad (2.1)$$

Ref. 4 describes the differential geometry of S^3, and the way in which its symmetry group reduces to that of flat space in the limit of large radius.

To define theories with global N=1 supersymmetry, we need the forms of the Lagrangians of fields of spin 0, ½ and 1. We will only consider massless fields henceforth; these are the ones for which index calculations are most troublesome in flat space.

The general Lagrangian for a free scalar field is

$$\mathcal{L} = \partial_0 A^* \partial_0 A - \partial_i A^* \partial_i A + 2\, e\, A^* i\partial_0 A + e' A^* A \tag{2.2}$$

The last two terms are of order ρ^{-1} and ρ^{-2}, and are allowed by the global symmetries of $R \times S^3$ for any real constants e and e . We will later find that these non-minimal terms are in fact required by supersymmetry. For the (Euclidean) action to be bounded from below, one requires that $\underline{e' \leq 0}$. This observation will prove to be crucial later.

For a spin 1 gauge boson, the (Abelian) electric and magnetic fields are given by

$$v_{io} = \partial_i v_o - \partial_o v_i \qquad\qquad v_{ij} = \partial_i v_j - \partial_j v_i - 2\,\epsilon_{ijk} v_k \tag{2.3}$$

and the Lagrangian is $\mathcal{L} = -\frac{1}{4} v_{mn} v^{mn}$, where m,n = 0,1,2,3.

We now turn to spin ½ fermions. We only consider Weyl fermions, since a Dirac fermion can be obtained by combining two Weyl fermions. To write down the Weyl equation, one usually follows the tetrad formalism [5]. A tetrad is a locally inertial frame of one-forms e^m , which, in terms of the coordinate differentials dx^μ, can be expanded as $e^m = e^m{}_\mu dx^\mu$. The torsion-less spin-connection $\omega^m{}_n$ is a one-form defined by $de^m = -\omega^m{}_n \wedge e^n$. Then the Weyl equation is

$$-i\,\bar{\sigma}^m e_m{}^\mu D_\mu \Psi = 0 \quad , \quad D_\mu = \partial_\mu + \frac{i}{2}\omega_{\mu mn}\sigma^{mn} \tag{2.4}$$

The corresponding action is invariant under general coordinate and local Lorentz transformations, and also conformal transformations of the metric.

If a different spin-connection is used (thereby introducing torsion), the action is not invariant under such local transformations. We find, in fact, that one has to allow some torsion in order to define theories with global supersymmetry in $R \times S^3$. Since our primary purpose is to study supersymmetry, we only keep those symmetries which are consistent with it. All other symmetries (namely, the local invariances mentioned above) must be given up if necessary. In $R \times S^3$, it turns out that the only symmetries of the action that are consistent with supersymmetry are the rigid rotations of S^3 , time translation, R-invariance and, of course, the discrete symmetries C, P and T.

The general Lagrangian for a Weyl fermion is

$$\mathcal{L} = \bar{\Psi}\,(-i\bar{\sigma}^m \partial_m + d\,\bar{\sigma}^0)\,\Psi \tag{2.5}$$

where d can be any real number, and the term containing it is again of order ρ^{-1}. We have followed the notation of [6] in equations (2.4) and (2.5).

From these Lagrangians, one can readily obtain the energy spectra of the various fields, and the degeneracy at each energy level. This will be shown later in a figure.

3. Supersymmetry, Superalgebra and R-invariance

We now turn to the simplest supersymmetric theory, the Wess-Zumino model. This model describes a complex scalar, a Weyl fermion and a complex auxiliary field, collectively called a scalar multiplet. In analogy with flat space, but mindful of all the non-minimal terms possible, we write down the following Lagrangian.

$$\mathcal{L} = \Psi(-i\bar{\sigma}^m\partial_m + d\bar{\sigma}^0) + A^*\Box A + F^*F$$
$$+ 2e A^* i\partial_0 A + e' A^* A \qquad (3.1)$$

We then discover that this action is supersymmetric only if d, e and e' satisfy a one-parameter set of relations

$$d = -r-1 \qquad e = r-1 \qquad e' = r(r-2) \qquad (3.2)$$

With the restriction on e' in sect. 1, r must lie in the interval [0,2].

The fermionic and bosonic spectra are shown in fig. 1. The figure indicates both the energy levels and the degeneracy of each level (on the right and left respectively of the vertical lines in the figure). For the scalar multiplet, we look at the two spectra on the left; for the vector multiplet, look at the two spectra on the right (and set r=0 for the fermion spectrum, as discussed shortly). A negative energy in the figure simply means that the quantum operator for that mode creates antiparticles, rather than annihilate particles.

The supersymmetry generators act on the fields to give [4]

$$[Q_\alpha, A] = \sqrt{2}\,\Psi_\alpha \qquad \{Q_\alpha, \Psi_\beta\} = \sqrt{2}\,\epsilon_{\beta\alpha} F$$
$$\{\bar{Q}_{\dot\alpha}, \Psi_\beta\} = -i\sqrt{2}\,\sigma^m_{\beta\dot\alpha}\partial_m A - \sqrt{2}\, r\,\sigma^0_{\beta\dot\alpha} A$$
$$[\bar{Q}_{\dot\alpha}, F] = \epsilon_{\dot\alpha\dot\beta}\left(i\sqrt{2}\,\bar{\sigma}^{m\dot\beta\alpha}\partial_m\Psi_\alpha + \sqrt{2}\,(r+1)\,\bar{\sigma}^{0\dot\beta\alpha}\Psi_\alpha\right) \qquad (3.3)$$

All other commutators and anticommutators give zero. From this we deduce

$$\{Q_\alpha, Q_\beta\} = \{\bar{Q}_{\dot\alpha}, \bar{Q}_{\dot\beta}\} = 0 \qquad \{Q_\alpha, \bar{Q}_{\dot\alpha}\} = 2i\sigma^m_{\alpha\dot\alpha}\partial_m - 2R\,\sigma^0_{\alpha\dot\alpha} \qquad (3.4)$$

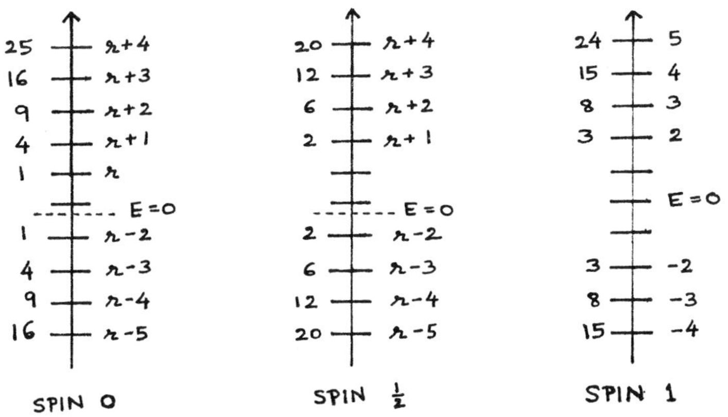

Fig. 1. The spectrum of a scalar multiplet as a function of r, and of a vector multiplet (with r set equal to zero for the fermion).

where R is an operator which acts on the fields as follows

$$[R , A] = r A \qquad [R , \Psi] = (r-1) \Psi \qquad [R , F] = (r-2) F \tag{3.5}$$

Note that the operator R has a ρ^{-1} hidden in it, so it does not appear in flat space.

The superalgebra (3.4) is completed by

$$[\partial_i , Q_\alpha] = -i(\sigma^i \bar{\sigma}^o Q)_\alpha \qquad [\partial_i , \bar{Q}_{\dot{\alpha}}] = i(\bar{Q}\bar{\sigma}^o \sigma^i)_{\dot{\alpha}} \tag{3.6}$$

All other commutators vanish. This describes supersymmetry in R x S^3.

It can be shown that the R-operator in (3.4) is identical to that introduced in [2], and that a theory is supersymmetric in R x S^3 only if it is also R-invariant. This can be understood directly from the superalgebra, since the anticommutator of two supersymmetry generators contains the R-operator. It can also be shown directly, by considering a typical interaction term coupling different scalar multiplets (of R-charges r_1 , r_2 , ...), and discovering that the term is supersymmetric only if the condition for R-invariance $\sum_i r_i = 2$ holds.

The necessity of R-invariance in R x S^3 may provide a rationale for restricting the kind of theories one considers in flat space. It is startling that curved space considerations, even in the limit that the curvature goes to zero, may impose restrictions on flat space theories. It is heartening that R-invariant theories have been found to be preferable on phenomenological grounds as well.

We turn next to the other important supersymmetric model, namely that describing a vector multiplet. A massless vector multiplet contains a spin-one gauge boson, a spin-half fermion called the gaugino, and a real auxiliary field. Considering an Abelian multiplet for simplicity, the Lagrangian is found to be

$$\mathcal{L} = -\frac{1}{4} v_{mn} v^{mn} + \bar{\lambda}(-i\bar{\sigma}^m \partial_m - \bar{\sigma}^o)\lambda + \frac{1}{2} D^2 \tag{3.7}$$

The superalgebra (3.4) holds, with the R-charges of the fields given by

$$[R , v_{mn}] = 0 \qquad [R , \lambda] = \lambda \qquad [R , D] = 0 \tag{3.8}$$

This is consistent with the known result that a gauge multiplet has R-number equal to zero.

We can generalize all this to nonabelian gauge groups, and also couple the gauge fields to scalar multiplets. All globally supersymmetric theories in four dimensions can be built out of scalar and vector multiplets, and interactions between them.

4. Supersymmetry Breaking and Witten index

We now come to the topic of super symmetry breaking. We will discuss symmetry breaking which is dynamic, rather than explicit.

In flat space, the concept of the index Tr $(-1)^F$ introduced by Witten [3] has been very useful. The index is defined to be the difference of the number of bosonic states and the number of fermionic states, in the subspace of physical states which have zero spatial momentum. The index is an integer and is usually independent of small changes (analytic) in the various parameters defining the theory. It is therefore sufficient to calculate it in perturbation theory, for weak coupling.

If the index is non-zero, it can be shown that the theory does not break supersymmetry. In the zero momentum subspace, there exists a hermitian operator Q, the square of which is the Hamiltonian H. Hence Q acting on a bosonic state of non-zero energy gives a fermionic state of the same energy, and vice versa. By pair counting, states of non-zero energy do not contribute to $\mathrm{Tr}\,(-1)^F$. All we need then is the contribution from zero energy states.

If supersymmetry is broken, Q acting on the vacuum does not annihilate it. So the expectation value of Q in the vacuum is positive. There are then no states of zero energy in the theory, and the index is zero. Conversely, if the index is non-zero, there must be states of zero energy and supersymmetry cannot be broken.

In $R \times S^3$, this argument needs to be modified, because the spatial derivatives do not commute with the supersymmetry generators, and the square of Q also contains the R operator. We therefore proceed as follows. Define the hermitian operator $Q = Q_1 + \bar{Q}_1$, so that

$$Q^2 = 2\,(H + R - i\,\partial_3) \tag{5.1}$$

Let us now restrict ourselves to the subspace of states which are annihilated by the operator $R - i\partial_3$ (which also commutes with Q and H). The vacuum can be shown to belong to this subspace. Then the previous index arguments go through. States of non-zero energy in this subspace come in pairs which contribute nothing to the index. The index only knows about zero energy states, which carries information about supersymmetry breaking.

For many theories, index consideration are useful only if they are formulated in $R \times S^3$, and not in flat space. For exmple, nonabelian gauge theories with scalar multiplets lying in a complex representation of the gauge group cannot be analyzed by the methods in [3], due to technical reasons having to do with massless fermionic modes. The same is true for scalar multiplets in a real representation of the gauge group (but with their masses set exactly equal to zero at tree level). But all such theories can be analyzed on the sphere as we will demonstrate.

Note that if a theory is discovered to have no supersymmetry breaking on the sphere for any value of the radius, then in the limit of infinite radius also the vacuum energy must be zero. Therefore the theory cannot break supersymmetry in flat space either.

For scalar multiplets which couple to gauge fields, the value of r (that is, the R-charge of the scalar field A) cannot be chosen arbitrarily. The current associated with R-invariance is like an axial $U(1)$ current, and is anomalous in general. The anomaly only vanishes for a particular value of r.

As an example, consider N_f flavors of N and $\bar{N}$ multiplets A_i and $\bar{A}_i$, $i=1,\ldots,N_f$, which are coupled to a $SU(N)$ gauge field. (For $N=3$, this describes supersymmetric QCD). The $U(1)_R$ current has a triangle anomaly, coming from the gaugino carrying R-charge $+1$, the A_i fermions of R-charge $r_i - 1$, and the $\bar{A}_i$ fermions of R-charge $\bar{r}_i - 1$. If we choose

$$r_i = \bar{r}_i = 1 - \frac{N}{N_f} \tag{5.2}$$

then the R-anomaly vanishes, and the quantum theory of massless quarks has the symmetry $SU(N_f) \times SU(N_f) \times U(1)_V \times U(1)_R$.

If $N_f \geqslant N$, there exists a non-anomalous $U(1)_R$ symmetry with r in the range $0 \leqslant r < 2$. $\mathrm{Tr}\,(-1)^F$ is then non-zero in perturbation theory, since both the gaugino and the scalar multiplet fermions have positive energy (fig. 1), and there is a bosonic vacuum at zero energy. Hence supersymmetry must remain unbroken. If $N_f < N$, and one requires that $0 \leq r \leq 2$, it is again true that the index is non-zero. However, $U(1)_R$ is then anomalous. The currents associated with R-invariance (and therefore with supersymmetry via (3.4)) are not conserved. Supersymmetry is then explicitly broken (not at tree level, but at one loop), and the argument relating $Tr\ (-1)^F$ to supersymmetry is no longer valid.

Thus index arguments in $R \times S^3$ tell us that supersymmetry is unbroken if $N_f \geq N$, but they have nothing to say if $N_f < N$.

Similarly, consider a supersymmetric grand unified theory based on SU(5), with M generations of scalar multiplets transforming as 10's and $\bar{5}$'s. Suppose we ignore the Higgs structure entirely. Once again, there are contributions to the R-anomaly from the SU(5) gauginos, the 10 fermions and the $\bar{5}$ fermions. The anomaly vanishes if $r_5 + 3\, r_{10} = \frac{2}{M}\,(2M - 5)$. One can satisfy this and simultaneously have $r_{\bar{5}}$ and r_{10} lie in the interval [0,2] only if $M \geq 3$.

We have considered two examples, one with chiral fermions and the other with non-chiral but massless fermions. In both cases, index theorems tell us that supersymmetry is unbroken if the number of generations satisfies a certain condition. Remarkably, these are exactly the same conditions that were derived earlier by completely different methods ([7] for QCD and [8] for SU(5)) based on instanton calculations. We see that index arguments yield the same result much more quickly. In all theories in which it is known by instanton computations that supersymmetry is unbroken, we have checked that index calculations say the same thing. The advantage of index arguments is that they work equally easily in all theories, whereas those based on instantons may become prohibitively difficult in complicated theories.

We conclude as follows. Taking the limit of infinite radius, we can definitely rule out supersymmetry breaking in flat space in theories in which (i) the action is R-invariant (since only such theories are supersymmetric in $R \times S^3$ to begin with, and (ii) all scalar multiplets which couple to non-abelian gauge fields have anomaly-free values of r lying in the range $0 \leq r < 2$. These conditions are sufficient, but may not be necessary in flat space. They are necessary only if one wishes to approach flat space from $R \times S^3$.

I thank R. Cutkosky and V. P. Nair for many stimulating discussions. This work was supported by the U.S. Department of Energy under contract DE-AC02-76ER0306.

References

1. R. E. Cutkosky, J. Math. Phys. 25, 939 (1984).
2. A. Salam and J. Strathdee, Nucl. Phys. B87, 85 (1975);
   P. Fayet, Nucl. Phys. B90, 104 (1975).
3. E. Witten, Nucl. Phys. B202, 253 (1982).
4. D. Sen, Carnegie-Mellon preprint CMU-HEP85-12, in press at Nucl. Phys. B.
5. S. Weinberg, Gravitation and Cosmology (Wiley, New York, 1972);
   T. Eguchi, P. B. Gilkey and A. J. Hanson, Phys. Rep. 66, 213 (1980).
6. J. Wess and J. Bagger, Supersymmetry and Supergravity
   (Princeton University, Princeton, 1983).
7. I. Affleck, M. Dine and N. Seiberg, Nucl. Phys. B241, 493 (1984).
8. Y. Meurice and G. Veneziano, Phys. Lett. 141B, 69 (1984).

P A R T I C I P A N T S

Alonzo-Sanchez, F.
Harvard University, U.S.A.

Belangar, G.
TRIUMF, Canada

Brink, L.
Göteborg, Sweden

Burgess, C.
Institute Advanced Study, U.S.A.

Campbell, B.
University of Alberta, Canada

Candelas, P.
University of Texas, U.S.A.

Chiu, T.W.
National Taiwan University, R.O.C.

Chou, K.C.
Institute of Theoretical Physics, China

Couture, M.
Chalk River Nuclear Laboratories, Canada

DeWit, B.
University of Utrecht, Netherlands

Dolan, B.
University of Glasgow, U.K.

Dolan, L.
Rockefeller University, U.S.A.

Douglas, R.
University of British Columbia, Canada

Draper, T.
University of California, U.S.A.

Elias, V.
University of Western Ontario, Canada

Englert, F.
University of Brussels, Belgium

Espriu, D.
Harvard University, U.S.A.

Evans, D.
University of Winnipeg, Canada

Floreanini, R.
Massachusetts Inst. of Technology, U.S.A.

Fogleman, G.
San Francisco State University, U.S.A.

Fujikawa, K.
Hiroshima University, Japan

Gegenberg, J.
University of New Brunswick, Canada

Goddard, P.
University of Cambridge, U.K.

Godfrey, F.
TRIUMF, Canada

Grossman, B.
Rockefeller University, U.S.A.

He, H.X.
Institute of Atomic Energy, China

Ho-Kim, Q.
Laval University, Canada

Hu, B.L.
University of Maryland, U.S.A.

Hull, C.
University of Cambridge, U.K.

Ibanez, L.E.
University of Madrid, Spain

Jackiw, R.
Massachusetts Inst. Technology, U.S.A.

Kafkalidis, M.S.
Columbus, Ohio, U.S.A.

Kim, D.Y.
Regina University, Canada

Kobes, R.
Memorial University, Canada

Kodaira, J.
Hiroshima University, Japan

Kunstatter, G.
University of Winnipeg, Canada

Labastida, J.M.F.
Princeton University, U.S.A.

Lam, C.S.
McGill University, Canada

Lancaster, D.
University of Sussex, U.K.

Lee, H.C.
Chalk River Nuclear Laboratories, Canada

Lee, S.C.
Institute of Physics, R.O.C.

Leivo, H.P.
University of Toronto, Canada

Li, D.X.
McGill University, Canada

Li, L.F.
Carnegie-Mellon University, U.S.A.

London, D.
TRIUMF, Canada

Louis, J.
University of Pennsylvania, U.S.A.

MacKenzie, R.
University of Cambridge, U.K.

Maillet, J.M.
University Pierre et Marie Curie, France

Mann, R.B.
University of Toronto, Canada

Mayrand, M.
Université de Montreal, Canada

McKeon, G.
University of Western Ontario, Canada

Migneron, R.
University of Western Ontario, Canada

Mukku, C.
Madras University, India

Myers, R.
Princeton University, U.S.A.

Nauman, C.
University of Guelph, Canada

Ng, J.
TRIUMF, Canada

Niemi, A.J.
LBL/Ohio State, U.S.A.

Olynyk, K.
Fermi Lab., U.S.A.

Patera, J.
Centre de Rech. Mathematique, Canada

Rabin, J.M.
Enrico Fermi Institute, U.S.A.

Rajpoot, S.
King's College, U.K.

Samant, S.
University of Western Ontario, Canada

Semenoff, G.
University of British Columbia, Canada

Sen, D.
Carnegie-Mellon University, U.S.A.

Sharp, R.T.
McGill University, Canada

Spence, W.J.
University of Southampton, U.K.

Steele, T.
University of Western Ontario, Canada

Strathdee, J.
ICTP, Trieste, Italy

Talon, M.
University Pierre et Marie Curie, France

Taylor, J.C.
Univeristy of Cambridge, U.K.

Toms, D.J.
University of Newcastle-upon-Tyne, U.K.

Van Proeyen, A.
Katholieke University, Belgium

Verma, R.C.
Punjab University, India

Vaillet, C.M.
University Pierre et Marie Curie, France

Viswanathan, K.S.
Simon Fraser University, Canada

Vogt, E.W.
TRIUMF, Canada

Walton, M.
McGill University, Canada

Weiss, N.
University of British Columbia, Canada

Woloshyn, R.M.
TRIUMF, Canada

Wong, B.
SLAC, U.S.A.

Wu, Y.S.
Universty of Utah, U.S.A.

Yao, Y.P.
University of Michigan, U.S.A.

Zanon, D.
Harvard University/
University of Pisa, Italy

INDEX

B

T